Land Degradation and Desertification: Assessment, Mitigation and Remediation

Pandi Zdruli · Marcello Pagliai · Selim Kapur ·
Angel Faz Cano
Editors

Land Degradation and Desertification: Assessment, Mitigation and Remediation

CIHEAM-Mediterranean Agronomic Institute,
Bari, Italy

Italian Society of Soil Science

International Union of Soil Sciences

European Commission, DG JRC Institute for
Environment and Sustainability

Editors

Pandi Zdruli
International Centre for Advanced
Mediterranean Agronomic
Studies (CIHEAM)
Mediterranean Agronomic Institute of Bari,
Land and Water Resources Management
Department
Via Ceglie 9
70010 Valenzano, Bari
Italy
pandi@iamb.it

Selim Kapur
Department of Soil Science & Archaeometry
University of Cukurova
Adana
Turkey
kapurs@cu.edu.tr

Marcello Pagliai
Consiglio per la Ricerca e la
Sperimentazione in Agricoltura
Centro di Ricerca per
Agrobiologia e la Pedologia
Piazza D'Azeglio, 30
50121 Firenze
Italy
marcello.pagliai@entecra.it

Angel Faz Cano
Depto. of Ciencia y Tecnología
Agraria
Universidad Politécnica de
Cartagena
Paseo Alfonso XIII, 52
30203 Cartagena
Spain
angel.fazcano@upct.es

ISBN 978-90-481-8656-3   e-ISBN 978-90-481-8657-0
DOI 10.1007/978-90-481-8657-0
Springer Dordrecht Heidelberg London New York

Library of Congress Control Number: 2010923830

© Springer Science+Business Media B.V. 2010
No part of this work may be reproduced, stored in a retrieval system, or transmitted in any form or by any means, electronic, mechanical, photocopying, microfilming, recording or otherwise, without written permission from the Publisher, with the exception of any material supplied specifically for the purpose of being entered and executed on a computer system, for exclusive use by the purchaser of the work.

Printed on acid-free paper

Springer is part of Springer Science+Business Media (www.springer.com)

Participants of the 5th International Conference
on Land Degradation

held at the

Mediterranean Agronomic Institute of Bari, Italy
18–22 September 2008

Foreword

Land degradation and desertification issues are now milestone pillars of the international environmental and development agendas. Not only because they affect the livelihoods of billions of people and have direct consequences on the well-being of entire societies, but also due to devastating effects on ecosystem's stability, functions and services, loss of biodiversity and an endless list of other ill-related severances. Problems are exacerbated when land degradation, mostly a human-induced process is combined with naturally occurring drought. It is for these reasons that the recent terminology adopted by the United Nations Convention to Combat Desertification (UNCCD) involves Desertification, Land Degradation and Drought (DLDD). That represents a major shift for the UNCCD itself covering thus the entire planet Earth and bringing it closer to similar UN Conventions like the Biological Diversity (CBD) and the UN Framework Convention to Climate Change (UNFCCC). However, for the UNCCD the major focus will still be placed on drylands and particularly in Africa.

The scientific community has invested more than half a century research in land degradation and desertification and much is known now compared with the time when Auberville for the first time in the 1950s coined the term "desertification", interesting enough not in the drylands but in tropical forests of Africa. Yet, the link between science and policymaking appears to be week and information flow among many stakeholders involved in the combat against land degradation and desertification is not moving fast either.

Results "on-the-ground" in the last decade are not yet convincing many local stakeholders that progress has been made, despite numerous excellent examples of sustainable natural resources management worldwide as documented also by this book. Recent trends ask for a paradigm shift in support of sustainable land management rather than simply focusing on combating land degradation. Climate change will continue to dominate the environmental agenda and its effects will impact also the land degradation-affected areas that will experience additional adversities. While recognising the needs for further mitigation actions to alleviate climate change effects, adaptation to the new climatic conditions will be the final unavoidable choice as the history of nature evolution has shown.

This book contains selected papers of the 5th International Conference on Land Degradation held at the Mediterranean Agronomic Institute of Bari, Italy (IAMB) in

18-22 September 2008. The event was sponsored also by the Italian Society of Soil Science (SISS), the International Union of Soil Sciences (IUSS) and the European Commission, Joint Research Centre, Institute for Environment and Sustainability (EC, JRC, IES). These institutions have been involved for long in land degradation research and mitigation studies.

We are delighted to offer this unique opportunity of presenting papers covering a wide range of topics and geographical areas, all of them serving the purpose of understanding better the cause-effect relationships of land degradation and desertification and to identify the best options for assessment, monitoring, mitigation, and remediation.

Cosimo Lacirignola
Bari, Italy

Marcello Pagliai
Florence, Italy

Stephen Nortcliff
Reading, UK

Luca Montanarella
Ispra, Italy

Editors' Note

The editors would like to express their gratitude to all the authors for their prompt response, hard work and professionalism in preparing their chapters. For some of them this was a first opportunity to write a chapter in a special Springer book, but this was exactly what we intended when we invited them to collaborate in this effort. We are sure this strengthens and enriches the book. We realised that it has been a great challenge to complete this book in a such short time. All of this was possible, thanks to authors' friendly cooperation and enthusiasm that made our tasks easier.

The scientific content of each chapter is the responsibility of individual author(s) and despite our continuous efforts to improve their content, there may be additional questions and comments. We thus invite the reader to kindly ask or write directly to each corresponding author for further clarification. Our editorial tasks were carried out in full respect of everyone's beliefs and research findings, and wherever necessary, to improve the content of each chapter. We did this without any prejudice for individual or professional gains.

A particular word of thank you goes to Ms. Margaret Deignan, Associate Editor at Environmental Sciences Unit of Springer who was the first to write and solicit us about the preparation of this book as she noticed the potential for this publication. We are thankful also to all the sponsors of the 5th International Conference on Land Degradation and in particular to the International Centre for Advanced Mediterranean Agronomic Studies (CIHEAM) and its Vice President Prof. Giuliana Trisorio Liuzzi, to the Director of the Mediterranean Agronomic Institute of Bari in Italy (IAMB), Dr. Cosimo Lacirignola to Deputy Director of IAMB, Dr. Maurizio Raeli, and to the Head of Land and Water Resources Management Department of IAMB, Dr. Nicola Lamaddalena for their enormous support in successfully organising this conference. Among us, an exceptional thank you is addressed to Dr. Pandi Zdruli in recognition of his endless and tireless efforts, scientific perseverance and scrutiny that have left their mark throughout this book.

Bari, Italy	Pandi Zdruli
Florence, Italy	Marcello Pagliai
Adana, Turkey	Selim Kapur
Cartagena, Spain	Angel Faz Cano
October 2009	

Contents

Part I Background Papers

1. **What We Know About the Saga of Land Degradation and How to Deal with It?** 3
   Pandi Zdruli, Marcello Pagliai, Selim Kapur, and Angel Faz Cano

2. **Moving Ahead from Assessments to Actions: Could We Win the Struggle with Soil Degradation in Europe?** 15
   Luca Montanarella

3. **Moving Ahead from Assessments to Actions by Using Harmonized Risk Assessment Methodologies for Soil Degradation** 25
   C.L. van Beek, T. Tóth, A. Hagyo, G. Tóth, L. Recatalá Boix, C. Añó Vidal, J.P. Malet, O. Maquire, J.H.H. van den Akker, S.E.A.T.M. van der Zee, S. Verzandvoort, C. Simota, P.J. Kuikman, and O. Oenema

4. **"Zero-Tolerance" on Land Degradation for Sustainable Intensification of Agricultural Production** 37
   Minh-Long Nguyen, Felipe Zapata, and Gerd Dercon

5. **A Methodology for Land Degradation Assessment at Multiple Scales Based on the DPSIR Approach: Experiences from Applications to Drylands** 49
   Raul Ponce-Hernandez and Parviz Koohafkan

6. **Global Warming, Carbon Balance, and Land and Water Management** 67
   Ahmet Ruhi Mermut

Part II Land Degradation and Mitigation in Africa

7. **The Use of GIS Data in the Desertification Risk Cartography: Case Study of South Aurès Region in Algeria** ... 81
   H. Benmessaud, M. Kalla, and H. Driddi

8 **Land Degradation and Overgrazing in the Afar Region, Ethiopia: A Spatial Analysis** 97
B.G.J.S. Sonneveld, S. Pande, K. Georgis, M.A. Keyzer, A. Seid Ali, and A. Takele

9 **Effects and Implications of Enclosures for Rehabilitating Degraded Semi-arid Rangelands: Critical Lessons from Lake Baringo Basin, Kenya** 111
Stephen M. Mureithi, Ann Verdoodt, and Eric Van Ranst

10 **Assessment of Land Desertification Based on the MEDALUS Approach and Elaboration of an Action Plan: The Case Study of the Souss River Basin, Morocco** 131
R. Bouabid, M. Rouchdi, M. Badraoui, A. Diab, and S. Louafi

11 **Assessment of the Existing Land Conservation Techniques in the Peri Urban Area of Kaduna Metropolis, Nigeria** 147
Taiye Oluwafemi Adewuyi

12 **The Use of Tasselled Cap Analysis and Household Interviews Towards Assessment and Monitoring of Land Degradation: A Case Study Within the Wit-Kei Catchment in the Eastern Cape, South Africa** 163
Luncendo Ngcofe, Gillian McGregor, and Luc Chevallier

13 **Environmental Degradation of Natural Resources in Butana Area of Sudan** 171
Muna M. Elhag and Sue Walker

14 **Land Suitability for Crop Options Evaluation in Areas Affected by Desertification: The Case Study of Feriana in Tunisia** 179
S. Madrau, C. Zucca, A.M. Urgeghe, F. Julitta, and F. Previtali

15 **Strategic Nutrient Management of Field Pea in South-Western Uganda** 195
P. Musinguzi, J.S. Tenywa, and M.A. Bekunda

Part III Land Degradation and Mitigation in Asia

16 **Effectiveness of Soil Conservation Measures in Reducing Soil Erosion and Improving Soil Quality in China Assessed by Using Fallout Radionuclides** 207
Y. Li and M.L. Nguyen

17 **Policy Impacts on Land Degradation: Evidence Revealed by Remote Sensing in Western Ordos, China** 219
Weicheng Wu and Eddy De Pauw

18	Assessment of Land Degradation and Its Impacts on Land Resources of Sivagangai Block, Tamil Nadu, India A. Natarajan, M. Janakiraman, S. Manoharan, K.S. Anil Kumar, S. Vadivelu, and Dipak Sarkar	235
19	New Approaches in Reclamation of Degraded Soils with Special Reference to Sodic Soil: An Indian Experience R.K. Isaac, D.P. Sharma, and N. Swaroop	253
20	Soil and Water Degradation Following Forest Conversion in the Humid Tropics (Indonesia) Gerhard Gerold	267
21	Relationships Between Land Degradation and Natural Disasters and Their Impacts on Integrated Watershed Management in Iran Hamid Reza Solaymani Osbooei and Mahnaz Bafandeh Haghighi	285
22	Modelling Carbon Sequestration in Drylands of Kazakhstan Using Remote Sensing Data and Field Measurements P.A. Propastin and M. Kappas	297
23	The Effectiveness of Two Polymer-Based Stabilisers Offering an Alternative to Conventional Sand Stabilisation Methods Ashraf A. Ramadan, Shawqui M. Lahalih, Sadiqa Ali, and Mane Al-Sudairawi	307
24	Mountainous Tea Industry Promotion: An Alternative for Stable Land Use in the Lao PDR Ayumi Yoshida and Chanhda Hemmavanh	323
25	Rehabilitation of Deserted Quarires in Lebanon to Initial Land Cover or Alternative Land Uses T.M. Darwish, R. Stehouwer, C. Khater, I. Jomaa, D. Miller, J. Sloan, A. Shaban, and M. Hamze	333
26	The Impact of Land Use Change on Water Yield: The Case Study of Three Selected Urbanised and Newly Urbanised Catchments in Peninsular Malaysia Mohd Suhaily Yusri Che Ngah and Ian Reid	347
27	Reclamation of Land Disturbed by Shrimp Farming in Songkla Lake Basin, Southern Thailand Charlchai Tanavud, Omthip Densrisereekul, and Thudchai Sansena	355

28 The Effect of Bio-solid and Tea Waste Applications on
    Erosion Ratio Index of Eroded Soils 367
    Nutullah Ozdemir, Tugrul Yakupoglu, Elif Ozturk, and
    Orhan Dengiz

29 Modern and Ancient Knowledge of Conserving Soils in
    Socotra Island, Yemen 375
    Dana Pietsch and Miranda Morris

Part IV Land Degradation and Mitigation in Europe

30 Content of Heavy Metals in Albanian Soils and
    Determination of Spatial Structures Using GIS 389
    Skender Belalla, Ilir Salillari, Adrian Doko, Fran Gjoka,
    and Majlinda Cenameri

31 Radioisotopic Measurements (^{137}Cs and ^{210}Pb) to Assess
    Erosion and Sedimentation Processes: Case Study in Austria ... 401
    L. Mabit, A. Klik, and A. Toloza

32 Development and Opportunities for Evaluation of
    Anthropogenic Soil Load by Risky Substances in the
    Czech Republic 413
    Radim Vácha, Jan Skála, and Jarmila Čechmánková

33 Land Degradation in Greece 423
    Sid. P. Theocharopoulos

34 Factors Influencing Soil Organic Carbon Stock Variations
    in Italy During the Last Three Decades 435
    M. Fantappiè, G. L'Abate, and E.A.C. Costantini

35 Monitoring Soil Salinisation as a Strategy
    for Preventing Land Degradation: A Case Study in Sicily, Italy .. 467
    Giuseppina Crescimanno, Kenneth B. Marcum, Francesco
    Morga, and Carlo Reina

36 Severe Environmental Constraints for Mediterranean
    Agriculture and New Options for Water and Soil
    Resources Management 477
    N. Colonna, F. Lupia, and M. Iannetta

37 Assessment of Desertification in Semi-Arid Mediterranean
    Environments: The Case Study of Apulia Region
    (Southern Italy) 493
    G. Ladisa, M. Todorovic, and G. Trisorio Liuzzi

38	**Spatial Variability of Light Morainic Soils** Michał Czajka, Stanisław Podsiadłowski, Alfred Stach, and Ryszard Walkowiak	517
39	**Studding the Impacts of Technological Measures on the Biological Activity of Pluvial Eroded Soils** Geanina Bireescu, Costica Ailincai, Lucian Raus, and Lazar Bireescu	529
40	**Achievements and Perspectives on the Improvement by Afforestation of Degraded Lands in Romania** Cristinel Constandache, Viorel Blujdea, and Sanda Nistor	547
41	**Investigating Soils for Agri-Environmental Protection in an Arid Region of Spain** C. Castaneda, S. Mendez, J. Herrero, and J. Betran	561
42	**Risk Assessment in Soils Developed on Metamorphic and Igneous Rocks Using Heavy Metal Sequential Extraction Procedure** S. Martínez-Martínez, Angel Faz Cano, Gerhard Gerold, J.A. Acosta, and R. Ortiz	569
43	**Assessing the Impact of Fodder Maize Cultivation on Soil Erosion in the UK** Mokhtar Jaafar and Des E. Walling	581

Part V   Land Degradation and Mitigation in the Americas

44	**Evolution and Human Land Management During the Holocene in Southern Altiplano Desert, Argentina (26°S)** Pablo Tchilinguirian and Daniel Olivera	591
45	**Metal Pollution by Gold Mining Activities in the Sunchulli Mining District of Apolobamba (Bolivia)** T. Teran, Angel Faz Cano, M.A. Munoz, J.A. Acosta, S. Martinez-Martinez, and R. Millan	605
46	**Areas Degradated by Extraction of Clay and Revegetated with *Acacia mangium and Eucalyptus camaldulensis*: Using Soil Fauna as Indicator of Rehabilitation in an Area of Brazil** Cristiane Figueira da Silva, Eliane Maria Ribeiro da Silva, William Robertson Duarte da Oliveira, Maria Elizabeth Fernandes Correia, and Marco Antonio Martins	619
47	**Conservation Tillage in Potato Rotations in Eastern Canada** M.R. Carter, R.D. Peters, and J.B. Sanderson	627

48	**An Assessment of Soil Erosion Costs in Mexico**	639
	Helena Cotler and Sergio Martínez-Trinidad	
49	**Predicting Winter Wheat Yield Loss from Soil Compaction in the Central Great Plains of the United States** . . .	649
	Joseph G. Benjamin and Maysoon M. Mikha	

Index 657

Contributors

J.A. Acosta Sustainable Use, Management, and Reclamation of Soil and Water Research Group, Department of Agriculture Science and Technology Department, Technical University of Cartagena, 30203 Cartagena, Spain, j.a.acostaaviles@uva.nl

Taiye Oluwafemi Adewuyi Department of Geography, Nigerian Defense Academy, Kaduna, Kaduna State, Nigeria, taiyeadewuyi@yahoo.com

Costica Ailincai University of Agriculture and Veterinary Medicine, 700490 Iaşi, Romania, ailincai@univagro-iasi.ro

Sadiqa Ali Kuwait Institute for Scientific Research, Safat 13109, Kuwait

Mane Al-Sudairawi Kuwait Institute For Scientific Research, Safat 13109, Kuwait

K.S. Anil Kumar National Bureau of Soil Survey and Land Use Planning, Regional Centre, Bangalore 560024, India

C. Añó Vidal CIDE-(CSIC, Universitat de València, Generalitat Valenciana), Cami de la Marjal, s/n, Apartado Oficial, 46470 Albal, Valencia, Spain, carlos.anyo@uv.es

M. Badraoui Institut National de la Recherche Agronomique, Rabat, Marocco, badraoui@inra.org.ma

M.A. Bekunda Department of Soil Science, Makerere University, Kampala, Uganda

Skender Belalla Centre of Agriculture Technology Transfer, Fushë Kruja, Albania

Joseph G. Benjamin USDA-ARS, Central Great Plains Research Station, Akron, CO 80720, USA, joseph.benjamin@ars.usda.gov

H. Benmessaud Laboratoire Risques Naturels et Aménagement du Territoire, Faculté des sciences, Université El Hadj Lakhdar, Batna, Algérie, ha123_m123@yahoo.fr

J. Betran Laboratorio Agroalimentario, DGA, 50059 Zaragoza, Spain

Geanina Bireescu Biological Research Institute, 700107 Iaşi, Romania, bireescugeanina@yahoo.com

Lazar Bireescu Biological Research Institute, 700107 Iaşi, Romania, lazarbireescu@yahoo.com

Viorel Blujdea Climate Change Unit, European Commission – Joint Research Centre, Institute for Environment and Sustainability, Ispra, Italy, viorel.blujdea@jrc.it

R. Bouabid Ecole Nationale d'Agriculture de Meknès, 50000 Meknès, Morocco, rachid.bouabid@gmail.com

Angel Faz Cano Sustainable Use, Management, and Reclamation of Soil and Water Research Group, Department of Agriculture Science and Technology, Technical University of Cartagena, 30203 Cartagena, Spain, angel.fazcano@upct.es

M.R. Carter Agriculture and Agri-Food Canada, Crops and Livestock Research Centre, Charlottetown, PE C1A 4N6, Canada, carterm@agr.gc.ca

C. Castaneda Soils and Irrigation Department (associated with CSIC), AgriFood Research and Technology Centre of Aragon (CITA), 50059 Zaragoza, Spain, ccastanneda@aragon.es

Jarmila Čechmánková Research Institute for Soil and Water Conservation, Prague 15627, Czech Republic

Majlinda Cenameri Polytechnic University of Tirana, Tirana, Albania

Luc Chevallier Council for Geoscience, Bellville 7535, South Africa, lchevallier@geoscience.org.za

N. Colonna Biotechnologies, Agroindustry and Health protection Department, ENEA, 00123 Rome, Italy, nicola.colonna@enea.it

Cristinel Constandache Forest Research and Management Institute Bucuresti, Focsani Station, 620018 Focşani, Romania, cicon66@yahoo.com

Maria Elizabeth Fernandes Correia Embrapa Agrobiologia, Seropédica CEP 23890-000, RJ, Brazil, correia@cnpab.embrapa.br

E.A.C. Costantini CRA-ABP, Research Centre for Agrobiology and Pedology, 50121 Florence, Italy, edoardo.costantini@entecra.it

Helena Cotler Instituto Nacional de Ecología, Coyoacán, México D.F., CP 04530, México, hcotler@ine.gob.mx

Giuseppina Crescimanno Dipartimento ITAF, Università di Palermo, 90128 Palermo, Italy, gcrescim@unipa.it

Michał Czajka Department of Mathematical and Statistical Methods, Poznan University of Life Sciences, 60-637 Poznan, Poland, michalczajka@gmail.com

Contributors

Cristiane Figueira da Silva Centro de Ciências e Tecnologias Agropecuárias, Universidade Estadual do Norte Fluminense Darcy Ribeiro, Campos dos Goytacazes CEP 28015-620, RJ, Brazil, cristiane@uenf.br.marcos@uenf.br

Eliane Maria Ribeiro da Silva Embrapa Agrobiologia, Seropédica CEP 23890-000, RJ, Brazil, eliane@cnpab.embrapa.br

T.M. Darwish National Council for Scientific Research-Remote Sensing Center, Mansourieh, Lebanon, tdarwich@cnrs.edu.lb

William Robertson Duarte de Oliveira Embrapa Agrobiologia, Seropédica CEP 23890-000, RJ, Brazil, eliane@cnpab.embrapa.br

Eddy De Pauw International Center for Agricultural Research in Dry Areas (ICARDA), Aleppo, Syria, e.de-pauw@cgiar.org

Orhan Dengiz Department of Soil Science, Agricultural Faculty, Ondokuz Mayis University, Samsun 55139, Turkey, odengiz@omu.edu.tr

Omthip Densrisereekul Faculty of Natural Resources, Prince of Songkla University, Hat Yai, Songkla, Thailand

Gerd Dercon Soil and Water Management & Crop Nutrition Subprogramme, Joint FAO/IAEA Division of Nuclear Techniques in Food and Agriculture, International Atomic Energy Agency, A-1400 Vienna, Austria, g.dercon@iaea.org

A. Diab Ministère de l'Aménagement du Territoire, de l'Eau et de l'Environnement, Rabat, Morocco

Adrian Doko Centre of Agriculture Technology Transfer, Fushë Kruja, Albania, adriandoko@hotmail.com

H. Driddi Laboratoire Risques Naturels et Aménagement du Territoire, Faculté des sciences, Université El Hadj Lakhdar, Batna, Algérie

Muna M. Elhag Faculty of Agricultural Sciences, University of Gezira, Wad Medani, Sudan, munaelhag@yahoo.com

M. Fantappiè CRA-ABP, Centro di ricerca per l'agrobiologia e la pedologia, 50121 Firenze, Italia, info@soilmaps.it

K. Georgis Ethiopian Institute for Agricultural Research (EIA), Addis Ababa, Ethiopia

Gerhard Gerold Department of Landscape Ecology, Geographisches Institute Georg-August-Universität Göttingen, Göttingen 37077, Germany, ggerold@gwdg.de

Fran Gjoka Agricultural University of Tirana, Tirana, Albania

Mahnaz Bafandeh Haghighi KAM Consulting Engineers Company, Tehran, Iran, mahnaz.haghighi@gmail.com

A. Hagyo Research Institute for Soil Science and Agricultural Chemistry, Hungarian Academy of Sciences, 1022 Budapest, Hungary

M. Hamze National Council for Scientific Research, Beirut, Lebanon, hamze@cnrs.edu.lb

Chanhda Hemmavanh Lanexang Avenue, Vientiane Municipality, Vientiane, LAO PDR, chanhda2006@hotmail.com

J. Herrero Estación Experimental de Aula Dei, CSIC, 50059 Zaragoza, Spain, jherrero@syrsig.mizar.csic.es

M. Iannetta Biotechnologies, Agroindustry and Health Protection Department, ENEA, 00123 Rome, Italy, massimo.iannetta@enea.it

R.K. Isaac Department of Soil Water Land Engineering Management, Faculty of Engineering, Allahabad Agricultural Institute, Deemed University, Allahabad, UP 211007, India, isaac_rk@hotmail.com

Mokhtar Jaafar Faculty of Social Sciences and Humanities, School of Social, Development and Environmental Studies, Universiti Kebangsaan Malaysia, 43600 Bangi, Selangor, Malaysia, mokhtar@eoc.ukm.my

M. Janakiraman National Bureau of Soil Survey and Land Use Planning, Regional Centre, Bangalore 560024, India

I. Jomaa National Council for Scientific Research-Remote Sensing Center, Mansourieh, Lebanon

F. Julitta DISAT, Università degli Studi di Milano-Bicocca, Milano, Italy

M. Kalla Laboratoire Risques Naturels et Aménagement du Territoire, Faculté des sciences, Université El Hadj Lakhdar, Batna, Algérie

M. Kappas Department of Geography, Georg-August University, Göttingen 37077, Germany, mkappas@uni-goettingen.de

Selim Kapur Department of Soils and Archaeometry, University of Çukurova, 01330 Adana, Turkey, kapurs@cu.edu.tr

M.A. Keyzer Centre for World Food Studies, VU University Amsterdam (SOW-VU), 1081 HV Amsterdam, The Netherlands

C. Khater National Council for Scientific Research-Remote Sensing Center, Mansourieh, Lebanon

A. Klik Institute of Hydraulics and Rural Water Management, University of Natural Resources and Applied Life Sciences, Vienna, Austria

Parviz Koohafkan Land and Water Division, Department of Natural Resources Management and Environment, Food and Agriculture Organisation, 00153 Rome, Italy, parviz.koohafkan@fao.org

P.J. Kuikman Alterra, Soil Science Center, 6700 AA Wageningen, The Netherlands, peter.kuikman@wur.nl

G. L'Abate CRA-ABP, Centro di ricerca per l'agrobiologia e la pedologia, 50121 Firenze, Italia, info@soilmaps.it

Gaetano Ladisa CIHEAM Mediterranean Agronomic Institute of Bari, Valenzano 70010, BA, Italy, ladisa@iamb.it

Shawqui M. Lahalih Kuwait Institute for Scientific Research, Safat 13109, Kuwait

Y. Li Institute of Environment and Sustainable Development in Agriculture, CAAS, Beijing 100081, China, yongli32@hotmail.com

Giuliana Trisorio Liuzzi CIHEAM/Department of Engineering and Management of Agricultural and Forest Systems, University of Bari, Bari 70126, Italy, giuliana.trisoriol@agr.uniba.it

S. Louafi Compagnie d'Aménagement Agricole et de Développement Industriel, Rabat, Morocco

F. Lupia INEA, National Institute of Agricultural Economics, 36 00187 Rome, Italy, lupia@inea.it

L. Mabit Soil Science Unit, FAO/IAEA Agriculture & Biotechnology Laboratory, IAEA Laboratories Seibersdorf, Vienna A-1400, Austria, l.mabit@iaea.org

S. Madrau DIT-UNISS, Department of Territorial Engineering, Università degli Studi di Sassari, 07100 Sassari, Italy, geopedol@uniss.it

J.P. Malet Institut de Physique du Globe de Strasbourg (IPGS), CNRS, Strasbourg, School and Observatory of Earth Sciences (EOST), F-67084 Strasbourg Cedex, France, jeanphilippe.malet@eost.u-strasbg.fr

S. Manoharan National Bureau of Soil Survey and Land Use Planning, Regional Centre, Bangalore 560024, India

O. Maquire Institut de Physique du Globe de Strasbourg (IPGS), CNRS, Strasbourg, School and Observatory of Earth Sciences (EOST), F-67084 Strasbourg Cedex, France, oliver.maquire@unicaen.fr

Kenneth B. Marcum Centre for Urban Greenery and Ecology, National Parks Board, Garden city, Singapore, kenneth_marcum@nparks.gov.sg

S. Martínez-Martínez Department of Agriculture Science and Technology, Technical University of Cartagena, 30203 Cartagena, Spain, silvia.martinez@upct.es.

Sergio Martínez-Trinidad Instituto Nacional de Ecología, Coyoacán, México D.F., CP 04530, México

Marco Antonio Martins Centro de Ciências e Tecnologias Agropecuárias, Universidade Estadual do Norte Fluminense Darcy Ribeiro, Campos dos Goytacazes CEP 28015-620, RJ, Brazil

Gillian McGregor Department of Geography, Rhodes University, Grahamstown 6140, South Africa, g.k.mcgregor@ru.ac.za

S. Mendez Réserve Naturelle Nationale Vallée d'Eyne, Ferme Cal Martinet, 66800 Eyne, France

Ahmet Ruhi Mermut University of Saskatchewan, Saskatoon, SK S7N, Canada; Department of Soil Science, Faculty of Agriculture, Harran University, Şanlıurfa 63200, Turkey, a.mermut@usask.ca

Maysoon M. Mikha USDA-ARS, Central Great Plains Research Station, Akron, CO 80720, USA

R. Millan Department of Environment, CIEMAT, E-28040 Madrid, Spain, rocio.millan@ciemat.es

D. Miller Center for Environmental Informatics, Penn Sate University, Philadelphia, PA, USA

Luca Montanarella Institute for Environment and Sustainability, European Commission, Joint Research Centre, I-21020 Ispra, VA, Italy, luca.montanarella@jrc.it

Francesco Morga Dipartimento ITAF, Università di Palermo, 90128 Palermo, Italy, francescomorga@gmail.com

Miranda Morris School of History, St. Andrews University, St. Andrews, Fife KY16 9AL, UK

M.A. Munoz Sustainable Use, Management, and Reclamation of Soil and Water Research Group, Department of Agriculture Science and Technology, Technical University of Cartagena, 30203 Cartagena, Spain

Stephen M. Mureithi Range Management Section, Department of Land Resource Management and Agricultural Technology, University of Nairobi, Nairobi, Kenya, stemureithi@yahoo.com

P. Musinguzi Department of Soil Science, Makerere University, Kampala, Uganda, patmusinguzi@agric.mak.ac.ug

A. Natarajan National Bureau of Soil Survey and Land Use Planning, Regional Centre, Bangalore 560024, India, athiannannatarajan@gmail.com

Mohd Suhaily Yusri Che Ngah Department of Geography, Faculty of Social Sciences and Humanities, Sultan Idris Education University, Tanjong Malim Perak 35900, Malaysia, suhaily@upsi.edu.my

Luncendo Ngcofe Council for Geoscience, Bellville 7535, South Africa, Lngcofe@geoscience.org.za

Minh-Long Nguyen Soil and Water Management & Crop Nutrition Subprogramme, Joint FAO/IAEA Division of Nuclear Techniques in Food and Agriculture, International Atomic Energy Agency, A-1400 Vienna, Austria, m.nguyen@iaea.org

Sanda Nistor Forest Research and Management Institute, Focsani Research Station, 620018 Focşani, Romania, icasvn2006@yahoo.com

O. Oenema Alterra, Soil Science Center, 6700 AA Wageningen, The Netherlands, oene.oenema@wur.nl

Daniel Olivera CONICET-INAPL-UBA, CP 1426 Buenos Aires, Argentina, deolivera@gmail.com

R. Ortiz Department of Agricultural Chemistry, Geology and Edaphology, University of Murcia, Campus of Espinardo, 30100 Murcia, Spain, rortiz@um.es

Hamid Reza Solaymani Osbooei Department of Watershed Management, Forest, Range and Watershed Organisation, Tehran 11445-1136, Iran, hrsolaymani@yahoo.com

Nutullah Ozdemir Department of Soil Science, Agricultural Faculty, Ondokuz Mayis University, Samsun 55139, Turkey, nutullah@omu.edu.tr

Elif Ozturk Department of Soil Science, Agricultural Faculty, Ondokuz Mayis University, Samsun 55139, Turkey, elifo@omu.edu.tr

Marcello Pagliai CRA-ABP Research Centre for Agrobiology and Pedology, 50121 Florence, Italy, marcello.pagliai@entecra.it

S. Pande Centre for World Food Studies, VU University Amsterdam (SOW-VU), 1081 HV Amsterdam, The Netherlands

R.D. Peters Agriculture and Agri-Food Canada, Crops and Livestock Research Centre, Charlottetown, PE C1A 4N6, Canada, rick.peters@agr.gc.ca

Dana Pietsch Institute of Geography, University of Tübingen, Tübingen 72070, Germany, dana.pietsch@uni-tuebingen.de

Stanisław Podsiadłowski Department of Agricultural Engineering, Poznan University of Life Sciences, 60-625 Poznan, Poland, stapod@au.poznan.pl

Raul Ponce-Hernandez Environmental and Resource Studies Program, Trent University, Peterborough, ON K9J 7B8, Canada, rponce@trentu.ca

Franco Previtali DISAT, Università degli Studi di Milano-Bicocca, Milano, Italy, franco.previtali@unimib.it

P.A. Propastin Department of Geography, Georg-August University, Göttingen 37077, Germany, ppropas@uni-goettingen.de

Ashraf A. Ramadan Kuwait Institute for Scientific Research, Safat 13109, Kuwait, aramadan@kisr.edu.kw

Lucian Raus University of Agriculture and Veterinary Medicine, 700490 Iaşi, Romania, rauslucian@univagro-iasi.ro

L. Recatalá Boix CIDE-(CSIC, Universitat de València, Generalitat Valenciana), Cami de la Marjal, s/n, Apartado Oficial, 46470 Albal, Valencia, Spain, luis.recatala@uv.es

Ian Reid Geography Department, Loughborough University, Leicestershire LE11 3TU, UK, ian.reid@lboro.ac.uk

Carlo Reina Dipartimento ITAF, Università di Palermo, 90128 Palermo, Italy

M. Rouchdi Ecole Nationale d'Agriculture de Meknès, 50000 Meknès, Morocco

Ilir Salillari Centre of Agriculture Technology Transfer, Fushë Kruja, Albania, ilirsalillari@gmail.com

J.B. Sanderson Agriculture and Agri-Food Canada, Crops and Livestock Research Centre, Charlottetown, PE C1A 4N6, Canada

Thudchai Sansena Geo-Informatics and Space Technology Development Agency, Bangkok, Thailand

Dipak Sarkar National Bureau of Soil Survey and Land Use Planning, Regional Centre, Bangalore 560024, India

A. Seid Ali Afar Pastoral and AgroPastoral Research Institute (APARI), Semerra, Ethiopia

A. Shaban National Council for Scientific Research-Remote Sensing Center, Mansourieh, Lebanon

D.P. Sharma Department of Soil Science and Agricultural Chemistry, Faculty of Agriculture, Allahabad Agricultural Institute, Deemed University, Allahabad, UP 211007, India

C. Simota ICPA, Bucharest, Romania, c.simota@icpa.ro

Jan Skála Research Institute for Soil and Water Conservation, Prague 15627, Czech Republic

J. Sloan Center for Environmental Informatics, Penn Sate University, Philadelphia, PA, USA

B.G.J.S. Sonneveld Centre for World Food Studies, VU University Amsterdam (SOW-VU), 1081 HV Amsterdam, The Netherlands, b.g.j.s.sonneveld@sow.vu.nl

Alfred Stach Quaternary Research Institute, Adam Mickiewicz University, 61-701 Poznan, Poland, frdstach@amu.edu.pl

R. Stehouwer College of Agricultural Sciences, Penn State University, Pennsylvania, PA, USA

N. Swaroop Department of Soil Science and Agricultural Chemistry, Faculty of Agriculture, Allahabad Agricultural Institute, Deemed University, Allahabad, UP 211007, India, narendra_swaroop2003@yahoo.com

A. Takele Ethiopian Institute for Agricultural Research (EIAR), Addis Ababa, Ethiopia

Charlchai Tanavud Faculty of Natural Resources, Prince of Songkla University, Hat Yai, Songkla, Thailand, charlchai.t@psu.ac.th

Pablo Tchilinguirian Institute of Archaeology, UBA, CP 1426 Buenos Aires, Argentina, pabloguirian@gmail.com

J.S. Tenywa Department of Soil Science, Makerere University, Kampala, Uganda

T. Teran Department of Agriculture Science and Technology, Technical University of Cartagena, 30203 Cartagena, Spain, tania.teran@upct.es

Sid. P. Theocharopoulos N.AG.RE.F. Soil Science Institute of Athens, Athens 14123, Greece, Sid_Theo@nagref.gr

Mladen Todoroviç Department of Land and Water Resources Management, CIHEAM Mediterranean Agronomic Institute of Bari, Valenzano 70010, BA, Italy, mladen@iamb.it

A. Toloza Soil Science Unit, FAO/IAEA Agriculture & Biotechnology Laboratory, IAEA Laboratories Seibersdorf, Vienna A-1400, Austria

G. Tóth European Commission, Joint Research Centre, I-21020 Ispra, VA, Italy, gergely.toth@jrc.it

T. Tóth Research Institute for Soil Science and Agricultural Chemistry, Hungarian Academy of Sciences, 1022 Budapest, Hungary, tibor@rissac.hu

A.M. Urgeghe Centro Interdipartimentale di Ateneo NRD-UNISS (Nucleo Ricerca Desertificazione), Università degli Studi di Sassari, 07100 Sassari, Italy

Radim Vácha Research Institute for Soil and Water Conservation, Prague 15627, Czech Republic, vacha@vumop.cz

S. Vadivelu National Bureau of Soil Survey and Land Use Planning, Regional Centre, Bangalore 560024, India

C.L. van Beek Alterra, Soil Science Center, 6700 AA Wageningen, The Netherlands, christy.vanbeek@wur.nl

J.H.H. van den Akker Alterra, Soil Science Center, 6700 AA Wageningen, The Netherlands, janjh.vandenakker@wur.nl

Sjoerd E.A.T.M. van der Zee Wageningen-UR, 6700 HB Wageningen, The Netherlands, sjoerd.vanderzee@wur.nl

Eric Van Ranst Laboratory of Soil Science, Department of Geology and Soil Science, Ghent University, B-9000 Gent, Belgium, eric.vanranst@rug.ac.be

Ann Verdoodt Laboratory of Soil Science, Department of Geology and Soil Science, Ghent University, B-9000 Gent, Belgium

S. Verzandvoort Alterra, Soil Science Center, 6700 AA Wageningen, The Netherlands, simone.verzandvoort@wur.nl

Sue Walker University of Free State, Bloemfontein 9300, South Africa, walkers.sci@mail.uovs.ac.za

Ryszard Walkowiak Department of Mathematical and Statistical Methods, Poznan University of Life Sciences, 60-637 Poznan, Poland, rwal@au.poznan.pl

Des E. Walling Department of Geography, School of Geography, University of Exeter, Exeter, Devon EX4 4RJ, UK, d.e.walling@exeter.ac.uk

Weicheng Wu International Center for Agricultural Research in Dry Areas (ICARDA), Aleppo, Syria, w.wu@cgiar.org

Tugrul Yakupoglu Department of Soil Science, Agricultural Faculty, Ondokuz Mayis University, Samsun 55139, Turkey, tugruly@omu.edu.tr

Ayumi Yoshida Zhejiang University, Hangzhou, Zhejiang Province 310029, PR China, ayumi_2180@hotmail.com

Felipe Zapata Soil and Water Management & Crop Nutrition Subprogramme, Joint FAO/IAEA Division of Nuclear Techniques in Food and Agriculture, International Atomic Energy Agency, A-1400 Vienna, Austria

Pandi Zdruli Department of Land and Water Resources Management, CIHEAM-Mediterranean Agronomic Institute of Bari, 70010 Valenzano, BA, Italy, pandi@iamb.it

Claudio Zucca NRD-UNISS, Desertification Research Group, Università degli Studi di Sassari, 07100 Sassari, Italy, clzucca@uniss.it

About the Editors

Dr. Pandi Zdruli over the last decade has been a Senior Research Scientist and Professor of Soil Science and Natural Resources with the International Centre for Advanced Mediterranean Agronomic Studies (CIHEAM), Land and Water Resources Management Department of the Mediterranean Agronomic Institute of Bari, in Italy. Prior to this position he was Visiting Scientist with the European Commission's Joint Research Centre in Ispra, Italy, Senior Fulbright Research Fellow at the United States Department of Agriculture, Natural Resources Conservation Service (USDA NRCS) in Washington DC, USA, and Chief of the Pedology Department of the Soil Science Institute of Tirana in his native Albania. Dr. Zdruli is author of over 46 scientific and technical papers, Editor-In-Chief of 10 books and distinguished member of a number of professional national and international organisations. He has over 28 years of experience in agriculture and rural development, soil science, sustainable land management, land degradation and desertification studies and integrated environmental impact assessments.

Dr. Marcello Pagliai has been for more than a decade Director of the Research Centre for Agrobiology and Pedology of the Italian Agricultural Research Council and President of the Italian Soil Science Society from 2002 to 2008. He was also Chairman of the Commission 2.1 (Soil Physics) of the International Union of Soil Sciences (IUSS) for the period 2002–2006. His main research efforts are centred in the fields of Soil Micromorphology and Soil Physics and particularly on: soil-conditioner interactions, effects of waste applications and organic materials on soil structure, evaluation of the impact of different tillage systems and management practices on soil quality, soil crusting, soil physical and biological degradation, soil compaction, sensibility and vulnerability and paddy soils research. Coordinator of several research projects on soil management and conservation, he is also author and co-author of 191 publications, Associate-Editor-In-Chief of the European Journal of Agronomy (Elsevier) and Member of the Editorial Board of Soil and Tillage Research (Elsevier).

Dr. Selim Kapur is Professor of Soil Science and Archaeometry at the University of Çukurova, in Adana, Turkey. He has organised various international meetings and events since the mid sixties when in Madrid, Spain in 1966 was held the first Meeting on Red Mediterranean Soils that later developed as Meetings of Soils with

Mediterranean Type of Climate; the tenth of such meetings was held in 2009 in Lebanon. Dr. Kapur has also organised the 1st International Conference on Land Degradation held in Adana in 1996 and latter on again in Adana he prearranged the 12th International Meeting on Soil Micromorphology. He is member of the European Soil Bureau Network and Secretary of the Working Group on Land Degradation of the International Union of Soil Sciences. He has been a Wageningen STIBOKA (now ALTERRA) and Hohenheim University fellow scientist. Dr. Kapur is author and also reviewer/editor of numerous papers and chapters published in national and international journals as well as books that have contributed on the interdisciplinary character of soil science, and in particular of soil micromorphology. Has contributed largely to the development of the Anthroscape context in relation to soils and has more than 35 years of research and experience in sustainable land management.

Prof. Angel Faz Cano received his Bachelor, Master and PhD degrees from the University of Murcia, in Spain. He is currently professor of Soil Science at the Agricultural Science and Technology Department, and Director of the Research Group on Sustainable Use, Management and Reclamation of Soil and Water, Technical University of Cartagena, Spain. Current national and international projects include soil usage and global change in semiarid areas: carbon cycle, agricultural application of pig slurries and organic residues from horticulture for crop production, mining and industrial polluted soils risk assessment, reclamation and landscape design. In addition to Spain, he has made extensive research on soils and water management in Latin America. Currently he is Vice chair of the Working Group on Land Degradation of Internatonal Union of Soil Sciences (IUSS), and just elected in 2008 as Vice chair of the Commission Soil Geography of IUSS.

Part I
Background Papers

Chapter 1
What We Know About the Saga of Land Degradation and How to Deal with It?

Pandi Zdruli, Marcello Pagliai, Selim Kapur, and Angel Faz Cano

Abstract The 5th International Conference on Land Degradation held at the Mediterranean Agronomic Institute of Bari, Italy in September 2008 brought together some 100 people from 37 countries worldwide. A number of international organisations, like FAO, IAEA, EC, and CIHEAM were also present. The conference was split into 8 sessions where 83 papers (43 oral) were presented. In total 235 abstracts were received. The main outcome was that the fight against land degradation and desertification could be successful if the right policy instruments are put in place and most importantly when local people are both authors and actors of the development process. Moreover, soil conservation and restoration should be one component of an integrated ecosystem management strategy that should include also water, biodiversity, livelihoods and human impacts on ecosystems. There are numerous positive results when dealing with land degradation worldwide. They should be used to emphasise the urgent needs for further actions to accelerate and scale up progress and not to induce complacency. Improved land resources management measures should build on scientific evidence, local innovation and knowledge and be locally tested and validated before being applied at larger scale. Natural resource base conservation should continue to be a priority for national governments and international organisations but Africa requires particular attention. The recent financial, economic and food global crisis should not overshadow the urgent needs to deal with natural resource management and conservation and mitigate climate change impacts.

Keywords Land degradation · Desertification · UNCCD · Sustainable land management · Mitigation · Remediation · Future perspectives · Possible solutions

P. Zdruli (✉)
Department of Land and Water Resources Management, CIHEAM-Mediterranean Agronomic Institute of Bari, 70010 Valenzano, BA, Italy
e-mail: pandi@iamb.it

1.1 Introduction

Bari, Italy, September 2008. Yet another international conference on land degradation (ICLD) organised by the Land Degradation Working Group of the International Union of Soil Sciences (IUSS) following four previous ones held in Adana, Turkey (1996), Khon Kaen, Thailand (1999), Rio de Janeiro, Brazil (2001), and Cartagena, Spain (2004). These were not the only events dealing with land degradation nationally and internationally as numerous similar ones have taken place over the last half-century in many places around the world. Most likely others will follow. Hence, a "common" and intriguing question would be: "do we really need another conference to discuss land degradation?"

To make it different from the previous ones, the Scientific Committee of this 5th ICLD decided to use a different approach, synthesised as: *Moving ahead from assessments to actions: Could we win the struggle with land degradation?* We think this conference gave some interesting answers to the above question, no matter how provocative, challenging or controversial that theme might have been.

In one of the discussions held during the conference Dr. Zdruli mentioned that: "in only one, out of many land degradation meetings I have attended over the last 2 decades around the world I heard someone to discuss erosion in a balanced manner, all the others identified only its negative aspects and consequences". Is there some truth from this rare case? What could have been Egypt without erosion or the Po River Valley in Italy, the largest fluvial deposit in Europe? What about landslides? Surely they devastate property, infrastructure, endanger public health and safety, but in some cases like in Papua New Guinea or Jamaica they are known for supporting better crops and creating new possibilities for cultivation (Stocking and Murnaghan, 2001). So are volcanic eruptions, as devastating as horrendous to destroy entire communities, but beneficial as well for the formation of the fertile Andosols. Reynolds (2008) analysed land degradation from various viewpoints and noticed for instance that an eroded landscape in Mexico is very attractive for the movie industry, thus providing additional jobs and income for the poverty stricken local population. Additionally, pastoralists could have far different views on degradation from farmers as they may benefit (at least in the short term) from overstocking their flocks in such areas. Hence, much depends on the angle one looks at the problem.

Land degradation is both a natural and human-induced process. It existed before the human race populated the earth and will continue to exist. However, humans have a two-sided effect on it: mitigate or accelerate. Devastating pictures of eroded landscapes and impoverished drylands are often used to show the "evils" of land degradation and desertification. We think that it is much easier to show the darker side of the story rather than the opposite. Thus, and to show that we are not fighting a lost battle, we asked participants of the 5th ICLD to bring forward some of these results. One could mention the millenary grape terraces of Cinque Terre in the north-western Ligurian coast of Italy that are a living example of human ingenuity to grow crops and preserve the environment. Many other good examples can be found around the world and have been well documented by the World Overview of Conservation Approaches and Technologies (WOCAT, 2007).

Despite the existence of the controversial definitions and confusion between "soil" and "land" degradation and desertification (this last is also land degradation but in well-defined climatic domains) the substance of the problem is the same: degradation of the resource base and reduced capacity for continued productivity and maintenance of global ecosystem services.

Measuring the extent and severity of land degradation has been quite challenging. Until very recently, there was only one global assessment of human-induced soil degradation, the GLASOD database (Oldeman et al., 1991). This has been debated over the years due to its main limitations being the qualitative assessments that produced disputable results and poor relationships between land degradation and policy-pertinent criteria (Sonneveld and Dent, 2007). Notwithstanding, GLASOD deserves credit for bringing the issue of land degradation to the world agenda.

Efforts to develop other global and regional assessments of land degradation continued (Eswaran et al., 2003; Holm et al., 2003; Prince et al., 2007; Bai et al., 2008). Safriel (2007) reports for an alternative method currently under development for detecting land degradation trends, using a surrogate called Residual Net Primary Production (RESTREND), which is based on an analysis of the residuals of the productivity-rainfall relationship during a defined time period. The GEF-UNEP-FAO sponsored LADA project (Land Degradation Assessment in Drylands) is presently engaged in development of standard methods to assess global land degradation (GLADA) and preliminary results are reported by Bai et al. (2008). Using the (Rural Urban Extent) RUE adjusted NDVI/NPP (normalised difference vegetation index/net primary production) index to globally detect significant biomass changes, they indicate that 23.54% of the Earth is degraded and 1.5 billion people are affected.

Further efforts have been devoted by Eswaran et al. (2003), Kapur and Akça (2004), Kapur et al. (2004) and Eswaran et al. (2005) to assess mitigation measures for land degradation within human-reshaped landscapes (i.e. Anthroscapes), based on combinations of appropriate indigenous technologies and scientific know-how. However, whatever the method, caution is needed when carrying out land degradation analyses as data could be collected using a wide range of approaches which inevitably contain assumptions that may not be comparable between different sites or regions.

It has been a matter of concern that even after more than a decade of the United Nations Convention to Combat Desertification (UNCCD), there is still considerable uncertainty on the global status of land degradation and desertification. Is desertification at global scale progressing, remaining stable or decreasing? Although the scientific community is struggling to provide a reliable response, there is not yet a clear, sound and scientific answer to this question. Also, even if such assessments were accurate and available, there is still considerable need for mitigation actions "on the ground" to alleviate the hardships of land degradation on vulnerable populations.

1.2 Topics of 5th ICLD

- Multidisciplinary assessment of land degradation and desertification at local, national, regional and global scales;
- Interaction between natural ecosystem components (land, water, biodiversity) and socio-economic indicators and their overall impact on land degradation;
- Impacts of poor land management on natural resources and examples of best management practices in reducing land degradation impacts;
- Promotion of income-generating activities that alleviate poverty through enhancement of sustainable crop production systems and valorisation of indigenous knowledge in sustainable ecosystem management;
- Participatory management of natural resources as a mean to sustain both productivity and environmental sustainability;
- Establishing the role and responsibilities of various stakeholders in reducing the negative impacts of land degradation and enhancing soil conservation measures;
- State and development of policy options, management strategies, and guidelines for sustainable natural resources use and management;
- Development of economically sustainable measures that match soil quality with environmental stability.

1.3 Discussions on the Controversies of Land Degradation

Stocking and Murnaghan (2001) describe the "land degradation wall" and its many biophysical "bricks", such as soil degradation, landscape alteration, water deterioration, soil erosion by water and wind, nutrient depletion, loss of biodiversity, climate change, reduced vegetation cover, pollution, drought, compaction, sedimentation, reduced organic matter and salinisation. Each of these contributes at various intensities to the land degradation process. They may be consistent for a specific area, but rarely simultaneous in the same area. Thus, careful analyses of various local conditions are required, when dealing successfully with land degradation and desertification analyses.

The recent outcomes of the Committee for the Review of the Implementation (CRIC) of the UNCCD Convention held in Istanbul, Turkey in November 2008, revealed a number of failures (UNCCD CRIC7, 2009) when dealing with desertification mitigation. Amongst others, they include: shortcomings in up-scaling good practices, disseminating available knowledge, and closing the gap between scientists, decision/policy makers and local communities. Moreover, there has been a failure in attempts to mainstream activities at the national level, mobilise resources and converge desertification, land degradation and drought (DLDD) from a global issue to a local one requiring immediate solutions. At the closure of the CRIC 7 of the UNCCD, the improved application and translation of National Action Programmes into science-based regional projects were identified as recommended procedures to mitigate regional land degradation and desertification. In addition, results based Monitoring and Evaluation (M & E) schemes were identified to monitor impacts on programming. For instance, the statement of the Annex IV countries (parties of the Northern Mediterranean Basin) was strongly in line with this resolution and for the future formats of the CRIC Sessions, urging other Annexes to discuss the issues related to the implementation of the Convention on the "Regional Level". Accordingly, the foreseen efforts and initiatives were conferred to the Dryland Science for Development (DSD) consortium as the strategy to support the Committee on Science and Technology (CST) COP-9 in a scientific conference format.

Land degradation processes, causes, intensities and effects, are well documented in the literature (Conacher and Sala, 1998; Rubio et al., 2002; Ryan, 2002; Stocking, 2003; Zdruli and Costantini, 2008). However, there are still many difficulties in distinguishing between human-induced and natural degradation processes and mitigation measures. There is a need to conduct specific assessments and test mitigation measure effectiveness prior to presenting them as the remedy or the solution to the problem. Moreover, in many countries there are shortages of reliable data that can be used to demonstrate the extent and the intensity of land degradation and desertification.

After years of regarding these processes as mainly biophysical ones and wrongly disregarding their socio-economic nature (Reynolds and Stafford Smith, 2002), the scientific community is increasingly acknowledging that land degradation often results from combined human-induced causes (unsustainable land use practices, such as overgrazing, deforestation, etc), as well as natural causes such as climate change, drought, etc (Adams and Eswaran, 2000; Cangir et al., 2000; Safriel, 2007). However, distinctions between areas already affected and those highly vulnerable to land degradation are not clearly distinguished in most assessments, and this complicates the impacts of the processes and creates confusion for decision-makers (Safriel, 2007). This underlines the need for a holistic and integrated approach, which takes into account not only biophysical aspects but also social, institutional, governance as well as economic and political dimensions of such processes. This was also pointed out in the well-documented article in Science Magazine, 11 June 2004 (Kaiser, 2004), with the headline: "Soils the last frontier". Many shortcomings in land degradation assessments are due to inadequate knowledge of cause-effect relationships between severity of degradation and agricultural productivity (Nachtergaele, 2003).

It is widely believed that the Green Revolution of the seventies largely succeeded in Asia and Latin America (Eswaran et al., 1997) because the genetic improvements of newly created cultivars were followed by improvements in land/soil and water management, but it did failed in Africa because these last actions were not taken. This is a lesson that yet remains valid especially for the sub-Saharan Africa that is characterised by poor-resource farmers and small landholdings (0.5–2 ha). Unless these farmers take actions to endorse sustainable soil and water management, results would be far less convincing.

The economic impacts of land degradation are still very uncertain. Wiebe (2003) estimates the economic effects of soil erosion globally at 0.05% per year of the total production value. Other authors admit similar values, but point out that off-site effects are much higher in economic terms. Controversially, studies in the mid 1990s predicted higher figures, reaching as much as 10% of the value of agricultural production each year according to a joint study of UNEP, UNDP and FAO (Pimentel et al., 1995). These estimates re-introduce the crucial question of data availability and their quality.

The criteria for designating classes of land degradation (i.e. low, moderate, high) are mainly based on land properties rather than on their impact on productivity or ecosystem functions and services. Even though the link between land degradation and productivity loss is well documented, there is still contradictory evidence on this. Studies show that crop productivity is a function of many variables and depends on soil and weather characteristics as well as on technological management. Thus, land degradation as a biophysical process cannot be separated from its socio-economic impacts.

New research identifies the processes whereby "temporary depletion" of land for increased income can be justified (the "profits" can be invested for education, health, etc., and once the land users receive increased income they are likely to re-invest in land improvement). However, where is the point of no return? Recent findings from Niger demonstrate that community level land rehabilitation activities achieved through agroforestry and reforestation combined with soil conservation (Pender and Ndjeunga, 2008) can be expected to yield high rates of return but with high variance. Other findings based only on soil conservation (*zai* planting pits; contour stone bunds; application of organic and inorganic fertilizer) show lower but still positive production impacts.

A workshop held at the FAO, Rome, December 2006, evaluated the cost of inaction. The workshop concluded that the rates of return from successful projects in arid areas could be as high as 30%, but the economic losses from continuing degradation without treatment could reach as high as several percentage points of the GDP per year, if such projects were not implemented (Global Mechanism, 2006). It is clear that the costs of amelioration of degraded lands are much higher than preventing them from becoming degraded in the first place: prevention is cheaper than cure (Zdruli et al., 2007).

Reynolds et al. (2007) suggest that dryland degradation can be confronted with renewed optimism if both ecosystem functions and livelihood needs are given equal importance. Additionally the Dryland Development Paradigm (DDP) approach (Reynolds et al., 2007) offers a comprehensive framework for integrated assessments. Thomas (2008) concludes recently that it is better to focus on Sustainable Land Management (SLM) rather than simply combating land degradation and desertification.

1.4 Conference Findings and Recommendations

The following are the major conclusions of the conference:

The implications of land degradation are of equal concern in arid and hyperarid drylands areas as well as in semiarid and dry subhumid regions. *Reduction of soil organic matter and soil biodiversity losses* reduce soil fertility and have direct consequences on crop productivity and other soil/ecosystem functions and services.

Despite their inherently poor organic matter content (Zdruli et al., 2004), further such losses in dryland soils could provoke irreversible degradation of the resource base. In view of increasing effects of resource constraints on the global economy (land, energy, water) and the recent financial, economic and food global crisis, *targeted research on ecological-economic interactions and application of adapted national policies and action plans* are recognized as instrumental in the fight to mitigate land degradation and prevent further losses of productive lands.

The international and national communities have been involved for decades in tackling these problems. It is encouraging to note that increasingly many success stories in sustainable management of natural resources are being identified, as shown also in this book. These case studies show that when an *enabling policy environment* is created, when the *enabling policy instruments* are put in place, and when *local stakeholders are both authors and actors* of the land management process, it is possible to make positive changes and to reverse the trend of land degradation and desertification. Environmental measures, which include interventions on land and water in the range of ecosystems spanning the agricultural, forest and livestock sectors, should be assessed in terms of *impacts on both productivity, ecological functions* and on the effects they have *on ecosystem stability, resilience, human livelihoods, and global life support systems.*

In addition, stronger links must be developed to ensure *implementation* of scientific information in the development of policies and programmes to mitigate land degradation and desertification. Application of M & E schemes, with land degradation indicators and analyses of trends and impacts of adapted remedial measures needs *proper transfer to decision making at policy levels.* This, however, will require improved coordination at national, regional and local levels.

We must now move to the next step: *implementing remedial measures to prevent and mitigate* land degradation and desertification, including *local adaptation measures also in terms of climate change that are essential.* Tackling the causes ensures permanence of positive changes and conservation measures. However, not all soil and water conservation measures work well as there are plenty of examples of failures due to being ill-adapted in mitigating constraints and/or in terms of limited impact. Improved land resources management measures should build on local innovation and knowledge and be locally tested and validated before being applied at larger scale or being transferred to other locations even in similar ecosystems.

The global agenda for *sustainable land management should remain a priority for national Governments and international organisations.* Africa in particular and some countries in Latin America and Asia require special attention due to limited resources, research capacities, and evidence of little progress in stimulating agricultural and economic growth.

Addressing the issue of land degradation requires the adoption of a holistic *approach to ecosystem management,* underlined by the concept of sustainability. The involvement and commitment of decision makers is crucial for the success of programmes that stimulate a transition to an era of innovation to achieve sustainable use of resources, development and growth. The Millennium Ecosystem Assessment

(MA) framework on ecosystem services is a useful methodology in this context and expectations are that the UNCCD in particular should apply an ecosystem approach in the broadest sense of its work.

Policies should be context specific. Not all work the same way in different ecological, economic and cultural environments, and they should be drafted and adapted to the local cultural values and knowledge. Rigid top down approaches have failed to halt land degradation as they have ignored issues such as empowerment of land users, land tenure, and the fundamental principles of sustainability. In this context, reliable procedures for scaling up and down both assessments and recommended practices at multiple scales are essential; being conscious that *solely bottom up methods* in natural resources management can also be detrimental (Zdruli et al., 2006). *Good coordination* and continuous interaction between *regional, national and local stakeholders* and across sectors is needed for putting in place *enabling policy and environmental programmes,* including the *required capacity and institutional building.* Included are also the issues of how to deal with *poverty reduction and forced migration* as driving forces in mitigation of land degradation. The endorsement of *harmonised bottom–up and top–down management and promotion of income generating activities,* are necessary to reverse trends and promote sustainable development of the affected areas.

Many policies and programmes have been formulated and are implemented to combat desertification and promote sustainable land management (SLM) under the auspices of the UNCCD and through the initiatives of governments, donor agencies, international, non-governmental organizations and local civil society groups. To date, such efforts have not succeeded in halting or reversing the problem on a large scale. In part, this is because the great magnitude of the problem, compared to the resources employed to address it. However, even where major policy efforts and large investments have been persuaded to promote sustainable land management, these have not always been effective.

One problem undermining the effectiveness of policy and institutional responses to land degradation and desertification is *inadequate diagnosis* of the underlying causes of the problem and insufficient links of prescribed remedies in the contexts where they are being pursued. Farmers and pastoralists may degrade the land on which their livelihoods depend on for many reasons. On one hand there is the lack of awareness of the problem or of a profitable and sustainable option to address it, along with the lack of resources or capability to implement equally profitable options, insufficient incentives to address the problems and implementation of solutions that have off-site effects (e.g., sedimentation caused by soil erosion or contributions to global climate change or biodiversity). On the other hand there is a lack of clear and secure property rights, missing or incomplete markets (e.g., poorly functioning output, land, labour or credit markets), social institutions and preferences (e.g., social norms preventing women from being able to make land improving investments in some countries), and difficulties of attaining effective coordination and collective action where it is necessary to improve land management (e.g., in managing rangelands or improving watershed management). The effectiveness of prescribed policy and programme approaches for the promotion of SLM depends on how well they address such underlying problems.

Combating land degradation and desertification thus should lead to an *increase in rural incomes,* which would allow people to get access to health and other public services. Similarly, efforts to force farmers to use prescribed soil and water conservation measures, build terraces or to plant unsuitable trees have sometimes led to increased land degradation as a result of poor maintenance or destruction of the measures due to farmers' opposition. Often *land use changes* are associated with increased erosion or salinity build up having thus negative effects on soil quality.

Despite of the obvious *importance of interactions between the policy and political environment and land degradation processes,* there is currently little work that has directly addressed how policies, and the political decisions influence and shape global land degradation and desertification. Although some work has shed important light on the role of actors, actor networks and stakeholders' political interests in both alleviating or exacerbating desertification, much work still remains to be done with regard to the specific role that policy and politics play in influencing land use decisions with potential negative effects on degradation processes. What is the impact of political conflicts on resource base degradation? There are many examples showing that increased political and social unrest and instability leads to increased poverty and degradation of natural resources.

Recent evidence suggests that soil erosion risk is decreasing in the EU countries (OECD, 2008). Positive results in increasing soil organic matter content in the US soils are reported as well for the areas under conservation reserve programmes or CRP. However, this should not be considered as the fight with land degradation is over in these regions or in these particular aspects. If we win one battle there are many others to be won. Thus there is the need more than ever for a *Directive for Soil Protection in Europe* and not to use the above positive examples to induce complacency.

There is a need to explore why Governments at various levels, take actions on environmental protection during certain periods and afterwards tend to forget either to follow them up or even to evaluate their effectiveness. This leads to argue about the *existing political, governance, and decision-making processes,* which amongst other things question the influence of international conventions – especially of the UNCCD as a main driving force, or do these actions, derive from more pragmatic reasons? The problem is that people can't wait until a tsunamy, hurricane, earthquake or any other form of natural disasters occurs so that decision makers could take action. History shows that in the 1930s it was not until the clouds of wind-eroded sands from Midwest USA reached the Congress at Capitol Hill in Washington DC, that action was taken to prevent from happening again this "great dust bowl". The USDA Soil Conservation Service (USDA SCS) was set up in that decade as a direct result of those powerful reminders. Since then, due to a concerted campaign based on conservation measures including strip contouring and (more recently) no-till farming promoted by the USDA Natural Resources Conservation Service (former SCS), the situation has been reversed and the Midwest is one of the most productive agricultural regions in the world. This is an extraordinary historical example of reversing land degradation: it is a lesson that must not be ignored!

So, can we win the struggle with land degradation? Yes we can (as it was shown in this conference), but we need to be aware first that sustainable land management can only be assured if all the components of the equation are given equal importance in a holistic and integrated manner. *We need to be also conscious that there are no "ready recipes" for each farm, nation, region, and beyond. They need to be "tailored" according to* specific conditions and we should be also prepared that this fight might be quite long.

Acknowledgment The corresponding author of this chapter wishes to thank Dr. Julian Dumanski, retired World Bank staff and former employee of Agriculture and Agri-Food Canada for kindly reviewing and commenting the chapter.

References

Adams, C.R. and Eswaran, H. (2000). Global land resources in the context of food and environmental security. In: S.P. Gawande et al. (eds.), Advances in Land Resources Management for the 20th Century. Soil Conservation Society of India, New Delhi, pp. 35–50, 655pp.

Bai, Z.G., Dent, D.L., Olsson, L. and Schaepman, M.E. (2008). Proxy global assessment of land degradation. Soil Use and Management 24:223–234.

Cangir, C., Kapur, S., Akca, E., Boyraz, D. and Eswaran, H. (June 2000). An assessment of land resources consumption in relation to land degradation in Turkey. Journal of Soil and Water Conservation 55:253–259.

Conacher, A. and Sala, M. (eds.). (1998). Land Degradation in Mediterranean Environments of the World: Nature and Extend, Causes and Solutions. Wiley, UK.

Eswaran, H., Almaraz, R., Reich, P. and Zdruli., P. (1997). Soil quality and soil productivity in Africa. Journal of Sustainable Agriculture 10:75–94.

Eswaran, H., Beinroth, F.H. and Reich, P. (2003). A global assessment of land quality. In: K. Wiebe (ed.), Land Quality, Agricultural Productivity, and Food Security: Biophysical Processes and Economic Choices at Local, Regional, and Global Levels. Edward Elgar Publishing Ltd, Northampton, MA.

Eswaran, H., Kapur, S., Akca, E., Reich, P., Mahmoodi, S. and Vearasilp, T. (2005). Anthroscapes: A landscape unit for assessment of human impact on land systems. In: J.E. Yang, T.M. Sa and J.J. Kim (eds.), Application of the Emerging Soil Research to the Conservation of Agricultural Ecosystems. The Korean Society of Soil Science and Fertilizers. Seoul, Korea, pp. 175–192.

Global Mechanism. (2006). International workshop on the cost of inaction and opportunities for investments in arid, semi-arid and dry sub humid areas. Rome, 4–5 December 2006. http://www.globalmechanism.org/dynamic/documents/document_file/final_report_en_120407.pdf [last accessed 31 March 2008].

Holm, A.M.S., Cridland, W. and Roderick, M.L. (2003). The use of time integrated NOAA NDVI data and rainfall to assess landscape degradation in the arid shrubland of Western Australia. Remote Sensing of Environment 85:145–158.

Kaiser, J. (11 June 2004). Wounding Earth's fragile skin. Special edition: Soil the final frontier. Science 304:1616–1622.

Kapur, S. and Akça, E. (2004). Environmentally Friendly Indigenous Technologies. Encyclopedia of Soil Science 1. Marcel Dekker, New York. doi: 10.1081/E-ESS 120006648.

Kapur, S., Zdruli, P., Akça, E., Arnoldussen, A., Gencer, O., Kapur, B., Öztürk, A. and Eswaran, H. (2004). Anthroscapes of Turkey: Sites of historic sustainable land management (SLM). Proceedings of the 2nd SCAPE Workshop, 13–16 June 2004. Cinque Terre, Italy, pp. 71–79.

Millennium Ecosystem Assessment (MA) http://www.millenniumassessment.org/en/index.aspx [last accessed 2 April 2009].
Nachtergaele, F. (2003). Land degradation assessment in drylands: The LADA project. In: L. Montanarella and R.J.A. Jones (eds.), Land Degradation, EUR 20688 EN. Office for Official Publications of the European Communities, Luxembourg, 324pp.
OECD. (2008). Environmental Performance of Agriculture in OECD Countries Since 1990. Paris, France.
Oldeman, L.R., Hakkeling, R.T.A. and Sombroek, W.G. (1991). World Map of the Status of Human-Induced Soil Degradation. 2nd edn. ISRIC, Wageningen.
Pender, J. and Ndjeunga, J. (2008). Assessing Impacts of Sustainable Land Management Programs on Land Management and Poverty in Niger. International Food Policy Research Institute. Mimeo, Washington DC.
Pimentel, D., Harvey, C., Resosudarmo, P., Siclair, K., Kurz, D., McNair, M., Crist, S., Shpritz, L., Fitton, L., Saffouri, R. and Blair, R. (24 February 1995). Environmental and economic costs of soil erosion and conservation benefits. Science 267:1117–1123.
Prince, S.D., Wessels, K.J., Tucker, C.J. and Nicholson, S.E. (2007). Desertification in the Sahel: A reinterpretation of reinterpretation. Global Change Biology 13:1308–1313.
Reynolds, J.(2008). From DDP to Atlas: A possible road. Presentation and personal communication at the 1st meeting on defining a roadmap for the development of a new World Atlas of Desertification, JRC – IES, 3–5 December 2008 Ispra, Italy.
Reynolds, J. and Stafford Smith, M. (eds.). (2002). Global Desertification: Do Humans Create Deserts? (Dahlem Workshop Report No. 88). Dahlem University Press, Berlin, 438pp.
Reynolds, J., Stafford Smith, M., Lambin, E., Turner, B., Mortimore, M., Batterbury, S., Downing, T., Dowlatabadi, H., Fernandez, R., Herrick, J., HuberSannwald, E., Jiang, H., Leemans, R., Lynam, T., Maestre, F., Ayarza, M. and Walker, B. (11 May 2007). Global desertification: Building a science for dryland development. Science 316(5826):847–851.
Rubio, J.L., Morgan, R.P.C., Asins, S. and Andreu, V. (eds.). (2002). Man and Soil at the Third Millennium. 2 Volumes. Geoforma Ediciones/Centro de Investigaciones sobre Desertificacion, Logrono.
Ryan, J. (ed.). (2002). Desert and dryland development: Challenges and potential in the new millennium. Proceedings of the 6th International Conference on the Development of Drylands, 22–27 August 1999. ICARDA Aleppo, Syria, Cairo, Egypt, xiv+655pp.
Safriel, U. (2007). The assessment of global trends in land degradation. In: R.V.K. Sivakumar and N. Ndiang'ui (eds.), Climate and Land Degradation. International Workshop on Climate and Land Degradation. Springer-Verlag, Berlin Heidelberg, pp. 2–36.
Sonneveld, B.G.J.S. and Dent, D. (2007). How good is GLASOD? Journal of Environmental Management. doi: 10.1016/j.jenvman.2007.09.008.
Stocking, M. (21 November 2003). Tropical soils and food security: The next 50 years. Science 302:1356–1359.
Stocking, M. and Murnaghan, N. (2001). Handbook for Field Assessment of Land Degradation. Earthscan, London.
Thomas, R. (2008). Critical review: 10th Anniversary review: Addressing land degradation and climate change in dryland agroecosystems through sustainable land management. Journal of Environmental Monitoring 10:595–603. doi: 10.1039/b801649f.
UNCCD CRIC7. (2009). Report of the seventh session of the Committee for the Review of the Implementation of the Convention. Istanbul, Turkey 3–14 November 2008. INCCD/CRIC(7)/5. http://www.unccd.int/cop/officialdocs/cric7/pdf/cric5-eng.pdf [last accessed 1 April 2009].
WOCAT. (2007). Where the Land is Greener – Case Studies and Analysis of Soil and Water Conservation Initiatives Worldwide. In: W. Critchley and H. Lineger (eds.). WOCAT, Switzerland.
Wiebe, K. (2003). Land quality, agricultural productivity and food security. In: K. Wiebe (ed.), Land Quality, Agricultural Productivity and Food Security: Biophysical Processes and Economic Choices at Local, Regional and Global Levels. Edward Elgar Publishing, Lyme, NH.

Zdruli, P. and Costantini, E. (eds.). (2008). Moving ahead from assessments to actions: Could we win the struggle with land degradation? Book of Abstracts. 5th International Conference on Land Degradation, Valenzano, Bari, Italy 18–22 September 2008. 2 Volumes, 976pp. CIHEAM/IUSS/ISSS/EC JRC.

Zdruli, P., Jones, R. and Montanarella, L. (2004). Organic matter in the soils of Southern Europe. European Commission, Joint Research Centre – European Soil Bureau; Expert Report prepared for DG ENV/E3 Brussels, EUR 21083 EN 16pp. Office for Official Publications of the European Communities, Luxembourg.

Zdruli, P., Lacirignola, C., Lamaddalena, N. and Trisorio Liuzzi, G. (2007). The EU-funded MEDCOASTLAND thematic network and its findings in combating land degradation in the Mediterranean region. In: R.V.K. Sivakumar and N. Ndiang'ui (eds.), Climate and Land Degradation. International Workshop on Climate and Land Degradation. Springer-Verlag, Berlin Heidelberg, pp. 422–434.

Zdruli, P., Liuzzi, G.T., Akça, E., Donma, S., Doğru, C., Kapur, B., Serdem, M., Pekel, M., Türkoğlu, G., Çelmeoğlu, N., Durak, A. and Kapur, S. (2006). From politics to policies: The must for the implementation of sustainable land management within the Mediterranean context. In: P. Zdruli and G. Trisorio Liuzzi (eds.), Managing Natural Resources Through Implementation of Sustainable Policies. MEDCOASTLAND publication 5, IAM Bari, Italy, pp. 167–178.

Chapter 2
Moving Ahead from Assessments to Actions: Could We Win the Struggle with Soil Degradation in Europe?

Luca Montanarella

Abstract The EU Thematic Strategy for Soil Protection has identified eight major threats to European soils. They include erosion, organic matter decline, compaction, salinisation, landslides, contamination, sealing and biodiversity decline. Yet a Framework Directive for Soil Protection as a legally binding document for all the EU member states has to become reality. This chapter emphasises the importance of soil functions and services in support of Europe's agricultural productivity and environmental sustainability. It draws conclusions also on the elaborated available soil legislation in the EU and explores the linkages between policy measures, implied agricultural soil conservation practices and soil degradation processes. It also emphasise the needs for additional soil research and awareness activities throughout Europe to further support soil conservation.

Keywords EU · Soil thematic strategy · Soil threats · Legislation · Research · Soil awareness

2.1 Introduction

The adoption of the EU Thematic Strategy for Soil Protection by the European Commission on 22 September, 2006 has given formal recognition of the severity of the soil and land degradation processes within the European Union and its bordering countries. The Strategy includes a communication (European Commission, COM(2006)231) outlining the strategy, a proposal for framework directive for soil protection (European Commission, COM(2006)232) as a legally binding instrument and an extended impact assessment (European Commission, SEC(2006)620) that has quantified soil degradation in Europe, both in environmental and economic terms.

L. Montanarella (✉)
Institute for Environment and Sustainability, European Commission, Joint Research Centre, I-21020 Ispra, VA, Italy
e-mail: luca.montanarella@jrc.it

This impact assessment is based mainly, but not exclusively, on reports (Van-Camp et al., 2004a, b, c, d, e, f) by the Joint Research Centre (JRC) of the Commission and the Working Groups set up to assist the Commission, and reports carried out for the Commission in assessing the economic impacts of soil degradation and economic, environmental and social impacts of different measures to prevent soil degradation.

Available information suggests that, over recent decades, there has been a significant increase in soil degradation processes, and there is evidence that these processes will further increase if no action is taken. Soil degradation processes are driven or exacerbated by human activity. Climate change, together with individual extreme weather events, which are becoming more frequent, will also have negative effects on soil.

Soil degradation processes occurring in the European Union include erosion, organic matter decline, compaction, salinisation, landslides, contamination, sealing and biodiversity decline.

The strategy proposed by the European Commission is based on four pillars: A binding legislative instrument (the proposed soil framework directive), integration of soil protection into existing legal instruments at European level, new and enhanced research activities related to soil protection and a renewed effort in awareness raising initiatives.

2.2 Framework Legislation for Soil Protection

The European Commission has proposed a draft soil framework directive as one of the essential elements of the soil thematic strategy. The proposed directive contains a large number of innovative approaches to soil protection that, if fully implemented, would lead to substantial reversal of the current negative trend in soil degradation in Europe.

At the core of the directive is the definition of soil as the full layer of unconsolidated materials from the surface down till the bedrock. Such a definition largely exceeds the traditional "pedological" definition of soils (WRB, 2006): .... *any material within 2 m from the Earth's surface that is in contact with the atmosphere, with the exclusion of living organisms, areas with continuous ice not covered by other material, and water bodies deeper than 2 m.* The directive therefore aims to comprehensive soil protection beyond the traditional views relating soil protection strictly to the protection of its agricultural function. Indeed the directive fully recognizes the multi-functionality of soils and aims towards the protection of these functions more then the actual protection of the soil per se.

The main functions recognized by the directive are:

(a) biomass production, including in agriculture and forestry;
(b) storing, filtering and transforming nutrients, substances and water;
(c) biodiversity pool, such as habitats, species and genes;
(d) physical and cultural environment for humans and human activities;

(e) source of raw materials;
(f) acting as carbon pool;
(g) archive of geological and archaeological heritage.

The list of functions identified within the directive do not precisely match the traditional six functions recognized by the soil science community (Blum, 1993), but have been selected as well for their relevance to stakeholders and decision makers in general. In particular the function of soils as major carbon pool has been singled out as a very crucial function within the current climate change debate (Lal, 2000).

In order to achieve the protection of the above functions of soils in Europe, an approach by priority areas is proposed: Member States are required to delineate priority areas for measures to combat the various soil threats as identified by the directive in Annex I: Erosion, loss of organic matter, compaction, salinization, landslides. A separate approach is proposed for soil contamination, addressing the issue of contaminated sites, their inventory and successive restoration measures. The criteria for the delineation of priority areas have been proposed in annex I to the directive and have been derived from the results of a specific working group of the European Soil Bureau Network (European Soil Bureau, 2006). The proposal by the European Commission received already positive opinions from the Committee of the Regions, the Economic and Social Committee and the European Parliament that adopted a favourable opinion both of the strategy and the proposed directive.

The Council still has not achieved a common position of the 27 EU Member States, despite substantial efforts by the Portuguese and French Presidencies to reach consensus. Five EU Member States have formed a blocking minority for different reasons: Germany, Austria and The Netherlands for reasons of subsidiarity, claiming the lack of competence of the European Union in legislating about an issue like soils, which are to be considered of strictly local, and therefore National competence. The United Kingdom and France have objections of proportionality and costs, claiming that the actual economic benefits of the proposed directive would not out weight the cost of implementation. This in clear contradiction to the extended impact assessment that has documented that the total costs of soil degradation just for erosion, organic matter decline, salinization, landslides and contamination, on the basis of available data, would be up to €38 billion annually for EU25.

Despite the current difficulties in the acceptance by the EU Member States of the soil framework directive, soil protection activities are increasing in Europe in the framework of the soil thematic strategy thanks to the initiatives within the other three pillars of the strategy: integration, research and awareness raising.

2.3 Integration

Several existing policies and legislative instruments at EU level already allow achieving extensive soil protection targets. Among them, certainly the Common Agricultural Policy (CAP) can be an instrument to achieve improved soil conditions in agricultural land. In the framework of the Cardiff-Process (European Commission, 1998, COM/98/0333), environmental objectives are to be integrated into EU sectoral policies, including the Common Agricultural Policy (CAP). Consequently, the protection of the environment is an important objective of the CAP. The CAP comprises two principal forms of budgetary expenditure: market support, known as Pillar One, and a range of payments for rural development measures known as Pillar Two.

Cross-compliance, a horizontal tool for both pillars and compulsory since the implementation of the CAP reform 2003 (Council Regulation (EC) No 1782/2003), plays an important role in soil protection, conservation and/or improvement. Under cross-compliance rules, the receipt of the single farm payment and payments for eight rural development measures under axis 2 is conditional on a farmer's compliance with a set of standards.

First, enforcement of implementation and control of EU environmental directives were promoted through compliance with the Statutory Management Requirements (SMR: Annex III). Second, the Good Agricultural and Environmental Conditions (GAEC: Annex IV) were introduced to prevent land abandonment that could result from the decoupling of direct aids from production. GAEC specifically include protection against soil erosion, maintenance or improvement of soil organic matter, and maintenance of a good soil structure. The fact that GAEC are defined at national level enables Member States to address soil degradation processes flexibly according to national priorities and local needs. Some Member States used GAEC to compensate for gaps in their existing national legislation on soil protection, while other Member States already had a legislative basis in place and merely adopted it for cross-compliance. This has resulted in the situation that national designs of GAEC are highly variable in scope and detail of describing measures.

Within the second pillar of the CAP, a wide range of measures can be supported under Council Regulation (EC) No 1698/2005. Member States and regions are obliged to spread their rural development funding across three thematic axes, (1) competitiveness; (2) environment and land management; and (3) economic diversity and quality of life, with minimum spending thresholds applied per axis (i.e. 10% for axes 1 and 3, and 25% for axis 2). "LEADER" is a horizontal axis (minimum spending of 5%; 2.5% in the new Member States) complementing the three thematic axes. Axis 2 measures are of particular interest within the scope of soil protection, since both environmental improvement and preservation of the countryside and landscape encompass soil degradation processes. Regarding this axis, Member States are encouraged to focus on key actions; of which some explicitly refer to soil, such as the delivery of environmental services, in particular water and soil resources; or stressing the role of soils in adapting to climate change.

Currently the most important pieces of environmental EU legislation with respect to soil quality are the Nitrates Directive (91/676/EEC) and the Water Framework Directive 2000/60/EC). Others also have beneficial effects but these are smaller, as a result of the specificity of their objectives.

- The Water Framework Directive 2000/60/EC), including its daughter directives such as the new Groundwater Directive has the objective to prevent and reduce pollution, promote sustainable water use, protect the aquatic environment, improve the status of aquatic ecosystems and mitigate the effects of floods and droughts. Because of the link between water and soil quality, measures taken under the Water Framework Directive may contribute to reducing soil contamination, with expected positive side-effects on soil biodiversity. Soil degradation processes (especially erosion and local and diffuse soil contamination) were identified as impacting on water quality, rather than being positively affected by improved water quality.
- The Nitrates Directive (91/676/EEC) is designed to protect the Community's waters against nitrates from agricultural sources, one of the main causes of water pollution from diffuse sources, and is thus primarily targeting water quality. However, it is expected to have positive effects on local and diffuse soil pollution by nitrates (and phosphates). Also in particular cases, soil compaction might be positively affected, as fertiliser spreading is banned in the winter period (with prevailing wet or water-saturated soils).
- Avoiding pollution or deterioration of agricultural soils is regarded as an implicit precondition for the protection or recovery of habitats under the Birds Directive 79/409/EEC) or the Habitats Directive (92/43/EEC). Soil biodiversity is likely to benefit from the (extensive) farm practices by the implementation of these directives. Positive effects on (local and) diffuse soil contamination are expected too. A coherent European ecological network known as "Natura 2000" is integrating the protected areas of both directives.
- The Sewage Sludge Directive (86/278/EEC) addresses the decline of organic matter and soil contamination, through regulating the use of sewage sludge on agricultural land, while encouraging its correct use (through the application of limits on the concentrations of certain substances, or outright bans where needed).
- The Plant Protection Products Directive (91/414/EEC) concerns the authorisation, placing on the market, and use and control within the Community of plant protection products in commercial use. It will be replaced with a Regulation once the Commission proposal (COM (2006) 388) is adopted by the Council and the European Parliament. This Regulation will be complemented by a Thematic Strategy on the Sustainable Use of Pesticides and its corresponding legislative proposal for a Framework Directive (COM(2006) 373 final) which address risks resulting from the actual use of pesticides (mainly plant protection products and biocides). Both pieces of legislation are expected to have repercussions for soil contamination and soil biodiversity.

The literature review and the policy implementation survey in the EU-27 show that there is a large spectrum of policy measures favouring soil protection throughout the EU (EC-JRC, 2008). These measures are implemented at the national and regional level and take account of the local conditions, and in doing so use the flexibility provided within the legislative EU frame. It is necessary to explore the linkages between available policy measures, implied agricultural soil conservation practice and soil degradation processes. This link can be either two-stage, by supporting or requiring a specific farming practice which positively affects soil quality, or one-stage with a direct link to soil quality. Especially with regard to voluntary incentive-based measures, it is important to monitor the uptake, as this provides an indication of their relevance to the social, economic and natural environment of farms and to their likely impact. Increasing awareness and advice have an important effect on levels of uptake and compliance with prescriptions.

2.4 Research

Research is the third pillar of the soil strategy. Soil research activities have been largely neglected in previous times by major research funding agencies (Hartemink, 2008). Indeed the large knowledge gaps still existing have been extensively recognized by the EU soil strategy and have been indicated as one of the main reasons for the lack of policy action in achieving better soil protection in Europe. Particularly the area of soil biology has been singled out as a topic for future research priorities at EU level.

Recently first signals of a reversal of the negative trend in soil science related research could be detected (Hartemink and Mc Bratney, 2008). A substantial increase of publication rates as well as of research funding dedicated to soil science could be observed.

At European level, important new research initiatives have been funded within the 7th Framework Programme for Research and Development (FP7) starting from 2007. Projects like DIGISOIL, ISOIL, eSOTER, SOILSERVICE and others will substantially contribute to filling the gap identified by the soil strategy. Direct research actions of the European Commission through the Joint Research Centre contribute as well, particularly developing the European Soil Data Centre (ESDAC) and further developing the European and Global soil databases and information systems.

2.5 Awareness Raising

The last of the four pillars of the soil thematic strategy is probably the most relevant: Without a substantial awareness raising effort at all levels it will be rather difficult to make substantial progress in soil protection in Europe. Soils are still largely neglected by the public opinion and are usually seen just a surface for building housing and infrastructure or as a dumping place for waste and other materials. The active role that soils play in the ecosystem in providing many key services to us, like clean water, healthy food, biodiversity, pollutants storage and filtering, raw materials, etc.... is largely neglected and hardly recognized by the average

European citizen. It is therefore of the highest importance to increase the investments and efforts towards increased awareness among the European citizens of the importance of soils and of the necessity of protecting this limited resource for future generations.

Educational activities in schools at all level should be further encouraged and increased, as well as the organization of media events and the compilation of communication materials. The European Commission, through its Joint Research Centre, has already launched several initiatives in order to raise awareness on soil protection needs. The European Summer School for Soil Survey, as well as the compilation of the Soil Atlas of Europe (Jones et al., 2005), are among successful initiatives to bring soils closer to the wider public. Still a lot needs to be done in order to get the awareness of the need of soil protections at levels comparable to the protection of air and water.

2.6 Conclusions

Can we win the struggle with soil degradation in Europe? In the previous paragraphs we have tried to highlight the actions proposed by the European Commission as well as their status of implementation. The lack of actual monitoring data for soils across Europe prevents us from providing hard evidence of current trends; nevertheless first signals of positive developments can be detected.

The recently published report by OECD (2008) reports that *overall for the OECD there has been some improvement or stability in soil erosion, from both water and wind. An increase in the share of agricultural land within the tolerable erosion risk class has been accompanied by a reduction in areas at moderate to severe erosion risk*. No clear explanation of the causes of such positive trend in OECD countries is provided; nevertheless these findings are also confirmed by recent results of the JRC, confirming that the overall trend in Europe is of a constant decline of soil erosion rates, mainly due to the growth of forest areas within the European Union, thus achieving better protection of the soils from water erosion processes.

Unfortunately only limited soil monitoring activities exist (Arrouays et al., 2008) within Europe, making any assessment of time trends of the various soil threats practically impossible at EU scale. Certainly the full implementation of the EU Soil Thematic Strategy, including the proposed Soil Framework Directive, would allow winning the struggle against land degradation in Europe. First signals of a reversed trend are there and an increased awareness of the importance of soils as an asset for future generations may create sufficient political consensus for reaching the ultimate goal of a sustainable soil management for Europe.

References

Arrouays, D., Morvan, X., Saby, N.P.A., Le Bas, C., Bellamy, P.H., Berényi Üveges, J., Freudenschuß, A., Jones, A.R., Jones, R.J.A., Kibblewhite, M.G., Simota, C., Verdoodt, A. and Verheijen, F.G.A. (eds.). (2008). Environmental Assessment of Soil for Monitoring: Volume

IIa Inventory & Monitoring, EUR 23490 EN/2A. Office for the Official Publications of the European Communities, Luxembourg, 180pp.

Blum, W.E.H. (1993). Soil and environment volume I. Integrated soil and sediment research: A basis for proper protection. In: H.J.P. Eijsackers and T. Hamers (eds.), Soil Protection Concept of the Council of Europe and Integrated Soil Research. Kluwer Academic Publisher, Dordrecht, pp. 37–47.

Council of the European Union. (1986). Council Directive 86/278/EEC of 12 June 1986 on the protection of the environment, and in particular of the soil, when sewage sludge is used in agriculture. Official Journal L 181, 04/07/1986.

Council of the European Union. (1991). Council Directive 91/414/EEC of 15 July 1991 concerning the placing of plant protection products on the market. Official Journal L 230, 19/08/1991.

Council of the European Union. (1991). Council Directive 91/676/EEC of 12 December 1991 concerning the protection of waters against pollution caused by nitrates from agricultural sources. Official Journal L 375, 31/12/1991.

Council of the European Union. (1992). Council Directive 92/43/EEC of 21 May 1992 on the conservation of natural habitats and of wild fauna and flora. Official Journal L 206, 22/7/1992.

Council of the European Union. (2000). Directive 2000/60/EC of the European Parliament and of the council of 23 October 2000 establishing a framework for community action in the field of water policy. Official Journal L 327, 22/12/2000.

Council of the European Union. (2003). Council Regulation 1782/2003/EC of 29 September 2003 establishing common rules for direct support schemes under the common agricultural policy and establishing certain support schemes for farmers and amending Regulations 2019/93/EEC, 1452/2001/EC, 1453/2001/EC, 1454/2001/EC, 1868/94/EC, 1251/1999/EC, 1254/1999/EC, 1673/2000/EC, 2358/71/EEC and 2529/2001/EC.

Council of the European Union. (2005). Council Regulation 1698/2005/EC of 20 September 2005 on support for rural development by the European Agricultural Fund for Rural Development (EAFRD). Official Journal L 277(1), 21/10/2005.

EC-JRC (European Commission, Joint Research Center). (2008). Sustainable Agriculture and Soil Conservation (SoCo Project). Interim Report, WP1 – Stock-taking of the current situation within an EU-wide perspective. unpublished report.

European Commission. (1998). Communication "Partnership for Integration" (COM(98)0333 C4-0410/98). Official Journal C 359, 23/11/1998 P. 0091.

European Commission. (2006). Proposal for a Directive of the European Parliament and of the Council establishing a framework for the protection of soil and amending Directive 2004/35/EC. COM (2006) 232.

European Commission. (2006). Report from the Commission on the Implementation of Directive 79/409/EEC on The Conservation of Wild Birds, COM (2006) 164 final.

European Commission. (2006). Thematic Strategy for Soil Protection, COM (2006) 231.

European Commission. (2006). Impact Assessment of the Thematic Strategy on Soil Protection, SEC (2006) 1165.

European Soil Bureau Research Report. (2006). European Soil Bureau Research Report No. 20, EUR 22185 EN. Office for Official Publications of the European Communities, Luxembourg, 94pp.

Hartemink, A.E. (December 2008). Soils are back on the global agenda. Soil Use and Management 24:327–330.

Hartemink, A.E. and McBratney, A. (2008). A soil science renaissance. Geoderma 148: 123–129.

IUSS Working Group WRB. (2006). World reference base for soil resources 2006.World Soil Resources Reports No. 103. FAO, Rome.

Jones, A., Montanarella, L. and Jones, R.J.A. (eds.). (2005). "Soil Atlas of Europe", European Soil Bureau Network, European Commission. Office for Official Publications of the European Communities, L-2995 Luxembourg, 128pp.

Lal, R. (2000). Soil Conversion and Restoration to Sequester Carbon and Mitigate the Greenhouse Effect. III International Congress European Society for Soil Conservation, Valencia.

OECD. (2008). Environmental Performance of Agriculture in OECD Countries Since 1990. Paris, France.

Van-Camp, L., Bujarrabal, B., Gentile, A.R., Jones, R.J.A., Montanarella, L., Olazabal, C. and Selvaradjou, S.-K. (2004a). Reports of the Technical Working Groups established under the Thematic Strategy for Soil Protection. Volume I Introduction and Executive Summary, EUR 21319 EN/1. Office for Official Publications of the European Communities, Luxembourg, 126pp.

Van-Camp, L., Bujarrabal, B., Gentile, A.R., Jones, R.J.A., Montanarella, L., Olazabal, C. and Selvaradjou, S.-K. (2004b). Reports of the Technical Working Groups established under the Thematic Strategy for Soil Protection. Volume II Erosion, EUR 21319 EN/2. Office for Official Publications of the European Communities, Luxembourg, pp. 127–309.

Van-Camp, L., Bujarrabal, B., Gentile, A.R., Jones, R.J.A., Montanarella, L., Olazabal, C. and Selvaradjou, S.-K. (2004c). Reports of the Technical Working Groups established under the Thematic Strategy for Soil Protection. Volume III Organic Matter, EUR 21319 EN/3. Office for Official Publications of the European Communities, Luxembourg, pp. 311–496.

Van-Camp, L., Bujarrabal, B., Gentile, A.R., Jones, R.J.A., Montanarella, L., Olazabal, C. and Selvaradjou, S.-K. (2004d). Reports of the Technical Working Groups established under the Thematic Strategy for Soil Protection. Volume IV Contamination and Land Management, EUR 21319 EN/4. Office for Official Publications of the European Communities, Luxembourg, pp. 497–621.

Van-Camp, L., Bujarrabal, B., Gentile, A.R., Jones, R.J.A., Montanarella, L., Olazabal, C. and Selvaradjou, S.-K. (2004e). Reports of the Technical Working Groups established under the Thematic Strategy for Soil Protection. Volume V Monitoring, EUR 21319 EN/5. Office for Official Publications of the European Communities, Luxembourg, pp. 653–718.

Van-Camp, L., Bujarrabal, B., Gentile, A.R., Jones, R.J.A., Montanarella, L., Olazabal, C. and Selvaradjou, S.-K. (2004f). Reports of the Technical Working Groups Established Under the Thematic Strategy for Soil Protection. Volume VI Research, Sealing and Cross-cutting Issues, EUR 21319 EN/6. Office for Official Publications of the European Communities, Luxembourg, pp. 719–872.

Chapter 3
Moving Ahead from Assessments to Actions by Using Harmonized Risk Assessment Methodologies for Soil Degradation

C.L. van Beek, T. Tóth, A. Hagyo, G. Tóth, L. Recatalá Boix, C. Añó Vidal, J.P. Malet, O. Maquire, J.H.H. van den Akker, S.E.A.T.M. van der Zee, S. Verzandvoort, C. Simota, P.J. Kuikman, and O. Oenema

Abstract Almost all developed countries use risk assessment methodologies (RAMs) for the evaluation of risks related to soil degradation, viz. soil organic matter decline, erosion, landslides, salinization and/or compaction. However and for various reasons, seldom the use of such RAMs seldom results in actual measures to combat soil degradation in practice. In this study the current status of RAMs in EU-27 was evaluated and factors hampering the implementation of action plans were explored. To do so we used a so-called risk assessment chain, which describes the five successive steps of any risk assessment for soil threats viz., (1) notion of the threat, (2) data collection, (3) data processing, (4) risk interpretation and (5) risk perception. Based on this assessment we identified three factors that hampered the execution of measures to combat soil degradation following the application of soil RAMs:

- Many RAMs are incomplete and focus on the first steps of the risk assessment chain, and ignore the decision for action to combat land degradation;
- Member states preferably monitor soil threats that are clearly present (e.g. landslides) and may overlook "slow killers" like compaction and soil organic matter decline.
- Different RAMs for the same threat provide different results for the same exposure. This undermines the scientific credibility of the RAMs and the plausibility of the severity of the threat and may result in loss of commitment to take remedial actions.

These factors may be overcome by harmonizing RAMs, i.e. by making results comparable and/or compatible. Therefore, complete RAMs, i.e. covering all aspects of the risk assessment chain, should be developed for each threat and different RAMs for the same threat should be made intercomparable, i.e. yield similar risk perceptions for a certain exposure to a threat. We recommend implementing a Tiered

---

C.L. van Beek (✉)
Alterra, Soil Science Center, 6700 AA Wageningen, The Netherlands
e-mail: christy.vanbeek@wur.nl

methodology, where the Tier 1 method is a standardized and uniformly applicable method across EU-27, at a relatively low spatial resolution and is used to identify areas at risk. The Tier 2 method is a regional-specific and more detailed assessment of the risk in the areas identified by the Tier 1 method, where the Tier 2 method is harmonized to the Tier 1 method. We urge to initiate this process timely considering that as long as different unharmonized soil RAMs are used simultaneously, the implementation of remedial measures will be frustrated by ambiguous results.

Keywords Erosion · Compaction · Landslides · Soil organic matter decline · Salinization

3.1 Introduction

In many countries risk assessment methodologies (RAMs) are used for the evaluation of risk related to soil degradation, e.g. soil organic matter decline, salinization, compaction, erosion and landslides. These soil RAMs generally consist of five successive steps that are visualized in Fig. 3.1. The notion of the threat refers to the definition of threat. Data collection refers to data derived from field measurements, remote sensing images and/or data statistics on land use, climate, etc. Data processing involves the quantification of a rate or state of the soil threat, using simulation modelling, empirical modelling, factorial assessment or expert evaluation of the data. Data interpretation refers to the comparison of the rate or state of the soil threat with previously defined threshold values. In the final step, the risk perception step and the risk of the soil threat is assessed in terms of the sense of urgency of actions and remedial measures. Based on this final step legal authorities may decide to adopt action plans to halt soil degradation.

Fig. 3.1 The risk assessment chain, starting with the definition (notion) of the soil threat (*below*) and ending with risk perception (*top*)

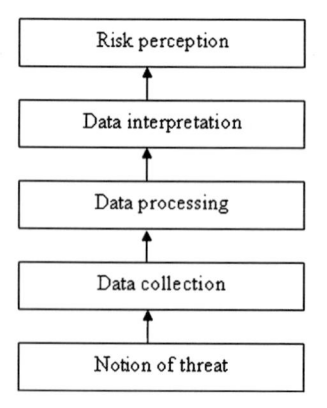

However, currently many different RAMs are used and the use of different RAMs for one and the same threat puts the plausibility of the results at stake. Kamrin (1997) reported an interesting case, which refers to fish consumption in the USA. In the USA several federal states bordering the Great Lakes used different RAMs to evaluate risks related to consumption of sport fish. In this case the use of different RAMs resulted in conflicting advices about consumption, notwithstanding that it concerns the same fish. Ultimately the use of different RAMs resulted in ambiguous interpretation of risk exposures and loss of public support to policy. This may hold equally well for policies meant to decrease soil degradation.

At present, there is a non-binding soil thematic strategy at force in EU-27 (EU, 2006). In the future a soil thematic directive is foreseen, which will be based in part on the soil thematic strategy, and which may result in more obligations towards the protection of soils in EU Member States. Provided that the soil framework directive does come into practice, Member States will be obliged to assign priority areas for all soil threats within 3 years following the ratification of the Directive. For the priority areas action plans to mitigate soil degradation have to be developed and executed. The designation of priority areas will likely be performed using RAMs. The current use of different RAMs is detrimental for reasons of incompatibility of results and hence possibly subjective identification of priority areas. Ultimately, the use of different RAMs may have consequences for equal market access throughout the EU-27 when these RAMs are used to define measures that will restrict certain economic activities.

The use of different soil RAMs is not necessarily detrimental, as long as results are comparable and/or compatible. This is the objective of harmonization, although the term harmonization is subject to quite some discussion as there are different perceptions and interpretations. Here, harmonization is defined as making results compatible or comparable, hence consistent, and thereby minimizes the differences between standards or measures with similar scope. Harmonization can be applied at all levels of the risk assessment chain. The most direct way of harmonization is by making risk perceptions of different RAMs comparable, i.e. harmonization at the highest possible level of the risk assessment chain. This could be achieved using conversion factors to calculate the outcome of one RAM into the other. However, such an approach is impossible when the notions of the soil threat differ and cause-effect relationships are non-linear. In these cases each level of the risk assessment chain has to be harmonized. An extreme form of harmonization is standardization in which all procedures and methodologies at all levels of the risk assessment chain are prescribed. The concepts of harmonization and standardization used in this study are schematically visualized in Fig. 3.2.

The current use of unharmonized RAMs for soil threats may result in different, and possibly conflicting, outcomes with regard to the severity of a soil threat. This puts the plausibility of soil RAMs at stake and may have consequences for the implementation of actions plans to combat soil degradation. Therefore, an inventory was made on the use of different RAMs for salinization, erosion, landslides,

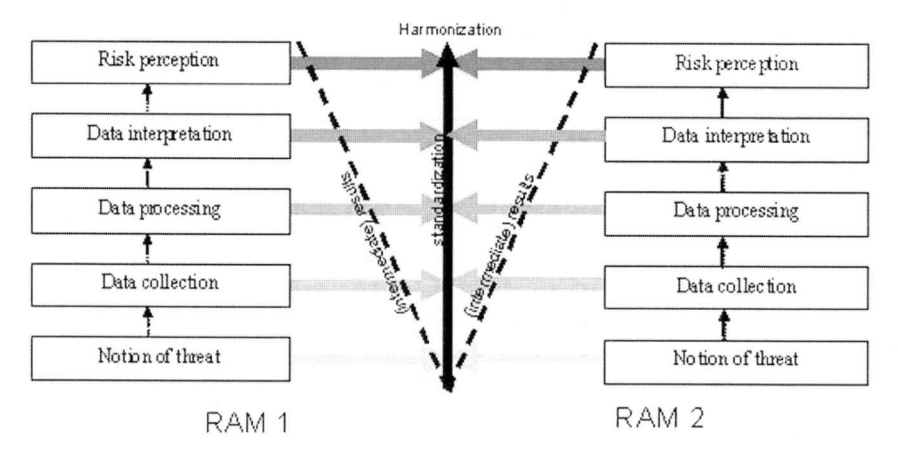

Fig. 3.2 Conceptual visualization of the meanings of harmonization and standardization of RAMs as used in this chapter. The *triangle* in between the two risk assessment chains represents the increasing divergence of (intermediate) results of two RAMs, from bottom to top. Standardization (*bold vertical arrow*) applies to prescribed procedures and activities at each level of the risk assessment chain, whereas harmonization (*horizontal arrows*) implies the use of conversion factors at the highest possible level (most direct way, indicated by *dark colour*) and possibly at other levels. Ultimately, both standardization and harmonization should result in comparable risk perceptions

compaction and soil organic matter decline in EU-27 and options for harmonization were explored so as to pave the way for consistent actions to halt soil degradation in EU-27.

3.2 Materials and Methods

To obtain an overview of RAMs currently used across EU-27, questionnaires were sent out to scientists and policy makers in all Member States of the EU-27. We made six different questionnaires, one for each soil threat (called "thematic questionnaire") and a general policy questionnaire. Each questionnaire was sent to national contact persons, or in case of decentralized governments, to regional contact persons. The policy questionnaire focused on the decision factors of policy makers to adopt and use RAMs (or not) in national or regional legislations. The policy questionnaire also inquired about the position of RAMs for each threat within the institutional structures and about the perception of urgency of the different threats. Thematic questionnaires focused on the scientific and technical details of the RAMs related to the steps in the risk assessment chain of Fig. 3.1. Questions referred to different fields of discipline, viz. policy relevance, responsiveness, analytical soundness, data availability and measurability, ease of interpretation and cost-effectiveness of the RAMs. More details about the questionnaires, the distribution of the questionnaires and the database can be downloaded from www.ramsoil.eu. Scientific literature reviews and web searches on the implementation on actions plans to combat soil degradation completed the assessment of the current status of soil RAMs in EU-27.

Additionally, two case studies were performed on the diversity of outcomes when using different RAMs for the same location. The case studies concerned (i) soil erosion in Romania and (ii) soil compaction in The Netherlands. For soil erosion in Romania two RAMs were used: the SIDASS-WEPP approach (Simota et al., 2005) and the PESERA approach (Kirkby et al., 2008). The following scenarios were used:

- PESERA modelling (JRC simulations) using the raster with 1 km grid for soil properties coming from the EU-soilGIS scale 1:1,000,000, raster from Corine Land Cover with 1 km grid and DEM with the grid space of 1 km.
- SIDASS modelling (WEPP methodology) with slope based on Slope index linked with each polygon in soil map of Europe at the scale of 1:1,000,000.

For soil compaction in The Netherlands the methods described by Jones et al. (2003) and the SOCOMO model (van den Akker, 2004) were compared. The "Jones" method uses FAO–UNESCO soil texture classes and pedotransfer functions for estimating subsoil densities. The subsoil densities are subsequently used to estimate the current packing density, which is considered as an indicator for susceptibility for soil compaction as shown in Table 3.1.

The SOCOMO model uses data from national soil maps and calculates the allowable wheel load based on texture classes as shown in Table 3.2.

More details about the case studies can be found in Tóth et al. (2009) and Hoogland and van den Akker (2009).

Table 3.1 Susceptibility to compaction depending on soil texture and packing density (After Spoor et al., 2003)

Texture class	Packing density		
	Low <1.4 g cm^{-3}	Medium 1.4–1.75 g cm^{-3}	High > 1.75 g cm^{-3}
Course	Very high	High	Moderate
Medium (<18% clay)	Very high	High	Moderate
Medium (>18% clay)	High	Moderate	Low
Medium fine (<18% clay)	Very high	High	Moderate
Medium fine (>18% clay)	High	Moderate	Low
Fine	Moderate	Low	Low
Very fine	Moderate	Low	Low
Organic	Very high	High	–

Table 3.2 Soil mechanical properties and allowable wheel loads of a Terra Tire 73 × 44.00 – 32 at pF 2.5 dependant on soil texture classes. F_Pv is the allowable wheel load based on compression strength (SS); F-MC is allowable wheel load based on shear strength (Mohr Coulomb equation with cohesion C and angle of internal friction ϕ)

Texture	Clay content	C (kPa)	ϕ (°)	SS (kPa)	Depth (cm)	F_Pv (kN)	F-MC (kN)
Course sand	< 8	10	32	240	32	125	29
Sand	< 8	12	28	198	32	103	30
Sandy loam	< 8	10	32	122	32	62	29
Sandy loam	8–18	10	32	140	27	66	29
Clay loam	18–25	14	31	79	27	36	
Light clay	18–35	26	36	118	22	49	
Medium clay	35–50	26	36	96	22	39	
Heavy clay	> 50	34	38	114	22	48	
Sandy silt	< 18	15	39	82	22	29	
Silt loam	< 18	26	37	110	22	47	

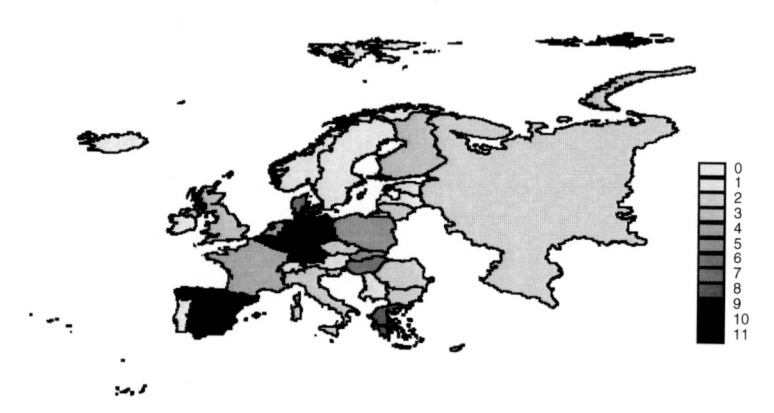

Fig. 3.3 Number of RAMs used to assess different aspects of soil degradation (all threats). The number of applied RAMs increases from *light* to *dark*

3.3 Results and Discussion

3.3.1 Inventory of Soil RAMs

Following the results of the questionnaires it appeared that all but one EU Member State used at least one soil RAM, and/or was working on the implementation of one or more soil RAMs (Fig. 3.3). Countries with federal or autonomous regional governments like Germany and Spain used different RAMs for different regions. For these countries the total number of applied RAMs may exceed the number of considered soil threats (5).

Off all reported RAMs more than 50% was still in development (Table 3.3). Of the remaining 50% that was already in practice, the majority concerned process

3 Moving Ahead from Assessments to Actions by Using Harmonized RAMs

Table 3.3 Countries that report one or more RAMs in practice (x) or in development (*). Underlined symbols indicate regional organization of RAMs. Numbers refer to publication of RAM in literature (i.e. no returned questionnaire). Data from all (policy + specific) questionnaires

	Erosion	SOM decline	Salinization	Compaction	Landslides
Austria					
Belgium	x*	*		*	x
Bulgaria					
Czech Republic	x	x	x	x	x
Denmark	*	*		*	
Estonia					
Finland	x*				
France					x
Germany	x*	*		x̲*	
Greece	*	*	*	*	*
Hungary	x	*	x	x	
Ireland					
Italy	*	1		x	2
Latvia					
Lithuania	*				
Luxembourg					
Malta					
Netherlands	x	x			*
Poland	x	*			
Portugal					
Romania	x				
Serbia	*	*	*	*	*
Slovakia					
Slovenia		*			
Spain	x̲	*	>1		x
Sweden					
UK		3			

quantifications, rather than risk assessments, i.e. only performed the first 3 steps of the risk assessment chain of Fig. 3.1. In other words, the majority of the RAMs yield a rate or state of the soil threat. However, a rate or state of a soil threat doesn't tell a policy maker whether action should be taken now, within months or somewhere in the future, let alone, what action should be taken. This incompleteness of RAMs puts the unambiguous interpretation of soil threats at stake, as outcomes at different levels in the risk assessment chain, e.g. results of the data processing step and the data interpretation step, cannot be compared.

There were some clear differences between soil threats with regard to development of RAMs. For instance, the developments of RAMs for landslides were ahead of the development of RAMs of other soil threats in terms of completion of the risk assessment chain and in terms of harmonization. This has several reasons: (1) landslides occur in a limited number of countries, (2) most landslides occur instantaneous and consequences are almost always catastrophic, which is a strong driver for policy makers and (3) external parties, e.g. insurance companies, demand for unequivocal risk assessments. In the data processing and data interpretation steps of the risk assessment chain, landslide RAMs combine expert judgment, empirical approaches and to a lesser extent mathematical simulations. Currently, a trend in harmonizing procedures and proposing standards is gaining ground in the landslide scientific community following the execution of various EU-funded projects, though differences in terminology may still hamper exchanges of information.

Most RAMs were found for erosion and SOM decline (Table 3.3), which reflects the widespread appearance of these phenomena. For salinization least RAMs were observed, which is probably related to the limited number of countries in which this threat occurs. Table 3.3 also shows that for SOM decline the majority of the RAMs was still in development, while for erosion and compaction the majority of the RAMs was already in practice. More details about the inventory of current RAMs in Europe can be found in van Beek et al. (submitted).

3.3.2 Case Studies

The case study on erosion in Romania showed that the use of different RAMs yielded differences in the estimation of spatial distributions and patchiness of erosion (Fig. 3.4). Moreover, the use of different RAMs resulted in different estimations of the affected areas when a certain threshold (in this case 1 t ha^{-1} y^{-1}) was applied. Differences in affected areas may yield up to 36% depending on the use of different soil RAMs and on the spatial scale of data input (not shown).

The case study on compaction in the Netherlands showed large discrepancies in spatial occurrence and severity of the threat (Fig. 3.5). However, a major part of the differences was caused by the differences in indicator used in the RAMs, viz. allowable wheel load in the SOCOMO approach and vulnerability to compaction in the "Jones" approach. The definition of the indicator is part of the first step of the risk assessment chain (Notion of threat, Fig. 3.1) and hence this case study demonstrates

Fig. 3.4 Soil loss (t ha^{-1} y^{-1}) in Romania evaluated using SIDASS-WEPP model and map of Europe scale 1:1,000,000 (*above, left*) and using PESERA model at 1 km grid (*above, right*). Source: Tóth et al. (2009)

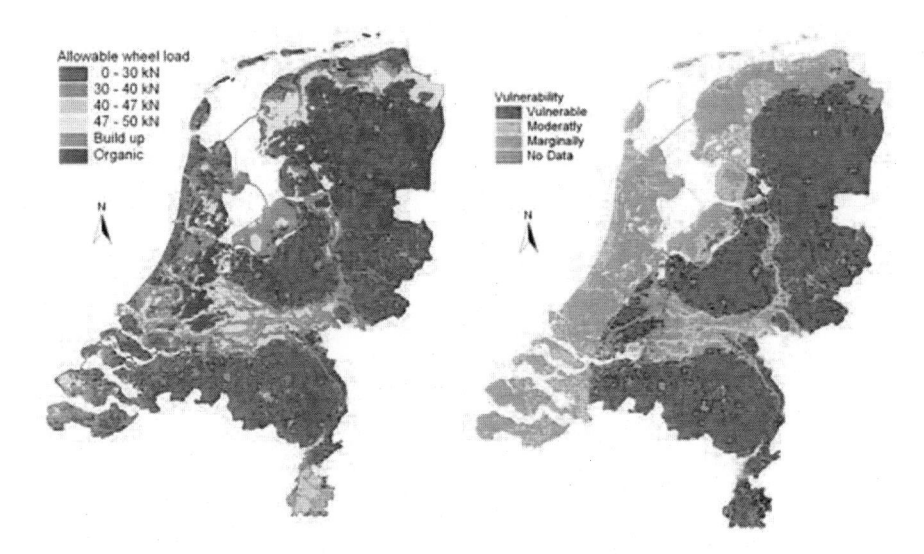

Fig. 3.5 Vulnerability to compaction based on susceptibility and climate (*left*) and maximal allowable wheel load of a Terra Tire 73 × 44.00 – 32 on soil with pF 2.5 (*right*)

that differences in the first step in the risk assessment chain results in incomparable outputs in the data processing step.

3.3.3 Implementation of Actions Plan

There are currently more than 200 treaties, agreements, conventions and protocols at force in the field of environment, but only a few of them are directed towards the protection of soil (EEA, 2000). Spain has a national action plan to combat erosion and desertification (MARM, 2008) and the UK has a national soil action plan (DEFRA, 2004). Furthermore Greece, Italy and Portugal have regional action plans to combat desertification, which includes aspects of soil erosion (www.MIO-ECSDE.org). These national initiatives are just a fraction of the RAMs that are currently in practice or in development in EU-27 (Table 3.3). This observation demonstrates the difficulties that are experienced by policy makers going from inventory of risk exposures towards action plans to prevent these risks.

3.4 Conclusions and Outlook

At present, the use of RAMs for soil degradation in EU-27 seldom results in the implementation of action plans to reduce soil degradation effects. This would suggest that soil degradation is not a serious issue in EU-27. However, various reports (e.g.) Greenland (2006), van Lynden (1995) and EEA (2000) indicate that soil degradation is a serious issue in EU-27. These conflicting results are based in part on the following facts:

– Incompleteness of RAMs for soil degradation assessment: the vast majority of the reported RAMs are process quantification as they yield a rate or state of a soil threat, but do not provide information about the actual risk perception (notably the final step of the risk assessment chain in Fig. 3.1). The risk perception should tell policy makers whether the risk is acceptable or not. However, to establish risk classes the contribution of policy makers to the development of RAMs is warranted as risk perception is not only a matter of scientific understanding, but also about social acceptance and political willingness.
– Uneven attention by policy makers between different threats: Our results showed that RAMs for "high impact" threats like landslides and erosion were much further elaborated than the "slow killers" like compaction and soil organic matter decline. However, "slow killers" are often predisposing factors for "high impact" threats and therefore should be part of an integrated assessment to combat soil degradation.
– The use of different RAMs result in different, and possibly conflicting, risk perceptions and inability to compare outputs of RAMs due to differences in notions of the threat. This may ultimately have consequences for the assignment of priority areas and the public support to policies.

Abovementioned shortcomings may be overcome when RAMs are harmonized. However, harmonization of soil RAMs is not an easy task as soil RAMs have often certain regionally specific characteristics that can not be easily incorporated in another soil RAM. For that reason we suggest a two-tier approach for the identification of geographical areas at risk for soil threats. The Tier 1 method is a standardized and uniformly applicable method across EU-27, at a relatively low spatial resolution and is used to identify areas at risk. The Tier 2 method is a regional-specific and more detailed assessment of the risk in the areas identified by the Tier 1. Hence, the Tier 1 method is similar for all Member States of EU-27, while the Tier 2 method is chosen on the basis of its specificity for areas/regions or Member States, by the Member States in question. The results of the Tier 2 approach should be compared and harmonized as far as possible with the results of the Tier 1 approach. The results of a number of recent explorative studies on the occurrence of soil threats in EU-27 may be used as a starting point for the identification of proper Tier 1 methods.

If such a two-tier approach appears to be not feasible, for whatever reasons, we recommend "generic harmonization", i.e. combining standardization and harmonization at all levels of the risk assessment chain for all RAMs in use. For instance, the notion of the threat, data collection and risk perception steps of the risk assessment chain are standardized (i.e. prescribed) whereas the data processing and data interpretation steps are harmonized, i.e. member states can use the models and threshold values that are most applicable to their (environmental) situation.

With regard to data collection several programmes/manuals are available that provide already standardized data inventories (Kibblewhite et al., 2008). However, this "generic harmonization" will be a major undertaking as we noticed that differences occur between RAMs at each level of the risk assessment chain. We urge to initiate this process timely considering that as long as different unharmonized soil RAMs are used simultaneously, the implementation of remedial measures will be frustrated by ambiguous interpretation of data.

Acknowledgments This study was part of the RAMSOIL project on harmonization of risk assessment methodologies for soil threats. The RAMSOIL project was funded by the European Commission, DG Research, within the 6th Framework Programme of RTD (Priority 8 – Specific Support to Policies, contract n 44240). The views and opinions expressed in this publication are purely those of the writers and may not in any circumstances be regarded as stating an official position of the European Commission. The RAMSOIL project was co-financed by the Dutch Ministry of Agriculture, Nature and Food Quality as part of the strategic research program "Sustainable spatial development of ecosystems, landscapes, seas and regions" (KB-01-001-005).

References

DEFRA (Department of Environment, Food and Rural Affairs). (2004). The first soil action plan for England: 2004–2006. DEFRA report PB9411. www.defra.gov.uk

EEA (European Environmental Agency). (2000). Down to earth: Soil degradation and sustainable development in Europe. EEA issue series No. 16.

EU. (2006). Communication fro the commission to the council, the European Parliament, The European Economic and Social Committee of the regions. Thematic Strategy for Soil Protection: Commission of the European communities Brussels.

Greenland, D.J. (2006). Soil management and soil degradation. European Journal of Soil Science 32(3):301–322.

Hoogland, T. and van den Akker, J.J.H. (2009). Comparison of two RAMs for compaction: A case study for The Netherlands. RAMSOIL project report 4.2. Available at www.ramsoil.eu

Jones, R.J.A., Spoor, G. and Thomasson, A.J. (2003). Vulnerability of subsoils in Europe to compaction: A preliminary analysis. Soil and Tillage Research 73:131–141.

Kamrin, M.A. (1997). Environmental risk harmonization: Federal/state approaches to risk assessment and management. Regulatory Toxicology and Pharmacology 25:158–165.

Kibblewhite, M.G., Jones, R.J.A., Baritz, R., Huber, S., Arrouays, D., Micheli, E. and Stephens, M. (2008). ENVASSO Final Report Part I: Scientific and Technical Activities. ENVASSO Project (Contract 022713) coordinated by Cranfield University, UK, for Scientific Support to Policy, European Commission 6th Framework Research Programme.

Kirkby, M.J., Irvine, B.J., Jones, R.J.A. and Govers, G. (2008). The PESERA coarse scale erosion model for Europe. I. Model rationale and implementation. European Journal of Soil Science 59(6):1293–1306.

MARM. (2008). Programa de Acción Nacional de Lucha contra la Desertificación. Ministerio de Medio Ambiente, Rural y Marino, Madrid.

Simota, C., Horn, R., Fleige, H., Dexter, A., Czyz, E.A., Diaz-Pereira, E., Mayol, F., Rajkai, K. and de la Rosa, D. (2005). SIDASS project – Part 1. A spatial distributed simulation model predicting the dynamics of agro-physical soil state for selection of management practices to prevent soil erosion. Soil and Tillage Research 82(1):15–18.

Spoor, G., Tijink, F.G.J. and Weisskopf, P. (2003). Subsoil compaction: Risk, avoidance, identification and alleviation. Soil and Tillage Research 73:175–182.

Tóth, T., Simota, C., van Beek, C., Recatalá-Boix, L., Añó-Vidal, C. and Hagyó, A. (2009). RAMSOIL project report 4.1. Available at www.ramsoil.eu

van Beek, C.L., Tóth, T., Hagyo, A., Tóth, G., Récatala Boix, L., Añó Vidal, C., Malet, J.P., Maquire, O., van den Akker, J.J.H., van der Zee, S.E.A.T.M., Verzandvoort, S., Simota, C., Kuikman, P.J. and Oenema, O. (submitted) Towards harmonization of risk assessment methodologies for soil threats in Europe. Soil Use and Management.

Van den Akker, J.J.H. (2004). SOCOMO: A soil compaction model to calculate soil stresses and the subsoil carrying capacity. Soil and Tillage Research 79:113–127.

van Lynden, G.W.J. (1995). European soil resources: Current status of soil degradation, causes, impacts and need for action. Manhattan Pub. Nature and Environment 74:100p.

Chapter 4
"Zero-Tolerance" on Land Degradation for Sustainable Intensification of Agricultural Production

Minh-Long Nguyen, Felipe Zapata, and Gerd Dercon

Abstract The demand to improve soil health, arrest land degradation, in particular desertification in agro-ecosystems and protect land and water resources for food production and sustainable agricultural and socio-economic development is expected to increase in the next 50 years as a result of the continuing worldwide population growth and the increased reliance on limited natural resource-based economy. Moreover, the intensive competition for land and water resources from industrial, urban and other sectors and the impacts of widespread soil degradation and global climate change will place increasing pressure on the need to improve sustainable land and water use and management. The objective of the Soil and Water Management & Crop Nutrition (SWMCN) Subprogramme of the Joint FAO/IAEA Division of Nuclear Techniques in Food and Agriculture is to assist Member States to use isotopic and nuclear-based techniques to diagnose constraints and pilot-test interventions to intensify crop production in a sustainable manner through the integrated management of soil, water and nutrient resources without land degradation. This objective is pursued through a range of activities including (a) co-ordinated research projects (CRP) which involve international networks of national agricultural research organizations from developing countries, advanced research institutes and CGIAR institutions, and (b) technical co-operation projects (TCP) that promote technology transfer through technical support and institutional capacity building in FAO and IAEA Member States. This chapter will report on the application of isotopic and nuclear techniques to unravel processes and factors that affect land degradation and major findings obtained from both CRPs and TCPs that were aimed to avoid and mitigate land degradation. Since land degradation includes not only soil erosion but also the decline in soil quality and their constituents (such as water and nutrients) with its subsequent reduction in crop production, projects that are

M.-L. Nguyen (✉)
Soil and Water Management and Crop Nutrition Subprogramme, Joint FAO/IAEA Division of Nuclear Techniques in Food and Agriculture, International Atomic Energy Agency, A-1400 Vienna, Austria
e-mail: m.nguyen@iaea.org

associated with improving soil health, minimizing nutrient mining, combating soil salinity, soil acidity and desertification and enhancing water use efficiency will also be briefly presented and discussed.

Keywords Nuclear and isotopic techniques · Land degradation · Erosion · Soil · Water management · Food security · IAIE

4.1 Introduction

The present world population of 6 billion is expected to reach 8 billion by the year 2020. Most of the population increases will occur in developing countries, where the largest fraction depends upon agriculture for their livelihoods. Against the background of projections on increased population growth and pressure on the availability of land and water resources worldwide, several developing countries will face major challenges to achieve food security in a sustainable manner, considering their available per capita land area, severe scarcity of fresh water resources and particular infrastructure and socio-economic conditions (Lal, 2000; Brown, 2009).

This is further compounded by severe global soil degradation, in particular Sub-Saharan Africa and South Asia, and increased risks of soil erosion, in particular desertification (Lal, 2007). Worldwide soil degradation is currently estimated at 1.9 billion hectares and is increasing at a rate of 5–7 million hectares each year (Lal, 2006). Soil degradation and food insecurity are intricately linked with long-term social, economic and environmental impacts resulting in human migration, social unrest, food crises and global instability (Doos, 1994; Alexandratos, 1995; Brown, 2009).

Enhancing sustainable food production will require the combined use of the following strategies for land and water resource management: (a) agricultural intensification on the best arable lands that are currently being farmed with minimum environmental degradation; (b) rational utilization of the marginal lands, and (c) arrest land degradation and restore degraded soils (Lal, 2000). Besides, these pressing issues, there are several other environmental problems that would also need to be addressed. These include: (a) Increasing risks and impacts of global warming and climatic variability; (b) rising energy demands, in particular renewable energy sources; (c) expanding urbanization and industrialization and related infrastructure development; and (d) deteriorating water and air quality. All of them will likely have negative impacts and induced changes on agro-ecosystems thus placing increased pressures on sustainable land and water resources to produce sufficient food, feed, fibre and fuel for the ever increasing world population (Lal, 2007; Verchot and Cooper, 2008).

This chapter reports on the objective, strategies and main project activities of the Soil and Water Management & Crop Nutrition (SWMCN) Subprogramme of the Joint FAO/IAEA Division of Nuclear Techniques in Food and Agriculture, with particular emphasis on the role of nuclear-based techniques in the development of integrated soil, water and nutrient management practices for sustainable intensification of agricultural production and conservation of the natural resource base without impact on soil health and land degradation.

4.2 Operational Strategy of the Soil and Water Management and Crop Nutrition Subprogramme

In 1964, two United Nations Organizations, the Food and Agriculture Organization (FAO) and the International Atomic Energy Agency (IAEA) established the Joint FAO/IAEA Division of Nuclear Techniques in Food and Agriculture at the IAEA Headquarters in Vienna, Austria, to strengthen capacities for using nuclear-based methods to develop technologies for sustainable food security and to disseminate these through international co-operation in research, training and outreach activities in Member Countries of FAO and IAEA (IAEA, 2008a; FAO, 2008). To achieve this mission, the Division has five disciplinary Sections, namely the Soil and Water Management and Crop Nutrition, Plant Breeding and Genetics, Animal Production and Health, Insect and Pest Control and Food and Environment Protection. Each Section is linked to a Laboratory Unit located at the Agriculture and Biotechnology Laboratory (ABL) in Seibersdorf. (FAO/IAEA, 2007, 2008a; FAO, 2008).

The strategic objective of the SWMCN Subprogramme, consisting of the SWMCN Section at IAEA Headquarters and the Soil Science Unit in Seibersdorf, is to develop and promote the adoption of nuclear-based technologies for optimising soil, water and nutrient management practices in targeted cropping systems (and agro-ecological zones), which support intensification of crop production and preservation of the natural resource base (FAO/IAEA, 2008b). Nuclear-based techniques (stable and radioactive isotopes, neutron moisture and gamma density probes) provide unique and quantitative data on nutrient and water dynamics in the soil-plant system, and therefore, have advantages over conventional techniques in providing essential or value-added information for properly defining and quantifying "land productivity" constraints and assessing the value of the interventions designed to alleviate them with the ultimate goal of enhancing sustainable intensification of agricultural production.

To achieve the strategic objective of the Subprogramme, the following activities are implemented:

- Development, evaluation and standardization of new nuclear and related methodologies for achieving sustainable intensification of crop production systems in Member States. This is done through the Research Contract Programme by promoting global and regional thematic research and development networks called Co-ordinated Research Projects (IAEA, 2008b). A list of completed and ongoing CRPs of the SWMCN Subprogramme can be found in Table 4.1.

Table 4.1 List of Coordinated Research Projects (CRP) implemented by the Soil and Water Management and Crop Nutrition Subprogramme of the Joint FAO/IAEA Division of Nuclear Techniques in Food and Agriculture (IAEA)

CRP code and title	Period	Publication
(a) Completed CRPs		
D1.50.05. The assessment of soil erosion through the use of Cs-137 and related techniques as a basis for soil conservation, sustainable agricultural production and environmental protection (Joint CRP with F3.10.01)	1995–2001	Handbook for the assessment of soil erosion and sedimentation using environmental radionuclides. Kluwer Ac. Publ., Dordrecht, the Netherlands (2002) Special issue STILL Research 69, 1–2 (2003)
D1.20.06. Management of nutrients and water in rainfed arid and semi-arid areas for increasing crop production	1997–2002	IAEA TECDOC 1468. IAEA, Vienna (2005)
D1.20.07. Use of nuclear techniques for developing integrated nutrient and water management practices for agroforestry systems	1998–2006	IAEA TECDOC 1606. IAEA, Vienna (2008)
D1.50.06. Development of management practices for sustainable crop production systems on tropical acid soils through the use of nuclear and related techniques	1999–2004	IAEA Proceedings series STI-PUB-1285. IAEA, Vienna (2006)
D1.50.07. Integrated soil, water and nutrient management for sustainable rice-wheat cropping systems in Asia	2001–2006	IAEA TECDOC (in preparation)
D1.50.08. Assess the effectiveness of soil conservation techniques for sustainable watershed management using fallout radionuclides	2002–2007	IAEA TECDOC (in preparation)
D1.20.08. Selection for greater agronomic water-use efficiency in wheat (drought) and rice (salinity) using carbon isotope discrimination	2003–2008	IAEA TECDOC (in preparation)
D1.50.09. Integrated soil, water and nutrient management in conservation agriculture	2004–2009	Research contractors from ARG, BRA, IND, MOR, PAK, TUR and UZB; technical contractor from CHI and research agreements from AUL and CIMMYT-Mexico

Table 4.1 (continued)

CRP code and title	Period	Publication
(b) Ongoing CRPs		
D1.50.10. Selection and evaluation of food (cereal and legume) crop genotypes tolerant to low nitrogen and phosphorus soils through the use of isotopic and nuclear-related techniques	2006–2011	Research contractors from BKF, BRA, CMR, CPR, CUB, GHA, MAL, MEX, MOZ and SIL; Technical contractors from USA and research agreements from AUL, WARDA, TSBF-CIAT, IITA and FRA
D1.20.09. Managing irrigation water to enhance crop productivity under water-limiting conditions: a role for isotopic techniques	2007–2012	Research contractors from CPR (2), BKF, MLW, MOR, PAK, TUR, VIE, and ZAM; technical contractors from USA (2) and research agreements from AUS and SPA
D1.20.10. Strategic placement and area-wide evaluation of water conservation zones in agricultural catchments for biomass production, water quality and food security	2008–2013	Research contractors from CPR, EST, IRA, LES, NIR, ROM, TUN and UGA Technical contractors from UK and USA and research agreements from FRA, UK and USA
D1.20.11. Integrated isotopic approaches for an area-wide precision conservation to control the impacts of agricultural practices on land degradation and soil erosion	2009–2014	Research contractors from CHI, CPR (2), POL, MOR, RUS, SYR and VIE; Technical contractors from GFR, NZE and UK and research agreements from AUL, CAN and UK

- Provision of assistance to developing Member States to build-up national capacity (human and infrastructure), transfer and apply nuclear-based technologies in sustainable agricultural development through Technical Co-operation Programme. Technical and scientific backstopping will be provided in identifying, formulating and implementing interregional, regional and national TC projects in topics related to sustainable intensification of crop production systems (IAEA, 2008c).
- Assist Member States in developing human resources through training courses, workshops, fellowship training, scientific visits and through development of training materials.
- Provide supportive research, training, and analytical services through the Soil Science Unit, Agriculture and Biotechnology Laboratory (ABL) at Seibersdorf, near Vienna, Austria (IAEA, 2008d).
- Synthesise and disseminate information from the SWMCN Sub-programme research activities through publications and databases and information exchange

through Newsletters, web pages, etc. to transfer technological packages to beneficiaries and end-users to enhance the impact of the projects.
- Promote the dissemination of nuclear technologies to the scientific community through the organization of international/regional meetings.

The effectiveness and impact of these activities is enhanced by further creating synergies between the Research Contract and Technical Co-operation Programmes to the benefit of the Member States. These activities are also implemented by establishing linkages to existing projects and collaborating partnerships with CGIAR Centres (e.g. CIMMYT, IITA, IRRI, TSBF-CIAT and WARDA) and Advanced Research Institutes (currently 20 ARIs participate in CRPs, coming from developed countries such as Australia, Canada, France, Germany, New Zealand, UK and USA).

4.3 Approaches, Strategies and Project Activities Related to Land Degradation

Combating or avoiding land degradation for sustainable intensification of crop production systems requires an integrated soil-water-plant nutrient approach at field and catchment level to improve the productivity of the systems and at the same time to restore and maintain soil fertility and enhance soil health and its resilience against degradation. Isotopes of nitrogen (^{15}N), phosphorus (^{32}P), carbon (^{13}C) and oxygen (^{18}O) together with soil moisture neutron probes have been used to develop integrated soil-plant approaches to ameliorate soil infertility and related soil problems/constraints; improve nutrient and water use efficiency and to provide sustainable intensification of crop production in agroforestry systems, dryland agriculture, tropical high phosphorus-fixing acid soils and rice-wheat cropping systems (Chalk et al., 2002; Nguyen and Zapata, 2006; FAO/IAEA, 2008b).

Soil organic matter (SOM) plays an important role in improving soil functions, promoting soil health and mitigating land degradation. The influence of soil and water conservation measures in SOM accumulation and the consequent impact of accumulated SOM on soil nutrient dynamics was investigated with the use of ^{15}N and ^{13}C based isotopic techniques in various agroecosystems. Data obtained under a CRP on conservation agriculture (CA) (Table 4.1, D1.50.09) indicates that zero tillage (ZT) could sequester up to 17 Mg C ha^{-1} more than conventional tillage in Ferralsols of the Brazilian Cerrado region over a 13-year period.

Carbon isotope signatures (isotopic ^{12}C/^{13}C ratios) in the SOM provided valuable information on the effects of tillage and type of crop rotation on soil C sequestration. A higher soil C accumulation was observed under ZT when the N-fixing legume vetch (*Vicia villosa*) with biological N fixation inputs of 127 kg N ha^{-1} as measured by the ^{15}N isotope dilution technique (Urquiaga et al., 2006) was included in the crop rotation involving root crops (Sisti et al., 2004).

Although CA may enhance C sequestration as shown in a review conducted by one of the agreement holders (Govaerts et al., 2009) of the CRP D1.50.09, the difference in soil C sequestration between CA and conventional tillage depends not only on a C input from crop residues but also on a net external input of N (Govaerts et al., 2009). Conventional tillage can diminish the input from a N-fixing green-manure because this N-input can be mineralised to soil mineral N (Alves et al., 2002), which in turn can be removed from the top soil depth by leaching or in gaseous forms. The use of ^{15}N-labelled green manure was able to quantify the relative importance of these N removal processes and hence management practices can be put in place to reduce N losses and enhance N retention for promoting C sequestration (Kirchmann and Bergström, 2001; Bergström and Kirchmann, 2004; Seo et al., 2006; Christopher and Lal, 2007).

Isotopic techniques are also playing an important role in quantifying the beneficial impacts of CA on mitigating soil erosion and the related loss of soil nutrients (Schuller et al., 2007). Fallout radionuclide (FRN) such as ^{137}Cs, ^{210}Pb and ^{7}Be are increasingly used by FAO/IAEA Member States through both CRP and TCPs (Table 4.1) as tracers to quantify soil erosion/sedimentation rates because the FRN technologies are cheaper and more effective at large-scale soil erosion evaluations taking into account of temporal and spatial variabilities, compared to costly conventioanl techniques using field plots and soil erosion pins (Zapata, 2003; Mabit et al., 2008; Zapata and Nguyen, In Press).

For example, Schuller et al. (2007) from the CRP on soil conservation measures (Table 4.1, D1.50.08) reported that 16 years after implementing zero tillage in southern Chile, there was a substantial reduction in the soil erosion rates as measured by Cs-137 (half-life of 30.2 years) of about 87% (from 11 t ha^{-1} year^{-1} to 1.4 t ha^{-1} year^{-1}). However such a beneficial effect can be readily lost if the mulch layer of old crop residues was removed by burning. Using the short-lived radionuclide Be-7 (half-life of 53.4 days), to measure a short-term (a 27-day period) erosion event occurred just after a dramatic burning event, Schuller et al. (2007) reported substantial soil losses of 12 t ha^{-1} over this 27-day of exceptionally wet (400 mm) period. This represents a dramatic acceleration of the average soil losses under CA from 1.4 t.ha^{-1} on a year basis to 12 t ha^{-1} over a short (27 days) runoff event.

The CRPs mentioned above have created an effective network of research scientists and national agricultural research institutes involving in the combined use of FRNs and stable isotopes such as ^{15}N to assess the relative impacts of different soil conservation measures on soil erosion and land productivity. The success of these CRPs has stimulated an interest in many Member States in the use of these methodologies to identify factors and practices that can enhance sustainable agriculture and minimize land degradation. At present there are 37 Member States through Technical Co-operation (TC) projects at both national and regional levels using FRNs to address issues relating to sustainable land management.

For example, one major regional project on "Sustainable Land Use and Management Strategies for Controlling Soil Erosion and Improving Soil and Water Quality" (RCA Project RAS/5/043) has recently been concluded. This project involved participants from the following 14 Member States in the East Asia and the Pacific region: Australia, Bangladesh, China, India, Indonesia, Republic of Korea, Malaysia, Mongolia, Myanmar, Pakistan, Philippines, Sri Lanka, Thailand, Vietnam. The FRN technology has been successfully used by the participating countries to assess soil erosion, evaluate soil conservation measures (e.g. forestation, terracing, contour cropping, contour hedgerow systems), and to better understand the link between soil redistribution and soil quality (e.g. Soil Organic Matter) in the landscape. The inter-institutional and multi-disciplinary approach (close collaboration between nuclear and soil science institutes) adopted by most participating Member States was one of the key factors of this success.

The expertise gained in the project can be used to further train scientists and technicians from the region. Participants were established with related policy-making and development-oriented institutions. These partnerships are an important vehicle for the dissemination of the FRN technology to assess soil redistribution and improve land management practices. The participants of the project and their partners in the countries have already used the information obtained in the regional project as a basis to formulate development projects for enhancing the adoption of improved soil conservation and water management practices. The following examples highlight the impact of this Regional TCP project (RAS/5/043) in Member States.

In China, the Office of the World Bank Project in Baota district, Yan'an has adopted the information obtained from the RAS/5/043 for selecting effective soil conservation measures to control soil erosion. The project area was 800 km^2 in the Yanhe River watershed of the Chinese Loess Plateau. Through the use of a cash forest area and the establishment of tree, shrub and grass cover in a structured framework, annual sediment delivery after a 6-year period (1998–2004) was reduced by 77% compared to that annually produced (8.32 million tons) before the project implementation (1998).

In the Philippines, the RAS/5/043 project has contributed towards the establishment of an integrated management plan for sustainable land and water use. Data on the spatial pattern and rates of soil redistribution under different land use and management practices were used in the assessment and recommendations for improving soil conservation measures in the watershed area and to make local farmers and communities aware of the importance of stopping erosion to ensure their agricultural productivity by adopting appropriate farming practices. The introduction of soil conservation measures in the Pilot Project area covering approximately 6,600 ha within the Angat watershed has been successful at mitigating soil erosion and improving crop production and farmers' income by about 15%.

The success outlined above has stimulated additional commitment from the Agency in responding to request from Member Sates in Latin America. This year a new regional Technical Co-operation project has been initiated on "Using Environmental Radionuclides as Indicators of Land Degradation in Latin American, Caribbean and Antarctic Ecosystems", RLA5051, which was approved for 5 years duration from 2009 until 2013.

Land degradation affects about 300 million hectares of land in the Latin American and Caribbean region, out of this 51% of agricultural land (180 million hectares). The ARCAL (Regional Cooperative Agreement for the Advancement of Nuclear Science and Technology in Latin America and the Caribbean) Regional Strategy Profile identified the unsustainable use of arable land and the resulting permanent loss of productive agricultural areas as one of the most significant environmental challenges to sustainable food production and water supply in Latin America and the Caribbean region.

The following 14 Member States are participating: Argentina, Bolivia, Brazil, Chile, Cuba, Dominican Republic, El Salvador, Jamaica, Haiti, Mexico, Nicaragua, Peru, Uruguay and Venezuela. The project aims to enhance soil conservation and environmental protection in Latin American, Caribbean and Antarctic environments in order to ensure sustainable agricultural production and reduce on and off-site impacts of land degradation. Soil redistribution rates will be determined to estimate erosion/sedimentation rates and assess the effect of human intervention on soil ecosystems in selected areas of the 14 participating countries in the region. The main expected outcome of this project is enhanced regional capacity for sound assessment of land degradation and improved national and regional policies for soil conservation and environmental protection in Latin America, Caribbean and Antarctic ecosystems by using fallout radionuclides.

In order to target appropriate soil and water conservation measures to reduce the area-wide impacts of different land uses within a catchment, a new CRP (D1.20.11, Table 4.1) entitled "Integrated Isotopic Approaches for area-wide precision conservation to control the impacts of agricultural practices on land degradation and soil erosion" has been initiated in 2009 with 14 participants from 12 Member States. This CRP aims at up scaling the use of the proven FRN based technique and the combined use of FRN with the compound specific stable isotopes to apportion the source of soil losses from different land uses within a catchment.

4.4 Looking Ahead to the Future

Land degradation can be influenced by land use as well as by extreme climatic events resulting from climate change. Adaptation to climate change will require the ability to identify the sources of soil loss at an area-wide scale for the purposes of remediation, rehabilitation and sustainable agricultural and economic development. The SWMCN Subprogramme will be increasingly involved in the application of a suite of nuclear techniques to monitor the impacts of climate change and variations in land use activities on soil health and water availability for crop and livestock production systems under rain-fed and irrigated conditions on the area-wide basis (i.e. catchment scale), as well as on a field-plot scale.

The losses of soil and its constituents that are expected due to climate change will become increasingly important elements of land management practices under climate change scenarios. In this context, nitrogen, oxygen, hydrogen and carbon stable isotope ratios will be increasingly used to trace the efficiency use of land constituents and external inputs for crop productivity under different climate change scenarios and soil salinity and acidity conditions that are expected to worsen under those scenarios.

Sources of soil loss/sediment production will be identified using compound specific stable isotopes (CSSI) of carbon, hydrogen and nitrogen as well as other fingerprints in plants, animal manure and soil samples. Information obtained through combined iso-source (IS) and other advanced modelling tools that use several isotopes to identify causes of soil loss will be used to assess the effectiveness of different land use practices and soil conservation measures in response to changes in climate.

The overall aim will be to provide holistic and innovative land and water management practices for (i) developing sustainable food production systems, (ii) arresting land degradation and enhancing land carbon sequestration, and (iii) rehabilitating degraded and marginal lands.

4.5 Conclusions

Nuclear-based techniques offer great potential in the development of integrated soil, water and nutrient management technologies for addressing land degradation issues and promoting conservation of the natural resource base. These techniques are widely used in essentially all developed and in an increasing number of developing countries in agronomic and related environmental research to enhance productivity and sustainability of agro-ecosystems.

The SWMCN Subprogramme of the Joint FAO/IAEA Division of Nuclear Techniques in Food and Agriculture strengthens national capacities for using these nuclear-based techniques methods and disseminate them through international co-operation in research, training and other outreach activities in FAO and IAEA Member States. In this way both UN organizations meet their mandates and contribute to the UN commitment towards achieving the Millennium Development Goals of Extreme Poverty and Hunger Reduction and Environmental Sustainability (FAO, 2008; IAEA, 2008a; United Nations, 2008).

References

Alexandratos, N. (ed.). (1995). World Agriculture: Towards 2010-An FAO Study, Rome.
Alves, B.J.R., Zotarelli, L., Boddey, R.M. and Urquiaga, S. (2002). Soybean benefit to a subsequent wheat cropping system under zero tillage. In: Nuclear Techniques in Integrated Plant Nutrient, Water and Soil Management. IAEA, Vienna, pp. 87–93.
Bergström, L. and Kirchmann, H. (2004). Leaching and crop uptake of nitrogen from nitrogen-15-labeled green manures and ammonium nitrate. Journal of Environmental Quality 33: 1786–1792.
Brown, L.R. (2009). Could food shortages bring down civilization? Scientific American, May issue:50–57.
Chalk, P.M., Zapata, F. and Keerthisinghe, G. (August 2002). Towards integrated soil, water and nutrient management in cropping systems: The role of nuclear techniques. In: CD-ROM, IUSS (ed.), Soil Science, Confronting New Realities in the 21st Century, Transactions of the 17th World Congress of Soil Science. Bangkok, Thailand, pp. 2164/1–2164/11.

Christopher, S.F. and Lal, R. (2007). Nitrogen management affects carbon sequestration in North American cropland soils. Critical Reviews in Plant Sciences 26:45–64.

Doos, R. (1994). Environmental degradation, global food production, and risk for large-scale migrations. Ambio 23(2):124–130.

FAO. (2008). http://www.fao.org

FAO/IAEA. (2007). Nuclear Technology Serving Agriculture. The Joint FAO/IAEA Programme of Nuclear Techniques in Food and Agriculture. IAEA, Vienna, Austria.

FAO/IAEA. (2008a). Joint FAO/IAEA Food and Agriculture Programme. http://www-naweb.iaea.org/nafa/index.html

FAO/IAEA. (2008b). IAEA. Subprogramme Soil and Water Management and Crop Nutrition. http://www-naweb.iaea.org/nafa/swmn/index.html

Govaerts, B., Verhulst, N., Castellanos-Navarrete, A., Sayre, K.D., Dixon, J. and Dendooven, L. (2009). Conservation agriculture and soil carbon sequestration: Between myth and farmer reality. Critical Reviews in Plant Science 28:97–122.

IAEA. (2008a). Our work. http://www.iaea.org/OurWork/index.html

IAEA. (2008b). Co-ordinated Research Activities, NACA Welcome. http://www-crp.iaea.org/html/welcome.html

IAEA. (2008c). Technical Co-operation Programme. http://tc.iaea.org/tcweb/default.asp

IAEA. (2008d). Soil Science Unit. http://www.iaea.org/OurWork/ST/NA/NAAL/agri/soi/agriSOImain.php

Kirchmann, H. and Bergström, L. (2001). Do organic farming practices reduce nitrate leaching? Communications in Soil Science and Plant Analysis 32:997–1028.

Lal, R. (2000). Soil management in the developing countries. Soil Science 165:57–72.

Lal, R. (2006). Encyclopaedia of Soil Science. 2nd edn. CRC Press, Boca Raton, Fl, USA.

Lal, R. (2007). Farming carbon. Soil and Tillage Research 96:1–5.

Mabit, L., Benmansour, M. and Walling, D. (2008). Comparative advantages and limitations of the fallout radionuclides ^{137}Cs, ^{210}Pb$_{ex}$ and ^{7}Be for assessing soil erosion and sedimentation. Journal of Environmental Radioactivity 99:1799–1807.

Nguyen, M.L. and Zapata, F. (July 2006). Use of nuclear techniques in addressing soil-water-nutrient issues for sustainable agricultural production. In: CD-ROM Abstracts-Session No.1-1, 18th World Congress of Soil Science. Philadelphia, USA.

Schuller, P., Walling, D.E., Sepúlveda, A., Castillo, A. and Pino, I. (2007). Changes in soil erosion associated with the shift from conventional tillage to a no-tillage system, documented using ^{137}Cs measurements. Soil and Tillage Research 94:183–192.

Seo, J.H., Meisinger, J.J. and Lee, H.J. (2006). Recovery of nitrogen-15-labeled hairy vetch and fertilizer applied to corn. American Society of Agronomy 98:245–254.

Sisti, C.P.J., dos Santos, H.P., Kohhann, R., Alves, B.J.R., Urquiaga, S. and Boddey, R.M. (2004). Change in carbon and nitrogen stocks in soil under 13 years of conventional or zero tillage in southern Brazil. Soil and Tillage Research 76:39–58.

United Nations. (2008). UN Millennium Development Goals. http://www.un.org/millenniumgoals/

Urquiaga, S., Jantalia, C.P., Zotarelli, L., Araujo, E.S., Alves, B.R. and Boddey, R.M. (2006). Nitrogen dynamics in soybean-based crop rotations under conventional and zero tillage in Brazil. In: IAEA Proceedings Series "Management Practices for Improving Sustainable Crop Production in Troicla Acid Soils". IAEA-STI-PUB 1285, Vienna, Austria, pp. 13–46.

Verchot, L.V. and Cooper, P. (2008). International agricultural research and climate change: A focus on tropical systems. Agriculture, Ecosystems and Environment 126:1–3.

Zapata, F. (2003). The use of environmental radionuclides as tracers in soil erosion and sedimentation investigations: Recent advances and future developments. Soil and Tillage Research 69:3–13.

Zapata, F. and Nguyen, M.L. (In Press). Soil erosion and sedimentation studies using environmental radionuclides, Chapter 7. In: K. Froehlich (ed.), Environmental Radionuclides-Tracers and Timers of Terrestrial Processes. Elsevier, Hamsterdam.

Chapter 5
A Methodology for Land Degradation Assessment at Multiple Scales Based on the DPSIR Approach: Experiences from Applications to Drylands

Raul Ponce-Hernandez and Parviz Koohafkan

Abstract A methodology for land degradation assessment based on indicators of drivers-pressures-state-impacts-responses (DPSIR), applicable at multiple scales is described in this chapter, and the key steps in methodological development are discussed. Procedures for observation, measurement and quantification of DPSIR indicators at each scale both, on the ground and from existent data are described together with the technological requirements for generating such databases. Issues pertaining to field sampling, up-scaling procedures and specially the integration of multi-thematic data of biophysical, socio-economic and land management indicators of DPSIR for land degradation are addressed and their application described. An algorithm for compiling a synthetic mapping legend that integrates such multi-thematic indicators is designed and its usefulness demonstrated with results in a set of case studies in drylands of different parts of the world. The overall assessment of the usefulness of the methodology proposed for assessment work in other locations and at a variety of scales and ecological conditions is also discussed.

Keywords DPSIR · Land degradation · Drylands · Multi-scale assessment · Mapping

5.1 Introduction

Drylands of the world are fragile ecosystems that represent the source of livelihood of about 1.2 billion people worldwide in over 100 countries. Yet, not enough is known about the nature, severity and extent of land degradation, its causes, and the responses from land users in such fragile ecosystems (UNCCD, 2004). Efforts

---

R. Ponce-Hernandez (✉)
Environmental and Resource Studies Program, Trent University, Peterborough, ON K9J 7B8, Canada
e-mail: rponce@trentu.ca

to characterize and quantify the type, intensity and extent of degradation processes at multiple scales from local to global are being carried out within the context of the LADA project (FAO, 2007). An important aspect of the LADA approach to assessment is that it attempts to link the states of degradation to its root causes (drives and pressures) through the driver-pressure-state-impact-response (DPSIR) approach. This approach is used in the global millennium ecosystem assessment – MEA – (World Resources Institute, 2005). A methodological framework for the assessment of land degradation at multiple scales (from local to global) is proposed in this chapter. The proposed framework builds on an earlier proposal by Ponse-Hernandez and Koohafkan (2004). The framework uses indicators of both, the state of land degradation and its causes (i.e. driving forces and pressures) as the vehicle for the assessment, thus integrating biophysical to socio-economic data on such indicators.

5.2 Methods

The essence of the proposed framework involves twelve (12) core sets of activities (Fig. 5.1), namely: 1: Area and scale definition; 2: Selection of Indicators of DPSIR; 3: Selection of procedures and tools to measure or estimate the values of indicators;

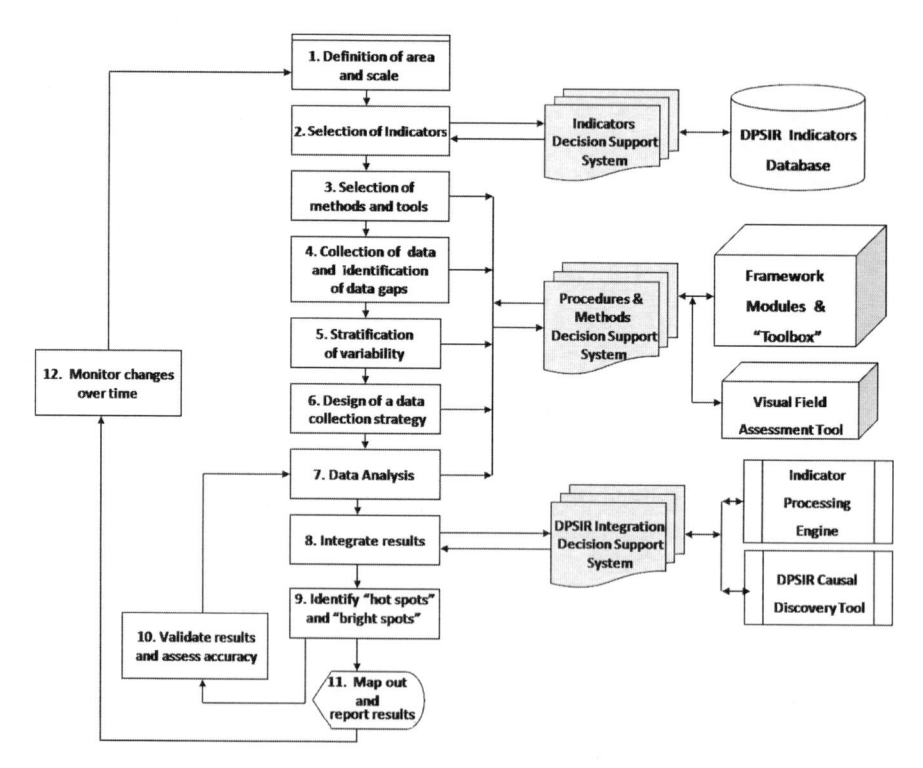

Fig. 5.1 Procedures in the DPSIR methodological framework for land degradation assessment and their decision support systems, databases and tools

4: Collection of existing data and identification of data gaps; 5: Partition variability through stratification; 6: Design of a statistically-reliable sampling scheme, based on the stratification and data collection through field forms and surveys of relevant indicators at the scale of the assessment; 7: Data analysis; 8: Integration of results; 9: Identification of "hot spots" and "bright spots"; 10: Validation of results and accuracy assessment; 11: Design of mapping legend integrating all DPSIR indicators and mapping of results; 12: Monitoring changes over time.

The structure of the framework is generic enough to lend flexibility to its application to dryland conditions anywhere in the world. Thus, its adaptability and usefulness are predicated on the ability to identify an appropriate set of locally relevant indicators at the chosen scale of the assessment, and on the selection and availability of the required tools and methods to achieve their measurement or estimation. The methodological framework proposes to support these important decisions with three decision support systems – DSS (Fig. 5.1). In turn, these DSS are to be supported by global databases of relevant indicators (e.g. the LADA indicator database), a set of procedural modules for guidance on the measurement, estimation or modelling of the selected indicators, including visual field assessment methods.

These make up a sort of "toolbox" from where the assessor can obtain guidance on procedural tools. Two "engines" for processing the measured/estimated indicators and for finding causality (i.e. connecting states of degradation to its relevant drivers and pressures) complete the framework tools. Prior work on an intensive compilation of indictors for the LADA project was carried out, amongst others, by Bot and Snell (2002) and yielded a relatively large number of indicators. These were augmented and refined for field application by Dixon (2003) and are the DPSIR indicators used in the methodology proposed by Ponce-Hernandez and Koohafkan (2004), and used in the proposed framework. The indicator databases together with the "toolbox" of methods, procedures and tools should allow for conducting an assessment at any scale and with any situation of data.

The procedures in the toolbox range from simple conventional procedures, visual field assessments to sophisticated modelling, including remote sensing tools (Ponce-Hernandez and Koohafkan, 2004). The development and computer automation of these DSS, supporting databases and toolboxes, at the time of writing this chapter, are work in progress. It is envisaged that they will be eventually ported to the Internet as a spatially explicit system for online access, relying on the availability of online global mapping tools of the kind of Google Earth^{TM} and/or Google Maps^{TM} or similar.

The results of case studies presented here are derived from the application of the proposed methodological framework whose procedures and tools, except from GIS functions, remote sensing image analysis and statistical and geostatistical estimation techniques, for the most, were not yet automated at the time of the study. Hence, conventional field measurement and observation procedures and data forms were employed, where necessary.

Geostatistical estimation, in particular the various forms of Kriging techniques (i.e. point and block estimates to generate continuous surfaces with ordinary, universal, and disjunctive Kriging and co-regionalization with co-kriging) underpins the upscaling and downscaling procedures for estimation of indicator values at a range of scales, together with conventional statistical procedures for data aggregation. Where the intrinsic hypothesis and assumptions of regionalized variable theory necessary for geostatistical estimation could not be met, even after trend removal or modelling, deterministic estimation techniques were used (e.g. bi-cubic splines and distance-based functions). Unfortunately, given the scope of this chapter, an in-depth account of upscaling and downscaling procedures used is beyond reach. The case studies illustrate the use of any of these techniques where appropriate.

5.2.1 Field Sampling

The field sampling scheme, in most case studies, consisted of a stratified stagewise random sampling, where the number of samples is proportional to the size of the stratum or terrain map unit to sample and its location. Constraints of terrain access, sampling effort and time were also determinants of the final sample size of indicators. Often, the sampling was achieved in three stages corresponding to hierarchical levels of strata (i.e. Satellite scene, land systems within the scene and land facets within land systems depending on scale and the terrain units identified (Ponce-Hernandez and Koohafkan, 2004)).

The stratification of the multivariate environmental variability in the assessment area (Step 5 in the framework illustrated in Fig. 5.1) was achieved with the intense use of remote sensing and GIS data. In particular, archival Landsat Thematic Mapper (TM) and Enhanced Thematic Mapper (ETM+), draped on top of GIS-driven digital terrain models (DTM) were used. When necessary, due to lack of available imagery, other global imaging software (e.g. Google Earth Pro^{TM} and NASA's World Wind) capable of displaying 3-D imagery of terrain, were also used to aid in the identification of the units for stratification and the landscape units for sampling and for the assessment.

The units used in the stratification process at the regional (i.e. sub-national level) were those part of the Land Systems Approach (Beckett et al., 1972; Mitchell, 1973; Moss, 1983). The land systems approach identifies, delineates and describes environmental units of uniform character. This approach was developed for rapid reconnaissance mapping with air photographs and satellite images. Inherent in the procedure is its integrated nature, since several environmental factors, notably the most salient attributes of the terrain, are surveyed simultaneously, creating a map of integrated terrain units. The land system and the land facet were the chosen units to be used in the assessment at sub-national and regional scales.

The land system is an area of a recurring pattern of topography, soils, vegetation, moisture regime and land use/cover, and with a relatively uniform climate. The land facet is an area within which, for most practical purposes, environmental conditions are uniform. Typically, a land system is a recurring pattern of land facets. The identification of land systems and land facets was greatly facilitated by the use

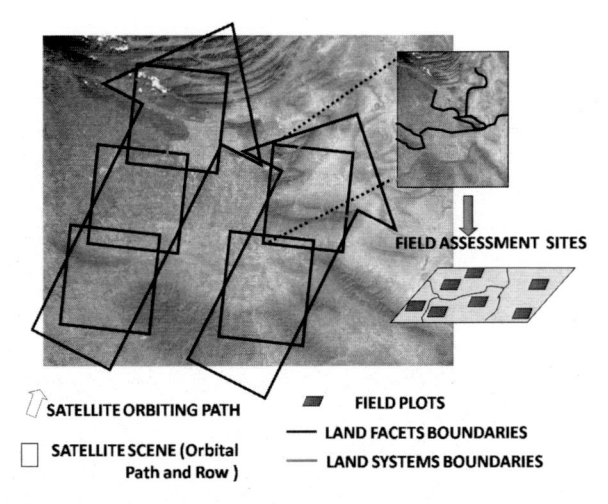

Fig. 5.2 Illustration of the stage-wise random sampling procedure involving 3 stages (satellite scene, land system, land facet and assessment sites within land facet)

of 3-D perspectives of terrain from the GIS-driven DTM and satellite image colour composites. The stage-wise field-sampling scheme is illustrated in Fig. 5.2.

5.2.2 National and Sub-national Scale

At the national and sub-national scales two main indicators of physical and biological land degradation are the decrease in net primary productivity (NPP) and vegetation cover recession over time. Since land degradation is understood as the decrease in productivity of the land over time, the dynamics of these two indicators could be readily analysed from computed band-ratio indices of vegetation (e.g. the Normalized Difference Vegetation Index – NDVI) derived from ratios of red and infrared spectral bands of satellite images (TM and ETM+). The estimation of NPP at a given time requires of an in-situ empirical function that relates NPP to NDVI and to photosynthetic active radiation (PAR). In none of the case studies presented here such function existed and therefore the analysis was restricted to the dynamics of NDVI through change detection over time.

To enable this, threshold values of NDVI corresponding to limits of vegetation health were identified and the NDVI images for different comparable times classified according to such thresholds. This allowed approximating real changes in vegetation as close as possible. Areas of with "healthy" vegetation cover for any given image-date were then computed and compared with areas from reclassified NDVI images of the same area at the latest date, thus enabling the computation of differential areas of either vegetation recession (i.e. degradation), or of (rarely) vegetation expansion (i.e. aggradations).

5.2.3 Regional and Local Scale

At the local scale (i.e. within a land facet of a given land system), the field assessment of the state, intensity and extent of physical, chemical and biological land degradation was carried out at the chosen sampling sites through observation/measurements of indicators by means of field forms and interviews with farmers and land managers using semi-structured questionnaires. The sampling site consisted of a quadrate of approximately 0.25 ha (50 by 50 m each side). The degree of intensity and extent were measured using the procedures in the procedures "toolbox". Appendix shows a sample of the indicators included in the field forms used at the regional and local scale.

5.2.4 Data Analysis, Integration and Mapping

Once measured or estimated in the field, the data analysis process (Steps 7 and 8 in Fig. 5.1) leads to the integration of indicators into an assessment. The integration process is not simple. The indicator processing "engine" offers a range of algorithmic possibilities, from a user-driven compilation of weighting factors based on perceived local importance to linear weighted combinations of selected relevant indicators, to computation of internal ratings to aggregate indicators of a given degrading process. The combination of indicators could be a fairly large number such as derived from a combinatorial calculation. Since the indicators are integrators themselves, therefore they are integrated into processes by type of degradation. This integration allows for the compilation of results into a mapping legend, which is applicable to each assessment site and unit (i.e. land facet). Box 5.1 shows the elements of the legend used for the integration of indicators of states and drivers and pressures into an assessment.

5.3 Case Study Results

The applicability of the methodological framework was tested in several countries. Here, for brevity, only three are shown representing different scales.

5.3.1 Lybia

At the national and sub-national scales, vegetation cover recession over time indicated by negative changes in NDVI was mapped as indicator of land degradation. The changes in NDVI were computed from Landsat TM satellite images over 12 years (Ponce-Hernandez and Mohammed, 2007). Although the threshold values of NDVI for vegetation are quite low, the state of vegetation cover degradation as indicated by negative changes in NDVI is quite evident as shown in Fig. 5.3.

5 Experiences from Applications to Drylands

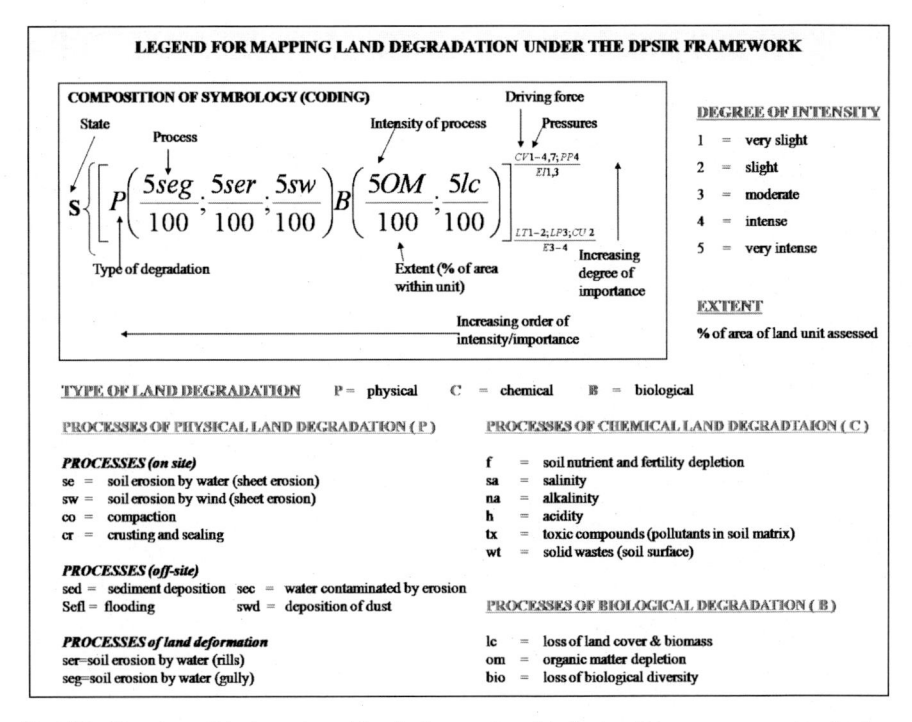

LEGEND FOR MAPPING LAND DEGRADATION UNDER THE DPSIR FRAMEWORK

COMPOSITION OF SYMBOLOGY (CODING)

$$S\left\{\left[\underset{\uparrow}{P}\left(\frac{5seg}{100},\frac{5ser}{100},\frac{5sw}{100}\right)B\left(\frac{5OM}{100},\frac{5lc}{100}\right)\right]\overset{\overset{CV1-4,7;PP4}{\underset{\uparrow}{E/1,3}}}{\underset{\underset{E3-4}{LT1-2;LP3;CU\,2}}{}}\right\}$$

- State
- Process (Intensity of process)
- Driving force / Pressures
- Type of degradation
- Extent (% of area within unit)
- Increasing degree of importance

Increasing order of intensity/importance →

DEGREE OF INTENSITY
1 = very slight
2 = slight
3 = moderate
4 = intense
5 = very intense

EXTENT
% of area of land unit assessed

TYPE OF LAND DEGRADATION   P = physical   C = chemical   B = biological

PROCESSES OF PHYSICAL LAND DEGRADATION (P)

PROCESSES (on site)
- se = soil erosion by water (sheet erosion)
- sw = soil erosion by wind (sheet erosion)
- co = compaction
- cr = crusting and sealing

PROCESSES (off-site)
- sed = sediment deposition   sec = water contaminated by erosion
- Sefl = flooding   swd = deposition of dust

PROCESSES of land deformation
- ser = soil erosion by water (rills)
- seg = soil erosion by water (gully)

PROCESSES OF CHEMICAL LAND DEGRADTAION (C)
- f = soil nutrient and fertility depletion
- sa = salinity
- na = alkalinity
- h = acidity
- tx = toxic compounds (pollutants in soil matrix)
- wt = solid wastes (soil surface)

PROCESSES OF BIOLOGICAL DEGRADATION (B)
- lc = loss of land cover & biomass
- om = organic matter depletion
- bio = loss of biological diversity

Box 5.1 Elements of the legend used for the integration of indicators into an assessment under the DPSIR methodological framework

5.3.2 Lebanon

At the sub-national and regional scales, in the Yammouneh Plateau and Bekaa Valley of Lebanon a regional land degradation assessment was conducted. The land systems and land facets were mapped (Fig. 5.4) and used as units for sampling and for reporting the assessment. Analysis and integration of data collected from field sampling of land facets allowed for the assessment of the state (i.e. type, intensity and extent) of land degradation (Ponce-Hernandez, 2005). The results of the assessment are shown in Fig. 5.5 in terms of state of degradation as per the DPSIR legend. It should be noted that although causality was established for each unit assessed, it is not shown in this legend due to space constraints in the map of Fig. 5.5.

5.3.3 Mexico

In "El Alegre" watershed, San Luis Potosi, Mexico, an assessment was conducted using the proposed DPSIR methodological framework at regional and local scales. Land Facets were mapped (Fig. 5.6) and field data on DPSIR indicators collected at specific sites within Land Facets (Fig. 5.7), following the statistical stage-wise

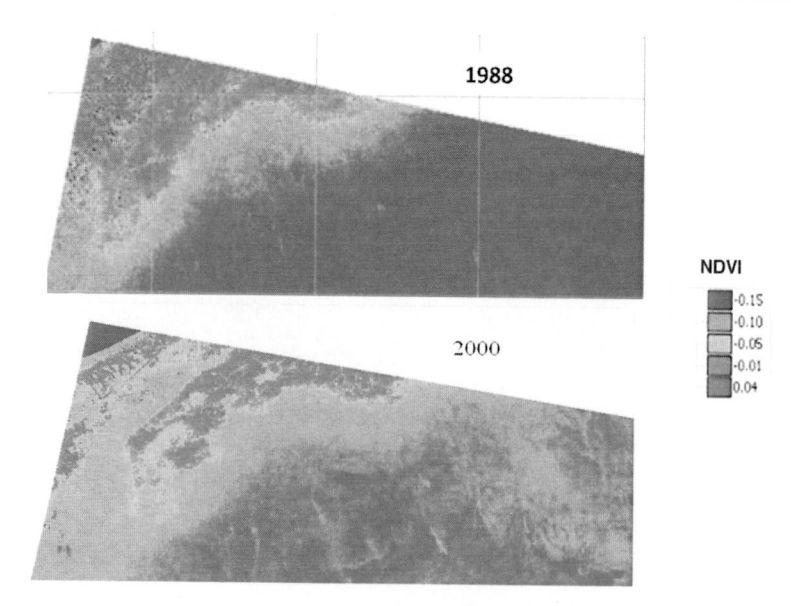

Fig. 5.3 Vegetation recession over 12 years as shown by computed NDVI images of Northeast Lybia (Al-jabal Al-akhdar region), depicting values of Chlorophylically-active vegetation cover (positive values closer to 1, and in *bright* and *intense red*). The recession of active vegetation in the 12-year period (1988–2000) is quite apparent (for colors, see online version)

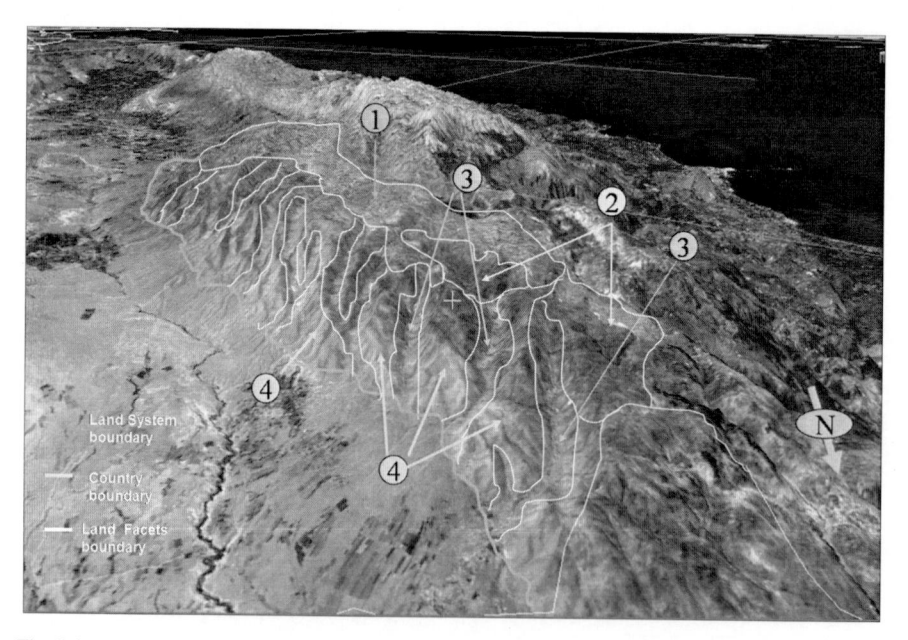

Fig. 5.4 Land Facets of the Land System "Yammouneh Plateau" identified and mapped from a 3D oblique perspective using NASA satellite imagery. Land Facets are identified by *numeral labels*

5 Experiences from Applications to Drylands

Fig. 5.5 Orthogonal view of land degradation assessment results in each of the land facets of the Land System "Yammouneh Plateau" and some of "Bekaa Valley" according to the proposed DPSIR legend. Only states of physical (p), biological (b) and chemical (c) degradation and their spatial extent are shown. No causality is shown in this legend due to space constraints

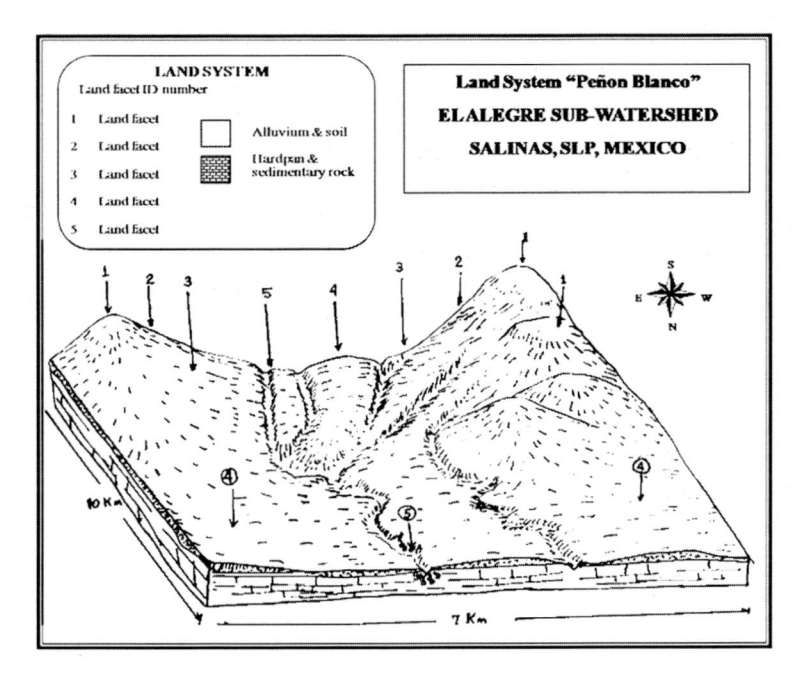

Fig. 5.6 Land facets for the assessment in "El Alegre" Watershed, San Luis, Mexico

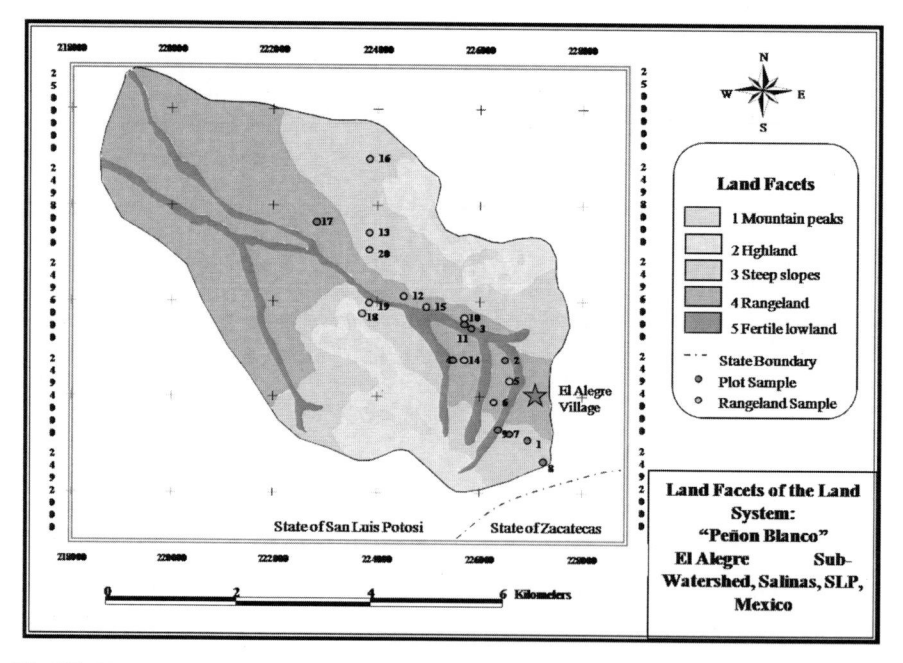

Fig. 5.7 Mapped Land Facets showing sampling sites for field data collection in San Luis, Mexico

sampling design. The indicators were observed, measured and recorded from field sampling on the ground, analysed and integrated. The results of the state (type, intensity, extent) of degradation were mapped (Fig. 5.8) and are shown in Table 5.1. The causes (drivers and pressures) were also integrated in the mapping legend for the same land units.

5.4 Discussion and Conclusions

The results obtained from applying the proposed methodological framework at three different scales allowed for the determination of biophysical states of degradation with different levels of detail. National scale assessments have as main indicator of land degradation negative changes in net primary productivity (NPP) of the land. Variables related to this indicator are derived typically from remote sensing satellite data aided by meteorological records. Long-term NPP decreases are a powerful indicator of persistent ecosystem degradation and the decrease of land productivity. The proposed framework includes procedures and tools for national-scale detection of such changes, as demonstrated with the Lybian example. But the availability of multispectral satellite data at the required dates becomes a very important factor.

At sub-national and regional scales not only the bio-physical states of degradation but their type, i.e. physical (P), chemical (C) or biological (B), their intensity

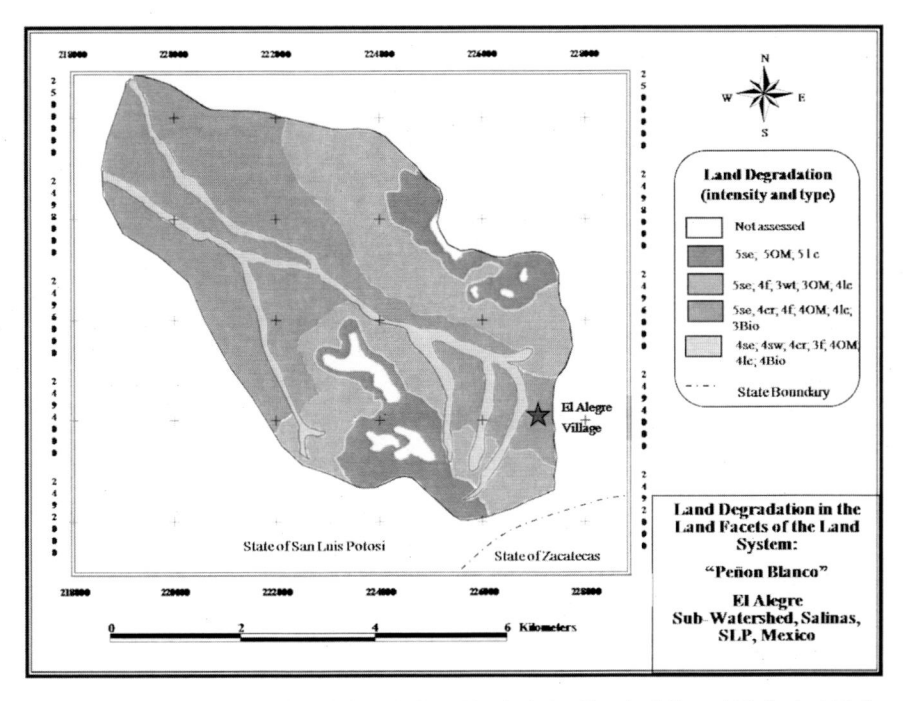

Fig. 5.8 Intensity and type of land degradation (Physical-P-, Chemical-C- and Biological-B) in the Land Facets mapped, as determined from the integration of field data of indicators based on the DPSIR approach in "El Alegre" Watershed, San Luis Potosi, Mexico

(in the scale of 0 – no degradation – to 5 – very high intensity) and extent (i.e. area affected from 100% of unit area evaluated and coded as denominator in the legend) are of significant importance. Just as important is the integration of such degradation states to their social, economic, cultural and management causes (drivers and pressures). For the drivers and pressures, indicator variables are also obtained from published work, databases, surveys and interviews and are linked one by one to the states of degradation. Networks of causal chains are established either manually (ad-hoc designed forms) or using computer models, e.g. Bayesian Networks for causal exploration (Ponce-Hernandez and Ahmed, 2008). In the results presented in this chapter causality was established from one-at-a-time linkage between drives and pressures to states of degradation using ad-hoc designed paper forms.

Once the networks of causal chains are identified the causes are coded and shown in the furthermost right position of the DPSIR mapping legend (last column in Table 5.1).

The results of the case studies are also valuable in that they demonstrated the flexibility of the framework in terms of its applicability to a range of scales in different dryland environments, and in terms of offering a fairly comprehensive database of scale-dependent indicators, paired to a wide range of procedures and tools for their observation, measurement and estimation in the field and from surveys. Perhaps

Table 5.1 Land degradation assessment in land facets, through the DPSIR legend, at regional/watershed scale in "El Alegre" watershed, San Luis, Mexico

Facet ID	Land form and rock	Soils	Land use type/land cover	Hydrology and moisture regime	Land degradation status (DPSIR legend)[a]
2	Steep mountain and hillside slopes Lower Cretaceous volcano-sedimentary rocks: flysch (siltstone, limestone)	Skeletal Eutrhric Lithosol and bare rock. Soils are gravely or stony and shallow	Sub-thorny and thorny brush and shrubs, crasicaule cacti	Contributing sites, excessive surface drainage, extremely low water retention capacity	$\left[P\left(\frac{5_{seg}}{100};\frac{5_{ser}}{100};\frac{5_{sw}}{100}\right) B\left(\frac{5_{OM}}{100};\frac{5_{lc}}{100}\right)\right]^{\frac{CV1-4,7;PP4}{Ef1,3}}_{\frac{Lf1-2;LP3;CU2}{E3-4}}$

5 Experiences from Applications to Drylands

Table 5.1 (continued)

Facet ID	Land form and rock	Soils	Land use type/land cover	Hydrology and moisture regime	Land degradation status (DPSIR legend)[a]
3	Extended slopes of alluvial fans and pediments and dip-slopes. Clastic volcanic debris, sediments and Quaternary alluvium with minor quantities of carbonates evaporites and scattered igneous rocks, andesites mixed with siltstones, limestones and marls	Predominantly Eutthric Lithosol, shallow and gravely with a petrocalcic phase, a calcareous hardpan within the top 20 cm of depth. Inclusions of Haplic Phaeozem in petro-calcic phase and presence of the near-surface calcareous hardpan and an impervious, surface-sealing clay crust	Natural pastures underlain a sub-thorny and thorny brush and shrubs mixed with scattered rosulifolious and crasicaule cacti. Small areas of annual permanent agriculture interspersed with natural pastures	Intermediate site, receiving considerable runoff contributions from steeper hillside slopes, thinly veneered with fluvial gravels of steep, seasonally draining streambeds, arroyos and gullies dissecting areas with considerable laminar surface runoff	$\left[P\left(\frac{5seg}{30};\frac{5ser}{60};\frac{4sed}{20}\right) C\left(\frac{4f}{100}\right) B\left(\frac{3OM}{100};\frac{3lc}{80}\right) \right] \frac{\overline{CV1-4,7;AP1-2;PP4}}{\frac{FI1-2;EI1,3}{LT1-2;A1,i3}}$

Table 5.1 (continued)

Facet ID	Land form and rock	Soils	Land use type/land cover	Hydrology and moisture regime	Land degradation status (DPSIR legend)[a]
4	Gentle extended slopes of Quaternary alluvial fans and pediments of alluvial valleys with clastic and alluvial deposits strongly dissected by seasonal streambeds	Predominantly Haplic Phaeozem with a petro-calcic phase consisting of a near-surface (<50 cm) calcareous hardpan, inbedded in deeper and fine alluvial material and an impervious surface sealing clay crust	Mainly annual permanent agriculture, corn and beans crops with small areas with oats. Agricultural areas surrounded by natural pastures and cacti of various kinds, and sub-thorny bush	Receiving sites, mostly through seasonal runoff of laminar kind, and through seasonal streambeds and gullies. Extensive presence of rills of various sizes and gullies. Medium water holding capacity	$\left[P\left(\frac{4se}{100}; \frac{4sw}{90}; \frac{4cr}{100}\right) C\left(\frac{3f}{100}\right) B\left(\frac{4OM}{100}; \frac{4c}{100}; \frac{4bio}{100}\right) \right] \frac{CV1-5;AP1-2;PP5-6}{\frac{ET1,4}{H_3C1=2;E4}}$
5	Flat or gently sloping Quaternary alluvial and fluvial valleys and banks near streambeds mixed with clastic volcanic and fluvial debris and gravels	Haplic Phaeozem with a deep petro-calcic phase of a calcareous hardpan, inbedded in deeper and fine alluvial silts and clays mixed with fluvial debris	Annual permanent agriculture with corn and beans and other crops, where not impeded by fluvial debris, with thorny shrubs and cacti edges	Receiving sites, medium to high moisture retention capacity and occasional seasonal flooding and sedimentation	$\left[P\left(\frac{4se}{100}; \frac{4sw}{90}; \frac{4cr}{100}\right) C\left(\frac{3f}{100}\right) B\left(\frac{4OM}{100}; \frac{4c}{100}; \frac{4bio}{100}\right) \right] \frac{CV1-5;AP1-2;PP5-6}{\frac{ET1,4}{H_3C1=2;E4}}$

[a]DPSIR Legend as described in Box 5.1 above and in Ponce-Hernandez and Koohafkan (2004).

the most striking feature of the methodology is its integration of the biophysical aspects of land degradation to their socio-economic, cultural and management causes identified in terms of driver forces and pressures on the land.

The methodology still requires much testing, improvement and refinement, particularly in the automation of its many aspects, and in the validation and verification of the accuracy of predicted assessment results against long-term monitoring data. On the other hand, automation would be particularly advantageous in the development of the decision support systems, the procedural toolboxes, the indicator processing engine, the causality explorer and an automated mapping and legend integrator. These developments, when achieved, could render a powerful tool to assist land assessors in their land degradation assessments, the identification of root causes (drivers and pressures) and in the design of policies to address them.

5.1 Appendix: Sample of indicator variables for field observation, measurement and recording

No	Indicator	Measurement
	Evidence of erosion	
1	Rills	Length, width and depth at all points where the width or depth changes significantly.
2	Gullies	Include the contributing area. Note the location of each width and depth measurement along the length
3	Pedestals	From pedestal surface to soil surface. Note the feature causing the pedestal to form (i.e. stone on soil surface)
4	Plant/tree root exposure	Prevalence and average depth
5	Fence post/other exposure	Distance from notch/mark to soil surface
6	"Waterfall" soil loss	Length, width and depth on down slope side of tree/post
7	Rock exposure	Estimate area with exposed rocks or bedrock on surface
8	Tree mound	Height of soil mound around shrub/tree
9	Build up against barriers/tree trunk/plant stem	Height and width of soil build up on upslope side
10	Sediments in drains	Depth
11	Enrichment ratio	Percenatge of fine particles downslope: Percenatge of fine particles upslope
12	Solution notches (indentation or discolouration)	Distance from notch to soil surface
13	Soil/rooting depth	Note if evident
14	Armour layer depth	Note if evident

5.1 Appendix (continued)

No	Indicator	Measurement
	Evidence of poor soil structure/compaction	
15	Structure	Massive, blocky, etc.
16	Consistency	Very hard, friable, etc.
17	Infiltration	Slow Med Fast
18	Water holding capacity	Low Med High
19	Differential crop growth pattern	Note and describe if evident
20	Crop root growth characteristics i.e. corkscrewing, thickening, etc.	Note and describe if evident
	Evidence of chemical degradation:	
23	pH <6 or >7.5	
23	Salinity – colour change, structure, "dusty" surface	Note and describe if evident
25	Sodicity – resistant species	Note and describe if evident
26	Alkalinity	
27	Toxicity/contamination – plant deficiencies	Note and describe if evident
28	Nutrient deficiencies	Note and describe if evident
	Evidence of biological degradation	
29	Percenatge of Land cover	Estimate using a representative 10×10 m plot
30	Soil temperature and aridity	Estimate
31	Above and below-ground soil biomass	Estimate – High Med Low
32	Soil Moisture	Estimate – High Med Low
33	Biodiversity of above and belowground flora/fauna	# of species observed
	Evidence of water quantity and quality degradation	
34	Change of permanent waters to seasonal/periodic	Land user's observation
35	Drought frequency	From climate station data/land user's observation
36	Increasing depth to water table	Land user's observation
37	Change in annual rainfall	From climate station data/land user's observation
38	Atmospheric contamination/effect on plants	Note and Describe if evident
40	Terrestrial carbon depletion	Note and Describe if evident
41	Change in annual air temperature	From climate station data/land user's observation
42	Decline of crop yield	Land user's observation

References

Beckett, P.H.T., Webster, R., McNeil, G.M. and Mitchel, C.W. (1972). Terrain evaluation by means of a databank. Geographical Journal 138(4), 430–456

Bot, A. and M. Snel (2002). Some suggested indicators for Land Degradation Assessment of Drylands. Draft paper. Food and Agriculture Organization, Rome. LADA Virtual Centre ftp://ftp.fao.org/agl/agll/ladadocs/paper_281102.doc

Dixon, R. (2003). Application of the LADA Framework Approach for Land Degradation Assessment in Drylands: Case Studies in Mexico. Consultancy Report. AGLL, FAO, Rome. ftp://ftp.fao.org/agl/agll/ladadocs/mexicocasestudies1.pdf

FAO. (2007). Land Degradation Assessment in Drylands. The LADA project. FAO, Rome. http://lada.virtualcentre.org/pagedisplay/display.asp?section=ladahome

Mitchell, C.W. (1973). Terrain Evaluation. Longmans, London.

Moss, M.R. (1983). Landscape synthesis, landscape processes and land classification, some theoretical and methodological issues. GeoJournal 7(2):145–153.

Ponce-Hernandez, R. (2005). Applied land degradation assessment methods: Methodological framework and training manual for the West Asia Region. United Nations Economic and Social Cooperation for West Asia (ESCWA). Training workshop December, 2005. Beirut, Lebanon.

Ponce-Hernandez, R. and Ahmed, O. (2008). A Bayesian network approach to finding the likely causes of land degradation based on DPSIR indicators and field observations. In: P. Zdruli and E. Costantini (eds.), Moving Ahead from Assessments to Actions: Could We Win the Struggle with land Degradation? Proceedings of the 5th International Conference on Land Degradation. Volume 2. September 2008. Bari, Italy.

Ponce-Hernandez, R. and Koohafkan, P. (2004). Methodological Framework for Land Degradation Assessment in Drylands (LADA) – Simplified Version – Report on a Consultancy as Visiting Scientist. Food and Agriculture Organization of the United Nations, Rome. http://ftp://ftp.fao.org/agl/agll/lada/LADA-Methframwk-simple.pdf

Ponce-Hernandez, R. and Mohammed, H. (2007). A methodology for the assessment of land degradation and monitoring the advance of desertification in drylands, based on the DPSIR indicators approach, remote sensing and GIS techniques: Application to Lybia. Proceedings of the Desert and Desertification Conference. Arab Centre for Saharian Research and Development of Saharian Societies. 19–21 March, 2007. Sebha, Lybia.

UNCCD. (2004). United Nations Convention to Combat Desertification http://www.unccd.int/

World Resources Institute. (2005). Millennium Ecosystem Assessment (MA). Ecosystem and Human Well-Being: Desertification Synthesis. World Resources Institute, Washington, DC.

Chapter 6
Global Warming, Carbon Balance, and Land and Water Management

Ahmet Ruhi Mermut

Abstract With more intense, longer droughts in larger areas since the 1970s, particularly in the tropics, one of the most important issues facing the world today is the need to ensure food security through the sustainable management of water and soil resources. There is a need to understand interactions between climate and land degradation through dedicated observations of the climate system, proper assessment and management of water and land resources, and advances in climate science. The debate on the complex issue of quantification of carbon stocks is still evolving. It is generally agreed that carbon sequestration, especially in soils, could be a highly cost effective and environmentally sound mitigation technique. Natural resources, particularly land and water are increasingly restricted both in quality and quantity in most parts of the world. There are still sufficient water resources however, to produce food and fibber for a growing population but that trends in consumption, production and environmental patterns, if continued, will lead to water crises. Consumption and pollution of water by agriculture are becoming serious concerns.

Keywords Drought · Climate variability · Food security · Carbon sequestration · SLWM · Conservation agriculture

6.1 Introduction

Global warming is defined as the increase in the average annual temperature of the Earth's near-surface, air, and oceans. This will cause an increase in extreme weather events such as storms and floods, changes in agricultural yield, and extinction of plant and animal species. The term *climate change* may also mean periods of overall temperature change including global cooling (US Department of Energy, 1999). The United Nations Framework Convention on Climate Change (UNFCCC) uses

---

A.R. Mermut (✉)
University of Saskatchewan, Saskatoon, SK S7N, Canada;
Department of Soil Science, Faculty of Agriculture, Harran University, Şanlıurfa 63200, Turkey
e-mail: a.mermut@usask.ca

P. Zdruli et al. (eds.), *Land Degradation and Desertification: Assessment, Mitigation and Remediation*, DOI 10.1007/978-90-481-8657-0_6,
© Springer Science+Business Media B.V. 2010

the term *climate change* for human-caused change, and *climate variability* for other changes. World needs dedicated action and resilience to climate change. Greenhouse gas, due to human activity, is just only one factor among others. Some scientists suggest that the change is due to solar activity (irradiation) amplified by cloud seeding via *galactic cosmic rays*. Effect of volcanic activities could be considerable. Burning of the fossil fuel has produced about three-quarters of the increase in CO_2 over the past 20 years. Most of the rest is due to land-use change, in particular deforestation.

The sustainable use and protection of the land and water resources requires careful management to maintain the environmental integrity and sustain agricultural production. Shortage in water can cause serious world crisis. At a time that we talk so much about climate change, we are also approaching the limits of human exploitation of land and freshwater resources, to grow essential food and fibber for human basic needs. Improper use of land resources causes degradation. Degraded soils, parched aquifers, polluted waters, and the loss of plant and animal species threaten food production in poor, heavily populated countries. Environmental degradation is one of the greatest risks to future world food security. This short synthesis attempts to assess the linkages between global warming, carbon balance, and land and water management.

6.2 Global Warming and Drought

Drought is a normal, recurring feature of the climate in most parts of the world. In general, means acute water shortage. Having adequate drought mitigation strategies in place could greatly reduce the impacts. Recurring or long-term drought can bring about desertification. Drought is a multi-dimensional phenomenon that needs to be understood and explained using insights developed by different disciplines (www.Drought.gov). Drought is not simply low rainfall. It disrupts crop production and animal breeding, encourages erosion and affects the productivity of farming enterprises resulting in the decline of the national economy. The loss of vegetation has long-term implications for the sustainability of the agricultural industries. Water quality suffers, and toxic algae outbreaks may occur; plants and animals are also threatened. Bushfires and duststorms often increase during dry times (Australian Government, Bureau of Meteorology. http://www.bom.gov.au/climate/drought)

Drought appears to be a cyclic process and human activities also play a significant role. Increases in global temperatures may in turn cause broader changes, including glacial retreat and worldwide sea level rise. Changes in the amount and pattern of precipitation may result in flooding and drought. Many evidences were reported about the fluctuation of climatic parameters in the literature.

There is growing scientific evidence that the human influence on the global climate began around 8000 years ago with the start of forest clearing to provide land for agriculture and 5000 years ago with the start of Asian rice irrigation. A marked increase in aridity at 2200 BC, wind activities, and volcanic eruptions in northern Mesopotamia have caused considerable degradation of the land use condition. Radiocarbon grain samples retrieved from the excavation site in Tell

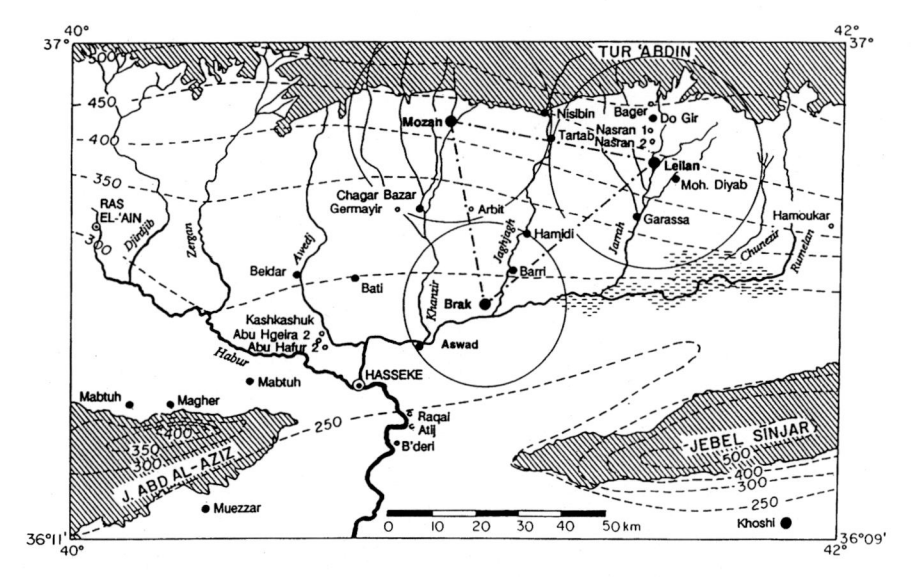

Fig. 6.1 Ancient urban settlements in the Habur Plains, Northeast Syria (Habur Plains, Northeast Syria, urban settlement system at 2600–2200 BC. *Circles* indicate probable areas of agriculture and herding sustain each city. *Diagonal lines* are elevation ≥500 m above sea level: *dotted lines* are modern rainfall isohyets (in mm). *Large circles* are sites of 75–100 ha. *Medium circles* are sites of 25–50 ha. *Small hollow circles* are sites less than 10 ha. Tephra deposits have been retrieved from Tell Leilan, Tell Bager, Tell Nasran1, Tell Nasran 2, Abu Hgeira 2, and Abu Hafur 2). Source: Weiss et al. (1993)

Leilan in Syria just 20 km from the Turkey-Syrian border provided calibrated dates of 2280–2040 BC (Forrest et al., 2004). Abrupt climate change caused abandonment of cities in the Habur plain in Northern Mesopotamia and collapse in adjacent regions suggested that the impact of the abrupt climatic change was extensive (Fig. 6.1).

This has caused regional desertification, and sudden collapse of the Akkadian empire (Weiss et al., 1993). A severe drought had occurred that even earthworms had not survived. This was among the earliest evidence that severe climate change could be responsible for the rise and fall of civilizations together with miss management of land resources.

During the AD 1990s the Waubay Lakes complex in eastern South Dakota experienced historically high water levels. A 1,000-year hydroclimate reconstruction was developed from local bur oak (*Quercus macrocarpa*) tree-ring records and lake-sediment cores. Lake shoreline analyses and drainage features provides late-Quaternary geomorphic context. Tree-ring width and shell geochemistry of the ostracode *Candona rawsoni* show marked coherence, indicating synchronous responses to moisture balance in vegetation and lake salinity. Geomorphic evidence suggests that buffering of lake-system expansion occurred during pluvial periods by evaporative dynamics (Shapley et al., 2005).

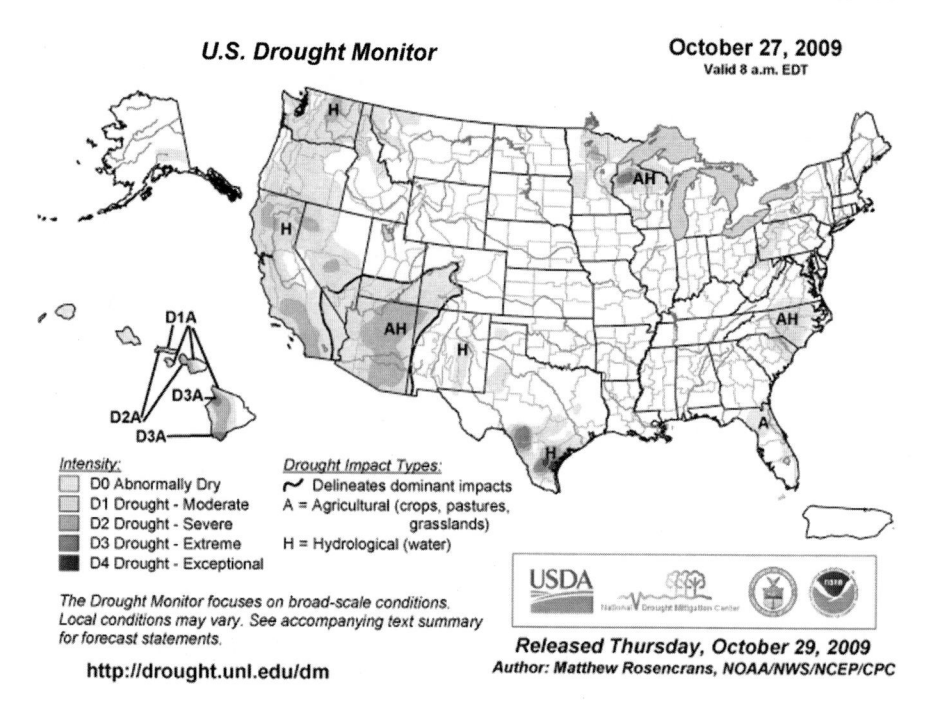

Fig. 6.2 US Drought map, based on the National Weather Service (NWS). National Centers for Environmental Prediction are depicting current areas of dryness and drought

Figure 6.2 shows a drought map, which is based on analysis of the data produced by the NWS forecast products. It utilized the HPC 5-day QPF and 5-day Mean Temperature progs, the 6–10 Day Outlooks of Temperature and Precipitation Probability, and the 8–14 Day Outlooks of Temperature and Precipitation Probability, valid as of late Wednesday afternoon of the USDM release week. The NWS forecast web page used for this section is: http://www.cpc.ncep.noaa.gov/products/forecasts

6.3 Carbon Balance

In the past decade, increasing awareness of CO_2 build-up in the atmosphere and the threat of global warming has instigated society to find means to reduce atmospheric CO_2. The concept of greenhouse gas reduction by sequestering carbon in different terrestrial ecosystems, or withdrawal of CO_2 from the atmosphere, has been extensively discussed over the last 2 decades. While the debate on the complex issue of quantification of carbon stocks is still evolving, it is generally agreed that carbon sequestration can be a highly cost effective and environmentally friendly sound mitigation technique. This would also be a response to commitments by signing parties under the conventions of:

6 Global Warming, Carbon Balance, and Land and Water Management   71

(i) Climate Change (Keyoto Protocol);
(ii) Biological Diversity; and
(iii) Combating Desertification.

Therefore, strategies that could lead the amelioration of these problems are likely to be of great global significance.

People interested in the carbon flux area have not yet succeeded to translate the vast current knowledge of carbon dynamics that scientists have produced, into real agronomic practices. There is a need to carry out practical works to design appropriate strategies for carbon sequestration. The research so far carried out show the multiple agricultural and environmental benefits. Funding for research and development to address the practical implementation of carbon sequestration is needed, if the potential of this new paradigm is ever to be realized.

There is a general agreement that with appropriate management technologies soil can function as a sink and contribute the process of CO_2 reduction in the atmosphere. This would mean drawing CO_2 out of the air and converting it to biomass (plants) or soil organic matter. By using water and energy from the sun, plants are naturally capable of converting CO_2 to carbohydrates or biomass and consequently organic matter in the soil. Preliminary estimates suggest that using appropriate management techniques ∼40–80 Pg C that would be produced through the combustion over the next 50–100 years could be sequestered in the cropland. This would mean that carbon sequestration offers a mean to control the CO_2 levels in the air to keep 550 ppm critical threshold level.

The total amount of carbon stored in terrestrial ecosystems is large. According to US Department of Energy (1999) it is ∼2,000 ± 500 Pg. The rate of the process is estimated to be ∼2 Pg C year^{-1} (∼0.1% of the current storage). About 75% of terrestrial carbon occurs in the soil, therefore, they are essential in terms of carbon sequestration. The potential for carbon sequestration appears to be large in comparison with current rate for terrestrial ecosystems (5–10 Pg C year^{-1}, when all terrestrial ecosystems are considered). What is the maximum capacity to sequester carbon is not yet known.

There are two fundamental approaches to sequestering carbon:

(i) Protection of ecosystem that store carbon so that sequestration can be maintained (increasing residence time), and
(ii) Manipulations of ecosystems to increase carbon sequestration beyond the current conditions.

All factors remaining the same, the rate of C sequestration in soils is higher for warm humid regions than dry cool regions. Great differences occur also between non-degraded soils as compared to severely degraded ones. On the other side it may seem unlikely to sequester large amounts of carbon in dryland regions, in comparison to other agro-ecological zones of the world. But, according to UNEP (1997) drylands store 60 times more carbon than the carbon added to atmosphere by fossil fuel. Drylands cover 450 million hectares area. A small change in the rate of carbon

Table 6.1 Estimates of carbon sequestration potential of some major land-use types with projected annual carbon storage and time frames (UNEP, 1997)

Option	Area (million hectares)	Rate (TC ha^{-1} year^{-1})	Period (year)	Cost ($ US tC)	Total (Mt C year^{-1})
Dryland crop management	450	0.3–1.0	5–20	1–5	135
Halophytes	130	0.5–5.0	Indefinite if harvested 5 year if not	170 (irrigated and harvested) 20 (dryland not harvested)	65
Bush encroachment	150	0.1–0.5	15–50	10–20	37
Energy crops	20 (5% of dryland crop area)	4–8	Indefinite	2–5	80
Domestic biofuel efficiency	Not applicable	Not applicable	Indefinite	2–5	75
Agroforestry (arid)	50	0.2	30	2–10	10
Agroforestry (semiarid)	75	0.5	20	2–10	38
Agroforestry (subhumid)	150	1.5	15	2–10	225
Improved pasture (semiarid Asia)	10 (2,500 degraded globally)	0.1	30	10	1
Savannah fire control	900 (globally)	0.5	30	1–5	450
Woodland management	400 (globally)	0.5	30	1–5	200

sequestered in dryland regions can have large impacts on CO_2 in the atmosphere. Over 1 billion people currently live in susceptible drylands and any effort to restore productivity of these eco-regions will be of benefit for their livelihoods. Table 6.1 shows carbon sequestration potential of some major land used types.

Technological options considered to sequester carbon in agricultural land are not many. Some of these were developed for temperate and tropical regions, others for developing economies. Application may differ from one area to another region. There is a strong need to do applied research thus to determine the actual values that can be used to calculate the economical benefits of carbon sequestration.

One of the fundamental arguments is that, about 50% of soil organic matter (SOM) is lost in the topsoil, due to intensive agricultural practices. Uncultivated soils were in equilibrium with the native vegetation and accumulated large soil organic carbon reserves (SOC), but cultivation has disrupted this steady state equilibrium (Lal et al., 1999). There are reliable estimates that many cultivated soils in North America have lost substantial amount of SOM due to crop cultivation (Acton and Gregorich, 1995; Bruce et al., 1999), which resulted in decline in production, increased soil erosion and soil degradation. It is estimated that lost carbon from the soil will take 25–50 years to store it back, with current technologies (Lal et al., 1998). With good management practices it may be possible to exceed the original native SOM content of many soils.

Lal et al. (1999) suggest that intensification of agriculture on good soils can be achieved through the widespread adoption of:

(i) conservation tillage and residue management,
(ii) irrigation and water management systems,
(iii) improved cropping systems, including agroforestry and sustainable fertiliser management

(i) *Conservation tillage and Residue management*: Conservation tillage (CT) is a method designed to keep most crop residues on the soil surface. This way soil is protected against erosion and water losses by runoff and evaporation are also reduced. Reliable data show that traditional intensive tillage decrease soil carbon as it encourages rapid mineralisation of soil organic matter. Fallow periods in rotation have been used in semiarid regions to conserve moisture for succeeding crop. However, fallow especially in combination with conventional tillage exposes the soil to erosion (especially wind) and creates temperature and moisture conditions that speed up the process of organic matter decomposition in the soil.

Conservation tillage stores or builds more organic matter into the soil and provides long-term productivity and sustainability by enhancement of soil quality and improvement of soil resilience (Reicosky et al., 1995; Grant, 1997). Technologies related to conservation tillage are well adopted by farmers in North America and currently gaining momentum by farmers elsewhere. More than 1/3rd of the land farmed in the USA is now managed with a CT system, including no till, minimum till, or ridge till (Lal et al., 1999). One of the encouraging advantages of this method is reduced farm input.

(ii) *Irrigation and water management systems*: Irrigation, especially in drought-prone soils can enhance SOC content. Experimental data on the impact of the irrigation on SOC dynamics are rare. Bruce et al. (1999) suggest that conversion of dryland to irrigated agriculture may increase the SOC content with an average rate of 100 kg ha^{-1} year^{-1}. Irrigation of soils in arid and semi-arid regions also affects the SOC pool and its dynamics. This is a complex issue and very little attempt is made to increase our understanding on this aspect of carbon cycle.

(iii) *Improved cropping systems, including agroforestry and sustainable fertiliser management*: It is natural that the application of fertilisers (N, P, and K) would increase the overall biomass production, including root biomass. Long-term experimental studies around the world have clearly proven this. Especially over the last 50–60 years fertilizers have increased notably food production. We should recognize that this is a good strategy for increased food production especially in the developing countries, which in turn can help to stabilize deforestation and reduce greenhouse gas emissions. More biomass production means increased chance of carbon sequestration. Other organic fertilizers and inputs, such as crop rotation, and agroforestry, can also do this.

A number of other techniques could used to increase and enhance soil carbon sequestration. They include:

Organic fertilisation and other organic inputs: This includes green manures with especially legume species, manure compost, manure sewage sludge, wood processing remains, and peat, beside crop residues. Adding organic matter on severely eroded soils reduces the risk of erosion by promoting the formation of aggregates that resist to soil degradation.

Crop Rotation: Forages and legumes have extensive rooting systems that leave large amounts of organic matter. When used with conservation tillage, crop rotation adds more organic matter to the soil.

Agroforestry: This is a rather new system of combination of fast growing trees with agriculture that also include feed to supports livestock (Mergen, 1986). It provides habitat for bio-diversity and produces goods and services (Winterbottom and Hazelwood, 1987). This system can increase carbon sequestration substantially (Unruh et al., 1993). It is a compromise solution to continuous crop production, supporting livestock and carbon sequestration. Extensive research is now going on around the world. ICRAF in Nairobi Kenya is established to specifically deal with agroforestry. Little is known however about carbon storage within a certain time frame (Schroeder, 1993).

One of the advantages of agroforestry systems is providing a more hospitable environment for biodiversity, both above and belowground. Sacnchez et al. (1996) suggest that substantial biodiversity benefits are likely if agroforestry covers large areas and maintained for relatively long time.

The flux of carbon among plants, soils, and the atmosphere is still poorly understood. It is important to recognize that carbon sequestration is an immature field of study. Understanding how to increase soil carbon stocks in agricultural lands is critical to increasing sustainability of food production and mitigation of degraded lands.

Interesting research results could be mention regarding Canada's carbon cycling two main national objectives (Acton and Gregorich, 1995). These are:

1. To determine weather Canadian agriculture is a source or sink of atmospheric CO_2,
2. To reduce uncertainties about the processes that determines the exchange of CO_2 between land and atmosphere.

It is calculated *that the amount of CO_2 released from the soil, when native Prairie grasslands were first cultivated, is equivalent to that released by about 10 years of fossil fuel consumption in Canada.* Currently there seems to be a balance established with the farming system used. Several methods were tested to sequester atmospheric CO_2 in soils and several management options were also identified in Canada.

The quantification of carbon storage requires many years of studies, as the altered new system attains a balance between soil carbon inputs and outputs after so many years. This is unfortunately one of the problems we face in carbon sequestration studies. Canada is interested in developing land management systems that maintain biodiversity, sustainability, and agricultural competitiveness. Canadian ecological condition allows only one crop a year and the growing seasons is short, due to low atmospheric and soil temperatures. The climate varies between semi-arid to humid. About 18 years of tillage treatments of soils from Eastern Canada under corn showed that no-till increased organic matter in the soil (both at the surface and throughout the soil profile). Table 6.2 shows clearly that organic matter has increased in no till treatment.

Table 6.2 Organic matter at two depths after 18 years of various tillage treatments of a soil from Ontario, Canada under corn

Tillage system	Soil organic matter (tonnes per hectare)		
	0–15 cm	15–30 cm	0–30 cm
No till	86	65	151
Chisel plough	73	52	125
Disc	74	58	132
Moldboard plough	66	64	130

6.4 Land and Water Management

The sustainable use and protection of our natural resources requires careful management to maintain the environmental integrity of the resource. Many countries are developing programmes to promote the best use of the resources for their social and economic benefits while protecting associated resource values, the environment, and public health and safety.

A good example in this context is the Australia's Commonwealth Scientific and Industrial Research Organisation (CSIRO) that tries to shape land and water resource management through finding new ways to use less water in agriculture and other industries, to re-use or recycle urban and industrial waters and wastes, while also protecting the rivers, catchments and groundwater reserves (http://www.clw.csiro.au). Multidisciplinary teams tackle land and water challenges with combined force. Most of their work is performed in partnership with government and research scientists. However, the complexity of the landscape still remains a challenge and limits our understanding. Experience, combined with the knowledge and expertise of the partners in industry, government, academia and the community, has made the CSIRO Land and Water Division an excellent example that can be used in other parts of the world.

Fraiture et al. (2009) in a recent study concluded that there are sufficient water resources to produce food for a growing population but that trends in consumption, production and environmental patterns, if continued, will lead to water crises in many parts of the world. Recent increase in food prices, partially caused by the increasing demand for agricultural products in non-food uses, underline the urgent need to invest in agricultural production, of which sustainable land and water management is a crucial part. The current situation and the long-term outlook require a fresh look at approaches that combine different elements such as the importance of access to water for the poor, providing multiple ecosystem services, rainwater management, adapting to efficient irrigation systems, enhancing water productivity, and promoting the use of low-quality water in agriculture.

6.4.1 Sustainable Water Use

The water available for agriculture becomes limiting due to population growth, competition from other water users, and drought and degradation of water quality. It is, therefore, important to ensure that every drop of water counts for crop water use. To manage sustainable water use the following measures are suggested: (1) Grow drought-resistant plants, (2) Apply water efficiently, (3) Manage soil and water to minimize water loss, (4) Conserve water for critical growth periods, (5) Use irrigation practices that enhance root growth, and (6) Minimize downstream environmental damage caused by irrigation runoff and deep percolation.

The European Environment Agency (EEA) (2001) provides an overview of the main natural and artificial causes and impacts of extreme hydrological events, such as floods and droughts, in European countries. This report also gives an overview on policy responses to prevent such events and reduce their damage. One of the main contributions of this report is the identification of driving forces, pressures, state, impacts and responses concerning floods and droughts. EEA has compiled information on extreme events in EEA member countries that includes also the Central and Eastern European countries.

Consumption and pollution of land and water by agriculture are becoming serious concerns. Water resources can be used much more efficiently in producing food and fibre, while minimizing pollution and supporting ecosystems. How to achieve this depends on mindsets and societal goals, as well as how the institutional systems and structures will be able to respond to such needs.

References

Acton, D.F. and Gregorich, L.J. (1995). The health of our soils, towards the sustainable agriculture in Canada. Agriculture and Agri-Food Canada, Research Branch. Center for Land and Biological Resources Publication. 1906/E

Australian Government, Bureau of Meteorology http://www.bom.gov.au/climate/drought/livedrought.shtml

Australian Government CSIRO http://www.clw.csiro.au/index.html

Bruce, J.P., Frome, M., Haites, E., Janzen, H., Lal, R. and Paustian, K. (1999). Carbon sequestration in soils. Journal of Soil and Water Conservation 54:382–389.

de Fraiture, C., Molden, D. and Wichelns, D. (2009). Investing in water for food, ecosystems, and livelihoods: An overview of the comprehensive assessment of water management in agriculture. Agriculture and Water Management (available on line).

European Environment Agency. (2001). Sustainable Water Use in Europe Part 3: Extreme Hydrological Events: Floods and Droughts. EEA, Copenhagen.

Grant, F.R. (1997). Changes in soil organic matter under different tillage and rotations: Mathematical modelling in ecosystems. Soil Science Society of America Journal 61: 1159–1175.

Lal, R., Follet, R.F., Kimble, J.M. and Cole, V.R. (1999). Managing US cropland to sequester carbon in soil. Journal of Soil Water Conservation 54:374–381.

Lal, R., Kimble, J.M., Follet, R.F. and Cole, V.R. (1998). The Potential of US Cropland to Sequestre Carbon and Mitigate the Greenhouse Effect. Sleeping Bear Press, Ann Arbor, MI, 128pp.

Mergen, F. (1986). Agroforestry-an overview and recommendations for possible improvements. Tropical Agriculture 63:6–9.

Forrest, Francesca deLillis, Mori, L., Guilderson, T. and Weiss, H. (2004). The Akkadian Administration on the Tell Lailan Acropolis. Imperialism and Cooptation on the Habur Plain. Yale University, New Haven. http://leilan.yale.edu/pubs/files/poster1/poster1.jpg; http://www.etana.org/abzu/abzu-displayentry.pl?RC=16588

Reicosky, D.C., Kemper, W.D., Langdale, G.W., Douglas, C.L., Jr. and Rasmussen, P.E. (1995). Soil organic matter changes resulting from tillage and biomass production. Journal of Soil Water Conservation 50:253–262.

Sanchez, P.A., Buresh, R.J. and Leakey, R.R.B. (1996). Trees, soils and food security. Paper presented at the Discussion Meeting on Land Resources: On the Edge of Malthusian Precipice? London, 5 December 1996.

Schroeder, P. (1993). Agroforestry systems: Integrated land use to store and conserve carbon. Climate Research 3:53–60.

Shapley, M.D., Engstrom, D.R. and Osterkamp, W.R. (2005). Late-Holocene flooding and drought in the Northern Great Plains, USA, reconstructed from tree rings, lake sediments and ancient shorelines. The Holocene 15(1):29–41.

UNEP. (1997). World Atlas of Desertification. 2nd edn. In: N. Middleton and D. Thomas (eds.), United Nations Environment Programme. Edward Arnold, New York.

US Department of Energy. (1999). Carbon sequestration, state of the science. A working paper for Road Mapping Future Carbon Sequestration Research and Development. US Department of Energy Office of Science, Office of Fossil Fuel.

Unruh, J.D., Houghton, R.A. and Lefebvre, P.A. (1993). Carbon storage in agroforestry: An estimate for Sub-Saharan Africa. Climate Research 3:39–52.

Weiss, H., Courty, M.-A., Wetterstrom, W., Guichard, F., Senior, L., Meadow, R. and Curnow, A. (20 August 1993). The genesis and collapse of third millennium north Mesopotamian civilization. Science, New Series 261(5124):995–1004.

Winterbottom, R. and Hazelwood., P.T. (1987). Agroforestry and sustainable development: Making the connection. Ambio 16:100–110.

Part II
Land Degradation and Mitigation in Africa

Chapter 7
The Use of GIS Data in the Desertification Risk Cartography: Case Study of South Aurès Region in Algeria

H. Benmessaud, M. Kalla, and H. Driddi

Abstract Risk cartography is a primordial step for the valuation and management of desertification process but it is a complicated one, which necessitates large amounts of spatial and statistical data. The use of Geographic Information Systems (GIS) permits to manage and use these data efficiently. The objective of our study was the compilation of desertification sensitivity maps of the area south of Aurès region, Algeria. We used a GIS system following the MEDALUS methodology (Mediterranean Desertification and Land Use), which use qualitative indices to define zones sensitive to desertification. The creation of the database consists of four information layers (soil, vegetation, climate and the socio-economic data) accomplished with ground validation. Once the database was completed, its elaboration helped identifying sensitivity areas through the use of various indices. The result is a middle scale risk map, which presents a synthesis of the desertification intensity. The map is an efficient tool to help decision makers endorse sustainable land management strategies for the protection of natural resources, especially to those greatly affected by increased aridity.

Keywords Desertification sensitivity · Indices · Risk cartography · Geographical Information System · MEDALUS concept · South Aurès · Algeria

7.1 Introduction

Having large parts of its territory under arid and semi-arid areas, Algeria is one of world's most affected countries by desertification. With 2 million square kilometres of desert and 382,000 km^2 dominated by semi-arid and dry sub-humid areas,

H. Benmessaud (✉)
Laboratoire Risques Naturels et Aménagement du Territoire, Faculté des sciences, Université El Hadj Lakhdar, Batna, Algérie
e-mail: ha123_m123@yahoo.fr

Algeria is the second largest African country as far as the territory is concerned (Abdessemed, 1981; Berkane and Yahiaou, 2007). Desertification has a strong presence in the country; however, the steppe areas are the most sensitive zones to desertification covering about 20 million hectares (Ansar, 2002). Combating thus desertification requires first the acquisition of data, such as soil, biodiversity, socio-economics, etc that are necessary for problem analysis. These data were derived from previous studies (Ballais, 1981; Benmessaoud et al., 2007) and using a GIS were elaborated for desertification impact analyses (Benmessaoud et al., 2007; Bensaid, 2006). The GIS proved to be a powerful tool for such purpose.

The region south of the Aurès has experienced over the last decades rapid degradation of natural resources (Dubois et al., 1997; PNUD/UNSO DGF/Algérie, 2001; Oussedik et al., 2003). The degradation process includes wind erosion and sand encroachment, animal and crop product reduction, and migration. Sustainable development of the area thus necessitates the establishment of digital databases and data elaboration for identification of desertification sensitivity areas.

Our work addresses a specific case study in the region south of the Aures (Eastern Algeria) that was finalised with the completion of a sensitivity to desertification map at medium scale using a GIS system and following the MEDALUS methodology (Mediterranean Desertification and Land Use).

7.2 Background Information

The Aurès constitute a geographical entity located East of the Saharan Atlas Mountain. This whole mountain is very steep and heavily exposed to the process of desertification (erosion), particularly in its southern part that is in direct contact with the Sahara.

Geographically, the study area is located (Fig. 7.1), between the meridians (6° 29' and 5° 36') East and the parallel (35° 15' and 34° 41') North.

The study area is located in a transition zone between the North Atlas and the flat desert expanding in the south. From the climatic point of view the area belongs to the Mediterranean climate and is characterized by hot and dry summers and cold wet winters in the highlands and mild in the plains, however remains generally affected by aridity (Fig. 7.2).

7.3 Working Methodology

The different types of Environmental Sensitivity Areas (ESA) to desertification could be analysed in relation to various parameters such as landform, soil, geology, vegetation, climate, and human action. Each of these parameters were grouped into various uniform classes with respect to the behaviour on desertification and then weighting factors were assigned to each class. The following four indicators were

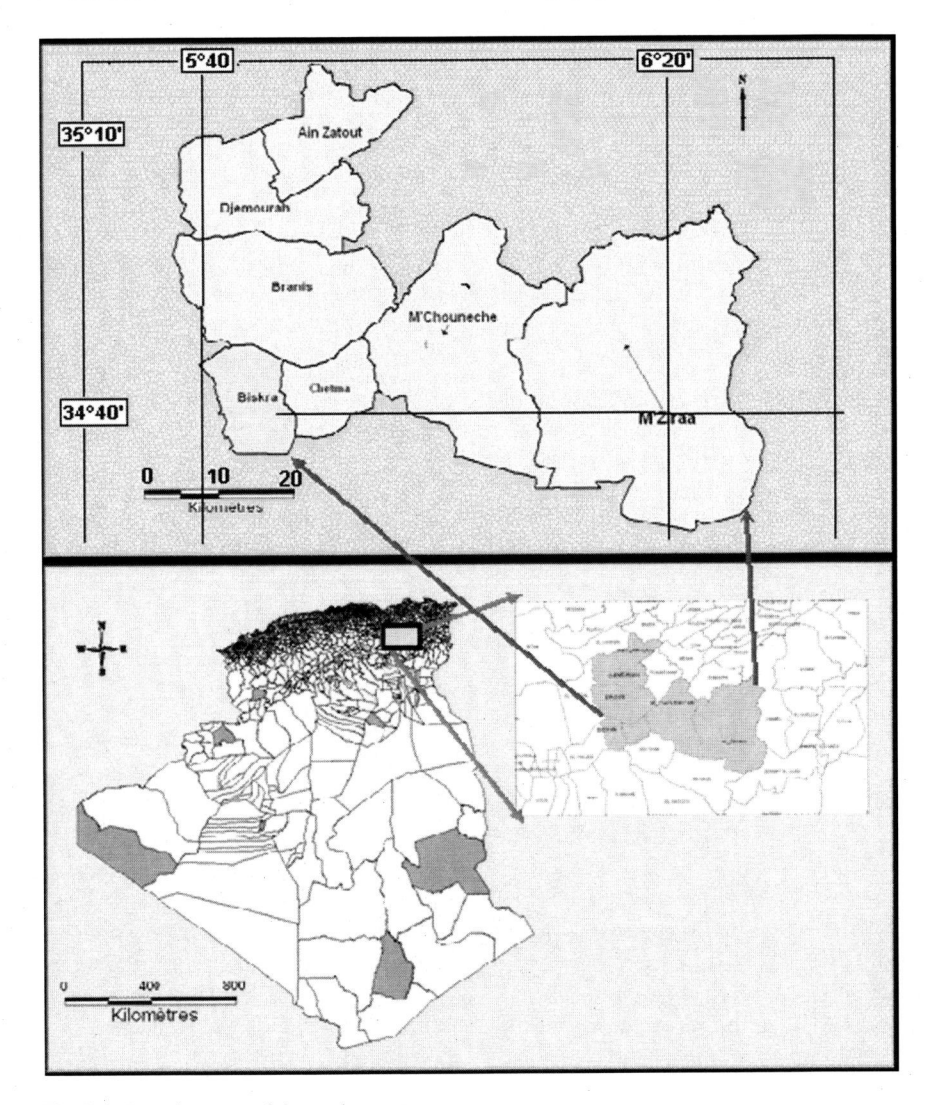

Fig. 7.1 Location map of the study area

evaluated (a) soil quality, (b) climate quality, (c) vegetation quality, and (d) management quality. After the computation of four indices for each indicator, the ESAs to desertification were defined by combining the abovementioned indicators. All the data were introduced in a regional GIS, and overlaid in accordance with the developed algorithms to finalise the compilation of ESAs to desertification. This approach includes parameters, which could be easily found in existing soil, vegetation, and climate reports of the area.

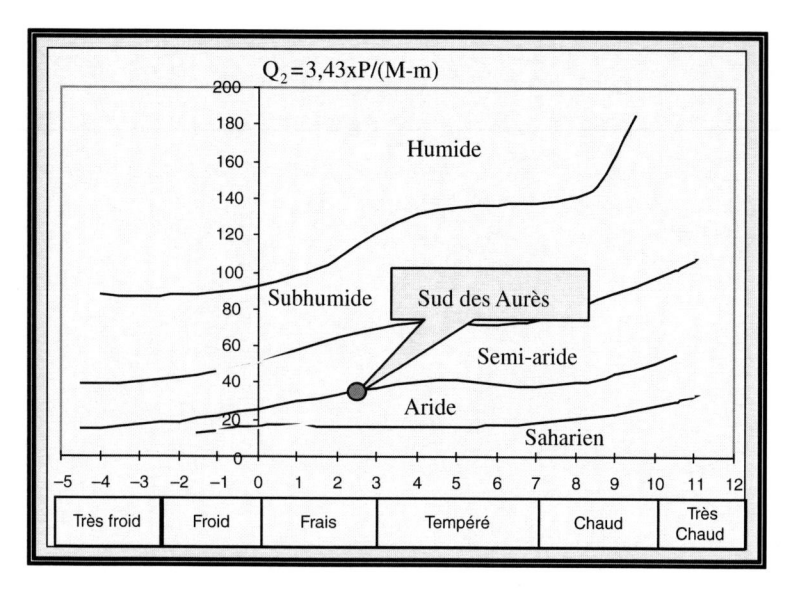

Fig. 7.2 Climagramme D'Emberger

7.3.1 Soil Quality Indicators

Soil is a dominant factor of the terrestrial ecosystems in the semi-arid and dry subhumid zones, particularly through its effect on biomass production. Soil quality indicators for mapping ESAs can be related to (a) water availability, and (b) erosion resistance. These qualities can be evaluated using simple soil properties or characteristics such as texture, parent material, soil depth, slope angle, drainage, stoniness, etc. available in regular soil survey reports The use of these properties for defining and mapping ESAs requires the definition of distinct classes with respect to the degree of land protection from desertification (Table 7.1).

Soil quality index (SQI) was then calculated as the product of the above attributes, namely soil texture, parent material, rock fragment cover, soil depth, slope grade, and drainage conditions using as the following algorithm (Table 7.2):

SQI = (texture × parent material × rock fragment × depth × slope × drainage) 1/6

7.3.2 Climate Quality

Climate quality was assessed using parameters that influence water availability to the plants such as the amount of rainfall, air temperature and aridity, as well as any climate hazards such as frost that might inhibit or even prohibit plant growth. Annual

Table 7.1 Classes, and assigned weighing indices for the various parameters used for soil assessment

SQ characteristics	Class	Description	Characteristic	Index
Texture	1	Good	L, SAL, SL, LS, LC	1
	2	Moderate	SC, SiL, SiCL	1.2
	3	Poor	Si, C, SiC	1.6
	4	Very poor	S	2
Parent material	1	Good	Shale, schist, basic, ultrabasic, conglomerates, unconsolidated	1
	2	Moderate	Limestone, marble, granite, Rhyolite, ignibrite, gneiss, siltstone, sandstone	1.7
	3	Poor	Marl, pyroclastics	2
Slope	1	Very gentle to flat	<6%	1
	2	Gentle	6–18%	1.2
	3	Steep	18–35%	1.5
	4	Very steep	>35%	2
Drainage	1	Well drained	–	1
	2	Imperfectly drained	–	1.2
	3	Poorly drained	–	2
Soil depth	1	Deep	>75	1
	2	Moderate	75–30	2
	3	Shallow	15–30	3
	4	Very shallow	<15	4

Table 7.2 Soil quality index description

	Class	Description	Index
Soil quality index (SQI)	1	High quality	< 1.13
	2	Moderate quality	1.13–1.45
	3	Low quality	>1.46

precipitation was classified in three classes considering the annual precipitation of 250 mm as a crucial value for soil erosion and plant growth (Table 7.3).

The most effective measure for assessing soil water availability is by calculating the difference between precipitation and evapotranspiration and run-off. However, this calculation requires relatively large amount of data such as soil moisture retention characteristics and vegetation growth characteristics. Therefore, the simple BAGNOULS-GAUSSEN aridity index was used. This index was grouped into six classes as shown in Table 7.3. Slope aspect was divided into two classes (a) NW

Table 7.3 Classes and weighing indices for climate quality assessment

	Class (mm)	Index
Rainfall	> 500	1
	250–500	2
	< 250	3
Aridity	< 50.0	1
	50–75	1.1
	75–100	1.2
	100–125	1.4
	123–150	1.8
	> 150	2

Table 7.4 Climate quality index

Class	Description	Range
1	Very favourable	<1.15
2	Favourable	1.15–1.81
3	Unfavourable	>1.81

and NE and (b) SW and SE assigning the indices 1 and 2, respectively. The above three attributes were then combined to assess the three climate quality index classes (CQI) shown in Table 7.4 using the following algorithm:

$$CQI = (rainfall \times aridity \times aspect)1/3$$

7.3.3 Vegetation Quality

Vegetation quality was assessed in terms of (a) fire risk and ability to recover, (b) soil erosion protection (c) drought resistance and (d) plant cover. The existing dominant types of vegetation in the Mediterranean region were grouped into four categories according to the fire risk. Four categories were used also for classifying the impacts of vegetation in regard to soil erosion. Four categories were used for the classification of vegetation with respect to drought resistance. Finally, plant cover was distinguished into three classes (Table 7.5).

The vegetation quality index (VQI) was assessed as the product of the above vegetation characteristics related to sensitivity to desertification using the algorithm below. The VQI index was classified into three classes defining the quality of vegetation with respect to desertification sensitivity (Table 7.6).

$$VQI = (fire\ risk \times erosion\ protection \times drought\ resistance \times vegetation\ cover)1/4$$

7   The Use of GIS Data in the Desertification Risk Cartography

Table 7.5 Relationship between type of vegetation and quality index

	Class	Description	Type of vegetation	Index
Fire risk	1	Low	Sand, and Chott	1
	2	Moderate	Sebkha	1.3
	3	Very high	Course, Hills, Culture, Forest	2
Erosion protection	1	Very high	Mountainous	1.3
	2	Moderate	Course, Hills, Culture	1.8
	3	Low	Sand, and Chott sebkha	2
Drought resistance	1	High	Sand, and Chott	1.2
	2	Moderate	sebkha	1.4
	3	Low	Course, Hills	1.7
	4	Very low	Cultures, Forest	2
Plant cover	1	High	>40	1
	2	Low	10–40	1.8
	3	Very low	<10	2

Table 7.6 Vegetation quality index

	Class	Description	Range
Vegetation quality index (VQI)	1	High quality	1.2–1.6
	2	Moderate quality	1.7–3.7
	3	Low quality	3.8–16

7.3.4 Management Quality or Degree of Human Induced Stress

To be able to assess the impacts of management quality and human induced stress the land was classified in the following categories according to the major land use patterns of the study area:

1. Agricultural land: Cropland and Pasture
2. Natural areas: Forest, shrubland and bare land
3. Mining areas (quarries, mines, etc.)
4. Recreation areas (parks, compact tourism development, tourist areas, etc.)
5. Infrastructure facilities (roads, dams, etc.)

After defining the main land use type in a certain piece of land, then the intensity of land use and the enforcement of policy on environmental protection were assessed for each particular type of land use for any specific area. The management

Table 7.7 Management quality index

	Class	Description	Range index
The management quality index (MQI)	1	High	1–1.25
	2	Moderate	1.26–1.50
	3	Low	>1.51

quality index (MQI) was assessed using the following algorithm and the results are presented in Table 7.7.

$$MQI = (\text{land use intensity} \times \text{policy enforcement})^{1/2}$$

7.4 Matching Results

The final step comprised the matching of the physical environmental qualities (soil, climate, vegetation) and the management quality for the definition of the various types of ESAs to desertification. The four derived indices are multiplied for the assessment of the ESAs index (ESAI) as following:

$$ESAI = (SQI \times CQI \times VQI \times MQI)^{1/4}$$

The ranges of ESAI for each of type of the ESAs (as they were defined above), included three subclasses in each type as they appear in Table 7.8. Each type of ESAs is defined on a three-point scale, ranging from 3 (high sensitivity) to 1 (lower sensitivity) in order that the boundaries of the successive classes of ESAs could be better integrated. It must be pointed out that the range for each type of ESAs has been adjusted in such a way that it could include the various types of ESAs resulted from the various studies conducted in the past in the target area of the south Aurès.

7.4.1 Results and Interpretation

Following are detailed analyses of obtained results for the study area.

Table 7.8 Types of ESAs and corresponding ranges of indices

Types of ESAs	Description	Ranges of indices
1	Non affected	0–1.22
2	Potential	1.23–1.30
3	Fragile	1.31–1.41
4	Critical	1.41–2

7 The Use of GIS Data in the Desertification Risk Cartography

7.4.1.1 Analysis of Soil Quality

SQL queries and simple queries were made and after each query was executed the table request was updated until finally the entire database was validated and these results were shown in three soil qualities classes for a total area of 2,501 km^2 that are distributed as follows (Table 7.9 and Fig. 7.3).

The high soil quality category occupies an area of 1,341 km^2 or 53.63% of the total area, with a quality index below 1.33. This class is mostly occupied by forest vegetation, is less subject to human pressure than the other two. This explains better soil stability and lower impacts of erosion. The moderate class covers an area of 595.47 km^2 or 23.81% and has an index between 1.33 and 1.45. It is mostly spread over the central part of the study area. The class of poor quality cover an area of 564.96 km^2 or 22.56% of the total area, with a quality index above 1.45 and it occupies the southern part of the case study.

Table 7.9 Distribution of soil quality areas

SQI	Description	Area (km^2)	Area (%)
<1.33	High quality	1,341	53.63
1.33 à 1.45	Moderate quality	595.47	23.81
>1.45	Low quality	564.96	22.56

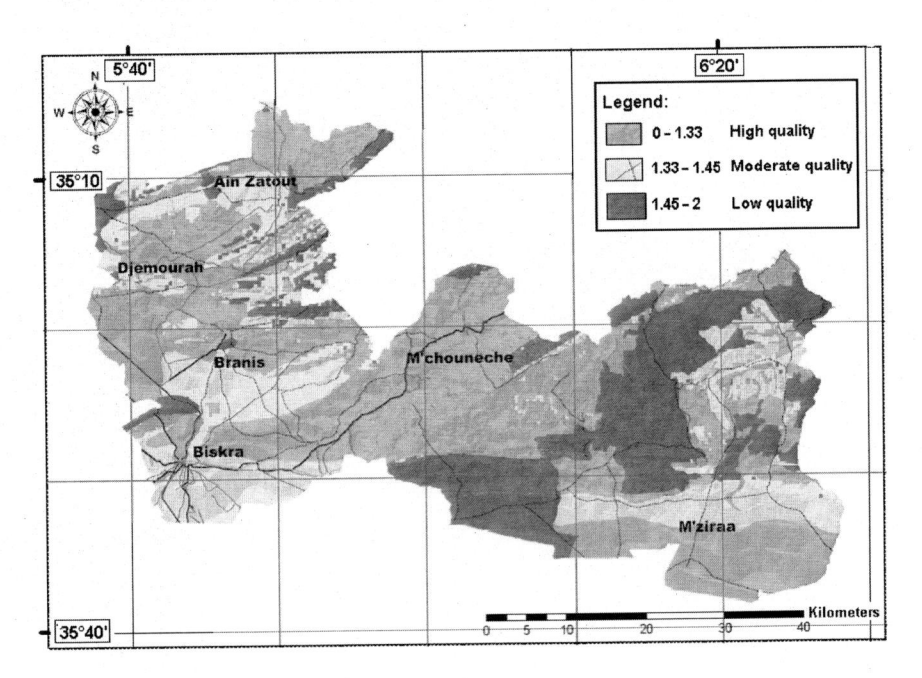

Fig. 7.3 Soil quality map of southern Aurès region

7.4.1.2 Analysis of the Climate Quality

Climate quality was shown in three classes and was rather a simpler SQL interpretation compared to the previous soil quality map. Table 7.10 and Fig. 7.4 show the results.

The very favourable climate class occupy an area of 747.92 km^2 or 29.91% of the total area of 2,501 km^2, with a quality index below 1.34. This class is located in higher altitudes with relatively heavy rainfall. The favourable class covers an area of 744.90 km^2 or 29.79% and has an index between 1.34 and 1.81. It is the biggest of the three categories as the region of south Aurés is located in the arid zone where rainfall does not exceed 200 mm. The unfavourable class extends to the rest of the area, with a quality index above 1.81 and it is spread over the southern part of the case study.

Table 7.10 Area of different climate quality

CQI	Description	Area (km^2)	Area (%)
<1.34	Very favourable	747.92	29.91
1.34–1.81	Favourable	744.90	29.79
>1.81	Unfavourable	1,007.73	40.30

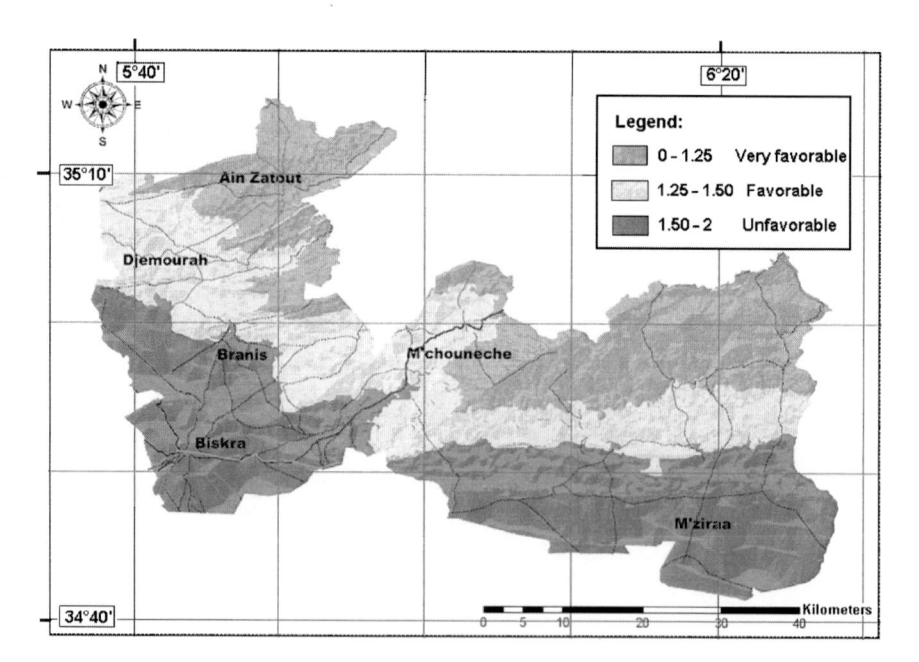

Fig. 7.4 Climate quality map of southern Aurès region

7.4.1.3 Analysis of the Vegetation Quality

An SQL analysis was used to establish the areas covered by three classes of vegetation quality to quantify their respective extension areas as they are shown in both Table 7.11 and Fig. 7.5.

The area with good vegetation occupies only about 24.16 km² or 0.97% of the total area, with a quality index below 1.13. More than 50% of this class is covered by forest vegetation (scrub oak green) (Schoenenberger, 1971). The moderate class quality covers an area of 520.35 km² or 20.82% and has an index between 1.13 and 1.38. It occupies the northern part corresponding to the degradation of forests and expansion of crops and pastures. The poor quality class extends to the rest of the area with a percentage of 78.21%, and a quality index above 1.38. It occupies the southern part.

Table 7.11 Vegetation quality classes and their respective areas

VQI	Description	Area (km²)	Area (%)
<1.13	High quality	24.16	0.97
1.13–1.38	Moderate quality	520.35	20.82
>1.38	Low quality	1,956.46	78.21

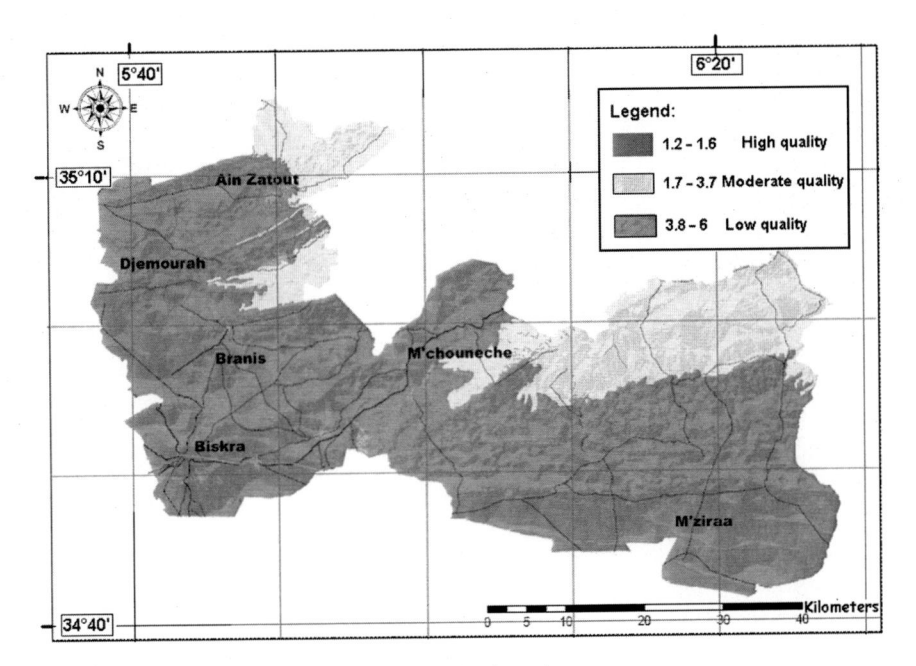

Fig. 7.5 Vegetation quality map of the southern Aurès region

7.4.1.4 Management Quality or Degree of Human Induced Stress

The map obtained after treatment with SQL and simple selection made possible to distinguish three socio-economic definition areas given in Table 7.12 and Fig. 7.6.

The class of good quality occupies an area de 178.55 km^2 or 7.14%, with a quality index between 1 and 1.25. It is localized mainly in the town of Ain zatout whose population has increased from 3,847 inhabitants in 1987 to 4,000 in 1998, or an average annual growth rate of 0.36%. This rate is far too low compared to the national average of 2.34%.

The moderate quality class occupies the largest area, or 2,211.44 km^2 (88.44%) of the total area, with an index between 1.25 and 1.50. This class occupies almost all municipalities in the study area. Vegetables, cereals, fruit trees and phoeniciculture represent the main agricultural production of the area. The poor quality class covers an area of 110.57 km^2 or 4.42% out of the total area, with a quality index above 1.5. These are predominantly pasturelands with dense population.

Table 7.12 Areas of the different management quality

MQI	Description	Area (km^2)	Area (%)
1–1.25	High	178.55	7.14
1.25–1.50	Moderate	2,211.44	88.44
> 1.50	Low	110.57	4.42

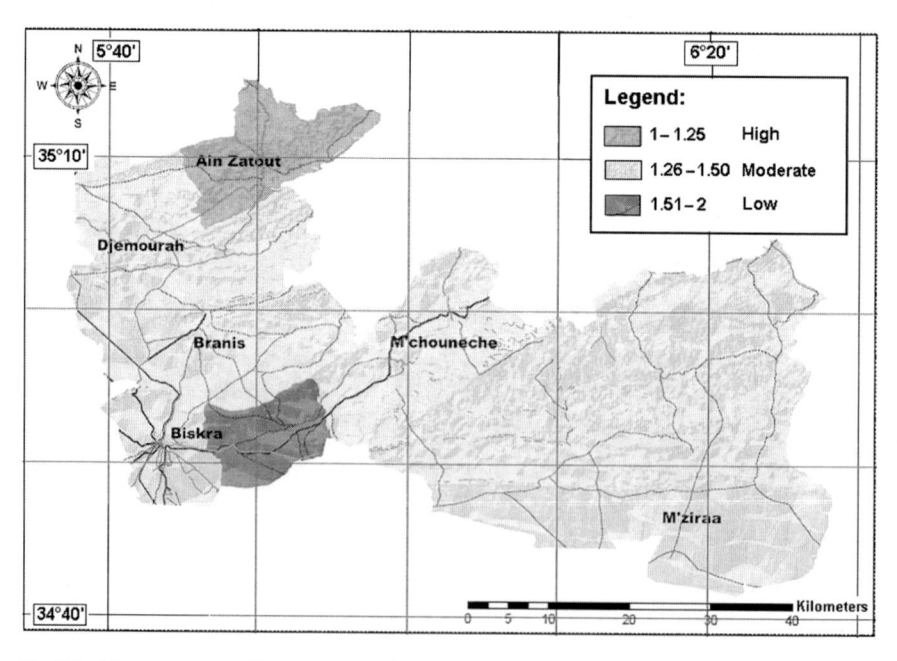

Fig. 7.6 Management quality map of southern Aurès

7.5 Analysis and Interpretation of the Desertification Sensitivity Map

The final map produced from the combination of previously described indicators enabled us to understand and classify areas sensitive to desertification in the region of south Aurès. We divided them in four classes i.e. (1) Highly sensitive, (2) Sensitive, (3) Insensitive, (4) Unaffected, as shown on Table 7.13 and Fig. 7.7.

Only 2.80% or 69.92 km² are included in the unaffected areas with an index of sensitivity between 0 and 1.22. This class of sensitivity is located north of the study area at the town of Ain zatout in a forest area that have a significant recovery capacity and good conditions for phoeniciculture despite covering a limited surface. The forests consist of Maquis and cover more than 50% of the territory. The relatively high rainfall reduces drastically the desertification risk.

The Insensitive areas with an extension of 218.40 km², or 8.73% of the total area, have a sensitivity index 1.22 and 1.30. They are located mainly on the mountain areas north of the case study including the thicket of M'ziraa and M'chouneche,

Table 7.13 Extension of sensitivity to desertification areas

Type	ESAs	Area (km²)	Area (%)
Very sensitive	1.40–2	1,696.16	67.83
Sensitive	1.30–1.40	516.07	20.64
Insensitive	1.22–1.30	218.40	8.73
Unaffected	0–1.22	69.92	2.80

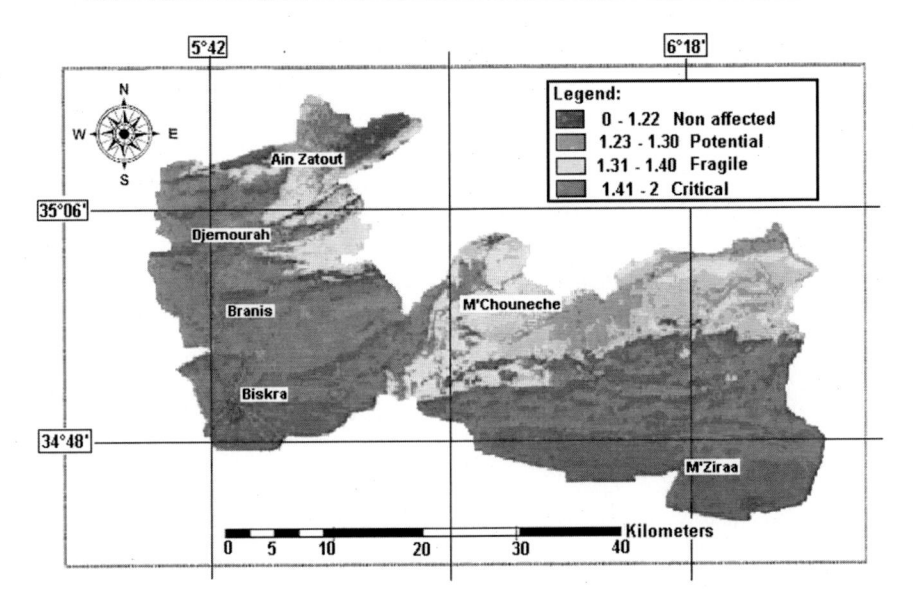

Fig. 7.7 Map of sensitivity to desertification of southern Aurès region

where goat grazing is largely practiced on old maquis accelerating thus the process of forest degradation.

The sensitive class is located in the northern communes of M'ziraa, M'chouneche and Djemourah. It covers an area of 516.07 km^2, or 20.64% of the total area and the sensitivity index vary between 1.30 and 1.40. This sensitivity affects mainly scrub and rangelands and is due to the poor quality of soil, climate and expansion of crops and pastures.

The very sensitive class stretches over an area of about 1,696 km^2, or 67.83% of the total area with an index ranging between 1.40 and 2. It is very sensitive and occupies the largest part of the case study. It is located mainly in the south and moderately on North West and affects all municipalities in the study area: M'ziraa, M'chouneche, Chetma, Biskra, Bran, and low Djemourah Ain zatout. The desertification process in this part is almost irreversible.

7.6 Conclusions

This work has clearly identified the main causes and problems related to desertification in the South Aurès region. It also quantified both in spatial and quantitative terms the intensities of desertification. The analysis of results shows that desertification threatens virtually the entire area and appears across various sensitivity classes. In fact over 88% of the area has been classified as sensitive to very sensitive, while only less than 12% is included into lower sensitivity classes.

The MEDALUS scientific approach responds well to the identification of desertification sensitivity areas. It could be used also as framework for upgrading results from regional to a larger scale. The approach is based on a number of indicators that need to be evaluated separately then included in thorough evaluation process. All of this has to be supported by the power of GIS. This process starts with data collection in a harmonized basis, which allows assessing the risk of the desertification process and allows developing tools for decision support systems.

References

Abdessemed, K. (1981). Le cèdre de l'Atlas (Cedrus atlantica Manetti) dans le massif de l'Aurès et du Belezma – Étude phytosociologique, problème de conservation et d'aménagement. Thèse Doctorat, Université d'Aix-Marseille.

Ansar, A. (Juin 2002). L'Aurès Oriental: Un milieu en dégradation. Journal Algérienne des régions arides. Revue semestrielle No. 01. Ed: C.R.S.T.R.A. Biskra (Algérie).

Ballais, J.L. (1981). Recherches géomorphologiques dans les Aurès (Algérie). Thèse de doctorat. Paris, Université Paris I.

Benmessaoud, H., Kalla, M., Dridi, H. and Arar, A.K. (2007). Utilisation des SIG pour la réalisation d'une carte de sensibilité à la désertification de la région Sud des Aurès,- Algérie. Actes du colloque GEOTUNIS 2007-Tunis du 15 au 17 Novembre 2007.

Bensaid, A. (2006). SIG et télédétection pour l'étude de l'ensablement dans une zone aride: le cas de la wilaya de Naâma (Algérie). Thèse de doctorat en géographie, Université Es.Senia, Oran–Algérie, 325p.

Berkane, A. and Yahiaou, A. (2007). L'érosion dans les Aurès, Sécheresse Vol. 18, No. 3, juillet-août-septembre 2007.

Direction générale des forêts. (2001). Document de projet d'appui au Plan d'action national Algérie. PNUD/UNSO DGF/Algérie, 88p.

Dubois, J.M. et al. (1997). La réalité de terrain en télédétection: Pratiques et méthodes. Actes des journées scientifiques de Sainte-Foy- Agence Universitaire de la Francophonie- AUPELF-UREF, 356p.

Oussedik A., Iftène T. and Zegrar A. (2003). Réalisation par télédétection de la carte d'Algérie de sensibilité à la désertification, sécheresse No. 02, Vol. 14, pp. 195–201.

Schoenenberger, A. (1971). Étude du couvert forestier de l'Aurès oriental et inventaire des espèces pastorales du massif des Beni Imloul, Projet Algérie 15, A.D.F., Constantine Agérie.

Chapter 8
Land Degradation and Overgrazing in the Afar Region, Ethiopia: A Spatial Analysis

B.G.J.S. Sonneveld, S. Pande, K. Georgis, M.A. Keyzer, A. Seid Ali, and A. Takele

Abstract Pastoralist societies in dryland areas anticipate the harsh climatic conditions with migration patterns that optimise the use of available forage and watering points. Yet, these traditional institutions are under increasing pressure due to a mounting population, encroaching of traditional grazing areas by sedentary agriculture and restrictions on transboundary movements. Indeed, the last decades witnessed an intensified use of these rangelands and the threat of overgrazing, a major cause of land degradation, should be taken seriously. This also motivates the current study where we analyse the relationship between grazing patterns and land degradation in the nomadic pastoralist areas of the Afar Region, Ethiopia. However, this is not an easy task because trekking patterns and concentrated grazing areas are not known in sufficient detail to engage in a fully spatial-temporal analysis. Therefore, we simulate the effect of migration by analysing land degradation-overgrazing relationships under various area accessibility scenarios, gradually releasing administrative boundary restrictions for pastoralists from district zone to state level. A grazing supply to demand ratio is applied to analyse the incidence of overgrazing whereas land degradation is estimated using time series analysis of the Rainfall Use Efficiency (RUE). The study shows that fodder shortages at district level in the western Afar are partly compensated at zonal level while the demand-supply ratio at state level is close to one. Significant negative trends in RUE are found in the north-eastern part of the Afar, in isolated pockets along the Awash River and near escarpments with the Highlands. A better understanding of the land degradation-overgrazing relationship requires more information on trekking patterns, including possible visits outside the study area.

Keywords Afar region · Ethiopia · Grazing · Overgrazing · Nomadic pastoral · Land conflicts

B.G.J.S. Sonneveld (✉)
Centre for World Food Studies, VU University Amsterdam (SOW-VU), 1081, HV Amsterdam, The Netherlands
e-mail: b.g.j.s.sonneveld@sow.vu.nl

8.1 Introduction

Sustainability of livestock management in pastoral areas has been hotly debated in the last decades. Many consider pastoral systems unsustainable as overstocking and overgrazing may lead to land degradation (e.g. Lamprey, 1983; Sinclair and Fryxell, 1985) and desertification (e.g. Homewood and Rodgers, 1987; Leach and Mearns, 1996; Dodd, 1994). These problems seem to echo "the tragedy of the commons" (Hardin, 1968) as individuals reap the benefits of overgrazing communal lands without sharing the costs with other land users. Yet, the view that pastoral systems are synonymous with unsustainability came under serious criticism (e.g. Behnke and Abel, 1996; Sullivan, 1996; Sullivan and Rohde, 2002) as inappropriate government orchestrated interventions limited the traditional strategies of nomadic pastoralists to deal with harsh dryland conditions (Sanford, 1983; Ellis and Swift, 1988). Indeed, many consider now nomadic[1] pastoralists as an epitome of sustainability (Desta and Coppock, 2004) that has been negatively influenced by external factors. For example, in arid lands population growth and appropriation of land for irrigation (Hundie, 2006) has caused pressure to mount, with overgrazing and violent conflicts among land users as most visible symptoms (Rass, 2006). Moreover, the development plans for massive expansion of biofuel plantations (Biopact, 2007) on "marginal" drylands can be expected to restrict further land's accessibility to pastoralists (Cotula et al., 2009).

Afar state in northeast Ethiopia (Fig. 8.1) is a good case in point. This arid region (Fig. 8.2) hosts 1.5 million people, 78% of which are pastoralists. Some of these pastoralists migrate along the seasons between a permanent base and distant grazing areas in semi-nomadic conditions, while the pure nomads among them move without a fixed homestead (CSA, 2003).

These movements are based on cautious decisions that capitalize on all perennial and current information by choosing trekking routes that can make best use of the forage expected to be available on the way. However, accessibility to rangelands and watering points has increasingly been hampered by new developments in the area. Indeed, expansion of sedentary agricultural settlements along the Awash River, implementation of the state-owned Tenaha sugar plantation and increasing incidence of contested territorial claims by different ethnic groups from outside the region (Rettberg, 2008) are prominent examples of new barriers to migration. The increasing pressures on the natural resource base is also a growing concern for policy makers (Pantuliano and Wekesa, 2008) who fear that land degradation might reduce grazing capacities and impoverish already fragile living conditions of the pastoralists.

The urgent calls for policy interventions are, therefore, to be taken seriously and justify a thorough analysis that explains land degradation patterns in its geographical dependence of grazing activities. Moreover, such an analysis needs to be conducted

---

[1] According to the type of trekking patterns pastoralists are called nomadic (irregular movements) or transhumant (regular movements between fixed locations).

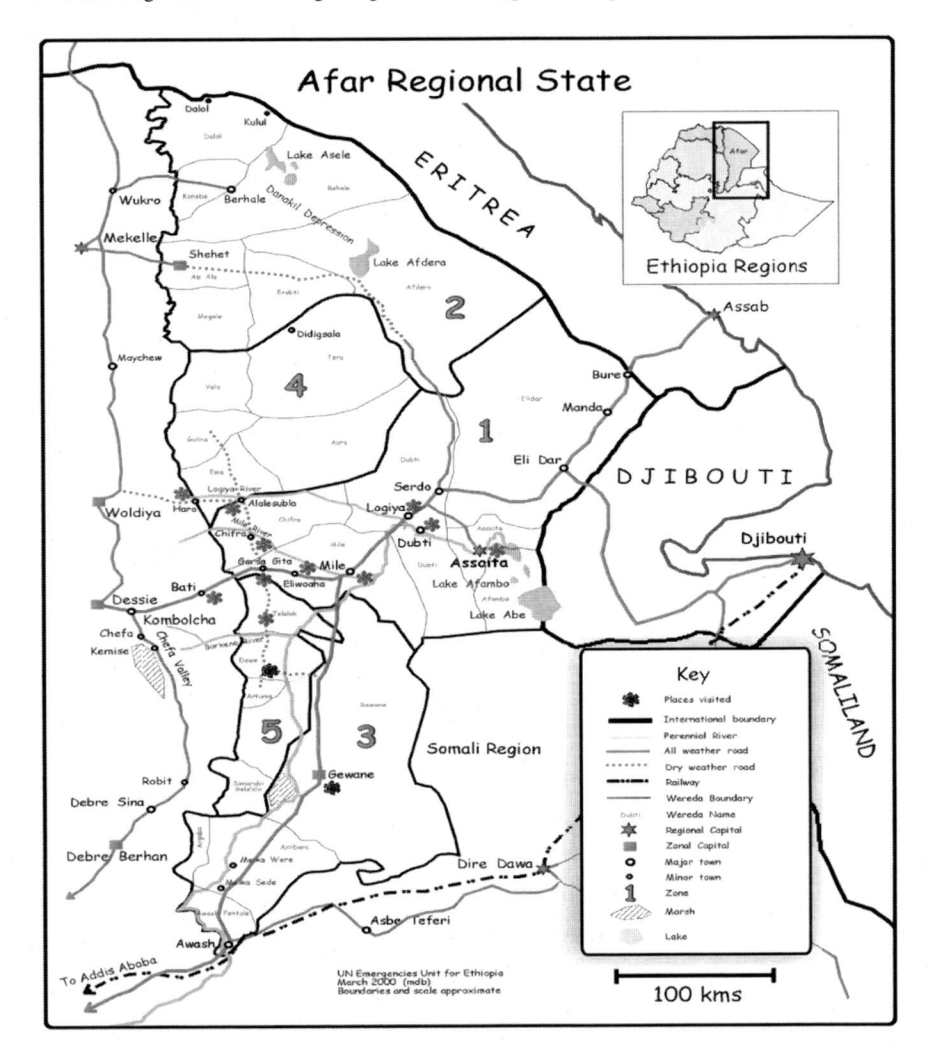

Fig. 8.1 Afar state

for the entire Afar area as it is at the state level where most important decisions concerning land use are taken.

This motivates the current study where we relate land degradation patterns to overgrazing in a spatially explicit manner for the entire Afar Region. However, this is not an easy task as detailed information on migration patterns, the main mechanism to mitigate the effect of overgrazing, remains largely absent and mapping of these trekking routes is still in an experimental stage (e.g. Sonneveld et al., 2009).

Certainly, a straightforward crossing between land degradation patterns and livestock does not account for flexibility and efficiency of trekking routes that manage relatively sparse vegetation in dryland areas. Therefore, we mimic the impact

Fig. 8.2 Average annual rainfall in mm

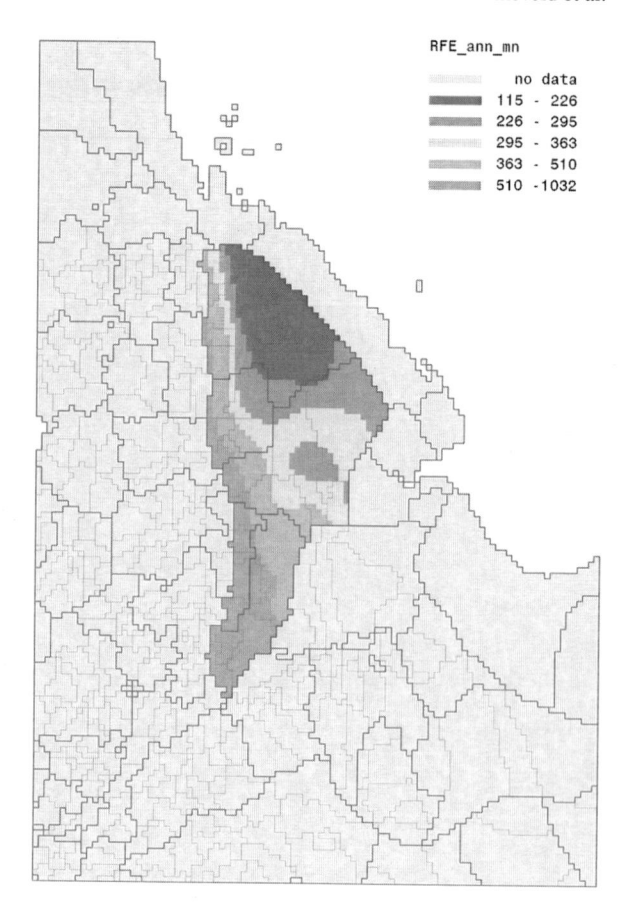

```
RFE_ann_mn
        no data
        115 -  226
        226 -  295
        295 -  363
        363 -  510
        510 - 1032
```

of migration patterns by analysing various area accessibility scenarios, gradually releasing cultural-administrative boundary restrictions for pastoralists from district (woreda), sub province (zone) to state (kililoch) level.[2] Interestingly, where production function analysis in sedentary agriculture aims to gain its explanatory power by respecting data at a fine resolution, information sources for studies on nomadic pastoral systems, necessarily have to do the opposite and aggregate their spatial-temporal dimensions to incorporate the supply-demand interactions in full.

The degree of overgrazing is based on a grazing supply to demand ratio that we calculate for the various aggregated administrative levels. A ratio below 1 shows a shortage of forage availability, equal to 1 represents a balanced situation and higher then 1 a surplus of forage. For the assessment of land degradation we use time series analysis of the Rainfall Use Efficiency (RUE) that divides annual sum of NDVI by

---

[2] Afar State consists of 5 administrative zones and 29 woredas. Farmer associations (326) and neighbourhood associations (32) (kebeles) are the most decentralized administrative units.

annual rainfall, separating rainfed driven vegetation dynamics from other factors. The use of RUE for land degradation assessment is based on Le Houérou's (1984) assumption (cited by Veron et al., 2006) that limiting resource use efficiency of plant traits, favoured by natural selection, is reflected in spatial patterns of the biophysical environment (soils and climate).

Departures from the average RUE, thus, result from anthropogenic influences. The RUE offers an attractive solution to the assessment of land degradation in dryland areas (Geerken and Ilaiwi, 2004; Prince et al., 1998) as availability of rainfall data and remotely sensed estimates of NDVI at adequate temporal and spatial scales ensures its applicability at various administrative levels up to the regional scale (Reynolds and Stafford Smith, 2002). Moreover, NDVI is linearly correlated to Above Ground Net Primary Productivity (Dent and Bai, 2008; Hall et al., 2006), which makes it a good estimator of ecosystem functioning (McNaughton et al., 1989) and, thus, of land degradation (UN, 1994).

This chapter is organized as follows. Section 8.2 presents the data sources and indicates how these were processed. Section 8.3 presents the results of this study and Section 8.4 concludes.

8.2 Data and Methodology

8.2.1 Livestock, Grazing Demand and Production

We use the pastoral livestock enumeration for the Afar Region (CSA, 2003) to analyse the herd composition of nomadic pastoralists per woreda (Annex). To compare grazing demand of the different species in common units, the body weight is converted into the Tropical Livestock Unit using coefficients from FAO, 2005. Figure 8.3 shows the TLU density as number of TLU's per hectare of the woreda.

We calculated grazing demand based on Boudet and Riviere (1968) and Minson and McDonald (1987) by assuming that livestock needs 2.5% of its body weight for a sustained growth, which in turn results in a consumption of 6.25 kg of forage dry matter daily for each TLU. Next, we used these figures together with the TLU inventory to calculate the total grazing demand (d_w) at each woreda w.

The grazing capacity (or fodder supply) is based on forage data of the regional PHYGROW model, Texas A&M University (Stuth et al., 2003) that was made available as 8 × 8 km grids for Eastern Africa. PHYGROW bases its calculations on: soil parameters, plant community characteristics, and livestock management rules, while the model is driven by satellite-based gridded weather data. Regular verifications in Ethiopia, Kenya, Tanzania, and Uganda show that PHYGROW's simulation output of available forage has a good correlation with observations in the field ($R^2 = 0.96$) (Jama et al., 2002). Grazing capacity is calculated for each pixel using PHYGROW results corrected for a utilization factor of 0.35, thus accounting for grass composition and temporal variability of productivity (Abule et al., 2007).

Fig. 8.3 TLU density per woreda

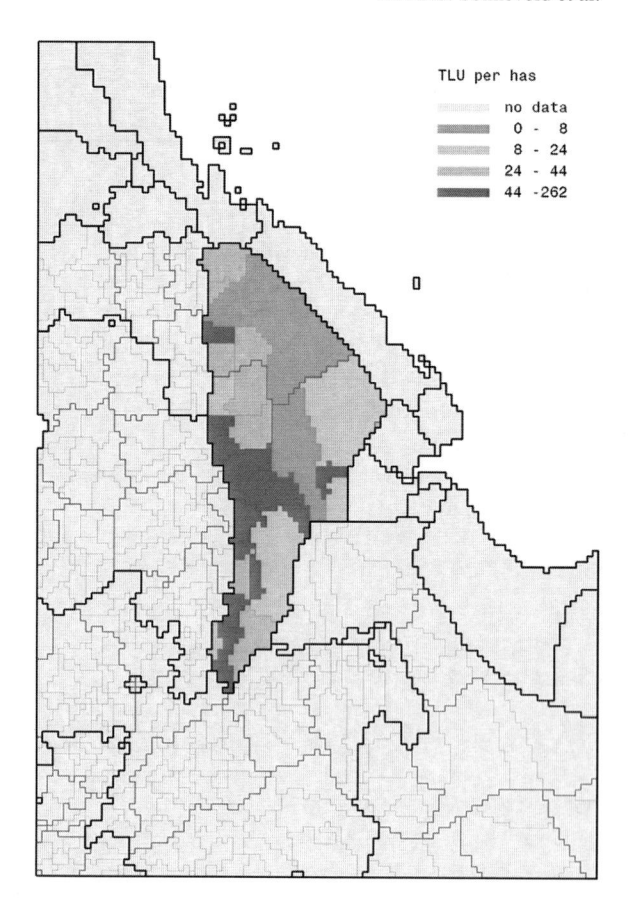

The production supply (f_n) for each pixel n is expressed in TLU ha^{-1}. The median of this statistic derived from all pixels that belong to the woreda area ($n \in w$) is used to express the production supply per woreda, f_w. We use the median to remove sensitivity of estimation from outliers. We then calculate supply to demand ratio at woreda level as $R_w = \frac{f_w}{d_w}$. Analogue we calculated supply, f_z, in zone z and the corresponding demand d_z as an area weighted mean of d_w. Zonal level supply to demand ratio is similarly calculated as $R_z = \frac{f_z}{d_z}$. Finally we repeated this exercise to get our statistic $R_s = \frac{f_s}{d_s}$ at state level.

8.2.2 RUE

The data for the RUE analysis is based on annual rainfall and NDVI data for the period 1980–2002. Rainfall data is obtained in grid format from Climate Research Unit of the University of East Anglia (Mitchell and Jones, 2005) at monthly time scale and 0.5° spatial resolution. NDVI data is obtained from FEWS-NET data set (Tucker et al., 2005) at 10 day time scale and 8 km spatial resolution. Rainfall and NDVI data are resampled to a common pixel size of 8 km, for a georeference that covers parts of Ethiopia, Eritrea, Djibouti and Somalia (as in Fig. 8.3). Finally, the RUE statistic is computed as a ratio of annual sum of NDVI to annual precipitation. The slope of the trend and its t-statistic are calculated for each of the pixels.

8.3 Results

In this section we start with the forage supply–demand ratio at the different spatial levels, followed by the RUE analysis; we end with a discussion on the spatial patterns of land degradation and overgrazing.

Figure 8.4a, b show the forage supply grazing demand ratio by woreda and zone, respectively. The analysis per woreda shows that deficits mainly occur in the western part near the escarpments of the Ethiopian Highlands, where also the highest TLU densities are reported. Surpluses are shown in the central and eastern woredas near the border with Eritrea and Djibouti. Concerning the supply to demand ratio for zones, it is interesting to note that deficits in the woredas that belong to Zone 1 and 2 are compensated by surpluses in the east and central Afar. Zones 4–5 do not have

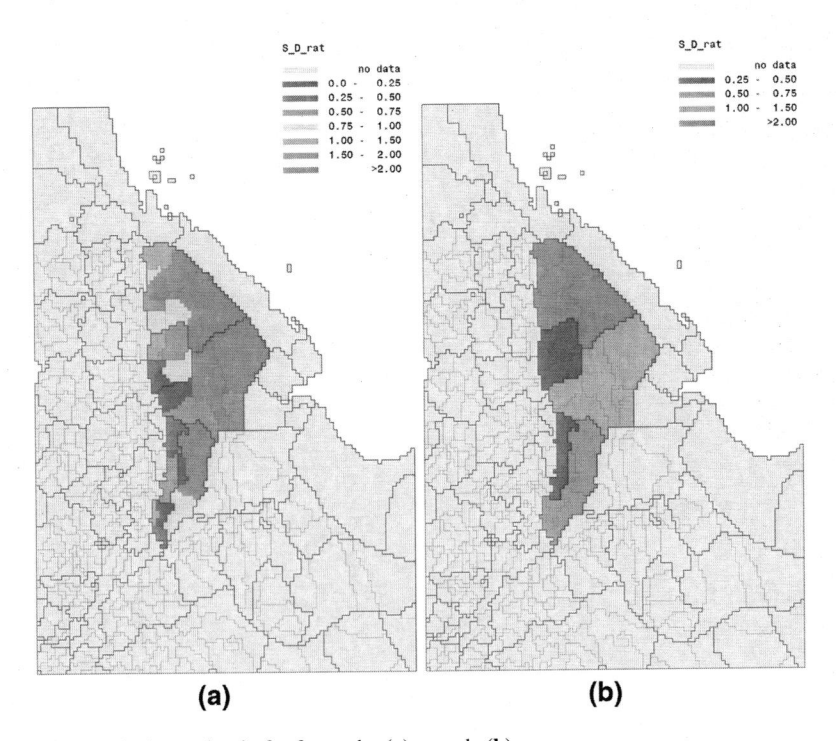

Fig. 8.4 Supply demand ratio for forage by (**a**) woreda (**b**) zone

many surplus areas and forage demand remains larger than its availability. For the entire Afar State we observe a supply to demand ratio of 0.95, indicating that current forage demand exceeds forage availability by a small amount at state level.

We now turn to the land degradation assessment. Figure 8.5 shows the slope of the RUE analysis, whereas Fig. 8.6 presents its t-statistic. In general, we observe a declining trend of the RUE in the north-eastern corner near the border with Eritrea, gradually becoming less pronounced towards the south-west direction and turning into positive values in the southern cone of Afar State. Exceptions of positive trends are found in pockets along the Awash River and in the North near the escarpments with the Highlands. Figure 8.6 shows the significance of the slope estimate, which is low to very low for most of the Afar; i.e. in most cases the trend does not significantly deviate from zero. Alarming is the situation in the Northern part of the Afar state where declining RUE trends are significant to very significant.

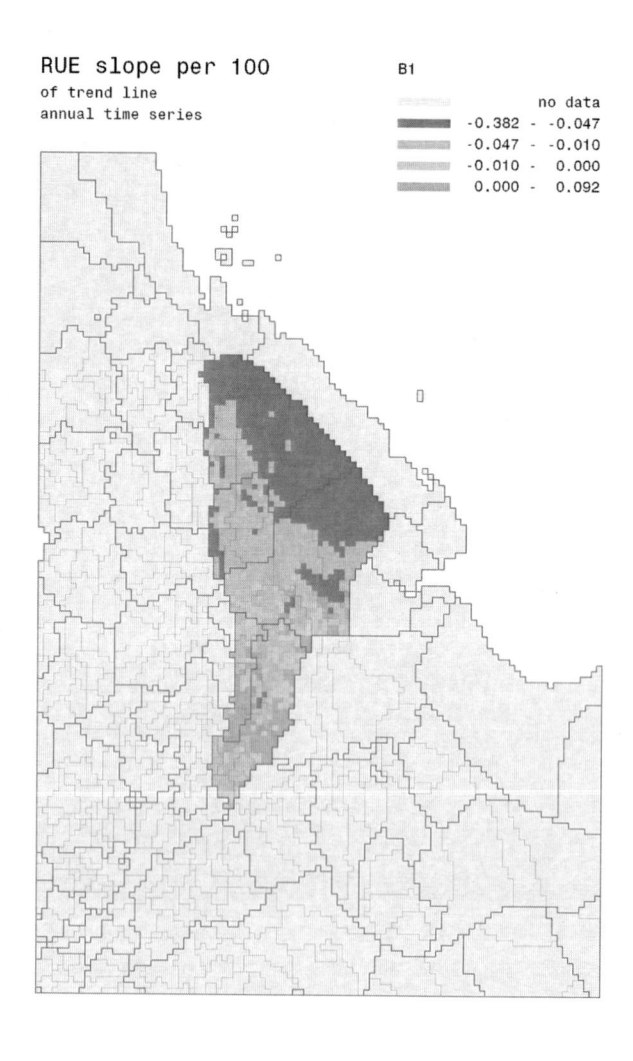

Fig. 8.5 Slope of RUE trend

Fig. 8.6 Student's t-statistics for slope

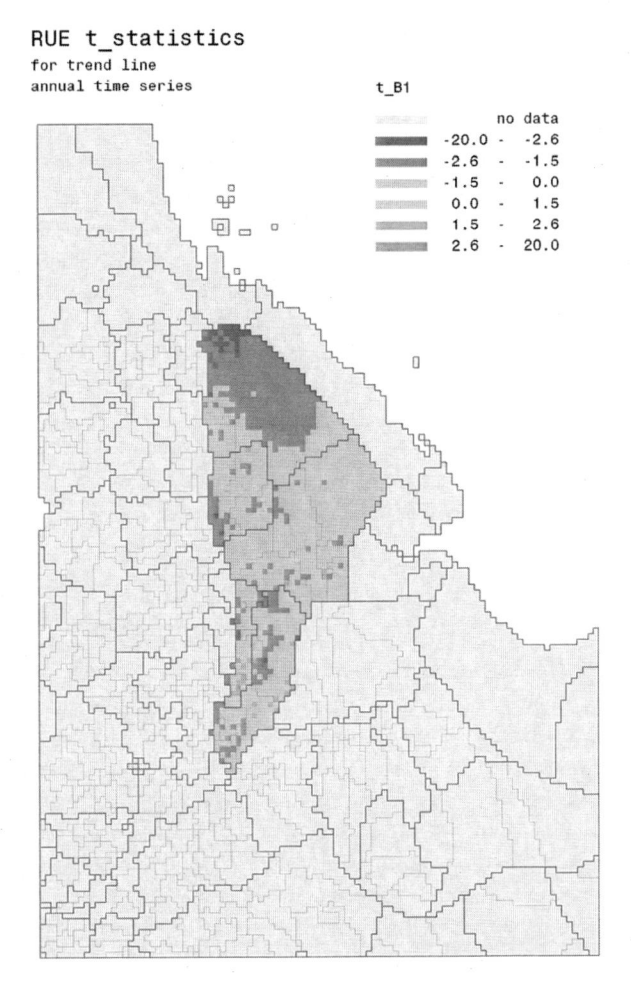

These negative RUE values in north Afar are explained by Tesfay (2004) who reported that in this area cultivation by Tigray nobility encroached onto prime grazing lands pushing the pastoralists to less potential lands with overgrazing as a consequence. Furthermore, the hazard of overgrazing in North Afar seems to be especially pronounced in times of drought when grazing areas are seriously limited. Yet we found in the same area with negative RUE trends that supply demand ratios are still positive, even at the woreda level. This would confirm that studies on nomadic pastoralism couldn't be confined to localized sites, as we have to account for migration patterns that cover extended areas.

We can conclude that at woreda level the land degradation trend does not correspond to the spatial patterns of overgrazing. This confirms that detailed information on trekking patterns is required in order to analyse the spatial incidence of overgrazing and land degradation. As part of such a database, information is required on conflict areas (that are now often avoided), as well as transboundary migration (for e.g. many Afar people live in Eritrea and might also visit the Afar state).

8.4 Conclusion

We found that most of Afar State shows stable RUE trends, which confirms that, at state level, with a supply-demand ratio near one, forage production more or less meets grazing demand. In the Northern part we found a significant degradation, most likely caused by the encroachment of cultivated areas into prime rangelands, which might have resulted in extended fallow periods without vegetative coverage. Despite the fact that local forage supply still exceeded the local grazing demand, but the visits of other clans could alter this balance and cause the incidence of land degradation.

The results may support the argument that if mobility of pastoralists continues unhampered, it results in sustainable land management, whereas restricted accessibility leads to overgrazing and land degradation. As such the results of this study contain an important message for the Afar authorities concerning new land developments like the Tenaha dam, expansion of sedentary cultivation along the Awash River and the potential to cultivate biofuel plantation, which could seriously interfere with traditional migration patterns. Clearly, such developments might also be beneficial for the Afar State and authorities should, therefore, aim to minimize the tensions between these initiatives and the pastoralist communities.

For example, authorities could guarantee safe passages through large-scale plantations to allow pastoralists unrestricted access to rangelands. Furthermore, research is urgently needed that supports the drought coping strategies of the pastoralists such as the design of the best spatial configuration of a system of groundwater pumps and forage storage points. When such "enclaves" are well regulated they could help pastoralists through dire periods and avoid overgrazing and land degradation of the few areas that are not yet affected by drought. However, to support these policy interventions research should avail of a consolidated data base that has detailed information on trekking routes, biophysical resources, land uses, market prices, conflict zones, household/pastoralist surveys, and narratives on coping strategies in appropriate spatial and temporal dimensions. The collection and organization of these data at the appropriate level and its organization in a dynamic modelling environment is one of the main scientific challenges in the coming years.

Annex: Livestock Herd Composition and TLU Density per Woreda

Woreda	Area (km²)	Cattle	Goats	Sheep	Camel	Equine	Total TLU	Pastoral/total	TLU density
Elidar	13,118	25,516	435,827	83,374	45,508	8,884	132,597	0.90	10
Asayit	1,357	60,823	21,423	15,840	2,922	454	57,516	0.86	42
Afambo	1,884	24,359	14,469	11,367	1,014	0	23,924	0.86	13
Dubti	8,670	34,593	48,929	43,771	5,320	1,498	45,164	0.88	5
Mille	4,766	115,817	282,105	232,337	62,018	8,349	224,391	0.89	47
Chifra	3,291	300,198	284,624	320,003	112,667	23,523	450,362	0.88	137
Dalol	3,245	48,905	253,666	60,688	9,075	15,481	92,079	0.90	28
Kuneba	675	13,575	65,715	30,533	3,178	4,420	27,417	0.89	41
Berahi	7,154	556	77,025	6,409	2,244	2,177	13,115	0.92	2
Afdera	12,938	10,219	42,912	17,787	9,764	3,278	27,613	0.89	2
Ab-Ala	1,282	28,918	138,684	31,921	19,679	7,028	67,881	0.89	53
Megale	1,968	33,224	159,230	51,403	15,111	3,975	69,014	0.89	35
Erabti	2,457	30,777	362,369	94,872	23,146	8,977	105,050	0.91	43
Bure M	1,104	137,062	125,647	160,608	48,597	11,673	203,946	0.88	185
Gewane	8,648	90,414	70,218	75,258	22,833	4,737	118,041	0.87	14
Amibar	3,916	88,581	113,699	44,915	35,664	3,662	132,164	0.87	34
Dulech	1,260	94,591	158,257	75,089	67,126	5,339	180,923	0.88	144
Argoba	471	12,799	9,126	966	574	2,638	13,673	0.88	29
Awash	1,042	73,351	77,662	33,085	36,779	2,501	114,839	0.88	110
Yalo	1,820	32,725	72,441	29,465	8,379	4,790	49,781	0.88	27
Teru	3,657	111,898	216,301	195,910	26,455	7,323	169,748	0.88	46
Awura	3,015	73,358	121,600	107,468	30,444	3,530	120,880	0.88	40
Golina	1,326	106,543	92,490	86,565	23,012	5,875	135,698	0.87	102
Ewa	1,205	192,167	134,759	183,252	42,632	11,336	245,821	0.87	204
Telala	1,391	82,396	170,849	91,954	43,710	7,266	148,813	0.88	107
Dawe	1,060	41,550	75,604	61,244	15,253	3,501	67,797	0.88	64
Artuma	374	52,360	109,460	83,099	18,709	5,343	87,518	0.88	234
Furisa	1,284	35,688	96,304	44,689	13,229	4,951	61,947	0.88	48
Semuro	1,247	37,888	129,118	29,375	14,711	5,787	67,784	0.88	54

Conversion factor TLU: Cattle = 0.79; Sheep =0.13; Goat = 0.10; Camel = 1.22; Equines = 0.63.
Source: CSA (2003)

References

Abule, E., Snyman, H.A. and Smit, G.N. (2007). Rangeland evaluation in the middle Awash valley of Ethiopia: I. Herbaceous vegetation cover. Journal of Arid Environments 70:253–271.

Behnke, R.H. and Abel, N.O.J. (1996). Revisited: The overstocking controversy in semi-arid Africa. World Animal Review 87:3–27.

Biopact. (2007). ICRISAT launches pro-poor biofuels initiative in drylands. Through: http://biopact.com/2007/03/icrisat-launches-pro-poor-biofuels.html

Boudet, G. and Riviere, R. (1968). Emploi pratique des analysis fourrageres pour l'appreciation des paturages tropicaux. Revue de levage et de medecine veterinaire des pays tropicaux 21(2): 227–266.

CSA. (2003). Pastoral Areas Livestock Enumeration, Results for Afar Region. The Central Agricultural Census Commission, Central Statistical Authority, Addis Ababa, Ethiopia.

Cotula, L., Vermeulen, S., Leonard, R. and Keeley, J. (2009). Land Grab or Development Opportunity? Agricultural Investment and International Land Deals in Africa. IIED/FAO/IFAD, London/Rome. ISBN: 978-1-84369-741-1.

Dent, D.L. and Bai, Z.G. (2008). Assessment of land degradation using NASA GIMMS: A case study in Kenya. In: E. Hartemink, A. McBratney and M. de Lourdes Mendonça-Santos (eds.), Digital Soil Mapping with Limited Data. Springer, Netherlands, p. 21. Assessment of Land Degradation Using NASA GIMMS: A Case Study in Kenya, pp. 247–258.

Desta, S. and Coppock, D.L. (2004). Pastoralism under pressure: Tracking system change in southern Ethiopia. Human Ecology 32(4):465–486.

Dodd, J.L. (1994). Desertification and degradation in sub-saharan Africa – the role of livestock. BioScience 44:28–34.

Ellis, J.E. and Swift, D.M. (1988). Stability of African pastoral ecosystems: Alternate paradigms and implications for development. Journal of Range Management 41:450–459.

Geerken, R. and Ilaiwi, M. (2004). Assessment of rangeland degradation and development of a strategy for rehabilitation. Remote Sensing of Environment 90(2004):490–504.

Hall, F., Masek, J. and Collatz, G.J. (2006). Evaluation of ISLSCP initiative II FASIR and GOMMS NDVI products and implications for carbon cycle science. Journal of Geophysical Research 111:D22S08. doi:10.1029/2006JD007438.

Hardin, G. (1968). The tragedy of the commons. Science 163:1243–1248.

Homewood, K. and Rodgers, W.A. (1987). Pastoralism, conservation and the overgrazing controversy. In: D. Anderson and R. Grove (eds.), Conservation in Africa: People, Policies and Practice. Cambridge University Press, Cambridge, 355pp.

Hundie, B. (2006). Property Rights among Afar Pastoralists of Northeastern Ethiopia: Forms, Changes and Conflicts. Presented at Survival of the Commons: Mounting Challenges and New Realities. Eleventh Conference of the International Association for the Study of Common Property, Bali, Indonesia, June 19–23, 2006.

Jama, A., Gibson, Z., Stuth, J., Kaitho, R., Angerer, J. and Marambii, R. (2002). Setting up a livestock early Warning System Monitoring Zone: Site selection, characterization, and sampling for the PHYGROW model. In: Livestock Early Warning Systems for East Africa. Proceedings of the Planning and Evaluation Workshop. Ethiopian Agricultural Research Organization, Addis Ababa, Ethiopia, May 5–7, 2002, pp. 33–39.

Lamprey, H.F. (1983). Pastoralism yesterday and today: The overgrazing controversy. In: F. Bourliere (ed.), Tropical Savannas. Ecosystems of the World. Volume 13. Elsevier, Amsterdam, 730pp.

Le Houérou, H.N. (1984). Rain use-efficiency: A unifying concept in arid-land ecology. Journal of Arid Environments 7:213–247.

Leach, M., Mearns, R. (eds.). (1996). The Lie of the Land: Challenging Received Wisdom on the African Environment. Cambridge University Press, Cambridge, 240pp.

McNaughton, S.J., Oesterheld, M., Frank, D.A. and Williams, K.J. (1989). Ecosystem-level patterns of primary productivity and herbivory in terrestrial habitats. Nature 341:142–144.

Minson, D.J. and McDonald, C.K. (1987). Estimating forage intake from the growth of beef cattle. Tropical Grasslands 21:116–122.

Mitchell, T.D. and Jones, P.D. (2005). An improved method of constructing a database of monthly climate observations and associated high-resolution grids. International Journal of Climatology 25:693–712.

Pantuliano, S. and Wekesa, M. (2008). Improving Drought Response in Pastoral Areas of Ethiopia Somali and Afar Regions and Borena Zone of Oromiya Region. Humanitarian Policy GroupOverseas Development Institute, London.

Prince, S.D., Brown de Colstoun, E. and Kravitz, L.L. (1998). Evidence from rain-use efficiencies does not indicate extensive Sahelian desertification. Global Change Biology 4:359–374.

Rass, N. (2006). Policies and Strategies to Address the Vulnerability of Pastoralists in Sub-Saharan Africa. PPLPI Working Paper No. 37. Pro-Poor Livestock Policy Initiative. Food and Agriculture Organization of the United Nations, Rome, Italy.

Rettberg, S. (2008). Der Umgang mit Risiken im Spannungsfeld zwischen Konflikten und Nahrungskrisen. Eine politisch-geographische Untersuchung in der Afar Region Ethiopien. PhD Thesis. Bayreuth University, Germany.

Reynolds, J.F. and Stafford Smith, D.M. (2002). Global Desertification: Do Humans Cause Deserts? Volume 88. Dahlem University Press, Berlin.

Sandford, S. (1983). Management of Pastoral Development in the Third World. John Wiley and Sons, New York, 316pp.

Sinclair, A.R.E. and Fryxell, J.M. (1985). The Sahel of Africa: Ecology of a disaster. Canadian Journal of Zoology 63:987–994.

Sonneveld, B.G.J.S., Keyzer, M.A., Georgis, K., Pande, S., Seid Ali, A. and Takele, A. (2009). Following the Afar: Using remote tracking systems to analyze pastoralists trekking routes. Journal of Arid Environments 73(11):1046–1050.

Stuth, J., Schmitt, D., Rowan, R.C., Angerer, J. and Zander, K. (2003). PHYGROW (Phytomass Growth Simulator) User's Guide. Technical Documentation. Ranching Systems Group Department of Rangeland Ecology and Management. Texas A&M University.

Sullivan, S. (1996). Towards a non-equilibrium ecology: Perspectives from an arid land. Journal of Biogeography 23:1–5.

Sullivan, S. and Rohde, R. (2002). On non-equilibrium in arid and semi-arid grazing systems. Journal of Biogeography 29:1595–1618.

Tesfay, Y. (2004). Pastoral natural resources management in Northern Afar. In: Y. Tesfay and K. Tafere (eds.), Indigenous Rangeland Resources and Conflict Management by the North Afar Pastoral Groups in Ethiopia. A Pastoral Forum Organized by the Drylands Coordination Group (DCG) in Ethiopia, June 27–28, 2003, DCG Report No. 31. Mekelle, Ethiopia.

Tucker, C.J., Pinzón, J.E., Brown, M.E., Slayback, D., Pak, E.W., Mahoney, R., Vermote, E. and El Saleous, N. (2005). An extended AVHRR 8-km NDVI data set compatible with MODIS and SPOT vegetation NDVI data. International Journal of Remote Sensing 26(20):4485–4498.

UN. (1994). UN Earth Summit. Convention on Desertification. UN Conference in Environment and Development, Rio de Janeiro, Brazil, June 3–14, 1992. DPI/SD/1576. United Nations, New York.

Verón, S.R., Paruelo, J.M. and Oesterheld, M. (2006). Assessing desertification. Journal of Arid Environments 66:751–763.

Chapter 9
Effects and Implications of Enclosures for Rehabilitating Degraded Semi-arid Rangelands: Critical Lessons from Lake Baringo Basin, Kenya

Stephen M. Mureithi, Ann Verdoodt, and Eric Van Ranst

Abstract The establishment of enclosures, denoting areas closed off from grazing for a specific period, is a well-known management strategy for restoring degraded semi-arid rangeland ecosystems. Range enclosure has profound ecological (biophysical) effects and a number of socio-economic implications that vary significantly, depending on local conditions. Understanding the consequences of the rising trend of rangeland enclosure is thus imperative for sustainable planning and management of these fragile ecosystems. Indeed, what administrators require is not a general policy for or against enclosure, but rather some understanding of the various effects of enclosure under different circumstances. Ultimately, researchers may be able to present policy-makers with a typology of different kinds of enclosure movements, and with a systematic discussion of the probable outcome of each kind of movement. Therefore, the spontaneous enclosure of the range by livestock owners may raise new problems, but may also permit new approaches to the development of the livestock industry in the arid and semi-arid areas in Africa. This paper seeks to highlight the effects and implications of using enclosures for rehabilitating degraded semi-arid rangelands and draw practical lessons to help us achieve increased restoration capability in the future.

Keywords Semi-arid rangelands · Land degradation · Rehabilitation · Enclosures · Baringo · Kenya

9.1 Introduction

Rangelands outside the protected areas – national parks, game reserves and private ranches and conservancies in Sub-Saharan Africa are under serious threat from land

---

S.M. Mureithi (✉)
Range Management Section, Department of Land Resource Management and Agricultural Technology, University of Nairobi, Nairobi, Kenya
e-mail: stemureithi@yahoo.com

degradation to a level where restoration may not be feasible. Most arid and semi-arid rangelands in Kenya for example have become degraded "wastelands" no longer able to support their diverse cultures, plants, and wildlife (UNEP/GoK, 1997). Severe rangeland degradation upsets the dynamics of these fragile ecosystems, affecting the energy flows, biogeochemical cycles, hydrological cycles, increased aridity (Dregne, 1992), and possibly all resulting into a downward spiral of ecosystem structure and function decline (King and Hobbs, 2006). The regenerative capacity is often compromised leading to loss of biodiversity. The ultimate effect of the foregoing is a livelihood crisis for the pastoral communities dependent on these ecosystems. The persistent menace of recurrent droughts, floods, and disease outbreaks leading to large losses of livestock and dryland crop failure are commonplace (UNEP, 2000). Food insecurity is a growing problem, and increasing poverty poses a major threat not only to the livelihoods of the pastoral communities that depend on these rangelands, but also to biodiversity. In these areas, rehabilitation and restoration of heavily degraded rangelands is often obligatory for pastoralist land use to be sustainable, even though implementing restoration projects in communally utilized rangelands is a complex endeavour (de Groot et al., 1992).

Rehabilitation and restoration of degraded drylands is a subject that at present times receives attention in many parts of the world especially in Sub-Saharan Africa. Increasing evidence exists to show that there has been a great deal of community based organisations, non-governmental organisations, local governmental and international efforts aimed at rehabilitating degraded rangelands using various approaches (GoK, 1997; UNEP/GEF, 2002; RAE, 2003). Yet, there exist very few cases of successful rehabilitation initiatives in the arid and semi-arid rangelands particularly in Eastern Africa (Makokha et al., 1999; RAE, 2003; Mengistu et al., 2005). One outstanding example is in Lake Baringo Basin, where significant areas of severely degraded semi-arid land have been successfully restored to productive grassland using enclosures. Starting with a number of communal enclosures that served as community demonstrations, private enclosures today form a mosaic in the entire Lake Basin.

This paper reviews the effects and implications of a growing trend of rehabilitating degraded semi-arid rangelands using enclosures. Critical lessons can be learnt from the on-going enclosure movement in the semi-arid rangelands of Lake Baringo Basin.

9.2 Rehabilitation of Degraded Semi-arid Rangelands

In the wake of increased land degradation in the past few decades particularly in arid and semi-arid rangelands, it has become increasingly necessary trying all available approaches and strategies to restore these ecosystems (Biamah, 1988; RAE, 2003). According to the UNCCD (1994), combating desertification includes activities, which are part of integrated development in arid, semi arid and dry sub-humid areas aimed at prevention and reduction of land degradation; rehabilitation of partly degraded land and reclamation of desertified land. Heady (1999) observed that rangeland rehabilitation or improvement implies implementing change to attain a particular economic value.

Degraded lands severely impacted by intensive and repeated disturbance may still provide a wide range of products (e.g. fuelwood, poles, cattle, sheep and goat grazing, herbal medicine) and valuable ecosystem services for the local community as cash income sources (Meyerhoff, 1991). Therefore, caution needs to be exercised in the identification of degraded land, its importance to local communities and the need for new methods of rehabilitation and management. Rehabilitation, if needed, should also seek to identify and enhance the ecological and socio-economic value of such lands to local communities and not deprive them of existing or foreseen benefits.

In other words, before embarking on a major rehabilitation programme, the objectives and implications should be carefully examined (Miller and Hobbs, 2007). According to Harris et al. (1996) and de Groot et al. (1992), the objectives may include enhancing pasture availability for the pastoralists, preventing further soil erosion hence saving the open water bodies from accelerated siltation, restoring scenic quality, or restoring a natural ecosystem. Milton et al. (1994) stressed the need to recognize and treat degradation at early stages, because management inputs and costs increase for every additional step in the degradation process. The choice of the approach and method also plays an important role. Mututho (1986) showed that because of high capital involvement in for example structural measures for range rehabilitation and low economic returns in grazing lands, most of them have produced only very little success.

9.2.1 Rangeland Rehabilitation in Kenya

Rehabilitation of degraded rangeland in Kenya dates back to as early as 1919 when the then District Commissioner of Machakos (quoted by Pereira and Beckeley, 1952) wanted to revegetate degraded grazing land through closure of the land for a couple of years. The method did not achieve its objectives. Other equally expensive measures have been tried in Machakos, Kitui, Baringo, Marsabit, and Turkana Arid and Semi-arid Lands (ASAL) districts (Pereira and Beckeley, 1952; Jordan, 1957; Pratt, 1964; Bogdan and Pratt, 1967; Lusigi, 1981; Smith and Critchley, 1983; Muhia, 1986). Measures used in these areas to restore productivity of degraded areas were: reseeding (seedbed prepared using hand-held tools, ox-plough or tractor-drawn cultivators); rotational grazing within paddocks; mass revegetation of the area using locally adapted grasses and using watering points to distribute grazing pressure thus giving degraded areas time for natural revegetation.

Other physical measures included installation of terraces and runoff harvesting structures, which, despite their apparent effectiveness, are potentially dangerous to the livestock under rangeland conditions. Almost all of these rehabilitation methods involved setting aside the degraded area in some form of closure, carrying out some restoration measures and allowing time for regeneration. However, many of these methods tended to place more emphasis on the physical and technical intervention than socioeconomic and cultural ones, leading to failure due to lack of embracing the needs and priorities of the pastoral communities.

In addition, the efforts and emphasis of combating land degradation in Kenya have in the past dominantly focussed on soil conservation measures for agricultural areas, aiming to curb the problem of soil loss and declining productivity of food crops. In the rangelands on the other hand, rehabilitation has targeted restoring the carrying capacity for livestock and wildlife species, by increasing revegetation and cover in overgrazed areas, and by preventing soil erosion (Harris et al., 1996; Pratt, 1964).

Such measures have to be designed in line with the changing nature of the soil component. Some of the common soils in semi-arid areas are particularly vulnerable to disturbances (Hudson, 1987), either because they have high susceptibility to erosion (high erodibility) or because of their chemical and physical properties. The sealing properties of Fluvisols in the semi-arid plains for example results in reduced infiltration rates and soil loss thereby rendering grazing lands bare, eroded and with soil crusts that inhibit seed emergence in the following wet season. Other soils (Vertisols) within the Njemps Flats area in Lake Baringo basin for instance have been described as unstable due to relatively high smectite clay content (Biamah, 1988). Hence, physical conservation measures are not effective in controlling erosion. Instead, a blend of both physical and biological conservation measures (i.e. making water harvesting micro-catchments and planting protective cover) would be the best options of conserving the soil and minimising erosion under the prevailing conditions.

Therefore, activities aimed at proper management and utilisation of rangelands needs to adopt a holistic and multidisciplinary approach (Harris et al., 1996), to restore the integrity of these fragile ecosystems. Any intervention however, has to reflect the needs of the local people (de Groot et al., 1992; Makokha et al., 1999; Meyerhoff, 1991), Gachimbi (1990). The most urgent need in the degraded arid and semi-arid lands is an effective, low-cost, and reliable system or measure of soil and water conservation, which will reduce soil erosion and soil sealing. The expected result is improved infiltration, which will allow revegetation under the prevailing soils and climate conditions. Consequently, for the range to recover from a degraded state, rehabilitation measures such as water-harvesting embankments, reseeding and tree planting amongst others become as necessary as the protection from further irrational exploitation. Many protection measures tend to take the form of temporary enclosure of the range to keep off ungulates.

9.3 Enclosures Approach for Rangeland Rehabilitation

The enclosure approach for rehabilitating degraded grazing lands involves closing off some part of the degraded open range from grazing for a given period usually not less than 3 years, to allow regeneration of vegetation (Behnke, 1986). This concept of the range enclosure that implies setting apart an area from grazing and wood harvesting, is not new. The traditional pastoralists used to set aside dry season grazing areas and kept livestock off such designated common pool resource areas, *which were not fenced*. Nevertheless, the breakdown of the traditional community leadership structures and governance, the influx of people with a different way of life from the neighbouring high potential areas has led to a collapse of the pastoral set aside policy in the rangelands. Thus setting aside areas for annual deferred grazing has become increasingly difficult, and in some cases, is no longer possible (Dietz, 1987). The result, akin to the case of Lake Baringo Basin, is many people with small herds, which graze *everywhere* whereas in the past, there were fewer people with large herds who *grazed certain areas and left others ungrazed* as dry season reserves (Meyerhoff, 1991).

These factors combined create a grazing *free-for-all* in the rangelands tending to the *tragedy of commons* scenario (Hardin, 1968; Anderson, 1980; Bonfiglioli, 1992). This led to overgrazing and degradation resulting to an ecosystem that can hardly maintain its stability, function, and structure. Moreover, the ensuing competition between livestock keepers for control of a diminishing range resource is fuelling the drive of range enclosure, as the pastoralists attempt to do something about their declining resource base.

In the Lake Baringo Basin, the severity of rangeland degradation has made the life of the pastoralists very harsh. Overstocking in the open range has undermined the economic welfare of local livestock keepers who face high levels of stock loss amidst other problems at the end of the dry season, especially if it is protracted. Their response has been to enclose a portion of their rangelands for their exclusive use, while emulating the communal rehabilitation enclosures set up by a community based Rehabilitation of Arid Environments (RAE) Trust after realizing that degraded land can be restored successively (RAE, 2003; 2004). Therefore, behind the fencing of the range to combat land degradation, lies the struggle to address the basic need of food security. With land rehabilitation, each farmer can provide for his own household and livestock, and successful enclosures mean that pastoral households could do without food aid (Makokha et al., 1999), thus giving them dignity and independence.

Prior to vegetation establishment, the fences play an important role of protecting the vulnerable tree and grass seedlings and regenerating remnants of existing vegetation (see Fig. 9.1). Harris et al. (1996) and Sands et al. (1970) suggested that well-adapted and hardy species should be selected as pioneer vegetation for reseeding during rehabilitation programmes. Biamah (1986) supported range reseeding for quick establishment of perennial grass species in the Lake Baringo Basin. Earlier reseeding experiments by Pratt showed *Cenchrus ciliaris* and *Eragrostis superba* as examples of the main grass species suited to semi-arid conditions in Baringo (Pratt, 1964).

Besides Baringo, enclosures have been used elsewhere for varying specific objectives. Examples include range restoration in Chepareria Division of West Pokot District in western Kenya where *Vi-Agroforestry Project* has been working with

Fig. 9.1 Enclosures establishment starts with fencing (Electric, Live Cacti, or Cut thorn-bush). Photo by SM Mureithi, 2005

the agro-pastoral community to establish enclosures in their fields (Makokha et al., 1999; Kitalyi et al., 2002). Other measures include revegetation of degraded dryland forest areas in Ethiopia (Mekuria et al., 2007; Cleemput et al., 2004; Mengistu et al., 2005), and in north-west Tanzania, where using indigenous knowledge, the local people are practising a natural resource management system called *Ngitili* – a Sukuma word meaning enclosure.

These enclosures help in conservation of grazing and fodder lands by encouraging vegetation regeneration and tree planting (HASHI, 2002). In South Darfur in Sudan (before the current crisis in the region), enclosure were established primarily to produce and sell fodder in the more profitable commercial fodder markets at Nyala, rather than to provide feed for local livestock (Behnke, 1985; 1986). According to (Behnke, 1985), Nyala as a railway-end town, contained numerous milk cows kept for household milk supplies, horses and donkeys used for the haulage of domestic water or commercial goods from the railhead, and a variable number of animals being held for marketing or shipment to Khartoum.

Enclosing is also famous as a basic measure for revegetation and stabilisation of wandering sand dunes and desertified lands. This has been done in both desert and semi-desert areas of China (IPALAC, 2006; Zhang et al., 2005), and in India amongst other countries. Sinha et al. (2006) reported that Thar Desert in India shows tremendous resilience for regeneration when it is protected by fenced enclosures for a certain period. In these cases, enclosing sandy areas helped the recovery of natural vegetation, hence stabilizing the sandy soil material.

Rangeland ecosystems are complex in ecological and socioeconomic dynamics. Enclosing one part of the range gives rise to more complexity, since the enclosed and unenclosed areas are parts of a unified whole.

9.4 Enclosure Movement in Lake Baringo Basin

In 1982, a community-based project was started in Baringo District in the mid-west of Kenya under the auspices of the Rehabilitation of Arid Environments (RAE) Trust (formally Baringo Fuel and Fodder Project-BFPP). It aimed to rehabilitate severely degraded areas around Lake Baringo and on the surrounding hills, which were subject to heavy grazing pressure (Meyerhoff, 1991; de Groot et al., 1992). The intention was to work with agropastoralist communities to achieve sustainable land management systems in arid and semi-arid areas. Rehabilitation commenced by enclosing areas of various sizes from 6 to 400 ha, preparing seedbed and water harvesting structures followed by reseeding with indigenous grass species alongside planting a variety of indigenous and exotic tree species.

Since 1982, a spontaneous rangeland enclosure trend has gained momentum in the Lake Baringo Basin, following successful rehabilitation of more than 1,430 ha of severely degraded land using reseeded communal enclosures (RAE, 2004). The Lake Basin pastoral inhabitants request cost shared-assistance from RAE Trust to rehabilitate their denuded land, which is still communally owned. Presently, more than 250 ha of "private" land has been rehabilitated in Baringo and the neighbouring Laikipia Districts. Furthermore, the RAE Trust by year 2006 received additional requests for assistance to reclaim more than 30,000 ha of communal degraded land (RAE, 2004). Successfully rehabilitated total area may seem insignificant compared to the spatial vastness of the degraded rangeland in Baringo and the rest of the arid and semi-arid districts in Kenya. However the positive impact of the enclosures to the agropastoralists' households and the environment is evident.

Presently, the Lake Baringo Basin is dotted by individual farmer's ("private") enclosures (Fig. 9.2), as locals try to emulate the communal enclosures and after realising that degradation can be combated successfully, hence giving them an opportunity to address their livelihood problems. Community leaders from other areas, including neighbouring Laikipia, Turkana, and Samburu districts have also requested the RAE Trust to expand its operations into their areas (RAE, 2004).

During our research in Baringo, many herd owners were using cut-thorn bushes (*Acacia sp.* and *Prosopis sp.*) and planting thorn cactus (*Opuntia ficus-indica*) to enclose their land. This trend is giving rise to two major categories of enclosures in terms of management and ownership. First category is the reseeded, communally owned, and – managed enclosures, while the second is the individual farmers'

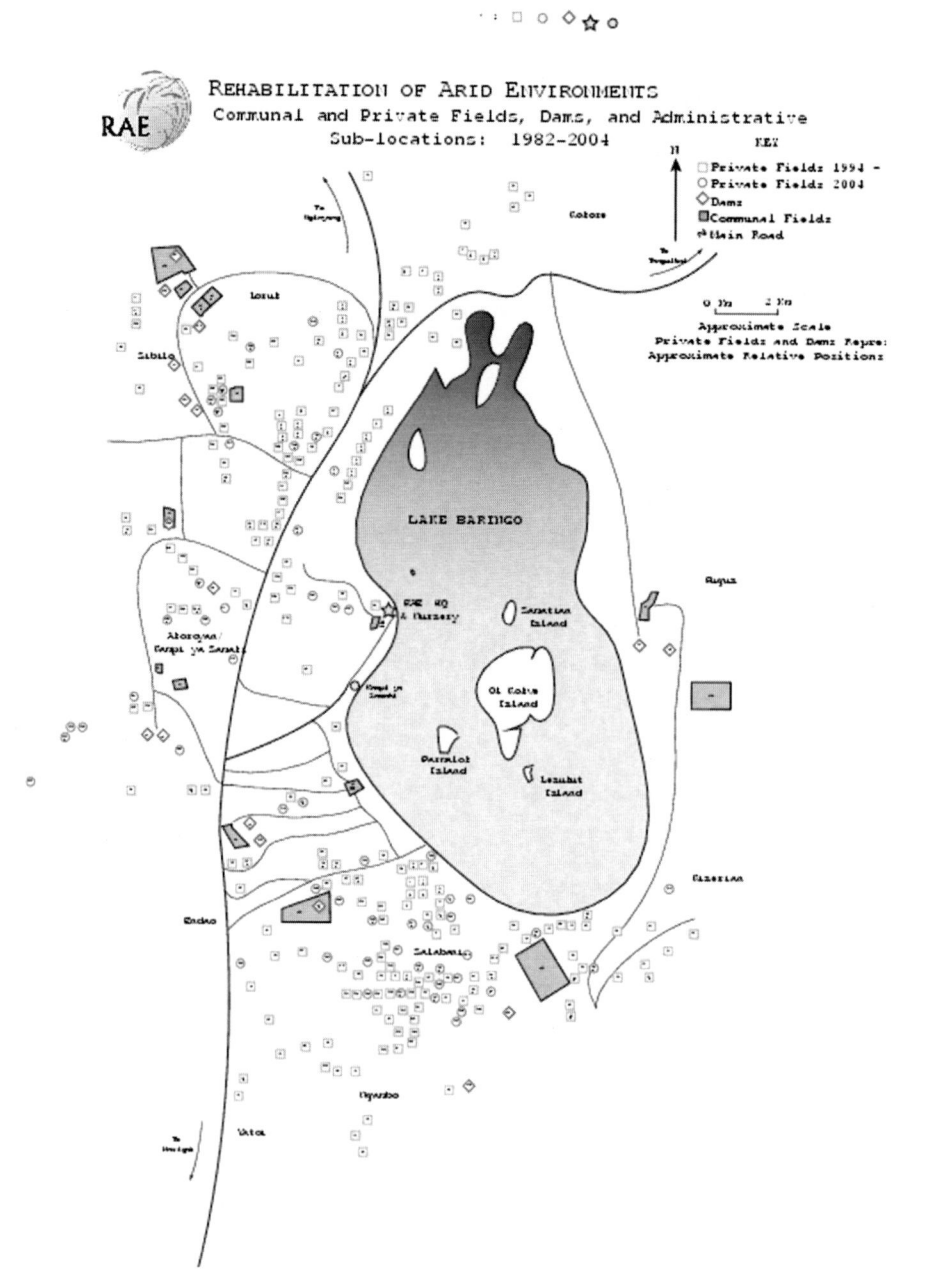

Fig. 9.2 RAE Trust-Working Field Map (not to scale) as at 2005, showing the communal (*orange blocks*) and private enclosures (*tiny green squares*) (for colors, see online version)

reseeded and privately managed enclosures. A third category comprises of naturally regenerated enclosures but their success rate has been low in the Lake Basin. A private enclosure is a piece of land fenced off by an individual household and planted with trees and grass or allowed to regenerate naturally (passive rehabilitation).

When an individual household head decides to enclose an area near their *boma* – (a Kiswahili word for cattle corrals found inside the cut-thorn bush fence enclosing a single pastoral household temporary settlement), one has to obtain consent from the community elders. Once the consent is granted and the household puts up a fence, other people in the community respect that household's negotiated rights for that specific area, as well as the adjacent area, and henceforth recognise that the land belongs to them (RAE, 1998). The result is an increasing trend towards respect for individual land tenure. Private enclosures owners do not have formal ownership of the fenced land as it is still owed communally.

Even though rehabilitated land flourishes with a diversity of woody and herbaceous species long lost in other un-rehabilitated areas, this trend seemingly welcome by the local communities raises major questions on the future of pastoralism and livestock husbandry in East African rangelands. Benhke (1986) identified major concerns on rangeland enclosure deserving attention. These include questions of technical efficiency and productivity; problems of range ecosystems conservation; and the related issues of economic equity and economic growth.

Behnke (1986) noted that "if the fencing of rangeland by livestock owners is likely to become more common, then administrators, rangeland scientists and policy makers will need to have some idea of the benefits and costs arising out of the shift from open-range to fenced forms of animal husbandry". The main drivers for this shift in Lake Baringo Basin for example, is the ardent need for rehabilitating severely degraded land, to attain pasture security for livestock, and hence food security for the local pastoralist communities during the dry seasons and drought. In addition, this would reduce soil erosion, reclaim gullies, and significantly limit the amount of topsoil and sediments being deposited into Lake Baringo (RAE, 2003, 2004). As early as 1974 the Government of Kenya (GoK, 1974) described this area estimated to be 2,115 km^2, as an "ecological emergency area".

9.5 Effects of Range Enclosures

Rangeland enclosure has profound ecological (biophysical) effects and a number of socio-economic implications that vary significantly, depending on local conditions. Understanding the consequences of the rising trend of rangeland enclosure has been shown to be imperative. Benhke (1986) and Graham (1988) argued that what administrators require is not a general policy for or against enclosure, but rather some understanding of the various effects of enclosure under different circumstances. Ultimately, researchers may be able to present policy-makers with a typology of different kinds of enclosure movements, and with a systematic discussion of the probable outcome of each kind of movement. Therefore, the spontaneous enclosure of the range by livestock owners may raise new problems, but may also permit new approaches to the development of the African livestock industry.

9.5.1 Ecological Effects of Enclosures

One consequence of the range enclosure is the notable difference inside and outside the fence after vegetation regeneration (Fig. 9.3). An enclosure in a severely denuded area tends to become an ecological island, and may thrive in desirable herbaceous and woody plants and overall biodiversity above- and belowground, from microbial to higher trophic levels (Verdoodt et al., 2009; RAE, 2004; Stelfox, 1986). This, however, depends on a number of factors. Biamah (1988) and Makokha et al. (1999) singled out the severity of the degradation and range condition before intervention, time allowed for restoration, and the enclosure size and management after range restoration as important factors influencing the success of the rehabilitation. Biamah (1988) further argued that the semi-arid rangeland areas typically are resilient and capable of regeneration even though the process of regeneration can be delayed by natural forces (1-year and multiyear droughts) or by the interference of human activities like grazing, time of grazing introduction, and stock densities.

The extent of degradation is an important factor determining whether the range will recover at all. The removal of vegetation cover from an ecosystem results in a compounding effect of degradation with the soil being the worst hit component. With absence of vegetation (Fig. 9.4), the soil is deprived of organic matter, which is the key to soil fertility and productivity especially in drylands (FAO, 2004), and is highly exposed to the agents of erosion.

Once the vegetation cover is restored, it improves the soil structure (Bronick and Lal, 2005), soil water balance (Hongo et al., 1995) chemical soil fertility (Jaiyeoba, 1995; Descheemaeker et al., 2006a; Mekuria et al., 2007), and restores the soil

Fig. 9.3 Inside–outside contrast in a communal enclosure. Hard setting is visible in the foreground-exposed soils. Photo by SM Mureithi, 2005

Fig. 9.4 (*Left*): Severely degraded rangeland in Njemps Flats being prepared for rehabilitation (*Right*): same field converted into productive pasture using enclosure approach. Source: RAE (2003)

biodiversity and ecosystem services (Su et al., 2005) through reduced soil erosion (Descheemaeker et al., 2006b). This clearly illustrates the linkages and feedback loops occurring between biotic and abiotic components of the rangeland ecosystem, capable of enforcing or reversing land degradation (Perrow and Davy, 2002a, b; King and Hobbs, 2006; Monger and Bestelmeyer, 2006).

Nevertheless, it is still questionable whether an eroded range can go back to is potential, as the probability of reversing grazing-induced change in the rangeland is inversely related to the amount of disturbance involved in the transition (SRM, 1995). O'Connor (1991) and Westoby et al. (1989) argues that severely degraded rangelands may never return to their original state, even when rested for decades.

9.5.2 Effects of Enclosures on Vegetation

Establishment of vegetation inside the enclosures is a relatively a slow process, which is dependent on the reliability of rainfall and the effectiveness of the seedbed and rainwater harvesting structures in place (Pratt, 1964). After initial reseeding and tree planting, only the well-adapted and hardy species thrive. Once established, such vegetation can support accumulation and recycling of nutrients by providing organic matter through litter fall and dead roots, improve soil structure and availability of nitrogen (when legumes and annual grasses are present).

Vegetation inside the enclosures established in a completely bare degraded area tends to follow the shrub-herbaceous plant theory of Gilad et al. (2004), where establishment of one form of plant life synergises that of another. In Baringo (RAE, 2004; 2005) and the neighbouring district of West Pokot (Makokha et al., 1999), some locally threatened species that had "disappeared" are reported to be present inside established enclosures. Range enclosure is thus providing a community-friendly way of restoring dryland biodiversity. Improved infiltration capacity inside the enclosures enhances the moisture available for the established plants (Ekaya and Kinyamario, 2003; Hongo et al., 1995), thereby increasing cover, and standing biomass (Cleemput et al., 2004; Makokha et al., 1999).

Higher herbaceous and woody species composition has been observed inside communal enclosures established in a degraded rangeland (Makokha et al., 1999; Cleemput et al., 2004; Mengistu et al., 2005; Verdoodt et al., 2009), compared to the outside. This was attributed to deliberate human influence such as the choice of herbaceous and woody species used for the rehabilitation. The enclosures have higher forage and browse value, which represents the high percentage of the vegetation palatable to the livestock (Milton et al., 1997). The high proportion of seedlings reported in the enclosures in Ethiopia by Mengistu et al. (2005), is an indicator of recruitment of the plants through germination, and implies the existence of a good potential for the restoration of woody communities. In contrast, the bulk of the species outside the enclosures is composed of a variety of invasive (usually noxious) species, most which are as disturbance and land degradation indicators.

9.5.3 Effects of Enclosures on the Soil

Vegetation in the rangelands has significant influence on the soil, which on the other hand, influences the kind of vegetation present, as influenced by the climatic factors and herbivory and human influence (McClanahan and Young, 1996). Protection of reseeded range is necessary to allow the initial increase of vegetation cover, which in turn plays crucial role in covering the soil. Biamah (1986, 1988) reported that increasing ground cover in Njemps Flats prevented soil sealing through raindrop interception and splash erosion. This sequentially encouraged infiltration thus reducing high runoff rates and soil erosion. Vegetation breaks up the falling raindrops so that they reach the soil surface as small droplets hence reducing their impact. Furthermore, trees utilize deep water table water, improve soil physical condition, and reduce the ground level wind speed and thus its erosive potential (Dregne, 1992).

In addition to protecting the soil from erosion agents, the tree canopy-herbaceous layer interaction improves soil fertility through addition of nitrogen and organic matter. In addition, the vegetation supplies plant litter, which decomposes to supply the soils organic carbon pools (Kellman, 1979). Dregne (1992) asserted that shrubs play an important role in maintaining a pool of soil nutrients in desert ecosystems by creating islands of fertility beneath their canopies through accumulation of organic matter.

In natural grasslands, Heady (1956) observed that any dead material above the soil surface is referred to as litter, mulch, or plant residues. Litter increases soil moisture through its effect on infiltration, evaporation, and runoff. It tends to stabilise soil moisture and temperatures, thus improving conditions for germination (Ekaya and Kinyamario, 2001). Part of the litter may be buried in the soil through animal activities, e.g. hoof activity through trampling or arthropod activity leading to development of good soil structure. Therefore, if not burned, litter can contribute significantly to the build-up of organic matter in the rangelands (Isichei and Sandford, 1980). Organic matter has many beneficial effects on the soil's physical, chemical, and microbiological properties. Examples include increasing water holding capacity, and cation exchange capacity, lowering the bulk density of the soil and increasing the microbial activity amongst others (Dumanski and Pieri, 2000). Litter also plays a crucial role in nutrient cycling in these ecosystems.

9.5.4 Effects Outside the Enclosures

Overuse of forage plants by the free ranging herbivores leads in the open grazing areas, to a shift of assemblages dominated by toxic and spinescent woody plants and numerous species of invading forbs (O'Connor, 1991). This change in composition may come about because unpalatable plant species that are usually ignored herbivores tend to thrive, out competing those preferentially selected (Milton et al., 1994). According to (Mureithi, 2006), this shift led to a highly patchy and heterogeneous rangescape, having range condition deteriorating from within, and consequently, resulted in an overall declining range trend. Eventual concentration of both wild and domestic herbivores not allowed utilizing the enclosures, in addition to human traffic, tend to compound the soil and land degradation problem on the other side of the fence. Given the fragility, stochasticity and ecological limitations of these non-equilibrium ecosystems, such a trend may eventually be ecologically disastrous.

Local people are well aware of the impact of their activities on the open range and of the negative implications of these activities. There are two sets of reasons why they continue to carry them out. First, there is usually no alternative means of making an income, and second, the open range is to all intents and purposes an uncontrolled resource since the breakdown of traditional structures of governance by pastoralist's elders. The state owns the majority of rangeland, but apart from the heavily protected areas such as nature reserves, most is de facto open access. With no rules for usage, or no enforcement of rules, each individual makes the most of his or her opportunity, because if not, someone else will – the tragedy of the commons (Hardin, 1968), or as it may more correctly be described, the tragedy of the *open access* resources ensues unabated.

9.5.5 Socio-Economic Implications of the Enclosures

A successfully regenerated enclosure becomes key resource area for the respective household or community in a harsh environment. The social and economic consequences of the range enclosure are varied, depending on the accessibility of the enclosure benefits to the pastoral households, and the environmental goods and services tapped either directly or indirectly by the society. The households that have access to the communal enclosures on one hand are enjoying improved livelihoods as a result of income generating activities that have enabled them to profit from the reclaimed land (Makokha et al., 1999; Kitalyi et al., 2002; RAE, 2004).

Examples of income generation from rehabilitated communal fields in Baringo enclosures include, amongst others, sale of various commodities from the enclosures (fattened livestock, cut grass for thatch, fodder or hay, grass seed), renting dry season grazing, and poles for fuelwood or other domestic uses. Bee-keeping using the Lungstroth hives has also been introduced and is picking up well. The individual enclosure owners can tap the same benefits depending on the enclosure time, treatment (whether reseeded or naturally regenerated) and the management.

On the other hand, there are households that neither have access to communal nor own private enclosures. The herd owners in this cluster are forced to graze their livestock in the open range, where the competition of scarce pasture resources is very high (Nyang, 1988). Benhke (1986 asserted that range enclosure will in the short term, exacerbate the problem of overstocking on the open range by withdrawing parts of the range from communal use, and by forcing more livestock into the remaining area. The pressure for further enclosure will therefore increase as individuals watch the commonage shrink and attempt to grab their piece of it before it is too late. In this way enclosure movements build within themselves pressures for their own expansion.

However, this scenario creates the haves and the haves-not situation in the range, and may eventually become a recipe for conflicts in these communally owned rangelands (Mureithi, 2006). Unless the policy is enacted to guide this changing land use and land ownership (addressing land tenure, access and land rights) in the rangelands, the likelihood of instability occurring remains imminent.

9.6 Policy Implications

Land ownership and access to resource, especially grazing land and water is indeed a very delicate issue in the African pastoral systems. For this reason, Makokha et al. (1999) suggested a very key point that, those households without enclosures should not be viewed as *non-adopters*, as most farmers have logical reasons for not establishing enclosures. The intervention programme should develop a set of interventions that could also assist these pastoralists to improve their livelihoods; they do not necessarily have to establish enclosures. Interventions on improved breeds, milk and stock marketing, nutrition education may be applicable to them as well as to those farmers with enclosures (Makokha et al., 1999). Speeding up the land adjudication process may also help some of these pastoralists, as individual owners are more likely to place their land under enclosure management, hence minimizing further land degradation, as well as breaking their poverty cycle (Meyerhoff, 2005 – personal communication to the first author).

The increasing trend of rangeland enclosure poses concerns on the long-term planning and policy implications of this process. It has been argued that range enclosure is producing a new and distinctly African system of range livestock management, in which animals alternate between both enclosed pastures and the open range (Benhke, 1986). In these cases it would be unrealistic to base development plans on the assumption that we are still dealing with traditional, fully or semi-nomadic pastoralists.

On the other hand, it would be equally unrealistic for administrators to press for the complete sedentarisation of pastoralists in the range, or immediate development of self-contained, fenced ranches, on the model of the standard group or individual ranch schemes common in Africa in the 1960s and 1970s. A more reasonable objective would be to devise suitable policy responses to the hybrid form of enclosed and open-range animal management, which is now developing, and to sponsor technical research, which will address the characteristic problems of this form of production (Benhke, 1986; Nyang, 1988).

9.7 Future Trends

Successful rehabilitation of severely degraded semi-arid rangelands in Lake Baringo Basin using communal enclosures has led to an upsurge of private enclosures on communal land. This trend is likely to be adopted in other semi-arid rangelands in Sub-Saharan Africa. Communal and individual farmer's enclosures form a mosaic in the severely degraded semi-arid rangelands of Lake Baringo Basin. Recent research shows rangeland enclosure has significantly improved the range condition of the enclosed areas, whereas degradation ensues in the open grazing areas (Verdoodt et al., 2009).

Enclosures with higher biomass production will support higher grazing capacity with good management, implying that less hectarage is required for 1 TLU in the enclosures, than that recommended in the open grazing rangeland. This translates into more economic gain to the enclosure owners, since livestock keeping is their main source of livelihood. A great task facing the RAE Trust today is to carry out stepwise community mobilisation and education to enable the resource users embrace the changes in natural resource management. Hopefully, this will enable them to adapt to the unfolding reality of managing the once open communal rangelands within a fence.

The sustainability of any positive rehabilitation work however, lies in a proper and extensive land policy and tenure reform in addressing the needs of pastoral communities. The communal use of resources has to great extent fuelled land degradation in the Lake Baringo Basin. Most people in this area indicated that they would care more for the land if they owned it, contrasting the present scenario where they own it communally and the free access of the common persist (Mureithi, 2006). The success of the rehabilitation may eventually lie in land adjudication and education to enable people to manage resources in a changing system, thus empowering them to meet themselves their fundamental needs (food, water and forage). The policy makers therefore, should look beyond indicators of land degradation, and quick fix technical solutions, to indicators of food insecurity that simultaneously degrade human health and nutrition and seek their possible long-term solution.

Acknowledgments This study was made possible through the financial support provided by the Flemish Interuniversity Council (VLIR) of Belgium. We are indebted to all the Njemps Flats communal and private enclosure farmers who warmly welcomed us to their fields. Special thanks to the staff Rehabilitation of Arid Environments (RAE) Trust in Baringo and the Range Management Section, Department of Land Resources Management & Agricultural Technology, University of Nairobi, for kind assistance during the fieldwork.

References

Anderson, D. (1980). Grazing, goats and government: Ecological crises and colonial policy in Baringo, 1918–1939. Department of History, Staff Seminar Paper No. 6, University of Nairobi.

Behnke, R.H. (1985). Rangeland Development and the Improvement of Livestock Production: Policy Issues and Recommendations for the Western Savannah Project, South Darfur, Sudan. Mokoro Ltd., Brill, Bucks, UK.

Behnke, R. (1986). The implications of spontaneous range enclosure for African livestock development policy. Network Paper No.12, September 1986. International Livestock Centre for Africa (ILCA), Addis Ababa, Ethiopia.

Biamah, E.K. (1986). Technical and socioeconomic considerations in rehabilitating and conserving an eroded/denuded catchment area: A case study of the Chemeron catchment area in Central Baringo. In: D.B. Thomas, E.K. Biamah, A.M. Kilewe, L. Lundgren and B.O. Mochoge (eds.), Soil and Water Conservation in Kenya. Proceedings of the 3rd National Workshop. Kabete, Nairobi. 16–19th September 1986.

Biamah, E.K. (1988). Environmental degradation and rehabilitation in central Baringo, Kenya. In: J. Rinamwanich (ed.), Land Conservation for Future Generations. Proceedings of the 5th International Soil Conservation Conference. Volume 1. 18–29 January 1988, Bangkok Thailand.

Bogdan, A.V. and Pratt, D.J. (1967). Reseeding denuded Pastoral Land in Kenya. Republic of Kenya. Ministry of Agriculture and Animal Husbandry, Nairobi, Kenya, pp. 15–30.

Bonfiglioli, A.M. (1992). Pastoralists at a Cross Road: Survival and Development Issues in African Pastoralism. UNICEF/UNSO Project for Nomadic Pastoralists in Africa.

Bronick, C.J. and Lal, R. (2005). Soil structure and management: A review. Geoderma 124:3–22.

Cleemput, S., Muys, B., Kleinn, C. and Janssens, M.J.J. (2004). Biomass estimation techniques for enclosures in a semi-arid area: A case study in Northern Ethiopia. Paper presented in Deutscher Tropentag, 2004, Berlin, Germany. Available at: http://www.tropentag.de/2004/abstracts/full/3.pdf Accessed 04/05/06.

de Groot, P., Field-Juma, A. and Hall, D.O. (1992). Reclaiming the Land: Revegetation in Semi-arid Kenya. African Center for Technology Studies (ACTS) Press, Nairobi kenya and Biomass Users Network (BUN), Harare Zimbabwe.

Descheemaecker, K., Muys, B., Nyssen, J., Poesen, J., Raes, D., Haile, M. and Deckers, J. (2006a). Litter production and organic matter accumulation in exclosures of the Tigray highlands, Ethiopia. Forest Ecology and Management. 233:21–35.

Descheemaecker, K., Nyssen, J., Rossi, J., Poesen, J., Haile, M., Raes, D., Muys, B., Moeyersons, J. and Deckers, J. (2006b). Sediment deposition and pedogenesis in exclosures in the Tigray highlands, Ethiopia. Geoderma 132:291–314.

Dietz, T. (1987). Pastoralists in dire straits: Survival strategies and external interventions in a semi-arid region at the Kenya-Uganda border, western Pokot, 1900–1986. ISBN 90 6983 0188.

Dregne, H.E. (1992). Degradation and Restoration of Arid Lands. International Centre for Arid and Semi-arid studies, Texas Tech University, Lubbock, TX.

Dumanski, J. and Pieri, C. (2000). Land quality indicators: Research plan. Agriculture, Ecosystems and Environment 81:93–102.

Ekaya, W.N. and Kinyamario, J.I. (2001). Production and decomposition of plant litter in an arid rangeland of Kenya. African Journal of Range and Forage Science 2001(18):125–129.

Ekaya, W.N. and Kinyamario, J.I. (2003). Herbaceous vegetation productivity in an arid rehabilitated rangeland in Kenya. In: N. Allsopp et al. (ed.), Proceedings of the 7th International Rangelands Congress, Durban, South Africa. 26th July-1st August 2003.ISBN: 0–958–45348–9.

FAO. (2004). Carbon sequestration in dryland soils. World Soil Resources Reports Series – 102, 129 pg. ISBN: 9251052301 Y5738/E. Available online at: http://www.fao.org/docrep/007/y5738e/y5738e00.htm#Contents

Gachimbi, L.N. (1990). Land Degradation and its control in the Kibwezi area, Kenya. Unpublished MSc Thesis, Faculty of Agriculture, University of Nairobi.

Gilad, E., von Hardenberg, J., Provenzale, A., Shachak, M. and And Meron, E. (2004). Ecosystem engineers. From pattern formation to habitat creation. Physical Review Letters 93:098105 (1–4).

GoK and Government of Kenya. (1974). Republic of Kenya National Development Plan, 1974–1978. Government Printers, Nairobi.

GoK and Government of Kenya. (1997). Development Policy for the Arid and Semi-arid Lands (ASALS). Government Printers, Nairobi.

Graham, O. (1988). Enclosure of the East African Rangelands: Recent Trends and Their Impacts. Pastoral Development Network – Overseas Development Institute, London. ISSN: 0951 1911.

Hardin, G. (1968). The tragedy of the commons. Science 162:1243–1248.

Harris, J.A., Birch, P. and Palmer, J. (1996). Land Restoration and Reclamation. Principle and Practice. Longman Publishers, Singapore.

HASHI, Hifadhi Ardhi Shinyanga. (2002). Greening the Desert – Tanzania. Available at http://www.HandsOnGreeningtheDesert - Tanzania.htm Accessed 04/05/06

Heady, H.F. (1956). Changes in California annual plant community induced by manipulation of natural mulch. Ecology 37:798–812.

Heady, H.F. (1999). Pespectives of rangeland ecology and management. Rangelands 21(5): 23–33.

Hongo, A., Matsumoto, S., Takahashi, H., Zou, H., Cheng, J., Jia, H. and Zhao, Z. (1995). Effect of exclosure and topography on rehabilitation of overgrazed shrub-steppe in the Loess Plateau of Northwest China. Restoration Ecology 3:18–25.

Hudson, N.W. (1987). Soil and water conservation in semi-arid areas. FAO Soil Bulletin 57.

IPALAC, International Program for Arid Lands Crops. (2006). Sand dune stabilization with plants in China. Available at http://www.ipalac.org/code/abstracts/ab_46.html Accessed 08/05/06

Isichei, A.O. and Sandford, W.W. (1980). Nitrogen losses by burning from Nigerian grassland ecosystems. In: T. Rosswal (ed.), Nitrogen Cycling in West African Ecosystems. SCOPE/UNEP International nitrogen unit, Royal Swedish Academy of Sciences, Stockholm, pp. 203–224.

Jaiyeoba, I.A. (1995). Changes in soil properties related to different land uses in part of the Nigerian semi-arid Savannah. Soil Use and Management 11:84–89.

Jordan, S.M. (1957). Reclamation and pasture management in the semi arid areas of Kitui district. East African Agricultural and Forestry 23:84–88.

Kellman, M. (1979). Soil enrichment by neo-tropical savannah trees. Journal of Ecology 67: 565–577.

King, E.G. and Hobbs., R.J. (2006). Identifying linkages among conceptual models of ecosystem degradation and restoration: Towards and integrative framework. Restoration Ecology 14(3):369–378.

Kitalyi, A., Musili, A., Suazo, G. and Ogutu, F. (2002). Enclosures to protect and conserve: For better livelihood of the West Pokot community. RELMA Technical Pamphlet Series No. 2. Nairobi, Kenya: Regional Land Management Unit (RELMA). Swedish International Development Cooperation Agency (Sida), Nairobi, Kenya, 24 p. + vi p.

Lusigi, W. (1981). Combating Desertification and Rehabilitating Degraded Production Systems in Northern Kenya. IPAL Technical Report No. A-4. pp. 126–128.

Makokha, W., Lonyakou, S., Nyang, M., Kareko, K.K., Holding, C., Njoka, T.J. and Kitalyi, A. (1999). We work together: Land rehabilitation and household dynamics in Chepareria Division, west Pokot District, Kenya. RELMA Technical Report No. 22. Nairobi Kenya: RELMA/SIDA. ISBN 9966-896-42-2. p. 81.

McClanahan, T.R. and Young, T.P. (eds.). (1996). East African Ecosystems and their Conservation. Oxford University Press, Inc., New York.

Mekuria, W., Veldkamp, E., Haile, M., Nyssen, J., Muys, B. and Gebrehiwot, K. (2007). Effectiveness of exclosures to restore degraded soils as a result of overgrazing in Tigray, Ethiopia. Journal of Arid Environments 69(2):270–284.

Mengistu, T., Teketay, D., Hulten, H. and Yemshaw, Y. (2005). The role of enclosures in the recovery of woody vegetation in degraded dryland hillsides of central and northern Ethiopia. Journal of Arid Environments 60:259–281.

Meyerhoff, E. (1991). Taking Stock: Changing Livelihoods in a Agropastoral Community. African Center for Technology Studies (ACTS) Press, Nairobi, Kenya and Biomass Users Network, Harare, Zimbabwe.

Meyerhoff, E. (2005). Personal Communication to the first author. RAE Trust Headquaters, Lake Baringo. Kenya. Date: 15th September 2005.

Miller, J.R. and Hobbs, R.J. (2007). Habitat restoration–do we knowwhat we're doing? Restoration Ecology 15(3):382–390.

Milton, S.J., Richard, W., Dean, J. and Roger, P.E. (1997). Rangeland health assessment: A practical guide for ranchers in arid Karoo shrublands. Journal of Arid Environments 39:253–265.

Milton, S.J., Dean, W.R.J., du Plessis, M.A. and Siegfried, W.R. (1994). A conceptual model of arid rangeland degradation: The escalating cost of declining productivity. BioScience 44(2):70–76.

Monger, H.C. and Bestelmeyer, B.T. (2006). The soil-geomorphic template and biotic change in arid and semi-arid ecosystems. Journal of Arid Environments 65:207–218.

Muhia, C.D.K. (1986). Grazing land management and improvements. In: D.B. Thomas et al. (ed.), Soil and Water Conservation in Kenya. Proceedings of a 3rd National Workshop, U.O.N./S.I.D.A. Nairobi, pp. 315–322.

Mureithi, S.M. (2006). The effect of enclosures on rehabilitation of degraded semi-arid land in Lake Baringo Basin, Kenya. Unpublished M.Sc. Thesis, Ghent University Belgium.

Mututho, J.M. (1986). Some aspects of soil conservation on grazing lands. In: D.B. Thomas et al. (ed.), Soil and Water Conservation in Kenya. Proceedings of a Third National Workshop, U.O.N./S.I.D.A. Nairobi, pp. 323–331.

Nyang, M. (1988). Vi Tree Planting Project: Supplementary Impact Assessment on Farmers Without Enclosures in Wet Pokot, July 1997. RSCU/Vi-TPP, Kitale, Kenya.

O'Connor, T.G. (1991). Local extinction in perennial grasslands: A life-history approach. The American Naturalist 137:735–773.

Pereira, H.C. and Beckeley., V.R.S. (1952). Grass establishment on eroded soil in a semi-arid African reserve. Empire Journal of Experimental Agriculture 21:11–14.

Perrow, M.R. and Davy, A.J. (2002a). Handbook of Ecological Restoration. Volume 1. Principles of Restoration. Cambridge University Press, Cambridge.

Perrow, M.R. and Davy, A.J. (2002b). Handbook of Ecological Restoration. Volume 2. Restoration in Practice. Cambridge University Press, Cambridge.

Pratt, D.J. (1964). Reseeding denuded lands in Baringo district, Kenya. 11-Techniques for dry alluvial sites. East African Agriculture and Forestry Journal 29:243–260.

RAE, Rehabilitation of Arid Lands Trust. (1998). Baringo Private Fields Handbook. RAE. Kampi ya Samaki. Marigat, Kenya.

RAE, Rehabilitation of Arid Lands Trust. (2003). URL: http://www.raetrust.org/ Accessed 16/12/03.

RAE, Rehabilitation of Arid Lands Trust. (2006). URL: http://www.raetrust.org/ Accessed 20th May 2006.

RAE, Rehabilitation of Arid Lands Trust. FactSheet. (2004). RAE. Kampi ya Samaki. Marigat, Kenya.
RAE, Rehabilitation of Arid Lands Trust. FactSheet. (2005). RAE. Kampi ya Samaki. Marigat, Kenya.
Sands, E.B., Thomas, D.B., Knight, J. and Pratt, D.J. (1970). Preliminary selection of pasture plants for the semi-arid areas in Kenya. East African Agriculture and Forestry Journal 36:49–57.
Sinha, R.K., Bhatia, S. and Vishnoi, R. (2006). Desertification control and rangeland management in the Thar desert of India. RALA Report No. 200. Available at http://www.rala.is/rade/rade-Sinha.PDF Accessed 08/05/06
Smith, P.D. and Critchley, W.R.S. (1983). The potential of runoff harvesting for crop productivity and range rehabilitation in semi-arid Baringo. In: D.B. Thomas and W.M. Senga (eds.), Soil and Water Conservation in Kenya. Nairobi, pp. 305–323.
SRM, Society for Range Management. (1995). Evaluating rangeland sustainability: The evolving technology. Rangelands 17(3):85–92.
Stelfox, B.J. (1986). Effects of livestock enclosures (Bomas) on the vegetation of the Athi plains, Kenya. African Journal of Ecology 24:41–45.
Su, Y.Z., Li, Y.L., Cui, J.Y. and Zhao, W.Z. (2005). Influences of continuous grazing and livestock exclusion on soil properties in a degraded sandy grassland, inner Mongolia, northern China. Catena 59:267–278.
UNCCD. (1994). Part 1 (Article 1a). UNEP, Nairobi.
UNEP, United Nations Environmental Programme. (2000). Devastating Drought in Kenya: Environmental Impacts and Responses. UNEP, Nairobi, 159p.
UNEP/GEF, UNEP's Action in the Framework of Global Environmental Facility. (2002). Protecting the Environment from Land Degradation. Division of GEF Coordination UNEP, Nairobi Kenya.
UNEP/GoK, United Nations Environmental Programme/Government of Kenya. (1997). National Land Degradation Assessment and Mapping. UNEP, Nairobi.
Verdoodt, A., Mureithi, S.M., Liming, Ye. and Van Ranst, E. (2009). Chronosequence analysis of two enclosure management strategies in degraded rangeland of semi-arid Kenya. Elsevier J. Agriculture, Ecosystems and Environment 129(1–3):332–339.
Westoby, M., Walker, B. and Noy-Meir., I. (1989). Opportunistic management for rangelands not at equilibrium. Journal of Range Management 42:266–274.
Zhang, J., Zhaob, H., Zhang, T., Zhao, X. and Drake, S. (2005). Community succession along a chronosequence of vegetation restoration on sand dunes in Horqin Sandy land. Journal of Arid Environments 62:555–566.

Chapter 10
Assessment of Land Desertification Based on the MEDALUS Approach and Elaboration of an Action Plan: The Case Study of the Souss River Basin, Morocco

R. Bouabid, M. Rouchdi, M. Badraoui, A. Diab, and S. Louafi

Abstract Following the definition of the UNCCD, desertification affects a major part of Morocco. A large extent assessment is usually difficult due to the lack of appropriate methods. The MEDALUS approach is one of the available approaches developed initially for the Mediterranean Europe for desertification sensitivity assessment based on four main indicators (soil, climate, vegetation and management) that are obtained from various parameters. This approach has been applied (with slight modifications) to the Souss river basin in west central Morocco as a case study to assist towards the implementation of the National Action Plan (NAP) to Combat Desertification. Remote sensing data coupled with field and other relevant data were integrated in a GIS database to produce individual maps depicting the four previously mentioned indicators. Such maps were then overlaid to derive a comprehensive Desertification Sensitivity Map (DSM). This map, as well as ground appraisal information, was used to propose an action plan comprising a list of potential interventions that may contribute to alleviating desertification problems in the region. The interventions were formulated with the consent of the local stakeholders in a participatory process. They included both direct interventions towards reducing land degradation, but also interventions that are linked to alleviating poverty, and offering alternative income sources to the local population in order to reduce land resources pressures. The DSM and the action plan proposed were adopted as a general guideline framework that would be translated into comprehensive detailed local/communal action plans.

Keywords Desertification · MEDALUS · GIS · Souss river basin · Morocco

R. Bouabid (✉)
Ecole Nationale d'Agriculture de Meknès, 50000 Meknès, Morocco
e-mail: rachid.bouabid@gmail.com

10.1 Introduction

Desertification is a global phenomenon resulting from combined effects of natural and human factors leading to land resource degradation with a subsequent reduction of its potentialities. Over the past 2 decades, the concept of desertification was set to *land degradation in arid, semi-arid and dry sub-humid areas resulting from various factors, including climatic variations and human activities* (UNCCD, 2000). In this context, desertification sensitivity is viewed as the degree of vulnerability or response of the environment/land to the impact of natural or anthropic activities. Consequently, governments were encouraged to establish their national action programmes and work towards implementing appropriate measures to combat this phenomenon.

Arid and semi-arid areas, covering about third of the earth surface are particularly threatened by desertification. Morocco, being located in a south Mediterranean environment, is also exposed to this threat, where climate severity, coupled with human pressures (both in extensive and intensive agriculture), have induced an acceleration of land degradation processes and lead to major environmental and socio-economic impacts. Desertification control should rely first on an assessment based on reliable data and approaches (factors and degree of degradation), and second on the use of the assessment outcome to trigger awareness and decision making towards desertification control implementations. The human factor is to be put in the centre of reasoning, from the early diagnostics to final implementation, in order to achieve reliable results.

Different models have been proposed to assess desertification at different scales with different approaches and parameters [Methodology for assessment of desertification (FAO-UNEP, 1984); Land quality indicators (FAO-UNEP, 1997); MEDALUS (Kosmas et al., 1999); DPSIR-framework (GIWA, 2001); Classification system for desertification in China (Jun Hou et al., 2003); Iranian Model of Desertification Assessment (Ahmadi and Nazari Samani, 2006), etc]; however, there is no consensus on the proper or best way, to assess desertification risk. Those integrating physical and anthropic factors and using spatial geo-information and tools are particularly preferred. The MEDALUS (Mediterranean Desertification and Land Use) methodology proposed by the MEDALUS project (EC DGXII Environment Programme) has been adopted in our study with minor modifications to adjust it to a basin context. This approach has been used in various areas with Mediterranean type of climate (Basso et al., 1999a, b; Sabir et al., 2005; Schall and Becker, 2007; Zehtabian et al., 2004; Vacca et al., 2009); however, most of these studies were limited to generating desertification risk maps, with limited cases going further to propose the necessary and appropriate interventions for combating desertification in the areas assessed.

The objectives of our work were (i) to apply the MEDALUS model to assess desertification risk in the Souss basin using GIS and remote sensing tools, and (ii) to use this information as a basis for proposing possible actions to combat this phenomenon in target priority areas.

10.2 Materials and Methods

10.2.1 Study Area

The Souss river basin is located in west central Morocco, east of the city of Agadir (Fig. 10.1). The Atlantic Ocean in the west, the High Atlas Mountains in the north, and the Anti-Atlas Mountains in the south border it. It lies over approximately 16,000 km^2 and is characterized by a semi-arid to sub-desert climate. The mean annual temperatures precipitations vary respectively from 14 to 20°C and 150 to 300 mm, from south to north, with great intra and inter annual variations. The geological formations are very diverse, including Jurassic limestone, Cretaceous marls, Triasic clays and Doleretic basals, in the north, Quaternary sediments in the plain, and Precambrian schists, sandstones, limestones granites and quarzites in the south.

10.2.2 Methodology

The methodology consisted of the following main steps:

- *Data collection and processing*: this step included collecting relevant information on the characteristics of the basin, baseline data, previous studies on land degradation, past and ongoing projects, etc.

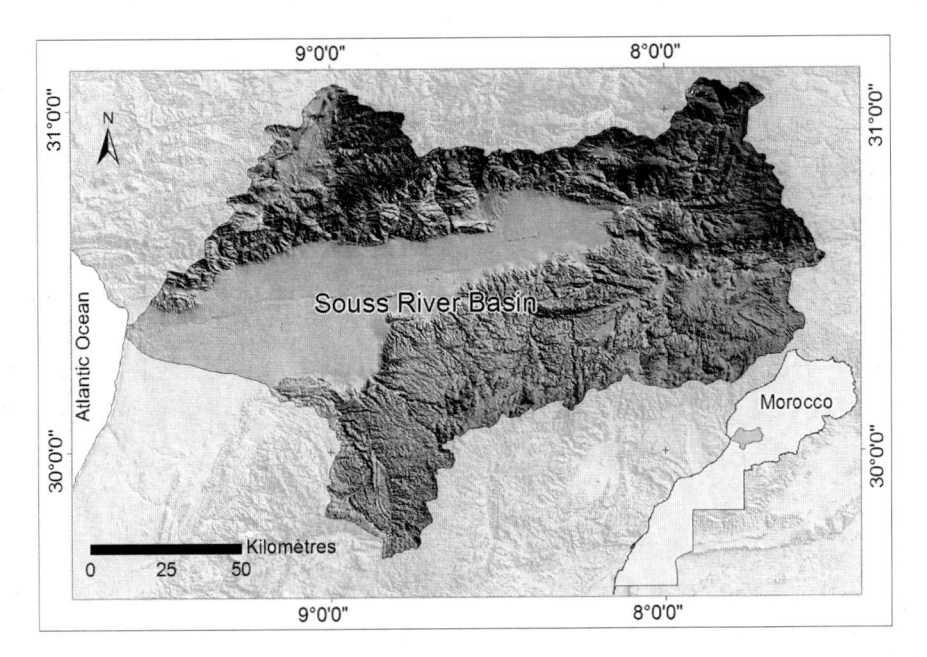

Fig. 10.1 Location and delineation of the study area

- *Field appraisals*: field transects and surveys were conducted with local population and other stakeholders to better understand the study area, including environment (physical and human components), the driving forces, as well as the various relationships that may help to interpret better the results and to propose appropriate interventions for desertification mitigation. The surveys were conducted using a participatory approach based on a semi-structured interview and questionnaires.
- *Implementation* of the MEDALUS model: the necessary data were processed and integrated into a GIS database in order to elaborate the various layer maps needed to develop the desertification sensitivity map for the basin. A desertification zoning and a spatial generalisation with respect to rural communal (RC) limits were performed as an aid for prioritisation in the action plan.
- The DSM in addition to the collected field information, were used to propose a *package of actions* that would enable combating directly or indirectly the desertification processes.

The MEDALUS model is based on the determination of 4 main indicators: soil quality, vegetation quality, climate quality, as well as management quality and human factors. These indicators are themselves determined based on various parameter maps (Fig. 10.2). The various parameter maps were determined as follows and were classified according to the manual proposed by the MEDALUS model (Kosmas et al., 1999):

SQI: Soil Quality Indicator
- PM: Parent material
- SL: Slope
- SD: Soil depth
- ST: Soil texture

$$SQI = (PM \times SL \times SD \times ST)^{0.25}$$

VQI: Vegetation Quality Indicator
- FR: Fire risk and ability to recover
- SEP: Soil erosion protection
- PDR: Plant drought resistance
- PC: Plant cover

$$VQI = (FR \times SEP \times PDR \times PC)^{0.25}$$

CQI: Climate Quality Indicator
- P: Precipitation
- AR: Aridity
- AS: Aspect

$$CQI = (P \times AR \times AS)^{0.33}$$

MQI: Management Quality Indicator
- ILUAA: Intensity of land use in agricultural area
- PP: Population Pressure
- AP: Animal Pressure

$$MQI = (ILLUA \times PP \times AP)^{0.33}$$

DSI: Desertification Sensitivity Indicator

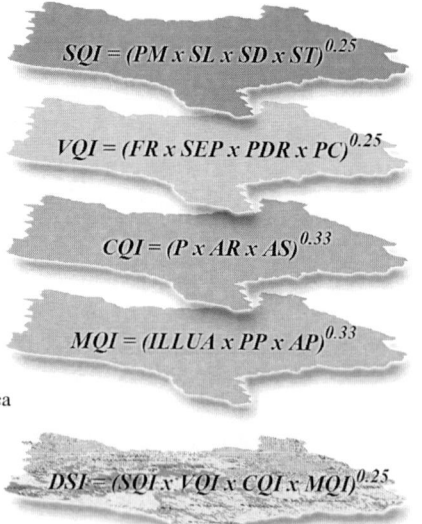

$$DSI = (SQI \times VQI \times CQI \times MQI)^{0.25}$$

Fig. 10.2 Illustration of the indicators and their sub-parameters used in the MEDALUS model adopted in this study

- *Soil Quality Indicator (SQI):* soil information was derived from available soil surveys and completed for unmapped areas by expert interpretation and ground truthing of various soil forming factors (litho-geology, climate, vegetation and relief). Slope was determined from a DEM of the area obtained by a Triangulation Irregular Network (TIN) prepared from contour lines of 1:100,000 scale topographic maps. Salinization was not considered in the study due lack of data.
- *Vegetation Quality Indicator (VQI):* The parameters needed for this indicator were derived from the land use map as well as from the field transects and surveys. The land use map was obtained from a mosaic Landsat ETM+ images using a supervised classification combined with ground truthing.
- *Climate Quality Indicator (CQI):* Climate data were collected from several weather stations covering the basin. The data were interpolated to determine the precipitation map. The Bagnouls-Gaussen aridity index (BGAI) was used for the aridity parameter. Aspect was generated from the DEM.
- *Management quality indicator and human factor (MQI):* the intensity of land use was derived from the land use map and ground knowledge of the agricultural practices in the area. Abandonment of terraced land was not used because of its limited extent with respect to the scale of work; "Fire" was considered to be already included in the VQI. "Population pressure" and "Animal pressure" parameters based on population and livestock censuses were included to take into consideration the effect of the anthropic factor.

Data for all the parameters was processed in a GIS environment to produce the four indicator maps as shown in Fig. 10.2. These parameters were then overlaid to produce the desertification sensitivity map (DSM). The weightings attributed to some of the parameter and subsequently to each indicator in the geometric

Table 10.1 Extent of the three classes in each indicator map

Indicator	Classes	Description	Area (ha)	Percentage of area
Soil	Satisfactory	< 1.13	719,762	44.6
	Moderate	1.13 to 1.45	873,075	54.1
	Low	>1.45	20,980	1.3
Vegetation	Satisfactory	< 1.13	263,052	16.3
	Moderate	1.13 to 1.45	703,624	43.6
	Low	>1.45	647,141	40.1
Climate	Satisfactory	<1.15	109,740	6.8
	Moderate	1.15 to 1.81	593,885	36.8
	Low	> 1.81	910,193	56.4
Management	Satisfactory	1 to 1.25	1,321,716	81.9
	Moderate	1.25 to 1.50	222,707	13.8
	Low	>1.50	69,394	4.3

means calculated were slightly modified from the original MEDALUS model in order to account for specific characteristics of the basin (Table 10.1). In addition, the term "satisfactory quality" was used instead of "high quality" for classifying the indicators.

10.3 Results and Discussion

10.3.1 Preliminary Appraisal

The field appraisals conducted over the basin allowed a better understanding of the major constraints and driving forces involved in the land degradation process. These included obviously natural and anthropic factors, which are linked to the socio-economic conditions prevailing in the various agro-ecosystems present in the study region. In an arid area like the Souss basin, the severe climate is one of the most important natural drivers of desertification, with more than 2/3 of the basin not exceeding 200 mm annual rainfall. Furthermore, during the last 3 decades, the basin has experienced very recurrent drought years with a trend towards decreasing annual precipitations, a phenomenon that can be attributed probably to the observed global climate changes. The severe conditions coupled with drought effects have lead to the reduction of land productivity and vegetation cover, with both direct and indirect effects on population's income and stability (accelerating migration). A large part of the basin is placed on steep slopes with fragile soils and parent materials leading to significant water erosion, dominantly rill and gully erosion. The Argane forest, native to Morocco and to the Souss basin, a major source of income for the population (argane oil extraction), is under a continuous degradation with limited and dispersed efforts for its regeneration.

The field surveys have also shown that all forms of land degradation are present and their degree of importance differs from one area to another. The northern High Atlas region, dominated by more or less degraded forest and rangeland areas are under extensive pressure of overgrazing and the south mountains are almost becoming bare of vegetation. The central irrigated plain, which has been under very intensive cropping (various cash crops), shows evidence of loss of soil fertility, soil compaction, and the trend for salinization. Ground water has also been severely affected by both the lowering of the water table (about 60 m during the last 30 years) as well as seawater marine intrusion along the coast (USAID, 2004; ABHAS, 2009; Choukrallah et al., 2007).

The human factor is a major driving force in desertification. However, its impact is not to be pejoratively incriminated for land degradation and desertification. Various scenarios could be found in our study area, among which (i) the lack of knowledge leading to miss-management and to a disequilibrium between land potentialities and uses, (ii) ignorance and lack of awareness, (iii) limited resources and poverty forcing the population to inappropriate practices, (iv) lack of alternative income to divert the local populations from pressure over the resources, and (v) the feel of abandonment leading to an overreaction, etc. Establishing a sense of harmony between the population in place and the available resources, means establishing a good strategy towards rural development, including: upgrading knowledge and practices, enhancing awareness for resource preservation, implementing socio-economic projects (infrastructure, roads, schools, hospitals, telecommunication, etc), creating thus new opportunities of alternatives for income sources, etc.

10.3.2 Desertification Indicators

Figure 10.3 shows the maps of the four indicators determined to produce the desertification sensitivity map, while Table 10.1 gives the extent of the classes for each indicator map.

The SQI map (Fig. 10.3a) shows that more than 50% of the basin area has soils with satisfactory quality (Table 10.1). These soils are located mainly on the southeastern part of the basin. The low quality of the soils in the central plain and northwestern mountains is due to the fine and loamy texture making them very sensitive to compaction, low infiltration, limited drainage, etc. In the northwestern mountain, water erosion is of concern and the soils with low quality are limited to steep slope areas with very fragile parent materials such as the clayey formations of Triasic age. However, this indicator need to be taken with precaution, in

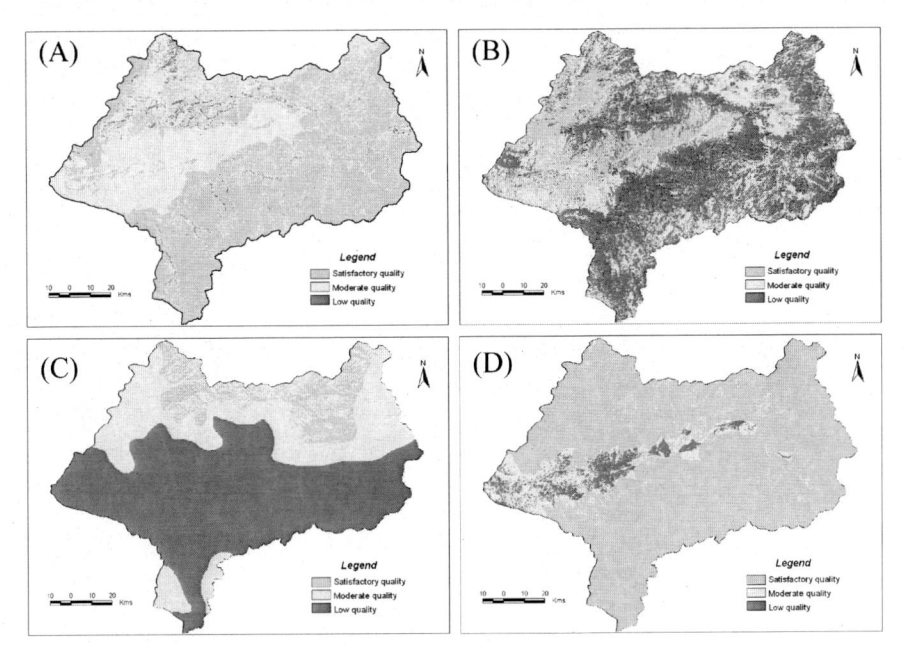

Fig. 10.3 Maps of the four indicators used to produce the desertification sensitivity map. (**a**): SQI; (**b**): VQI; (**c**): CQI and (**d**): MQI

that, satisfactory quality does not mean "agronomic quality" (productivity) of the soil, but rather its low degree of sensitivity to degradation based on the parameters used by the approach (parent material, depth, texture and slope). The dominance of coherent parent materials and the favorable texture classes in these areas made these two parameters somewhat dominant in the geometric average leading to the SQI.

The VQI map (Fig. 10.3b) indicates that the areas having low to moderate vegetation quality cover respectively 40 and 44% of the basin (Table 10.1) that are located mainly on the southeastern part of the basin. This trend reflects the effect of the harsh climate characterizing these areas leading to a poor and to a low protection of the soils. The result may appear somewhat contradictory with the SQI, in that, land with low quality vegetation should be of low soil quality. As discussed previously, in areas where the parent material is coherent, the resulting soil depth and soil texture do not reflect the real state of soil degradation.

The CQI map (Fig. 10.3c) indicates that the low and moderate climate quality area occupy a large part of the basin with respectively 56 and 37%, while the satisfactory quality areas are only about 7% (Table 10.1). This means that almost the whole basin is vulnerable to desertification due to its arid to arid-subdesertic climate, and that the climate will be a very determining factor in the desertification sensitivity assessment.

The MQI map (Fig. 10.3d) indicates, based on the parameters used, that besides the central plain area, a large part of the basin is of satisfactory management quality. The plain is particularly vulnerable because of the pressure from an intensive cropping as well as from the high population density with limited sustainable land management practices applied for soil and water conservation. In the low population density and animal charge were in favor of a moderate management quality in the mountainous areas; though, in small areas not depicted at the scale of the basin, the impact from animal grazing, wood extraction, etc, are still very visible. The integration of other parameters in the MQI, such as water management, soil fertility (organic matter), soil conservation, etc, may help to better express variations of this indicator.

10.3.3 Desertification Sensitivity Map

The desertification sensitivity map obtained from the overlay of the four previously discussed indicator (Fig. 10.4) shows that a large part of the basin (72%) is critically vulnerable to desertification. These areas are mainly located in the Anti-Atlas Mountains, the central plain and their surroundings. The high desertification sensitivity in the south is attributed to the severity of climate, being influenced by the desert currents. Contrary to what might be expected, the plain area (both the intensive and semi-intensive zones) are also dominantly fragile to critical, a condition largely endorsed to the vulnerability of the soils, owed to a fragile soil texture and its consequences on compaction, low infiltration, and salinity risk. The intensification for cash crops has impacted both soil fertility as well as ground water (quality

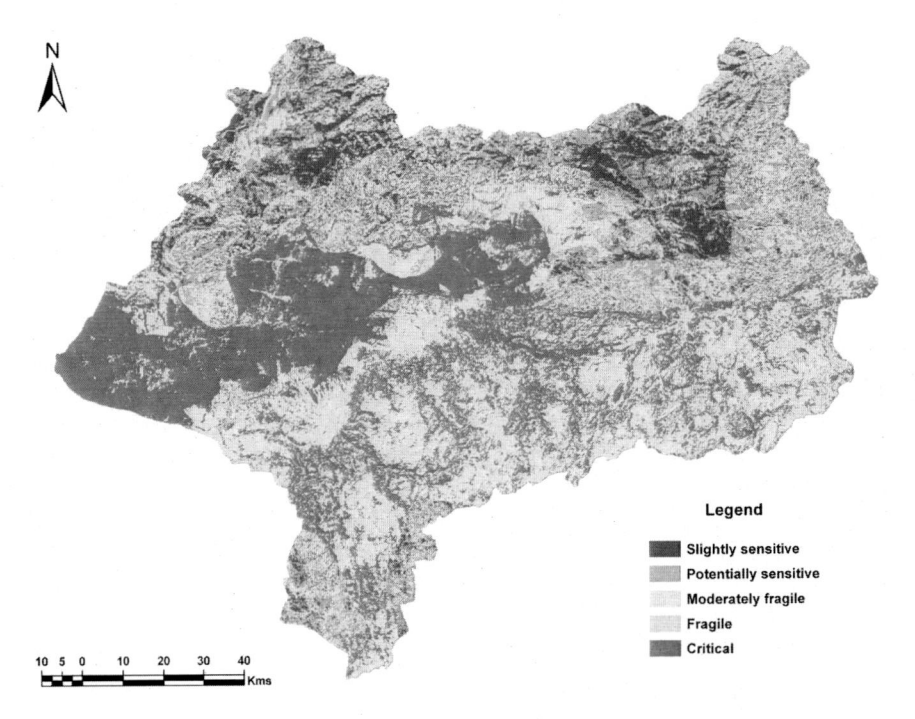

Fig. 10.4 Desertification sensitivity map of the Souss river basin

and quantity). The High Atlas areas are rather moderately sensitive to desertification risk due to the mild climate and the persistence of a vegetation cover.

Areas that are potentially sensitive represent about 15%, and areas slightly sensitive cover only 13% (Table 10.2). In the context of the study area, it was preferred to use the description "slightly sensitive" instead of "not affected" for the class with a desertification index <1.17 as suggested originally by the MEDALUS methodology. In the south Moroccan conditions, almost all areas are exposed to some degree of desertification, and therefore, the criterion "not affected" was considered not appropriate for our conditions.

In general, the variability observed follows the trend of the CQI, indicating that the climate is a very influencing factor in this case. Similar results have been

Table 10.2 Extent of the desertification sensitivity classes in the basin

Classe	Type	DSI	Area (ha)	Area (%)
1	Critical (C)	> 1.37	452,164	28
2	Fragile (F)	1.30 to 1.37	309,739	19
3	Moderately fragile (M)	1.23 to 1.30	406,101	25
4	Potentially sensitive (P)	1.17 to 1.22	237,826	15
5	Slightly sensitive (S)	< 1.17	207,987	13

reported in other studies (Farajzadeh and Egbal, 2007) using the same methodology. This could be also explained by the fact that the model gives higher weight indices to the climate for the "low" and "moderate" quality classes; a very suitable approach in the case of areas with harsh climates.

Having a significant part of the basin classified as not very affected by desertification, does not mean that it should not receive any form of attention. On the contrary, the less sensitive areas are those that need particular attention in order to maintain their natural soil fertility status with minimal interventions. The critical zones are already in an advanced state of degradation that interventions to combat desertification would be either too costly or with minimal impact or success. The prioritization in this case might be in the opposite sense, that is, the less sensitive areas are those to be targeted first in order to ensure their sustainability by preserving their quality. Prevention is better than mitigation.

If a DSM, like the one obtained above, is supplied to decision makers, it would be very difficult to apprehend as such because the driving forces leading to desertification in each part of the basin, are not well elucidated, especially considering that the area is very large and the processes may differ from one ecosystem to another and even within each ecosystem. Therefore, the other layer maps such as the ones represented in Fig. 10.3 should always accompany a DSM. In the case of the present study, three detailed reports were made available, one dedicated to the initial appraisal, the second to the implementation of the MEDALUS approach and the production of the DSM, and the third to the interventions to be considered in the action plan.

10.4 Proposition of an Integrated Action Plan to Combat Desertification in the Souss Basin

The National Action Plan for combating desertification (NAP-CD) (MADRPM, 2001) recommended that any local action plan should be based on a participatory approach involving the various stakeholders and partners. The present study aimed to be compliant with this recommendation and therefore has undertaken the following steps in order to propose a general action plan that can be later detailed at the communal level following a "communal action plan approach". The action plan definition relied on three main directives:

- *Identify* from the desertification sensitivity map the priority zones that really need immediate and specific interventions, given large extent of the basin and that actions cannot target simultaneously the whole area.
- *Use* the information collected in the initial phase of the study along with the knowledge gathered during the field transects and surveys to propose a package of specific actions suitable to the various agro-ecosystems.
- *Validate* the actions proposed in a stakeholder workshop to have the consensus of the various actors in the basin in order to have good chances for it to be adopted and implemented.

It is very important to note that at the scale of the study area (16,138 km^2 ~ 1.6 million hectares), a detailed and comprehensive action plan would require thorough ground studies with important funding and qualified human resources; therefore, the action plan proposed in this work comprised a package of potential specific actions to the various priority zones and recommended an approach that can be adopted for the local comprehensive communal action plans (CAP). In compliance with the NAP-CD, the package comprises *direct* actions targeting land degradation phenomena, as well as *indirect* actions aiming at indirectly reducing the pressure on the resources by improving the different agricultural systems, offering alternatives of income sources to assist the livelihoods and the well being of the local population.

The spatial entity for intervention can differ from one project to another (basin, watershed, province, commune, village, etc). The choice of either one is usually justified by the objectives of the work. In the context of the basin, the "Rural Communal" (county) entity was considered to be appropriate for local action plans since various other projects worked or are working at this direction (Goldnick and Moumadi, 2004; HCEFLCD, 2009; ABHS, 2009; ADS, 2008). The local desertification action plan could be included in the "Communal Development Plan" for a more integrated strategy at the local level. Based on the DSM a spatial generalization (average) was performed using the communal limits to determine a zoning in terms of the importance of desertification at the communal level, which helped better orienting the interventions proposed in the action plan. The three main zones identified are:

- Zone I: the least sensitive to desertification, located mainly in the north, with a fair climate, an appreciable vegetation canopy and low population pressure. However, it is believed that this zone should deserve as much attention, yet even more attention than the other two zones, in order to sustain its environmental quality.
- Zone II: critical; located mainly in the plain area with intensive and semi-intensive agriculture with high population pressure, both rural and urban.
- Zone III: very fragile to fragile, covering almost the 2/3 of the southern part of the basin. The vulnerability of this zone is due mainly to the aggressiveness of the climate and the vegetation degradation. The environment is being deserted and agriculture is limited to confined spaces and oasis. Nomadic extensive grazing is dominant.

The actions proposed in the plan to respond to desertification problems (natural, human and socio-economic) were categorized in six categories:

- *Direct actions:* this group includes direct actions aiming at reducing or alleviating land resource (soil water and vegetation) degradation using appropriate measures adapted in the context of the area and taking into consideration the local knowledge. This group includes interventions to be implemented in several sites, targeting soil erosion control, water harvesting, ground water conservation and recharge, forest regeneration, pasture improvement, wastewater treatment and reuse, etc

- *Indirect actions:* this group includes actions that will contribute indirectly to the preservation of the resources through improvement of productivity of the existing agro-systems, energy saving techniques, and creating alternatives of income that will reduce the dependence on the available resources and consequently reduce the pressure. Actions of this category include interventions targeting land productivity and crop production improvement, diversification of cropping systems, improvement of animal production and sedentarization of the livestock, introduction of energy saving techniques for firewood, proximity advisory and outreach, etc.

- *Income generating actions:* this category includes actions that will promote additional incomes to the local population and therefore divert it from overusing the available resources. They concern both existing activities that require enhancement, as well as alternative new activities. They include the valorization (individually or in groups) of local agricultural and non-agricultural products such as argane and olive oil, honey, medicinal and aromatic plants, prickly pears, craft and other artisanal, rural and eco-tourism activities, etc.

- *Local development actions*: this type of actions aims at contributing to the human and socio-economic development in the area, creating favorable conditions in terms of infrastructures, public and private services, in order to help reduce poverty and ensure favorable conditions for the success of the other actions. These actions include, among others, education (especially for girls), professional trainings for farmers and youth, roads and transportation for enclaved zones, electrification, health, potable water, etc.

- *Transversal actions:* this type of actions will accompany the other categories of actions previously discussed and aims at contributing to training, increasing awareness, capacity building, promoting societal and professional organization, empowerment, developing tools for information dissemination. This type of actions should the various stakeholders (local rural populations, mainly farmers, all genders considered; NGOs; technical services, agriculture, forestry, range, etc; administrators; decision makers, elected communal representatives, local authority representatives, government agencies; etc). These types of actions are to be reasoned according to the specificities of each target group and use adapter tools for better communication.

- *Coordination, monitoring and evaluation:* a desertification action plan would not be successful without the consent of all stakeholders and partners, both at the basin and at the communal levels. To be compliant with the NAP-CD, it should be coordinated by a body that would foster it, take the lead for harmonizing and consolidating the efforts, capitalizing on previous experiences, fund raising, establishing mechanisms and indicators for monitoring and evaluation, disseminating results and information. A workshop organized in this context led to the consent to attribute the coordination to the Regional Directorate of Water, Forest and Combating Desertification as this task is supposed to be among its mission.

Table 10.3 Example of a summarized action sheet proposed in the action plan

ACTION no: 1.2.1	
Title	Soil and water conservation in mountainous areas
Description	This type of action comprises measures for soil and water conservation, both on agricultural and forest areas that will contribute to reducing soil loss, improving soil fertility and productivity, conserving water, and preventing sedimentation in the reservoirs downstream.
Target zones	Zone 1
Priority sites/ communes	• Areas upstream of Abdelmoumen, Aoulouz and Chakouken dams • Tiqqui, Argana and Admine forests • SIBE (sites of biological and ecological interest) sites • Other sites to be defined by the local communal action plans
Components	• Appropriate soil tillages practices on crop lands • Soil and water conservation techniques (catchment ditches, stone breaks, cuvettes, etc) • Water harvesting techniques • Strip planting • Mechanical and biological correction of ravines and gullies • Plantations of fruit trees adapted to water scarcity and harsh climate • Costal sand dune stabilisation
Beneficiaries	Farmers (all gender), local associations, Argane forest right-users, herder associations, water user associations,...
Coordinating body	Regional directorate for Water and Forest
Partners	Regional Agricultural Office, local NGOs, Regional Service for Environment, others
Approximate cost	10,000 to 20,000 MAD[a]/hectare

[a] Moroccan currency.

The action plan comprised a list of 37 types of actions that target several sites or communes. As a general guideline, the actions proposed were reported as instructive sheets, each with a brief description, target zones, target sites, beneficiaries (if any), components, leading body, partners, and an approximate cost when possible (Bouabid et al., 2007). A summarized example of an action sheet targeting soil and water conservation in mountainous areas is given in Table 10.3.

10.5 Conclusions

This work is to be considered as case study in methodological and technological approaches for the assessment and proposing actions to combat land degradation and desertification towards promoting sustainable land management. Desertification risk evaluation was accomplished using the MEDALUS model supported by field participatory appraisals. The results showed that the Souss Basin is in general critical to fragile to desertification especially in the southern part. The climate is a very determining factor, aggravated by both physical and anthropic factors. The desertification sensitivity map along with the other collected information was used to propose an action plan specific to the basin, taking into consideration the driving forces in the various agro-ecosystem. The maps produced, (individual parameters, indicators, and desertification sensitivity), offer good pictures for better grasping land degradation, while the action plan gives general guidelines for possible interventions to combat the various aspects of desertification. Comprehensive local plans are to be derived from the proposed plan using in-depth appraisals and participatory communal approaches.

Acknowledgment Study initiated and supported by the Secretary of State for Water and Environment, and contracted by ADI-ONA, Morocco.

References

ABHS (Agence de bassin Hydraulique du Souss Massa) (2009) Etudes et travaux réalisés entre 2000–2007. Services/Etudes et Travaux. www.abhsm.ma. [Last accessed: 15 Apr 2009]
ADS (Agence de développement Social). (2008). Programme d'Appui à l'amélioration de la situation et de l'emploi de la femme rural et gestion durable de l'Arganeraie dans le sud-Ouest du Maroc. Projet Arganier. Rapport Annuel. Coopération Maroc-EU. Convention No MAR/AIDCO/2002/0521. ADS, Agadir Maroc.
Ahmadi, H. and Nazari Samani, A.A. (2006). Introducing Iranian model of desertification potential assessment (IMDPA) for desertification. Proceedings of the 1st International Conference on Water Ecosystems and Sustainable development. October 9–15, st2006. Xinjiang University, Urumqi, China.
Basso, F., Belloti, A., Faretta, S., Ferrara, A., Mancino, G. and Quaranta Psante, G. (1999b). Application of the MEDALUS methodology for defining ESA in the Lesvos Island. European Commission.
Basso, F., Bellotti, A., Faretta, S., Ferrara, A., Mancino, G., Pisante, M., Quaranta, G. and Taberner, M. (1999a). Application of the proposed methodology for defining ESAs: The Agri basin, Italy. In: C. Kosmas, M. Kirkby and N. Geeson (eds), The Medalus project: Mediterranean desertification and land use. Manual on key indicators of desertification and mapping environmentally sensitive areas to desertification. European Commission, Project ENV4 CT 95 0119 (EUR 18882), pp. 74–79.
Bouabid, R., Rouchdi, M. and Badraoui, M. (2007). Elaboration d'un Programme d'action de Lutte Contre la Désertification pour le Bassin du Souss. Rapport Phase III. Secrétariat d'Etat Chargé de l'Environnement/Société ADI-ONA. Rabat Maroc.
Choukrallah, R., Bellouch, H. and Baroud, A. (2007). Gestion intégrée des ressources en eau dans les bassins du Souss Massa. In: F. Karam, K. Karaa, N. Lamaddalena and C. Bogliotti (eds), Harmonization and Integration of Water Saving Options: Convention and Promotion of Water Saving Policies and Guidelines, pp. 67–69. Options Méditerranéennes, Série B, No. 59.
FAO-UNEP. (1984). Provisional Methodology for the Assessment and Mapping of Desertification. FAO, Rome.
FAO-UNEP . (1997). Land Quality Indicators and Their Use in Sustainable Agriculture and Rural Development. FAO, Rome.
Farajzadeh, M. and Egbal, M.N. (2007). Evaluation of MEDALUS model for desertification hazard zonation using GIS; Study area: Iyzad Khast plain, Iran. Pakistan Journal of Biological Sciences 10(16), 2622–2630.
GIWA (Global International Water Assessment). (2001). DPSIR framework for State of Environment Reporting (Driving Forces-Pressures-State-Impacts-Responses). European Environmrnt Agency (EEA).

Goldnick, K. and Moumadi, H. (2004). La lutte contre la désertification et la pauvreté pour un développement local durable. Projet appui au PAN-LCD. GTZ (Gesellschaft für Technische Zusammenarbeit), Service de Coopération Allemande. Rabat Maroc.

HCEFLCD (Haut Commissariat aux Eux et Forêt et à la lutte contre la désertification). (2009). Projet Protection de la Nature et Lutte Contre la Désertification (PRONA-LCD): Notre de Présentation. HCEFLCD, Rabat Maroc.

Jun Hou, W., Zhi Qiang, Z., Bao Quan, J. and Fan Rong, M. (2003). Classification system for desertification and its quantitative assessment methodology in China. Forest Studies In China 2003(5), 3.Chinese Academy of Forestry, Beijing, China.

Kosmas, C., Kirkby, M. and Geeson, N. (1999). The MEDALUS project: Mediterranean desertification and land use. Manual on key indicators of desertification and mapping environmentally sensitive areas to desertification. European Commission, Project ENV4 CT 95 0119 (EUR 18882).

MADREF (Ministère de l'Agriculture du Développement Rural et des Eaux et Forêts). (2001). Programme national de lutte contre la désertification: Rapport principal. [http://www.unccd.int/actionprogrammes/africa/national/2002/morocco-fre.pdf. Last accessed 15 Apr 2009].

Sabir, M., Qarro, M., Badraoui, M. and Rochdi, M. (2005). Mapping desertification vulnerability in Morocco. Proceedings of the 1st Global International Studies Conference: Fourth AFES-PRESS Workshop on Reconceptualising Security: Security Threats, Challenges, Vulnerabilities and Risks. University of Istanbul, Turkey, 24–27 August 2005.

Schall, N. and Becker, M. (2007). Identification of desertification prone areas, a practioner guide: Example from Lebanon. CoDeL project. GTZ/Ministry of Agriculture of Lebanon/NCRS/ACSAD.

UNCCD (United Nations Convention for Combating Desertification). (2000). Assessment of the Status of Land Degradation in Arid, Semi-arid and Dry Sub-humid Areas. United Nations Convention to Combat Desertification, Bonn, Germany.

USAID (United States Agency for International Development). (2004). Souss integrated water management Project. Final Report. USAID, Washington, DC.

Vacca, A., Loddo, S., Serra, G. and Aru, A. (2009). Soil Degradation in Sardinia Italy: Main Factors and Processes. Options Méditerranéennes, Série A, No. 50.

Zehtabian, G.H., Ahmadi, H., Khosravi, H. and Rafiei Emam, A. (2004). An approach to desertification mapping using MEDALUS methodology in Iran. Desert Journal 10, 205–223.

Chapter 11
Assessment of the Existing Land Conservation Techniques in the Peri Urban Area of Kaduna Metropolis, Nigeria

Taiye Oluwafemi Adewuyi

Abstract This study assessed the effectiveness of the existing land conservation techniques in the peri-urban area of Kaduna metropolis with the aim of ascertaining if the existing conservation methods have helped to alleviate land degradation, and provide sustainable land use. Random sampling method was used to collect data from field observation, measurement and semi-structured interviews, which are analysed using descriptive statistics. The results revealed that there exist local conservation techniques along with the modern ones, some are physical and others are biological methods, even though techniques such as agroforestry, which is known to be the best method of farming is present but is yet to take root in the area. Some of these conservation methods are not standardized neither are they implemented in a scientific manner to ensure effectiveness and efficiency without causing further damage to the land, and there may be no end to land degradation in Kaduna if the current approaches to conservation are not improved on. It is suggested that improved water management, improved farming techniques, economic empowerment and education of the land users be employed in refining existing techniques, through which poor management practices such as bush burning, mono-cropping and overgrazing will be avoided while farmers may easily embrace new practices such as agroforestry, which provides farmers with income and food all year round as well as protect the environment from further degradation.

Keywords Agroforestry · Land conservation · Land degradation · Sustainable land use · Techniques

T.O. Adewuyi (✉)
Department of Geography, Nigerian Defense Academy, Kaduna, Kaduna State, Nigeria
e-mail: taiyeadewuyi@yahoo.com

11.1 Introduction

Mariko (1991) stated that the Earth is humanity's life support system and any society must find a way to use its resources in an intelligent, economical and rational way. In turn, it is also important to enrich the land whose resources are not inexhaustible. However, in order to manage and use the products and by-products of the land and the natural environment in a rational way, the soil must be worked intelligently as its fertile surface is exhaustible.

Mortimore (1998) noted that conservation has different meanings to different people. For some, it implies the exclusion of humans from protected natural reserves and to others it entails the protection of threatened species or habitats in ecosystems that are already occupied or exploited by human populations. The United Nations (1994) consider land conservation those activities that are part of the integrated development of land in the arid, semi-arid and dry sub-humid areas for sustainable development and which are aimed at the prevention and/or reduction of land degradation, the rehabilitation of partly degraded land, and the reclamation of desertified land.

The causes of declining biodiversity and land degradation are often multiple and complex and usually involve a combination of human and natural factors. The impacts of land degradation are also multiple in effects and range from natural to socio–economic considerations. From field observation, direct and indirect relationships between the state of natural resources (soil, vegetation, water, and ecosystem), the biological diversity at species level (animal, plant and microbial species) and the ecosystem level (habitats, interactions, and functions) and the management of those resources have been discovered. The management practices directly or indirectly affect the capacity of land users to conserve and sustain resources. It also provides goods and ecological services such as timber, herbs and eco-tourism. The assessment and monitoring of biodiversity and associated ecosystem services, therefore, require an integrated suite of biological and socio-economic indicators.

There are three major principles and direction of strategy for combating land degradation (Hamorouni et al., 2001). First is the sustainable use of water, soil and vegetation resources by ensuring their protection and conservation, and at the same time stimulating proper social and economic development. Second, ensuring land development by encouraging livestock farming and regeneration of natural vegetation coupled with a better use of soil and water resources. The third is the integration of farmers into all development and protection actions, by providing them with logistical support, efficient advice and enabling them to pass from unreliable types of agriculture to more regular ones while ensuring a reliable source of subsistence.

From these principles, various studies summarized the technical measures for land degradation as follows: soil and water conservation (Ben Hassine et al., 1998; Ogunwole et al., 2002), water collection and saving (Adewumi and Kolawole, 2002), sand invasion control, regeneration of forest and reforestation of bare land (IRA, 1991), development and rehabilitation of small irrigated areas, combating soil salinization, the re-use of drainage water in agriculture, re-use of treated waste water in agriculture, agriculture and pastoral development and the improvement of degraded soils (UNEP, 1985; Mtimet et al., 2002).

However, conservation techniques are not limited to technical measures but include a range of economic, social and institutional measures, which vary from place to place. Some major examples of these measures are as follows: appropriate fertilizer management (Rayar, 1995), supporting research, training of farmers, community and extension officers on how to use natural resources (Gadzama, 1995), creation and improvement of local infrastructure, support to small scale farmers and women, building awareness, continuous monitoring, institutional policy and re-orientation, improving trade, improving the economy and reducing poverty, developing microeconomic reforms vis-à-vis international, regional and bilateral cooperation (Harou, 2002).

In the monitoring and dealing with land degradation, the effectiveness of the existing land conservation methods is very important. Therefore examining the adequacy of the current measures is a very important step in mitigating land degradation. It is equally important from the point of view that some land conservation techniques employed by land users may have caused or aggravated the rate of land degradation.

Consequently, it is important to examine the roles of various land conservation techniques and their contributions to sustainable land use and development. As a result, this study aimed at examining the existing land conservation techniques and its effects on land degradation in the study area. The specific objectives are: firstly, to examine the existing land conservation techniques, secondly, to evaluate their effectiveness, thirdly, to suggest how to improve the existing land conservation techniques if necessary.

11.2 Methodology

11.2.1 Study Area

The study area is the entire peri-urban area of Kaduna metropolis. The peri-urban areas lie within a 500 m corridor from the outskirts of the city. These zones are transition areas from rural to urban and they lack adequate infrastructure in comparison to the main city. The study area therefore circled the city and forms an irregular shape. It falls within latitudes 10° 22′ 00″–10° 40′ 00″ N and longitudes 7° 20′ 00″–7° 28′ 00″ E with the elevation ranging from 600 to 650 m above mean sea level.

The approximate size of the study area is 24,000 m^2 (24 km^2). It falls within Igabi, Chikun, Kaduna North and South local government areas of Kaduna State, Nigeria (Fig. 11.1). A larger percentage of the northern part of the study area belongs to the Nigerian Defence Academy, the Nigerian Airforce, the old airport and the Nigerian Army. The area being a military zone is excluded from the recent wave of physical development in the city. However, it is not free from agricultural use (Dogo, 2006).

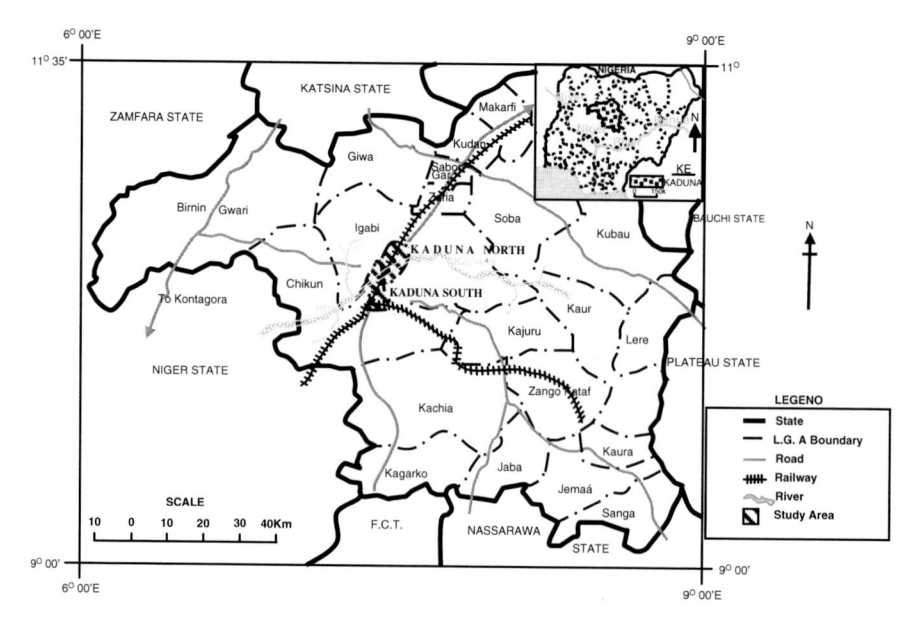

Fig. 11.1 Kaduna metropolis in Kaduna state

The existing land use is predominantly agriculture, while the land cover is dominated by natural vegetation (Plate 11.1). The area's original vegetation is Guinea Savannah, which has been replaced by cultured vegetation that is characterized by tall and short grasses with medium height trees interspersed within shrubs and herbs.

River Gora with six tributaries drains the northern area. The remaining parts do not fall in restricted environment like most of the northern part. As a result, both farming and grazing go on hand in hand. The vegetation is the same for both the northern and southern parts.

River Kaduna and its tributaries drain the southern area. However, there is a section along the eastern part where the river creates a sharp boundary between urban land use and rural land use due to the absence of a bridge. However, a year after the fieldwork for this research had been completed, a bridge was being built over the river along the axis, but this has not altered the land use situation.

The trees are generally moderate in size, ranging from 5 to 15 m in height and 15 to 100 cm in trunk diameter. The crops grown are mainly tubers (yams and potatoes), cereal (maize, guinea corn and millet) and vegetables (spinach, tomatoes, cabbage, onions etc). Cattle, goat and sheep grazed the vegetation from time to time.

11.2.2 Data and Methods

Studies on land conservation are both biophysical and socio-economic in nature. As a result the methodology of data collection was designed to reflect these two factors.

11 Assessment of the Existing Land Conservation Techniques

Plate 11.1 SPOT satellite imagery of Kaduna metropolis and its Peri-Urban area

Since the aim of the study is to examine the existing land conservation measures and its effects on land degradation in the area, first hand information was collected through observation, measurement and interviews after a thorough reconnaissance survey.

The sampling design for the study was the random sampling method. The sampling technique was random in order to allow transects that form the study site to be chosen only by chance so as to avoid any kind of influence for a proper representation of the study area. A base map was produced during the reconnaissance survey that served as the sample frame. From the sample frame, 96 portions of 500 m wide were created. These 96 portions were numbered from 1 to 96. Twelve portions were then selected randomly. At each portion, transects of 50 m wide and 500 m long were randomly demarcated for the field observations and measurements (Fig. 11.2).

The field measurements/observations were carried out using quadrant method and observation techniques accepted for standard fieldwork. On these selected transects, observation and measurement of land use, land cover, conservation techniques, content of manure, soil type, texture and colour are carried out. The semi-structure interviews on people's opinions on the existing conservation techniques and their effectiveness was conducted randomly on farmers found in the

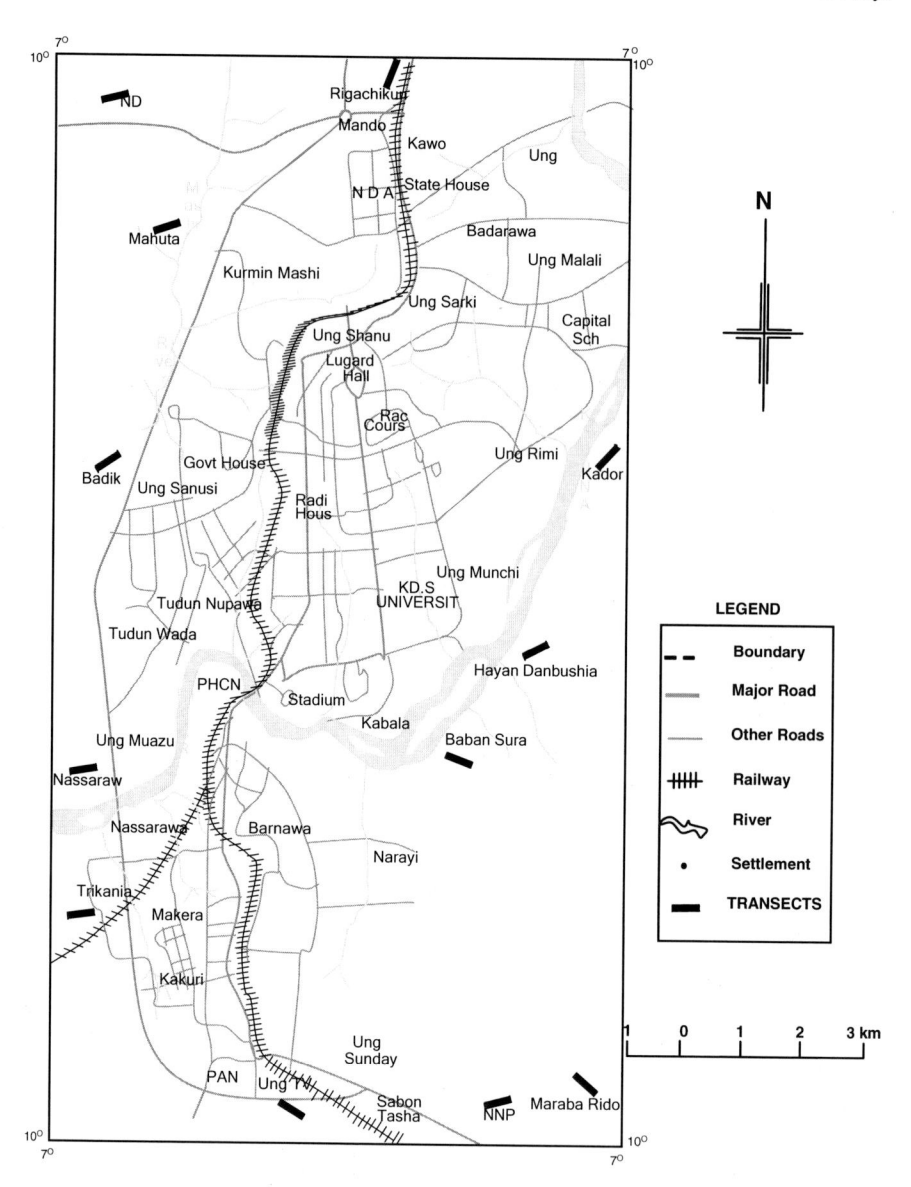

Fig. 11.2 The study (Source: Fieldwork, 2005)

area. The interview was utilized to gather opinion from people who live and work in and around the study area. Some of the information that was sought for includes crop yield, types of land degradation, farming system, crop grown, land use, land cover, vegetation type, species and diversity, conservation techniques and problems of farming.

Data processing involved various descriptive statistical tools such as mean and percentages. The analysis entails using comparison and inferences while the discussion focused on relating the result of the analysis with other study's result and their implication on sustainable environment.

11.3 Results and Discussion

11.3.1 From Field Observation

Observation data from the fieldwork, which are presented in Table 11.1 revealed that the biological techniques comprise the use of fertilizers, manure, crop residue, crop rotation farming method and mixed cropping method. The use of fertilizer is the most common agricultural technique. It was practiced in six out of the twelve transects selected (50%) namely TP2, TP3, TP4, TP5, TP7 and TP9 closely following the use of fertilizer and the use of crop residues. Crop residues from previous season's harvest are intentionally allowed to remain on the farm to decay. These crop residues consist of leaves, stems, branches, stalks, chaff and shell of grains such as the shell of groundnuts. In some cases where the grains are processed on the farm, it may include waste from such processing. This practice occurred in four transects (33%) namely TP1, TP2, TP9 and TP10.

The application of manure as a method of land conservation was used in three transects (25%) namely TP4, TP7 and TP8. The content of the manure varies a lot. In some places cow dung and chicken waste constitute the manure while others use household refuse.

The result is that most manure applied to the farm contains several components with no specific proportion as shown in Plate 11.2 from TP4.

Other observations in the field include crop rotation, a farming system where the crops planted are rotated between seasons to allow depleted/extracted soil nutrients by plants to recover. This is a common practice in the study area as noticed during fieldwork. It was also observed that some farmers employed mixed cropping method for land conservation. Under this approach instead of rotating different crops on a piece of land, many types of crops are simultaneously planted over some years. This allows the replacement of crop nutrient being depleted by a crop to be fixed by another thereby conserving the land. This mixed cropping is shown in Plate 11.3 from TP1.

Three physical conservation techniques observed during the fieldwork are ridging across the slope, mounding raised beds and construction of water channels. These physical conservation techniques are not popular in the area as each of these techniques was only employed in one transect. Ridges are made across the slope direction in order to reduce the rate of soil erosion. This is a common practice where the slope is moderate and the soil is at risk of erosion if cultivated.

The use of raised beds is employed primarily to distribute and retain water for plant consumption. This approach is commonly used along the flood plain called

Table 11.1 Summary of results from field observation

Transect	Land use	Land cover	Soil textural class	Soil texture	Soil colour	Type of degradation	Erosion type and number	Farming system	Major crop planted	Conservation techniques
TP 1 Rigachuku	Crop farming	Trees and crops	Silty clay	Fine	Brown	Sheet erosion and deforestation	Sheet – numerous	Crop and animal – rain fed	Maize and guinea corn	Crop residue left on farm
TP 2 NDA	Crop farming	Trees, crops, Grasses and litters	Sandy clay	Coarse	Reddish	Sheet & gully erosion, deforestation and badland	Numerous sheet and one gully	Crop – rain fed	Maize and guinea corn	Use of fertilizer and crop residue
TP 3 Mahuta	Crop farming	Trees and grasses	Silty clay	Fine	Fine brown	Deforestation	NIL	Crop and animal – rain fed	Maize and millet	Use of fertilizer
TP 4 Badiko	Crop farming	Trees and crops	clay	Fine	Dark brown	Flooding, Gully, pollution and refuse dumping	One gully	Intensive crop farming – Irrigation	Maize, rice, cabbage, tomatoes and lectus	Use of fertilizer, manure, raised beds and construction of water channels
TP 5 Nasarawa	Crop farming	Trees, crop and litters	silty	Fine	Brown	Gully and excavation pits	Two gully	Crop – rain fed	Maize	Use of fertilizer
TP 6 Trikania	Crop farming and refuse dump	Shrubs, grasses and refuse	Silty sand	Medium Fine	Light brown	Erosion, soil crusting, mining and refuse dump	Three gully	Crop – rain fed and irrigation	Maize, sugar cane, cocoa yam	NIL

Table 11.1 (continued)

Transect	Land use	Land cover	Soil textural class	Soil texture	Soil colour	Type of degradation	Erosion type and number	Farming system	Major crop planted	Conservation techniques
TP 7 NNPC	Forest, crop and animal farming	Trees and grasses	clay	Fine	Dark brown	Soil crusting and dumping of refuse	NIL	Crop and animal – rain fed	Maize, millet and guinea corn	Use of fertilizer and manure
TP 8 Maraba Rido	Crop farming	Trees, grasses and litters	silty	Fine	Brown	Soil crusting and deforestation	NIL	Crop – rain fed	Maize and guinea corn	Use of manure
TP 9 Baban Sura	Crop farming	Trees	silty	Fine	Light brown	Erosion, badland and sandy deposit	Five sheet and rill	Crop – rain fed	Maize and guinea corn	Use of fertilizer and crop residue
TP 10 Hayan Danbushia	Crop farming	Trees, grasses and litters	Silty clay	Fine	Dark brown	Deforestation	NIL	Crop and Plantation – rain fed	Maize, millet, guinea corn and soya-beans	Use of crop residue
TP 11 Television	Crop farming Soil mining Refuse dump	Grasses and refuse	Silty clay	Fine	Orange brown	Soil mining, erosion, Refuse dumping and Bush burning	One gully and 42 burrow pits	Crop – rain fed	Maize, guinea corn, sugar cane and cocoa yam	Ridging across slope
TP 12 Kadore	Forest	Trees, grasses and litters	Silty clay	Fine	Dark brown	NIL	NIL	NIL	NIL	Litters allow to remain on the ground

Plate 11.2 Preparation for manuring in one of the farms

Plate 11.3 Mixed cropping in one of the farms in the study area

fadama by dry season irrigation farmers. It requires water to be pumped either through a pumping machine or manually on to the highest point on the farms from where it is then circulated through the force of gravity. The construction of a water channel observed in the field served three purposes: to transfer water from one point to another, for maximum water utilization and to prevent gully erosion and water logging along the flood plain. An example of a water channel constructed to prevent gully erosion is shown in Plate 11.4 from TP3.

Plate 11.4 A constructed water channel

11.3.2 From Administered Questionnaires

The existing conservation techniques reported by respondents are the same with the ones observed previously except for the inclusion of bush burning. Bush burning method was observed but not considered along with the methods mentioned above because it was seen as a method of farm clearing. However, some farmers claimed that they use the ash derived from the burning as a soil-enriching component. A good example is shown in Plate 11.5 where bush burning is used as a conservation technique. This technique was observed in all the transects.

11.3.3 Problems of Existing Land Conservation Techniques

The effectiveness of the existing land conservation techniques was examined by considering the roles of existing conservation methods on biophysical features such as soil, vegetation, water and biodiversity on one hand, and on socio-economic variables such as crop yield, return on investment and security of tenure on the other hand. The conservation methods employed are rendered ineffective by destructive practices of the people. An example is the practice of outright cutting down of trees and uncontrolled pruning. From investigation (interview) the farmers agreed that trees are either cut or pruned in preparation of the land for farming. The percentage vary from 62.5% for pruning, 22.9% for cutting down trees while 14.6% were for neither prune nor cut. Statistics collected for wood collection shows that 60.4% is used for fuel wood, 22.9% for ash, 12.7% for animal feeds and 4% for manure, resulting in total that only 27% being used for ash and manure to improve the fertility of the soil while 13% as animal feed and the remaining 60% is used as fuel wood. This confirms the finding of Mortimore (1998) that most people in Nigeria still depend on fuel wood for energy.

Some level of awareness exists on conservation techniques, which has led to local practices of conservation. However, these practices are not scientifically proven. For example, baseline data on soil condition are mostly not available to determine the type and level of intervention that will improve the soil. Consequently, some of the measures taken are abused thereby aggravating degradation in the area in the long

Plate 11.5 Bush burning as a conservation technique

run. This is evident from the arbitrary use of refuse and animal dung as a measure of improving soil nutrient in several places within the study area.

It seems that the effort on grazing control is yielding desired results because overgrazing contributed only 2.1% to land degradation in the area as revealed by the result of the interview. Consequently controlled grazing is a good conservation measure for soil improvement in the area. However, 35.4% of the farmers interviewed reported that a loss in soil fertility implies, either the farming systems employed by the farmers reduce soil quality or the conservation techniques employed by the farmers are inadequate, or both. However evidence tends towards inadequacy of the conservation techniques. It is also obvious, that the flood control measures adopted by the farmers are not very effective. Hence, 41.7% of them claimed that flood is their major problem.

11.4 Implications and Recommendations

The failure of the conservation measures to achieve the desired goals has led to a number of problems for the human and natural environment. Such problems have led to many complains by farmers, which include increase in cost of farming (through increase in labour, time and finance), increase in the use of fertilizer, reduction in crop yield, loss of soil fertility and loss of agricultural land. The overall effect is low return on investment which farmers claimed is a major concern and that it has lead to lack of social and economic security and over exploitation of marginal resources. To correct these problems and ameliorate there implications on the environment, the following suggestions are provided based on the experience gathered from the field.

11.4.1 Education of the Land User

In view of the expected role of the local people in the success of any conservation technique, the first step towards the improvement of land conservation practices in the study area is education, particularly, mass literacy of the land users. Education will be particularly helpful in the areas where the introduction of new land conservation techniques is in the process of refining the existing ones. This will facilitate effective communication, dedication, trust and the building of farmers' confidence in the new methods. For instance, the arbitrary dumping of urban refuse on farms as manure will be mitigated because appropriate enlightenment will teach the farmers how to sort the waste in order to separate them into toxics/non toxics and biodegradable/non biodegradable substances before their. In a similar manner, education will expose the farmers to more information and knowledge on various types of soil, their characteristics, need and best use in a sustainable way.

11.4.2 Economic Empowerment

The second suggestion on how land degradation can be controlled in the study area is through economic empowerment of the people. If all farmers in the area are given the best education and information on how to protect the environment without economically empowering them, land degradation will be inevitable. This is so true because struggling poor people would not really care about the consequences of their actions on the environment unless they meet their basic needs. Therefore, they will farm marginal land, practice farming systems that degrade the environment and apply agricultural techniques that introduce other environmental problems. This fact is evident in the studies of Warren et al. (2001) and Butterbury (2001) where people who depend solely on the land for their livelihood degrade the land to get whatever that is possible to be able to survive. Bielders et al. (2001) in his study also confirmed that those with other sources of income are not desperate and their actions are not so destructive.

Thus, any intervention that excludes economic empowerment may fail. That is why an approach such as the agroforestry farming system is highly recommended to be practiced in the area. This is the only farming system that encourages a broad base diversification into crop farming, animal husbandry and forestry. It is not only that each of these units generates income in different ways but also they do so at different times therefore providing the farmers with regular income throughout the year. It is the suggestion of this study that when animal husbandry and tree planting is combined with crop farming the environment will be better utilized in a sustainable way. The resultant effect would be better environmental resources conservation and reduction in the rate of land degradation.

11.4.3 Improved Farming Techniques

The third suggestion on how land degradation can be reduced is by using improved farming techniques by all farmers. Namely, improved tillage methods (e.g. zero tillage, ploughing along contour lines, terracing), rotating crops, intercropping, the use of legumes (which biologically fix nitrogen) as biofertilizers, mulching, composing and rainfall harvesting. These farming methods can improve crop production, use less water and reduce pressure on the marginal land for food or cash crop production. The emphasis is that all farmers should adopt using these techniques irrespective of what they grow. This is required to have a consistent and generally spread sustainable land use management.

This has being found in different studies, for instance, Ogunwole et al. (2003) and Ogunwole and Raji (2001) confirmed that better tillage practices increases the yield of rain fed crop. Adeoye (1990) also shows that grass mulch improved soil condition (soil temperature and profile water storage) and grain yield, while Bodunde and Ogunwole (2000) have shown that fruit yield of tomato varieties grow better when soil moisture is not stressed. This study in particular revealed an average of 50% increase in production.

11.4.4 Better Water Management

Alongside is the need for better water management practices since water is the key factor to land conservation. Although this study assumed that the available quantity and distribution of rain in Kaduna peri urban area is adequate, however there is the need to utilize this resource in a sustainable manner so that excess or lack of it will not aggravate the factors of land degradation. To this end and in agreement with Adewuyi small barrages should be built across the area to harvest rainwater. These barrages will help to store excess rainwater during the rainy season and provide water to the farmers for irrigation farming during the dry season. They will also help to control flooding in the area thereby reducing land degradation; these barrages will help to stabilize the microclimate condition by lowering the atmospheric temperature, and improve the land cover distribution and biodiversity of the area.

11.5 Conclusions

In conclusion, we state that the issues of environmental protection require the support of all citizens as stakeholders. When the environment is neglected, the country loses valuable resources that lead to bigger problems that may eventually become difficult to reverse like the case of desertification and big gullies that now threaten various parts of the country. Solving these problems may demand important revenues that otherwise could be used for other essentials needs of life. Therefore, this is the time for action to be taken and not time to play politics with issues concerning the environment and the future of our children.

References

Adeoye, K.B. (1990). Effects of amount of mulch and timing of mulch application on maize at Samaru, Northern Nigeria. Samaru Journal of Agricultural Research 7:57–66.

Adewumi, J.K. and Kolawole, A. (2002). Water harvesting for dryland farming. The Zaria Geographer 15(1):24–33.

Ben Hassine, H., Aloui, H. and Amdouni, M. (1998). Evolution mensureelle des reserves en azote mineral dans deux types de sols cerealiers, sous climat sub-humidedu word-quest de la Tunisie, campanes agricoles 96-97 et 97-98 Tunis; direction des sols, ministere de sols, minstere de l' agriculture.

Bielders, C.L., Alvey, S. and Cronyn, N. (2001). Wind erosion: The perspective of grass-root communities in the Sahel. Land degradation and development 12(1):57–70.

Bodunde, J.G. and Ogunwole, J.O. (2000). Productivity of tomato (*Lycopersicon Esculentum mill.*) genotypes grown on residual soil moisture. The plant scientist 1(1&2):115–125.

Butterbury, S.P.J. (2001). Landscapes of diversity: A local political ecology of livelihood diversification in south-western Niger. Ecumene 8(4):437–464.

Dogo, B. (2006). Problems and prospects of agriculture as a secondary occupation amongst military personnel in Ribadu Cantonment-Kaduna. The academy journal of defence studies 12:125–136.

Gadzama, N.M. (1995). Sustainable development in the arid zone of Nigeria, monograph series, No 1, centre for arid zone studies. University of Maiduguri, pp. 1–32.

Hamrouni, H., Mtimet, A., Morel, C. and Moutonnet, P. (2001). Acquisition de valeurs de references pour raisonner la fertilite phosphate de deux sols calcaire de Tunisie central cultives en ble. Tunis: Direction des sols and Vienna: International Atomic Energy Agency (IAEA).

Harou, P.A. (2002). What is the role of markets in altering the sensitivity of arid land systems to perturbation. In: J.F. Reynolds and S. Stafford (eds.), Global desertification: Do humans cause deserts? Workshop Report 88. Dahlem University press, Berlin, pp. 253–274.

IRA (Instit des Regions Arides). (1991). Seminaires national sur la lutte centre la desertification 4–6 decembre 1989. Revue des regions arides. Numero special. Medenine, Tunisia.

Mariko, K.A. (1991). Reforming land tenure and restoring peasants' rights: Some basic conditions for reversing environmental degradation in the Sahel, IIED, Dryland networks programme, issue paper No. 24.

Mortimore, M. (1998). Roots in the African Dust: Sustaining the Sub-Saharan Drylands. Cambridge University press, UK.

Mtimet, A., Attia, R. and Hamrouni, H. (2002). Evaluating and assessing desertification in arid and semi-arid areas of Tunisia: Process and management strategies. In: J.F. Reynolds and S. Stafford (eds.), Global desertification: Do humans cause deserts? Workshop report 88. Dahlem university press, Berlin, pp. 197–214.

Ogunwole, J.O., Alabi, S.O. and Onu, I. (2003). Evaluation of three long staple lines of cotton to levels of fertilizer under moisture stressed and unstressed conditions. Crop Research 25(1): 50–57.

Ogunwole, J.O. and Raji, B.A. (2001). Tillage methods and fertilizer effects on soil phosphorus distribution and yield of late sown cowpea in a Typic Haplustalf. Nigeria Journal of Biological Sciences 1:43–49.

Ogunwole, J.O., Yaro, D.T., Bello, A.L. and Lawal, A.B. (2002). Soil conservation technologies in the sustenance of soil productivity in northern Nigeria. The Zaria Geographer 15(1):103–112.

Rayar, A.J. (1995). Desertification and soil degradation in arid and semi-arid regions of Nigeria. Inaugural Lecture Series 55:1–30.

United, N. (1994). UN Earth summit convention on desertification. UN conference on environment and development, Rio De Janeiro, Brazil, June 3–14, 1992.DPI/SD/1576. New York.

United Nations Environmental Programme. (1985). Desertification control in Africa. Desertification control programme activity centre. UNEP actions and directory of institutions. Nairobi 1:126 pp.

Warren, A., Batterbury, S.P.J. and Osbahr, H. (2001). Soil erosion in the Sahel of West Africa: A review of research issues and an assessment of new approaches. Global environmental change-human and policy dimensions 11(1):79–95.

Chapter 12
The Use of Tasselled Cap Analysis and Household Interviews Towards Assessment and Monitoring of Land Degradation: A Case Study Within the Wit-Kei Catchment in the Eastern Cape, South Africa

Luncendo Ngcofe, Gillian McGregor, and Luc Chevallier

Abstract Land degradation is a global problem affecting many countries. In South Africa extensive degradation can be related to a history of unjust land policies, which resulted in over-exploitation of the land. According to Hoffman and Todd (Journal of Southern African Studies 26:743–758, 2000) the problem is most severe in the communal districts of the Limpopo Province, Eastern Cape, and Northwest Province. Our study used a combination of GIS and Remote Sensing techniques together with field visits and household interviews to determine the spatial characteristics, history and nature of land degradation in the Wit-Kei catchment, in the Eastern Cape Province, South Africa. Vegetation cover and bare-ground change were selected as land degradation indicators. Using time-series analysis of Landsat images over an 18-year period (1984, 1993, 1996, 2000 and 2002), the rate and nature of change was assessed. Results from the Tasselled Cap Analysis technique showed an unexpected overall vegetation cover increase as well as a bare-ground increase in other parts of the study area. Based on field visits and interviews, the vegetation increase was explained by the presence of the invasive Euryops shrub. Bare-ground increase occurred mainly in former cultivated lands where erosion features in the form of gullies and dongas have become problematic. Landholders commented on the decline in food production over time, increase in dongas, and replacement of grassland by Euryops. The occurrence of erosion features on bare-ground and the increase of alien vegetation shown by GIS and Remote Sensing techniques was corroborated by the field and household survey, which added a further dimension to the underlying causes. The study demonstrates the value of using a multidisciplinary approach to obtain a holistic view of degradation, from which better-informed management decisions may be made.

Keywords Remote sensing · GIS · Land cover · Grassland · Invasive shrubs · Land policies · South Africa

---

L. Ngcofe (✉)
Council for Geoscience, Bellville 7535, South Africa
e-mail: lngcofe@geoscience.org.za

12.1 Introduction

Land degradation is an important global issue due to its negative impact on the environment and quality of life (Eswaran et al., 2001). According to Hoffman and Ashwell (2001), 50% of the land in South Africa is affected by land degradation, with the Limpopo Province, Eastern Cape and Northwest Province as the most affected Provinces (Fig. 12.1). Hoffman et al. (1999) argue that the extensive land degradation in South Africa can be related to a history of unjust land policies and over-exploitation of land. Two categories of degraded land were recognised: communal lands, which consisted of former self-governing territories also known as homelands (predominantly populated by Black South Africans) and commercial areas, which consisted of land owned by white farmers. Land degradation occurring in communal areas was attributed to poor land management according to the former apartheid government. However, land degradation nonetheless occurred on commercial lands too despite apartheid government aid to white farmers.

Bertram and Broman (1999) suggested that assessment of the past and present land degradation factors needs to be completed in order to address the issue of land degradation and to achieve sustainable land use. In line with this philosophy, we adopted a time series analysis for the assessment and monitoring of land degradation (with vegetation cover and bare-ground change as chosen land degradation indicators) in the Wit-Kei sub-catchment in the Eastern Cape Province of South Africa.

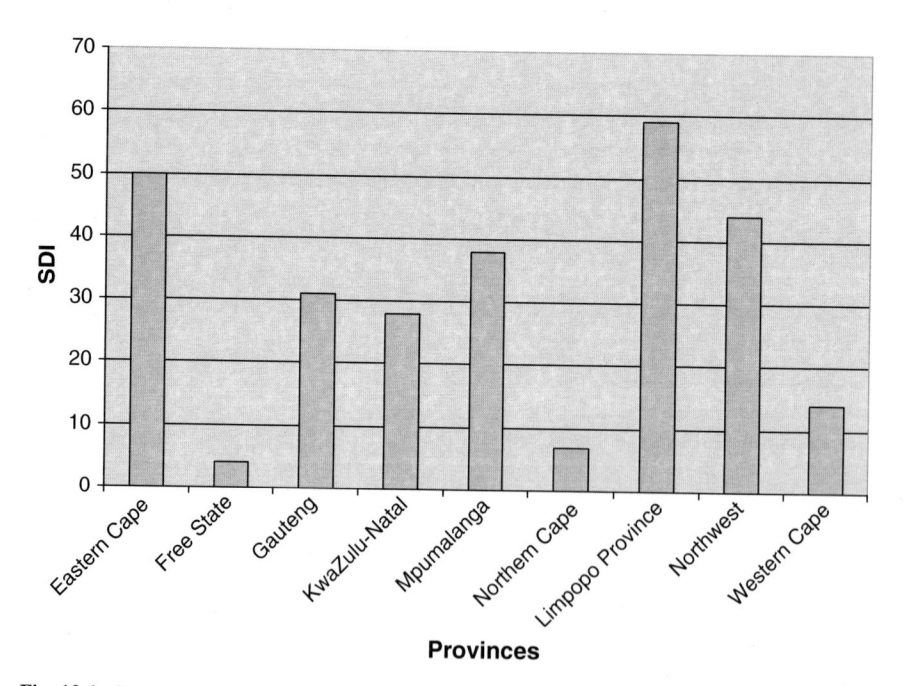

Fig. 12.1 Soil degradation index (SDI) in the croplands of South Africa (Hoffman and Ashwell, 2001)

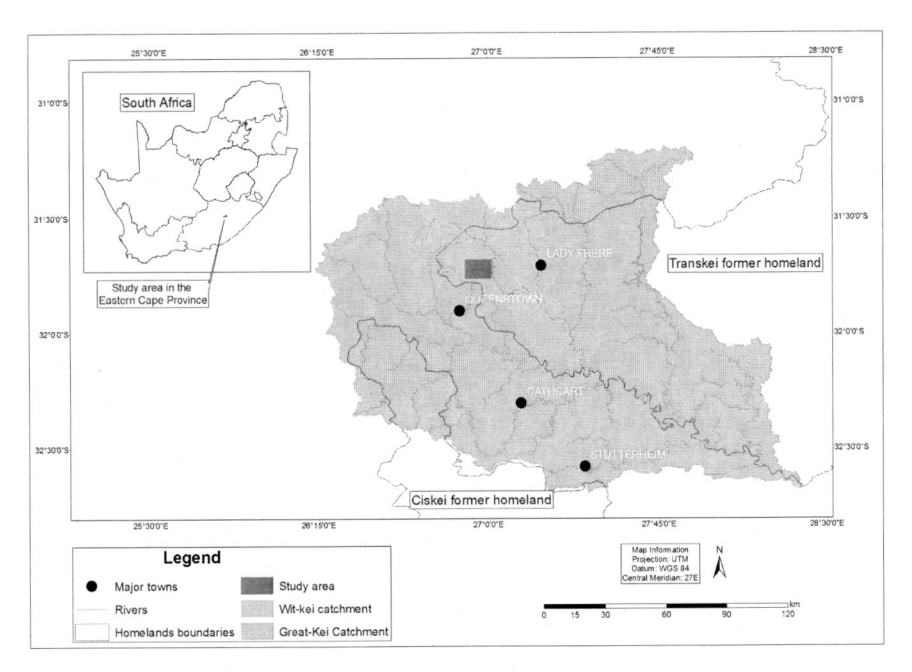

Fig. 12.2 Study area within the Great Kei Catchment, Eastern Cape Province, South Africa

The Wit-Kei is the sub-catchment of the Great Kei catchment within the Transkei former homeland (Fig. 12.2).

12.2 Methodology

Tasselled Cap Analysis (TCA) of Landsat images (1984, 1993, 1996, 2000 and 2002) and household survey were the chosen techniques for this study. According to Mather (2003) and Jensen (1996) TCA clearly differentiates between vegetation and bare-ground by reducing original data of Landsat image into 3 bands namely: brightness (which is designed to highlight bare-ground), greenness (which is designed to highlight vegetation) and wetness (which is designed to highlight wetlands). The result of TCA for each band is a grey scale image where brightness represents area of interest. The brightness and greenness TCA bands were further analysed for vegetation and bare-ground monitoring for the study period through density slicing.

Density slicing is the mapping of a range of contiguous grey scale levels of a single band image (Mather, 2003). These values for density mapping are chosen based on the histogram analysis of each image separately. The histogram being defined as a graphical representation of the brightness values that comprise an image (Mather, 2003). Histograms are automatically divided into 3 ranges, those with low

Fig. 12.3 Histogram analysis of Tasselled Cap

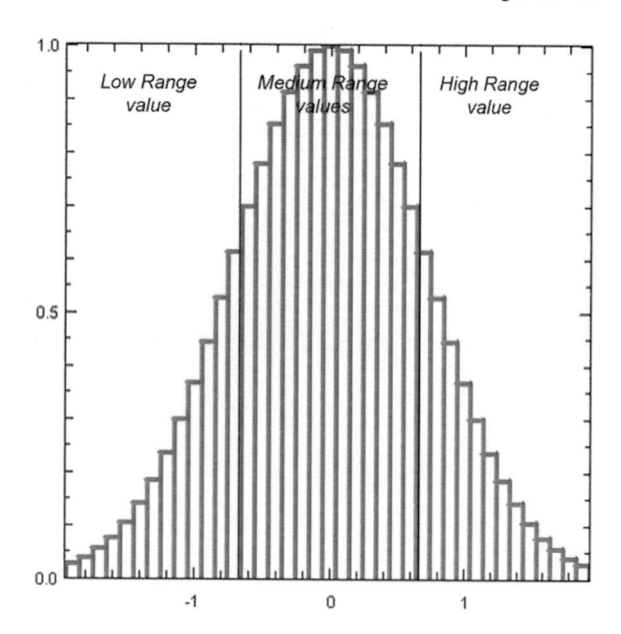

range value (representing dark pixels in an image), medium range value (representing grey pixels in an image) and high range values (representing bright pixels in an image) (Fig. 12.3). The density slice was computed to select only the high range values of the TCA of each image. The areas for the selected high range values in each year (for brightness and greenness TCA bands) were calculated and analysed.

Household surveys through focus group interviews together with field visits of the area were also conducted to add a human dimension to the research. Robson (2002) argued that the focus group technique enriches the discussion as people with different background view the topic differently. Photographic evidence was also employed in assessing the extent and nature of land degradation in the area. A limited number of 32 people were interviewed due to time constraints and homogeneity of the study area (all people were amXhosa). De Vos et al. (1998) argue that selecting a minimum sample number minimises the need of a larger sample in order to accommodate different groups. In cases where there is homogeneity, he recommends a sample size of 30 people as enough.

12.3 Results and Discussion

Density slicing was conducted in order to define threshold values from the TCA for the greenness and brightness components which highlighted only vegetation cover and bare-ground. The average annual rate of change of vegetation and bare-ground were then analysed. The results showed somewhat unexpected overall vegetation

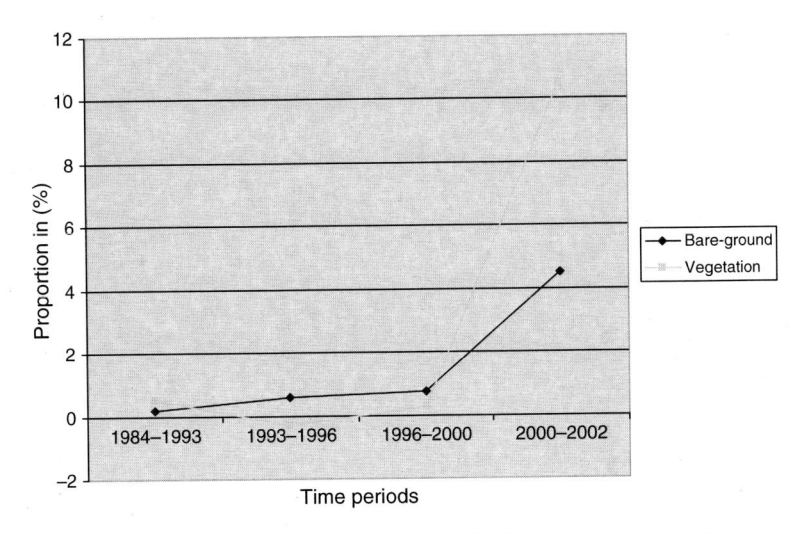

Fig. 12.4 Average annual rate of change of bare-ground and vegetation cover

cover increase together with the increase of bare-ground (Fig. 12.4). The increase of vegetation within the study area was in fact a further indication of land degradation rather than recovery. Palmer and Tanser (1999, Hoffman et al. (1999), Evans and Geerken (2004) and Kakembo et al. (2007) also acknowledges that although vegetation may increase this may be a sign of land degradation due to invasion of indigenous vegetation by alien species. TCA cannot differentiate between alien species and indigenous vegetation, demonstrating the vital role played by ground observations, which complement the remote sensing studies.

Interviews with local communities confirmed that the indigenous vegetation in the study area is being depleted while the alien *Euryops* (also known as "Lapesi"

Fig. 12.5 Observed land degradation features

by the local community) have shown a notable increase. Dongas and rills are also recognised as land degradation indicators by the interviewees. Photographs captured during field visit in the area also reveal the status of land degradation in area (Fig. 12.5).

12.4 Conclusions

Kakembo et al. (2007 suggested that an increase in vegetation cover does not necessarily mean that land degradation is being rehabilitated, a finding confirmed by our study. We observed that the increase in vegetation in the study area is due to the encroachment on indigenous vegetation by the alien shrub species *Euryops,* which is constitutes another form of land degradation. TCA cannot be used alone to assess the rate of land degradation, since the method cannot distinguish between indigenous and alien vegetation. TCA needs to be complemented by household interviews and field visits and photographic evidence to accurately assess the rate and extent of land degradation.

The effectiveness of the combination of Remote Sensing and household interviews is also noted by Kinlund (1996), Dzivhani (2001), and Florencia (2003) who contend that human memories, experience and perceptions in tandem with remote sensing techniques provide a holistic view in land degradation studies.

References

Bertram, S. and Broman, C.M. (1999). Assessment of soil and geomorphology in central Namibia, Swedish University of Agricultural Sciences; available: www.envi-impact.geo.uu.se// mfsbetram.pdf; [Last accessed 04/10/2006]

De Vos, A.S., Strydom, H., Fouche, C.B., Poggenpoel, M., Schurink, E. and Schurink, W. (1998). Research at Grass Roots: A Primer for Catering Professions. National Book Printers. Goodwood, Western Cape.

Dzivhani, M.A. (2001). Land degradation in the Northern Province: Physical manifestations and local perceptions. Unpublished Masters thesis. Stellenbosch University, Cape Town.

Eswaran, H., Lal, R. and Reich, P.F. (2001). Land degradation: An overview. In: E.M. Bridges, I.D. Hannam, L.R. Oldeman, F.W.T. Pening de Vries, S.J. Scherr and S. Sompatpanit (eds.), Responses to land degradation. Prac 2nd International Conference on Land Degradation and Desertification. Oxford Press, New Delhi.

Evans, J. and Greerken, R. (2004). Discrimination between climate and human induced dryland degradation. Journal of Arid Environments 57:535–554.

Florencia, F. (2003). Land cover change in rural areas of the forest-steppe ecotone of Andean Patagonia, Argentina: Utilising landsat data for the detection and analysing the change. Unpublished Masters thesis. Lund University, Sweden.

Hoffman, M.T. and Ashwell, A. (2001). Nature Divided: Land Degradation in South Africa. UCT press, Cape Town.

Hoffman, T., Todd, S., Ntshona, Z. and Turner, S. (1999). Land degradation in South Africa; Prepared for the Department of Environmental Affairs and Tourism. Available: http://www.nbi.ac.za [Last accessed 04/11/2005]

Jensen, R.J. (1996). Introductory Digital Image Processing. Prentice Hall Publishers, New Jersey.

Kakembo, V., Rowntree, K. and Palmer, A.R. (2007). Topographic controls on the invasion of Pteronia incana (Blue bush) into hillslopes in Ngqushwa (formally Peddie) district, Eastern Cape, South Africa. Catena 70:185–199.

Kinlund, P. (1996). Does Land Degradation Matter? Perspectives on Environmental Change in North-Eastern Botswana. Almqvist and Wiksell International, Stockholm.

Mather, P.M. (2003). Computer Processing of Remote Sensed Images: An Introduction. 2nd edition. John Wiley and Sons, New York.

Palmer, A.R. and Tanser, F.C. (1999). The application of a remotely-sensed diversity index to monitor degradation patterns in a Semi-Arid heterogeneous South African landscape. Journal of Arid Environment 43:477–484.

Robson, C. (2002). Real World Research. Blackwell publishers, Oxford.

Chapter 13
Environmental Degradation of Natural Resources in Butana Area of Sudan

Muna M. Elhag and Sue Walker

Abstract Environmental degradation has become a very serious problem in Africa since the Sahelian drought. It refers to the diminishment of local ecosystem or the biosphere as a whole due to human activity or the climate factors. Butana, in the north-eastern part of Sudan is known by many nomadic tribes as a good palatable grazing area during and after the rainy season. Rainfall plays a dominant role in the vegetation growth of the area. The goal of this study is to monitor the extent and severity of environmental degradation in relation to climate variability and change. The rainfall time series (1940–2004) for four weather stations were examined on monthly and annual bases to investigate any possible trends. The analysis of rainfall showed gradual decrease in the rainfall for the whole duration of the study at three out of the four stations. The progressive decline in the rainfall since late 1960s was significant and cannot be considered random for the northern part of the area. A significant increase in temperature, in autumn, is partly due to dry conditions observed since the late 1960s. Satellite image was used for routine natural resource monitoring and mapping land degradation. The Moving Standard Deviation Index (MSDI) increased considerably from 1987 to 2000 and the Bare Soil Index (BSI) for the degraded sites increased from 0–8 in 1987 to 32–40 in 2000. The BSI image difference indicated that the index increased between 14 and 43 over the 13 years. It is therefore, observed that different ecosystems in Butana area were subjected to various forms of site degradation, which led to sand encroachment, acceleration of dunes development and increased water erosion in the northern part of the area. The area has also, been subjected to vegetation cover transformation that made the pastures to deteriorate seriously in quality and quantity; however, in many parts of the area, the degradation is still reversible if land use and water point sites are organized.

M.M. Elhag (✉)
Faculty of Agricultural Sciences, University of Gezira, Wad Medani, Sudan
e-mail: munaelhag@yahoo.com

Keywords Environmental degradation · Moving Standard Deviation Index · Bare Soil Index · Sudan

13.1 Introduction

Environmental degradation refers to the diminishment of a local ecosystem or the biosphere as a whole due to human activity or climate factors. It occurs when nature's resources (such as trees, plants and other habitat, earth, water, air) are being consumed faster than nature can replenish (Ehrlich et al., 2000). Drylands cover about 5.2 billion hectares; a third of the land area of the globe (UNEP, 1992) and occupies roughly one fifth of the world population. Many of these lands appear to be undergoing various processes of degradation.

Since time immemorial the Butana northeastern part of Sudan has been know to have excellent pastures (Akhtar, 1994) and the best grazing land in Sudan. The grasses are palatable with high nutritional value for animals. Thus many nomadic tribes from adjacent as well as far away regions use it as grazing land during and after the rainy season. This study aims to quantify the causes of environmental degradation and their impact on soil and vegetative cover and socio-economic aspects. The specific objectives were to analyse the rainfall variability during the period 1940–2004 and to quantify the area affected and extent of environmental degradation in the area.

13.2 Materials and Methods

13.2.1 Study Area

The Butana is region lies between latitude 13° 50′ and 17° 50′ N and longitude 32° 40′ and 36° 00′ E. Figure 13.1 shows the location of the area, which is roughly kidney-shaped. Rainfall is the most important single determining factor in the climate because the temperature is high all year around.

13.2.2 Data Collection

The rainfall data used in this study were the monthly time series from 1940 to 2004 for three weather stations, namely (Shambat, Wad Medani, El Gadaref stations), and from 1970 to 2004 for the station of New Halfa. The Normalized Difference Vegetation Index (NDVI) for period from 1981 to 2003 was also used. The boundary of the area was adjusted according to the availability of the Landsat images for 1987–2000. Figure 13.2 shows the study area and the Landsat scenes. Limited field collection of data has been carried out during March to October 2005. Auxiliary

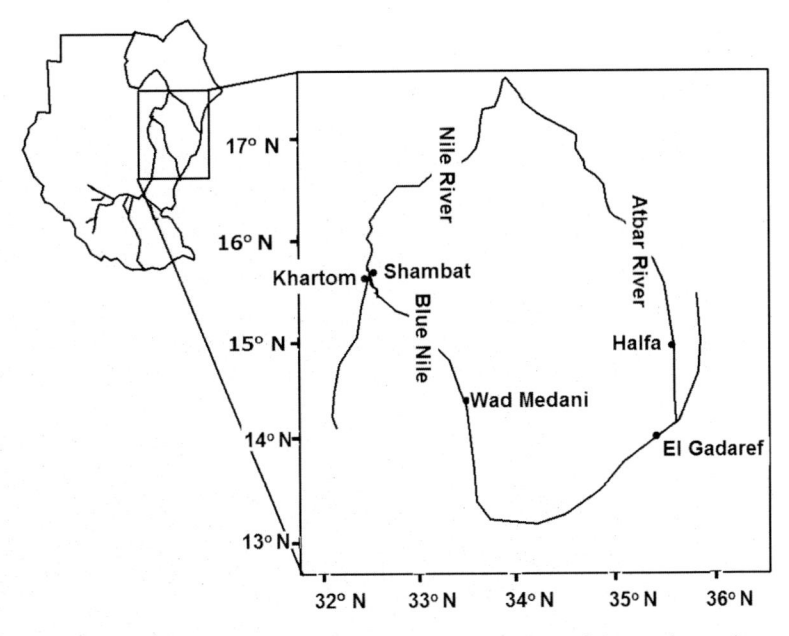

Fig. 13.1 Map of Sudan showing the studied area and the locations of the weather stations

Fig. 13.2 Map of the studied area showing the Landsat scenes (path/row) in World Reference System 2 (WRS2)

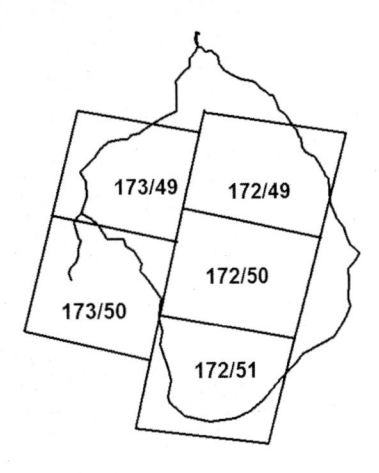

sources of information include topographic maps (scale 1:250,000). The Landsat data covering the same area from different dates were geometrically corrected to each other in order to cut out areas of interest and get the exact size of the area. The image-processing tool used in this study is ERDAS IMAGINE 8.5, and ArcMap9.1 software.

13.2.3 Landscape Pattern Index to Monitor Degradation

13.2.3.1 Moving Standard Deviation Index (MSDI)

The land degradation in semi-arid area may increase as result of increasing runoff and soil and water redistribution (Miles and Johnson, 1990). This leads to an increase in landscape heterogeneity or variability. The underlying assumption of the heterogeneity index (MSDI) is that a healthy landscape is less variable than a degraded landscape (Tanser, 1997). MSDI images were calculated by passing a 3 × 3 moving filter across the image (Elhag, 2006). The moving window calculates the standard deviations for nine pixels and assigns that value to the middle pixel. The standard deviation is then placed into a new map at the same location as the target pixel.

13.2.3.2 Bare Soil Index (BSI)

BSI is normalized index that separates two vegetation with different background viz completely bare, sparse canopy and dense canopy. The bare soil areas, fallow lands and vegetation with marked background response are enhanced using this index. The Bare Soil Index (BSI) was used for mapping bare soil and thus differentiating it from vegetation cover (Jamalabad and Abkar, 2004; Wessels, 2001). BSI was calculated according to the following formula

$$BSI = \frac{(B_5 - B_3) - (B_4 - B_1)}{(B_5 + B_3) + (B_4 + B_1)} * 100 + 100$$

Where:
B_1, B_3, B_4, B_5 is band 1, 3, 4, 5 respectively.

13.3 Results and Discussion

13.3.1 Rainfall Analysis

Time series analysis was conducted using long-term monthly and annual rainfall for the four stations. The result showed that there has been a significant decrease in the trend of monthly and annual rainfall during the period of 1940–2004 for Shambat and Wad Median. However, for the period of 1970–2004 the trend was not significant for New Halfa, which may be due to the limited period of the data (1970–2004) and does not give the patterns of the rainfall before the Sahelian drought (1969). El Gadaref station showed a gradual increase for monthly and annual rainfall but was not significant. These stations represent the south part of the study area (Table 13.1).

Table 13.1 The trend of the monthly and annual rainfall and their significant level at $P = 0.05$

Station	Monthly rainfall Trend	P	Annual Rainfall Trend	P
Shambat	−0.009	0.028*	−1.23	0.025*
New Halfa	−0.012	0.574	−2.22	0.379
Wad Medani	−0.016	0.044*	−2.15	0.002*
El Gadaref	0.012	0.425	1.12	0.101

* = Significant (at $P = 0.05$).

13.3.2 Vegetation Cover Changes

The condition of the vegetation cover in the area was examined using the peak NDVI (occurring at the end of August and the beginning of September). The departure from the long-term average of peak NDVI for each pixel was calculated using the Departure Average Vegetation Method. Figure 13.3 show that the area had a high percentage of departure from long-term average which reached > 30% during the drought years. Figure 13.3 also, showed increase in NDVI trends during the period from 1992 to 2003. Despite the fact that there were few years with high departure (e.g. 2000) but this departure are still less than that during the period from 1981 to 1991. This indicate that the area start to recover from the effect of the Sahelian drought.

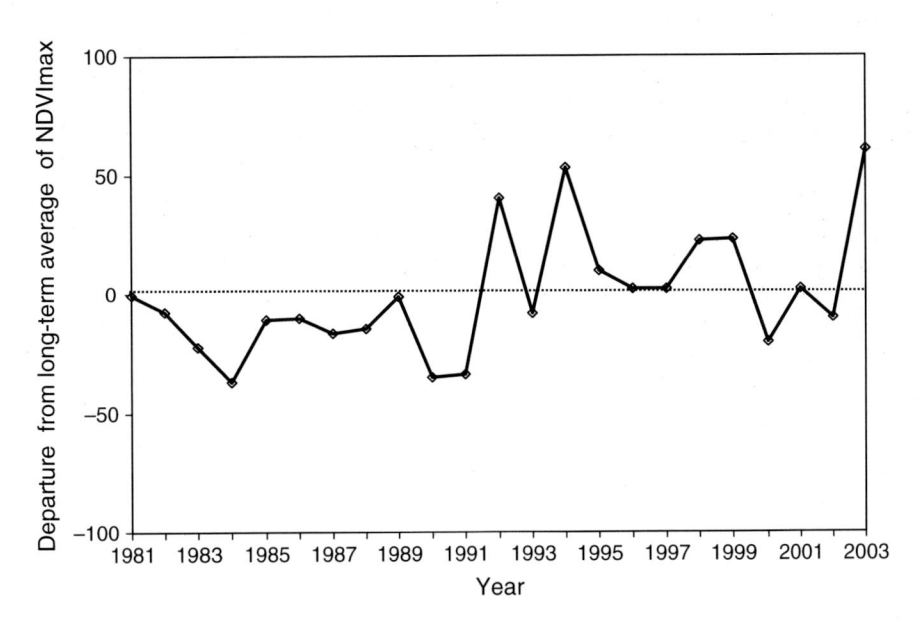

Fig. 13.3 Departure from the long-term average of peak NDVI for the studied area

13.3.3 Landscape Pattern Index for Degradation

13.3.3.1 Moving Standard Deviation Index

The MSDI shows that the degraded areas exhibited the highest MSDI values, but undegraded areas showed low MSDI values (Fig. 13.4), whoever, the study area shows lower MSDI values in 1987 than in 2000. These areas have been severely eroded with little or no vegetation around Sufeiya, Sobagh and Banat sites they show higher values in 2000. The image difference for MSDI shows that most of the areas have exhibited high MSDI during the last 13 years, indicating considerably increase in degraded areas from 1987 to 2000, especially, around Banat, Sufeiya and Sobagh.

13.3.3.2 Bare Soil Index

The BSI highlights areas that were potentially affected by erosion, possibly requiring intervention to counteract the severe degradation. Most of the study area had low BSI in 1987 except few sites which had high index (Fig. 13.5a), however, during 2000 most the areas exhibited highest BSI values except for irrigated and rainfed

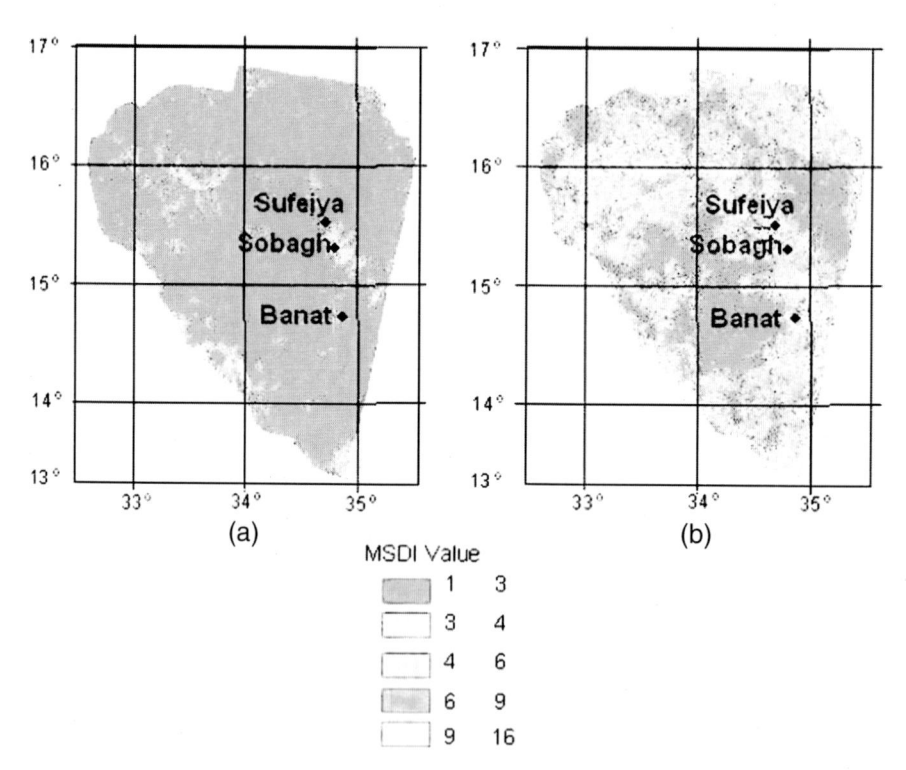

Fig. 13.4 Moving Standard Deviation Index (MSDI) for the study area (**a**) 1987 and (**b**) 2000 respectively

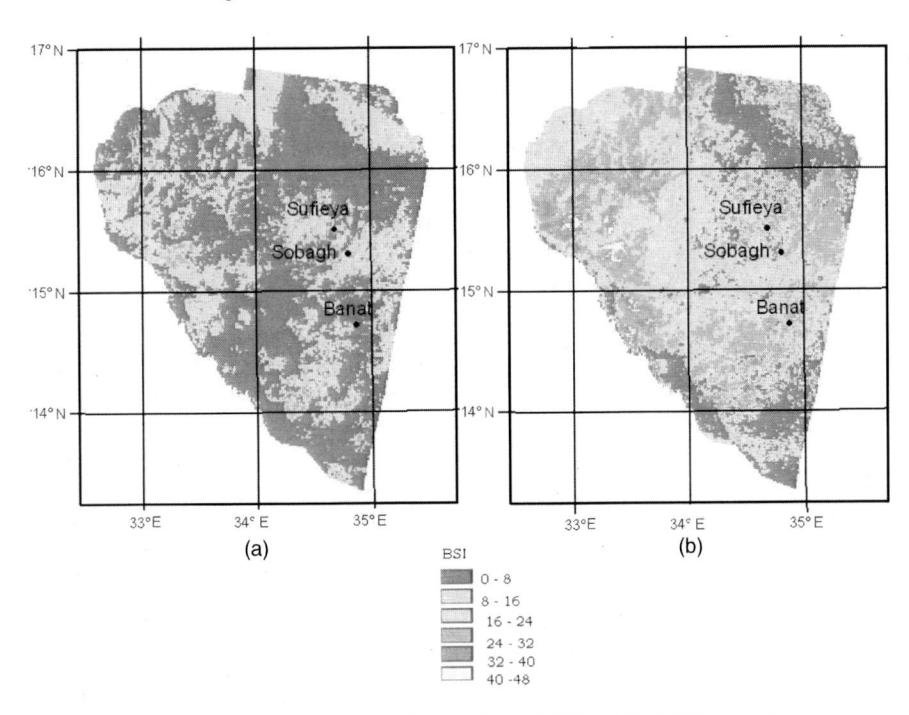

13 Environmental Degradation of Natural Resources in Butana Area of Sudan 177

Fig. 13.5 Bare Soil Index (BSI) for the study area for (**a**) 1987 and (**b**) 2000 respectively

areas, which had low values (Fig. 13.5b). Comparing the MSDI and BSI images it can be noticed that the degraded sites (e.g. Sufeiya, Sobagh and Banat) exhibited high values of both indices, which can be interpreted that, the degraded area is more heterogeneous and there is more bare land than undegraded area. Further on was

Photo. 13.1 Wind erosion around Banat village (photo courtesy S. Walker, August, 2005)

indicated that the bare soil in the irrigated scheme (south west and north east part) remains constant (–6 to 8) while in the west and central part of the study area the bare soil increased by about 14–43%. It can be noticed that the bare soil increased about 22–43% around Banat, as illustrated in Photo 13.1, this indicate the area is severely affected by wind erosion.

13.4 Conclusions

The extensive spatial, regular temporal coverage and reasonable cost of satellite imagery provides an opportunity to undertake routine natural resource monitoring. It can be noticed that the different ecosystems in the area are subject to various forms of site degradation. The environmental degradation has led to sand encroachment and to accelerated development of dunes and also increased the water erosion in the northern part of the area. Pastures have deteriorated seriously in quality and quantity. But in many parts the degradation is still reversible if organized land use and water points could be introduced. The MSDI proved to be powerful indicator of landscape condition for the studied area. The MSDI values increase considerably between 1987 and 2000, especially in Sufeiya, Sobagh and Banat areas. The BSI for these sites increased from 0–8 in 1987 to 32–40 in 2000. The image difference of the BSI indicated that the index increased about 14–43 over the last 13 years. It is concluded that the Remote sensing data can provide solid bases for evaluation and monitoring the environmental degradation but the result must be supported by the ground truthing as well as perceptions of the people living in the target areas.

References

Akhtar, M. (1994). Geo-ecosystem and pastoral degradation in the Butana. Animal Research Development 39:17–26.
Ehrlich, A.H., Gleick, P. and Conca, K. (2000). Resources and environmental degradation as sources of conflict. In: 50th Pugwash Conference on Science and World Affairs "Eliminating The Causes Of War". Queens' College, Cambridge. http://www.pugwash.org/reports/pac/pac256/WG5draft.htm .
Elhag, M.M. (2006). Causes and impact of desertification in the Butana area of Sudan. Ph.D. Thesis, University of the Free State, South Africa, p. 171.
Jamalabad, M.S. and Abkar, A.A. (2004). Forest Canopy Density Monitoring, Using Satellite Images. ISPRS Congress, Istanbul. www.isprs.org/istanbul2004/comm7/papers/48.pdf
Miles, R.L. and Johnson, P.W. (1990). Runoff and soil loss from four small catchments in the Mulga lands of southwest Queensland. 6th Australian Rangeland conference. Perth, pp. 170–184.
Tanser, F.C. (1997). The application of a landscape diversity index using remote sensing and geographical information systems to identify degradation pattern in the Great Fish River Valley, Eastern Cape province, South-Africa. M. Sc. Thesis. Rhodes University, South Africa, p. 167.
UNEP. (1992). Status of desertification and implementation of the United Nations Plan of Action to combat desertification. Report of the executive Director. United Nations Environment Program, Nairobi.
Wessels, K.J. (2001). Gauteng Natural Resource Audit. ARC-ISCW, Pretoria.

Chapter 14
Land Suitability for Crop Options Evaluation in Areas Affected by Desertification: The Case Study of Feriana in Tunisia

S. Madrau, C. Zucca, A.M. Urgeghe, F. Julitta, and F. Previtali

Abstract The study was carried out in the Tunisian site of a Cooperation Project, aiming to implement techniques for combating desertification. The approach of the project is based on promoting drought-resistant fodder shrubs as an alternative crop, in areas where the expansion of traditional cereal and olive trees in not suited rangelands constitute a land degradation factor. A land evaluation procedure was implemented based on a Map of Pedo-morphologic Units purposely created and on GIS based suitability models specifically adapted to the local conditions. The suitability maps obtained showed that about 18% of present rainfed cereal crops and 12% of olive plantations are located in not suited areas. The results also highlight the areas where further crop development could be more or less advisable and fodder shrubs could be introduced as an alternative and sustainable income generating option.

Keywords Crop suitability · Land evaluation · Drought-resistant shrubs · Soil survey · GIS · Tunisia

14.1 Introduction

The present work was carried out in the frame of a Euro-Mediterranean Cooperation Project, (MEDA-SMAP; Short and Medium Term Priority Environmental Action Programme). The Project (2002–2007) aimed to apply and demonstrate at local level and on an extensive scale, some techniques for combating desertification in the drylands, based on restoration of vegetation cover and production of fodder resources for the local breeders.

C. Zucca (✉)
NRD-UNISS, Desertification Research Group, Università degli Studi di Sassari, 07100 Sassari, Italy
e-mail: clzucca@uniss.it

In the target areas of Morocco (Rural Municipality of Ouled Dlim, Marrakech) and Tunisia (Imada de Skhirat, Feriana, Kasserine) drought-resistant fodder shrubs were used, over some thousands of ha, for restoration of strongly degraded pasturelands (Zucca et al., 2005; Bellavite et al., 2009). The species used are mainly *Atriplex nummularia*, (Morocco) and *Opuntia ficus indica* (Tunisia).

The research results reported in this paper derive from the Tunisian site. Rainfed cereals (barley in particular, *Hordeum vulgare*) play an important role in the human diet and breeding in the area, while olive crops (*Olea europea*) recently underwent relevant expansion driven by favourable international market factors, as in much of the "high steppe" parts of the country. This expansion, often carried out in rangelands, and the introduction of the associated agronomic practices such as tillage, contributed to intensify wind erosion and, in some cases, sand encroachment, both in plantations and in downwind areas. In areas threatened by desertification, as the sites studied by the present project, this activity is considered as highly sensitive by local authorities (NAP-TUN, 1999).

For this reason, it is of strategic importance to develop an adequate land planning process, able to evaluate the specific suitability (and vulnerability) of the land to the considered crops and to suggest alternatives in fragile areas. The fodder shrubs plantation, such as *O. ficus indica*, is considered an effective alternative in unsuited, degraded rangeland. The present study, associated with the cooperation project mentioned above, had the final goal of testing reliable criteria to support a suitable land zoning and in particular at providing the local decision makers with cartographic tools to best plan land use options, including establishment of new plantations. Such an approach requires a suitable cartographic support (land/soil maps) based on which land evaluation can be carried out.

Soil survey always involves a combination of theoretical and practical problems. The main one is the identification of the physical limits between the different soil bodies, especially when the variation of their attributes and the distribution patterns are continuous (Velásquez et al., 2004). Many interpretation strategies of the landforms have been conceived in order to infer from them a likely distribution of the underground parts of the soils. For a long time, laws and rules of variation of soil nature and distribution, connected and/or depending on laws and rules that govern the action of one or several soil forming factors (lithology, landforms and geomorphic dynamics, surface drainage, vegetation cover and, where existing, land use), have been investigated (Comolli and Previtali, 1999).

In particular, the concept of soil-landscape-(geo)morphological units is introduced as a retroaction between (geo)lithologic and geomorphic features, landform evolution, parent materials composition, drainage network and plant cover. In fact, the characteristics and quality of soils, their distribution and bio-geo-chemical processes are the result of interactions and retroactions between the above-mentioned factors (Wysocki et al., 2000). Especially soils and landforms develop together and "this development is a two-way street. Soils are affected by landforms, and through their developmental accessions and features, they in turn influence geomorphic evolution" (Schaetzl and Anderson, 2005).

14 Land Suitability for Crop Options Evaluation in Areas Affected by Desertification

This is why it is more appropriate to reason – spatially and at process level – in terms of "morphopedology" and not simply "pedology" (Steiner, 1991). This approach is particularly effective in areas where geomorphic processes are active and can strongly influence land capability and soil vulnerability to degradation: geomorphic dynamics should thus be given relevance in defining mapping units, beside soil features, in order to enhance the usefulness in land management and planning.

Hence, a correct application of this approach requires, as a first step, an iterative procedure (Legros, 1996), consisting of geomorphological "reading" of the land, composition of mapping units, survey of typology of the existing soils, feedback to the refinement of mapping units. As a second step second, a more accurate verification of soil distribution and its actual relationship with the soil forming factors are recognized as dominant, then reaching finally to the composition of Morphopedological Units (Fig. 14.1). The last ones, compared with maps of the existing land use, will guide the programmes of agro-ecological management and land evaluation (Piorr, 2003).

The evaluation of the suitability of a territory to a given crop is a complex multidisciplinary procedure. Many of suitability evaluation procedures in use are adaptations to the local conditions of the Framework for Land Evaluation (FAO,

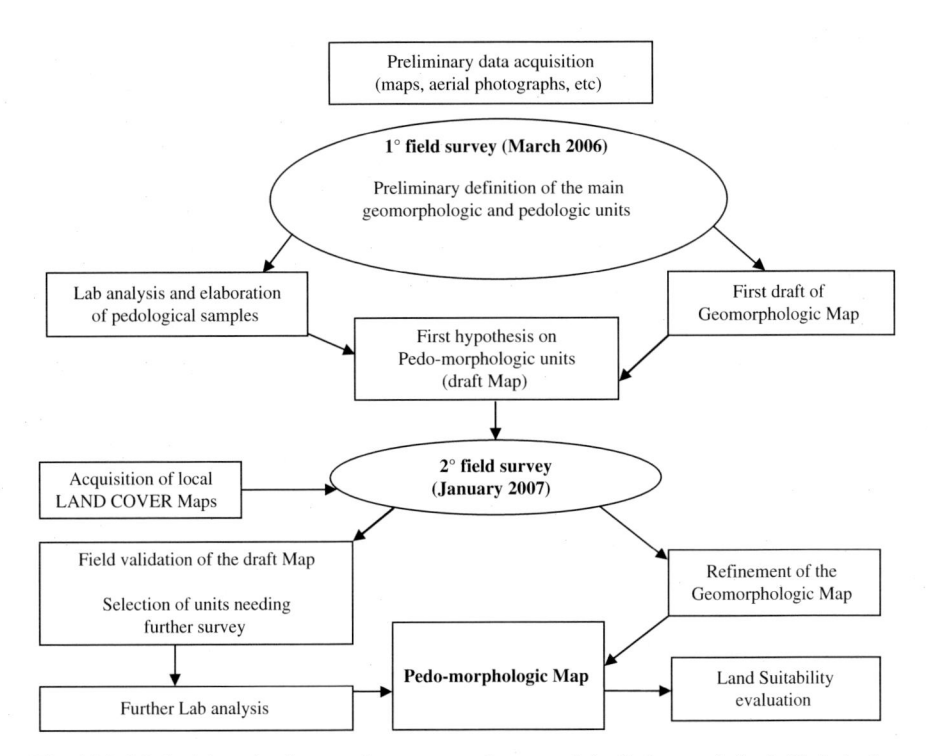

Fig. 14.1 Methodology implemented to generate the map of the Pedo-morphologic Units in the study area

1976), and focus on the severity of land limitations related to crops and land use. The distinction between the classes is based on the increase of the costs for the reduction or elimination of these limitations.

The olive tree (*Olea europea*) is a typical Mediterranean tree crop. It is relatively not exigent in terms of nutritional elements and water requirements, for these reason, up to a recent past, it was typically grown in marginal areas. Its wide diffusion gave rise to the selection of a high number of *cultivars* and forms of breeding. This variability does not allow to easily adopting a single suitability model. In Italy the work done on the suitability evaluation for olive crops has been reported and summarized by Franchini et al. (2006).

Other models have been recently proposed in Syria (Cools et al., 2003), Egypt (Wahba et al., 2007), Pakistan (Del Cima and Urbano, 2008), Iran (Lake et al., 2009) and Spain (de La Rosa et al., 2004). FAO (1983) proposed an approach for rainfed crops in general that had several local applications. Concerning barley, apart from the studies carried out by FAO-IIASA (1991), some recent research was done in Iran (Jalalian et al., 2007; Jafarzadeh et al., 2008; Behzad et al., 2009). These studies emphasise the marked relevance of salinity and alkalinity as limiting factors. As will be explained below, these factors were not included in the simplified model implemented by the present study, although some areas were in fact classified as not suitable a priori due to their salinity.

The main constraint of the models derived from the FAO Framework is the need for economic and socio-economic data, not always available. An alternative approach, called *VCs-Vocazione Colturale Specifica* (literally, "specific crop vocation"), was proposed by Danuso et al. (2001). The VCs only considers the agronomic requirements. This simplifies the collection of the necessary data, and allows a greater flexibility in results applications at different local contexts. This model was further adapted for rainfed olives and cereals by Sarria (2006), Urgeghe (2006), and Piras (2006). The results of previous evaluations showed a high reliability in spite of the relative simplification.

Finally, concerning *O. ficus indica*, the high degree of adaptation of the species to aridity and to shallow, marginal soils is very well documented (Mulas and Mulas, 2009). In particular, in the high steppe land in inner Tunisia, it proved to be productive and a possible strategic option for the marginal areas, which the local Government is actively promoting. For this reason no specific suitability model was here developed for the species, but it was considered as a "residual" option able to valorise soils not suitable for the traditional crops of the area, as proved by the cooperation project mentioned.

14.2 Materials and Methods

The study was conducted in an area covering about 27,500 ha (Fig. 14.2; left), located in Central Western Tunisia (Feriana, Kasserine), on the borderline with Algeria. The elevation of the area ranges between 750 and 1,100 a.s.l. Average

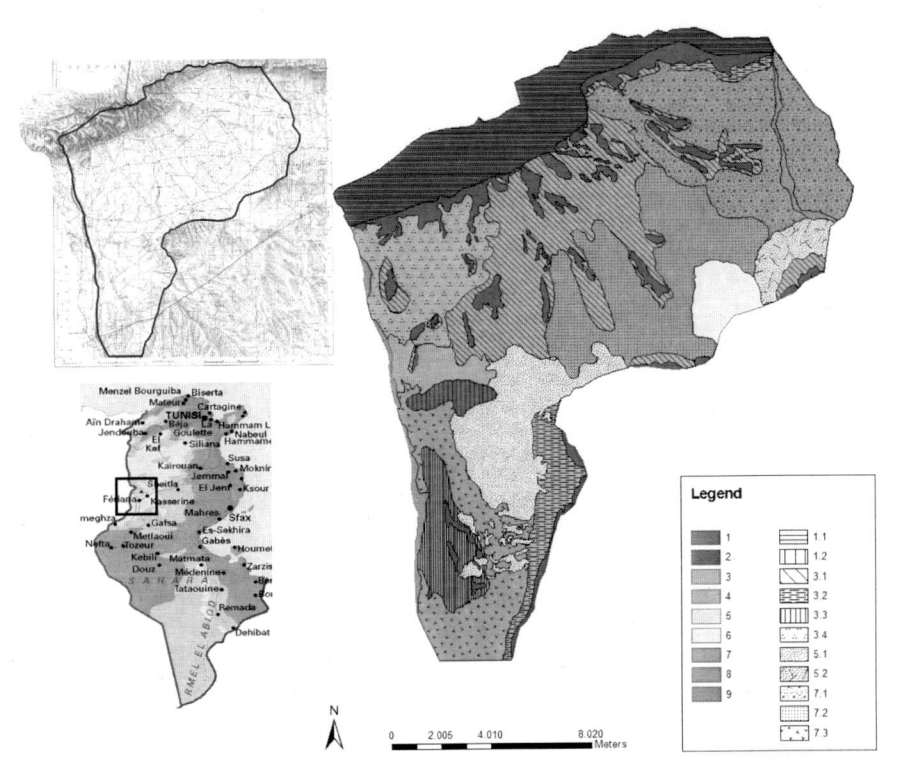

Fig. 14.2 *Left*: Location of the study area. The town of Feriana is located outside the area, a few km to the East. *Right*: Map of the *Pedo-morphologic Units*. 1. Rock outcrops; 2. Pediments; 3. Transition surface between pediments and alluvial fan; 4. Hydromorphic and salinized areas; 5. Areas of aeolian accumulation; 6. Areas of ancient endorheism and fossil sand dunes; 7. River terraces; 8. Bed and bank of the river Oued Saboun; 9. Bed and bank of the river Oued Saf Saf. Second order units are also shown, as listed in Table 14.1

annual rainfall (at Feriana for the period 1975–2005) is 314 mm, with maximum in September and average annual temperature (at Kasserine, some km away; for the period 1975–2005) is 15.9°C (2.8°C, average in January and 35.3°C in July). The northern relieves shown in Fig. 14.2 date back to the Cretaceous period and are mainly made of hard limestone and marl. The plain is characterised by Neogene (mainly constituted by highly incoherent greenish clay and yellowish sands) and Quaternary sediments. The Neogene sediments are particularly vulnerable to wind erosion.

According to the information included in the "Carte Agricole" (PNCA, 2004) of Tunisia, 66% of the surface is used for agriculture, mainly linked to small ruminants breeding. Rainfed cereal crops (barley in particular) cover 37% of the area and Alfa steppe (Stipa *tenacissima*) is spread in about 90% of the rangeland surface.

Table 14.1 Criteria for land evaluation. Average values and classes of the selected variables for each Pedo-morfologic sub-unit and associated scores for rainfed olive (O) and herbaceous (H) crops (from 1 to 0 the suitability ranges from high to low, as detailed for each variable in Table 14.3). Representative soils are also listed for most of sub-units

Pedo-morphologic unit	Sub unit	Soils (WRB, 2006)	Soil depth (d) Values (cm)	O	H	Texture (t) Class (USDA)	O	H	Skeleton (Sk) Value (%)	O	H	Available water reserve (wr) Values (mm)	O	H
1 Rock outcrops	Morpho tectonic landforms	Rendzic Leptosol	24	0	0	Loam	1	1	35	0.7	0	20-50	0.4	0
2 Pediments	Sandstone	–	24	0	0	Loam	1	1	35	0.7	0	20-50	0.4	0
	–	Epipetric hypocalcic Calcisols (Aridic)	18	0	0	Loamy sand	0.7	0	30	1	0.7	<20	0.4	0
3 Transition surface	Ondulated surfaces	Haplic Arenosols (Gypsiric, Eutric); Cutanic Luvisols (Hypereutric)	89	1	1	Sandy clay loam	1	0.7	10	1	1	>200	1	1
	Inherited erosion	Luvic Calcisol (Aridic);	180	1	1	Sandy loam	1	0	15	1	1	100-200	0.85	0.7
	Calcrete	Petric Calcisol (Aridic)	6	0	0	Loamy sand	0.7	0	70	0	0	<20	0.4	0
	Aeolian erosion		20	0	0	Loamy sand	0.7	0	40	0.7	0	<20	0.4	0

Table 14.1 (continued)

Pedo-morphologic unit	Sub unit	Soils (WRB, 2006)	Soil depth (d) Values (cm)	O	H	Texture (t) Class (USDA)	O	H	Skeleton (Sk) Value (%)	O	H	Available water reserve (wr) Values (mm)	O	H
4 Hydromorphic and salinized areas	–	Hypersalic Calcic Solonchak (Sodic Aridic)	160	1	1	Sandy loam	1	0	20	1	0.7	100–200	0.85	0.7
5 Aeolian accumulation	Aeolian sedimentation areas	Haplic Arenosol (Aridic)	30	0	0	Loamy sand	0.7	0	5	1	1	20–50	0.4	0
	weakly cemented aeolianites	Haplic Luvisol	110	1	1	Sandy loam	1	0	5	1	1	100–200	0.85	0.7
6 Ancient endorheism and fossil sand dunes	–	Luvic Calcisols	230	1	1	Sandy loam	1	0	2	1	1	100–200	0.85	0.7
7 River terraces	Downstream alluvial deposits (low/high energy)	Calcic Luvisol (Fragic); Hyperskeletic Leptosol (Aridic)	180	1	1	Sandy clay loam	1	0.7	5	1	1	> 200	1	1
	Endorheic areas	Cutanic Calcic Luvisol (Clayic)	190	1	1	Clay	0	0	3	1	1	> 200	1	1

Table 14.1 (continued)

Pedo-morphologic unit	Sub unit	Soils (WRB, 2006)	Soil depth (d)			Texture (t)			Skeleton (Sk)			Available water reserve (wr)		
			Values (cm)	O	H	Class (USDA)	O	H	Value (%)	O	H	Values (mm)	O	H
	Alluvial area with aeolian sheet	Haplic Arenosol (thapto luvisolic)	300	1	1	Sandy loam	1	0	35	0.7	0	100–200	0.85	0.7
8–9 Bed and bank of *Oueds* Saboun and Safsaf		–	50	1	0.7	Sand	0.7	0	1	1	1	20–50	0.4	0

In some areas the Alfa cover is protected and collected for transformation in local paper factories. In other places it is more or less degraded due overgrazing and excessive uptake.

Limited information exists on the soils of the area from the Carte Agricole but not suited for the aims of the present study. A specific soil survey was thus carried out to fill this gap and to support the following evaluations. Particular importance was given to geomorphologic aspects, as discussed above, due to the relevance of the past and present local geomorphic processes in influencing the landforms and the soils.

The coherent *Pedo-morphologic Units* terminology was adopted for the definition of mapping units in the study area. The land mapping phases were implemented as represented in Fig. 14.1. The geomorphologic and pedological surveys were carried out in two phases and 60 soil profiles were described and classified according to the WRB system (IUSS Working Group WRB, 2006).

Land Suitability evaluations related to rainfed olive crops and to rainfed cereals crops were carried out according to Danuso et al. (2001), modified and fitted to local conditions based on expert judgments, as described below.

The following variables were selected as basic parameters for land evaluation:

- soil depth to the "lithic or paralithic" contact (d);
- texture, described in terms of USDA classes (t);
- skeleton, as percentage of rock fragments (larger than 2 mm) on the total volume (sk);
- available water reserve, in mm, ranging from 20 to 200 (wr).

Additionally, the land parameter slope angle, in % (sl) was integrated. For each variable, an average value (or class, in case of texture), was estimated for each second level unit of the Pedo-morphologic Map, based on all profiles and observations data available for the Map units (Table 14.1).

The slope gradient map was derived from a Digital Elevation Model. Slope classes (Table 14.2) are established in accordance with the Guidelines for Soil Profile Description by FAO (2006). Average values were translated into parametric scores ranging from 0 to 1, as showed in Table 14.2. The suitability was then calculated using a GIS-algebra procedure (in ESRI-ARCGIS environment), through the following equation:

For rainfed herbaceous crop:

$$VA = sl^*(d + t + sk + wr)/4$$

For rainfed olive crops:

$$VA = sl^*wr^*(d + t + sk)/3$$

The results were grouped into four classes: (SC1) poorly suited; (SC2) moderately suited; (SC3) well suited; (SC4) highly suited. Finally, suitability maps were

Table 14.2 The scores assigned to ranges of values for each soil variable and slope angle classes (from 1 to 0 the suitability ranges from high to low)

Variable	Range	Score rainfed olive crops	Score rainfed cereals
Soil depth(d)	< 30 cm	0	0
	30–60 cm	1	0.7
	60–120 cm	1	1
	>120 cm	1	1
Texture(t)	Loam	1	1
	Sandy loam	1	0.7
	Sandy clay loam	0.7	1
	Loamy sand	0	0.7
	Clay	0	0
Skeleton (Sk)	>30%	0.7	0
	15–30%	1	0.7
	5–15%	1	1
	<5%	1	1
Available water reserve (wr)	<50 mm	0.4	0
	50–150 mm	0.65	0
	150–200 mm	0.85	0.7
	>200 mm	1	1
Slope angle classes (%)	0–2	1	1
	2–6	0.9	0.9
	6–13	0.75	0.75
	13–30	0.2	0.2
	30–50	0	0
	>50	0	0

overlaid to a land cover map produced by the project, to highlight relationship between suitability and current land use patterns.

14.3 Results and Discussion

Nine *Pedo-morphologic Units* were defined and mapped. Second level units were also mapped and typical soils are listed in Table 14.1. The Map is shown in Fig. 14.2 (right). The Map Units can be used to highlight specific "vulnerability" factors, mainly related to active geomorphic processes. These are not accounted for by the model, but of course should be considered once the agronomic and land limitations have been put into evidence by the model. Table 14.3 reports, in quantitative terms, the results of Land Suitability evaluation, while the land suitability maps obtained are shown in Fig. 14.3.

It can be noted that all areas assigned to SC1 are the same for both crops, because this class includes severe limitations, not compatible with agricultural crops. For this reason in Table 14.3 a single column reports the results obtained.

If the results are analysed according to the Map Units (MU), it could be observed that MU1 (Rock outcrops) and 2 (Pediments) were assigned 100% to SC1 for both

14 Land Suitability for Crop Options Evaluation in Areas Affected by Desertification 189

Table 14.3 Percentage of each Pedo-morphologic Unit (MU) assigned to each Suitability Class (SC), for both olive (O) and herbaceous (H) rainfed crops (SC1, not suitable; SC2, poorly suitable; SC3, moderately suitable SC4, very suitable)

Map unit (MU)	SC 1 (%) Both crops	SC 2 (%) O	H	SC 3 (%) O	H	SC 4 (%) O	H	Total hectares Per MU
1	100	0	0	0	0	0	0	3,490
2	100	0	0	0	0	0	0	2,117
3	34	0	0	5	14	60	51	8,261
4	100	0	0	0	0	0	0	538
5	0	81	81	1	19	18	0	2,809
6	0	0	0	2	100	98	0	625
7	1	1	21	48	38	50	40	9,263
8	0	100	100	0	0	0	0	164
9	2	98	98	0	0	0	0	269
								Tot 27,534
Total hectares and percentage per SC	9,058 (33)	2,750 (10)	4,673 (17)	4,962 (18)	5,856 (21)	10,764 (39)	7,947 (29)	

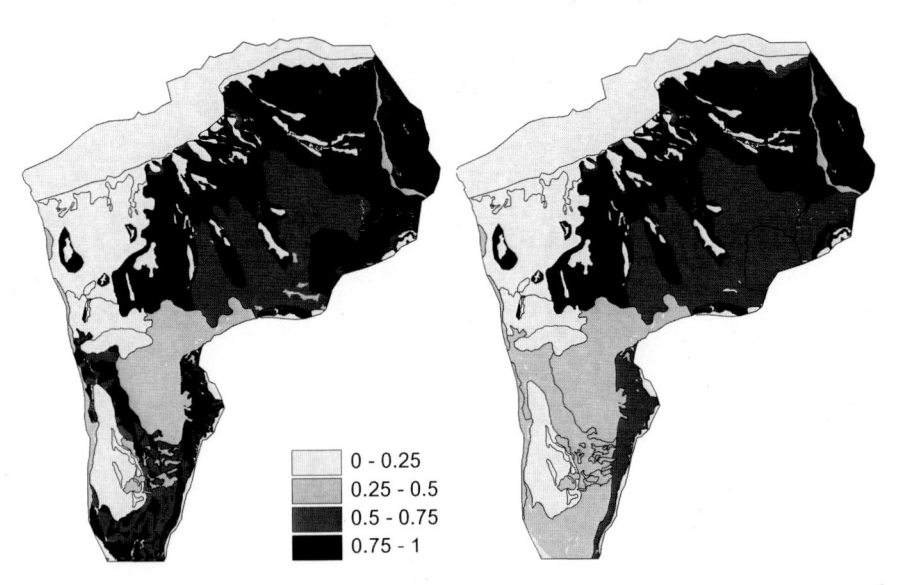

Fig. 14.3 Land Suitability Maps to rainfed olive (*left*) and cereal (*right*) crops. The ranges of reported values correspond to the four classes (higher score means higher suitability)

crops. In the case of MU2, this is mainly due to the shallow soils and to the steep slopes of the unit.

The same suitability class was assigned to MU4 (saline areas), for different reasons. In fact it was considered as not suitable a priori, by expert judgement, due to the high soil salinity, but actually the present model does not include salinity as a variable. This decision was taken in order to avoid designing a model much relying on data often not available.

MU3 yielded much more variable results. It presents a marked difference between sub-Unit 3.1, very suited in average and 3.3/3.4. The latter are poorly suited due to the thick calcrete crusts (3.3) and the shallow, quite light soils (3.4). A great part of these three sub-units were assigned the same suitability class (SC4 and SC1 respectively), for both crops. Thus sub-unit 3.2 in fact determines a large part of the difference in evaluation between the two crops. It should be added that MU3.4 is also characterised by a specific "vulnerability" related to the present intensity of wind erosion. In this case, the status of soils is degraded enough to suggest not to cultivate the area based on agronomic considerations alone; otherwise, considerations related to geomorphology would do so.

The evaluation of Map Unit 5 reflects quite well the different characteristics of the two sub-units 5.1 and 5.2. The former is poorly suited (SC2) for both crops, mainly due to the very sandy, shallow soils. The latter is characterised by weakly cemented aeolianites, originated by past Aeolian deposition. These soils are deeper and texture is much coarser, yielding SC3 for herbaceous and SC4 for olive. This evaluation is not completely satisfying if the vulnerability of the weakly structured soils of the sub-unit to water erosion is considered. Tillage should be done very carefully and SC2 for herbaceous would probably be more prudent. Similar considerations could be done for MU6, needing conservative agronomic practices.

The areas characterised by the highest suitability are mostly located in the western and central sectors, (alluvial terraces, MU7.1/7.2). These areas are almost flat and characterised by deep fine soils, with relevant water reserves. However, MU7.2 (endorheic areas) has some limitations due to the very fine texture and relatively poor drainage and that explains the high percentage values in SC3 class for both crops. MU7.3, in the southwestern sector, shows instead limitations related to texture and skeleton affecting herbaceous crops only, resulting in a higher proportion of SC2 values for such crops. Map Units 8 and 9 are relatively small and homogeneous and almost completely classified as SC2.

14.4 Conclusions

The work carried out has been focused on land suitability evaluation as a tool to analyse the land degradation risk related to specific land uses in an area already affected by desertification. Land degradation risk could be seen as a consequence of irrational allocation of land use, especially when unsuitable land is also vulnerable to degradation processes such as erosion. The approach implemented for producing the necessary cartographic support, or the *Pedo-morphologic* Map, was intended to meet the information needs by providing information on both agronomic properties and active geomorphic processes. Land evaluation also contributes to define possible mitigation strategies, by indicating where conservative crops such as fodders shrubs could best become an alternative to impacting crops.

The results obtained could be made clearer if considered in the light of present land use patterns. By overlaying the land suitability maps and the present land use map, the following considerations were made. About 18% of present rainfed cereal crops and 12% of olive plantations are located in not suited areas, some of them strongly affected by wind erosion. On the other hand, up to about 60% of the SC4 (highly suited) areas for both crops are already correctly exploited. The remaining SC4 surface is still in part (about 15% in average) not used, being classified as "open space, scarcely vegetated" (most often degraded rangeland). This is potentially available for crop development.

The poorly suited areas across different land units are generally associated with rock outcrops, salinity, active wind erosion or aeolian accumulation dynamics, thick calcrete crusts. In many cases they have already been successfully planted with fodder shrubs. As an example, Atriplex nummularia developed very well on the saline soils of the study area. *Opuntia ficus indica* proved to be effective on very sandy soils of MU5, even in presence of thick surface sand cover, but also on shallow soils subjected to severe wind erosion. Social acceptance has been high too, also due to the income generated. Further plantations could be preferentially made in these areas. Areas having intermediate suitability classes could be considered for crop development, but on a case by case basis, because in some situations conservative practice need to be adopted.

It can thus be concluded that, if integrated in the local land planning processes, the studies carried out, the methods developed and the Maps produced could significantly contribute to valorise possible crop complementarities, supporting sustainable rural development and reducing desertification risk.

Acknowledgments The authors warmly thank Dr. Nabil Gasmi, for collaboration on geomorphology and TELEGIS Laboratory of Prof. A. Marini for land cover mapping.

References

Behzad, M., Albaji, M., Papan, P., Boroomand Nasab, S., Naseri, A.A. and Bavi, A. (2009). Qualitative evaluation of land suitability for principal crops in the Gargar Region, Khuzestan Province, Southwest Iran. Asian Journal of Plant Sciences 8(1):28–34.

Bellavite D., Zucca C., Belkheiri O., Saidi H. (eds.). (2009). Etudes techniques et scientifiques à l'appui de l'implémentation du projet démonstratif SMAP de Lutte Contre la Désertification. NRD, Université des Etudes de Sassari, Italie. (In press).

Comolli, R. and Previtali, F. (1999). Il paradigma suolo-paesaggio nel rilevamento e nella cartografia pedologica. In: G. Brigati and G. Orombelli (eds.), Studi Geologici e Geografici in onore di Severino Belloni. Glauco Brigatti Publisher, Genova, pp. 207–222.

Cools, N., De Pauw, E. and Deckers, J. (2003). Towards an integration of conventional land evaluation methods and farmers soil suitability assessment: A case study in north-western Syria. Agriculture, Ecosystems and Environment 95 (2003), Elsevier:327–342.

Danuso, F., Giovanardi, M. and Donatelli, M. (2001). Applicazioni agronomiche delle conoscenze pedologiche. Bollettino Società Italiana Scienza del Suolo 50(2):251–280.

De la Rosa, D., Mayol, F., Diaz-Pereira, E., Fernandez, M. and De la Rosa, D., jr. (2004). A land evaluation decision support system (MicroLEISS DSS) for agricultural soil protection. With special reference to the Mediterranean region. Environmental Modelling and Software 19, Elsevier:929–942.

Del Cima, U. and Urbano, F. (2008). Selection of suitable areas for olive growing in Pakistan. In: Relazioni e monografie agrarie tropicali e subtropicali, Ministero Affari Esteri – Istituto Agronomico per l'Oltremare, Firenze Nuova serie, No. 119.

FAO. (1976). A framework for land evaluation. Soil Bulletin 32, Roma.

FAO. (1983). Guidelines: Land evaluation for rainfed agriculture. Soil Bulletin 52, Roma.

FAO. (2006). Guidelines for Soil Description. 4th edition. FAO, Roma.

FAO-IIASA. (1991). Agro-ecological land resources assessment for agricultural development planning. A case study of Kenya. Resources data base and land productivity. Technical Annex. 3. Agro-climatic and agroedaphic suitabilities for barley, oat, cowpea, green gram and pigeonpea. In: World Soil Resources Reports 71/3, Roma.

Franchini, E., Cimato, A. and Costantini, E.A.C. (2006). Attitudine dei suoli alle colture arboree: Colture da frutto. Olivo. In: E.A.C. Costantini (ed.), Metodi di valutazione dei suoli e delle terre, Valutazioni attitudinali per la gestione aziendale e la programmazione a livello locale. Min. Pol. Agric. Alimen. E Forestali – Osserv. Naz. Pedologico per la qualità del suolo agricolo e forestale – Centro Ric. Agricoltura – Collana metodi analitici per l'agricoltura. volume 7. Cantagalli and Siena Eds, pp. 579–642.

IUSS Working Group WRB. (2006). World reference base for soil resources. World Soil Resources Reports 103. FAO, Rome.

Jafarzadeh, A.A., Alamdari, P., Neyshabouri, M.R. and Saedi, S. (2008). Land suitability evaluation of Bilverdy Research Station for wheat, barley, alfalfa, maize and safflower. Soil and Water Resources 3(2008) (Special issue 1):S81–S88.

Jalalian, A., Givi, J., Bazgir, M. and Ayoubi, S. (2007). Qualitative, quantitative and economic evaluation of land suitability for barley, wheat, and chickpea in rainfed area of Talandasht district (Kermanshah province). Journal of Science and Technology of Agriculture and Natural Resources, Institute for Sustainable Futures, University of Technology, Iran. 10 4(A).

Legros, J.P. (1996). Cartographies de sols. Presses Polytechniques et Universitaires Romandes, Lausanne, p. 321.

Mulas, M. and Mulas, G. (2009). The strategic use of atriplex and opuntia to combat desertification. In: D. Bellavite, C. Zucca, O. Belkheiri and H. Saidi(eds.), Etudes techniques et scientifiques à l'appui de l'implémentation du projet démonstratif SMAP de Lutte Contre la Désertification. NRD, Université des Etudes de Sassari, Italie. (In press).

NAP-TUN. (1999). Programme d'Action Nationale de Lutte contre la Désertification. Ministère de l'environnement et de l'aménagement du Territoire. www.unccd.int .

PNCA. (2004). Programme National Carte Agricole. Société Tunisienne d'Ingénierie Informatique. Ministère de l'Agriculture et des Ressources Hydrauliques, Tunisie.

Piorr, H.P. (2003). Environmental policy, agri-environmental indicators and landscape indicators. Agriculture, Ecosystems and Environment 98:17–33.

Piras, F. (2006). La pianificazione del territorio quale strumento di programmazione. La Vocazione colturale specifica alla vite e all'olivo. Il caso del comune di Siniscola (NU). Degree thesis, Università degli Studi di Sassari, Facoltà di Agraria, A.A. 2005–2006 (in Italian).

Rahimi Lake, H., Taghizadeh Mehrjardi, R., Akbarzadeh, A. and Ramezanpour, H. (2009). Qualitative and quantitative land suitability evaluation for olive (Olea europaea L.) production in Roodbar region, Iran. Agricultural Journal 4(2), Medwell Publishing:52–62.

Sarria, I. (2006). Applicazioni di metodologie GIS nella pianificazione territoriale per l'utilizzo agricolo nel comune di Ittiri. Degree thesis, Università degli Studi di Sassari, Facoltà di Agraria, A.A. 2005–2006 (in Italian).

Schaetzl, R.J. and Anderson, S. (2005). Soils Genesis and Geomorphology. Cambridge University Press, West Nyack, NY, p. 817.

Steiner, F. (1991). The Living Landscape: An Ecological Approach to Landscape Planning. McGraw Hill, New York, p. 356.

Urgeghe, A.M. (2006). Caratterizzazione pedologica e studio di valutazione della attitudine del territorio all'utilizzo agricolo di un'area a rischio di desertificazione in Tunisia: L'Imada di Skhirat. Degree thesis, Università degli Studi di Sassari, Facoltà di Agraria, A.A. 2005–2006 (in Italian).

Velásquez, J., Ochoa, G., Oballos, J., Manrique, J. and Santiago, J. (2004). Metodología para la Delineación Cartográfica de Suelos. Revista Forestal Latinoamericanavii 36:15–34.

Wahbba, M.M., Darwish, K.M. and Awad, F. (2007). Suitability for specific crops using Micro Leis program in Sahal Baraka, Farafra Oasis, Egypt. Journal of Applied Sciences Research 3(7) INSInet Publication:531–539.

Wysocki, D.A., Schoeneberger, P.J. and LaGarry, H.E. (2000). Geomorphology of soil landscapes. In: M.E. Sumner (ed.), Handbook of Soil Science. CRC Press, New York, E8–E10.

Zucca, C., Lubino, M., Previtali, F. and Enne, G. (2005). The Euro-Mediterranean partnership: A participatory demonstration project to fight desertification in Morocco and Tunisia. In: P. Zdruli and G. Trisorio Liuzzi (eds.), Determining an Income-Product Generating Approach for Soil Conservation Management. IAM Bari, Bari, pp. 317–325.

Chapter 15
Strategic Nutrient Management of Field Pea in South-Western Uganda

P. Musinguzi, J.S. Tenywa, and M.A. Bekunda

Abstract The highlands of south-western Uganda account for the bulk of field pea (*Pisum sativum* L.) produced and consumed in the country. The crop fetches a stable price, which is as high as that of beef, but it has remained outside the mainstream of the research process. Low soil fertility, unfortunately, is poised to eliminate the crop. Nitrogen, phosphorus and potassium have variously been reported as deficient on the bench terraces where crop production is primarily done. Strategic nutrient management requires that the most limiting nutrient is known in order to provide a foundation for designing effective and sustainable soil fertility management interventions. A study was conducted on upper and lower parts of the bench terraces on the highlands in south-western Uganda to identify the most required macronutrient(s) in field pea production. Treatments included: 0 and 25 kg N ha^{-1}, 0 and 60 kg P ha^{-1}, and 0 and 60 kg K ha^{-1}, all applied factorially in a randomized complete block design. Parameters assessed included nodulation, nodule effectiveness for BNF and dry weight, shoot dry weight (SDW), and grain yield. Nutrient applications that resulted in the highest crop responses were considered as most required, and hence, most limiting to plant growth and yield. Phosphorus based nutrient combinations gave the highest increments in total and effective nodule numbers, as well as dry weight, irrespective of terrace position. On the other hand, N based combinations led to the highest shoot dry matter at flowering (39% higher over the control). The superiority of N was carried over up to final harvesting, with stover and grain yields edging out the other treatment regimes on either terrace positions.

Keywords Nutrients · Nodulation · Biomass · Grain Yield

P. Musinguzi (✉)
Department of Soil Science, Makerere University, Kampala, Uganda
e-mail: patmusinguzi@agric.mak.ac.ug

15.1 Introduction

Field pea is a legume crop, reportedly among the key components of human diets worldwide (Duke, 1981). It is cultivated for fresh green seed, tender green pods, dried seed and forage. In sub-Saharan Africa (SSA), field pea is a primary source of protein for many communities, in spite of the fact that it is hitherto largely understudied among legume crops. It is commonly grown in the cool highlands of Africa like in Ethiopia and Kenya.

In Uganda, field pea is a staple as well as a major income earner for most small-scale farmers in the highlands of southwestern (SW) region, where the agro-ecology is most suited for its production. The crop fetches a stable price, which is as high as that of beef, yet it has remained outside the mainstream of the research process. Apart from the local demand, which is far from satisfaction, the crop presents great potential for export to European countries where it is heavily consumed and forms a significant component of the diets and yet the local supply in Uganda is still very low.

The crop is traditionally grown in rotation with other crops on terrace benches constructed on the steep hills of SW Uganda. These terraces have been continuously cropped for decades without nutrient management attention hence; soil fertility depletion is among the major factors constraining field pea and crop production in general in this region (Lindblade et al., 1996; Siriri, 1998). In fact, on-farm yields currently stand at a paltry 400 kg ha^{-1}, contrasting with 2 t ha^{-1} obtained in other countries (FAO, 2000). Literature on soil fertility management is still hard to trace in the entire eastern Africa.

Field pea is a renowned heavy consumer of N and P (Poulain, 1989 as a legume, the crop is endowed with potential for replenishing soil N through biological N fixation (BNF)). The continuous production process results in depletion of the other nutrients, especially P and K that are required in fairly large quantities by the crop. Besides, the viable performance of BNF depends on the adequacy and balanced proportions of the spectrum of the other nutrients (Pulong, 1994).

In order to elevate the productivity of field pea to profitable levels in Uganda, it is prudent that renewed attention is focused on sustainable soil fertility management. The objective of this study, therefore, was to identify the most limiting nutrient to production of the crop as the strategic entry point to effective management and subsequent formulation of rationalized packages.

15.2 Materials and Methods

An on-farm experiment was conducted on bench terraces of SW Uganda, in Kabale district during two rainy seasons of 2005 (September–December) and 2006 (March–July). The area is characterised by mountainous relief of flat topped hills and ridges, underlain by partly granitised and metamorphosed Precambrian rock formations of the Karagwe–Ankolean system (Harrop, 1960). Elevation ranges between 1,300 and 2,400 m above sea level, with bi-modal annual rainfall pattern. The annual minimum and maximum means are 1,092 and 1,500 mm, respectively, while respective temperatures range between 10 and 23°C (KDMD, 2005). In 1999 the soils of the region were mainly classified by Wortmann and Eledu (1999) as Ferralsols.

Our study involved ten farm fields and these were in effect the replications. Treatments included two terrace positions, namely upper (UTP) and lower (LTP); N, P and K applied as 0 and 25 kg N ha^{-1}, 0 and 60 kg P ha^{-1}, and 0 and 60 kg K ha^{-1}. The experiment was factorial and was laid out in a randomized complete block design. N, P and K were applied in the form of urea, triple super phosphate and muriate of potash (KCL), respectively. The choice of the rates was based recommendations for a number of annual legumes cultivated in the country (NARO, 1996).

The experimental sites' soil (0–20 cm depth) was pre-analysed for pH (water), total N and organic matter content, available P, and exchangeable K$^+$ and Ca^{2+} using procedures outlined by Okalebo et al. (2002). A locally known field pea variety, which is "white and black eyed" called *Meisho* was used. Because available stockists did not carry certified field pea seed, farmers' stocks off the previous season were used. As such, seed viability was tested using routine procedures. In plots of 4 m by 4 m, seeds were planted by broadcasting and incorporated into soil. The experiment was entirely rain-fed, and was managed following farmer practices.

Parameters assessed included nodulation, nodule effectiveness for BNF and dry weight, shoot dry weight (SDW), and grain yield. Nodule and other associated evaluations were done 30 days after planting (DAP). For this purpose, 10 pea plants were randomly selected from each plot for destructive sampling. They were carefully dug out and the roots freed of adhering soil and debris using distilled water, prior to the evaluation. Nodules were removed, counted and sliced cross-sectionally for assessment for potential BNF using the leghaemoglobin pigment indicator Thereafter, each sample's nodules gravimetric moisture content determination and eventual dry nodule dry weight was determined. The process involved oven-drying at 65°C for 48 h prior to dry weight determination. All nodule data were expressed per plant as the unit. Immediately after nodule removal, the shoots were severed from the roots with a knife, bulked per sample and subjected to gravimetric moisture dry weight determination as described above.

At approximately 60% flowering, 10 plants were again randomly collected from each plot for further shoot dry weight (SDW) assessment using the gravimetric procedure. Finally, at physiological maturity (120 DAP for this variety), the remaining plants were uprooted, and root parts severed from the shoots with a knife. Fresh weight of the shoots with pods was recorded per plot before ten plants with seed were sub-sampled for dry weight assessment. The remaining plants of the harvested sample were air-dried to 14% moisture content, threshed and winnowed. The clean seeds were weighed and a portion sub-sampled for gravimetric moisture content seed dry weight determination. Seed yield values were extrapolated to yield per hectare prior to statistical analysis using analysis of variance (ANOVA) using GenStat. Significant treatment means were separated using Fisher's Least Significant Difference (LSD) test at 5% probability level.

15.3 Results and Discussions

15.3.1 Site Soil Characteristics

Soil pH was generally below the limit of 5.5 (Table 15.1) widely considered optimal agronomically. On the other hand, total organic matter and exchangeable bases (K^+ and Ca^{2+}) were in the adequate range, total N and plant available P were marginally adequate and critically low, respectively (Table 15.1). In fact, total N, soil organic matter, and exchangeable bases concentrations varied across the terrace, with higher concentrations in LPT than UPT. This observation is in conformity with the one made by Siriri (1998) a decade ago. Phosphorus values in the range of 3–7 folds less than the critical limit are suggestive that the element has great potential for limiting plant growth. Davies et al. (1985) and Pulong (1994) similarly reported considerable negative effects of sub-optimal soil P supply on field pea performance. On the other hand, the spatial gradient in soil chemical properties across the terraces suggests that soil management within the terrace ought to be disaggregated accordingly in order to achieve realistic crop responses. However, how management of such micro-site level resource variations fits into the small-scale farmers' biophysical and socio-economic frameworks needs further investigation.

15.3.2 Nodulation and Associated Parameters

Phosphorus based treatment outstandingly ($p<0.05$) increased nodule number and weight (Figs. 15.1 and 15.2). Application of P alone resulted in a nodule number increment of 54% over the control (no application), while application of N+P+K and P+K increased the numbers by 53 and 49%, respectively (Fig. 15.1). Furthermore, the highest increment in nodule weight of 66% was realised with P alone, followed

Table 15.1 Chemical characteristics of the site soil in SW Uganda

Terrace position	pH	Organic Matter (%)	Total N	Bray 1 P (mg kg^{-1})	K$^+$ Exchangeable (cmol. kg^{-1})	Ca^{2+}
Upper	5.2	3.8	0.21	1.76	0.61	4.96
Lower	5.3	4.4	0.27	5.84	0.66	5.84
[a]Critical values	5.5	3	0.2	15	0.4	4.5

[a]Critical values for most crops in East Africa (Okalebo et al., 2002).

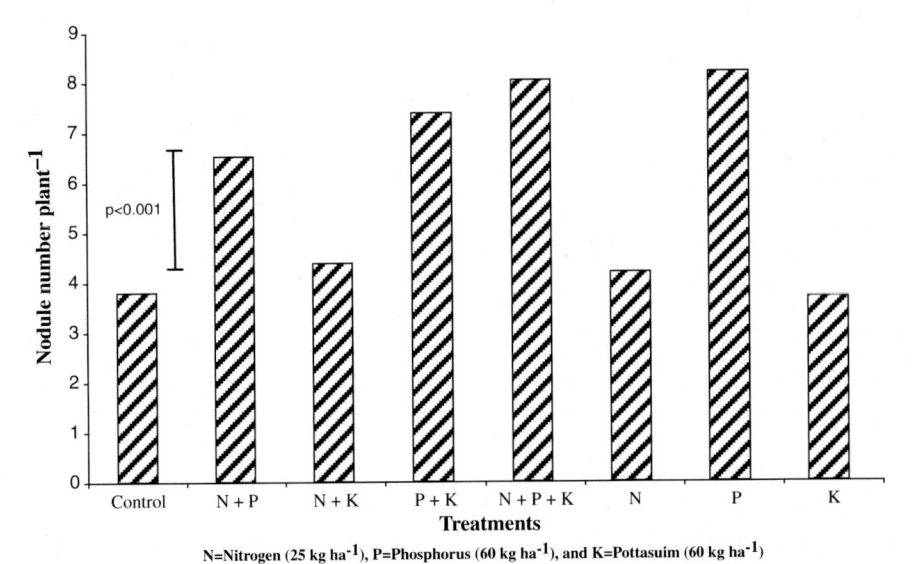

Fig. 15.1 Nutrient application effects on total nodule number in field pea in SW Uganda

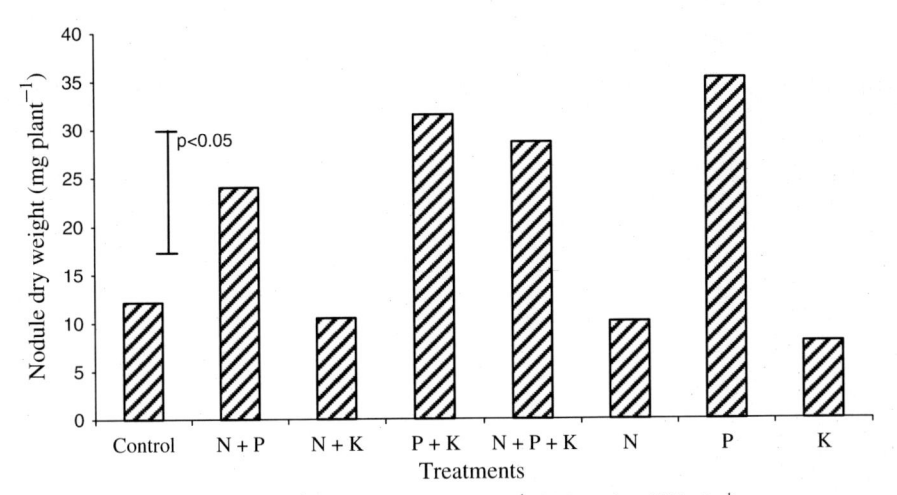

Fig. 15.2 Nutrient application effects on nodule dry weight in field pea in SW Uganda

by P+K with an increase of 62% (Fig. 15.2). On the other hand, sole application of N and K had no significant effect ($p>0.05$). This sequence of observations on nodule parameters demonstrates that phosphorus was most required nutrient relative to N and K. Phosphorus plays a key role in nodule activity through increased formation and availability of Adenosine Tri-phoshosphate (ATP), a resource material for the energy intensive for the N reduction process via nitrogenase enzyme

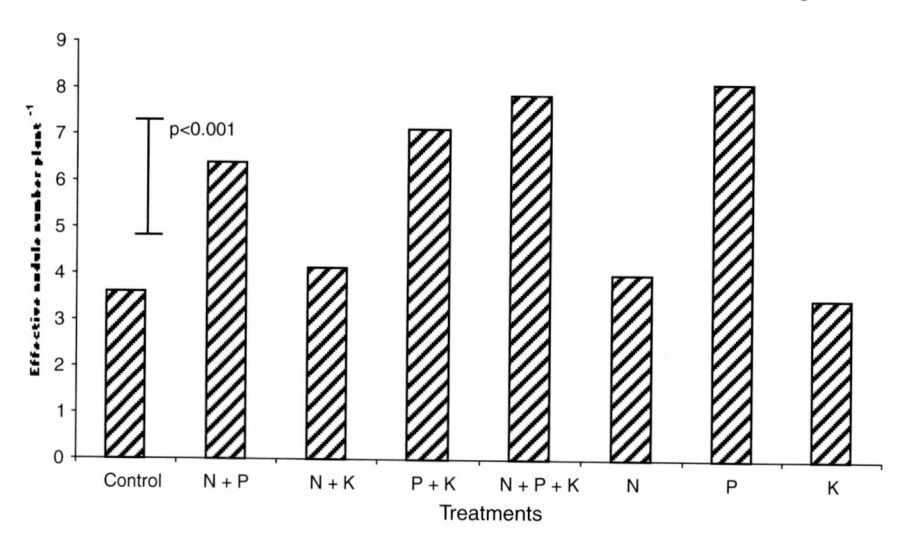

Fig. 15.3 Nutrient application effects on effective nodule number in field pea in SW Uganda

activity (Slinkard and Drew, 1988). In fact, this was evident in the P applied cases where the number of potentially BNF-effective nodules was the highest (Fig. 15.3). Surprisingly, there were a few nodules in the control treatment with the pigment indicative of BNF effective at such low available P concentrations levels. This could be explained by the possibility that the plant has evolved mechanisms for P access at such low levels. A possible candidate mechanism in this case is mychorrhyzae formations, which are known to enhance the plant's ability to access the higher plant's access to otherwise unavailable soil resources (Clark and Zeto, 2000). This speculation requires an independent investigation.

15.3.3 Shoot Dry Matter at 60% Flowering

Shoot dry weight (SDW) response was at variance with the nodule parameters discussed above (Fig. 15.4). In this case, it was the N based treatments that led to the highest SDW over the control. Generally, P and K based combinations trailed. Furthermore, plant sensitivity to the treatments was way superior on the UPT to that of the LPT, with SDW values several folds greater ($p<0.05$). In fact, on the LTP, only N resulted in significant increment in this parameter (Table 15.2).

The disappearance of the outstanding effect of P application on the crop and the emerging superior effect of N, suggest that the latter became limiting in the later stages of plant growth. This could be explained by the fact that the rate of N applied (25 kg ha^{-1}) was rather too low to support the crop through its requirement. In a study done elsewhere by McKenzie et al. (2001) observed high field pea yield responses only at rates as high as 60 kg N ha^{-1}. The observation also implies that

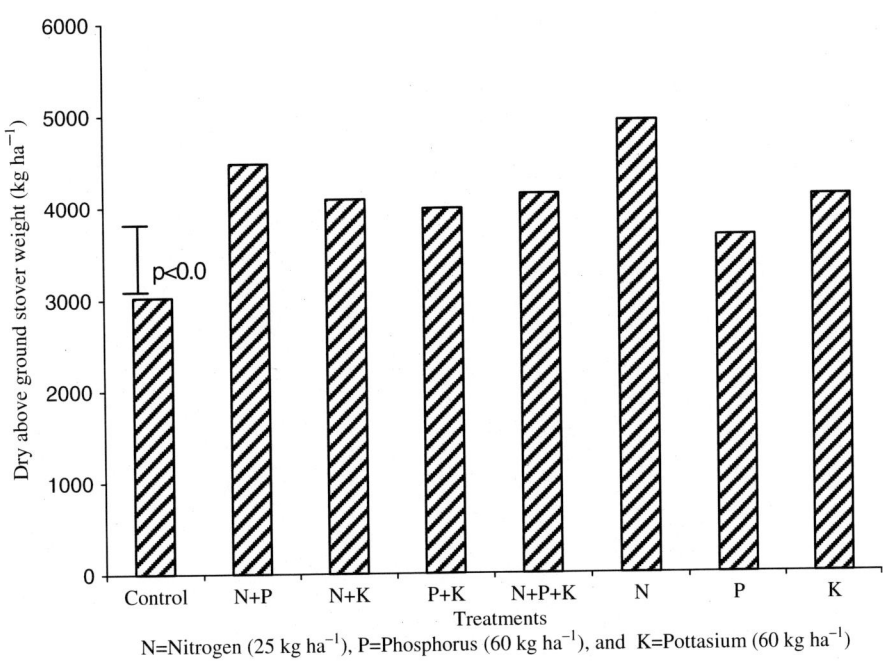

Fig. 15.4 Nutrient application effects on shoot dry weight at flowering in SW Uganda

N=Nitrogen (25 kg ha^{-1}), P=Phosphorus (60 kg ha^{-1}), and K=Pottasium (60 kg ha^{-1})

Table 15.2 Shoot dry matter response to nutrient application on bench terraces at flowering

	Dry above ground matter (kg ha^{-1})	
Nutrient treatments	Upper terrace position	Lower terrace position
Control	2,789	3,256
N+P	5,298	3,654
N+K	4,573	3,714
P+K	4,821	3,147
N+P+K	4,318	3,961
N	5,336	4,538
P	3,852	3,531
K	4,343	3,940
LSD$_{0.05}$	1,208	840

the observed soil test total N (Table 15.1) did not mineralise adequately to influence plant available N concentrations to sufficiently supplement applied N. This suggests that soil total N is not a dependable index for soil N availability to a crop. Furthermore, the emergent superior influence of N implies that biologically fixed N by the crop was equally inadequate to additionally cater for the crop needs. At this stage, this study has demonstrated that P and N are closely related as limiting nutrients in this region for field pea production; which one of them is the most limiting is a rather circumstantial phenomenon.

15.3.4 Stover and Grain Yield

Stover and grain yield at harvest were not significantly ($p>0.05$) influenced by nutrient application within each terrace position though the LTP generally registered significantly ($p>0.05$) higher stover dry weight (3,906 kg ha^{-1}) than its UTP counterpart (3,318 kg ha^{-1}). However, grain yield gave totally contrasting performance, with the upper part of terrace significantly ($p<0.01$) higher (958 kg ha^{-1}) than the lower part (730 kg ha^{-1}).

This contradiction could have been due to heavy rains that fell (>150–180 mm/month) during podding and grain filling stages, that caused lodging and injury of the plants including the rotting of the pods; such cases were observed more on the LTP than the upper part. According to Duke (1981), lodging is a common phenomenon that may significantly affect field pea yields. However, it was not clear as to why there was more pronounced lodging on the LTP than its counterpart position. The logical explanation is that higher fertility on the former promoted more vigorous plant growth, which in turn predisposed the crop free-fall. On the other hand, there are also physiologically based factors that could lead to the reversed observation. For instance, imbalance in N supply in the soil in excess of a certain level tends to over-favour vegetative growth at the expense of grain production. In this connection, less of the assimilates accumulated during the vegetative phase and eventually translocated into the reproductive apparatus of the plants. In a nutshell, these observations further underscore the need for targeted soil fertility management interventions for each of the terrace position in order to optimise as well as achieve sustainable crop production.

15.4 Conclusions

The most constraining nutrient in the soils of SW Uganda is P; its deficiency particularly manifests in terms of the intensity of BNF indicators (nodule numbers and effectiveness, and dry weight). However, the effect of P is closely followed by and eventually surpassed that of N, especially in terms of stover and grain yield. The LTP is generally more fertile than the UTP, but the former favours stover production, while the latter favours grain yield. Heavy rains and their attendant lodging effects appear to affect grain yields on the LTP more than the upper positions.

Acknowledgement Authors thank the Regional Universities Forum for Capacity Building in Africa (RUFORUM) for financing the research and the farmers of Kabale District for their participation.

References

Clark, R.B. and Zeto., S.K. (2000). Mineral acquisition by arbuscular mycorrhizal plants. Journal of Plant Nutrition 23:867–902.

Davies, D.R., Berry, G.J. and Dawkins., T.C.K. (1985). Pea (Pisum sativum L.). In: R.J. Summerfield and E.H. Roberts (eds.), Grain Legume. Crops. William Collins & Sons Co, London, pp. 147–198.

Duke, J.A. (1981). Hand Book of Legumes of World Economic Importance. Plenum Press, New York, pp. 199–265. Food and Agriculture Organization of the United Nations.1994- Production Year Book, Rome, Italy.

FAO. (1961–2000). FAO Production Yearbook (1961–2000). Rome, Italy. (http://museum.agropolis.fr/english/pages/expos/aliments/legumineuses/prodconso)

Harrop, J.F. (1960). The Soils of the Western Province of Uganda. Memoirs of the Research Division, 1/6. Department of Agriculture, Uganda. http://www.fertilizer.org/ifa/publicat/html/pubman/fpea.pdf.

KDMD (Kabale District Meteorology Department). (2005). Monthly and Annual Weather Data for Kabale District. Uganda. Ministry of Local Government. Government of Uganda, Kampala, Uganda.

Lindblade, K., Tumuhairwe, J.K., Carswell, G., Nkwiine, C. and Bwamiki, D. (1996). More people, more fallow: Environmentally favourable land use changes in south western Uganda. A Report. The Rockefeller Foundation, New York, USA, 67p.

McKenzie, R.H., Middleton, G., Clayton, G. and Bremer, E. (2001). Response of pea to rhizobia inoculation and starter nitrogen in Alberta. Canadian Journal of Plant Science 81:637–643.

NARO. (1996). National Agricultural Research Organisation. Ground-nut Grower Guide Edition, 12p. Ministry of Agriculture, Animal Industry and Fisheries. Government of Uganda, Kampala, Uganda.

Okalebo, J.R., Gathna, K.W. and Woomer., P.L. (2002). Laboratory Methods for Soil and Plant Analysis. A Working Manual. 2nd edition. Tropical soil fertility and Biology program, Nairobi Kenya. TSBF-CIAT and SACRED Africa, Nairobi Kenya. 128p.

Poulain, D. (1989). Nitrogen content and mineral composition of protein crops; Peas and lupins. Journal Proceedings. ATOUT POIS, Paris, France.

Pulong, M.A. (1994). Effect of fertilizer rates on yield, productive efficiency of pea on brown Podzolic soil. Acta Horticulturae 369:306–310.

Siriri, D. (1998). Characterization of the spatial variations in soil properties and crop yields across terrace benches of Kabale. MSc. Thesis. Makerere University, Kampala, 96p.

Slinkard, A.E. and Drew., B.N. (1988). Dry pea production in Saskatchewan. Ag. Dex. 140/10 (rev.). University of Saskatchewan, Saskatoon.

Wortmann, C.S. and Eledu., C.A. (1999). Uganda's Agro-Ecological Zones. A Guide for Planners and Policy Makers. Centro International de Agriculture Tropical, Kampala, Uganda, 29–31.

Part III
Land Degradation and Mitigation in Asia

Chapter 16
Effectiveness of Soil Conservation Measures in Reducing Soil Erosion and Improving Soil Quality in China Assessed by Using Fallout Radionuclides

Y. Li and M.L. Nguyen

Abstract China experiences the most severe water and wind erosion problems in the world. Using fallout radionuclide techniques (FRN), our objectives were to assess the extent of soil erosion and to quantify the beneficial effects of soil conservation measures at four sites extending from South West (SW) to North East (NE) China. At the Xichang site of SW-China, the combined use of FRN ^{137}Cs and ^{210}Pb$_{ex}$ measurements demonstrated that the effectiveness of vegetation species in reducing soil erosion decreased in the following order: shrubs > trees with litter layer > grasses > trees without litter layer. At the Yan'an site of Loess Plateau, sediment production estimated by ^{137}Cs declined by 49% due to terracing and by 80% due to vegetated (with grass and forest) compared to the cultivated hillslopes. Vegetated hillslope with grasses and forest increased soil organic matter (SOM) by 255%, soil available N (AN) by 198%, and soil available P (AP) by 18% while terracing increased SOM by 121%, soil AN by 103%, and soil AP by 162% compared with the entire cultivated hillslope. Both terracing and vegetating hillslopes were found to enhance soil porosity as shown by a decrease in soil bulk density (1.6 and 6.4%, respectively). At the Baiquan site in NE-China, soil loss as measured by ^{137}Cs tracer, decreased by 14% due to terracing and by 34% due to contoured tillage. At the Fengning site, data from ^{7}Be measurements indicated that 4 years of no tillage with high crop residues (50–56 cm depth) reduced soil erosion by 44% and no tillage with low residues (25 cm depth) reduced soil erosion rates by 33% when compared with conventional tillage practices.

Keywords China · FRN (fallout radionuclides) · Soil conservation measures · Soil erosion · Soil quality

Y. Li (✉)
Institute of Environment and Sustainable Development in Agriculture, CAAS, Beijing, 100081, China
e-mail: yongli32@hotmail.com

16.1 Introduction

Over the last decades, soil conservation measures, such as re-vegetation, terracing, and conservation tillage practies throughout the northern and western regions of China, have been extensively used for control of water erosion and sedimentation. These regions experience the most severe water and wind erosion problems in the world. According to the Chinese Water Resources Ministry, China has spent billions of yuans since 2001 planting trees, converting marginal farmland to forest and grasslands and fighting soil erosion. Over more than half a million sq km of the lands on hillsides have been set aside for afforestation or pasture. A total of 350,000 km^2 of plants and grassland have been restored and conserved. In addition, 11 major state-level water and soil conservancy projects are currently in operation, covering over 500 counties and cities. But few quantitative assessments have been made on the effecttiveness of such soil consevation measures in reducing soil erosion at regional and national scales because of lack of a large-scale assessment technique. A regional evaluation on the effectiveness of these soil conservation programmes is urgently needed for the development of integrated land and water management practices.

Fallout radionuclides (FRNs), including cesium-137 (^{137}Cs), excess lead-210 (^{210}Pb$_{ex}$), and beryllium-7 (^{7}Be), are an effective way for studying soil erosion and sedimentation within the agricultural landscape (Wallbrink and Murray, 1993; Zapata, 2003; Li et al., 2003). Cesium-137 from atmospheric nuclear weapon tests in 1950s and 1960s is a unique tracer for erosion and sedimentation, since there are no natural sources of ^{137}Cs in the environment. Lead-210 is a naturally occurring radionuclide from the ^{238}U decay series. It is derived from the decay of gaseous ^{222}Rn. Some ^{222}Rn in the soil diffuses into the atmosphere and decays to ^{210}Pb and subsequent fallout of ^{210}Pb to the landscape surface provides an input that is not in equilibrium (excess) with its parent ^{226}Ra (Robbins, 1978). By measuring ^{210}Pb and ^{226}Ra in the soil, ^{210}Pb$_{ex}$ can be calculated and used to measure soil movement (Joshi, 1987). Berylium-7 is also a naturally occurring radionuclide produced by the bombardment of the atmosphere by cosmic rays causing spallation of oxygen (O) and nitrogen (N) atoms in the troposphere and stratosphere. Production of ^{7}Be is relatively constant, producing a constant fallout deposition on the landscape (Wallbrink and Murray, 1994, 1996a).

When FRNs reach the soil surface, they are quickly and strongly adsorbed by exchange sites and are essentially non-exchangeable in most environments (Cremers et al., 1988; Robbins, 1978; Olsen et al., 1986). Biological and chemical processes move little of FRNs while physical processes of water, wind and tillage are the dominant factors moving FRN-tagged soil particles in the landscape surface. Accurately measuring ^{137}Cs, ^{210}Pb, and ^{7}Be in environmental samples is relatively easy using gamma ray spectrometer.

Measured patterns of the distribution of ^{7}Be, ^{137}Cs, and ^{210}Pb on the landscape provide information on short-term (<30 days), medium-term (~40 years) and long-term (~100 years) average soil redistribution rates and patterns, respectively. The basis of FRN technique involves comparing the measured inventories (total activity in the soil profile per unit area, horizontal distance) at study sites with an estimate of the total atmospheric input obtained from a reference site (Zapata, 2002). By comparing FRN measurements of the study site with the reference site, one can determine whether erosion (less FRN present than at the reference site) or deposition (more FRN than at the reference site) has occurred.

Against this background, studies were conducted in areas extending from south west (SW) to north east (NE) of China. The objective was to assess the effectiveness of soil conservation measures in reducing soil erosion using FRNs (^{137}Cs, excess ^{210}Pb, and ^{7}Be).

16.2 Materials and Methods

16.2.1 Study Sites

Four sites that represent diverse condtions in China were considered in this study (Fig. 16.1). They are (i) the Xichang site (27°43́N, 102°13É) in the purple soil region of SW-China, (ii) the Yan'an site (36°42′N, 109°31′E) on the Loess Plateau, (iii) the Baiquan site (47°30′N, 125°51′E) in NE-China, and (iv) the Fengning site (41°12′N, 116°38′E) in N-China. At the Yan'an and Baiquan sites, severe water erosion is the major cause of soil loss resulting from intensive tillage opreations

Fig. 16.1 Eroded landscapes at the 4 study sites: Yan'an, Xichang, Baiquan and Fengning

and inappropriate crop rotations on hillslopes. For the Fengning site, wind erosion is the main cause of soil erosion, resulting from conventional deep tillage operations without surface cover. At the Xichang site, over-grazing is considered to be the major cause of accelerated soil erosion and associated land degradation.

Cesium-137 (^{137}Cs) and excess ^{210}Pb (^{210}Pb$_{ex}$) were used to quantify the effectiveness of soil conservation measures which included terrace, vegetation and contour tillage to reduce soil erosion and sedimentation on hillslope landscapes at the Xichang, Yan'an and Baiquan sites. Beryllium-7 (^{7}Be) was used to assess the short-term (3–5 years) impacts of soil conservation tillage practices on soil erosion at the Fengning site.

16.2.2 Soil Sampling

At the Xichang site of SW-China, five typical species of plants were selected from the lower slope position on a 40-year restoration hillslope to evaluate effectiveness of trees, shrubs, and grasses to entrap sediments delivery from the upslope. They were: *Eucalyptus robusta Smith* (tree without litter layer), *Pinus massoniana Lamb* (tree with litter layer), *Camellia oleifera Abel* (shrub), *Dodonaea viscose* (shrub) and *Eulaliopsis binata* (grasses). Soil samples for determination of ^{137}Cs and ^{210}Pb$_{ex}$ were taken to a depth of 30 cm using a 6.74 cm diameter hand-operated core sampler at both the downslope and upslope for selected trees and shrubs, and at the growing site for the grasses.

At Yan'an site of Loess Plateau, the topography can be divided into hillslopes of 10–30° (inter-gully area) and gully slopes greater than 35° (gully area). Three hillslopes and two gully slopes were selected to assess spatial patterns of sediment production under specific land use and land management practices at hillslope scale. These hillsllopes were: (a) a cultivated hillslope, (b) a terraced farm hillslope, and (c) a vegetated hillslope, i.e. grassland on the top portion, forestland on the upper and middle portion, and grassland on the lower portion of the backslope. Two gully slopes were: one gully slope was cultivated and the other was grassland where natural grasses were sparsely distributed. Soil samplings for ^{137}Cs and soil quality analyses were undertaken using a 6.74 cm diameter hand-operated core sampler, with samples taken at 10 m intervals along cultivated and vegetated hillslopes and gully slope transects, and 3–5 m intervals along the terraced farm hillslope transect including all portions of the terrace. Three cores were collected at each sampling point to a depth of 50 cm and were then bulked to make a composite sample.

At the Baiquan site of NE-China, field sampling was collected on two pairs of sloping farmland. The first pair of slopes included a terraced slope less than 4° and a gentle slope of 6–10° without terraces. The relative elevation of the sloping farmland was about 15 m, and the horizontal length of slope was about 50 m. The second pair of slopes were managed by downslope farming and contour farming practices, respectively. The relative elevation of the slope was about 16 m, and the horizontal length of slope was about 102 m. Soil samples for determination of FRN inventory were taken to a depth of 15 cm at the upper, middle positions of slopes, and 60 cm at a lower slope position along three parallel transects in downslope direction. In total, five samples were collected at the upper, middle and lower positions of the slopes, respectively.

At the Fengning site of northern China, the major objective was to assess the possibility of using ^{7}Be to assess short-term changes in soil redistribution rates and soil quality parameters due to conservation tillage practices. There were five treatments: (i) Ref (Reference) – undisturbed never-cultivated grassland; (ii) CT – conventional tillage, spring wheat – fallow rotation; (iii) NT + HR – no-till for 3 years, spring wheat – surface residue cover with 50–56 cm high wheat straw fallow rotation; (iv) NT + LR – no-till for 3 years, spring wheat –surface residue cover with 25 cm wheat straw fallow rotation, and (v) T-G-conversion of cultivated land to grassland for 5 years. Soil samples for determining ^{7}Be were taken to a depth of 20 mm for all treatments immediately after wind erosion event in 2003 and 2004, respectively.

Reference sites for determining ^{137}Cs, ^{210}Pb$_{ex}$, and ^{7}Be were established at undisturbed, non-eroded, and uncultivated grassland in the four study areas. For the assessment of soil conservation measures on soil quality parameters, soil bulk density (BD) were determined over the 40-cm sampling depth, while soil available nitrogen (AN), soil available phosphorus (AP) and soil organic matter (SOM) were for the surface 10-cm only at Yan'an site of Loess Plateau (Li and Lindstrom, 2001).

16.2.3 Sample Analysis

Soil samples were air-dried and passed through a 2-mm sieve and weighed for the measurement of ^{137}Cs, ^{210}Pb$_{ex}$, and ^{7}Be specific activities. Soil samples for measuring ^{210}Pb were sealed in containers and stored for 28 days to ensure equilibrium between ^{226}Ra and its daughter ^{222}Rn (half-life 3.8 days). The amounts of ^{210}Pb$_{ex}$ in soil samples were calculated by subtracting ^{226}Ra-supported ^{210}Pb from the total ^{210}Pb concentration. Measurements of ^{137}Cs, ^{210}Pb, and ^{7}Be activities were conducted using Laboratory Gamma Spectrometer with Lab SOCS (Laboratory Sourceless Calibration Software) (BE5030, CANBERRA). Cesium-137 was detected at 661.7 kev peak while total ^{210}Pb was detected at 46.5 kev, and the ^{226}Ra-supported ^{210}Pb and ^{7}Be were detected at 609.3kev and 477.6 kev, respectively. Analytical precision for all FRNs was between ±5% and ±10%. Soil BD (Mg m^{-3}) calculation were based on volume of bulked soil cores and oven dried mass determinations (Pennock et al., 1994). Soil available N (mg kg^{-1}) was determined by using microfussion (Bremner, 1965) and soil AP (mg kg^{-1}) was determined using the method described by Olsen and Sommers 1982. Organic matter (percentage by weight) was measured by wet combustion (Nelson and Sommers, 1982).

16.3 Results and Discussion

At the Xichang site in SW-China, results through a combined use of fallout ^{137}Cs and ^{210}Pb$_{ex}$ measurements suggested that the effectiveness of plant species in entrapping sediment from upslope decreased in the following order: shrubs > trees with litter layer > grasses > trees without litter layer (Table 16.1). Based on the ^{137}Cs reference value at the permanent cover site consisting of a mixture of tree with litter layer (*Pinus massoniana Lamb*) and re-growth grasses (*Eulaliopsis binata*), a gain in ^{137}Cs was more than 39% under *Dodonaea viscose* (shrub) and *Camellia oleifera Abel* (shrub), compared with 13% under *Pinus massoniana Lamb* (tree with litter layer). In contrast, loss in ^{137}Cs was more than 33% under *Eucalyptus robusta Smith* (tree without litter layer) and about 7% under *Eulaliopsis binata* (re-growth grasses) due to over-grazing.

Table 16.1 Entrapment of sediment by different species of vegetation planted over the last 40 years at the Xichang site in SW-China

Plant species	Inventories (Bq m^{-2}) ^{137}Cs	^{210}Pb$_{ex}$	Loss or gain (%) ^{137}Cs	^{210}Pb$_{ex}$
Pinus massoniana Lamb (tree with litter layer)	951 ± 63	1,0794 ± 1,978	12.55	−14.14
Eucalyptus robusta Smith (tree without litter layer)	563 ± 51	5,379 ± 1,835	−33.35	−57.21
Dodonaea viscose (shrub)	1,182 ± 66	10,014 ± 1,984	39.94	−20.34
Camellia oleifera Abel (shrub),	1,198 ± 77	11,268 ± 2,234	41.82	−10.36
Eulaliopsis binata (re-growth grasses)	781 ± 48	8,897 ± 1,936	−7.62	−29.22
Pinus massoniana Lamb mixed with Eulaliopsis binata (permanent vegetation cover used as a reference site)	845 ± 53	12,571 ± 2,247	0	0

At the Yan'an site, results throuth using ^{137}Cs measurements indicated that landscape location had the most significant impacts on sediment production for cultivated hillslopes in the Loess Plateau, followed by the terraced hillslope, and to a lesser degree for the vegetated hillslope (Table 16.2). Sediment production increased in the following order: top < upper < lower < middle for the cultivated hillslope, and top < lower < upper < middle for the terraced hillslope. The mean value of sediment production declined by 49% for the terraced hillslope and by 80% for the vegetated hillslope compared with the cultivated hillslope. The vegetated gully hillslope reduced sediment production by 38% compared with the cultivated gully slope. This data demonstrated the effectiveness of terracing and perennial vegetation cover in controlling sediment delivery on a hillslope scale.

16 Effectiveness of Soil Conservation Measures in China

Table 16.2 Mean sediment production caused by water erosion and its standard deviation on hills and gully slopes under different land uses at the Yan'an site, derived from the measurement of ^{137}Cs

Location	Hillslope (t ha^{-1} a^{-1})			Gully slope (t ha^{-1} a^{-1})[c]	
	Cultivated	Terraced farmland	Vegetated (grass/forest)	Cultivated	Vegetated (grasses)
Top	17 ± 6	13 ± 10 (24)[a]	11 ± 6 (35)[a]	87 ± 41	54 ± 18(38)[b]
Upper	49 ± 30	27 ± 16 (45)	2 ± 8 (96)	87 ± 41	54 ± 18 (38)
Middle	71 ± 25	37 ± 36 (48)	17 ± 15 (76)	87 ± 41	54 ± 18 (38)
Lower	63 ± 16	25 ± 11 (60)	11 ± 21 (83)	87 ± 41	54 ± 18 (38)
Mean	50 ± 20	26 ± 19 (49)	10 ± 13 (80)	87 ± 41	54 ± 18 (38)

[a] reduction in soil losses due to terracing or vegetated on hillslope (%).
[b] reduction in soil losses due to vegetated on gully slopes (%).

According to Li et al. 2003, the amounts of ^{137}Cs and averaged ratios of ^{210}Pb to ^{137}Cs in the 0–5 cm surface soil (2.2–4.7 Bq kg^{-1} and 20.7:1 – 22.1:1, respectively) and in the 5–30 cm subsoil (2.6 Bq kg^{-1} and 28.6:1, respectively) on the cultivated hills and gully slopes were close to those of the deposited sediment in the reservoir (3.4 Bq kg^{-1} and 29.1:1, respectively). These results suggest that the main sediment sources in the catchment were from the cultivated slopes and from the gully hillslopes. Changes in land use types can greatly affect sediment production from gully hillslopes. An increase in grassland and forestland by 42% and a corresponding decrease in farmland by 46%, reduced sediment production by 31% in the catchment (Li et al., 2003).

Concerning the role of soil conservation measures in improving soil quality parameters, changes in ^{137}Cs, SOM, soil AN, soil AP and soil BD were specifically compared between the terraced and vegetated hillslope and the cultivated hillslope at the Yan'an site. The magnitude of ^{137}Cs, SOM, soil AN and soil AP were significantly lower on the steep cultivated hillslope than on the terraced and vegetated hillslope (Table 16.3). Vegetated hillslope increased SOM by 255%, soil AN by 198%, and soil AP by 18% while terracing increased SOM by 121%, soil AN by 103%, and soil AP by 162% when compared with the entire cultivated hillslope. Concentration of SOM, soil AN and soil AP in the terraced hillslope were 64, 68, and 223% of those in the vegetated hillslope, respectively. In contrast, terracing and vegetation resulted in a 1.6 and 6.4% decline in soil BD, respectively compared to an averaged value of 1.26 Mg m^{-3} for the entire cultivated hillslope. The changes in soil quality parameters are in agreement with redistribution pattern in ^{137}Cs inventory. Over the last 40 year, the loss in ^{137}Cs was much lower on terraced and vegetated hillslopes than on the cultivated hillslope based on the ^{137}Cs reference value (2,390 Bq m^{-2}) basis, which is further evidenced by the sedimentation data listed in Table 16.2.

Results from the Baiquan site in NE-China showed that the upper position contained the highest amount of ^{137}Cs whereas the lower position had the lowest ^{137}Cs

Table 16.3 Summary statistics for measured ^{137}Cs inventories and soil quality indicators on cultivated, terraced, and vegetated hillslopes at Yan an site

Soil variables		Cultivated hillslope			Terraced hillslope			Vegetated hillslope		
		Upper	Mid	Lower	Upper	Mid	Lower	Upper	Mid	Lower
^{137}Cs	Range, Bg m^{-2}	150–1,031	348–937	452–1,143	314–828	1,425–2,183	1,761–4,059	1,328–1,756	1,713–3,011	745–2,315
	Average, Bg m^{-2}	521	527	755	571(9.6)*	1,868(254)*	2,874(281)*	1,476(183)*	2,288(334)*	1,428(89)*
	SD, Bg m^{-2}	290	255	258	363	395	940	242	564	640
SOM	Range, %	0.25–0.38	0.39–0.49	0.33–0.40	0.45–0.92	0.74–1.01	0.74–1.05	1.05–1.55	0.89–1.98	0.46–1.22
	Average, %	0.33	0.43	0.36	0.68(106)	0.87(102)	0.93(158)	1.36(312)	1.52(253)	0.99(175)
	SD, %	0.05	0.04	0.03	0.33	0.13	0.14	0.07	0.41	0.32
AP	Range, mg kg^{-1}	1.01–1.99	1.62–2.20	1.50–1.92	3.68–4.59	2.41–3.96	2.20–14.06	1.77–2.61	1.32–2.27	1.31–2.64
	Average, mg kg^{-1}	1.39	1.96	1.70	4.13(197)	2.98(52)	6.14(261)	2.15(55)	1.79(−8.7)	2.01(18)
	SD, mg kg^{-1}	0.33	0.24	0.15	0.64	0.85	5.43	0.43	0.43	0.50
AN	Range, mg kg^{-1}	9.56–18.22	14.44–17.76	14.22–17.37	22.13–31.80	27.21–34.61	24.59–40.39	37.93–52.58	31.99–74.49	23.25–55.38
	Average, mg kg^{-1}	14.07	15.67	15.69	26.97(92)	31.29(100)	33.84(116)	45.44(223)	48.22(208)	41.62(165)
	SD, mg kg^{-1}	3.51	1.35	1.41	6.84	3.75	6.92	7.20	15.59	12.40
BD	Range, Mg m^{-3}	1.20–1.32	1.25–1.31	1.21–1.27	1.29	1.19–1.28	1.16–1.21	1.10–1.24	1.01–1.23	1.17–1.28
	Average, Mg m^{-3}	1.28	1.27	1.23	1.29(0.8)	1.25(−1.6)	1.18(−4.1)	1.16(−9.4)	1.16(−8.7)	1.22(−0.8)
	SD, Mg m^{-3}	0.05	0.03	0.02	0	0.05	0.02	0.07	0.08	0.04

* Numbers in brackets indicated changes in ^{137}Cs inventory and soil quality parameters due to terracing on hillslope (%).

Table 16.4 Effectiveness of terracing farmland on soil erosion reduction as compared with farmland without soil conservation measures at the Baiquan site (NE-China)

Slope position	^{137}Cs inventory (Bq m^{-2}) Terraced farmland	Uncertainty	No soil conservation	Uncertainty	Soil erosion rate (t ha^{-1} year^{-1}) Terraced farmland	No soil conservation	Reduction in soil losses by terraces (%)
Upper	1,184	95	1,079	84	22	24	8
Middle	948	78	862	66	26	27	4
Lower	832	65	296	27	28	36	22
Mean	988	79	746	59	25	29	14

Table 16.5 Effectiveness of contouring farm practice on soil erosion reduction as compared with downslope farm without soil conservation measures at the Baiquan site (NE-China)

Slope position	^{137}Cs inventory (Bq m^{-2}) Contour	Uncertainty	Downslope	Uncertainty	Soil erosion rate (t ha^{-1} year^{-1}) Contour	No soil conservation	Reduction in soil losses by contour cultivation (%)
Upper	1,151	97	809	71	23	33	30
Middle	1,191	116	612	53	22	40	45
Lower	1,197	96	929	72	23	32	28
Mean	1,179	103	784	65	23	35	34

for the sloping farmland (Table 16.4 and Table 16.5). The sediment budget, which was calculated using ^{137}Cs inventories, showed that by terracing the field, soil erosion rates were reduced by 14% for the entire slope, compared to downslope farming land (Table 16.4). On the contoured cropping farmland, different slope positions were found to have similar amounts of ^{137}Cs (Table 16.5), suggesting that contoured cultivation measures could effectively control soil erosion, especially in the middle position where serious soil erosion occurs. Compared with the downslope farmland (i.e., no soil conservation), contoured cultivation reduced soil erosion rates by 30, 45 and 28% on the upper, middle and lower slope positions (Table 16.5). Altogether, contoured cultivation reduced soil erosion rates by 34% over the entire slope. These results suggested that contour farming practices might be more effective in reducing soil losses than terracing in NE-China.

The data from ^{7}Be measurements suggests a significantly positive effect from soil conservation tillage practices in reducing wind erosion in N-China (Table 16.6). As compared with CT (conventional tillage), the mean fallout ^{7}Be inventory was increased by 56.8% for TG, 44.4% for NT+HR, and 32.5% for NT+LR.

Table 16.6 Summary statistics for beryllium-7 inventories under different soil conservation practices at Fengning site (N-China)

Treatments	Mean (Bq m^{-2})	SD (Bq m^{-2})	CV (%)	Change with CT (%)	Change with reference (%)
Ref	169.3	209.6	123.8	104.5	0.0
T – G	129.8	146.9	113.2	56.8	−23.3
NT + HR	119.6	152.2	127.2	44.4	−29.4
NT + LR	109.7	97.1	88.5	32.5	−35.2
CT	82.8	111.3	134.4	0.0	−51.1

By comparison with the NV (undisturbed, non-cultivated grassland) site, the conventional tillage operation resulted in more than a 50% loss of the total ^{7}Be amount, which was much higher than the soil conservation tillage practices. These indicate that soil losses were more substantial under CT than under conservation practices.

16.4 Conclusions

Fallout radionuclides were shown to be effective at assessing the beneficial effects of soil conservation measures in reducing soil erosion in four study sites. The combined use of fallout ^{137}Cs and ^{210}Pbex measurements suggested that shrubs are more effective in reducing soil erosion on eroded hillslopes than grasses, while planting trees without litter layer did not effectively control soil erosion at the Xichang site of SW-China. At the Yan'an site of the Chinese Loess Plateau, soil losses, estimated by the fallout ^{137}Cs measurements, declined by 49% due to terracing and by 80% for the vegetated hillslopes compared with the cultivated hillslopes. Vegetated hillslope increased SOM by 255%, soil AN by 198%, and AP by 18% while terracing increased SOM by 121%, AN by 103%, and AP by 162% when compared with the entire cultivated hillslope. In contrast, soil bulk density decreased by 1.6% due to terracing and by 6.4% due to vegetation.

A net soil erosion rate, measured by the fallout ^{137}Cs tracer, decreased by 14% on terraced farmland compared with a 34% reduction in soil loss due to contoured tillage in the Baiquan site of NE-China. The data from ^{7}Be measurements indicated that 4 years of no tillage with high (50–56 cm depth) residues reduced soil erosion by 44% and no tillage with low (25 cm depth) residues reduced soil erosion rates by 33% as compared with conventional tillage practices in the Fengning site of N-China.

Acknowledgements This study was supported by National Natural Science Foundation of China (No. 40671097 and No. 40701099), and the International Atomic Energy Agency (IAEA) (Research Contract No. 12323 and TC Project CPR5015), and National Key Basic Research Special Foundation Project of China (2007CB407204). Field sampling and soil analysis were assisted with X.C. Geng, J. Li, R. Funk, X.C. Zhang, F.H. He, D.H. Liu, L. Li, L. Sun, R. Li, Q.W. Zhang, and H.Q. Yu.

References

Bremner, J.M. (1965). Inorganic forms f nitrogen. In: C.A. Black et al. (eds.), Methods of Soil Analysis. Part 2. Agron.Monogr.9. ASA and SSSA, Madison, WI, pp. 1179–1237.

Cremers, A., Elsen, A., De Preter, P. and Maes, A. (1988). Quantitative analysis of radio caesium retention in soils. Nature 335:247–249.

Joshi, S.R. (1987). Non-destructive determination of lead-210 and radium-226 in sediments by direct photon analysis. Journal of Radioanalysis and Nuclear Chemistry Articles 116:169–182.

Li, Y. and Lindstrom, M.J. (2001). Evaluating soil quality-soil redistribution relationship on terraces and steep hillslope. Soil Science Society of America journal 65:1500–1508.

Li, Y., Poesen, J., Yang, J.C., Fu, B. and Zhang, J.H. (2003). Evaluating gully erosion using ^{137}Cs and ^{210}Pb/^{137}Cs ratio in a reservoir catchment. Soil and Tillage Research 69:107–115.

Nelson, D.W. and Sommers., L.E. (1982). Total carbon, organic carbon, and organic matter. In: A.L. Page et al. (eds.), Methods of Soil Analysis. Part. 2. 2nd edition. Agron.Monogr.9. ASA and SSSA, Madison, WI, pp. 539–580.

Olsen, C.R., Larsen, I.L., Lowry, P.D. and Cutshall, N.H. (1986). Geochemistry and deposition of 7Be in river-estuarine and coastal water. Journal of Geophysical Research 91:896–908.

Olson, S.R. and Sommers., L.E. (1982). Phosphorus. In: A.L. Page et al. (eds.), Methods of Soil Analysis. Part 2. 2nd edition. Agron. Monogr. 9. ASA and SSSA, Madison, WI, pp. 403–430.

Pennock, D.J., Anderson, D.W. and de Jong., E. (1994). Landscape scale changes in indicators of soil quality due to cultivation in Saskatchewan, Canada. Geoderma 64:1–19.

Robbins, J.A. (1978). Geochemistry and Geophysical application of radioactive lead. In: J.O. Nriagu (ed.), The Biochemistry of Lead in the Environment. Elsever, Amsterdam, pp. 285–393.

Wallbrink, P.J. and Murray, A.S. (1993). The use of fallout radionuclide as indicators of erosion processes. Hydrological Processes 7:297–304.

Wallbrink, P.J. and Murray, A.S. (1994). Fallout of 7Be over south eastern Australia. Journal of Environmental Radioactivity 25:213–228.

Wallbrink, P.J., and Murray, A.S. (1996a). Distribution of 7Be in soils under different surface cover conditions and its potential for describing soil redistribution processes. Water Resources Research 32:467–476.

Zapata, F. (ed.). (2003). The use of environmental radionuclide as tracers in soil erosion and sedimentation investigation: Recent advances and future developments. Soil and Tillage Research 69:3–13.

Chapter 17
Policy Impacts on Land Degradation: Evidence Revealed by Remote Sensing in Western Ordos, China

Weicheng Wu and Eddy De Pauw

Abstract This paper presents a multi-temporal monitoring and assessment of biomass dynamics in response to land cover change in Western Ordos, one of the most important dry areas in China, aiming to reveal the impacts of governmental land management policies on the biomass production of the rangeland ecosystem and on land degradation. Multi-temporal Landsat images (MSS 1978, 1979; TM 1987, 1989, 1991, 2006 and 2007; ETM+ 1999, 2001, 2002, 2004) were used in this research. An integrated processing algorithm, indicator differencing and-thresholding and post-classification differencing, was applied to reveal the land biophysical change and rangeland degradation, and a relevant biomass estimation model was developed for the rangeland ecosystem based on other researchers' work. Meteorological data since the 1960s were incorporated in the analysis to avoid false signals of degradation, as could arise from normal climatic variability. The results show that to some extent land management policies have been instrumental in the protection and recovery of grasslands biomass production. On the other hand, in the non-controlled and weakly monitored zones land degradation, in the form of biomass loss due to desert extension, vegetation degradation, salinisation and water-table decline has continued. This could be attributed to a combination of both natural and human factors, such as lack of protection against strong winds, collective grazing in the permitted rotation areas and previously controlled zones, and over-pumping for agricultural and sand control activities. From this case study, it seems that the effectiveness and rationality of land use policy depend on whether it can coincide with the interests of the local people while conserving the environment. Where there is a conflict between economic viability and environmental sustainability, land degradation is inevitable.

Keywords Biomass dynamics · Land use change and land degradation · Land use policy · Multi-temporal remote sensing · Ordos · China

---

W. Wu (✉)
International Center for Agricultural Research in Dry Areas (ICARDA), Aleppo, Syria
e-mail: w.wu@cgiar.org

17.1 Introduction

Since they can directly and indirectly influence land resources exploitation, creation of market and economic opportunities and the impact of policies on the ecosystem degradation has been increasingly recognized as some of the main social drivers. Although a certain number of researches have dealt with human-environmental interaction by linking land use change revealed by remote sensing with human activity (Serneels and Lambin, 2001; Veldkamp et al., 2001; Verburg et al., 2002; Wu et al., 2002; Wu, 2003b; Xie et al., 2005), few analyses have focused specifically on the impacts of policies on land degradation. The objective of this paper is to conduct a study, taking the Western Ordos Region in China as an example, to monitor the biophysical response of ecosystems to the implementation of different land policies through remote sensing using vegetation indices trajectories and biomass dynamics.

As a part of the Ordos Plateau and bordering the Loess Plateau on the south, the Ordos region is administratively located in Inner Mongolia and adjacent with Shaanxi Province on the southeast and Ningxia Province on the southwest (Fig. 17.1). The region is mainly sandy rangeland interleaved with desert patches

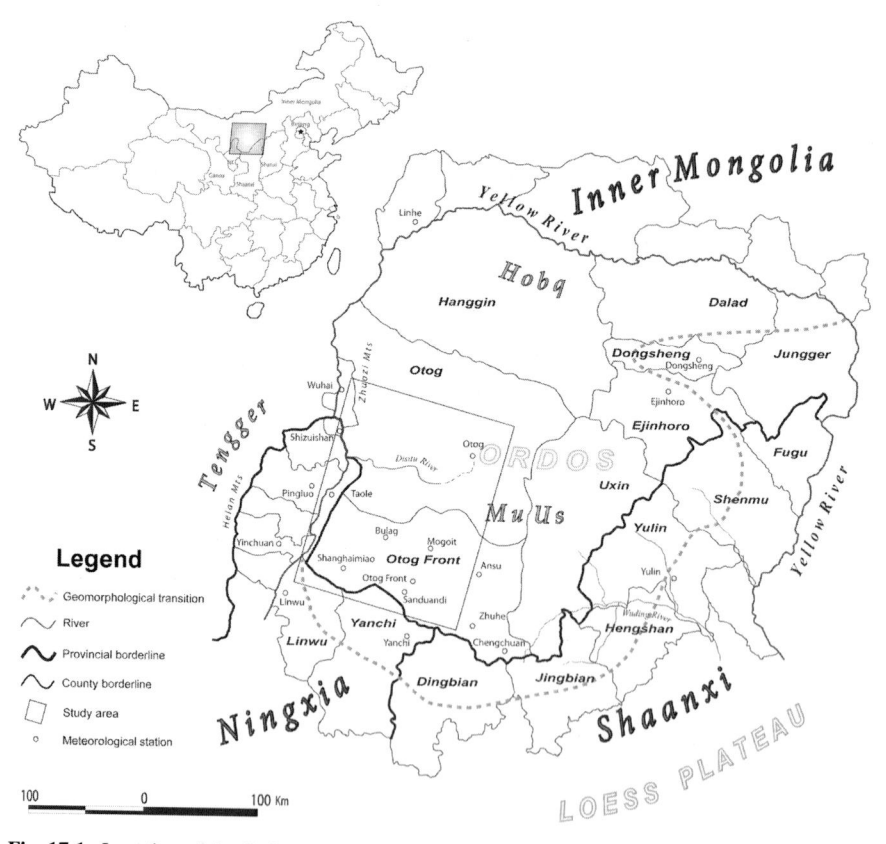

Fig. 17.1 Location of the Ordos Region and its administrative units. Note: The basic administrative unit shown in this figure in Inner Mongolia is Banner, which is equivalent to County in other provinces

17 Policy Impacts on Land Degradation

and locally some pieces of cropland, and thus named the *Mu Us Sandy Land*. The field investigations by Huang and Zhang (2006) identified as the main herbaceous and shrub species in this sandy land *Artemisia ordosica, Stipa bungeana, Juniperus vulgaris, and Caragana intermedia*.

The average annual precipitation is around 279 mm in the Banner Otog, of which 85% is concentrated in the period June–September. The wind blows mainly from the northwest (230 days) and its speed exceeds 17 ms^{-1} during more than 40 days per year. The highly concentrated rainfall and strong winds provoke soil erosion and water loss by runoff. Due to an abundant good-quality coal resource under the Plateau, as well as natural gas and oil reserves, Ordos has become in recent years one of the national energy bases under the mid-to-long term national strategy "To Develop the West". The fossil fuel exploitation driven by this development, combined with long-time human activities in grazing, deforestation and land reclamation for agriculture driven by a number of different local and national policies, together with medicinal herbs and fuel wood collection, have caused significant land use change and land degradation (Jiang et al., 1995; Zhang and Wang, 2001; Wu et al., 2005; Xu, 2006).

Among hundreds of national and local government policies, those related to, or having influenced land use and management in Western Ordos are listed in Table 17.1. Going back to their implementation dates, multi-temporal satellite images were acquired for revealing land cover change and land degradation and projecting biomass dynamics in time to understand the biophysical response of the rangeland ecosystem to the implementation of these policies. The images dated

Table 17.1 Multi-temporal satellite images used in this study

Captors	Scene	Acquisition dates	Spatial resolution	Mean haze	Policy implementation period
Landsat 5 TM	Path-Row: 129-33	2007 Aug 10	30 m	28.90	Period 3: 2000–2001, "Herbs collection forbidden" and "Grazing-forbidden and -rotation policy"
Landsat 5 TM		2007 July 09	30 m	30.07	
Landsat 5 TM		2006 Aug 07	30 m	35.28	
Landsat 7 ETM+		2004 Aug 25	30 m	13.18	
Landsat 7 ETM+		2002 Aug 20	30 m	13.78	
Landsat 7 ETM+		1999 Aug 12	30 m	31.06	
Landsat 5 TM		1991 Aug 30	30 m	27.87	Period 2: 1987–1988, Deng's "Open and reform" and "Legalization of the private economy"
Landsat 5 TM		1989 Sep 17	30 m	38.11	
Landsat 5 TM		1987 Sep 20	30 m	28.24	
Landsat 3 MSS		1979 Oct 09	56 m		
Landsat 3 MSS		1978 Aug 21	56 m	13.13	Period 1: 1979–1985, Nationwide implementation of Deng's "Household land tenure policy" and issues of the Decree of Grassland

Note: the haze values in digital count (DC) derived from the 4th Tasseled Cap feature are used for atmospheric correction.

1978, 1987, 1999 are considered to represent the initial state of the land at the beginning of each policy implementation period, and those of 2007 as representing the current state (see Table 17.1).

17.2 Data and Methods

17.2.1 Data

Multi-temporal Landsat images and the initial implementation dates of different policies concerning land use and management were compiled (Table 17.1) as well as meteorological data, especially monthly and annual rainfall from 1960 to 2007 (station locations shown in Fig. 17.1).

17.2.2 Method for Biophysical Change Extraction

Among a number of available change detection approaches, the post-classification differencing (Wu, 2008) and indicator differencing-and-thresholding algorithms were selected. Image pre-processing included image-to-image rectification (RMS error of 0.23–0.58 pixels), atmospheric correction using the COST model (Chavez, 1996; Wu, 2003b), transformation of the Enhanced Vegetation Index (EVI) developed by Huete et al. (1994) and the Normalized Difference Vegetation Index (NDVI) proposed by Rouse (1973) and Tucker (1979). After pre-processing a differencing and thresholding technique was applied to the EVI for the periods 1987–1999 and 1999–2007, and a post-classification differencing for the period 1978–1987 (overall classification accuracy >95%). For more details on this change detection technique is referred to (Wu et al., 2008). Here the emphasis is laid on the approaches for grassland biomass estimation.

17.2.3 Biomass Estimation Models

To project the biomass dynamics in response to land use change and land degradation in time, it is necessary to build up first biomass estimation models for the corresponding land use/cover type, in this case, rangeland interleaved with desert patches and croplands. In order to select an appropriate model, a comprehensive review was undertaken of the available estimation approaches, which is summarized in the following paragraphs.

Since 1980s a number of researchers have undertaken remote sensing-based biomass estimation for rangeland, grassland and savannah in different regions, such as Sahelian Africa (Tucker et al., 1983, 1985; Devineau et al., 1986; Justice and Hiernaux, 1986; Prince and Tucker, 1986, Diallo et al., 1991; Prince, 1991; Wylie et al., 1991 and 1995; Bénié et al., 2005), the grasslands in North America (Everitt et al., 1989; Merrill et al., 1993; Todd et al., 1998; Reeves, 2001; Reeves et al., 2001; Wylie et al., 2002; Butterfield and Malmstrom, 2004), in South America (Flombaum and Sala, 2007), in Inner Mongolia in China, and in Mongolia (Xiao et al., 1997, Kawamura et al., 2003; Akiyama et al., 2005; Kawamura et al., 2005; Akiyama and Kawamura, 2007; Ichiroku et al., 2008).

Tucker et al. (1985) found a strong correlation between the integrated satellite data (e.g. ΣNDVI) of the growing season and end-of-season aboveground herbaceous biomass for the Western Sahelian region where tree cover is less than 10%. Based on field measurements Devineau et al. (1986) obtained a non-linear relationship between herbaceous grassland biomass and NDVI, in the form B = 0.00216 NDVI$^{1.7}$ (t ha^{-1}, $R^2 = 0.927$), where NDVI is in fact NDVI × 100. Bénié et al. (2005) applied this equation to investigate the spatial-temporal dynamics of the herbaceous biomass in Burkina Faso. Buerkert et al. (1995) noticed that weed biomass is linearly correlated with NDVI in Niger in the form of B = 1.0417NDVI-0.2177 (t ha^{-1}, $R^2 = 0.777$).

Todd et al. (1998) used Tasseled Cap features (Greenness, Brightness and Wetness), NDVI and TM band 3 (Red) to investigate the above-ground biomass of the short-grass steppe of Eastern Colorado. They found that all of these indicators are well correlated to standing biomass ($R^2 = 0.62$–0.67) for grazed grassland but that the Red indicator is more responsive ($R^2 = 0.70$) than other indicators for ungrazed grassland. Frank and Karn (2003) obtained non-linear relationships between grassland biomass and NDVI in the Northern Great Plain in the form of B = 2.698 + 3,709.449 NDVI^3 (kg ha^{-1}, $R^2 = 0.83$, $p \leq 0.05$) and found that NDVI has good potential for use in predicting biomass and canopy CO_2 flux rates for grassland. Butterfield and Malmstrom (2004) also reported that NDVI was strongly correlated with above-ground green biomass of grasslands in California throughout the growing season ($R^2 = 0.78$) and that a single NDVI-biomass function may be applied to the grasslands up to the period of peak greenness.

While quantifying herbaceous biomass in rangeland ecosystems in western North Dakota, Reeves (2001) worked out the relationship between biomass and NDVI, resulting into the equation $B = NDVI(65.0112) + \left(\sum P(0.9) - \left(\sum Th\right)^2 (0.0013)\right)$, where B is the estimated biomass within each Thiessen polygon, NDVI is the average NDVI for a given polygon, $\sum P$ is the summation of precipitation from 1 January to the date of ground sampling, and $\sum Th$ is the summation of thermal time (TAVGdaily – 0) from 1 January to the date of ground sampling where TAVGdaily is the daily average temperature. Reeves (2001) considered that the relationship between grassland biomass and LAI is strongest when the biomass is at its peak in July ($R^2 = 0.78$).

Yu et al. (2004) analysed the relationship between above-ground net primary production and annual rainfall in Inner Mongolia, China and found that peak aboveground biomass (PAB) is positively correlated with the annual rainfall (PAB = 0.5515Rainfall-25.631, $R^2 = 0.684$), in which the slope exceeds those obtained from other dry regions in Africa and South America implying a higher rain-use efficiency in Inner Mongolia. Kawamura et al. (2003) conducted biomass estimation in the same area employing AVHRR NDVI and later these authors (Kawamura et al., 2005) used MODIS vegetation indices to undertake a similar study by establishing both linear and exponential relationships. For the live and total biomass (including live and dead), these relationships are summarized in the regression equations of Table 17.2. It is clear from their study that for both live and total biomass MODIS NDVI and EVI are of higher predictive power than the same indices derived from AVHRR, and that NDVI gives a better result than EVI. Akiyama et al. (2005) obtained a similar level of correlation between the live biomass and MODIS EVI in the same region, with equation B = 18.722 exp (5.698EVI) (R^2 = 0.744).

Table 17.2 Regression analysis results between biomass and vegetation indices (After Kawamura et al., 2005)

Coefficient of determination	Explanatory variable	Function type	a	b	R^2	Average error (g m^{-2})[a]
Total biomass (dry matter g m^{-2}) n = 30	AVHRR-NDVI	Linear	−58.23	571.37	0.53	± 54.24
		Exponential	11.13	6.07	0.64	± 36.65
	MODIS-NDVI	Linear	−160.02	628.08	0.75	± 40.08
		Exponential	16.31	4.26	0.83	± 33.16
	MODIS-EVI	Linear	−87.01	797.67	0.69	± 44.24
		Exponential	24.56	5.71	0.77	± 38.28
Live biomass (dry matter in g m^{-2}) n = 30	AVHRR-NDVI	Linear	−47.74	454.43	0.54	± 42.97
		Exponential	11.13	6.07	0.64	± 36.65
	MODIS-NDVI	Linear	−127.00	495.98	0.74	± 32.31
		Exponential	12.15	4.36	0.83	± 25.70
	MODIS-EVI	Linear	−73.39	644.10	0.71	± 33.75
		Exponential	18.35	5.86	0.80	± 28.48

Note: Linear-type: B = a + bX; Exponential-type: B = a∗Exp (bX); [a]Average error is calculated from the original "error sum of squares" by Kawamura et al. (2005).

Among the above mentioned models, the exponential one derived from MODIS NDVI for total biomass, by Kawamura et al. (2005) in Inner Mongolia, B = 16.31 exp(4.26∗NDVI) (R^2 = 0.83) produced the best fit with the field biomass data measured in Ordos: (1) desert steppe for the period 2002–2005 (9.5–175.1 g m^{-2} with a mean of 56.6 g m^{-2}) by Ma et al. (2008); (2) shrub-grassland within a range of 28–236 g m^{-2} in Mu Us Sandy Land in 2006 by Cheng et al. (2007); (3) 50–100 g m^{-2} in the Banners of Otog and Otog Front in the period July–August 2007 by the local government[1]; and 68–195 g m^{-2} (mainly 124–140 g m^{-2}) in the enclosed grasslands in the north Yanchi near Sanduandi of Otog Front by Shen et al. (2007). In addition, Hu et al. (2007) investigated the spatio-temporal dynamics of aboveground net primary productivity (ANNP) in Inner Mongolia and reported the ANNP varying from 28.53 to 157.78 g m^{-2} a in the western part of the Mu Us Sandy Land. Hence this model was selected for estimating the rangeland biomass for all other observation years.

---

[1]Ordos Weather Bureau, 2008: http://www.imwb.gov.cn/qxinfo/stinfo/200803/773.html

17.3 Results

The land degradation detection and multi-temporal biomass estimation results are shown in Figs. 17.2 and 17.3. Several types of degradation were observed. The first one is the southeasterly expansion of desert patches, especially in the non-controlled zones, at a rate of 11–21 m year^{-1}. In more detail, deserts and small patches of sand dunes expanded depending on location by 120–240 m, 90–180 m, and 60–150 m, or 240–570 m in total, in the periods 1978–1987, 1987–1999 and 1999–2007 (Fig. 17.2) and swallowed the grassland on their southeast margins where there were not enough shrubs (e.g., *Caragana Korshinskii Kom* and *Salix gracilior*) to block sand movement. This extension is a result of wind blowing from northwest, which occurs about 230 days per year. The second type is vegetation

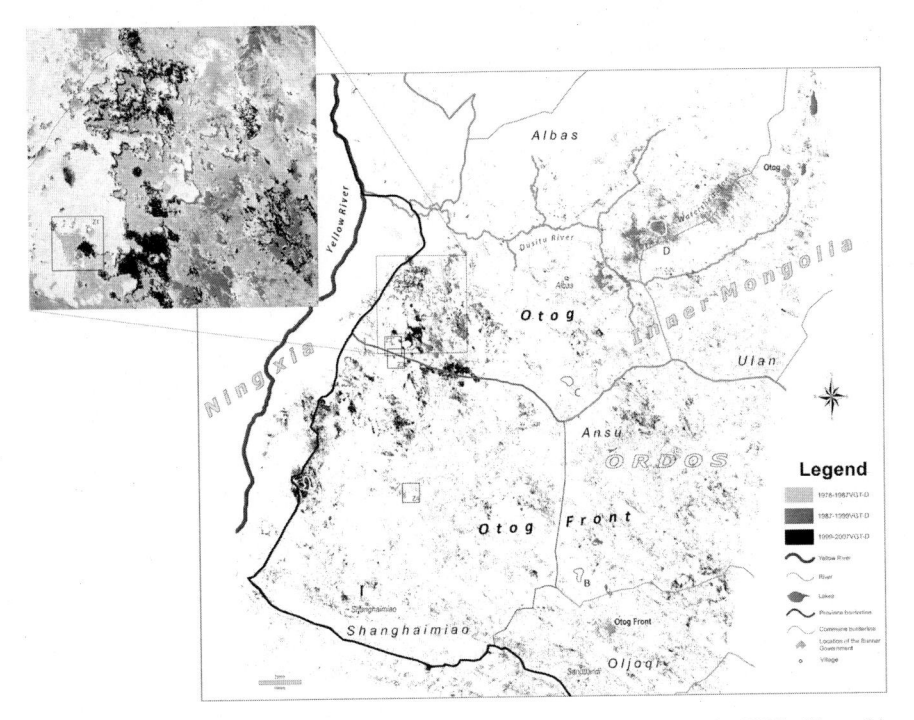

Fig. 17.2 Land degradation in the Western Ordos (modified from Wu et al., 2008). Note: this Figure shows (1) desert patches extending to the southeast (see the *up-left* zoom) along the dominant wind direction from NW to SE and (2) grassland in degradation in the observed periods 1978–1987, 1987–1999 and 1999–2007. Sites A, B, C and D were selected to check and calibrate the relationship between the biophysical feature changes (e.g., NDVI and biomass) and the annual rainfall variation. According to our previous work (Wu et al., 2008), Site A experienced degradation in 1978–1989 but controlled in 1991–1999 and again degradation after 2002; Sites B and C are protected or enclosed areas from grazing, no evident degradation was observed; site D suffered degradation in 1987–1999 but recovered after 2002. Zooms Z1, Z2 and Z3 are examples showing vegetation cover degradation around water points/settlements and Z4 a recovery (see Wu et al., 2008 for detail)

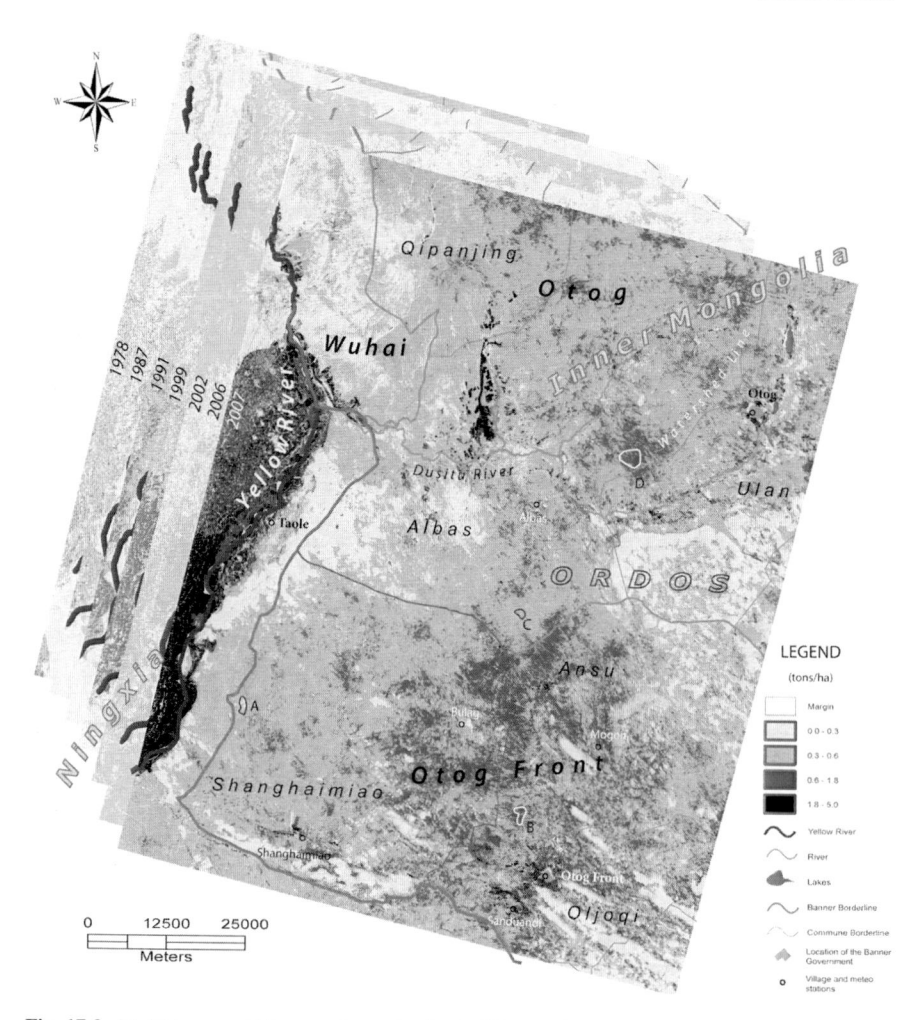

Fig. 17.3 Multi-temporal biomass dynamics in the Western Ordos

degradation around water ponds/settlements (with patch diameter varying from 300 to 1,600 m) and in some permitted rotation grassland, of which a part had previously been controlled. This kind of degradation is not stationary: in one period it was observed in one place, and in another period in another place.

While discerning vegetation degradation in some places, we also observed significant increase in vegetation vigor and cover (see Wu et al., 2008), especially in the recent decade. This greening trend is attributed to the conversion from grassland to agricultural land (including farmland, plantations of economic plants such as ephedra, licorice, etc.), and from natural grassland into pasture/forage land irrigated with underground water. Sand control in the sandy land and desert patches, by planting grasses, shrubs and trees in a grid pattern, has also increased the greenness of land cover. This practice has to some extent restored a number of degraded

17 Policy Impacts on Land Degradation

patches around water ponds/settlements, although the greenness in these patches has not yet reached the same level as the surrounding grassland after 20–30 years recovery.

In order to uncover the spatio-temporal variability of the biophysical features related to these land cover changes, and to understand the importance of human intervention in provoking land degradation, four typical sites, marked A, B, C and D (see Figs. 17.2 and 17.3 for locations), were extracted for checking biomass dynamics against annual rainfall (Fig. 17.4). Of these, sites A and D suffered degradation, whereas B and C experienced no significant change (see note of Fig. 17.2).

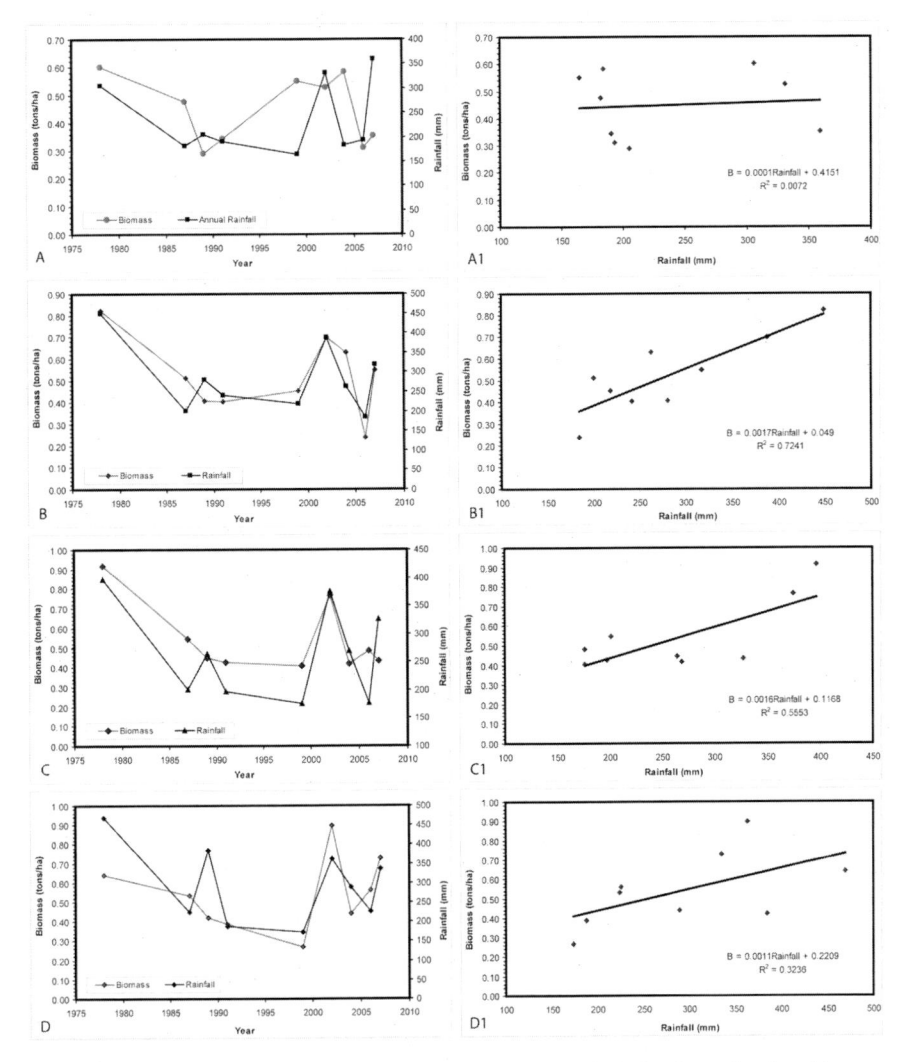

Fig. 17.4 Temporal biomass dynamics in the observed sites. Note: *Left side* graphs show biomass dynamics against annual rainfall in the Sites A, B, C and D, *right side* ones reveal their relationships

It was found that the average biomass production sensed by satellites in the sites A and D is not well correlated to their annual rainfall (A1 and D1 in Fig. 17.4, $R^2 = 0.007$–0.324). However, in the enclosed or protected sites B and C a strong positive relationship exists between biomass density and the annual rainfall (B1 and C1 in Fig. 17.4) with R^2 values of 0.555–0.724, which are close to the correlation ($R^2 = 0.684$) obtained by Yu et al. (2004) in Inner Mongolia.

17.4 Discussion and Conclusions

Land degradation as influenced by unfavourable land cover changes, is a complicated phenomenon related to both natural and human factors. After weighing the importance of the two groups of factors, Zhu and Liu (1989) concluded that the anthropogenic factors account for 94.5% of the responsibility in provoking desertification in China. It is thus essential to analyse the impacts of human activities on the environmental change and land degradation.

Heretofore, recognized human factors include overgrazing, population growth, land reclamation, institutional weaknesses, irrational policies, land tenure, market economy, water overuse, over-collection of fuelwood, over-excavation of wild medicinal and edible herbs, exploitation of fossil fuel (coal, oil and gas), over-hunting, culture and lack of education (Jiang et al., 1995; Erdunzhav, 2002; Enkhee, 2003; Wu, 2003a; Gai, 2007). However, not all of these socio-economic and cultural factors had the same importance in effecting land degradation in history. Policies are the underlying forces driving other kinds of socio-economic activities (proximate causes) which directly lead to land use change, development of market economy and new enterprises, and exploitation of natural resources in Ordos (Wu et al., 2008).

In the past centuries, land reclamation from grassland for agriculture was, despite unsuccessful outcomes, undertaken again and again, driven by different national policies such as "Consolidating the frontier with immigrants from the interior of the country for reclamation" in Dynasty Qing and "Giving prominence to agricultural food production" in the period 1956–1974. Of little productivity under the semi-arid and arid climate conditions, rainfed cropland was, after 2–3 years of use (Enkhee, 2003; Wu, 2003a), often abandoned and exposed to soil erosion and desertification. New land reclamation was conducted elsewhere for food production. These practices constituted a vicious cycle "Reclamation-Cultivation-Abandonment-Reclamation" leading to land degradation in the rangelands (Wu, 2003a). As Enkhee (2003) analysed, land reclamation, regardless of natural conditions, might have been the major cause producing the initial state of the Hobq Desert in the north and the desert patches in the Mu Us Sandy Land (see the up-left zoom in Fig. 17.2 and the patches with biomass density < 0.3 t ha^{-1} in Fig. 17.4) in the south in Ordos since the Dynasty Tang (A.D. 618–907). Unfortunately, neither remote sensing images nor maps are available to back up this kind of historical land degradation analysis.

In the recent decades the impacts of policies can be analysed in a more tangible way. In the period 1979–1984, Deng's policy "Household responsibility for agricultural production" and the promulgation of the "Decree of Grassland" in 1985 had greatly aroused the enthusiasm of peasants and raised the agricultural production, but left the grassland in a situation "collective grassland and private cattle" which lasts up to today. To gain more personal profits and income, each herdsman had an incentive to raise as many animals as possible on the public land, inevitably leading to a Chinese variant of "the Tragedy of the Commons" (Hardin, 1968) and the "institutional defect" (Erdunzhav, 2002), which is the direct consequence of "indefinite land property" (Gai, 2007). During this period, land degradation, despite its localized character, occurred throughout the study area (1978–1987 VGT-D in Fig. 17.2).

In 1987–1988, under the development strategy "Invigorating the domestic economy and opening to the outside world", Deng's "Open and reform" policy, and the decree on "Legalization of the private economy", hundreds of rural enterprises and companies were established in Ordos, based on the region's agricultural and pastoral products like food, wool and natural resources (coal, oil, gas, medicinal and edible wild herbs). Widespread collection of herbs for providing materials to these enterprises and for increasing family income induced local people to overturn the fragile sandy soils in search for licorice roots (*Glycyrrhiza uralensis*) and *Nostoc commune* var. *Flagelliforme,* leading to a large reduction in biomass production in some areas and land degradation (e.g., see 1987–1999VGT-D and Site D in Fig. 17.2 and Fig. 17.4d). One bright spot during this period was the spontaneous establishment of an ecological construction enterprise – a non-governmental sand-control team – composed of the local peasants and shepherds. The objective of this team was to combat desertification by planting ephedra, licorice, *Hedysarumleave, Caragana korshinski, Artemisia sphaerocephala* and *Artemisia ordosica*, sea-buckthorn, etc., in a grid pattern to restore the degraded land and protect the sandy land from degeneration, and simultaneously bring economic benefits for the local people. No doubt this activity has greatly produced positive impacts on the environment, as evidenced by the biomass increase in the period of 1989–1999 in Site A (Fig. 17.4a).

After 1999, with the inauguration of the national middle-to-long term strategy "To Develop the West" in 1999, Ordos has become one of the National Energy Bases, thanks to its abundant fossil fuel resources. With the exploitation of coal, oil and natural gas and overuse of water in mining and agriculture, new forms of land degradation took place, particularly oil pollution, cropland destruction, water-table decline.[2] Aware of this serious land degradation, the central government promulgated a national order to "Forbid collection of herbs in grasslands" in 2000. Complementing this national policy, the local governments of Otog and Otog Front implemented a "Grazing-forbidden and -rotation policy with a subsidy system" in 2001 to treat grazing differently in different zones. The policies were implemented by closure of large pieces of grassland for recovery and by conversion of parts of

---

[2] http://yudefu186.bokee.com/viewdiary.15081810.html

highly productive grassland into pastures, cultivated with some aridity- and cold-resistant forage grasses such as alfalfa (*Medicago sativa*) and *Astragalus adsurgens* for animal breeding. As a result the previously open grazing became indoor dry-lot feeding. With the added boost of favourable rains biomass has since increased (e.g., Fig. 17.4). An interesting fact is that with the implementation of these policies, not only the vegetation vigour and biomass but also the cattle numbers have increased. This has led to an improvement of household income of the local peoples and their livelihood. The average per capita income of the rural people has increased by 60.9 and 119.1% respectively in the Banners Otog and Otog Front from 2000 to 2005 (Wu et al., 2008). However in areas where communal grazing was permitted or in the protected zones where surveillance was less effective, the grasslands have suffered even more grave destruction due to overgrazing (1999–2007 VGT-D in Fig. 17.2 and Fig. 17.4a), although the annual rainfall was normal in 2007.

In summary, the impacts of policy are complex and often entail positive and negative aspects, in terms of whether the policy can bring profits to the herdsmen and farmers while protecting grassland from degradation. The evidence from Ordos suggests that if there is a contradiction, land degradation is unavoidable. The impact assessment should be dialectically conducted from multiple dimensions, although it is difficult to unravel the effects of overlapping policies.

This study attempted to assess the impacts of policy from a viewpoint of biophysical change, as revealed by remote sensing in the Western Ordos rangeland. Despite its immaturity this technology provides interesting possibilities to look into the interaction between changes in socio-economic activities, especially those that are policy-driven, and the ecological system. In the absence of human intervention, biomass productivity is completely associated with the natural conditions, such as rainfall, temperature and radiation. When there is human intervention, this productivity is not any more related only to natural conditions but also to land use practices and exploitation of land resources. This difference makes it possible to discern the contribution of human impact on the rangeland productivity. This is one of the advantages that remote sensing technology has brought us. Another fact revealed in this case study is that it is possible for decision-makers to work out sustainable grazing and rangeland use policy by controlling grazing intensity through carrying capacity analysis based on biomass productivity estimation.

Land degradation is not an irreversible biophysical degeneration. It is produced by implementation of unwise policies (e.g., land reclamation in the early of 1970s Zhang and Wang, 2001; Enkhee, 2003; Wu, 2003a) or by exploitative activities under rational policies (e.g., overgrazing in the permitted areas under the "grazing forbidden and rotation" policy; overuse of underground water for planting trees on the sand dunes to combat desertification). At the same time land degradation can also be mitigated and reversed by reasonable execution of rational policies. For example, the legalization of the private economy in 1988 not only promoted land degradation but also gave rise to the sand control enterprise in the region, which contributed significantly to the ecological recovery of the region. Similarly the "grazing forbidden" policy implemented with subsidies to herdsmen has truly restored and protected some grassland.

References

Akiyama, T. and Kawamura, K. (2007). Grassland degradation in China: Methods of monitoring, management and restoration. Grassland Science 53:1–17.

Akiyama, T., Kawamura, K., Yokota, H. and Chen, Z. (2005). Real-time monitoring of grass and animal for steppe management in Inner Mongolia, China. Proceedings of the AARS (Asian Association of Remote Sensing) 2005 Conference. Mongolia, China. Available at: http://www.aars-acrs.org/acrs/proceeding/ACRS2005/Papers/EEC-1.pdf.

Buerkert, A., Lawrence, P.R., Williams, J.H. and Marschner, H. (1995). Non-destructive measurements of biomass in millet, cowpea, groundnut, weeds and grass swards using reflectance, and their application for growth analysis. Experimental Agriculture 31:1–11.

Butterfield, H. and Malmstrom, C. (2004). Phenological changes in the relationship between NDVI and aboveground biomass in California annual grasslands. Proceedings of Ecological Society of America (ESA) 2004 Annual Meeting. Portland, Oregon, USA (abstract available at: http://abstracts.co.allenpress.com/pweb/esa2004/document/37735).

Bénié, G.B., Kaboré, S.S., Goïta, K. and Courel, M-F. (2005). Remote sensing-based spatio-temporal modeling to predict biomass in Sahelian grazing ecosystem. Ecological Modelling 184:341–354.

Chavez, P.S., Jr. (1996). Image-based atmospheric correction – revisited and improved. Photogrammetric Engineering and Remote Sensing 62:1025–1036.

Cheng, X., An, S., Chen, J., Li, B., Liu, Y. and Liu, S. (2007). Spatial relationships among species, above-ground biomass, N, and P in degraded grasslands in Ordos Plateau, north-western China. Journal of Arid Environments 68:652–667.

Devineau, J.L., Fournier, A. and Lamachere, J.M. (1986). Programme d'évaluation préliminaire SPOT. PEPS No. 149 – SPOT Oursi. Centre ORSTOM de Ouagadougou, 52 pp.

Diallo, O., Diouf, A., Hanan, N.P., Ndiaye, A. and Prevost, Y. (1991). AVHRR monitoring of savanna primary productivity in Senegal, West Africa: 1987–1988. International Journal of Remote Sensing 12:1259–1279.

Enkhee, J., (2003). A historical rethink about the grassland desertification: The cultural dimension of development, Friends of Nature, No.7 (in Chinese, available at: http://old.fon.org.cn/enl/content.php?id=64).

Erdunzhav. (2002). Reflections on institutional deficiency that account for grassland desertification. Journal of Inner Mongolia University (Humanities and Social Sciences) 34:8–12.

Everitt, J.H., Escobar, D.E. and Richardson, A.J. (1989). Estimating grassland phytomass production with near-infrared and mid-infrared spectral variables. Remote Sensing of Environment 30:257–261.

Flombaum, P. and Sala, O.E. (2007). A non-destructive and rapid method to estimate biomass and aboveground net primary production in arid environments. Journal of Arid Environments 69:352–358.

Frank, A.B. and Karn, J.F. (2003). Carbon dioxide flux, biomass, and radiometric reflectance of northern Great Plains Grasslands. Journal of Range Management 56:382–387.

Gai, Z. (2007). Grassland property right and its eco-environmental protection, Globalization Forum (in Chinese, available at: http://www.china-review.com/gath.asp?id=18883).

Hardin, G. (1968). The tragedy of the commons. Science 162:1243–1248.

Hu, Z., Fan, J., Zhong, H. and Yu, G. (2007). Spatiotemporal dynamics of aboveground primary productivity along a precipitation gradient in Chinese temperate grassland, Science in China, Series D. Earth Sciences 50:754–764.

Huang, Y. and Zhang, M. (2006). Temporal and spatial changes of plant community diversity on the Ordos Plateau. Biodiversity Science 14:13–20.

Huete, A.R., Justice, C. and Liu, H.Q. (1994). Development of vegetation and soil indices for MODIS-EOS. Remote Sensing of Environment 49:224–234.

Ichiroku, H., Kawada, K., Kurosu, M., Batjargal, A., Tsundeekhuu, T. and Nakamura, T. (2008). Grazing effects on Floristic composition and above ground plant biomass of the grasslands in the Northeastern Mongolian Steppes. Journal of Ecology and Field Biology 31:pp. 115–123.

Jiang, H., Zhang, P., Zheng, D. and Wang, F., (1995). The Ordos plateau of China. In: J.X. Kasperson, R. E.Kasperson and B. L. Turner II (eds.), Regions at Risk: Comparisons of Threatened Environments. United Nations University Press, Tokyo (available at: http://www.unu.edu/unupress/unupbooks/uu14re/uu14re11.htm).

Justice, C.O. and Hiernaux, P.H.Y. (1986). Monitoring the grasslands of the Sahel using NOAA AVHRR data: Niger 1983. International Journal of Remote Sensing 7:1475–1497.

Kawamura, K., Akiyama, T., Watanabe, O., Hasegawa, H., Zhang, F.P., Yokota, H. and Wang, S. (2003). Estimation of aboveground biomass in Xilingol Steppe, Inner Mongolia using NOAA/NDVI. Grassland Science 49:1–9.

Kawamura, K., Akiyama, T., Yokota, H., Tsutsumi, M., Yasuda, T., Watanabe, O. and Wang, S. (2005). Comparing MODIS vegetation indices with AVHRR NDVI for monitoring the forage quantity and quality in Inner Mongolia grassland, China. Grassland Science 51:33–44.

Ma, W., Yang, Y., He, J., Zeng, H. and Fang, J. (2008). Temperate grassland biomass and its relation with environmental factors in Inner Mongolia (in Chinese), Chinese Science (C). Life Science 38:84–92.

Merrill, E.H., Bramble-Brodahl, M.K., Marrs, R.W. and Boyce, M.S. (1993). Estimation of green herbaceous phytomass from Landsat MSS data in Yellowstone National Park. Journal of Range Management 46:151–157.

Prince, S.D. (1991). Satellite remote sensing of primary production: Comparison of results for Sahelian grasslands 1981–1988. International Journal of Remote Sensing 12:1301–1311.

Prince, S.D. and Tucker, C.J. (1986). Satellite remote sensing of rangelands in Botswana II: NOAA AVHRR and herbaceous vegetation. International Journal of Remote Sensing 7:1555–1570.

Reeves, M.C., (2001). Quantifying Herbaceous Biomass in a Rangeland Ecosystem Using MODIS Land Products (available at: http://www.ntsg.umt.edu/projects/rangeland).

Reeves, M.C., Winslow, J.C. and Running, S.W. (2001). Mapping weekly rangeland vegetation productivity using MODIS algorithms. Journal of Range Management 54:A90–A105.

Rouse, J.W., Haas, R.H., Schell, J.A. and Deering, D.W. (1973). Monitoring vegetation systems in the Great Plains with ERTS', Third ERTS Symposium, NASA SP-351 I, pp. 309–317.

Serneels, S. and Lambin, E.F. (2001). Proximate cause of land-use change in Narok District, Kenya: A spatial statistical model. Agriculture, Ecosystem and Environment 85:65–81.

Shen, Y., Zhang, K., Du, L. and Qiao, F. (2007). Impacts of exclusion region on vegetation feature and diversity in Yanchi County, Ningxia. Ecology and Environment 16:1481–1484.

Todd, S.W., Hoffer, R.M. and Milchunas, D.G. (1998). Biomass estimation on grazed and ungrazed rangelands using spectral indices. International Journal of Remote Sensing 19:427–438.

Tucker, C.J. (1979). Red and photographic infrared linear combinations for monitoring vegetation. Remote Sensing of Environment 8:127–150.

Tucker, C.J., Vanparet, C.L., Boerwinkel, E. and Gaston, A. (1983). Satellite remote sensing of total dry matter production in the Senegalese Sahel. Remote Sensing of Environment 13:461–474.

Tucker, C.J., Vanpraet, C.L., Sharman, M.J. and Van Ittersum, G. (1985). Satellite remote sensing of total herbaceous biomass production in the Senegalese Sahel: 1980–1984. Remote Sensing of Environment 17:232–249.

Veldkamp, A., Verburg, P.H., Kok, K., De Koning, G.H.J., Priess, J. and Bergsma, A.R. (2001). The need for scale sensitive approaches in spatially explicit land use change modeling. Environmental Modeling and Assessment 6:111–121.

Verburg, P.H., Veldkamp, W.S.A. et al. (2002). Modelling the spatial dynamics of regional land use: The CLUE-S model. Environmental Management 30:391–405.

Wu, W. (2003a). Evaluation on land use and land cover changes in north Shaanxi, China. Photo-Interpretation 36:15–29.

Wu, W. (2003b). Application de la géomatique au suivi de la dynamique de l'environnement en zones arides, PhD dissertation, Université de Paris 1, France, p. 217.

Wu, W. (2008). Monitoring land degradation in drylands by remote sensing. In: A. Marini and M. Talbi (eds.), Desertification and Risk Analysis Using High and Medium Resolution Satellite Data. Springer, Berlin, pp. 157–169.

Wu, W., De Pauw, E. and Zucca, C. (2008). Land degradation monitoring in the West Mu Us, China. Proceedings of ISPRS 2008, Part B 8:847–858.

Wu, W., Lambin, E.F. and Courel, M-F. (2002). Land use and cover change detection and modeling for North Ningxia, China. Proceedings of Map Asia 2002, Bangkok, Thailand, Aug.6–9, 2002 (available at: http://www.gisdevelopment.net/application/environment/overview/envo0008.htm).

Wu, W., Zucca, C. and Enne, G. (2005). Land degradation monitoring in the Ordos region, China. Proceedings of the International Conference on Remote Sensing and Geoinformation Processing in the Assessment and Monitoring of Land Degradation and Desertification (RGLDD), Trier, Germany, pp. 618–625.

Wylie, B.K., Dendra, I., Piper, R.D., Harrington, J.A., Reed, B.C. and Southward, G.M. (1995). Satellite-Based herbaceous biomass estimates in the pastoral zone of Niger. Journal of Range Management 48:159–164.

Wylie, B.K., Harrington, J.A., Jr., Prince, S.D. and Denda, I. (1991). Satellite and ground-based pasture production assessment in Niger: 1986–1988. International Journal of Remote Sensing 12:1281–1300.

Wylie, B.K., Meyer, D.J., Tieszen, L.L. and Mannel, S. (2002). Satellite mapping of surface biophysical parameters at the biome scale over the North American grasslands, A case study. Remote Sensing of Environment 79:266–278.

Xiao, X., Ojima, D.S., Ennis, C.A., Schimel, D.S. and Chen, Z.Z. (1997). Estimation of aboveground biomass of the Xilin River Basin, Inner Mongolia using Landsat TM imagery. In: Inner Mongolia Grassland Ecosystem Research Station, Chinese Academy of Sciences (ed.), Research on Grassland Ecosystem 5, Science Press, Beijing, pp. 130–138.

Xie, Y., Mei, Y., Tian, G. and Xing, X. (2005). Socio-economic driving forces of arable land conversion: A case study of Wuxian City, China. Global Environmental Change (Part A) 15:238–252.

Xu, J. (2006). Sand-dust storms in and around the Ordos Plateau of China as influenced by land use change and desertification. CATENA 65:279–284.

Yu, M., Ellis, J.E. and Epstein, H.E. (2004). Regional analysis of climate, primary production, and livestock density in inner Mongolia. Journal of Environmental Quality 33:1675–1681.

Zhang, F. and Wang, L. (2001). Analysis on land use situation in the Banner Ejinhoro, Research Report of UNDP Project: Construction of the capacity to implement the UNCCD in China (available at: http://nic6.forestry.ac.cn/sts/tdly/tdly.html).

Zhu, Z. and Liu, S. (1989). Desertification and Its Control in China (in Chinese). Science Press, Beijing, China.

Chapter 18
Assessment of Land Degradation and Its Impacts on Land Resources of Sivagangai Block, Tamil Nadu, India

A. Natarajan, M. Janakiraman, S. Manoharan, K.S. Anil Kumar, S. Vadivelu, and Dipak Sarkar

Abstract A detailed cadastral level survey of land resources occurring in Sivagangai block of Tamil Nadu state, India, covering an area of about 44,600 ha, was carried out during the period 2006–2007. Based on this, 18 soil series were identified and 103 phases mapped at 1:12,500 scale. The study revealed severe sheet erosion on the uplands, heavy siltation of tanks and development of salinity/sodicity at the lowlands as major causes for the drastic decline in productivity. The study warrants systematic and timely efforts to arrest soil erosion on the uplands, proper maintenance of tanks to increase the storage capacity and recharge of the aquifers and providing drainage facilities to reclaim the lowlands and prevent the development of salinity/sodicity in the study area.

Keywords Detailed soil survey · Land degradation assessment · Salinity development · Sheet erosion · Soil series

18.1 Introduction

Tamil Nadu is the southernmost state of India. It is an agrarian state with more than 60% of its people still depending on agriculture for their livelihood. Out of the total geographical area of 13 million hectares, only about 50% is available for cultivation. This limited cultivable area is under severe strain due to increasing population pressure and competing demands of various land uses. Because of this, there is significant diversion of farmlands and water resources for non-agricultural purposes. Further, degradation due to soil erosion, salinity/alkalinity, water logging and

A. Natarajan (✉)
National Bureau of Soil Survey and Land Use Planning, Regional Centre, Bangalore 560024, India
e-mail: athiannannatarajan@gmail.com

depletion of nutrients has already affected about 6 million hectares of land in the state (Natarajan et al., 1997).

The degradation and diversion is continuing every year without any check and needs to be corrected urgently to maintain the sustainability of the ecosystem. For this, a thorough understanding of the factors and processes responsible for the types of degradation observed at field level is very essential. Detailed site-specific databases can help in assessing and treating the degraded lands in an effective manner. The farm-specific database can be obtained by carrying out detailed characterization and mapping of the available land resources such as soil, water, climate, minerals and rocks, vegetation, crops, land use pattern, socio-economic conditions and infrastructural facilities by using a suitable base map.

The land resources of Sivagangai block, like in other parts of the state, are facing serious problems of degradation like severe soil erosion and nutrient loss in the uplands, salinity, sodicity and water logging in the low lying and tank irrigated areas. Earlier investigations carried out in the block were on a smaller scale (SS and LUO, 1994; Natarajan et al., 1997) and provided only general information on the nature of the soils and other resources. Because of this, the extent and severity of degradation could not be assessed properly. Thus, in order to assess the exact nature and extent of various forms of degradation and its impact on land resources, cadastral level survey was carried out in Sivagangai block during 2006 and 2007.

18.2 Materials and Methods

18.2.1 Study Area

The Sivagangai block is located on the northwestern part of Sivagangai district in Tamil Nadu, India and lies between 9°45′ and 10°05′ North latitude 78°19′ and 78°33′ East longitude (Fig. 18.1). The total area of the block is 44,660 ha, which constitutes about 10.7% of the total geographical area of the district. The block has 51 revenue villages.

Geologically, the block can be considered as an extension of the larger Indian peninsular shield, which is composed of diverse rock types belonging to Archaean period. Charnockite and granites (Igneous rocks), hornblende biotite gneiss, garnetiferous-quartzo-feldspathic gneiss, quartzite (Metamorphic rocks) and Cuddalore sandstone, conglomerate sandstone with shale, boulder bed conglomerate shale and sandstone, alluvium and laterites (Sedimentary formations) are the major rock types observed. Hornblende biotite gneiss occurs extensively while all the others are limited in extent.

The elevation ranges from 80 to 120 m above MSL and the area is gently sloping with a slope percentage ranging from 1 to 4. The general direction of the slope is from northwest to southeast. Many small streams with a vast network of tanks drain the block. All the tanks are seasonal and dependent only on rain. The block forms part of the Inland Lateritic Plain in Tamil Nadu. Isolated hillocks, ridges,

18 Assessment of Land Degradation and Its Impacts on Land Resources

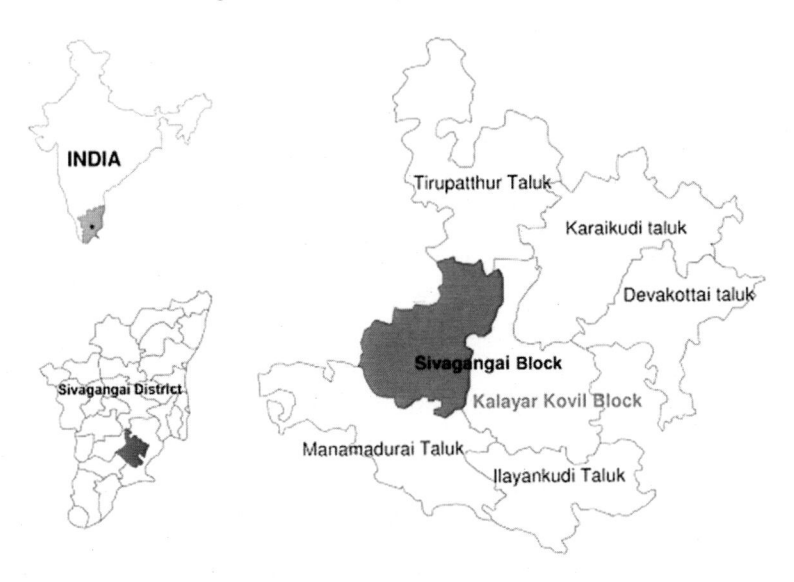

Fig. 18.1 Location map of Sivagangai block in Sivagangai district, Tamil Nadu

gently to very gently sloping uplands, lowlands and narrow valleys are the major landforms identified in the area. Very gently sloping uplands occur extensively in all the villages.

The climate is semi arid tropical and monsoonic type. The mean annual rainfall is 786 mm. The rainfall is erratic and not normal in many years, which varies from less than 400 mm to more than 1,300 mm in the area. Major part of the rainfall (51%) is received during the monsoon period. October, November and December are the rainy months and May and June are the hotter months of the block (Fig. 18.2). The soil moisture is dry in some or all parts of the soil pedon for 90 or more cumulative days in most years. The soil temperature at 50 cm depth is more than 22°C and the difference between mean summer and mean winter soil temperature is less than

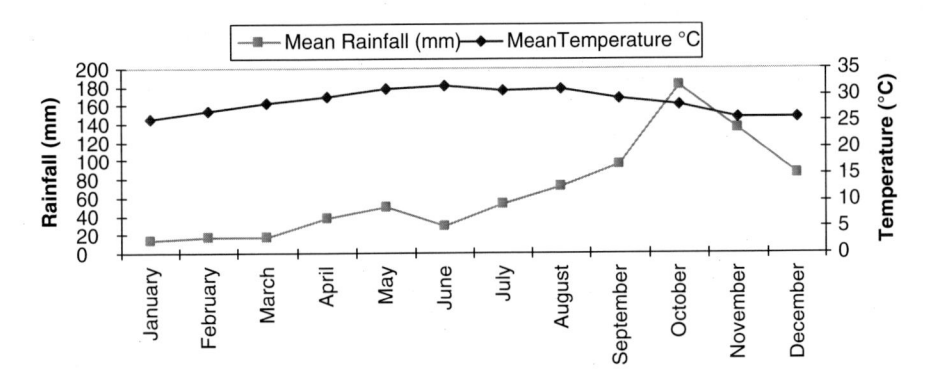

Fig. 18.2 Mean monthly rainfall and temperature of Sivagangai block

5°C. The growing period is about 4 months in a year. However, the probability of getting normal rainfall is only about 40% and hence, the growing period becomes shorter in 6 out of 10 years in the area.

Large tracts of cultivable lands are lying fallow due to non-profitability of agriculture. The net area sown is reducing every year (<20% at present) and fallow lands occupy large parts in all the villages. Cultivation is confined mostly in the tank-irrigated areas. Paddy, sugarcane, groundnut, banana, pulses, coconut and chillies are the major crops cultivated in the lowlands and short duration cereals, pulses and oilseed crops in the upland areas.

18.2.2 Methodology for Database Generation

The detailed soil survey of the villages was carried out by using cadastral maps of 1:12,500 scale as a base in conjunction with remote sensing data products of the same scale. Based on geology, drainage pattern, surface features, slope characteristics and land use, landforms and physiographic units were identified and profiles (755) were studied in transects for all the units (Natarajan et al., 2002). The soil and site characteristics were recorded on a standard proforma (Soil Survey Staff, 1993).

Based on the soil-site characteristics, the soils were grouped into 18 different soil series. Soil depth, texture, colour, amount and nature of gravel present, calcareousness, presence of limestone, nature of substratum and horizon sequence were used to identify different soil series occurring in the area. Based on variations observations on surface texture, slope, erosion, presence of gravels, salinity, sodicity etc., phases of soil series were identified and mapped. Soil samples were collected from representative pedons and characterized for particle size separates, soil reaction, electrical conductivity, free calcium carbonates, organic carbon, cation exchange capacity and extractable bases by standard recognised procedures (Jackson, 1973).

The soil maps were finalized separately for each village and the block map was prepared later by subjecting them to both cartographic and categorical generalizations (Fig. 18.3). The soil maps of the villages and other inputs collected during the survey were used to identify the constraints like shallow soil depth, erosion, gravelliness, calcareousness, salinity and sodicity in the area. Based on the interpretation, various land degradation maps were prepared by using GIS software.

18.3 Results and Discussion

Based on the survey, 18 soil series with 103 phases were identified and mapped. Out of this, eight soil series occur in the lateritic uplands and 10 in the lowlands. Out of the 103 phases mapped, 61 occur in the uplands and 42 phases in the lowland areas of the block (Natarajan et al., 2007). As per Soil Taxonomy (Soil Survey Staff, 2003), all the upland soils are grouped into Alfisols and lowland soils under

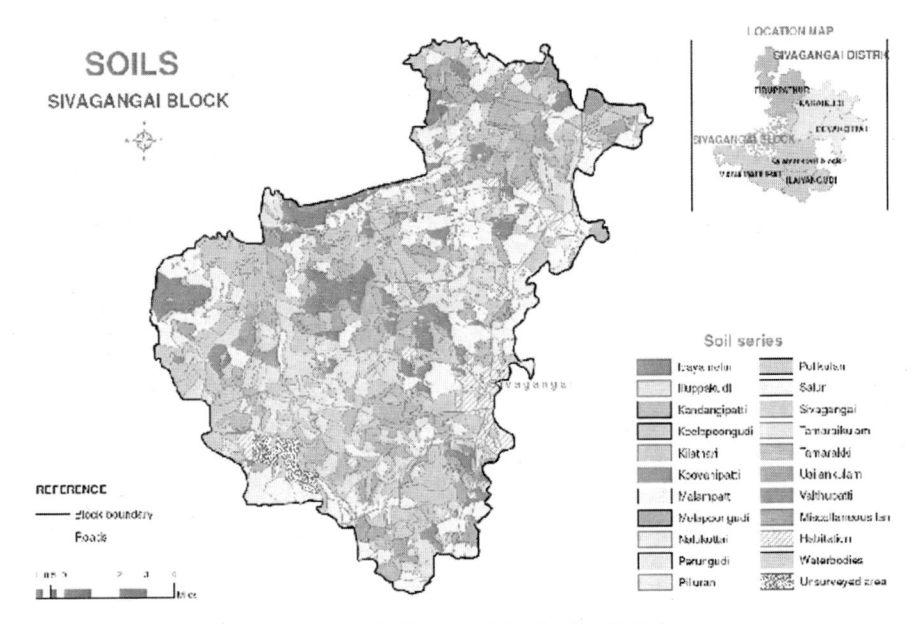

Fig. 18.3 Soil map of Sivagangai block, Sivagangai district, Tamil Nadu

Inceptisols (Table 18.1). Profile development is very well expressed in the upland soils.

In the lateritic uplands, out of the eight soil series mapped, Sivagangai, Malampatti, Idayamelur and Kandangipatti series occur extensively, occupying about 90% (15,264 ha) of the upland area (Table 18.1). Tamarakki, Tamaraikulam, Salur, Melapoongudi, Nalukottai and Kilathari series are the major lowland soils (14,224 ha) of the block. The phsico-chemical properties of major soils of the block are given in Table 18.2. Brief description of the major soil series identified in the block is given below.

18.3.1 Soils of Lateritic Uplands

Sivagangai series consists of very deep (>150 cm), dark red, well-drained, non-calcareous soils with gravelly clay texture. The gravel content ranges from 35 to 70% and increases with the depth of the soil. These soils occur normally on the lower part of the lateritic uplands. Thin patchy to thick continuous clay skins are noticed in all the subsoil horizons. This series is severely affected by sheet and rill erosion and in many villages surface layer and major part of the subsoil is eroded (Fig. 18.4).

Malampatti series consists of deep (100–150 cm), reddish brown, well-drained, non-calcareous soils with gravelly clay texture. The gravel content (>35%) increases with depth in the subsoil. These soils occur on the middle and lower part of the uplands with a slope gradient of 1–3%. The texture of the surface soil is dominantly

Table 18.1 Classification of the soils identified in Sivagangai block

Soil series	Family or higher taxonomic class	Area (ha)
Lateritic uplands		
1. Koovanipatti (Kv)	Clayey skeletal, mixed, isohyperthermic Rhodic Paleustalfs	304.75
2. Kandangipatti (Kp)	Clayey skeletal, mixed, isohyperthermic Rhodic Paleustalfs	1,575.46
3. Idayamelur (Im)	Clayey skeletal, mixed, isohyperthermic Rhodic Paleustalfs	2,822.06
4. Malampatti (Mp)	Clayey skeletal, mixed, isohyperthermic Rhodic Paleustalfs	5,802.38
5. Sivagangai (Sv)	Clayey skeletal, mixed, isohyperthermic Rhodic Paleustalfs	5,063.85
6. Keelapoongudi (Kl)	Fine, mixed, isohyperthermic Rhodic Paleustalfs	413.10
7. Illuppakudi (Ip)	Fine, mixed, isohyperthermic Rhodic Paleustalfs	452.93
8. Usilankulam (Uk)	Fine, mixed, isohyperthermic Typic Haplustalfs	442.71
Lateritic lowlands		
1. Pillurani (Pl)	Fine, mixed, isohyperthermic Typic Haplustepts	59.77
2. Valthupatti (Vp)	Fine, mixed, isohyperthermic Typic Haplustepts	252.07
3. Tamaraikulam (Tk)	Fine, mixed, calcareous, isohyperthermic Aquic Haplustepts	3,053.84
4. Tamarakki (Tm)	Fine, mixed, calcareous, isohyperthermic Vertic Haplustepts	3,295.15
5. Salur (Sl)	Fine, mixed, calcareous, isohyperthermic Oxyaquic Haplustepts	3,153.65
6. Perungudi (Pg)	Fine, mixed, calcareous, isohyperthermic Typic Haplustepts	436.09
7. Pulikulam (Pk)	Fine, mixed, isohyperthermic Typic Haplustepts	772.65
8. Melapoongudi (Mg)	Fine, mixed, isohyperthermic Typic Haplustepts	1,796.68
9. Nalukottai (Nk)	Fine, mixed, calcareous, isohyperthermic Aquic Haplustepts	1,917.78
10. Kilathari (Kt)	Fine, mixed, calcareous, isohyperthermic Aquic Haplustepts	1,007.16

loamy with less than 15% gravel. The texture of the subsoil is either sandy clay or clay with thin patchy to thick continuous clay skins. It is the dominant upland soil of the block and occurs in all the villages. Sheet erosion is severe in many places in this soil.

Idayamelur series consists of moderately deep (75–100 cm), dark red or dark reddish brown, well-drained upland soils with gravelly clay texture. The gravel content (35–70%) increases with depth of the soil. Surface texture is dominantly loamy with gravels observed in many places. Thin patchy to thick continuous clay skins are noticed in the subsoil. Severe sheet erosion is observed in many areas of this series.

Kandangipatti series consists of moderately shallow (50–75 cm), dark red or dark reddish brown, moderately well drained, calcareous (only in the subsoil)

Table 18.2 Physico-chemical properties of major soils series mapped in Sivagangai block, Tamil Nadu, India

Horizon	Depth (cm)	Sand (%)	Silt (%)	Clay (%)	pH	EC (dS m^{-1})	OC (%)	CaCO$_3$ (%)	CEC	Ca^{2+}	Mg^{2+}	Na$^+$	K$^+$	ESP	BSP
										(cmol (+) kg^{-1} soil)					
Sivagangai series: Survey no. 185, Malampatti village, Sivagangai block															
Ap	0–10	73.0	6.5	20.5	5.8	Tr	0.46	–	7.9	3.25	1.75	–	0.07	–	64
Bt1	10–24	58.4	6.6	35.0	5.9	Tr	0.43	–	9.5	4.25	2.00	–	0.07	–	67
Bt2	24–42	65.4	8.6	36.0	6.2	Tr	0.32	–	9.2	4.25	1.75	–	0.13	–	67
2Bt3	42–70	46.3	10.2	45.5	6.4	Tr	0.09	–	13.8	6.75	2.50	–	0.13	–	68
2Bt4	70–96	40.6	11.9	47.5	6.5	Tr	0.09	–	17.4	9.25	3.25	–	0.20	–	73
2Bt5	96–138	41.5	9.5	49.0	6.8	Tr	0.09	–	19.9	10.75	3.75	0.25	0.26	1.2	75
2Bt6	138–164	40.2	9.3	50.5	6.8	Tr	0.09	–	14.7	8.25	2.75	0.25	0.26	1.7	78
Malampatti series: Survey no. 11, Salur village, Sivagangai block															
Ap	0–16	63.4	13.3	23.3	6.2	0.04	0.46	–	9.4	5.0	2.0	0.2	0.3	2.1	80
Bt1	316–30	57.6	11.3	31.5	6.3	0.08	0.32	–	12.9	7.0	2.3	0.2	0.4	1.6	77
2Bt2	30–63	43.0	11.5	45.5	6.2	0.05	0.14	–	15.8	8.3	3.1	0.2	0.5	1.3	77
2Bt3	63–88	32.2	19.8	48.0	6.3	0.05	0.09	–	17.4	10.0	3.2	0.2	0.4	1.1	79
2Bt4	88–108	37.5	15.2	47.3	6.5	0.04	0.09	–	16.3	8.4	2.7	0.2	0.4	1.2	72
2Bt5	108–124	25.8	24.7	49.5	6.5	0.04	0.09	–	18.2	10.8	2.4	0.2	0.4	1.1	76
Idayamelur series: Survey no. 302, Malampatti village, Sivagangai block															
Ap	0–9	82.0	1.2	16.8	6.2	0.01	0.35	–	5.7	3.5	0.4	0.05	0.3	0.9	75
Bt1	9–19	71.0	1.0	28.0	6.3	0.01	0.33	–	8.2	3.2	2.2	0.15	0.4	1.8	73
2Bt2	19–36	53.3	2.5	44.3	6.2	0.01	0.17	–	12.2	5.7	2.7	0.20	0.4	1.6	74
2Bt3	36–62	43.4	7.8	48.8	6.2	0.01	0.14	–	15.2	6.7	4.6	0.25	0.3	1.6	78
2Bt4	62–85	44.2	6.8	49.0	5.9	0.01	0.08	–	13.9	6.6	3.8	0.20	0.4	1.4	79

(continued)

Table 18.2 (continued)

Horizon	Depth (cm)	Sand (%)	Silt (%)	Clay (%)	pH	EC (dS m^{-1})	OC (%)	CaCO$_3$ (%)	CEC	Ca^{2+}	Mg^{2+}	Na$^+$	K$^+$	ESP	BSP
										(cmol (+) kg^{-1} soil)					
Kandangipatti series: Survey no. 268, Kandangipatti village, Sivagangai block															
Ap	0–14	89.5	0.8	9.7	6.9	0.04	0.45	–	7.6	2.7	2.3	0.15	0.3	1.9	72
Bt1	14–36	64.0	0.9	35.1	6.7	0.02	0.34	–	16.2	9.8	1.9	0.30	0.4	1.8	77
2Bt2	36–69	56.0	8.3	35.7	6.6	0.04	0.22	1.1	16.4	9.9	2.5	0.25	0.4	1.5	80
Tamaraikulam series: Survey no. 200, Keelapoongudi village, Sivagangai block															
Ap	30–20	60.4	5.4	34.2	8.1	0.14	0.56	0.85	18.1	13.0	2.75	1.00	0.26	5.5	94
Bw1	20–38	55.7	7.5	36.8	8.4	0.98	0.41	1.94	19.6	14.0	3.25	1.25	0.30	6.3	96
Bw2	38–62	52.4	8.2	39.4	8.7	1.02	0.38	2.43	19.9	12.75	2.50	3.50	0.40	17.5	96
Bw3	62–99	52.7	7.1	40.2	8.7	0.96	0.33	3.58	23.2	14.75	3.00	4.00	0.50	17.2	96
Bk	99–128	46.8	8.4	44.8	8.9	1.17	0.21	5.61	23.7	15.0	2.75	4.50	0.60	18.9	96
Tamarakki series: Survey no. 296, Tamarakki Vadakkur village, Sivagangai block															
Ap	0–10	54.4	10.0	35.6	7.9	0.20	1.10	0.66	14.3	6.1	5.50	1.00	0.4	6.9	91
Bw1	10–27	55.8	8.5	35.7	8.3	0.13	0.62	1.09	14.4	8.6	3.00	1.25	0.4	8.7	92
Bw2	27–46	57.5	6.8	35.7	8.4	0.10	0.57	1.54	19.4	8.9	7.70	2.00	0.3	10.3	97
Bw3	46–62	57.8	9.3	35.9	8.6	0.12	0.48	1.58	20.0	9.0	7.80	2.50	0.3	12.5	98
Bw4	62–100	53.1	7.8	39.1	8.8	0.14	0.40	2.06	22.9	11.1	7.60	3.25	0.7	14.2	99
Bw5	100–112	51.5	9.2	39.3	9.0	0.21	0.34	2.41	26.5	13.6	8.20	4.25	0.3	16.0	99
Bk1	112–126	52.6	7.5	39.9	9.0	0.32	0.31	4.45	27.7	13.3	9.40	4.50	0.3	16.2	99

Table 18.2 (continued)

Horizon	Depth (cm)	Sand (%)	Silt (%)	Clay (%)	pH	EC (dS m^{-1})	OC (%)	CaCO$_3$ (%)	CEC	Ca^{2+}	Mg^{2+}	Na$^+$	K$^+$	ESP	BSP
										(cmol (+) kg^{-1} soil)					
Nalukottai series: Survey no. 106, Arasani village, Sivagangai block															
Ap	0–14	58.1	8.5	33.4	9.5	1.8	0.56	2.43	13.7	8.50	1.75	2.8	0.4	20.4	98
Bw1	14–39	49.4	10.4	40.2	9.7	3.2	0.41	5.30	21.8	13.75	2.50	4.9	0.6	22.4	100
Bw2	39–79	48.2	10.9	40.9	9.7	3.9	0.18	10.10	24.9	15.50	2.75	5.9	0.7	23.6	100
Kilathari series: Survey no. 265, Tamarakki Vadakkur village, Sivagangai block															
Ap	0–15	54.3	10.5	35.2	9.1	1.34	0.52	0.81	24.5	16.50	3.00	4.5	0.4	18.3	100
Bw1	15–34	44.4	10.2	43.4	9.4	1.51	0.44	1.54	26.8	17.50	2.50	6.1	0.5	22.7	99
Bw2	34–49	46.9	9.4	43.7	9.6	1.59	0.38	1.54	27.2	17.50	2.25	6.8	0.5	25.0	99
Bw3	49–80	46.6	10.3	43.1	9.4	2.10	0.32	2.42	28.4	18.25	2.50	6.9	0.7	24.2	100
Bw4	80–106	46.4	9.6	44.0	9.4	1.79	0.11	2.49	28.7	18.50	2.50	7.1	0.5	24.7	100
Bw5	106–150	45.1	10.2	44.7	9.5	2.20	0.09	3.62	28.9	19.25	2.25	6.7	0.5	23.1	99

Fig. 18.4 Severe soil loss due to sheet and rill erosion in Sivagangai series, S.No. 11, Cholapuram village, Sivagangai block

gravelly to extremely gravelly clay soils occurring on the lower part of the uplands, bordering lowlands. Few to common calcium carbonate nodules are observed in the subsoil of this series.

18.3.2 Lowland Soils

Tamaraikulam series consists of deep (100–150 cm), greyish brown, somewhat poorly drained, calcareous, clay soils occurring in almost levelled lowlands. Surface cracks, mottlings, calcium carbonate nodules, angular blocky structure and pressure faces are commonly observed. These soils are subjected to flooding for a short period. The organic carbon ranges between 0.1 and 0.7% and Ph from 8.4 to 8.9, which increases with the depth of the soil. The calcium carbonate nodule also increases with depth and below 100 cm the nodules range from 20 to 40%. The ESP is less in the first two layers (about 6%) and increases sharply to about 18% from third layer onwards (Table 18.2).

Tamarakki series consists of very deep (>150 cm), greyish brown, poorly drained, calcareous, clayey soils occurring in almost level low lands. These soils develop deep wide cracks in summer months. Reduction mottlings, angular blocky structure and pressure faces are noticed in the lower part of the subsoil. Calcium carbonate nodules occur after 100 cm depth. Reaction with dilute Hydrochloric acid is slight to violent in the surface soil. In the "Bk" horizon, the calcium carbonate nodules range between 20 and 40%. The ESP ranges from 7 to 16% and increases with depth of the soil.

Salur series consists of very deep (>150 cm), greyish brown, somewhat poorly drained, calcareous, clayey soils occurring in low land areas. Reduction mottlings and pressure faces are noticed in the subsoil. Calcium carbonate nodules occur after 100 cm depth. These soils are flooded for a short period.

Melapoongudi series consists of very deep (>150 cm), dark yellowish brown, moderately well drained, non-calcareous, clayey soils occurring in the low lands.

Fig. 18.5 Degraded sodic soils with columnar structure, Nalukottai series in S.No. 77, Arasani village, Sivagangai block

The surface soil contains less than 15% gravels and subsoil less than 35% gravels. Calcium carbonate nodules are absent. In the lower horizons, common and many iron gravels are noticed. The organic carbon (ranges from 0.1 to 0.8%) decreases regularly with depth. The EC is less than 1 dS m^{-1}

Nalukottai series consists of moderately deep (100–150 cm), dark greyish brown, poorly drained, strongly calcareous, very strongly alkaline, sodic soils with clayey texture. Many calcium carbonate nodules are noticed from 50 cm onwards. The soils are flooded frequently during the rainy season. It is one of the most degraded soils of this tract. It is affected by severe sodicity with well-developed columnar structure (Fig. 18.5). The pH is 9.5 in the surface and 9.7 in the subsoil. The EC is 1.8 dS m^{-1} in the surface and increases to 3.9 dS m^{-1} in the subsoil. The ESP ranges from 20 to 24%, which is the highest among all the lowland soils of the block.

Kilathari series consists of very deep (>150 cm), greyish brown, poorly drained, calcareous, very strongly alkaline, sodic, clayey soils occurring in nearly level low lands. Reduction mottles, pressure faces and columnar structures are noticed in the subsoil. Calcium carbonate nodules increase with depth. These soils are frequently flooded. The pH ranges from 9.1 to 9.5, EC from 1.3 to 2.2 dS m^{-1} and ESP from 18 to 23, showing an increasing trend with the depth of the soil. Severe sodicity and salinity phases of this series were mapped in many villages.

18.4 Causes for the Degradation of Resources

The detailed survey has brought out clearly that severe soil loss in the upland areas, heavy siltation of the tanks, invasion of prosopis (*Prosopis juliflora*) and other weeds in the tank beds and development of salinity, or alkalinity or sodicity in the lowlands are the major causes responsible for the severity of the degradation observed in the block.

18.4.1 Degradation Due to Soil Erosion

Among the major types of degradation identified, soil erosion is the most serious problem and a major cause for the loss of productivity in the upland areas of the block. Practically no parcel of land in the villages surveyed is left intact by this scourge. The erosion is either severe or moderate in most of the areas and slightly eroded soils occur only in patches.

In many soils, the intensity of erosion is so severe that it is difficult to see even a thin layer of soil left intact. But yet in many places there is no visible mark or sign of erosion seen at the surface. This is because of the nature of erosion, which is predominantly of sheet wash type. Since sheet erosion leaves no visible marks like rills or gullies, its impact is not assessed properly and often they are missed or ignored in the field, which generally results in under estimation of soil loss at field conditions.

Sheet erosion removes mostly the finer particles like silt, clay, organic matter and nutrients from the surface. If it is not checked, particularly in the initial stages, this leads to the complete removal of the surface soil first and then the subsoil from its place of occurrence and resulting, ultimately, in the accumulation of only coarser particles like sand, gravel, pebbles or stones at the surface (Fig. 18.4). The presence of large amounts of gravel at the surface of many upland soils is a clear indication of the extent of damage already caused by sheet erosion in the block.

Sheet erosion is also responsible for the reduction of soil depth in many upland soils. The slow removal of finer soil materials over a period of time is responsible for the occurrence of relatively shallow soils with coarser texture in many places. But this effect cannot be noticed easily at the surface. Only careful observation and comparison of similar soils at different locations can reveal this fact. The presence of thin soil layer followed by gravelly layer or completely exposed gravelly layer at or near the surface in many phases of the soils mapped is due to the prolonged removal of soils by sheet erosion in the block.

Generally, most of the upland soils in the block are characterized by the presence of large amounts of iron gravel or gravelly layer at some depth in the solum. These iron gravels form generally at deeper layers, due to the prevalence of alternate oxidation and reduction conditions. Even a short period of reduced environment is sufficient to bring the iron into circulation. At the interface between the oxidation and reduction zones, the reduced iron is oxidized and deposited as nodules. The continuation of this process leads to the formation of a thick nodular layer in the sub soil. The occurrence of distinct iron nodular layer after 100 cm depth in Illuppakudi, Keelapoongudi and many other upland soils is due to the existence of the conditions elaborated earlier. These nodules once formed are generally non reactive and resistant to any change in the soil.

Due to the existence of uncontrolled sheet erosion, in many areas, the soil material is completely eroded leaving only the dense and compact iron gravelly layer exposed at the surface. Since this process of erosion occurs over a period of time and there is no obvious mark or sign seen at the surface, we fail to realize the seriousness of the damage caused by this type of erosion in time. Only when all the soil material is eroded and the gravelly layer is exposed completely, we realize, suddenly, the extent of damage caused. But the damage is already done and it becomes too late to initiate any effective conservation measures in the field. The presence of large quantities of gravel at the surface, occurrence of exposed iron nodular layer or ironstones in many areas and dark coloured manganese coated indurated material near or at the surface are due to the effect of large scale uncontrolled sheet erosion prevalent in the block.

Generally, neglect of sheet erosion leads to the development of rills and gullies later. Next to the sheet erosion, significant amount of soil is eroded from the uplands due to rill erosion. Gully erosion is not common in this block and confined mainly to the lower part of the uplands. Since rills and gullies are noticed easily in the field, more attention is given in the various conservation programmes to check their spread, but maximum soil loss is due to sheet erosion, which needs to be tackled urgently.

Due to long neglect and lack of conservation, erosion has become a serious problem of this. If timely measures were initiated to control the effect of erosion, it would not have caused such a damage and soil loss. Actually, controlling erosion is not a difficult task in a terrain like this. The gently to very gently sloping landscape looks almost like a plain area with the slopes ranging from less than 1–3%. With such gentle and long slopes, even a simple structure like farm bunding is sufficient to arrest the runoff and minimize the loss of soil considerably. Construction of costly conservation structures is needed only in areas having deep rills or wider gullies and such gullied lands occur only in a limited extent in the block.

18.4.2 Degeneration of Tanks

Tank irrigation is an age-old practice in this area. The numerous tanks dotting this landscape have formed the lifeline of agriculture for centuries. The well-developed and intricately linked tank irrigation system is in total decay due to the neglect and misuse. Almost all the tanks are heavily silted at present due to unchecked erosion from the uplands. Silt accumulation not only reduces the storage capacity but also the percolation of water downwards. This is because, once the silt settles down at the bottom of the tank it closes almost all the pores and channels thus preventing the movement of water further into the ground water or any other aquifer nearby. Due to this, the ground water recharge is very much affected and whatever little water is stored in the tank is also lost due to evaporation. Almost all the tanks are in the same state.

Added to this is the misuse of tank beds for social forestry, brick making and illegal encroachment. The ill-conceived idea of planting trees in the tank bed reduces not only the storage capacity, but also completely prevents taking up of any desiltation work. Most of the tree species planted have a very high evapotranspiration rate, which results in huge loss of stored water. During poor rainfall years, the situation becomes much worse and whatever little water stored in the tank is not sufficient even to meet the evapotranspiration needs of the trees grown in the tank bed.

Another serious threat is the encroachment of the tank bed by Prosopis. Almost all the tanks in the block are encroached by this fast spreading species. It is a heavy feeder with very deep root system, which extracts large amounts of water for growth. This depletes the water level very fast from the tank bed and also from the surrounding areas. If the tanks have to serve the twin purposes of irrigation and recharge of the aquifers in the future, then controlling erosion from the uplands, desilting of the tank beds, removal of all the encroachments in the tanks, removal of social forestry plantations and complete eradication of prosopis from tank beds are to be undertaken urgently in this area.

18.4.3 Development of Salinity/Alkalinity in the Lowlands

The widespread occurrence of both salinity and alkalinity in the tank irrigated lowland areas of Sivagangai block, is mainly due to the neglect of the irrigation and drainage system and consequent degeneration of the tank network over a period of time. Due to this, more than 85% of the lowland soils have developed moderate to severe alkalinity and the remaining soils are already showing signs of alkalinity in the block. As per the survey, out of the total area of about 15,745 ha in the lowlands, about 3,000 ha have become already barren due to severe alkalinity problem, approximately 10,650 ha area is affected by moderate to strong alkalinity and another 2,100 ha area is showing mild alkalinity condition.

The development of salinity in the lowlands is due to the slow build up of salts in the soil, rise in ground water level and poor water management techniques followed in the command areas. Weathering and soil formation is a continuous process in any area. Normally, the salts released by weathering and carried by the runoff water is either deposited in the low-lying areas and depressions or leached further down in the profile. The leached salts may be completely removed from the soil or it may be deposited at some depth in the profile depending upon the amount and intensity of the rainfall.

In Sivagangai area, the total amount of rainfall (about 800 mm) is quite sufficient to leach most of the salts formed from the uplands. The occurrence of well-drained soils in the uplands clearly indicates the extent of leaching in the upland areas. Out of the eight soils identified in the uplands, all are well drained except one series, which has calcium carbonate in the substratum and occurs in the transition zone.

In the lowlands, the leached salts are either deposited at the surface or leached into the soil, ground water or into the streams and rivers downstream. All this depends on the amount of rainfall, slope of the land, texture and permeability of the soils and provision of free drainage in the area. Since the slope is negligible and permeability of the soil is low due to clayey texture, complete leaching and removal of salts from the soil is not possible from the lowland area. Due to this, salt concentration is always more in the lowland soils compared with upland soils. So, salt accumulation is a continuing process in the lowlands of the block, albeit slowly. Continued accumulation of salts has led to the formation of a compact layer/substratum at some depth as seen in some of the lowland profiles.

This is reflected by the high pH, EC values and calcium carbonate content of the lowland soils when compared with the upland soils (Table 18.2). All the lowland soils have moderate to very high pH (7.9 to more than 10), slight to moderate EC values (up to 4 dS m^{-1}) and moderate to high calcium carbonate content, indicating the extent of salt accumulation in the lowland areas of the block.

Mere build up or increase in the concentration of salts in the soil will not make the soil saline or will automatically lead to the development of salinity in any area. Salinity will not develop as long as the drainage of the area is free. So, the mechanism that triggered the formation of saline soils in the area is the rising ground water table. In the lowlands of Sivagangai block, poor drainage, heavy texture and presence of hard substratum at some depth in the soil hamper the free flow of water and help in the build-up of the ground water level. Due to this rise, the salts present in the soil becomes mobile and move along with the rising ground water to the surface and gets deposited after precipitation, resulting in the formation of saline soils, which is very widespread in this block. In these soils, the pH is less than 8.5, EC around 4.0 dS m^{-1} and ESP less than 15.

In many lowland soils, the salinity is moderate to severe. It is very widespread in the lowlands of Panaiyur, Pillur, Ponnakulam, Sendi Udayanathapuram, Kovanur, Arasani and Arasanur villages. The occurrence of salt affected soils in the block can be clearly seen from the Resourcesat imagery of the area around Panaiyur village (Fig. 18.6). The white tone of the imagery indicates vast tracts of salt affected areas in the block. Adopting proper water management techniques in the field can prevent salinisation.

18.4.4 Formation of Sodic Soils in Sivaganga Block

The removal of salts and replacement of other cations in the exchange complex by sodium has resulted in the formation of sodic soils in the area. In saline soils mapped, calcium and magnesium are the dominant ions and the exchange complex is saturated with either calcium or magnesium ions, even though sodium ions are also present in the soil. When the salts are removed, particularly from the surface

Fig. 18.6 Salt affected soils seen as a *white tone* on the FCC imagery from Resourcesat LISS-IV/P6, Panaiyur area, Sivagangai block, Sivagangai district, Tamil Nadu

layers, either by leaching due to irrigation or any change in the surface or subsurface hydrology, the exchange complex is slowly replaced by sodium ions. This leads to the dispersion of clay because of the highly deflocculating property of the sodium ions.

Once the clay is dispersed, it starts moving from the upper layers to the subsoil below. In the process, the pores present in the lower layers are slowly filled with the finer clay material from the surface, resulting in the formation of a dense, almost an impermeable layer in the subsoil. At this stage, the properties of both surface and subsoil are completely altered. Due to the removal of clay and other finer materials, the surface texture becomes coarse and development of columnar structure becomes prominent (Fig. 18.5) Below this depth, the structure is completely destroyed and the horizon becomes massive and almost impregnable even when it is wet. At this stage, the soil becomes almost sterile and cannot support any vegetation. Only few halophytes, which are resistant to high concentration of salts, can thrive in such soils. Continuation of this process has led to the widespread occurrence of sodic soils in the block.

In Sivaganagi block, such sodic soils are seen in many villages, particularly in Panaiyur, Pillur, Ponnakulam, Sendi Udayanathapuram, Kovanur, Arasani, Muthupet and Arasanur villages. Out of the ten soil series identified in the block, Kilathari and Nalukottai soils, occupying about 3,000 ha in the block (about 19% of the lowland area) are seriously affected by this problem (Fig. 18.7) These soils are commonly seen in association with saline soils in the surveyed villages. In many

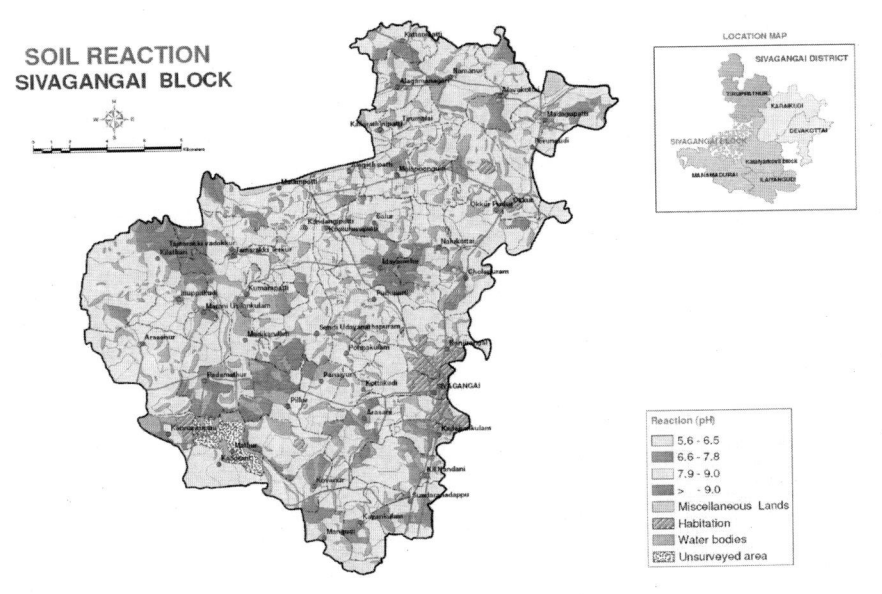

Fig. 18.7 Map showing the areas affected by salinity and sodicity in Sivagangai block

places, it is difficult to separate sodic soils from that of saline soils in the field due to their close association.

Normally sodic soils are observed in depressions or low-lying areas when compared with the occurrence of saline soils in the field. Due to high pH and ESP values, whatever little amount of organic matter present in these soils are also dissolved, giving rise to dark colouration to the surface soils in many places. Even the colour of the irrigation water is changed to dark brown due to the dissolution of the organic matter from these soils. These soils are lying almost barren and sodicity is emerging as a serious problem in the remaining lowland areas of the block.

18.5 Conclusions

The large scale mapping of the land resources occurring in the rainfed semi arid plains of Sivagangai block, covering an area of 44,660 ha, was carried out to identify the nature and exact extend of land degradation occurring at each and every parcel of land in the area. The survey helped in identifying site-specific problems of the area, which was not possible from the earlier small-scale surveys. In regard to the database, severe soil erosion in the uplands, heavy siltation of the tanks and development of large-scale salinity and sodicity in the lowlands are the major causes for the severity of the degradation observed. The degradation has drastically affected the productivity of the resources and because of this the cultivable area has come down to less than 10% of the total area in the block. Since the problems of the area are linked with each other, only an integrated management strategy will help in restoring the degraded resource base to its original glorious past.

References

Jackson, M.L. (1973). Soil Chemical Analysis. Printice Hall of India Pvt Ltd, New Delhi.

Natarajan, A., Janakiraman, M., Manoharan, S., Balasubramaniyan, V., Murugappan, K., Udayakumar, J., Ramesh, M., Chandramohan, D., Niranjana, K.V., Krishnasamy, M., Sennimalai, P., Thanikody, M., Kamaludeen, M., Krishnan, P., Rajamannar, G., Natarajan, S., Jagadeesan, P. and Vadivelu, S. (2007). Land Resources of Sivagangai Block and Land Resource Atlas of Sivagangai Block, Sivagangai District, Tamil Nadu. Soil Survey and Land Use Organization, Coimbatore and Tirunelveli and National Bureau of Soil Survey and Land Use Planning, Bangalore.

Natarajan, A., Krishnan, P., Velayutham, M. and Gajbhiye, K.S. (2002). Land Resources of Kudankulam, Vijayapati and Erukkandurai Villages, Radhapuram Taluk, Tirunelveli District, Tamil Nadu.

Natarajan, A., Reddy, P.S.A., Sehgal, J. and Velayutham, M. (1997). Soil Resources of Tamil Nadu for Land Use Planning. Executive Summary. NBSS Publications, Nagpur, 46b, 88 pp + 4 Sheets of Soil Map.

Soil, S. and Land Use Organisation. (1994). Reconnaissance Soil Survey report of Sivagangai Tk, Sivagangai district, Department of agriculture, Government of Tamil Nadu.

Soil Survey Staff. (1993). Soil Survey Manual, Agriculture Handbook No. 18. US Department of Agriculture, Washington, DC, USA, 437 pp.

Soil Survey Staff. (2003). Soil Taxonomy, Agricultural Handbook, Title 436. 3rd edn. US Department of Agriculture, Washington, DC, USA, 869 pp.

Chapter 19
New Approaches in Reclamation of Degraded Soils with Special Reference to Sodic Soil: An Indian Experience

R.K. Isaac, D.P. Sharma, and N. Swaroop

Abstract The study was conducted in sodic soil (alkali soil) having pH 10.5 and ESP 35.0 for reclamation with graded doses of gypsum applied with press mud organic matter and bio amelioration by planting forest tree species along with forage grasses. The highest yield was obtained at 50% gypsum ($CaSO_4$, $2H_2O$) application, which was comparable with gypsum applied at 25% gypsum requirement (GR) value with press mud in the rate of 10 t ha^{-1}. It is inferred that bio-amelioration resulted not only in decreasing pH but also in substantial improvement in organic matter and available P in the soil. This technique was found viable and eco-friendly, meaning that could be expanded on large scale in the country with the participation of farmers. Because of impeded drainage conditions, moisture retention at low tension near saturation assumes importance in alkali soils. Leaching of excess soluble salts by addition of water through frequent irrigation is helpful for reclaiming sodic soils.

Keywords Sodic saline soil · ESR · Acid-sulphate · Soil quality · Hydraulic conductivity · Indo-Gangetic plain

19.1 Introduction

Land degradation is a key environmental challenge that has the potential to threaten future viability of the agricultural system In India. Soil degradation, conversion of agricultural use, and urbanization has decreased the per capita arable land (Mishra, 2005). Land degradation as a function of natural environments, land use and its management includes evaluation of different land use categories for data collection

---

R.K. Isaac (✉)
Department of Soil Water Land Engineering Management, Faculty of Engineering, Allahabad Agricultural Institute, Deemed University, Allahabad, UP 211007, India
e-mail: isaac_rk@hotmail.com

and assessment that need to be confronted with traditional land use policies. Limited water availability, climate variation and injudicious use of agricultural resources threaten the future of India's agriculture and increase the risks of soil degradation.

The soil resources of India are enormous as nine out of twelve soil orders of USDA Soil Taxonomy occurs in the country (Mishra, 2005). Table 19.1 shows the soil resources and soil related constraints. However, soil related constraints are especially critical in arid, semi-arid and hilly region. Important constraints are low soil fertility and nutrient depletion, physical degradation and accelerated soil erosion. Application of ecological principles like land use capability classification involving ecological factors such as soil depth, texture, slope, production potential, population pressure, and water availability provide important inputs for effective land management alternatives. The limitation in land use affects the adoption of agricultural practices, water application methods and soil and water conservation measures for optimum crop production and sustainable land use.

Due to unabated wide spread land degradation of different kinds almost 187.7 million hectares in India are affected at unacceptable level. Among these, the salt-affected soils encompass millions of hectares with largest area in the state of Uttar Pradesh. More than 50% of the salt affected soils in the world are sodic, and their larger extension occurs in Australia. The expansion of irrigation and especially the use of poor quality irrigation water have aggravated the situation.

In earlier reports Abrol and Bhumbla (1971) reported 7 million hectares of salt affected soils occurring in different parts of the country. Bhumbla (1975) classified the distribution of these soils into six different classes viz, sodic, saline, potentially

Table 19.1 Soil resources of India and soil related constraints

Soil order	Land area (million hectares)	Percentage of the total area	Soil-related constraints
Alfisols	44.29	13.5	Weak soil structure, crusting, compaction, erosion by water
Aridisols	14.07	4.3	Drought stress, nutrient depletion, wind erosion, desertification, secondary salinisation
Entisols	92.13	28.0	Erosion, nutrient depletion, low soil organic matter
Inceptisols	130.37	39.8	Erosion, low soil organic mater nutrient imbalance
Mollisols	1.32	0.4	(no constraints)
Ultisols	8.25	2.5	Erosion by water, nutrient imbalance, acidification, P fixation
Vertisols	27.96	8.5	Massive structure, poor tilth, drought stress, water erosion
Histosols	0.002	–	High organic matter
Other	9.67	2.95	
Total	328.06	100	

Source: Velayutham and Bhattacharya (2000)

saline, coastal saline, deltaic saline and saline acid sulphate soils. It is to be noted and emphasized that Abrol and Bhumbla (1971) reported only barren uncultivated saline/sodic soils while projecting figures of 7.0 million hectares salt-affected soils, but they did not include 2.2 million hectares area of Rann and Kutch, which is an extensive saline marsh area where wild life exist, salt manufacturing is the main industry and cattle grazers abound. In case they had included this area, the total extent of salt effected soil in India would have been 9.0 million hectares.

The situation is rather critical. Out of the large arable area of 329 million hectares, 187.7 million hectares are degraded at different categories and threats and this is of major concern. About 95.65 million hectares of cultivated land suffers from physical degradation, out of which shallow soils cover 25.02 million hectares, hard soils cover 20.35 million hectares and highly permeable soils cover 10.77 million hectares. Soils with high mechanical impedance at shallow depth cover 10.63 million hectares, slowly permeable soils cover 9.43 million hectares and area with other physical constraints is 9.45 million hectares (Yadav, 1996)

Table 19.2 shows the states of Haryana, Punjab, Uttar Pradesh and Bihar that are largely the most salt-affected states. This was verified by remote sensing methodology adopted for the purpose of delineation of various types of salt-affected areas. Survey shows that large fertile land areas are being affected by salinization and alkalization due to expansion of furrow irrigation and water mismanagement greatly contributes to soil degradation through increasing salinization, sodification and water logging. (Estimates indicate that world wide 20% of irrigated land suffers from salinity/sodicity and water logging). Suri (2007) states that chemical degradation by salinization or alkalinization is reported to extent to 10.1 million hectares

Table 19.2 Severity and extent of soil degradation in the states of Haryana, Punjab, Uttar Pradesh and Bihar (million hectares)

Degradation process	Severity of degradation				Total area
	Low	Medium	High	Very high	
Water erosion					148.9
(a) Loss of top soil	27.3	99.8	5.4		132.5
(b) Terrain	–	11.8	–	4.6	16.4
Wind erosion					13.5
(a) Loss of top soil	0.3	5.5	0.4	–	6.2
(b) Loss of top soil/terrain deformation	–	4.6			
(c) Terrain deformation/ over-blowing			2.7		2.7
Chemical deterioration					13.8
(a) Loss of nutrients		3.7		–	3.7
(b) Salinisation	2.8	7.3		–	10.1
Physical deterioration					11.6
(a) Water logging	6.4	5.2			11.6
Total area	36.8	137.9	8.5	4.6	187.8

Source: Sehgal and Abrol (1994)

Table 19.3 Land degradation statistics in India

S No.	Agency/Organisation	Year	Area affected by land degradation (million hectares)	Criteria for delineation
1	National Commission on Agriculture (NCA)	1976	175.00	Based on secondary data only
2	Ministry of Agriculture, GOI	1985	173.64	Based on the land degradation statistics for the states
3	National Bureau of Soil Survey and Land Use Planning (NBSS&LUP)	1994	187.70	Mapping on 1:4.4 million scale based on Global Assessment of Soil Degradation (GLASOD) guidelines
4	Ministry of Agriculture, Department of Agriculture and Cooperation	1994	107.43	Based on the land degradation statistics from the States
5.	National Bureau of Soil Survey and Land Use Planning	2004	147.80	Mapping in 1:250,000 scale and GLASOD method

(Table 19.2) and the problem is increasing at an alarming rate in the canal irrigated areas. The industrial effluent, sewage sludge and agro-chemical residues, either in-situ or through the ground water resources are also degrading the soils.

Table 19.3 shows the differences that could be found when various sources of reporting and information are used to assess the extension of land degradation. The great differences are a source for concern and point out the need for unified methodologies to evaluate accurately the extent the of land degradation.

19.2 Characteristics of Sodic Soils

Sodic soils not only have adverse chemical and physical properties for plants but also poor tilth due to high soil loss (> 9.0 pH and high ESP >15). The virgin sodic soils depict advanced state of deterioration with distinct crust of alkaline earth carbonates. Sodic soils, truly called non saline-sodic soils exhibits water-soluble salts less than 4 dS m^{-1} and in absence of water soluble salts, are highly dispersed and impermeable to both water and air. The infiltration rate of sodic soils remain less than 0.5 cm day^{-1} due to sealing of water conducting pores which decreases steeply at an ESP of 15 attaining maximum fall up to 40 ESP, beyond which only marginal decrease in hydraulic conductivity occurs. Soil surveys in India in sodic soils have indicated the presence of loamy texture up to about 1 m depth below which lay a calcic horizon. This creates problems for tillage that could be possible over a narrow moisture range. Due to sodium saturated illuvation the formation of clay argilic and/or nitric horizons in such soils of Indo-Gangetic plains are common.

19.2.1 Genesis of Sodic Soils

Several studies have confirmed the occurrence of sodic soils (Table 19.4) in micro depressions in parts of Indo-Gangetic Plains (IGP). These have developed primarily through weathering of plagioclase feldspar through hydrolysis and carbonation processes (Bhargava and Rajkumar, 2004). Feldspar happens to be a common constituent of sodic as well as non-sodic soils. Although the IGP has very gentle slope gradients, there exist micro depressional position all over the plain with net difference in elevation in the range of 30–60 cm. The synthesis of Na_2CO_3 and $NaHCO_3$ primarily occur through weathering followed by several reactions like cation-exchange leading to Na saturation of most parts of the soil profile, precipitation of Ca and Mg as their carbonates and hydrolysis of exchangeable Na during infiltration of rain water, further augmenting the formation of sodium carbonate. The study conducted at the experimental farm in Karnal (Haryana state) established clearly that all alkali (sodic) areas occurred specifically in low lying depressions, frequently receiving runoff wash and passing through repeated cycles of wetting and drying, while the soils occurring at elevated sites not influenced by wetting and drying cycles remained relatively free of sodicity problem, although situated in close proximity with the ones affected.

Table 19.4 Recent estimated of sodic soils in Indo-Gangetic plain (India)

State	Area of sodic soils (.000 ha)
Jammu & Kashmir	20.0
Punjab	190.0
Haryana	255.7
Rajasthan	9.5
Uttar Pradesh	5,000.0
Bihar	229.0
Total	5,705.1

Formation of sodic soils has taken place in micro-depressions but is not related to greater in-situ weathering of plagioclase. It is, in fact, related to accumulation of surface runoff water charged with alkali carbonates in these depressions. These bio carbonates (HCO_3^-) turn to carbonate (CO_3^{2-}) as the soils passes through the drying phase. The hydrolysis of feldspar produces alkali bicarbonates and carbonates. It was further established on the basis of studies conducted at Central Soil Salinity Research Institute, Karnal, Haryana state in India that watering of aluminium-silicate minerals and dual factors of micro-relief and climate facilitate seasonal flooding and evaporation of the accumulated runoff wash play a critical role in the formation of sodic soils.

19.2.2 Properties of Salt-Affected Soils

Salt-affected soils in India occur in three different categories (1) saline soils (2) saline-sodic soils and (3) Non saline-sodic soils. Saline soils occur in several parts but most typically in the coastal area. This type of salt-affected soils contain high amount of water soluble salts of Cl, SO_4, NO_3, of Na, K, Ca and Mg. Due to high content of water soluble salts, the osmotic pressure of soil water show high electrical conductivity (>4.0 dS m^{-1}) and pH <8.2. The main problem in such soil is that due to excessive pH, water-soluble salts do not permit normal growth for many crop plants.

Non-saline-sodic soils are characterised by high pH (>8.5) and low water soluble salts (low electrical conductivity EC <4.0 dS m^{-1}). The salts are alkaline in chemical properties. The exchangeable sodium percentage is greater than 15, which make the soil to be in dispersed conditions when irrigation water is applied. In some parts of India, non-saline alkali (sodic) soils occur in patches with dark grey colour that has been found due to solubilization of soil organic matter content by strong alkali produced due to hydrolysis of sodium and potassium. The experiments conducted in non-saline sodic soils in Phulpur tehsil of Allahabad district of UP (India) and Kumarganj Faizabad in the vicinity of agricultural university farm and farmers field (1978) showed dark grey to whitish colour identified as non-saline sodic soil high in pH (>9.9). We describe below in detail the saline sodic soils.

19.2.2.1 Physical Parameters – Diagnostic Criteria for Saline Sodic Soils

Physical soil attributes as diagnostic criteria for sodicity problems were described as early as 1916 (Gedroiz, 1916) based on the concept of soil base exchange. Sodic soils occurring in IGP are highly dispersed in nature and on the degree of dispersion that is positively correlated with ESP ($r = 0.896$) and dispersion coefficient of 6.23 against ESP (= 15). The excess of exchangeable Na deteriorates soil structure through the process of swelling, dispersion and slumping of soil aggregates through wetting and drying (Mullins et al., 1990). Clay dispersion causes the blocking of soil pores at more than 50% (Gupta and Narain, 1971), reduces conductivity and soils exhibit poor soil water relations. Sodic soils are dense and compact having a bulk density of 1.6 g cc^{-1}, which may be 0.1–0.2 units higher than the normal soils. Reclamation treatments have been found to reduce the bulk density in the surface but insignificantly in the subsurface soils (Acharya and Abrol, 1978; Gupta and Jha, 2000).

A positive correlation was found between ESP/pH and bulk density with $r = 0.627$ and a value 1.54 g cc^{-1} has been found corresponding to sodicity of pH 8.5. The most favourable bulk density for wheat growth was recorded as 1.35 g cc^{-1} (Singh and Gupta, 1971; Acharya and Abrol, 1978). Due to impeded drainage conditions prevalent in alkali soils, moisture and low tension near saturation assumes importance. Moisture at higher suction (0.05 Mpa and above) assumes little importance in view of the extremely low hydraulic conductivity of alkali soils at these suctions (Acharya and Abrol, 1975).

pH-ESP Relationship

Sodic soils cause more deleterious effect on plants due to high pH (pH >9.0) and high exchangeable sodium percentage (ESP >20). In the state of UP, defective use of ground water has caused aggravating water logging problem with the fluctuating water table within 1.0 and 1.5 m depth. Deteriorations due to high pH, salt content and degree of sodium saturation values show gradual decrease with depth (Table 19.5).

Table 19.5 Typical characteristic properties of a sodic saline soil in Indo-Gangetic Plains (India)

Soil depth (cm)	Sodic soil				Saline soil			
	pHe	EC (dS m^{-1})	ESP	Texture	pHe	EC (dS m^{-1})	ESP	Texture
0–10	10.0	15.0	85	Sandy	7.2	56.0	3	Sandy
10–25	10.1	11.6	76	Loam	7.4	24.0	4	Loam
20–50	9.7	4.8	60	(SL)	7.6	6.0	9	(SL)
50–75	9.6	1.5	50		7.9	5.7	7	
75–90	9.3	2.0	30		7.3	5.0	3.0	

Under natural conditions a good correlation has been found between pH and ESP in the sodic soils (Govinda Iyer et al., 1963; Kanwar et al., 1963; Singh et al., 1971; Agarwal and Tripathi, 1974). Normally, ESP of 15, electrical conductivity of 4.0 dS m^{-1} or less and pH of 8.5 and above are considered the critical limits for distinguishing a sodic from non-sodic soil. In the recent past criteria for distinguishing sodic soil from non-sodic soil, ESP of 15 and pH of 8.5 has been advocated by some scientists. Abrol et al. (1980) also advocated these criteria with slight modification in pH (8.2) and ESP (15–20) but it must not be forgotten that pH 8.2 is not alkaline as even neutral salts could produce pH values as high as 8.5. Even in calcareous soil the abundance of lime (CaCO$_3$ >8.0%) does not permit the pH to go more 8.5. Hence, the research data obtained by Sharma (2004) support the criteria of ESP of 15 and pH 9.2 for distinguishing sodic soils from non-sodic soils.

Gupta and Narain (1971) provided the theoretical basis for the relationship between pH and ESP for sodic soils (Fig. 19.1). They observed that the exchangeable sodium ratio (ESR) of sodic soils containing CaCO$_3$ could be defined by P CO$_2$, selectivity pH coefficient, ionic strength and soluble sodium concentration. Using a selectivity coefficient of 0.345 (moles L^{-1})$^{-1/2}$ for Na-Ca exchange, ESR could be calculated for degraded sodic soils as follows:

$$\log ESR = pH - 5.236 + 1/2 \log CO_2 + \log (Na) + 0.51\sqrt{u}$$

Sodic soils occurring in the Indo-Gangatic plains of India are highly dispersed in nature and the degree of dispersion is positively correlated with ESP ($r = 0.896$) and dispersion coefficient of 0.23 against ESP 15 (Khosla et al., 1973; Sadhu et al., 1980).

Fig. 19.1 Relation between the (saturation paste) and ESP

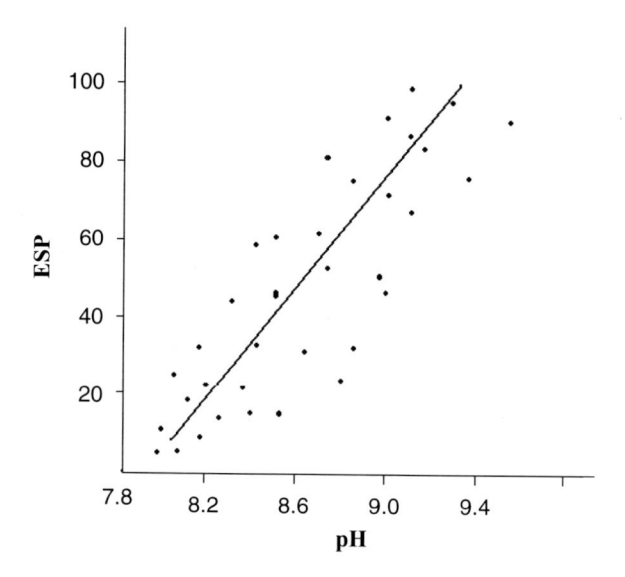

19.3 Reclamation Measures and Technologies for Improvement of Sodic Soils

Reclamation measures adopted for amelioration of sodic soils in India include the use of gypsum ($CaSO_4, 2H_2O$) and Pyrite (FeS_2). These reclaiming materials have been applied based on specific soil gypsum requirements. The graded doses of gypsum alone and in combination with organic materials such as compost, farmyard manure (FYM), Pressmud, sugarcane trash, water hyacinth, and wheat straw have better potential effects in the reclamation of both sodic and saline sodic soils.

The horizontal approach for increasing crop production through reclamation of large areas of sodic soils in the country to make them suitable for cultivation seems to be inevitable for food security due to the fast growing population of India. Being a costly and non-renewable natural resource, gypsum inputs necessitates to be the supplemented with local available organic amendments. The decomposing power of sodic land is lower than that of normal cultivated soils (Bajpai and Gupta, 1979). Quite often, due to low yield, the efficiency of added chemical fertilizer is very low and unprofitable in such soils. Reclamation of alkali/sodic soils basically requires neutralization of alkalinity and replacement of most of the sodium ions ($Na+$) from the exchange complex by the most favourable fertilizer use efficiency. Besides, chemical amendments such as gypsum/pyrite, organic amendments and green manures could be used to reclaim these lands.

19.3.1 Amendment Requirements

The quality of an amendment required for reclamation of an alkali soil depends on the exchangeable Na to be replaced and depth of the soil to be reclaimed. This quality is often referred to as gypsum requirement (GR) of the soil. Equivalent amount of any amendment can be calculated on the basis of conversion factors. While each amendment has a place in reclamation, effectiveness under different soil conditions is governed by several factors, the principal ones being the alkaline earth carbonates ($CaCO_3$) and the pH. Acid or acid formers react immediately with limestone naturally present in the sodic soils to provide soluble calcium. Materials such as sulphur or iron pyrite must first be oxidized to produce acid, which slowly dissolves native calcium carbonate present in the soil.

19.3.1.1 Relative Reclaiming Efficiency of Chemical and Organic Amendments

The reclamation of sodic soils with abundant Na_2CO_3 and $NaHCO_3$ that upon hydrolysis produce alkalinity leads to high pH and ESP causing poor physical properties. The reduction replacement of exchangeable Na by Ca ion and the Na ion is thus leached out of the soil root zone. Sodic soils developed in the Indo-Gangetic plains invariably contain free $CaCO_3$ concretions (locally called Kankar) throughout the soil profile. For reclaiming a sodic soil, $CaCO_3$ concretions play an important role. Reclamation can be accomplished by chemical and organic amendments. The result of the experiments conducted at experimental stations has shown that graded doses of different amendments viz. sulphuric acid, nitric acid, pressmud (carbonation and sulphitation processes) gypsum, aluminium sulphate, ferrous sulphate and FYM were used for the reclamation of sodic soils.

Experiments have showed that gypsum applied at the rate of 12–15 Mg ha^{-1} (50% GR for 0–15 cm soil), is sufficient to initiate the reclamation process in rice-based cropping systems. Field studies have also shown also that gypsum dose could be reduced from 50 to 25% GR when organic matter viz. FYM@ 20 Mg ha^{-1} was also applied. Reduction in pH and ESP changes in various treatments indicated that soil improvement was better in the surface layers in rotation having rice as one of the crops. The use of iron pyrite (FeS_2) was also found to be equally effective and is more beneficial in soils having calcium carbonate concretions in the topsoil. India has large reserves of pyrite (FeS_2) of sedimentary origin containing 10–40% sulphur in the form of FeS_2. Native calcium carbonate in sodic soils reacts with the acid formed to generate soluble Ca^{2+}, which replaces exchangeable Na^+ from the exchange complex of the soil.

The technology developed for reclaiming sodic soil depends on the proper choice of amendments. Experimental results have shown that both amendments proved effective in increasing grain yield of both rice and wheat but the grain yield after application of pyrite was lower than those after gypsum application. Since the northern state of Bihar adjacent to Uttar Pradesh has large reserves of iron pyrites, this amendment could be used to accelerate the reclamation process of the sodic and saline-sodic soils of this region. Pyrite of 5 mm size gave significantly higher yield of rice and wheat as compared to 10 mm size, but it needs to be applied under moist conditions for its fast reaction or oxidation.

The role of organic materials was more pronounced at lower quality of gypsum requirement (25% + FYM 10 t ha^{-1}) followed by gypsum applied at the rate of 50% GR. Both FYM and pressmud (by product of sugar industry) were found better in comparison to rice straw, wheat straw and fly ash with respect to physical properties of soil as well as yield of crops.

Bio-inoculants organic amendments integrated with gypsum were tried for the reclamation of sodic soils and their efficiency was tested on the yield of rice and wheat at Etawah district of Uttar Pradesh (Gupta and Jha, 2000). These were eight treatments consisting of rice straw, pressmud, water hyacinth with bio-enounlant and with gypsum applied @ 25% GR value. Rice and wheat yield increased significantly due to bio-inoculation with press mud and water hyacinth.

19.3.2 Acid – Sulphate Soils

These soils occur in coastal areas of India and in other areas with similar conditions having high organic matter content influenced by the intrusion of seawater or by the presence of sulphates under reduced condition that favour their formation. These soils occur in Kerela, Sunderbans area of West Bengal, Coastal Orissa and Andman and Nicobar Islands. These soils are characterized by very low pH, plant growth is poor due to less availability of Ca, N_2, Mg, micro nutrients and toxic elements like Al and Fe. These soils are rather difficult to manage due to intense degradation.

19.4 General Conclusions on Management and Improvement of Sodic Soils in the Indo-Gangetic Plains

Soils of Indo-Gangetic plains of India (IGP) by large have been assessed for their high productivity and still this alluvial plain constitutes the country's granary basket. Of the total 17.1 million hectares canal irrigated area, 1.38 million hectares falls in Haryana, 1.36 million hectares in Punjab, 3.08 million hectares in UP, 1.10 million hectares in Bihar and 1.5 million hectares in Rajasthan. Water from the Himalayan rivers was harnessed by the British government to augment food production and to ward off recurrence of famines. But due to faulty planning in its wake, the canal irrigation caused secondary salinisation/sodification under certain typical environmental situations.

Salinity has become a very serious problem for India. The change in the hydrology with the addition of extra irrigation water has disturbed the hydrologic balance of the fertile land in several parts of Uttar Pradesh state inside the Indo-Gangetic plains. When subsurface water fills the soil profile and the water comes to the soil surface, it enters a stream, or forms a wetland. Where evaporation is the only outlet to a wetland, the evaporating water leaves behind all the salt. Salts in the soil continue to accumulate over time if they are not continuously or periodically removed. The repeated application of irrigation water adds the salt concentration.

Management of saline soils includes leaching of excess soluble salts, lowering of water table depth, selection of suitable crops and improved land management practices. Leaching reduces the root zone salinity to a desired level through removal of excess salts. Subsurface drainage is also an effective tool for lowering the water table, removal of excess salts and prevention of secondary salinization.

Drip and pitcher methods are very useful for saline soils as they add water directly into the root zone at controlled rates and even saline water can be used under these methods without any detrimental effect on crop growth (Bandopadhyay et al., 2001) owing to dilution of salts at the root zone. Yield reduction in saline soils can be minimized through selection of salt tolerant crops and different planting techniques that can reduce the salt stress on plant growth. A better salinity control can be achieved by using sloping bed seeds planted on the sloping side just above the water line. Alternate furrow irrigation is advantageous as the salts can be displaced beyond the single seed row (Michael, 1978).

Sodic soils are characterized by low intake of water or normally less than 5 mm day^{-1} and therefore these soils are suited to lowland rice cultivation without puddling. Reclamation was found to improve infiltration but much below the level of normal soil when the subsoil was dense and equally alkaline (Gupta and Jha, 2000). Since such soils have very slow permeability, their hydraulic conductivity (HC) is 1/10–1/50th of normal soils. After 4 years of reclamation, saturated hydraulic conductivity of the surface soil was found to have increased significantly but improvement in sub-soil was nominal (Gupta and Jha, 2000) and HC was found to be negatively correlated with pH/ESP. Only a slight decrease in saturated HC was noticed as ESP increased from 5 to 14 but any further increase in ESP brought a sharp decline in HC.

Shallow rooting, greater moisture extraction from the surface layers, low upward water flux, etc result in rapid depletion of moisture in the surface in sodic soils affecting plant growth and the frequent demand of water to soak the shallow root zone. Light and more frequent irrigation is therefore needed in alkali soils. Heavy irrigation leads to water stagnation and lower yield. If the soil gets improved in deeper layers (i.e. from drainage), the crop is relieved of the greater moisture stress by downward improvement in water flow (Acharya et al., 1979).

Though the effects of sodic waters on soil degradation and crop yields have been reported extensively, only a few selected waters were tested and these did not interacted the effects of water quality parameters. It is therefore, imperative that adverse water qualities containing alkali salt ions be mixed with good quality water before used for irrigating sodic reclaimed land. Additionally, poor water quality could be improved with additions of soluble Ca^{2+} by dissolved gypsum ($CaSO_4.2H_2O$) to lower the high SAR value of water since dissolved electrolytes do help in the reclamation of sodic soil without any dispersing effect on soil properties. High irrigation frequency with low irrigation depth has been found beneficial to improve yields of most of the crops grown in the sodic soils.

Rice/wheat improving system for 1–2 years after reclamation has been found quite effective for sodic soils. The water requirement for rice cultivation even under continuous submergence is appreciably lower. It is, thus, appreciated that sustainability of crop yield is more effective on long term basis with low water requirement with rice-wheat/rice-barley legume cropping system. Rice is preferred as first crop in alkali soils as it can grow under submergence, can tolerate fair extent of ESP and can influence several microbial processes in the soil, which can improve soil physical and chemical properties along with additional returns on long-term basis.

Long-term experiments of planting forest tree species and forage grasses such as *Brachiaria mutica, Leptochloa fusca and Setaria sphacelata* have shown positive results in improvement of sodic soils. Planting of salt tolerant cultivars, drainage, leaching out of salt, graded use of gypsum, pyrite application at gypsum requirement values on land irrigated with sodic water have proved to be viable and an efficient approach with huge potentials for exploiting sodic land for crop production. Application of pressmud (sugarcane by product) with graded doses of gypsum proved beneficial for production of rice and wheat (Table 19.6). Pressmud is available in abundance in India.

Grain yield of rice and wheat showed significant increase in the yield at various doses of gypsum and pressmud. The highest yield was at 50% gypsum requirement which was comparable with gypsum at 25% GR + Pressmud @ 10 t ha^{-1}. The yield performance was at par with the application of organic matter at the same level of pressmud.

In a long-term experiment, bio-ameliorative effects of agro-forestry on sodic soil were studied. Soil samples were taken from a 5-year-old agro-forestry plantation field for determination of pH, organic carbon and available P (Jackson, 1973). The experiment showed a sharp decline in pH and alkalinity and conspicuous increase in organic carbon and available P content in the soil (Table 19.7).

The highest reduction in ESP was found with *Acacia nilotica* + forage grasses followed by *Prospis juliflora* and *Eucalyptus tereticornis*. Survival of all species was excellent which ranged from 72 to 100%. The maximum height was attained by Eucalyptus tereticornis, followed by prospis juliflora and Acacia nilotica indicating their high adaptation ability in such environments.

Some grasses like *Deplache fusca* (Karnal grass), *Bracharia mutica* (Para grass) and *Cynodon dactylon* (Bermuda grass) have been reported to produce 50% yield

Table 19.6 Relative efficiency of graded doses of gypsum with press mud on yield of rice and wheat grain (Qha^{-1})

Treatment combination	Rice	Wheat
Control	10.5	08.5
Gypsum @ 50% GR	34.40	26.50
Gypsum @ 25% GR + PM @ 10 t ha^{-1}	34.90	26.45
Gypsum @ 25% GR + PM @ 7 t ha^{-1}	29.50	22.80
Gypsum @ 25% GR + PM @ 5 t ha^{-1}	28.00	19.00
Gypsum @ 25% GR + OM @ 10 t ha^{-1}	34.42	25.80

19 New Approaches in Reclamation of Degraded Soils with Special Reference

Table 19.7 Changes in soil properties

Treatment	pH (1:2)	EC (dS m^{-1})	OC (g kg^{-1})	Av. P (kg ha^{-1})	ESP
T$_0$	Control	10.5	2.8	6.0	35.0
T$_1$	Prospis Juliflora	9.5	7.0	10.5	12.0
T$_2$	Eucalyptus tereticornis	9.6	8.0	11.2	14.0
T$_3$	Acacia nilotica + forage grass	9.0	12.0	12.2	10.2

at ESP level above 30, while *Leptochloa fusca* can grown in sodic soil with ESP as high as 80–90 (Kumar and Abrol, 1988). Among the various forest species evaluated, *Casuarina equisetifolia* (Soru), *Eucalyptus tereticarnis* (Safeda), *Leucaena leucocephala* (Subabul), *Dalbergia sissoo* (Shisam) were found to be most suitable for fuel, fodder and timber, respectively. For the reclamation of salt affected and water logged lands, aforestation with tree species like *Eucalyptus* spp. *Prospopis juliflora*, *Acacia nilotica*, *Azadirachtra indica*, is proven to be effective (Sharma, 2004).

Jain et al. (2002a) reported that growing two salt-tolerant species, viz., *Albizia procera* and *Acacia auriculiformis* for a year reduced the pH (9%) exchangeable Na (39%) and ESP (36%) and increased the soil organic carbon (88%), exchangeable Ca (58%), Mg (37%) and CEC (24%). The reduction in Na, Ca was more under *Acacia auriculiformis* (55%) than in *Albizia procera* (46%). Similar trend was observed with *Azadirachta indica* and *Pongamia pinnata* showing a better efficacy on a degraded silty clay loam soil with high sodicity (Jain et al., 2002b).

References

Abrol, I.P. and Bhumbla, D.R. (1971). Saline and alkali soils in India: Their occurrence and management. World Soil Resources, FAO, Report 41:42–51.

Acharya, C.L. and Abrol, I.P. (1975). A comparative study of soil water behaviour in sodic adjacent normal soil. Journal of Indian Society of Soil Science 23:391–401.

Acharya, C.L. and Abrol, I.P. (1978). Exchangeable Na and Soil water behaviour under field condition. Soil Science 125:310–319.

Acharya, C.L., Sandhu, S.S. and Abrol, I.P. (1979). Effect of exchangeable Na on the rate and pattern of water uptake by raya (Brassica Juncea L.) in the field. Agronomy Journal 71: 936–941.

Agrawal, G.K. and Tripathi, B.R. (1974). Journal of Indian Society of Soil Science 22:43.

Bajpai, P.D. and Gupta, B.R. (1979). Effect of salinity and alkalinity of soil on some important imerrbial activities. Journals of the Indian society of soil science 27:197–198.

Bandopadhyay, B.K., Sen, H.S., Maji, B. and Yadav, J.S.P. (2001). Saline and Alkaline Soils and Their Management. ICAR Monograph 1, ICAR, CSSRI, West Bengal, 72 pp.

Bhargava, G.P. and Rajkumar (2004). Genesis, characteristics and extent of sodic soils of the indo-gangetic alluvial plain. Extended Summaries, International Conference on Sustainable Management of Sodic Lands, pp. 15–22.

Bhumbla, D.R. (1975). Map of India, Salt Affected Soils. CSSRI, Karnal.

Gedroiz, K.K. (1916). The Absorbing Capacity of Soil and Soil Eolitic Basis. Translated from Russian by S.A. Wakeman, quoted from W.P. Kelley cation exchange in soil. Rain hold publications Corp, New York.

Govinda Iyear, T.A., Krishnamurthy, V.S., Ramadass, C. and Sathiadass, N. (1963). Madras Agricultural Journal 50:261.

Gupta, R.N. and Jha, C.K. (2000). Report of hydraulic conductivity, infiltration and bilk density studies in sodic soils. Sodic land reclamation project. Remote Sensing Applications Centre, UP, Lucknow, pp. 1–70.

Gupta, R.N. and Narain, B. (1971). Investigations on some physical properties of alluvial soils of Uttar Pradesh related conservation and management. Journal of the Indian Society of Soil Science 19:11–22.

Jackson, M.L. (1973). Soil Chemical Analysis. Prentice Hall of India, New Delhi.

Jain, R.K., Singh, B., Srivastava, N., Tripathi, K.P. and Behl, H.M. (2002a). Influence of tree growth on reclamation of sodic soil. Indian Journal of Agricultural Sciences 72:39–41.

Jain, R.K., Singh, B., Tripathi, K.P. and Srivastava, N. (2002b). Reclamation of a sodic soil through afforestation and Azadirachta indica and Pongamia pinnata. Journal of the Indian Society of Soil Science 50:147.

Kanwar, J.S., Sehgal, J.L. and Bhumbla, D.R. (1963). Journal of Indian Society of Soil Science 1:39.

Kumar, A. and Abrol, I.P. (1988). Bulletin No. 11, CSSRI, Karnal, p. 95.

Michael, A.M. (1978). Irrigation Theory and Practice. Vikas publication House Pvt Ltd, New Delhi, p. 725.

Mishra, B. (2005). Managing soil quality for sustainable agriculture. Journal of Indian Society of Soil Science 53(4):529–536.

Mullins, C.E., Machle, D.A., Narth kole, K.H., Tisdall, J.M. and Young, I.M. (1990). Hard Selling Soil Irrigation Advances in Soil Science II Soil Degradation. Springer, New York.

NBSSLUP (2004). Soil Resource Management Reports. National Bureau of Soil Survey and Land Use Planning, Nagpur.

Sehgal, J.L. and Abrol, I.P. (1994). Soil Degradation in India: Status and Impact. Oxford and IBH Publications Co Pvt Ltd, New Delhi.

Sharma and Ambekar. (2004). Efficient on-farm water management in sodic lands. International Conference on Sustainable Management of Sodic Lands, February 9–14, Lucknow (India).

Sharma, P.K. (2004). Emerging technologies of remote sensing and GIS for the development of spatial data infrastructure. Journal of the Indian Society of Soil Science 52(4):384–406.

Singh, Y.P. and Gupta, R.N. (1971). Fertilizer response to the physical effects of soil compaction. Journal of Indian Society of Soil Science 19:345–352.

Singh, K.S., Lal, P. and Singh, M. (1971). Indian Journal of Agricultural Research 5:115.

Suri, V.K. (2007). Perspectives in soil health management – a looking glass. Journal of Indian Society of Soil Science 55(4):436–443.

Velayutham, M. and Bhattacharyya, T. (2000). Soil recourse management. In: J.S.P. Yadav and G.B. Singh (eds.), Natural Resource Management for Agricultural Production in India. Indian Society of Soil Science, New Delhi, pp. 1–135.

Yadav, J.S.P. (1996). Land degradation and its effect on soil productivity, sustainability and environment. Journal of soil and water conservation 40:660–674.

Chapter 20
Soil and Water Degradation Following Forest Conversion in the Humid Tropics (Indonesia)

Gerhard Gerold

Abstract Indonesia's annual deforestation rate of −1.2% is dramatic compared for example to Brazil's 0.4% loss of forest cover. Sulawesi region of Indonesia still has a forest cover of 48%, but as population has been growing by 66% over the past 2 decades, massive land cover transformations are going on that have changed the land cover pattern and consequently soil and water resources of the region are altering as well. Since 2001, we investigate the impact of forest conversion on the water balance, nutrient losses and soil erosion of a small mesoscalic tropical catchment, which is integrated into the long-term interdisciplinary collaborative project STORMA (www.storma.de). The study was conducted in a small mountainous catchment, which is located at the north-eastern border of the Lore Lindu National Park Central of Sulawesi. The Nopu catchment (51S 01757231E, 9867683 N) covers an area of approximately 2.6 km^2. Since 2001 the catchment is monitored with three weirs (automatic stage recorder), one climate station and six rain gauges. Traditional *slash & burn* cultivation is predominant and heavy machinery is not used. At two sub-catchments (rainforest weir 3, slash & burn weir 2) changes in river discharge and nutrient outputs with time and differences were analysed since 2002. As indicators for land degradation we studied the changes of infiltration rate, water flow path (increasing interflow and surface flow) and soil nutrient output with river discharge. In the Nopu catchment slash & burn activities increased mainly since 2003/2004 with forest reduction until 2007 in the sub-catchment weir 2 from 87% (2001) to 26% forest cover (2007). Due to forest conversion, river discharge increased from 9 to 17% for the period 2002/2003–2005/2006 mainly driven by increase in overland flow and quick interflow. Three scenarios (forest/cacao/slash & burn (corn with cassava)) were simulated with the application of the water balance model WASIM-ETH. Simulation results supported the experimental results showing

G. Gerold (✉)
Department of Landscape Ecology, Geographisches Institut Georg-August-Universität Göttingen, Göttingen 37077, Germany
e-mail: ggerold@gwdg.de

an increase from 8% to 17% annual discharge for cacao and slash & burn scenario and give insight into the changing discharge components. The increase in overland flow leads to an increase in soil erosion with doubling of suspended sediment output from 2003 to 2005. Also, higher soil nutrient leaching and increase in quick interflow has caused an increase by ratios of 1.5–1.9 of the main cations (Ca, Mg, K) and nitrogen (TNb) in the sub-catchment partially deforested compared to the natural forest sub-catchment. After 2005/2006, land degradation indicators of suspended sediment output and soluted nutrient output indicate a "stabilizing effect" with decrease due to land use change to cacao agroforestry in the mid slopes of the catchment.

Keywords Rainforest conversion · Tropical catchment · River discharge simulation · Solute nutrient output · Suspended sediment output · Sulawesi-Indonesia

20.1 Introduction

Indonesia's annual deforestation rate of −1.2% is dramatic compared for example to Brazil, which exhibits only a 0.4% loss of forest cover (FAO, 2003). Sulawesi still has a forest cover of 48%, but as population has been growing by 66% over the past 2 decades massive land cover transformations are going on that are changing the land cover pattern. Consequently soil and water resources of the region are altering as well. Rainforest conversion, predominately into annual cultures and cacao-systems, was intensified in Central Sulawesi during the last decade. Since 2001, we investigated the impact of forest conversion on the water balance, on nutrient losses and on soil erosion of a small mesoscalic tropical catchment, which is integrated into the long-term interdisciplinary collaborative project STORMA (www.storma.de).

Besides many plot related studies on soil erosion for different land use types in the tropics (Van Dijk and Bruijnzeel, 2004) long-term catchment studies on changes of water balance components and nutrient losses with forest conversion are rare. Considering the increasing stress for water resources of developing countries in the humid tropics due to rapid deforestation rates and climate change, there is an urgent global research need for humid tropical hydrological and associated mesoscalic catchment processes (Leemhuis et al., 2007). According to Abbot and Reefsgard (1996) spatially distributed and time-dependent hydrological modelling is of outermost importance ("*the conditio sine qua non*") for investigations in this area.

Therefore the main questions to be answered are: (1) how does forest conversion affect main water balance components and river discharge, and (2) how does forest conversion effect suspended material and soluble nutrient output of a small mesoscale tropical catchment.

20.2 Material and Methods

20.2.1 The Study Site

The study was conducted in a small mountainous catchment, which is located at the north-eastern border of the Lore Lindu National Park, Central Sulawesi, Indonesia. The Nopu catchment (51S 01757231E, 9867683 N) covers an area of approximately 2.6 km^2, of which 86% are situated within the borders of the National Park. The catchment area receives an annual mean precipitation of around 2,500 mm without any clear dry season and the altitude ranges from 600 to 1,400 m a.s.l. with steep slopes at the upper catchment (Fig. 20.1).

Based on soil mapping for the whole catchment, Eutric and Dystric Cambisols (FAO, 1998) with fairly high cation exchange capacity (7–14 cmol kg^{-1} in B-horizon) and base saturation of 45–90% are the dominant soil types formed on andesite and granite parent materials (Kleinhans, 2003; Mackensen et al., 1999). The catchment was divided into three sub-catchments, each of them representing different stages of land use. About 20 years ago smallholders started to establish cacao plantations in the lowlands (subcatchment A, weir 3). The middle part (B,

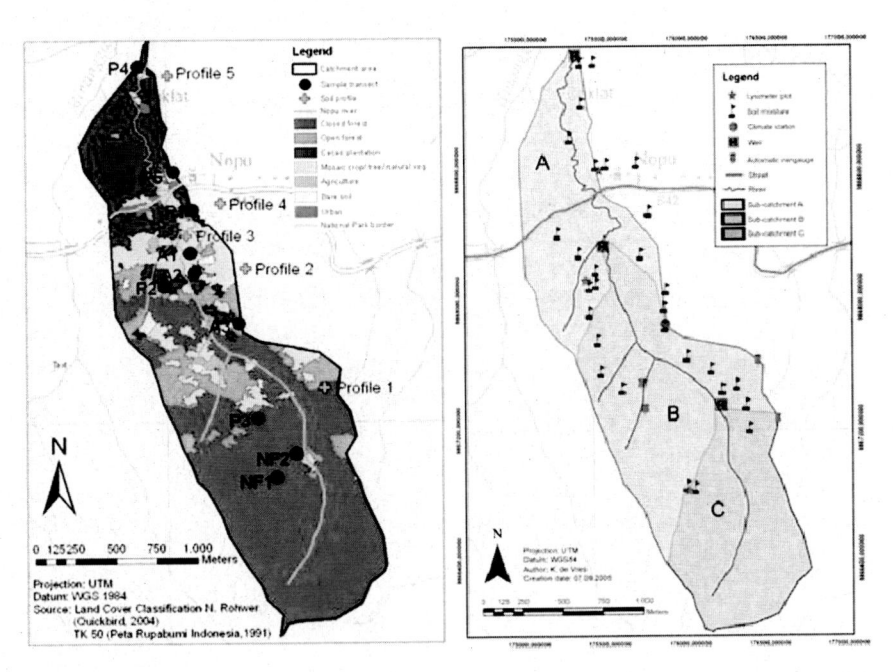

Fig. 20.1 Nopu catchment with instrumentation and land cover classification 2004. Sub-catchment B with weir 2 (W2) is forest conversion catchment, sub-catchment C with weir 3 (W3) is the reference forest catchment (natural rainforest)

weir 2:650 m–950 m.a.s.l.) is dominated by ongoing forest conversion with patches of annual crops, pasture and agro-forestry. Rainforest is found for most of the upper area (C, weir 3) of the catchment (Fig. 20.1, Gerold et al., 2004).

Since 2001, the catchment is monitored with three weirs (automatic stage recorder), one climate station and six rain gauges. Data are recorded regularly on a 10 min basis. Furthermore, soil moisture measurements (TDR system) are used for the main land use types such as cacao and slash & burn (maize) and natural forests, 4 times week^{-1} at 29 field points. Additionally lysimeter stations with measurements of seepage nutrient concentrations in the soil depth of 30/60/90/120/150/180 cm are installed (weekly samples). Since 2007, thorough rates and nutrient fluxes in two cacao- and one natural forest site in relation to open area input (incident rainfall) are investigated. Plot measurements on soil erosion were conducted for the main land use types in 2005 and 2006. Traditional slash & burn cultivation is predominant in the area and heavy machinery is not used.

20.3 Methods

Since 1999/2000 the middle part of the catchment is dominated by ongoing forest conversion with patches of annual crops, agro-forestry (cacao) and secondary forest. The lowest part is used for cacao plantations for the last 15–20 years but rainforest still remained dominant in the upper part of the catchment (above 900 m a.s.l.). Increasing forest conversion in the catchment was studied by satellite image detection (Quick Bird images from 2001, 2004, 2007, Fig. 20.1) validated with ground check and quantitative change detection for the main land cover types (object based image classification with eCognition V.30). For two sub-catchments (rainforest C, weir 3, slash & burn B, weir 2) we compared the changes in soil moisture, river discharge and nutrient output.

In 2001 in the context of the hydrological project, two automatic weather stations (AWS) with rain gauges (Friedrich Germany, 0.1 mm resolution, 5-min data interval) and four rain-gauge stations (tipping bucket system, Friedrich Germany, 0.2 mm resolution, 10-min data interval) were installed in the catchment from 620 until 1,075 m a.s.l., to investigate the pattern of spatial and temporal rainfall distribution (Fig. 20.1).

For the river discharge and hydro-chemical composition measurements, three composite (rectangular-rectangular-trapezoidal) cross section weirs were constructed for each sub-catchment (Fig. 20.1). They were built up at the outlet of the total catchment (weir 1, A), slash & burn sub-catchment (weir 2, B) and natural forest sub-catchment (weir 3, C) (Fig. 20.1). For each weir, water levels were recorded with automatic stage recorders since November 2001. Additionally, flow velocity measurements were carried out to establish a stage-discharge relationship for each weir (for details see Kleinhans, 2003; Lipu, 2007). At weir 2 (forest conversion sub-catchment) and weir 3 (forest sub-catchment), multi-parameter sensor (MPS-D SEBA Hydrometry, Germany) including measurements of temperature, electric conductivity, pH, O_2-content, water level and turbidity (in NTU) every 10 min were installed in November 2001. The turbidity sensor has a sensitivity of 0.5 NTU and the scale of measurement ranged from 0 to 1,070 NTU, at weir 2, and from 0 until 975 NTU at weir 3. The water stage data was used for corresponding time-intervals as a reference to control the turbidity value (Lipu, 2007).

For the determination of dissolved nutrient concentration (hydro-chemical parameter) manual weekly samples were taken at weir 2 and 3 from river discharge and analysed for the main cations, anions, pH, EC and DOC. Event based flood samples were taken manually.

For the measurements of suspended load concentration it was possible to install at weir 2 an automatic water sampler (ISCO 7200, USA, 24 bottles) with regular water samples for normal river discharge (base flow), 2 times week^{-1} (Monday and Thursday, 24 h) and for flood events (storm flow, water level $\geq$10 cm at weir 2, $\geq$5 cm at weir 3) with 15 min time interval (6 h) (October 2004 until December 2005). At weir 2 and 3, manual water samples for base flow and storm flow (same day and hour as weir 2) were launched from May 2003 until December 2005 (Lipu, 2007). Liquid suspended sediment samples were filtered (0.45 μm) and then weighed as oven dry sediment (mg l^{-1}) at the STORMA laboratory. Since November 2001 weekly and event triggered manual river discharge samples (1 l bottles) were taken for soluble nutrient output calculation. After filtration (0.45 μm) all chemical analysis were done by the STORMA-laboratory at Tadulakko University, Palu, Central Sulawesi, Indonesia. Ca; K; Mg; Na; Si were analysed using an ICP-OES Optima 2000 DV (Perkin Elmer). Minimum concentration for calibration was 0.1 mg l^{-1}. PO_4-P; NO_3-N and NH_4-N were analysed using an AA3 Autoanalyzer (Bran & Luebbe). The detection limit was 0.1 mg l^{-1} for NO_3-N and NH_4-N. For PO_4-P it was 0.05 mg l^{-1}.

To study the impact of forest conversion on the water balance of the Nopu River catchment we applied the Water Flow and Balance Simulation Model WASIM-ETH (Schulla and Jasper, 1999). WASIM-ETH is a process-based fully distributed catchment model. A grid gives the spatial resolution and the time resolution can vary from minutes to days. The main processes of water flux, storage and the phase transition of water are simulated by physically-based simplified process descriptions (Schulla, 1997). The meteorological input data of the model are interpolated for each grid cell and data output provide the simulation for the main hydrological processes like evapotranspiration, interception, infiltration and the separation of discharge into direct flow, interflow and base flow.

These calculations are modularly built and can be adapted to the physical characteristics of the catchments area. The model was applied on a daily basis at a resolution of 30 m pixels. The model was calibrated using measurement data from the field and missing parameters were used from literature (Kleinhans, 2003; Leemhuis, 2005). Water balance and discharge dynamics of the Nopu catchment were modelled from 19 September 2000 until 19 February 2003. The calibration period was from November 2001 to March 2002 and the validation period run from April 2002 to February 2003. Both periods include a rainy and a dryer season and the statistical quality of the model results were r^2 0.83 and 0.86 for calibration and validation, respectively (Nash and Sutcliffe, 1970). The difference between model and discharge measurement were low, i.e. for 2002 only 20 mm and for the total run 6 mm.

Different river discharges regression between suspended sediment concentration (SSC, g l^{-1}) and turbidity (NTU 100^{-1} units) were calculated (weir 2: $r^2 = 0.56$; weir 3: $r^2 = 0.72$) and used for the calculation of suspended sediment output (Lipu, 2007). Soluted nutrient output was calculated for the yearly mean and median value as well as sum of the daily nutrient output (daily discharge × nutrient conc.), multiplied with 365 n^{-1} ($n =$ sampling days) (Tables 20.5 and 20.6).

As indicators for land degradation we investigated the changes on infiltration rate, water flow path (increasing interflow and surface flow), suspended sediment output and dissolved nutrient output with river discharge.

20.4 Results

The comparison of the measured annual river discharge and the analysed land cover changes from 2001 until 2007 deliver a first general overview on the impact of forest conversion on the hydrology of the observed watershed (Table 20.2). In the Nopu catchment, slash & burn activities increased mainly since 2003/2004 with forest reduction until 2007 in the sub-catchment with weir 2 from 86% (2002) to 37% forest cover (2007). Land use classification based on satellite imagery shows a reduction of the natural forest in the lower catchment area from 82.1% in 2001 to 38.7% in 2004 and 15.1% in 2007 (Table 20.1).

Table 20.1 Land cover change in Nopu subcatchment 3 and 2 from 2001 until 2007 (weir 3 catchment with 80.6 ha, weir 2 catchment with 101.9 ha)

Land use	2001 W3 ha %	2001 W2 ha %	2004 W3 ha %	2004 W2 ha %	2007 W3 ha %	2007 W2 ha %
Closed forest	80.6	83.7	76.5	39.4	62.8	15.4
	100	**82.1**	94.9	**38.7**	77.9	**15.1**
Open forest		5.0	3.0	30.5	8.5	22.0
	4.9	3.7	29.9	10.5	21.6
Mosaic[a]	3.5	1.0	21.1	3.0	27.5
	3.4	1.3	20.7	3.7	27.0
Cocoa	7.4	0.03	7.0	1.4	23.0
plantation	7.3		6.9	1.7	22.6
Agriculture[b]	2.2	0	1.9	0.8	10.5
	2.2		1.9	1.0	10.3

[a]Mixture of cocoa, crops, natural forest patches and secondary forest
[b]Mixture of annual crops – most maize, cassava, peanuts.
Source: IKONOS 7th Jan. 2001, Quick Bird 23rd June 2004, Quick Bird 27.05.2007.

20 Soil and Water Degradation Following Forest Conversion in the Humid Tropics

Table 20.2 Water balance components areal precipitation (P), discharge (Q), evapotranspiration (aET), hydrological characteristics (discharge coefficient Qc, Q5) and specific suspended load output for the sub-catchments W3 (natural forest) and W2 (forest conversion)

Year/weir	Areal P (mm)	Q (mm)	aET (mm)	Qc (% of P)	Q5 (mm d^{-1})	Specific suspended load (kg ha^{-1} mm^{-1} Q)
2002/W3	2,550	929	1,826	28.4	9.2	0.45
2002/W2	2,542	1,373	1,169	54.0	5.0	0.32
2003/W3	2,734	814	1,920	29.8	7.1	1.34
2003/W2	2,579	1,378	1,201	53.4	4.4	1.40
2004/W3	2,090	351	1,739	16.8	2.2	0.37
2004/W2	2,081	691	1,390	30.7	2.8	2.56
2005/W3	2,942	1,029	1,913	35.0	10.8	1.04
2005/W2	2,645	1,642	1,003	62.1	5.2	1.77
2006/W3	2,843	1,325	1,518	46.6	10.6	0.81
2006/W2	2,682	1,506	1,176	56.2	5.3	1.86
2007/W3	3,193	1,352	1,841	42.3	10.0	0.81
2007/W2	3,013	1,633	1,380	54.2	6.8	1.04

Q5 = 5% of days equal or greater of this discharge (mm d^{-1}).

In the first step, an increase was identified for the area classified as mosaic (a mixture of annual crops, cocoa plantation, natural forest patches and secondary forest) from 3.4% in 2001 to 20.7% in 2004. The second increase revealed was that of cocoa plantation area from 6.9% in 2004 to 22.6% in 2007. Smallholders through slashing and cutting trees do the conversion manually. Slash & burn, which was still common until 2004, is hardly practiced anymore in the area. Before planting the cocoa trees, the ground is usually prepared for 1 or 2 years by planting annual crops (mainly maize).

By monitoring the discharge since 2002 from a changing (weir 2, deforestation) and a mainly natural sub-catchment (weir 3), we tried to investigate the effect of land use change on river discharge. The yearly discharge development and flood characteristic suggests a development of the lower sub-catchment discharge (W2) to a slightly more extreme runoff regime. For years with similar rainfall amounts (2002, 2003, 2005, 2006) runoff coefficient increased slightly in sub-catchment W2 from 2002/2003 to 2005/2006 by +3–8% (+ 260 mm) and peak discharges (Q5) increased from 4.4–5.0 mm d^{-1} to 5.2–5.3 mm (Table 20.2).

Despite high rainfall, Q5 is stable at about 10 mm d^{-1} in the forest sub-catchment (W3, 2005–2007). For both sub-catchments, discharge increased with increasing rainfall amounts. The forest sub-catchment (W3) has a more extreme behaviour (higher flood discharge Q5, higher total discharge in rain-laden years) due to higher steepness and lower water storage capacity (groundwater storage capacity 57.5 mm, in sub-catchment W2 = 82.4 mm, Kleinhans, 2003). Despite higher rainfall in 2005, discharge at W3 increased in 2006 by 300 mm. Hydrograph separation with digital filter method (TSPROC and Doherty, 2003; Kleinhans, 2003) indicated direct flow components (surface runoff and quick interflow) over the years of 25–31% for weir 2 and 35–40% for weir 3, except for drier years like 2004 revealing only 13%.

However, during storm flow events the rate of direct flow reached a maximum of 88% (W2) and 93% (W3) on an hourly base. Over all years (2002–2007) an average of 30% days year^{-1} was identified with direct flow components at both sub-catchments, particularly during the main rainy season (December–January, March–May) having high soil moisture content. The calculation of the surface runoff coefficient (SF) shows no significant differences between the sub-catchments with a variance of 2–13% (W3) and 5–12% (W2) of yearly discharge (driest and wettest year). According to Kleinhans (2003) surface runoff increased by 20% in the forest conversion sub-catchment (W2) compared to the "natural" forest sub-catchment (W3).

Based on the weekly river discharge samples (2002–2007), the specific yearly dissolved nutrient output was calculated for weir 2 and weir 3 (Table 20.5). For the cation Ca no significant differences ($p = <0.001$) exists from year to year (exception only for dry year 2004). However, between weir 2 and 3 there are significant differences from 2005 until 2007. For TNb (total nitrogen bound) main differences occur in 2005 and 2007. The converted sub-catchment (W2) exhibits an increase in dissolved nutrient output from 2002 (82% closed forest, Table 20.1) until 2006 (15% closed forest 2006, Table 20.1) for the main cations and TNb, but a decrease from 2006 to 2007 for Ca and an increase for K, TNb and phosphorus (Table 20.5) was recorded despite much higher precipitation in 2007.

In regard to phosphorus in both catchments from 2006 to 2007 there is a remarkable increase. With the overall dataset for base flow and storm flow, the dissolved nutrient concentration between weir 2 and weir 3 was tested for statistical significance (Wilcoxon, Mann & Whitney). The biogenic cation K, which is highly enriched in the forest and cacao sites (Nicklas, 2006) shows, parallel to the macronutrient N (TNb), a significant increase during storm flow conditions and significant higher concentrations at weir 2 for both discharge conditions. Pedogenic and geogenic cations (Ca, Mg, Si) show typical dilution effects during storm flow conditions (Table 20.4).

The development of suspended sediment load delivers an impression of soil erosion and sediment output (without bedload!) from the two sub-catchments. According to Lipu (2007) the volume weighted specific suspended load output increased by factors of 2–8 (weir2/weir3 kg ha^{-1} mm^{-1} runoff) from 2002 to 2004 and 2006 (Table 20.2). It is remarkable, that suspended load decreased after 2005/2006 to 1.04 kg ha^{-1} mm^{-1} Q in 2007!

Three scenarios (forest/cacao/slash & burn (corn with cassava)) were simulated based on the main processes of land use change in the catchment and related to land use pattern of 2002 (Table 20.3). The results of the three-water balance model exhibit clear differences to the real land use. The increase in discharge for annual crops is completely related to overland flow. Interflow and base flows are reduced compared to present land use.

The causes of increased surface run-off are linked to reduced interception and reduced infiltration. Research on saturated hydraulic conductivity clearly shows the

Table 20.3 Water balance of 2002, modeled water balance for different land use scenarios (forest, cacao, and traditional agriculture (corn with cassava)) and relative changes of the components compared to the present land use mixture (Kleinhans, 2003)

Simulated water balance [mm year^{-1}]	Present land use	Forest		Cacao		Annual crops	
Precipitation	2,360	2,360	–	2,360	–	2,360	–
Interception	409	480	17.4%	309	–24.4%	114	–72.1%
Throughfall	1,951	1,880	–3.6%	2,051	5.1%	2,246	15.1%
Transpiration and soil evaporation	1,027	1,046	1.9%	910	–11.4%	1,232	20.0%
Evapotranspiration	1,436	1,526	6.3%	1,219	–15.1%	1,346	–6.3%
Discharge	987	923	–6.5%	1,153	16.8%	1,064	7.8%
Direct flow (overland flow)	227	181	–20.3%	318	40.1%	414	82.4%
Interflow	489	476	–2.7%	542	10.8%	403	–17.6%
Base flow	271	266	–1.8%	293	8.1%	247	–8.9%
(Stock) storage changes	–63	–89	41.3%	–12	–81.0%	–50	–20.6%

reduced conductivity for older cacao fields, *imperata* grassland and annual cultures (corn) of k_{sat} 6 mm h^{-1} compared to k_{sat} 210 mm h^{-1} of forest and younger cacao fields. Additionally, the runoff coefficient increased from 2.2% for forest to 6–9% for corn and old cacao fields as revealed by the soil erosion plots (Wischmeier-plot-size, Gerold et al., 2010).

For the cacao scenario, all discharge components are contributing to the increase in discharge. The increase of overland flow is much lower than for the agricultural (annual crops) scenario due to the relatively high leaf area index of cacao with much higher interception values. With same soil parameters for all simulation run for the cacao scenario show an increase for all three discharge components. The forest scenario leads to the reduction of all discharge components. The reduction of base flow and interflow, however, are low compared to the surface run-off, which is reduced by 20.3%.

20.5 Discussion

Confirmed by several tropical paired catchment studies (Critchley and Bruijnzeel, 1996; Bonell, 2005) river discharge in general increases with forest conversion, but seasonal variation, peak floods and dry weather discharge depend on rainfall intensity, catchment characteristics and type of land use (e.g. pasture and annual crops versus agroforestry). The main characteristics of the Nopu catchment include the forest conversion with slash & burn activities that increased mainly since 2003/2004 and with reduction of closed forest in the sub-catchment of weir 2 from 82% (2002)

to 22% (2007). It is assumed that the natural forest in the sub-catchment area B (W2) will eventually be lost completely and that the major part of the mosaic land use in 2007 (27%) will also be converted into cocoa plantation.

From 2005 to 2006 one farmer opened the natural forest at the south-western border of the sub-catchment from weir 3 (about 2 ha with maize cultivation, own observation 2006). This was part of the beginning of the human impact with a decrease of closed forest from 94.9 to 77.9% in the sub-catchment 3 (Table 20.1). In consequence, changes in water balance components (interception – 10–15%, transpiration – 6–20% Table 20.3, Falk et al., 2005; Oltchev et al., 2008) led to an increase +11% of discharge in 2006 without increase in annual rainfall from 2005 to 2006. With comparable rainfall amount the discharge in sub-catchment B (weir 2) from 2003 to 2005 increased from 200 to 250 mm. In this period, closed forest decreased from 80 to 39%. Hilbert (1967) pointed out that a rough generalization of 1% forest conversion leads to an increase of discharge by 2.5–4.5 mm. Following this assumption the theoretical increase at weir 3 is between 43 and 77 mm (measurement +300 mm) and at weir 2 from 100 to 180 mm (measurement +200). This indicates, that forest conversion in the humid rainforest and in a mountainous steep area (average slope angle 24.5°) with high percentage of quick interflow (27–35% of total discharge, Kleinhans, 2003) could lead to quick response of total discharge. In the sub-catchment B, a much deeper unsaturated soil zone (ground water level 1–3 m during the year) with higher groundwater recharge (max. 140 mm month^{-1}) causes higher dry weather discharge (Q95 3–4 fold higher at weir 2 compared to weir 3).

Therefore, the discharge regime is less extreme. Applying the water balance method aET values vary between 4.2–5.2 mm d^{-1} (weir 3) and 2.7–3.8 mm d^{-1} (weir 2) (Table 20.2). The calculated evapotranspiration according to Penman-Monteith is 3.9–4.1 mm d^{-1} (Kleinhans, 2003). Real evapotranspiration (aET, mm year^{-1}) for tropical rainforests in SE-Asia show amounts between 1,440 mm (East Malaysia Kuraji and Paul, 1994) and 1,835 mm (East Malyasia Malmer et al., 2005 and water balance method). The difference in daily aET between sub-catchment 2 and 3 varies between 0.9 mm d^{-1} (2006) and 2.0 mm d^{-1} (2003), which is in line with the measured differences of rainforest and cacao/mixed agriculture (maize-cassava) (Hölscher, 1995; Oltchev et al., 2008).

Deforestation alters hydrologic processes and water quality in catchments worldwide. Experimental controlled catchment studies with paired watersheds and the use of modelling for evaluating and predicting hydrological consequences of land-use change had advanced in the last years (De Vries and Eshleman, 2004). But for the humid tropics consequences of land-use change on hydrology and mainly hydro-geochemistry have received little attention. Forested catchments in the humid tropics with soils having high base cation saturation release up to one order of magnitude more cations in stream water than those with cation-poor soils (Bruijnzeel, 1991). In south-western Brazilian Amazon basin for example, Biggs et al. (2002) analysed, only once, the wet and dry season of 49 river sites for streamwater hydrochemistry (catchment areas between 18 and 12,500 km^2) considering soil exchangable cation content and percentage of deforested watershed. They found that concentrations of potassium (K), sodium (Na) and chloride (Cl) increase with deforestation extended and the ratios of disturbed to undisturbed stream concentrations range from 1.2 to 3.1 (K) and 0.8–2.2 (Na). But they also found higher stream water cation concentrations in watersheds originating from soils with higher base cation stocks.

Markewitz et al. (2004) found higher stream water concentrations of Ca, Mg, K and Na during the wet season in a partially deforested eastern Amazonian watershed. Ballester et al. (2003) found similar result for larger rivers in Rondonia, but the relation of stream chemistry to land use was complicated by the fact that most of forest clearing in the region occurred on soils with higher original cation stocks.

The advantage of our longterm experimental catchment study is, that base flow and storm flow nutrient concentration were analysed and that the underlying geology and soil characteristic is similar for both sub-catchments (weir 2 and 3).

Median nutrient concentrations for base flow differ for all years between weir 2 and weir 3 referring to Mg, K, Na and TNb. For storm flow conditions a significant difference additionally exists for Ca and a general dilution happens with the exception of TNb (Table 20.4). This indicates, that higher soil solution concentrations of biogen K and nitrogen in the upper soil horizon (0–100 cm), originating from the slash & burn slopes, contributes to interflow input to river discharge. For Ca we measured highest concentration in 30 cm depth of the soil profile (lysimeter plot with maize/cacao) compared to forest and cacao lysimeter station (median 13 mg l^{-1} vs. 5 mg l^{-1}, locations, Fig. 20.1). Therefore, during storm flow, the saturated upper soil horizons contribute to higher calcium leaching and quick interflow transport to the river. Vertical gradient of cation concentration in soil solution exist for K with maximum of median K-concentration in 60–90 cm depth (7 mg l^{-1}) (data evaluation for 2002–2007). For K leaching to groundwater (base flow conditions) and additionally subsurface stormflow contributes to K export to the river.

Applying a conceptual model of water and solute transport pathways to a pasture hillslope in South-western Brazilian Amazon for 10 rainfall events (October–November 2002) Biggs et al. (2006) identified that near stream zones (equally to subsurface storm flow at downslopes in Nopu) controlled the export of K and total dissolved N and that Na and Si export was coming via groundwater from upslope areas (saturated saprolite). Box plots of vertical cation soil water solution from 30 to

Table 20.4 Median values (mg l^{-1}) and standard deviation (s) for the main soluble cations and TNb at the outlets of weir 3 (natural forest) and weir 2 (forest conversion) for base flow (BF) and storm flow (SF) (2002–2007)

Weir/element	Ca	Mg	K	Na	Si	TNb
BF 3	17.1 (3.7)	5.1 (1.2)*	2.8 (1.2)*	3.1 (1.3)*	7.0 (3.2)	0.20 (0.27)*
BF 2	17.6 (4.3)	5.9 (1.5)	3.3 (1.4)	6.0 (2.4)	8.5 (3.9)	0.24 (0.23)
SF 3	12.9 (3.2)*	3.7 (1.1)	3.4 (1.4)*	3.1 (1.2)*	4.9 (2.8)	0.52 (0.39)*
SF 2	15.0 (3.5)	4.4 (1.3)	4.1 (1.7)	4.7 (1.7)	6.3 (2.7)	0.65 (0.53)

Source: Since 2002 water samples weekly for BF, for SF event based; statistical significance between weir 2 and 3 at P = <0.001 with*.

180 cm depth for the three lysimeter plots (Fig. 20.1) with a constant increase of Na concentrations from 30 cm (median 2 mg l^{-1}) to 180 cm (median 5 mg l^{-1}) and for Si concentrations with 10 mg l^{-1}–15 mg l^{-1} and a dilution effect (storm flow concentrations lower than base flow concentrations, Table 20.4) support the assumption of geogenic export (Na, Si) with the slope groundwater to the stream zone.

Biggs et al. (2002) reported a significant increase of K and Na stream concentration with increasing deforestation, whereas Ca and Mg concentration is mainly influenced by the soil conditions (soil exchangeable cation content). This may explain that for Ca and Mg the human deforestation impact at weir 2 is overlapped by the main soluble soil cation stock (75–95% Ca + Mg saturation of CEC).

Chemical fingerprints of flow paths in a native small rainforest catchment in western Amazonia (Elsenbeer and Lack, 1995) dominantly covered by Ultisols showed very low soil water cation concentrations from 30 until 90 cm depth (<1 mg l^{-1} Ca, Mg, K) and low base- and storm flow river concentrations (<2 mg l^{-1}), which are typical for *terra firme* lowland streams of a clear water type of Amazonia (Brinkmann, 1985). Based on young relief and soil development in Central Sulawesi, even for natural mountainous rainforest catchments, nutrient concentrations belong even more to the category of "white water rivers".

The converted sub-catchment B (W2) has a significant increase from 2002/2003 onward for TNb-concentration and specific dissolved nitrogen output, indicator for the intensive new slash & burn practise in 2003 and 2004 (Table 20.5). Comparing the year 2002/2003 with 2006/2007, a significant increase also exists for Ca and K. The increase of specific TNb-output from 2006 to 2007 in sub-catchment C indicates the beginning influence of forest opening since 2006 associated with higher mineralisation rates of the litter- and humus horizon. Exhibiting higher discharge, the absolute nutrient output per ha and year from sub-catchment B (Table 20.2)

Table 20.5 Specific soluted nutrient output (kg ha^{-1} year^{-1} mm^{-1} Q) for the subcatchments W3 (natural forest) and W2 (forest conversion)

Year/weir	Ca	Mg	K	Na	Si	TNb	P
2002/W3	0.17	0.05	0.03	0.03	0.08	$0.13*10^{-2}$	$0.47*10^{-2}$
2002/W2	0.19	0.08	0.05	0.09	0.14	$0.21*10^{-2}$	$0.69*10^{-2}$
2003/W3	0.18	0.05	0.02	0.03	0.03	$0.07*10^{-2}$	$2.32*10^{-2}$
2003/W2	0.19	0.05	0.03	0.07	0.04	$0.08*10^{-2*}$	$1.50*10^{-2}$
2004/W3	0.11*	0.03	0.02	0.02	0.05	$0.26*10^{-2*}$	$0.03*y10^{-2}$
2004/W2	0.13*	0.04	0.02	0.04*	0.08	$0.27*10^{-2*}$	$0.06*10^{-2}$
2005/W3	0.17	0.05	0.03	0.04	0.05	$0.24*10^{-2*}$	–
2005/W2	0.21	0.06	0.04	0.07	0.07	$0.38*10^{-2*}$	–
2006/W3	0.18	0.05	0.03	0.04	0.07	$0.39*10^{-2*}$	$0.04*10^{-2}$
2006/W2	0.22*	0.06	0.04	0.07	0.09	$0.40*10^{-2*}$	$0.05*10^{-2}$
2007/W3	0.15	0.04	0.03	0.03	0.07	$0.47*10^{-2*}$	$0.12*10^{-2*}$
2007/W2	0.19	0.06	0.05*	0.08	0.10	$0.76*10^{-2*}$	$0.17*10^{-2*}$

– not enough data for calculation.
* significant difference between year and 2002/2003 at $P < 0.01$.

Table 20.6 Development of soluted nutrient output in sub-catchment W3 (forest) and W2 (forest conversion) between 2003 and 2006/2007 (kg ha^{-1}, average a. standard deviation)

Year	Weir/areal P (mm)	Ca	Mg	K	Na	Si	TNb	P
2003	W3 2734	148.4 (27.5)	38.1 (6.4)	19.7 (2.1)	25.7 (7.8)	25.0 (33.3)	0.6 (0.4)	4.4 (8.6)
	W2 2579	258.4 (35.1)	75.5 (14.4)	41.9 (8.4)	95.7 (24.7)	47.7 (67.0)	1.1 (0.7)	9.6 (2.0)
2006	W3 2843	235.6 (31.2)	65.2 (12.4)	35.4 (6.7)	51.1 (22.5)	94.7 (27.3)	5.2 (2.7)	5.3 (7.5)
	W2 2682	329.8 (36.3)	97.5 (7.9)	53.5 (13.9)	111.7 (11.7)	131.2 (26.7)	6.0 (3.2)	7.5 (2.6)
2007	W3 3193	202.3 (40.3)	57.1 (10.0)	40.5 (13.6)	41.6 (7.0)	87.9 (39.7)	6.4 (4.0)	1.6 (0.3)
	W2 3013	312.6 (75.8)	95.4 (19.6)	71.3 (18.6)	128.4 (3.0)	159.4 (54.1)	12.5 (4.9)	2.7 (0.3)

is higher by a factor of 1.5–2.2 in the sub-catchment B (W2) in relation to sub-catchment C (W3) (Table 20.6). The output relation between weir 2 and 3 from 2003 to 2007 increased tremendously for nitrogen! For the main cations, only for K a significant difference exists at weir 2 for the specific dissolved output comparing 2003 with 2007.

These findings correspond with paired catchment studies and broad watershed studies (Biggs et al., 2002). Neill et al. (2006) found significant higher concentrations only for K and Na in pasture streams compared to rainforest streams (factor 1.3–2.2), but Ca and Mg concentrations did not differ. Biggs et al. (2002) concluded that the lack of a strong land use signal for Ca and Mg might be due to the duration of the impact of deforestation on soil water solute concentration. But our findings, with no time lag between stream nutrient concentration analysis and deforestation, are more consistent with the geo-chemical model of mobility of elements and cation exchange capacity of the soil (Ludwig et al., 1999). Most soluble elements (K, Na, NO_3) have better detectable stream responses to vegetation disturbance.

Comparing ratios of disturbed (forest conversion) to pre-disturbance (natural forest) stream dissolved nutrient output (Table 20.6, 2007), ratios range from 1.5 to 1.7 (Ca, Mg) up to 3.0 (Na) and 2.0/1.8 for TNb and K. This is comparable to the range of enrichment reported by Neill et al., 2006, forest to pasture) which were Ca (1.2), Mg (1.2), K (2.3) and Na (1.3) and Biggs et al., 2002, forest to mixed agriculture) reporting rations for K (1.2–3.1) and Na (0.8–2.2).

Stream solute concentration and output of biogenic K (highest through fall enrichment by forest and agroforest systems (e.g. cacao) as well as nitrogen deriving from mineralisation of the A-horizon can be used as good indicator of human disturbances in natural watersheds.

Our results indicate along with the development of median nutrient concentrations in base flow and storm flow, that the patchiness of land use and ongoing conversion to cacao plantation after 1–2 years of slash & burn agriculture do not result in high increase of specific dissolved nutrient output for most cations. The absolute increase mainly results from the increase in discharge for both weirs. Between 2003 and 2005 with comparable annual rainfall in sub-catchment B there is the highest increase in discharge (+9% Qc), which correlates with the first intensive slash & burn impact in the sub-catchment (Table 20.1). Development of discharge coefficient and dissolved nutrient outputs (exception TNb) after 2005 indicates a "stabilizing effect" due to the increase in young cacao plantation. The same effect exists for the development of suspended load outputs from sub-catchment B with a remarkable decrease from 2005/2006 to 2007.

20.6 Conclusions

Additionally to the fact that higher soil erosion with higher sediment yield is induced by rainforest conversion to different land use types in the inner tropics the water balance is significantly disturbed as well. This also holds true in the case of small-scale agricultural land use. Type and magnitude of changes for individual water balance components in small mesoscalic catchments depends on land use types, age and kind of field practice. For our case of traditional shifting agriculture followed by cacao plantation, the quickest responses after forest conversion were found for soil nutrient leaching and dissolved nutrient export to the river for potassium and nitrogen. Also soil erosion with suspended load and sediment yield react quickly in years with highest deforestation impact. Hydrological responses react with time lag, depending on the rate of deforestation. For our case, forest cover decreased from 82 to 15% from 2001 to 2007. Consequently total runoff and peak discharges increased significantly from 2003 to 2005 in the forest conversion sub-catchment. But land use change to more cacao plantation (cacao boom in Sulawesi) decreased the discharge coefficient after 2005. The same development exists for the specific suspended load.

The main results are:

1. Due to forest conversion river discharge increase by 9–17% from 2002/2003 to 2005/2006 mainly driven by increase in overland flow and quick interflow (Gerold and Leemhuis, 2008). The application of the water balance model WASIM-ETH with simulation results for the Nopu catchment (forest, cacao-plantation, slash & burn with maize and cassava) support these experimental results and gives insight into the changing discharge components.
2. The increase in overland flow leads to an increase in soil erosion, so specific suspended load increased from 2002 to 2005 (maximum) and then decreased with extension of cacao agroforestry.
3. Specific soluted nutrient output mainly increased following deforestation for potassium and nitrogen and indicate the importance of these biogenic elements for the analysis of human disturbances in watersheds.

However, extensive land use practice with cacao plantations is less critical and can maintain soil fertility and water resources better than annual crops.

Acknowledgements This study was supported by the German Research Foundation (DFG – SFB 552). Special thanks to the Institute Pertanian Bogor (IBP), Java, Indonesia and the Universitas Tadulako (UNTAD), Palu, Central Sulawesi, Indonesia for the productive research collaboration.

References

Abbott, M.B. and Reefsgaard, J.P. (1996). Terminology, modelling protocol and classification of hydrological model codes. In: Abbott M.B., Refsgaard J.P. (eds.), Distributed Hydrological Modelling. Kluwer Academic Publishers, Dordrecht, Boston, London, pp. 17–37.

Ballester, V.M., Victoria, D.C., Krusche, A.V., Coburn, R., Victoria, R.L., Richey, J.E., Logsdon, M.G., Mayorga, E. and Matricardi, E. (2003). A remote sensing/GIS based physical template to understand the biogeochemistry of the Ji-Paraná river basin. Remote Sensing of Environment 87:429–445.

Biggs, T.W., Dunne, T., Domingues, F. and Martinelli, L.A. (2002). Relative influence of natural watershed properties and human disturbance on stream solute concentrations in southwestern Brazilian Amazon basin. Water Resources Research 38:1–16.

Biggs, T.W., Dunne, T. and Muraoka, T. (2006). Transport of water, solutes and nutrients from a pasture hillslope, southwestern Brazilian Amazon. Hydrological Processes 20:2527–2547.

Bonell, M. (2005). Runoff generation in tropical forests. In: Bonell M., Bruijnzeel L.A. (eds.), Forests, Water and People in the Humid Tropics. University Press, Camebridge, pp. 314–406.

Brinkmann, W.L.F. (1985). Studies on Hydrobiogeochemistry of a tropical lowland forest system. GeoJournal 11:89–101.

Bruijnzeel, L.A. (1991). Nutrient input-output budgets of tropical forest ecosystems: A review. Journal of Tropical Ecology 7:1–14.

Critchley, W.R.S. and Bruijnzeel, L.A. (1996). Environmental impacts of converting moist tropical forest to agriculture and plantations. IHP Humid Tropics Programme Series No. 10, UNESCO p. 48

De Vries, R. and Eshlemann, K.N. (2004). Land-use change and hydrologic processes: A major focus for the future. Hydrological Processes 18:2183–2186.

Doherty, J. (2003). PEST-model independent parameter estimation. pp. 1–9.

Elsenbeer, H. and Lack, A. (1995). Chemical fingerprints of hydrological compartments and flow paths at La Cuenca, western Amazonia. Water Resources Research 31:3051–3058.

Falk, U., Ibrom, A., Oltchev, A., Kreilein, H., June, T., Rauf, A., Merklein, J. and Gravenhorst, G. (2005). Energy and water fluxes avove a Cacao agroforestry system in Central Sulawesi, Indonesia, indicate effects of land use change on local climate. Meteorologische Zeitschrift 14:219–225.

Food and Agriculture Organisation of the UN. (1998). World reference base for soil resources, by ISSS-ISRIC-FAO. World Soil Resources Report no. 84. Rome.

Food and Agriculture Organisation of the UN. (2003). The State of the World's Forests. FAO, Rome, p. 151.

Gerold, G., Fremerey, M., Gudhardja, E. (eds.). (2004). Land Use, Nature Conservation and the Stability of Rainforest Margins in Southeast Asia. Environmental Science. Springer, Berlin, p. 533.

Gerold, G. and Leemhuis, C. (2008). Effects of "ENSO-events" and rainforest conversion on river discharge in Central Sulawesi (Indonesia) – problems and solutions with coarse spatial parameter distribution for water balance simulation. In: M. Sànchez-Marré, J. Béjar, J. Comas, A. Rizzoli, G. Guarisco (eds.), Proceedings of the International Congress on Environmental Modelling and Software Integrating Sciences and Information Technology for Environmental Assessment and Decision Making (iEMSs), Barcelona, 553–565.

Gerold, G., Murtilaksono, K., Monde, A., De Vries, K. and Lipu, S. (2010). Consequence of rainforest conversion on soil erosion, river discharge and suspended load in Central Sulawesi (Indonesia). Zschr. f. Geom. (submitted, in revision).

Hölscher, D. (1995). Wasser- und Stoffhaushalt eines Agrarökosystems mit Waldbrache im östlichen Amazonasgebiet. Göttinger Beiträge zur Land- und Forstwirtschaft in den Tropen und Subtropen 106:133.

Kleinhans, A. (2003). Einfluss der Waldkonversion auf den Wasserhaushalt eines tropischen Regenwaldeinzugsgebietes in Zentral Sulawesi (Indonesien). Dissertation, University of Göttingen.

Kuraji, K. and Paul, L.L. (1994). Effects of rainfall interception on water balance in two tropical rainforest catchments, Sabah, Malaysia. In: Proceedings of the International Symposium on Forest Hydrology. Tokyo, Japan, pp. 291–298.

Leemhuis, C. (2005). The impact of El Nino Southern Oscillation Events on water resource availability in Central Sulawesi, Indonesia. EcoRegio 21:150.

Leemhuis, C., Erasmi, S., Twele, A., Kreilein, H., Oltchev, S. and Gerold, G. (2007). Rainforest conversion in Central Sulawesi, Indonesia – Recent development and consequences for river discharge and water resources. Erdkunde 61:284–293.

Lipu, S. (2007). Rainforest conversion consequences on the suspended material load and output in the Nopu catchment in Central Sulawesi, Indonesia. EcoRegio 22:122.

Ludwig, B., Khanna, P.K., Hölscher, D. and Anurugsa, B. (1999). Modelling changes in cations in the topsoil of an Amazonian Acrisol in response to additions of wood ash. European Journal of Soil Science 50:717–726.

Mackensen, J., Ampt, J., Garrelts, A. and Kortekas, M. (1999). Report on reconnaissance soil survey in the Napu and Sopu valley, Central Sulawesi. Göttingen, pp. 45.

Malmer, A., van Noordwijk, M. and Bruijnzeel, A. (2005). Effects of shifting cultivation and forest fire. In: Bonell M. and Bruijnzeel L.A. (eds.), Forests, Water and People in the Humid Tropics. University Press, Cambridge, pp. 533–560.

Markewitz, D., Davidson, E., Moutinho, P. and Nepstad, D. (2004). Nutrient loss and redistribution after forest clearing on a highly weathered soil in Amazonia. Ecological Applications 14: 177–199.

Nash, J.E. and Sutcliffe, J.V. (1970). River flow forecasting through conceptual models, Part 1 – A discussion of principles. Journal of Hydrology 10:282–290.

Neill, Ch., Deegan, L.A., Thomas, S.M., Haupert, C.L., Krusche, A.V., Ballester, V.M. and Victoria, R.L. (2006). Deforestation alters the hydraulic and biogeochemical characteristics of small lowland Amazonian streams. Hydrological Processes 20:2563–2580.

Nicklas, U.G. (2006). Nährstoffeintrag durch Bestandsniederschlag und Streufall in Kakao-Agroforstsystemen in Zentral-Sulawesi, Indonesien. Diplomarbeit, Göttingen, pp. 134.

Oltchev, A., Ibrom, A., Priess, J., Erasmi, S., Leemhuis, C., Twele, A., Radler, K., Kreilein, H., Panferov, O. and Gravenhorst, G. (2008). Effects of land use changes on evapotranspiration of tropical rain forest margin area in Central Sulawesi (Indonesia): modeling study with a regional SVAT-model. Journal of Ecological Modelling 212:131–137.

Schulla, J. (1997). Hydrologische Modellierung von Flussgebieten zur Abschätzung der Folgen von Klimaänderungen. Züricher Geographische Schriften. Verlag Geographisches Institut ETH, Zürich.

Schulla, J. and Jasper, K. (1999). Model Description WASIM-ETH. ETH, Zürich.

Van Dijk, A.I.J.M. and Bruijnzeel, L.A. (2004). Runoff and soil loss from bench terraces. An event based erosion process model. European Journal of Soil Science 55:317–334.

Chapter 21
Relationships Between Land Degradation and Natural Disasters and Their Impacts on Integrated Watershed Management in Iran

Hamid Reza Solaymani Osbooei and Mahnaz Bafandeh Haghighi

Abstract Integrated watershed management entails coordination and cooperation as well as the management of soil and water for the attainment of several objectives. There are numerous examples and regulations to be cited from countries such as the US and Australia, which clearly depict measures and activities directed at the management of soil, water and other environmental resources. Reports indicate that discharges from groundwater resources, fountains, and Qanats in vast areas of Iran have drastically declined so that supplying drinking water in some areas is facing serious difficulty. Water table decline, gradual salinization of the underground reservoirs and their declining quality lead to more soil salinization and increasingly lower harvest yields. Combined with the low irrigation efficiency (around 30–35%), these problems indicate that less than 10% of the annual precipitation is being used for productive use. Increased salinity in both surface and ground water resources can be ascribed to overexploitation, depletion of the vegetation cover, deforestation, range degradation, and increasing agricultural developments. All in all, these factors have led to more runoff and to rising water level in low-lying areas. From a management perspective, considering the potential capacity of watersheds as well as supervising land use in watersheds with regard to the importance and functions it can have various governmental and strategic aspects of watershed management. Among the first steps to be taken in implementing a management system are strengthening the intersectional coordinating institutions, identifying critical and strategic watersheds (e.g., in terms of their importance for drinking water supply), defining types of land uses according to definite standards, and supervising the general trend of activities in the watershed. For solving the above problems, the participation of local communities in watershed management must be taken more seriously than ever. Mechanisms

H.R.S. Osbooei (✉)
Department of Watershed Management, Forest, Range and Watershed Organisation, Tehran 11445-1136, Iran
e-mail: hrsolaymani@yahoo.com

must be developed to remove undertakings by the government in favor of a privatized system of watershed executive management so that the public sector will be in a position to take its fundamental role of steering and mentoring the implementation of projects. Along these lines, it is essential for banks and the related systems to be empowered in order to offer financial support and backing.

Keywords Land degradation · Natural disaster · Iran · Watershed management

21.1 Introduction

The importance of soil, water, and natural resources as cornerstones and essential factors in national production are universally recognized in both developing and industrialized countries. However, the exploitation and management of these resources do not follow a unique pattern across the world. Whereas in some parts of the world proper management practices are employed, in others these vital resources are experiencing fatal devastation due to either natural or human-induced factors or both. The emerging crises in most regions of the world as a result of serious water shortage are the immediate results of population increase and reduced water quality due to imprudent water management by man and increased water pollution. Formulation of region-specific practices in watershed management gains increasing importance in the face of the global variations in climatic and demographic conditions.

Iran had a high birth rate in the decade from 1976 to 1986. The dire consequences of this rather high population explosion and the war from 1980 to 1988 are currently unfolding in all social and environmental aspects. Planners, researchers and managers in their concerted efforts to preserve the natural environment and its resources for future generations are busily engaged in finding practical ways to prevent the degradation inflicted by the growing demands of this large population on soil, water, and natural resources.

Compared with other sources such as petroleum, *water* cannot be easily transported. It flows under gravity and easily escapes access. Under our climatic conditions, it is one of the most scarce and most vital resources that have witnessed drastic quantitative fluctuations in recent years. Plant and animal resources in the region have been undergoing corresponding variations. This calls for a firm planning for prudent water consumption in the years to come. Water inflicts great damages to human beings and property during floods and endangers social life during droughts. It is, therefore, essential to establish a balance between supply and demand prior to, and as a preparation for, the emergence of water crises. The next limiting factor is *soil,* whose interactions with water and plant life are essential. Human intervention in this cycle leads either to production or to regression and deterioration of these resources. It is evident that human intervention in natural cycles can have both negative and positive effects. Just as population increase, overexploitation and the employment of heavy machinery can destroy natural resources, well-defined and thought-out management and resource utilization practices can bring with them higher yields, more productivity and sustainability. It is the coexistence of soil, water, and human resources that requires a sound management to create a balance in life, production, vegetation cover and forest; without the human element, nature will find a balance of its own.

21.2 Integrated Watershed Management

Management normally deals with establishing coordination among disparate entities and/or components with casual reference to the role of the manager. This approach, however, fails to distinguish the role played by the society and its culture in management. Water resources management is of special importance because of its vast temporal and spatial dimensions, its relationships with both natural and social laws, its interactions with governmental and non-governmental organizations, and its role in food security, social services, and community infrastructures. Water resources management involves the management of the geo-system, the bio-system, and the human community system. That is why water resources management must of necessity include the management of the whole water cycle. Such a management system will consist in coordination and cooperation in the management of the soil and water to achieve such objectives as harvesting reliable and pollution-free water; allocating timely water for irrigation purposes; supplying adequate drinking water; supplying for industrial, energy, and environmental water demands; soil erosion control in catchments; and flood and sediment control and increasing the effective life of dam structures. There are different aspects to this management that involve not only the various sectors and ministries but also the community and its culture. This vast and inclusive phenomenon will necessarily call for a national and comprehensive approach.

More than 130 years has passed since J.W. Powell first defined the term "watershed" or "catchments" in 1896. Watershed is defined as a natural unit for the management and preparation of land. The management of this natural unit involves the management of environmental resources aimed at maximum productivity at the least social, environmental, and economic damages. Integrated watershed management is a coordinated and well-behaved management of economic, social, biological and physical systems with the least negative effects on the resources while also securing and supplying for the benefits of the community.

Integrated watershed management entails coordination and cooperation as well as the management of soil and water for the attainment of several objectives. There are numerous examples and regulations to be cited from countries like USA and Australia (Burton, 1988) that clearly depict measures and activities directed at the management of soil, water and other environmental resources.

There is universal consensus among all those involved that, on a large scale, it is essential to exercise governmental authority in watersheds so as to regulate and define land use in such a way to observe the overall capacity of the watershed, to control water quality and quantity, and to minimize soil erosion and sediment deposition. However, most experts and practitioners maintain that, on a smaller scale, operations such as erosion and sedimentation control in smaller watersheds must be implemented as executive and corrective measures in those regions (IECA, 2000). Salinity control and supervision for optimal use of the resources may also be included under the rubric of corrective measures.

Integrated watershed management, as a new concept, calls for the collaborative activities of an assortment of expert groups including hydrologists, hydraulic engineers, soil conservationists, natural resources experts, geologists and a multitude of other engineers, experts and social scientists in a concerted effort assisted by local communities to seek ways of preventing further losses and degradation of natural resources.

Recognizing the watershed as the best unit for planning, development and resource management encourages appropriate and proper institutional arrangements in the watershed for a comprehensive management system. The different stages of management in the watershed require different institutional arrangements. The different stages include production, balance, and sustainability of the resources; harvesting, distribution, utilization and consumption. There are numerous factors involved in resource management that make watershed management a quite complicated task during the production and sustainability stages, requiring the coordination and collaboration among many organizations and sectors. Although initially managed easily with the erection of control infrastructures and water transfer systems, the stage of development, harvesting and distribution faces more and more management complexities with increasing control on resources, increasing prices, emergence of environmental problems, pollution of resources, and overexploitation of groundwater resources. When these issues are integrated with consumption and reuse management into a unified whole, then the need for more coordinated arrangements among the sectors and executive bodies involved becomes evident while the differences and conflicts for meeting the dynamic demands are also disclosed.

21.3 The Need for an Integrated Management

The gradual salinisation of soil, the devastation of vegetation cover, reduced soil productivity, increased erosion and sedimentation, and increasing chemical pollution were first noticed in Australia where, in 1985, the integrated watershed management was adopted and watershed was recognized as a management unit. In the United States, the change in the course of the Kimi River located in the Mississippi drainage basin and the reduction of its length by half resulted in the devastation of 18,000 ha of wetland with the subsequent extinction of valuable fish species, increased environmental pollution, and loss of beautiful natural sites. This fatal event led to Food Security Act of 1985. In 1987, a Resource Conservation and Development Program (RC&D) was also passed by the legislature, which enabled the government to control not only soil erosion but also farm production. According to this program, 40–45 million acres of land that had escaped farming due to excessive erosion was brought under tree and forage cultivation within a 10-year plan (Journal of Soil & Water Conservation, 1988).

Decades have passed now since the idea of watershed management was first put forth in countries like USA, Australia, and New Zealand. In others such as Canada, UK and US, it has been found that drinking and irrigation water supply must be left to the private sector while watershed management must remain within the hands of the government assisted by public participation. There are still other countries that are in the process of reviewing their policies with regards to the management of their strategic resources.

Although agriculture and livestock production have a history of over 5,000 years in Iran, evidence shows that the major trend in soil erosion and flood flows is of a recent history of no more than 100 years. Like in other parts of the world, there have been great efforts in Iran put to the development and proper utilization of soil and water resources. Introduction of mechanization and machinery in these areas and easy access to new equipment have enabled individuals and local communities to exercise a more rapid influence on nature. However, some have brutally invaded nature either for their base living or merely for more profits.

Official statistics indicate that Iran once had an estimated forest area of 19.5 million hectares. Today, this has dropped to around 12.4 million hectares while there has also been a drastic qualitative decline in forest areas. During the years 1985–1995, the poor, moderated and good quality of forest and ranges was decreased from 14, 60, and 16 million hectares to 9.3, 37.3, and 43.3 million hectares, respectively (Statistical Yearbook, 1997, Iran Statistics Center, 1999 and Ministry of Budget and Planning of I.R. of Iran).

It can thus be claimed that forests and ranges in Iran have paved a regressive path. This is only why Iran is more fragile and susceptible than other parts of the world. Despite having less than 1% of the world population, Iran has at its disposal less than 0.36% of world fresh water, and for worst also suffers from an uneven distribution. The per capita water 40 years ago was 4,000 m^3 this has currently been reduced to 1,200 m^3, and will further reduce to 400 m^3 in 40 years' time (Statistical Yearbook, 1986, Iran Statistics Center, Ministry of Budget and Planning of I.R. of Iran).

Iran has a long tradition in civilization. More than 80,000 operational and non-operational Qanats (underground water galleries), dams and weirs and a lot more of historical water structures have been reported by historians to exist along with a multitude of lively forests. But sadly today, we are witnessing an imbalance in our environment. The British Consulate in Iran reports in its travel accounts the names of the then thick forests in Sarakhs region, a region that is currently barren and dry (Rahbari and Roshani, 1986). The elders in many parts of the country such as Fars, Kurdistan, and Azerbaijan have also reported of forests and plant covers so thick that they were impassable.

Soil erosion is a major problem in Iran's watersheds. Soil erosion will occur wherever the soil is bare and water flows across it. The loss of topsoil from the land limits the growth of vegetation. Soil washed into rivers can cause flooding and further erosion. This disturbs the river habitats and the sedimentation downstream affects the capacity of streams, lakes and reservoirs.

Sediment in the river and reservoir causes environmental problems that have undesired effects on water quality, and create large losses and dangers. On the basis of measured data and monitoring in a number of reservoirs, the country will annually be facing 250 million cubic meter of sedimentation in the reservoirs and 400 million cubic meter of sedimentation to downstream of dam reservoirs and irrigation networks.

Population growth and need for food from the main sources of income in watershed area (agriculture and animal husbandry) over exploit the land. A decrease of vegetation cover and deforestation due to overgrazing and cutting the trees for fuels or cash (or to increase arable land) results in more flooding and landslides and destroy infrastructures. The climate change due to the human activities has also changed rainfall pattern and has increased the risk.

Infrastructures play an important role in reducing poverty, and have a direct impact on watershed management. Either lack or not having access to it by being destroyed with natural disasters could affect the rural communities and their lives.

Under natural conditions in the catchments, livestock and plant life establish an environment in balance for their interactions. If man succeeds to conserve this balance through his proper management, it will be possible to produce and supply the necessary food while also conserving soil, water, and plant resources.

Soil erosion and low range productivity follows range degradation. This will, in turn, lead to larger range area requirements by people to feed their livestock, which only means invasion on neighbouring range and involvement in tribal conflicts across the region.

Once the vegetation cover on watershed is degraded, rainwater will not infiltrate into soil and forms fast runoff flowing in small streams, which ultimately join each other to create destructive floods. The floods may then demolish and wash away all facilities such as roads, bridges, downstream villages, farmlands, and other infrastructures.

Implementation of watershed management practices and rehabilitation of vegetation cover will provide the conditions for a proper resource exploitation led by scientific management. It is essential to know that rehabilitation is not final job and exploitation must be guided by technical and scientific considerations. Otherwise, degradation and desertification will occur again and the investments will be wasted. It is even more essential to prevent rather than cure; i.e., technical consultation must be sought in the maintenance and conservation of catchments prior to their degradation.

The relationship between environment change, land degradation, natural disasters and the poverty is evolving and gradually being better understood. This chain is the major threat to the civilization, and specifically to the development of the poorest and most marginalized people. This chain if not controlled creates a moving vicious cycle, which will be growing, and expanding and more people will be affected by it. The disruptions in the natural environment not only lead to desertification and drying but also influence the constant and variable features of the strata in the biosphere as well. The air, soil and subsoil strata; plant and animal communities; and even the human legacy on the earth are all endangered by this disruption. The only way to reverse or to slow down this dangerous trend is a well-defined management system and a set of proper activities.

The establishment of Ministry of Jihad Construction in 1979 initiated the establishment of large-scale rural development activities Parallel to the infrastructure development, the shortcomings and constraints in water resources development in the country have led the decision makers to select agricultural developments as the key and pivotal indicator of development in general. Thus, a large-scale program of dam construction and water resource development has been initiated to meet the demands of agricultural development. Unfortunately, the activities conducted in some area could not stop land degradation and migration and in some cases accelerated the process. Ignorance of proper methods of watershed management and over exploitation by people in some marginal areas causes these watersheds to soon change into dry deserts devoid of life and vitality.

Lack of a comprehensive or holistic approach, improper interventions, and inadequate legislation with varied conflicting regulations are among the major causes for the natural resource degradation occurring in Iran. These have culminated in a feeling of possession of watersheds by local settlers, which has led to acquisition and devastation. As an example of this devastating trend, rather than developing development plans proportionate to watershed capacity, a plan was developed to secure self-sufficiency in meat production that inevitably increased livestock numbers from 3 to 7 times the carrying capacity of ranges. This was while another plan was developed for self-sufficiency in agricultural products that changed more range land into dry land farms. The two conflicting plans had drastic effects on the degradation of rangelands in Iran (but fortunately, a Livestock Balancing Act has been recently developed and approved that will hopefully reverse the trend). Other destructive activities include the application of excessive machinery on land, eradication of bushes, deforestation, and improper land use changes. These have cumulatively resulted in more flood occurrences, by 10 times, with their subsequent droughts and groundwater level decline in most areas.

Water erosion has gained an increasing momentum over the past half century. Studies indicate that 50% of maize yields are lost if 60 cm of topsoil is washed away and that in cases where 80 cm of the topsoil is washed away, the resulting reduction will be 100%. Studies on less deep soils (Pierce et al., 1984); show when 20 cm of topsoil is washed away, 50% of the crop is lost; when 25 cm of the topsoil is washed away, the losses will be 100% (Giam Pietro and Pimental, 1993). Studies on forest areas based on direct measurement of seed growths indicate that a loss of only 3 cm of the forest soil causes an average reduction of 80% in seed trees.

Over the past 3 decades, the sedimentation process in catchments areas has witnessed an increasing trend due to intensified soil erosion and flood events (Sharifi et al., 2002). The sediments have not only inflicted destructive damages on different water storage and transfer facilities but have caused adverse geomorphologic changes in rivers as well. This will naturally result in flooding risks in agricultural lands and industrial and municipal facilities in the neighbouring areas due to sediment deposition and reduced flow capacity. The environmental impacts due to sedimentation such as adverse effects on water quality and aquatic habitats a well as rising of wetland beds and natural lakes are alarming signals of enormous nationwide economic havoc. During only one flood event, approximately 10% of the Golestan dam reservoir was deposited with sediments, the economic damage being estimated at around $150 million (Pierce et al., 1984).

Reports indicate that discharges from groundwater resources, fountains, and Qanats in vast areas of the country have drastically declined so that supplying drinking water in some areas is facing serious difficulty. Water table decline, gradual salinization of the underground reservoirs and their declining quality lead to more soil salinization and increasingly lower yields. Combined with the low irrigation efficiency (around 30–35%), these problems indicate that only less than 10% of the annual precipitation is being used for productive use.

Another emerging aspect of the problems associated with improper management of watersheds is land salinization. Increased salinity in both surface and ground water resources can be ascribed to overexploitation, depletion of the vegetation cover, deforestation, range degradation, and increasing agricultural developments. All in all, these factors have led to more runoff and to rising water level in low-lying areas.

Salinity is a threat to soil productivity and public health in rural and urban communities living in catchments areas. Rural farms, urban development, infrastructures (bridges and roads), water users, and the environment are adversely affected by salinity. Although salinity may be a natural phenomenon in some areas, increased salinity in other parts is the direct result of uprising water table which is itself caused by changes in land use such as depletion of the vegetation cover, urban development, river flow regulation, irrigation and cropping. Such factors result in what is commonly called "secondary salinization".

If left unattended to, not only will salinity lead to decreasing harvests, as its first and foremost consequence, but sustained salinization will also result in alkalinity of the land and its ultimate desertification. When this happens, it will be too late to take any corrective measures and we should only witness the expansion of deserts, immigrations, rapid livestock mortality, and loss of our national legacy.

21.4 The Methods to Be Applied in Watershed Management

Among the first steps to be taken in implementing a watershed management system are strengthening the intersectional coordinating institutions, identifying critical and strategic watersheds (e.g., in terms of their importance for drinking water supply), defining types of land uses according to definite standards, and supervising the general trend of activities in the watershed.

Any sectoral approach must be abandoned in favour of an integrated, intersectoral approach. In sectoral approaches, it is only the economy of the cubic meter of harvested water that is given first priority in investments at the expense of environmental and community relocation costs. It must be borne in mind that communities are the backbones to all development. In such profit-seeking, limited approaches, even the costs for implementing upstream watershed management operations and the costs of constructing water transfer and distribution facilities are not taken into account. This is while in an integrated management priority is given to supplying water for communities settled over the years, abatement of economic stresses on these communities, decentralized rather than centralized investments in environmental plans and artificial recharge schemes, utilization of precipitation, and increasing soil moisture to assist range and forest reclamation.

In (semi-)arid countries, more than 90% of water resources is often used in irrigation. This is while precipitation in these regions undergoes considerable variations through the year and from year to year. The quantity of precipitation in different regions is directly related to the altitude of the watershed. In (semi-)arid zones, evaporation is also rather high, usually several times the quantity of precipitation. This combination of climatic and demographic conditions have led to the development of decentralized water utilization organizations through which participatory systems such as the Qanat, local water storage systems (Abanbar), Khooshab, and Ice water systems have evolved through the centuries.

In some areas in Iran, the population may not exceed 600 people km^{-2}. The population density varies at upstream and downstream watershed. Land farms in these areas are normally small and water consumption has naturally been organized in participatory systems. Single cropping and centralized water use is rarely ever found in (semi-) arid regions and is associated with numerous problems in cases where such instances exist. For instance, in situations where the population density is low and where population is concentrated along coastal areas or riverbanks, there is usually no concern for water evaporation and/or wastage so that no human intervention is generally needed to optimise on consumption or to increase the water available. The only concern in such areas would be transfer of water to population centres.

In contrast, in areas with high population density and high evapotranspiration rates, watershed management is concentrated on water transfer and, even more important, on reducing evaporation and on salvaging a percentage of the precipitation before it is lost to evaporation. In most rural watershed areas, the main difficulty lies in supplying drinking water, combating dry spells, and providing water for supplementary irrigation for dry land farming. Centralized investments by the government in constructing large dams to procure drinking water for downstream large cities or establishing centralized agricultural sites may not be considered as solutions to rural community problems. Instead, these communities always look for ways to mitigate dry spells and periodical droughts through water storage in underground reservoirs, through preservation of soil moisture, or through storage of water in places as near to their residence as possible.

There is no doubt that the Water Council Act of 2000 passed by the Iranian Parliament has triggered an important movement in non-sectoral integrated watershed management, The legislation has as its objective the formation of a supreme watershed management body to serve as the highest decision-making body in defining national policies and guidelines in this area. The need for the formation of a national water council had already been emphasized and reiterated by various authorities over the years. It is certain that the Council will establish different sub-commissions to address its agenda items including the major aspects of defining policies, major allocations, inspection and information dissemination, as well as the financial aspects of water concerning water supply and use, and priorities in investments. It is expected that the duties and liabilities of each sub-sector will be defined after the sub-commissions are set up. The National Water Council will be expected to establish, in a later stage, regional watershed management councils as management-executive bodies as well as local watershed management councils, which will be in charge of daily affairs and activities within the watershed. These lower councils will consist of political managers, executive managers and users' representatives.

A next step by the Council will be to define an upper uptake limit based on the base water year, water rights, and environmental considerations. Independent mechanisms will have to be developed for water allocations to environmental projects (watershed management, desertification control, range and forest reclamation...). The water pricing must become realistic in order to economize on water use and make water supply economical. Special exemptions or subsidies need to be given to small landowners and environmental projects. For agricultural and industrial establishments, the pricing system should be based on water consumption rates and total production rates for every cubic meter of water used. Volumetric measurement installations will allow exchange and trading on a larger scale between the downstream and upstream communities. In order to conserve the physical, chemical and biological quality of water, a quantified system of quality indices must be defined so that every salinity unit can be commercialised for industrial and agricultural users according to a legal regime. Privatization and services provided in irrigation management and watershed management must be organized within a governmental structure.

21.5 Concluding Remarks

A new era has definitely started in Iran in the integrated watershed management that so far pursued disparate and independent objectives such as supplying irrigation or drinking water, neglecting issues such as flooding, erosion, sedimentation or the various environmental changes. It is certain that this trend has become a thing of the past, as watershed areas cannot be maintained intact. They must be managed for various objectives and beneficial uses. It is also definite that industrial, mining, recreational, agricultural, and range uses of watersheds will continue even at a more intensified level due to the growing demands.

The development operations of the recent past concentrated on single objectives such as water harvesting or its use, thus pooling all investments on a single objective. However, while water economy must be regarded as a major concern, it is necessary to define new policies for such areas as natural resources and livestock breeding, the environment, erosion, sedimentation, flooding, etc. Thus, creating coordination among various bodies and agreement on the policies and planning must form an essential objective of the integrated watershed management.

An integrated and demand driven strategy for sustainable development of watersheds with empowerment of people and active participation and partnership of local communities is necessary to ensure productivity and sustainability in the watersheds (Sharifi and Haydarian, 1999). Any successful approach should closely be involves community sectors at grass-root levels, including NGOs, women and youth, in formulation, planning and implementation as well as launching awareness to sensitise all stakeholders to understand the impacts and to identify their roles.

In most regions, comparing with other natural resources, water is the scarcest substance. As a result, production relies heavily on underground and surface water, which become increasingly scarcer. Water will remain a barrier to the achievement of poverty alleviation and food security. Most of renewable water resources have already been committed by conventional method of dam construction but the demand for water is exceeded renewable water supplies. Therefore the future emphasis must be directed towards increasing the efficiency of water management and increasing water productivity and producing more crops per cubic meter. Unless there is an increase in watershed investment to generate higher employment, income, productivity and production opportunity for local inhabitants the trend of watershed degradation will continue to exist.

A second question to be addressed is whether or not the existing legislations are adequate for our long-term objectives. The answer is clearly negative. Currently, a major problem in the way of integrated management is the lack of a set of comprehensive laws that can serve as the foundation and supporter of an integrated watershed management.

The next issue of concern is the planning and financing system. Most of the organizations involved are entangled and choked by the national bureaucracy inherent in the planning and financing system. They normally face resistance against reforms and improvements even in cases where the sources of the problem are identified and/or disclosed. Lack of financial and human resources most often discourage them. Measures must be taken to lift these obstacles.

Public participation by the local communities in watersheds must be taken more seriously than ever. Mechanisms must be developed to remove executive undertakings by the government in favour of a privatised system of watershed executive management so that the public sector will be in a position to take its more fundamental role of steering and mentoring the implementation of projects. Along these lines, it is essential for banks and the related systems to be empowered in order to offer financial facilities.

References

Burton, J.R. (1988). Catchment management in Australia. Civil Engineering Transaction CE 30(4):145–152.

Giam Pietro, M. and Pimental, D. (1993). The NPG Forum. Negative Population Growth, Teaneck, NJ.

IECA. (2000). Journal of Erosion Control. International Erosion Control Association, September/October 2000, Santa Barbara, CA, pp. 68–75, 104.

Pierce, F.J., Larson, W.E. and Dowdy, R.H. (1984). Evaluating soil productivity in relation to soil erosion.

Rahbari, M. and Roshani, Gh. (Translators) (1986). Khorasan and Sistan (by Yate E and Sir Charles Edward). Yazadan Publisher, Tehran.

Sharifi, F. and Haydarian, A. (1999). On the natural resources management strategy in Iran. Proceedings of the Regional Workshop on Traditional Water Harvesting Systems, Tehran-Iran, UNESCO, pp. 345.

Sharifi, F., Saghafian, B. and Telvari, A. (2002). The great 2001 flood in Golestan Province, Iran: causes and consequences. International Conference on Flood Estimation, 6–8 March, 2002, Berne, Switzerland.

Statistical Yearbook. (1986). Iran Statistics Center, Ministry of Budget and Planning of I.R. of Iran, p. 391.

Statistical Yearbook. (1997). Iran Statistics Center (1999). Ministry of Budget and Planning of I.R. of Iran, p. 174.

Chapter 22
Modelling Carbon Sequestration in Drylands of Kazakhstan Using Remote Sensing Data and Field Measurements

P.A. Propastin and M. Kappas

Abstract Landsat ETM+ data were related with field measurements of carbon stocks residing in the vegetation and in the soil along a 230 km west-east transect within Kazakhstan's grassland zone. The biomass carbon was correlated to the mid-infrared corrected Normalized Difference Vegetation Index (NDVIc). Soil carbon responded most strongly to the multi-spectral features of the Kauth-Thomas transformation (KT) and was modelled using the non-linear multiple regression. Retrieved models were applied to the Landsat ETM+ image and carbon stocks were mapped over Shetsky district of Karaganda province at the pixel-by-pixel scale. Total terrestrial carbon stocks estimated for the area of Shetsky district account to 97.2 million tons, whereas most part of it (97.6%) is residing in soil.

Keywords Carbon stocks · Kazakhstan · Grassland · Remote sensing

22.1 Introduction

Drylands cover more than 30% of the earth land area and comprise a variety of ecosystems, such as different variations of grasslands, which are large reservoirs of carbon as well as potential carbon sinks and sources to the atmosphere (Heimann, 2001). In Kazakhstan, grasslands, expanding from north to south (600–1,000 km) and from west to east (about 3,000 km), cover the largest part of the total territory and represent the major pool for carbon absorption and are believed to offset significant proportion of carbon emissions associated with fossil fuel combustion

P.A. Propastin (✉)
Department of Geography, Georg-August University, Göttingen 37077, Germany
e-mail: ppropas@uni-goettingen.de

(Lal, 2004). Information on carbon stocks within landscapes is an important topic for annual reports for the Kyoto Protocol. The Government of the Republic of Kazakhstan signed the Kyoto Protocol in 2003. Obtaining of quantitative information on carbon stocks over the huge area of the country is a great challenge for the researchers and land use planners of this young independent state. However, a detailed investigation of the stocks changes in grasslands of Kazakhstan has not yet been done. Thus, monitoring carbon sequestration in the grasslands is of great interest in relation to understanding the current status of the global carbon cycle and to meeting requirements of the Kyoto Protocol.

Carbon stocks of an ecosystem, in general, include the carbon residing in soil and biomass. Biomass includes the aboveground and belowground living mass, and the dead mass of litter. Several different approaches to estimating larger-scale carbon stocks may be undertaken, with each approach having both advantages and disadvantages. A standard concept is used by the Intergovernmental Panel on Climate Change (IPCC, 1997), which establishes ecological regions within an area to be studied, quantifies historical land use and land use change within those regions, establishes current soil and biomass carbon concentrations for each land use unit, and calculates and then aggregates the total carbon within each unit (Eve et al., 2001; Woomer et al., 2004).

Another concept, providing a capability for wide-area monitoring biomass production and carbon stocks, employs analysis of biomass inventory data obtained by field survey using the regression models between remote sensing data and measured biomass data (Lu, 2006; Zheng et al., 2004). In order to scale biomass from plot estimates to landscape and regional levels, the estimates have to be linked with spectral reflectance of remote sensing data. Grasslands regions are characterized by relatively high response of the vegetation cover to remote sensing-derived vegetation indices such as simple ratio (SR), normalized difference vegetation index (NDVI), and the corrected normalized difference vegetation index (NDVIc) (Tieszen et al., 1997; Paruelo and Lauenroth, 1998; Diallo et al., 1991). Biomass is frequently calculated from linear and non-linear regression models established between these indices and field measurements (Lu, 2006).

Recent studies have also shown that various characteristics of soil cover such as moisture, organic carbon, iron, nitrogen, salt and sodium content can be effectively estimated from spectral reflectance of remote sensing data particularly in regions with transparent vegetation cover resembling grasslands and shrublands (Ben-Dor et al., 1999; Bartholomeus et al., 2007). Remote sensing of soil carbon is based on the existence of a relationship between spectral reflectance and the carbon content in air-dried soil of the upper horizon. Organic carbon in soils is strongly related to total organic matter, iron oxides and other soil constituents that can successively estimated from remotely sensed data (Vinogradov, 1981; Leone et al., 1994; Jarmer et al., 2003).

In this study, we estimate the carbon stocks in a grassland-dominated region in Central Kazakhstan and describe the effects of land use changes. An inventory of terrestrial carbon stocks was conducted employing regression models between fine-resolution satellite data and field measured carbon stocks occurring in soil and plants. The approach allowed the estimation of total carbon stocks for the year 2004 for each pixel over the 250 × 250 km study area. Results were considered in relation to design of algorithms for satellite-based monitoring of the Kazakhstan-wide carbon stocks.

22.2 Study Area

The study area is located in the middle part of Kazakhstan between 48°20′ and 49°30′ northern latitude and 72° and 74°10′ eastern longitude and encompasses the southern margin of the Kazakh Hills (Fig. 22.1). It comprises the northern area of the Shetsky district in Karaganda province. The climate of the region is dry, cold and high continental. Average annual precipitation is about 250–300 mm. The most part of precipitation falls during warm period from March to October. Inter-annual rainfall variation has a coefficient of variation of 20–35%. The temperature amplitude is relative high: average January temperature is below −12°C and average July temperature is about 26–28°C. The growing season starts in April and continues till October.

Steppe grassland and short grassland are the two vegetation types covering the study region. The steppe grassland is dominated by genera *Festuca* and *Stipa*. Few euryxerophilous forbs occur; the co-dominants are dwarf shrubs of the genus *Artemisia* and sometimes of other genera, particularly *Anabasis* and *Salsola*. Species diversity is about 12–15 species in a square metre. The height of the canopy decreases from 30–40 cm in the north to 15–20 cm in the south, while vegetation cover decreases from 50–70% to 20–30%, and even less. The vegetation growth in the study area is strongly dependent on precipitation dynamics. Grasses and shrubs in the vegetation cover grow during the whole vegetative period, but the vegetation growth is most rapid during May and early June (the period of greatest precipitation) in the southern part, and during June in the northern part of the study area. During droughty months in summer (July and early August) their rate of development is hindered. This period of semi-dormancy occurs throughout the study region (Propastin et al., 2007).

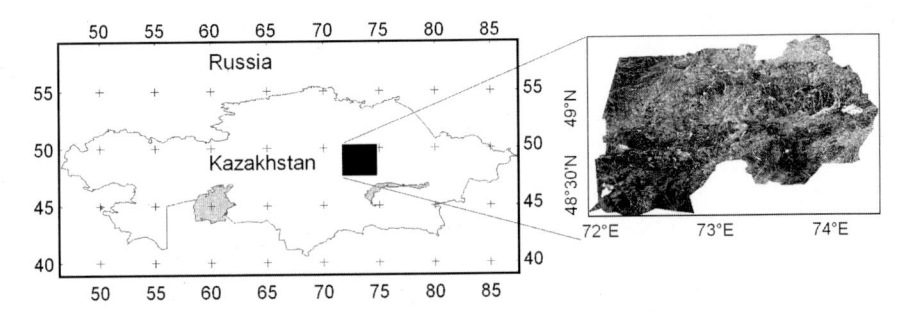

Fig. 22.1 Location of the study area (*black rectangle*) on a map of Kazakhstan (*left*) and a subset of Landsat ETM+ band 4 for Shetsky district (*right*)

22.3 Data

22.3.1 Carbon Data

Carbon data were collected from 14 georeferenced locations along a 230-km transect representing the entire range of ecological conditions and land uses in the study region. Fieldwork was carried out at the peak of growing season in June 2004. Data were sampled from four replicates within each test site. Each replicate occupied 0.0625 ha. Total carbon was defined as the sum of the herbaceous biomass, root, litter, and soil carbon pools, with biomass assumed to contain 0.47C. The peak-season living aboveground biomass and litter was measured by destructive sampling of 1.0 m^2, with samples weighted, sub-sampled, and dried at 65°C to constant weight to correct for moisture content. Roots were collected by excavating a square of 1.0 × 1.0 m to a depth of 50 cm with a narrow, flat-bladed shovel and handsaw. Coarse roots were hand sorted and washed. The remaining sample was dispersed in tap water, passed through a 2-mm sieve, and fine roots were collected without differentiation between live and dead roots. Roots were washed of gross mineral contamination, dried at 65°C to constant weight, and weighted. The proportions of litter and roots to aboveground biomass were calculated. The original values of dry matter were converted to carbon.

Total soil organic carbon to 50 cm was calculated from measurements of the total soil organic carbon concentrations of the 0–10, 10–20, 20–30, 30–40 and 40–50 cm soil layers and soil bulk density at 10, 20, 30, 40 and 50 cm depths. Soils were recovered in these five layers using a narrow, flat-bladed shovel. Organic carbon was determined by sulphuric acid and aqueous potassium dichromate mixture with external heating and then absorbance measured at 600 nm using a colorimeter (Nelson and Sommers, 1975). Samples for soil bulk density were collected by driving a thin-walled metal cylinder of known volume into the vertical face of the excavation with a wooden mallet at the five depths, withdrawing the filled cylinder, trimming soil protrusions with a knife, and storing the sample in a plastic bag for later soil moisture and bulk density determination. Bulk density was measured by oven drying soil cylinders of known weight and volume for moist soils.

22.3.2 Satellite Data

Mapping carbon sequestration relied on a fine-resolution satellite data set: a Landsat enhanced thematic mapper plus (ETM+) image acquired on 19 June 2004. The image has very good quality and is devoid of clouds over the study area. The terrain in the study area is predominantly flat, hence, atmospheric effects can be regarded as uniform. The data were first geometrically corrected using a set of ground control points extracted from 1:100,000 topographic maps. Then the data were radiometrically normalized using the absolute correction approach (Song et al., 2001). The Landsat ETM image of the study area was then subset spatially and spectrally for further data transformation and modelling.

22.3.3 Remotely Sensed Explanatory Variables for Carbon Stocks

After pre-processing the satellite data, a variety of vegetation indices and image transformations were calculated. Best results were produced by utilization of the near-infrared corrected normalized difference vegetation index (NDVIc) and features of the Kauth-Thomas transformation (KT). These indices and image transformations are to be briefly described here (Table 22.1). The NDVIc is a derivation from NDVI and was developed to incorporate middle-infrared information while at the same time accounting for background effects in the observed reflectance (Brown et al., 2000). The advantage of NDVIc over other vegetation indices such as NDVI is that the difference between cover types is very much reduced so that the accuracy for retrieval of biophysical variables for mixed cover types can be improved and a single algorithm can be developed without the use of a co-registered land cover map.

The Kauth-Thomas (KT) transformation (Kauth and Thomas, 1976) is commonly used for image transformation and enhancement. The KT transformation not only provides a mechanism for reducing data volume with minimal information loss but its spectral features are also directly related to the important physical parameters of the land surface (Crist and Cicone, 1984; Jin and Sader, 2005). In this study, the KT transformations of the six non-thermal ETM bands were performed to produce six multi-spectral features. All of them were potentially differentiated in terms of stability and change in a multi-spectral data set. We were primarily interested in the first three KT features (brightness, greenness and wetness), because they seem associated strongly with observed differences in vegetation and soil characteristics (Healey et al., 2005).

Table 22.1 Image variables used in the research

Index/image transformation	Formula*	Source
Corrected normalized Difference vegetation Index (NDVIc)	$(NIR - R)/(NIR + R)*(1 - \frac{MIR1 - MIR1_{min}}{MIR1_{max} - MIR1_{min}})$	Brown et al. (2000)
Tasselled cap transform Brightness (KT1) Greenness (KT2) Wetness (KT3)	$0.304*B + 0.279*G + 0.474*R + 0.559*NIR + 0.508*MIR1 + 0.186*MIR2$ $-0.285*B - 0.244*G - 0.544*R + 0.704*NIR + 0.084*MIR1 - 0.180*MIR2$ $0.151*B + 0.197*G + 0.328*R + 0.341*NIR - 0.711*MIR1 - 0.457*MIR2$	Kauth and Thomas (1976)

*NIR, near-infrared band (760–900 nm); R, red band (630–690 nm); B, blue band (450–520 nm); G, green band (520–600 nm); $MIR1$, middle-infrared band (1,550–1,750 nm); and $MIR2$, middle-infrared band (2,080–2,350 nm).

22.3.4 Integration of Field Measurements and Landsat Data to Produce Maps of Carbon Stocks

All the sample data have accurate coordinates derived from GPS devices and were located on the pre-processed Landsat image. These sample data were linked to the vegetation indices or the image transformations derived from the Landsat images to extract the value for each sample site. A window size of 3 by 3 pixels was placed over each individual test site to extract the mean value of each vegetation index or image transformation. Carbon parameters, such as standing crop carbon, total plant carbon, and soil carbon at different layers, from each site were associated with the satellite data using linear and non-linear regression approach. The leave-one cross-validation method was used for evaluation of modelling results. Values of root mean square error (RMSE) and determination coefficient (R^2) served as guides for accuracy assessment. Best accurate regression models were then employed to the Landsat ETM image in order to estimate carbon stocks for each pixel and to create final maps of carbon stocks over the entire study area (Table 22.2).

Table 22.2 Statistic models used for calculation carbon stocks from landsat data (g C m^{-2})

Parameter	Regression model	R^2	RMSE, g C m^{-2}
Standing crop	340.51*NDVIc-2.401	0.90	22.81
Total biomass	876.72*NDVIc-1.949	0.71	43.61
Soil 0–10 cm	exp(8.0033 – 0.0023*KT1 + 0.0135*KT2 – 0.0078*KT3)	0.58	352.35
Soil 0–20 cm	exp(8.8527 – 0.0032*KT1 + 0.0120*KT2 – 0.0064*KT3)	0.55	716.83
Soil 0–30 cm	exp(9.3184 – 0.0040*KT1 + 0.0089*KT2 – 0.0047*KT3)	0.51	962.22
Soil 0–40 cm	exp(9.4578 – 0.0034*KT1 + 0.0073*KT2 – 0.0029*KT3)	0.50	1,174.59
Soil 0–50 cm	exp(10.59 – 0.0078*KT1 – 0.0072*KT2 + 0.0101*KT3)	0.48	1,425.38

22.4 Results

Remote sensing-derived variables were useful predictors for carbon stocks in the living aboveground biomass, total biomass, and soil. Particularly the living green biomass responds very strongly to the remote sensing-derived NDVIc. The overall model for standing crop explained 90% of variance (p <0.0001). The model for total biomass has significantly lower explanation power with $R^2 = 0.71$, however, it was sufficiently for mapping this variable from Landsat data. The lower explanation power of the total biomass model is caused by enclosure of the vegetation parts that have no response (biomass of roots) or very weak response (biomass of litter) to remote sensing data. For carbon stocks in soil, the model's explanation power was significantly weaker: the percentage of explained variance ranged from 58 to 48%, depending on the soil layer. However, these models were also statistically significant at the $P <0.01$ level and showed sufficient accuracy in terms of their RMSE values for the application at regional scale with the Landsat data.

By employing the models from Table 22.1 both to Landsat TM and ETM images, we generated estimates of carbon stocks in biomass (separately in living aboveground biomass, and total biomass) and soil at different horizons for each pixel within the study area. The total carbon stocks were calculated by summing estimations of total biomass carbon and soil carbon. The final maps are presented in Fig. 22.2. The spatial distribution of carbon stocks residing in different ecosystem compartments shows clear pattern within the study area. There is a southwest-northeast gradient of carbon stocks, which is associated with general distribution of vegetation types across the study area driven by patterns in precipitation and temperature. It is well known that, because precipitation is the major limiting factor for vegetation growth in drylands, it controls strongly the spatial patterns in vegetation production (Robinson et al., 2002; Propastin and Kappas, 2008).

Total carbon stocks ranged from 3,837 g C m^{-2} in the southern section to 9,829 g C m^{-2} in the northern section (Fig. 22.2c). Soil is the most important carbon pool in the study area. The carbon residing in biomass ranged from 132 to 294 g C m^{-2} (Fig. 22.2b), whereas its overall part in the ecosystem total carbon stocks is less than 6% (Fig. 22.2d). The percentage of the carbon residing in biomass is significantly higher for pixels representing agriculturally used areas where it scores values >4.5%. In 90% of other pixels, it ranges from about 0.5 to 2.5%. The approach resulted in estimated total carbon stocks of about 97 million tones over the whole

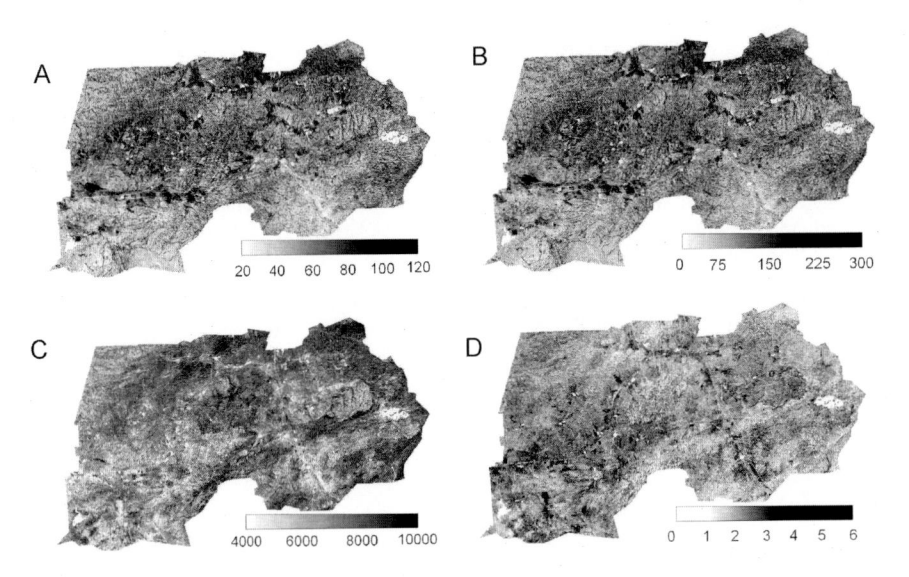

Fig. 22.2 Carbon stocks (g C m^{-2}) for 2004 in standing crop (**a**), in plant biomass (**b**) and total (**c**). *Panel* (**d**) shows the percentage of carbon stocks residing in biomass from the total carbon stocks

Table 22.3 Estimated terrestrial carbon stocks in Shetsky district (Karaganda province) for 2004

	Area (10⁶ ha)	Mean carbon stocks (t C ha⁻¹)	Area carbon stocks (t C 10⁶)	Percentage from total
Biomass	1.368	1.755	2.327	2.395
Soil	1.368	67.033	94.875	97.605
Total	1.368	68.787	97.203	100

study region with a mean value of 68.787 t C ha⁻¹. From the total system carbon, 97.6% is residing in soils (Table 22.3).

22.5 Conclusions

Based on Landsat ETM+ data and in situ measurements of carbon residing in different compartments of ecosystem, carbon stocks were mapped for a large grassland region in Central Kazakhstan. Remotely sensed data provided effective explanatory variables for predicting spatial variance in carbon across the study area. In the results presented here, carbon residing in biomass, including its aboveground, belowground parts and dead litter, could be modelled with high accuracy from the mid-infrared corrected vegetation index using linear regression approach. Soil carbon showed the strongest relationships with the multi-spectral features of the Kauth-Thomas transform (brightness, greenness, and wetness) and could be predicted by a non-linear regression model. The study suggests that terrestrial carbon stocks in Shetsky district (1.368 million hectares) was 97.2 million tons, with >97% of this residing in soil (0–50 cm). The findings of the study serve to a better understanding of carbon cycle in dry lands of the interior Eurasia and should play an important role in the establishing of an appropriate model for calculation of carbon assimilation in grasslands of Kazakhstan for annual reports for the Kyoto Protocol.

Acknowledgments The work of the first author was supported by a grant from the Space Research Institute at the Science Academy of the Republic of Kazakhstan. The authors gratefully acknowledge all colleges from the Space Research Institute and particularly N.R. Muratova for the field data provided and their helpful recommendations and approvals.

References

Bartholomeus, H., Epema, G. and Schaepman, M. (2007). Determining iron content in Mediterranean soils in partly vegetated areas, using spectral reflectance and imaging spectroscopy. International Journal of Applied Earth Observation and Geoinformation 9: 194–203.

Ben-Dor, E., Irons, J.R. and Epema, G.F. (1999). Soil reflectance. In: Rencz A.N. (ed.), Remote Sensing for the Earth Sciences: Manual of Remote Sensing. John Willey & Sons Inc., New York, pp. 111–118.

Brown, L.J., Chen, J.M., Leblanc, S.G. and Cihlar, J. (2000). Shortwave infrared correction to the simple ratio: An image and model analysis. Remote Sensing of Environment 77:16–25.

Crist, E.P. and Cicone, R.C. (1984). A physically-based transformation of Thematic Mapper data – the TM Tasseled Cap. IEEE Transactions on Geoscience and Remote Sensing 22: 256–263.

Diallo, O., Diouf, A., Hanan, N.P., Ndiaye, A. and Prevost, Y. (1991). AVHRR monitoring of savanna primary production in Senegal., West Africa: 1987–1988. International Journal of Remote Sensing 12:1259–1279.

Eve, M.D., Paustian, K., Follett, R. and Elliott, E.T. (2001). A national inventory of changes in soil carbon from national resources inventory data. In: Lal R., Kimble J.M., Follett R.F., Steward B.A. (eds.), Assessment Methods for Soil Carbon. Advances in Soil Science. Lewis Publishers, Boca Raton, FL, pp. 593–610.

Healey, S.P., Cohen, W.B., Yang, Z. and Krankina, O.N. (2005). Comparison of tasselled cap-based landsat data structures for use in forest disturbance detection. Remote Sensing of Environment 97:301–310.

Heimann, M. (2001). Zonal distribution of terrestrial and oceanic carbon fluxes. Max-Planck Institute für Biogeochemie. Technical Report 2.

IPCC. (1997). Revised 1996 IPCC guidelines for national greenhouse gas inventories reporting instructions. In: Houghton J.T. et al. (eds.), Intergovernmental Panel on Climate Change. Volume 1. Meteorological Office, Bracknell.

Jarmer, T., Udelhoven, T. and Hill, J. (2003). Möglichkeiten zur Ableitung Bodenbezogener Größen aus multi- und hyperspektralen Fernerkundungsdaten. Photogrammetrie, Fernerkundung & Geoinformation 14:115–123.

Jin, S. and Sader, S.A. (2005). Comparison of time-series tasselled cap wetness and the normalized difference moisture index in detecting forest disturbances. Remote Sensing of Environment 94:364–372.

Kauth, R.J. and Thomas, G.S. (1976). The tasselled cap – a graphic description of the spectral-temporal development of agricultural crops as seen by LANDSAT. Proceedings of the Symposium on Machine Processing of Remotely Sensed Data, Purdue University of West Lafayette, Indiana, pp. 4B-41–4B-51.

Lal, R. (2004). Carbon sequestration in soils of Central Asia. Land Degradation and Development 15:563–572.

Leone, A.P., Wright, G.G. and Corves, C. (1994). The application of satellite remote sensing for soil studies in upland areas of Southern Italy. International Journal of Remote Sensing 15: 1087–1105.

Lu, D. (2006). The potential and challenge of remote sensing-based biomass estimation. International Journal of Remote Sensing 27:1297–1328.

Nelson, D.W. and Sommers, L.E. (1975). A rapid and accurate method for estimating organic carbon in soil. Proceedings of the Indiana Academy of Science 84:456–462.

Paruelo, J.M. and Lauenroth, W.K. (1998). Interannual variability of NDVI and its relationship to climate for North American shrublands and grasslands. Journal of Biogeography 25:721–733.

Propastin, P. and Kappas, M. (2008). Reducing uncertainty in modelling NDVI-precipitation relationship: A comparative study using global and local regression techniques. GIScience and Remote Sensing 45:47–68.

Propastin, P., Kappas, M., Erasmi, S. and Muratova, N.R. (2007). Remote sensing based study on intra-annual dynamics of vegetation and climate in drylands of Kazakhstan. Basic and Applied Dryland Research 2:138–154.

Robinson, S., Milner-Gulland, E.L. and Alimaev, I. (2002). Rangeland degradation in Kazakhstan during the Soviet-era: Re-examining the evidence. Journal of Arid Environments 53:419–439.

Song, C., Woodcock, C.E., Seto, K.C., Lenney, M.P. and Macomber, S.A. (2001). Classification and change detection using Landsat TM data: When and how to correct atmospheric effects. Remote Sensing of Environment 75:230–244.

Tieszen, L., Reed, B.C. and Dejong, D.D. (1997). NDVI, C3 and C4 production, and distributions in the Great Plains grassland cover classes. Ecological Applications 7:59–78.

Vinogradov, B.V. (1981). Remote sensing of the humus content of soils. Soviet Soil Science 11:114–123.

Woomer, P.L., Toure, A. and Sall, M. (2004). Carbon stocks in Senegal's Sahel transition zone. Journal of Arid environments 59:499–510.

Zheng, D., Rademacher, J., Chen, J., Crow, T., Bresee, M., Le Moine, J. and Ryu, S.R. (2004). Estimating aboveground biomass using Landsat ETM+ data across a managed landscape in northern Wisconsin, USA. Remote Sensing Environment 93:402–411.

Chapter 23
The Effectiveness of Two Polymer-Based Stabilisers Offering an Alternative to Conventional Sand Stabilisation Methods

Ashraf A. Ramadan, Shawqui M. Lahalih, Sadiqa Ali, and Mane Al-Sudairawi

Abstract Sand stabilisation techniques have largely been utilised on the basis of trial-and-error at various locations in Kuwait. However, no specific technique has been identified as the most suitable so far. Short-term solutions, though attractive, are more costly and their effect is short-lived. Mechanical, chemical and biological dune and encroaching sand control techniques have been attempted, however, in isolation, leading to limited success with financial penalties sometimes. A common shortcoming of mechanical methods is the long-term, high-cost sand clearance work required to maintain efficiency, e.g., cost of clearing 415,000 m^3 of accumulated sand from As-Salmi Road-Kuwait reached KD108,166 during 5 months in 1993/1994. Chemical methods have their drawbacks too, mainly the pollution caused to the environment. For biological methods, it is well known that the revegetation of sand dunes of heights above 1.5 m is biologically non-viable without the initial use of mechanical and/or chemical techniques. Only after the soil has been stabilised and the threat of seedling burial and seedling root exposure has been eliminated, can one turn to biological techniques. In this paper we report on the preliminary findings of a series of laboratory experiments with the objective of assessing the stabilisation characteristics of two stabilisation chemicals, namely SUMF and SF-C, on sand taken from As-Salmi and Al-Atraf area in Kuwait. The series of the experiments conducted covered: grain size and chemical composition analyses as well as unconfined compression and water runoff erosion tests. The results obtained demonstrated the superiority of SUMF stabiliser when it comes to the mechanical strength. However, the water runoff erosion tests showed the SF-C treated samples to have higher resistance to water erosion than those treated with SUMF. Also, SF-C was superior as far as the thickness of the stabiliser layer is concerned.

Keywords Chemical techniques · Reforestation · Sand encroachment · Soil fixation · Kuwait

A.A. Ramadan (✉)
Kuwait Institute for Scientific Research, Safat 13109, Kuwait
e-mail: aramadan@kisr.edu.kw

23.1 Introduction

The climatic conditions in the State of Kuwait are hot, arid, with scanty rainfall. Hot and dry summer winds are usually experienced for the greater part of the year. Average annual rainfall and temperature are 110 mm and 26°C, respectively. In general the naturally available water in Kuwait is scarce and it varies in its quality, Al-Naser (1978). The factors of hot and dry weather, soil characteristics, wind speed and availability of local and regional sources of sand conspire to make the desert in Kuwait a suitable source for sand particles that can be easily entrained by the wind. The desert of Kuwait has a fragile environmental ecosystem balance between its natural elements including climate, water resources, soil, vegetation, and animal life (Kuwait National Report on the Implementation of the UNCCD, 2000). Man-related activities are the primary source disturbing this fragile balance. Examples include: overgrazing, successive sand and gravel quarrying, camping, off-road vehicle movement and military activities. As an example, Al-Dabi et al. (1996) reported an increase in the rate of sand dune formation in northwest Kuwait from 31 dunes year^{-1} during 1985–1989 to 321 dunes year^{-1} during 1989–1992 and related this to military activities. Human activities similar to those mentioned above, in the presence of some natural factors like high evaporation, low rainfall rates, sandy soils containing negligible amounts of organic matter and weak cohesion and high permeability between particles have initiated naturally-irreversible phenomena like soil migration and loss of its productive ability, increase in dust and sand storms, and the spread of mobile sand dunes as well as creation of new ones. Add to that the long drought periods, e.g., from 1958 till 1976 (Kuwait National Report on the Implementation of the UNCCD, 2000), which have affected Kuwait, producing the perfect conditions for a never-ending source of sand particles feeding storms and making the building blocks of active sand dunes.

The best and long-term solution to the problem of mobile sand dunes is to establish suitable, self-sustaining vegetation on them. The anticipated high cost is readily justified if one considers the direct threat posed by sand dunes due to burial of valuable infrastructure or blockage of stretches of main roads and the resulting serious social and economic problems. Add to that the savings on the cost of the continuous clearing of encroaching sands from roads, airport runways (civil and military), oil operational areas and towns. A study conducted by Ahmed et al. (1996) revealed that KD[1] 108,166 was spent on removing 415,000 m^3 from As-Salmi Road during 5 months only (26th June 1993 until 25th September 1993 and 19th July 1994 until 16th September 1994). Mechanical methods, e.g., fences, fences combined with ditches, sand ridges occasionally stabilised with oil, and paving have been tried with some success in Kuwait, Gharib and Al-Hashash (1985), Gharib et al. (1985), Anwar et al. (1987), Abdullah (1988), Khalaf et al. (1990) and Al-Sudairawi (1995). These techniques, however, require continuous maintenance and can become ineffective in a relatively short period of time. In this paper we report the results obtained from tests on two promising sand stabilisation chemicals, one of which

---

[1] Kuwait dollar.

is commercially available and the other has been developed and patented by KISR (Lahalih et al., 1988; Lahalih, 1997, 1998). The SF-C is prepared from concentrated solution of polymer modified styrene acrylic copolymer resin with a specially designed plasticiser while SUMF is a polyanionic Sulfonated Urea-Melamine Formaldehyde condensate. Both of the chemicals are reported to help reestablishment of the native vegetation cover and ultimately control migration of sand dunes.

23.2 Area of Concern

As-Salmi Road runs from As-Salmi border station in the south west to Al-Jahra city in the north east. The segment of As-Salmi Road of interest to this study runs close to Ali As-Salem Airbase, which is in a playa depression called Al-Atraf. Water catchments in this area collect water and sediments and act like a local source of sand when it dries during the summer season. A total distance of 50 km of the road is affected by mobile sand in the form of sand dunes or active sand sheets. According to Ahmed et al. (1996), near to Al-Atraf area, a 20 km stretch of As-Salmi Road, which is the subject of this study, is affected by dunes migrating from Ali As-Salem Airbase in the south-eastern direction (see Figs. 23.1 and 23.2).

The stretch of the road of concern to this project is located within the main path of the wind corridor and hence is subjected to the direct influence of Al-Huwaimliyah active mobile sand belts with various modes of Aeolian sand deposits (sand dunes, dust fallout and mobile sand). In their extensive study, Al-Ajmi et al. (1994), classified Kuwait into zones according to the severity of sand encroachment. The region surrounding Ali As-Salem airbase was considered to suffer from severe sand encroachment. The fact that about 30% of the sand cleared from the main roads in Kuwait comes from this Road, Ahmed et al. (1996), strengthens

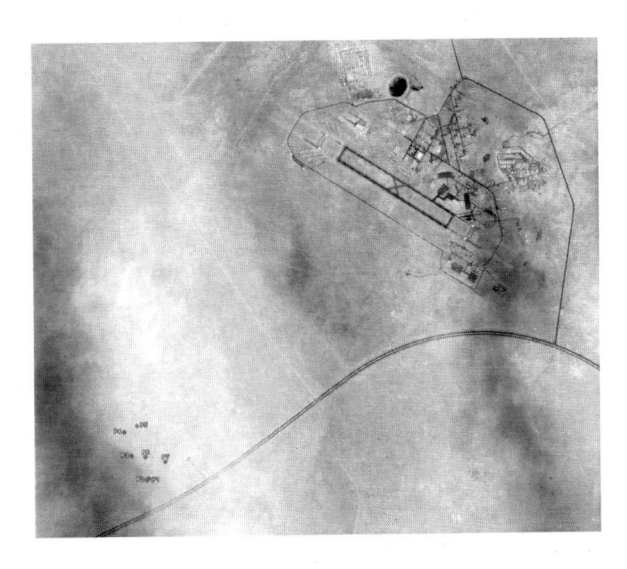

Fig. 23.1 The locations of the sand samples near As-Salmi Air Base

Fig. 23.2 The locations of the sand samples used in the study

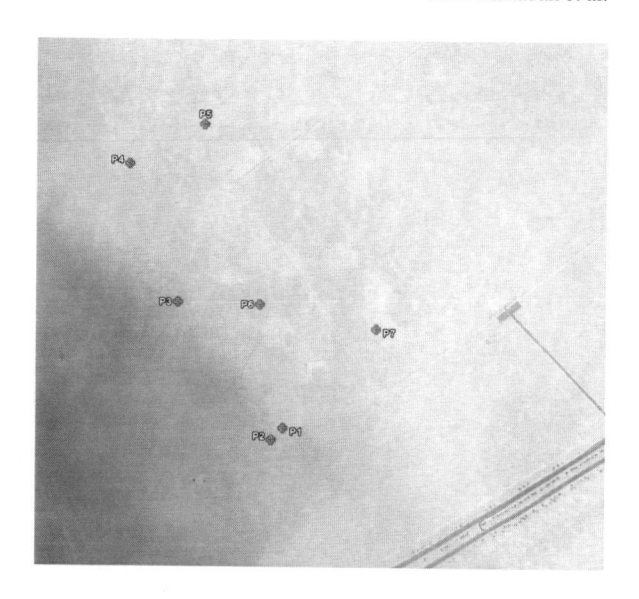

the previous argument. The axes of the existing sand dunes, which can be as close as tens of metres from the main road, are aligned with the prevailing wind direction in Kuwait, i.e. from northwest to southeast, and they have heights ranging from 0.5 to 4.5 m, Anwar et al. (1987). The rapid development of these dunes has been known for a long time and it is anticipated to continue unless preventive or control measures are implemented.

23.3 Materials, Equipment and Procedures

23.3.1 Grain Size Analysis

Two samples each weighing 10 kg were collected from 7 locations near As-Salmi Road (shown in Fig. 23.2). The samples were dried in an oven (70°C for 24 h). The dried disaggregated sediments of the samples were mixed thoroughly and split in two methods, i.e. (a) coning and sorting and (b) John Splitter Method, to obtain the desired weight of 0.2 kg for the grain size analysis using a nest of nine sieves.

23.3.2 Chemical Analysis

The chemical analysis was performed at the Petrography Laboratory of Kuwait Institute for Scientific Research (KISR) using an X-Ray Florescence spectrometry technique (XRF). A BRUKER-S4 PIONEER-Germany analyzer with

23.3.3 Unconfined Compression Tests

The compressive strength tests were conducted for different application rates as listed in Tables 23.1 and 23.2. For the control experiment, 640 g of sand was mixed with 112 ml of water to obtain a homogeneous mix. The mix was then cast in a cylindrical mould (50 mm diameter and 100 mm high). A 1 kg weight was used to compact the sample by 15 free falls from suitable height (30 cm). The moulds were left to cure in room temperature for 24 h before it was put in a preheated 70°C oven and left for another 24 h to cure. For the chemical stabilisers, the following procedure was used to make the samples:

Table 23.1 The different treatment levels of the sand with SUMF and the relevant solid content to sand weight ratio

Exp. No.	Mixture A (ml)	Water (ml)	Solution (ml)	Solution solid conc (%)	Solid content weight/sand weight (%)
1	102.82	681.18	784	2.86	0.5
2	205.5	578.5	784	5.71	1
3	308.26	475.74	784	8.57	1.5
4	411.01	372.99	784	11.43	2
5	513.76	270.24	784	14.29	2.5
6	616.51	167.49	784	17.14	3
7	719.27	64.73	784	20	3.5

Table 23.2 The SF-C to water mixing ratios for the SF-C compressive strength tests

(SF-C/water) mixing ratio	SF-C (ml)	SF-C (g)	Water (ml)	Total volume (ml)	Solid content in SF-C/sand Weight (%)
1/6	16	15.02	96	112	2.35
2/5	32	30.05	80	112	4.70
3/4	48	45.07	64	112	7.04
4/3	64	60.10	48	112	9.39
5/2	80	75.12	32	112	11.74
6/1	96	90.15	16	112	14.09
ALL	112	105.17	0	112	16.43

1. For the SUMF experiments, the below ingredients were mixed to get 2,570 ml of mixture A as shown below, the right amount of mixture A was then mixed with the right amount of water to reach the required concentration as given in Table 23.1. Seven experiments were conducted for each sand sample.

2. For the SF-C experiments, the right amounts of the SF-C concentrate solution were mixed with distilled water (<20,000 ppm salt) to make a diluted solution of volume = 112 ml as shown in Table 23.2. The seven experiments below covered mixing ratios from 1:6 (recommended) to pure SF-C.

	SNF	UF	PVA	H_2O	Mixture A	Solid content weight (g)
ml	128.5	385.5	1799	257	2570	449.08

3. 112 ml of the final mixture was mixed with 640 g of sand.
4. The mix was then cast in the cylindrical mould on three equal layers and compacted as mentioned above following Lahalih (1998).
5. The samples were left (4 for each concentration) to cure in room temperature for 24 h.
6. The samples were placed in a preheated 70°C oven for 24 h to cure.
7. Three of the four samples were used in the unconfined compressive test (one sample was kept for reference).
8. The samples were crushed at 1 mm min^{-1} rate.

The compressive strength tests were conducted in accordance with the international standards (American Society for Testing and Materials-ASTM C39). The compressive strength of the specimens was tested on a compressive test device (Testometric-AX FS-150KN).

23.3.4 Water Runoff Tests

In this task, the dried treated sand was subjected to water flow to simulate flooding in the water runoff test rig shown in Fig. 23.3. The steps for these tests were as follows:

1. Dry sand was wetted by 10% of its weight of distilled water to obtain a homogeneous wet mix.
2. The mix was compacted in (300 × 300 × 50 mm) trays with perforated bottom surface (1 mm holes) to allow water infiltration.
3. The compacted dry samples were sprayed with the SUMF (mixing ratios shown in Table 23.3) at the rate of 3.51 m^{-2} (recommended rate), except for the first experiment where the application rate is 7.01 m^{-2}. The application rates for SF-C are listed in Table 23.4.
4. Samples were allowed to dry under atmospheric conditions for 24 h and then in an oven at 70°C for another 48 h.
5. Samples were left under atmospheric conditions for 40–50 days.
6. The thickness of stabilised layer was measured.

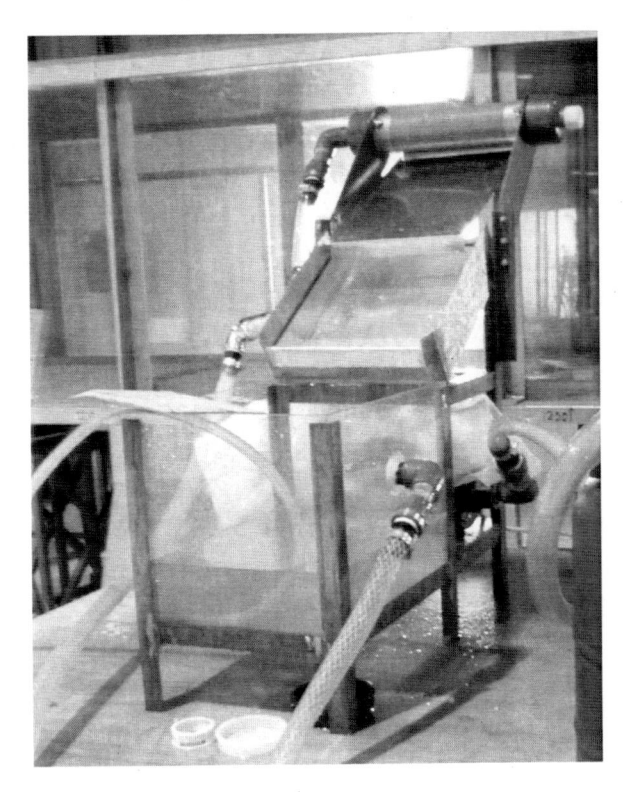

Fig. 23.3 The water runoff test rig

Table 23.3 The different application rates of the sand with SUMF and the relevant solid content to surface area ratio for the water runoff tests

Water runoff test (SUMF)

Exp. no.	SNF (ml)	UF (ml)	PVA (ml)	H$_2$O (ml)	Total volume (ml)	Solid content weight (g)	Solid content weight/area (g m^{-2})
1	24.77	74.31	346.79	184.13	630	108.00	120.00
2	12.39	37.16	173.39	92.06	315	54.00	60.00
3	6.19	18.58	86.70	203.53	315	27.00	30.00
4	3.10	9.29	43.35	259.27	315	13.50	15.00

7. Dried samples were weighed.
8. Trays of dried treated sand were inclined by 30° and then subjected to water flow from the top to simulate flooding. The water runoff test rig, shown in Fig. 23.3, was used for this experiment.
9. Water flow-rate was fixed at 6 l min^{-1} and the run lasted for 6 h.

Table 23.4 The different application rates of the sand with SF-C and the relevant solid content to surface area ratio for the water runoff tests

Water runoff test (SF-C)

Exp. no.	SF-C (ml)	SF-C (g)	H$_2$O (ml)	Total volume (ml)	(SF-C/water) mixing ratio	Solid content weight/area (g m^{-2})
1	8.57	8.05	51.43	60	1/6	8.94
2	17.14	16.10	42.86	60	2/5	17.89
3	25.71	24.15	34.29	60	3/4	26.83
4	34.29	32.19	25.71	60	4/3	35.77
5	42.86	40.24	17.14	60	5/2	44.72
6	51.43	48.29	8.571	60	6/1	53.66
7	60	56.34	0	60	ALL	62.60

10. At the end of each run, the sample was dried in a preheated 70°C oven for 48 h and then weighed to calculate the percentage of eroded material.
11. Once the sample has dried completely under atmospheric conditions it underwent another cycle of testing.
12. Each sample underwent three cycles of testing.

23.4 Results and Discussion

23.4.1 Grain Size Analysis

The grain size analysis was described in details in Ramadan and Lahalih (2007). In brief, the sand used can be categorised as fine to medium ranging between poorly-sorted to moderately-well-sorted. Its mean grain size and sorting range from 1.812 to 2.428 phi and 0.618 to 1.252 respectively. The skewness ranges from – 0.740 to 0.310 whilst the kurtosis values of the sand range from 1.747 (platykurtic) to 3.783 (mesokurtic) with an average of 2.320 (mesokurtic). Figure 23.4 shows the grain size distribution of the sand used.

23.4.2 Chemical Analysis

The results obtained are tabulated in Table 23.5. In general, SiO$_2$ represented 84% of the oxides in the sample.

23.4.3 Unconfined Compression Tests

The compressive strength of the dune sand from all locations improved by a factor of 177–266 and 19–31 due to the treatment with 2.5% (solid content/sand weight ratio %) SUMF and SF-C respectively. For the SF-C treated samples, the peak stress-solid

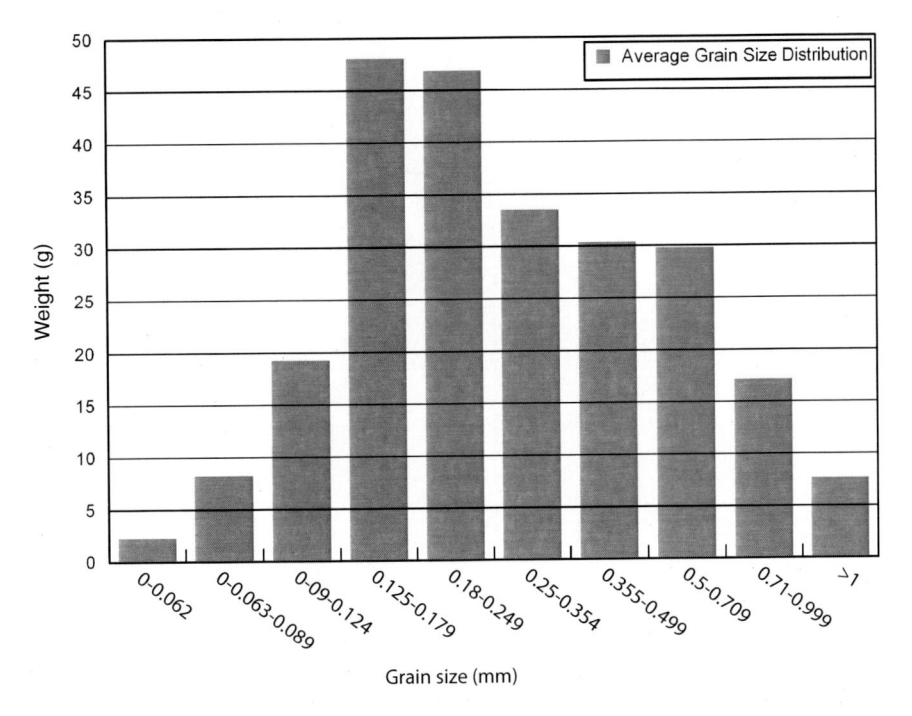

Fig. 23.4 Average grain size distribution for sand used

content weight ratio curve is linear in general (refer to Fig. 23.5) whilst for those samples treated with SUMF, the curve was parabolic, as shown in Fig. 23.6. The compressive strength tests revealed a proportional relationship between the solid content weight ratio and the compression strength of the sand treated with the two stabilisers. However, for the SF-C tests, the overall trend of monotonic increase in the peak stress values with the dose reached its maxima at 14% treatment and drops for higher treatment dosages, i.e. when sand is mixed with SF-C only. Treating the sand with the SF-C concentrate without water increases the compressive strength of the sand by a factor of 34, however this increase is similar to that achieved with 8% solid content weight ratio. Hence the maximum compressive strength is reached for an SF-C solid content weight ratio of 14%.

23.4.4 Water Runoff Test

For the SF-C treated sand, the samples show resistance to water runoff erosion starting from an application rate of 17.89 g m^{-2}, i.e. Experiment (Exp. 2), while for the SUMF, the resistance appears only at 60 g m^{-2} application rates, i.e. Exp. 2, and hence the SF-C is more suitable for stabilising surfaces which are prone to water runoff. Figure 23.7 shows that for the first cycle of the experiment, the weight of the sand eroded is nearly constant for the 17.89 and 26.83 g m^{-2} application rates

Table 23.5 The chemical composition of the dune sand from different locations

Sample\oxide (%)	CaO	Al$_2$O$_3$	FeO$_3$	MgO	K$_2$O	SiO$_2$	SO$_3$	Na$_2$O	ZnO	TiO$_2$	SrO	BaO	Mn$_2$O$_3$	L.O.I
A1	3.3	6.67	1.42	0.65	1.3	82.62	0.009	0.48	<0.001	0.19	0.01	0.01	0.02	3.035
A2	3.86	5.72	1.42	0.66	1.2	82.29	0.008	0.35	<0.001	0.12	0.01	0.01	0.02	3.435
A3	2.26	4.88	0.95	0.43	1.38	87.02	0.007	0.16	<0.001	0.14	0	0.01	0.01	1.984
A4	3.17	5.29	1.22	0.56	1.27	84.31	0.008	0.27	<0.001	0.18	0.01	0.01	0.02	2.901
A5	2.99	5.09	1.11	0.52	1.25	84.84	0.007	0.52	<0.001	0.17	0.01	0.01	0.02	2.689
A6	2.81	6.18	1.21	0.58	1.33	83.55	0.009	0.44	<0.001	0.16	0.01	0.01	0.02	2.685

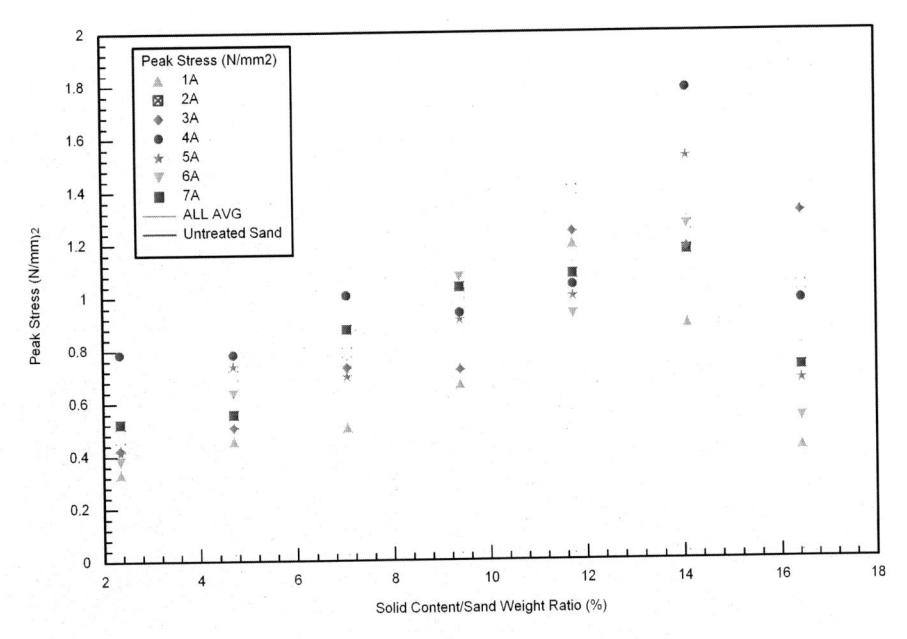

Fig. 23.5 Results of the unconfined compression test for sand treated with SF-C

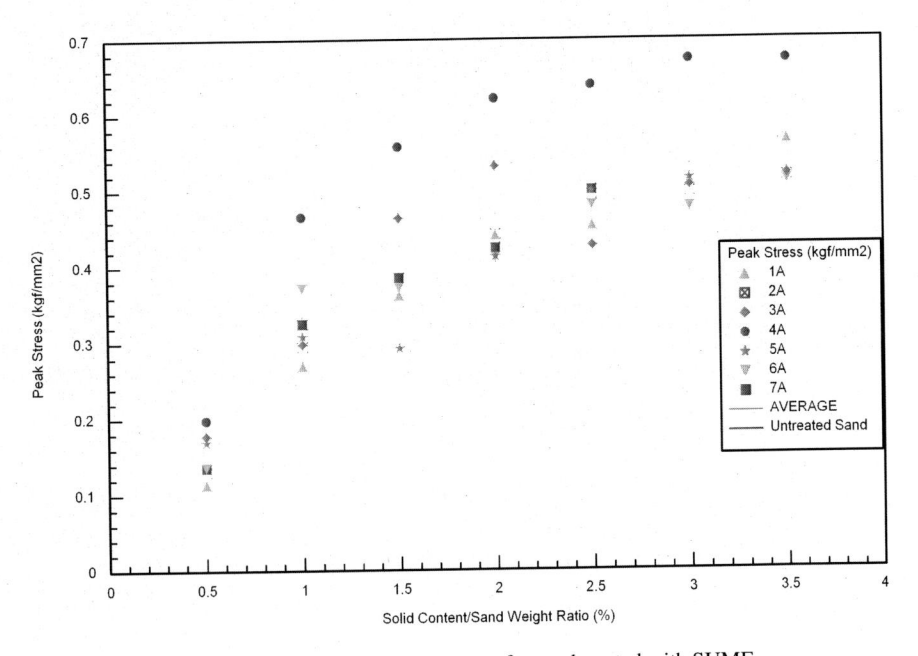

Fig. 23.6 Results of the unconfined compression test for sand treated with SUMF

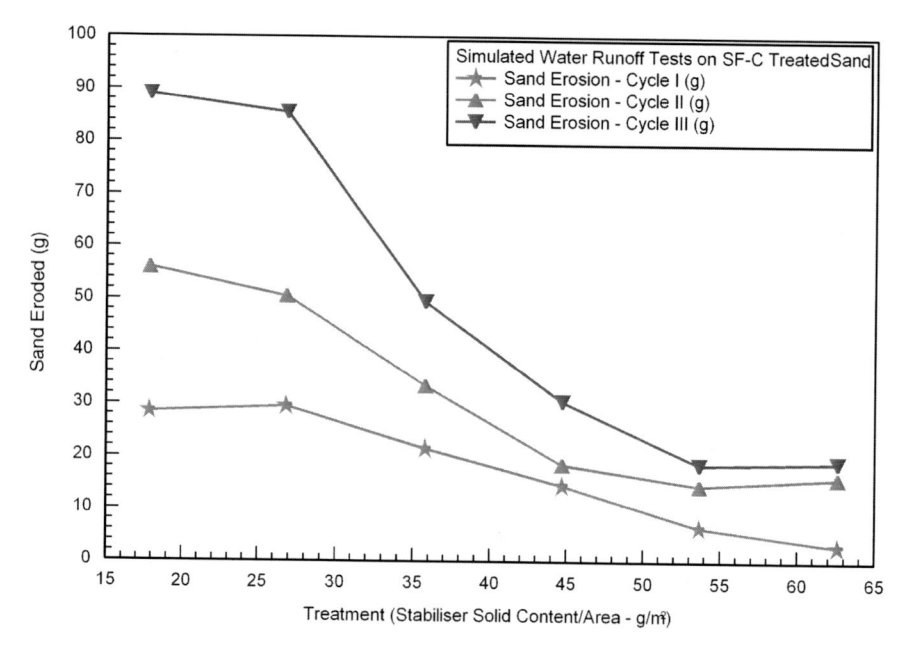

Fig. 23.7 The weight of eroded sand due to water runoff vs. application rate by SF-C

(Exp. 2 and 3). As the application rate increases, the weight of the eroded sand due to water runoff decreases linearly (linear inverse proportionality). The weight of the sand eroded due to the second cycle is nearly double that from the first cycle, i.e. 28.5 g compared to 56 g for Exp. 2. As the application rate increases, the difference between the weights of the sand eroded from the first and second cycles decreases to reach 4 g at Exp. 5, i.e. 14.5 g compared to 18.5 g from the 1st and 2nd cycles respectively. This difference increases for Exp. 6 and 7 (53.66 and 62.60 g m^{-2} respectively) so that for Exp. 7 the ratio of the weight of the sand eroded form the 2nd cycle to that from the 1st cycle is nearly 5:1. For the 3rd cycle, the weight of the eroded sand is nearly triple that eroded during the 1st cycle, 89 g compared to 28.5 g for Exp. 2. As the application rate increases, the difference between the weights of the sand eroded from the 2nd and 3rd cycles decreases to reach 3 g at Exp. 5, i.e. 16 g compared to 19 g from the 2nd and 3rd cycles respectively. In general, the weight of the sand eroded due to water runoff decreased as the application rate increased.

In contrast to Figs. 23.7, 23.8 is much simpler while the relationship between the weight of the eroded sand and the application rate is reversed. Here, this relationship is linearly directly proportional, i.e. the weight of the sand eroded increases with the application rate, which is unfavourable. For application rates less than 60 g m^{-2}, the samples failed after exposure to water runoff for 2 h and 40, 50 and 37 min for

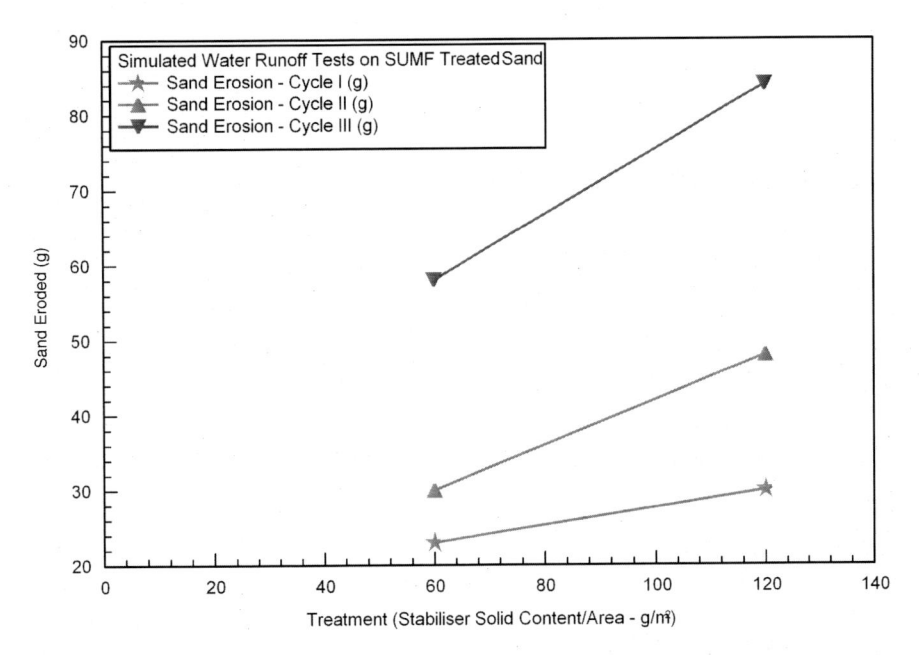

23 The Effectiveness of Two Polymer-Based Stabilisers 319

Fig. 23.8 The weight of eroded sand due to water runoff vs. application rate by SUMF

application rates of 30, 15 and 7.5 g m^{-2} respectively. The ratio between the weight of the sand eroded during the 1st cycle to that eroded during the 2nd cycle is 5:8 for Exp. 2 (60 g m^{-2}) and 3:4 for Exp 1 (120 g m^{-2}).

The ratio between the weight of the sand eroded during the 1st cycle to that eroded during the 3rd cycle is 5:14 for Exp. 2 (60 g m^{-2}) and 4:10 for Exp. 1 (120 g m^{-2}). As a quick comparison between the weights of the sand eroded from samples treated with the same amount, i.e. 60 g m^{-2}, of SF-C and SUMF, the weights of the sand eroded during the 1st, 2nd and 3rd cycles were 23, 30 and 58 g for the SUMF experiment compared to 4, 16 and 19 g for the SF-C experiment revealing the superiority of SF-C in increasing the surface resistance to water runoff erosion. Figures 23.9 and 23.10 show the weight of the sand eroded due to the water runoff tests versus the time of the experiment. In these two figures, each experiment is represented by a vertical line (not drawn), while in Figs. 23.7 and 23.8 each line represents one whole experiment, e.g. three runs of 6 h each. For a constant flow-rate of 6 l min^{-1}, the rate of loss per time can be treated as a straight line for the SF-C experiments but not for the SUMF experiments.

For SF-C the thickness of the stabiliser layer increased with stabiliser solid content weight per area, whilst for SUMF, the thickness of the stabiliser layer was nearly fixed at 8 mm, refer to Fig. 23.11. pH value varied between 7.35–7.55 and 7.5–7.85 for sand treated with SF-C and SUMF respectively.

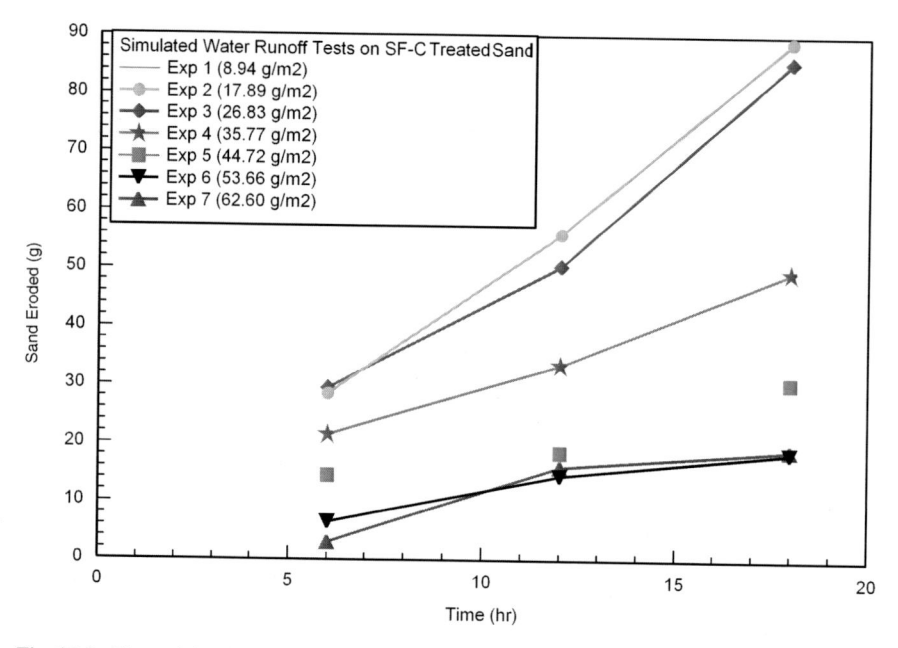

Fig. 23.9 The weight of eroded sand due to water runoff vs. exposure time for SF-C

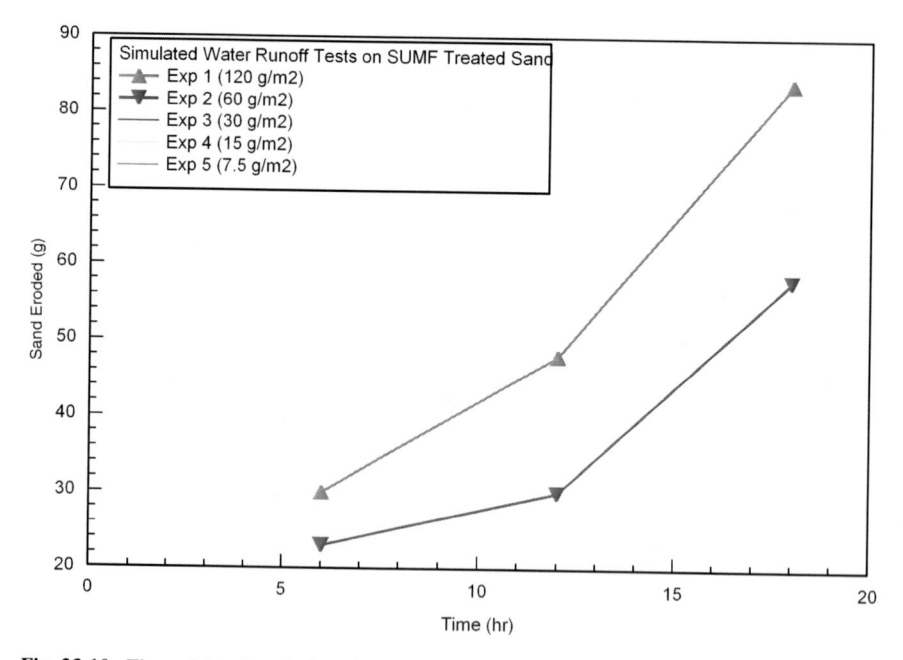

Fig. 23.10 The weight of eroded sand due to water runoff vs. exposure time for SUMF

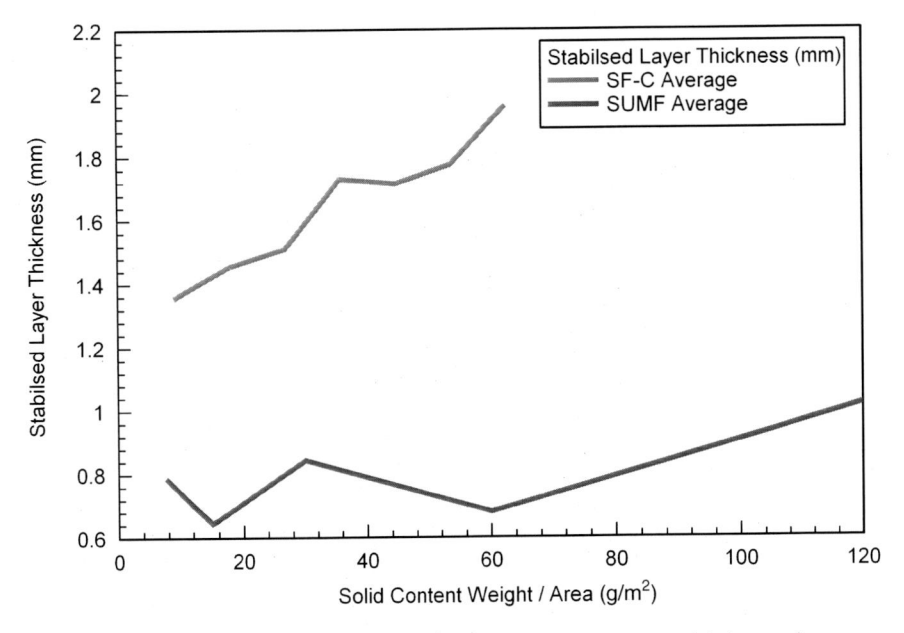

Fig. 23.11 Variation of stabilised layer thickness with the solid content weight/area ratio

23.5 Conclusions

The grain size analysis revealed that the sand could be categorised as fine to medium ranging between poorly-sorted to moderately-well-sorted. The SUMF treated sand had a peak stress value which was about 11 times the corresponding one for sand treated with SF-C at 3.5% stabiliser solid content to sand weight ratio. The SF-C treated sand had a compressive strength of 13.26 kg cm^{-2} when treated with 14% stabiliser solid content to sand weight ratio, while SUMF treated sand had a compressive strength of 55.06 kg cm^{-2} when treated with 3.5% stabiliser solid content to sand weight ratio. The above shows that SUMF is more suitable for stabilising surfaces, which are prone to mechanical forces (due to human/animal activities, e. g. grazing, camping and off-road driving).

For the SF-C treated sand, the samples showed resistance to water runoff erosion starting from an application rate of 17.89 g m^{-2}, i.e. Exp. 2, while for the SUMF, the resistance appeared only at 60 g m^{-2} application rates, i.e. Exp. 2, and hence the SF-C is more suitable for stabilising surfaces which are prone to water runoff. For the SF-C treated samples, the weight of the eroded sand reduced as the application rate increased, however, for the SUMF treated samples, the opposite was true.

In brief SF-C outperformed SUMF when erosion was caused by water while SUMF outperformed SF-C in compressive strength. When one takes into account that SUMF can act as a fertiliser (which helps the growth of native plants), and the other fact that water erosion is not a significant problem in Kuwait, one would prefer SUMF in conditions like those of Kuwait.

References

Abdullah, J. (1988). Study of control measures of mobile Barchan Dunes in the Umm-Al-Eish and West Jahra areas. KISR, Final Report. EES-98.KISR 2580. Kuwait.

Ahmed, M., Al-Dosari, A.M., Al-Otaibi, Y., Al-Awadi, L., Al-Mutairi, A. and Al-Rashed, A. (1996). A study of the cost and history of sand encroachment problems in Kuwait. KISR, Technical Report. VD002G. KISR 4939. Kuwait.

Al-Ajmi, D., Misak, R., Khalaf, F., Sudairawi, M. and Al-Dousari, A.M. (1994). Damage assessment of the desert and coastal environment of Kuwait by remote sensing. KISR, Final Report. K1SR 4405. Kuwait.

Al-Dabi, H., Koch, M., El-Baz, F. and Al-Sarawi, M. (1996). Mapping and monitoring sand dune patterns in northwest Kuwait using Landsat TM images. Proceedings Sustainable Development in Arid Zones. Kuwait.

Al-Naser, S. (1978). The role of science and technology for development in Kuwait. In: Behbehani K., Girghis M. and Marzouk M.S. (eds.), Proceedings, Symposium on Science and Technology for Development in Kuwait. Longman, London, pp. 109–116.

Al-Sudairawi, M. (1995). Study on sand control at KOC's operation areas of South-East, West and North Kuwait. KISR, Amendment/Extension "B" Report. EES-139. KISR 4622. Kuwait.

Anwar, M., Gharib, I., Zaghloul, N. and Hashash, M. (1987). A study of mobile sand control measures at Kuwaiti army camps. KISR, Final Report. EES-77. KISR 2220. Kuwait.

Gharib, I. and Al-Hashash, M. (1985). A preliminary assessment of mobile sand encroachment problems at Kuwaiti army camps. KISR, Final Report. K1SR 1792. Kuwait.

Gharib, I.M., Foda, M.A., Al-Hashash, M.Z. and Marzouq, F.Z. (1985). A study of control measures of mobile sand problems in Kuwaiti air bases. KISR, Final Report. KISR 1696. Kuwait.

Khalaf, F., El-Sherbiny, S., A1-Ajmi, D., Gopal, T. and Hashash, M. (1990). Study of sand control at KOC's operation areas of South-East, West and North Kuwait. KISR, Report. EES-139. KISR 3379. Kuwait.

Kuwait National Report on the Implementation of the United Nations Convention to Combat Desertification (UNCCD). (2000). Presented to the 4th Conference of the Parties. Kuwait, September 2000.

Lahalih, S.M. (1997). Method and composition for stabilizing soil and process for making the same. US Patent 5,670,567.

Lahalih, S.M. (1998). Method and composition for stabilizing soil and process for making the same. US Patent 5,824,725.

Lahalih, S.M., Absi-Halabi, M., Al-Awadhi, N. and Shuhaibar, K. (1988). Method for improving the mechanical properties of soil, a polymeric solution therefore, and a process for preparing polymeric solutions. US Patent 4,793,741.

Ramadan, A. and Lahalih, S. (2007). Laboratory assessment of two chemical sand stabilisers with potential use near As-Salmi Road – Kuwait. Proceedings of Desertification Control Conference. Kuwait, 12–14 May, 2007.

Chapter 24
Mountainous Tea Industry Promotion: An Alternative for Stable Land Use in the Lao PDR

Ayumi Yoshida and Chanhda Hemmavanh

Abstract Through fieldwork, analysis and planning of a model for a tea project, it has been found that the Phongsaly tea industry could be developed through co-operation with China's Yunnan province. The authors' tea industry model for Phongsaly province has been proposed to the authorities of Phongsaly and Yunnan as a co-operation project. Chinese investors have been invited to Phongsaly to set up tea businesses and to exchange agricultural technology.

Keywords Tea industry · Lao PDR · China PR · Sustainable land use · Policies · Trade · Income generating · Poverty alleviation

24.1 Introduction

Phongsaly is a mountainous province of the Lao PDR, located on the southern border of Yunnan province of China. The province's seven districts are considerably diverse in population, being home to 28 ethnic groups. This diversity, and the rugged topography of the area, makes the provincial government's task of managing and developing the local economy a difficult one.

With nearly 90% of the population depends on shifting cultivation. During the period 1997–1998 Phongsaly was classified as the third poorest province in the country: 64.2% of its people were officially below the poverty line (GoL, 2003). In 1992 forest area covered nearly 44% of the total land area, but by 2002 this figure had fallen to under 24%, a loss of 325 ha year^{-1} (DoF, 2005). As in other parts of the world, degradation of forest is occurring at an alarming rate.

Phongsaly province has a rich history of tea growing. The Phounoi ethnic group has, like the Blang and the Hani (usually known as the Samtao and the Ahka

---

A. Yoshida (✉)
Zhejiang University, Hangzhou, Zhejiang Province 310029, PR China
e-mail: ayumi_2180@hotmail.com

in Laos) minorities who live in Yunnan China, planted tea for centuries (Huang Guishu, 2005). Ancient horse and mule tracks for trading tea with China and other countries still run through Phongsaly. One of these connects from Yiwu Xishounbanna Yunnan China through Phongsaly then Hanoi-Haiphong Vietnam and then connecting Hong Kong by boat (Yunnan Daily, 2006).

Research of literature on this area reveals papers relating to anthropology, the environment, and poverty. Many papers concentrate on surveys of opium growing, shifting cultivation, and poverty alleviation but without supplying remedies for these problems (e.g. UNODC, 2002–2005). Chinese investors have proposed to develop tea plantations in Phongsaly as means of helping the local people there. This paper attempts to analyse the methods and possible impacts of tea plantation promotion to help develop a tea project model that will have the best chance of successfully growing tea and bringing socio-economic benefits to Phongsaly people.

24.2 Study Area: Khomen Village

Fieldwork was conducted in the village of Khomen in January 2007. The authors interviewed and held discussions with villagers and local government staff. Secondary research has been restricted by the fact that the written historical records on rural areas of Laos are often very limited (Sanderwell, 1999), hence much of the data used for this paper are based on professional experience and fieldwork.

Ban Khomen is located about 10 km south of Phongsaly district town and is around 270 km from the Lao-China border post at Boten in Luang Namtha. The village area covers 1,701 ha at elevations of 450–1,800 m. Annual rainfall is 1,500–2,000 mm. The villagers are Sinsily, known as Phounoi in Lao. This ethnic group moved from Burma since the year 1600 (LNFC, 2005). The population of the village is 313 people living in 80 households: about six people per household at a density of 18 persons km^{-2}. The population aged less than 17 years old or more than 64 years of age is 20% while those aged between 18–64 years make 80% of the total (Fig. 24.1)

24.3 Before the Tea Project

Prior to 1999 tea was not a commercial activity in Phongsaly. Tea plantations were small as local people used the leaves for their own consumption or as gifts and sold only to the local market. Livelihoods in Ban Khomen, as in most parts of the province, depended on the shifting cultivation of rice and several other crops. Khomen was accessible only by a narrow track, which was very difficult to use during the rainy season. Houses remained traditional, with thatched roofs that needed replacing every 3 years, and there were few motorbikes or other forms of transport in the village.

Fig. 24.1 The study area.
Source: Lao Urban Planning and Research Institute (2007)

24.4 The Tea Project

In 1999 a Chinese company established a tea station in the area. They contracted local people to develop their tea plantations. According to these contracts, the company sold 2,500 tea bush seedlings to each farmer at US$0.0052 per seedling along with a tea leaf cutting tool at $2 per tool. The farmers planted the bushes on their own land and agreed to sell tea produce to the company at the Phongsaly market price.

In 2005 another company from Malaysia came to Phongsaly district to invest in tea production. The local government allocated this firm a tea development zone near Ban Khomen, where the Chinese company had already started operations with local people. This move upset the Chinese company and certainly did not encourage it to increase its investment. This situation in indicative of the local government's lack of a clear policy on tea development and also a dearth of experience in promoting foreign direct investment (Phongsaly Planning and Investment Office, 2005). If the authorities were able to find good markets for tea they would be able to support several investors and so develop a stable tea industry.

It is certain that Phongsaly government and the villagers of Khomen support the project. However, support from the provincial authorities comes only in the form of information and advice – not money. All project funds come from foreign investors. Phongsaly provincial government's total agricultural development budget is very small and must be shared among various agricultural activities including forestry, irrigation, and livestock. The budget covers only agricultural monitoring and personnel cost, not extension (Phongsaly Planning and Investment Office, 2005).

24.4.1 Tea Plantation Areas

Following the implementation of these foreign tea projects, tea plantation area increased rapidly in Ban Khomen, rising from 63 ha in 1996 to 92 ha in 2006. Tea plantations have also been developed in 33 other villages in Phongsaly district, covering a total of about 1,022 ha in 2007 (Phongsaly Agriculture and Forestry Office, 2007; MAF, 2005) (Fig. 24.2).

Fig. 24.2 Tea plantations area. Sources: Authors' field survey, 2007

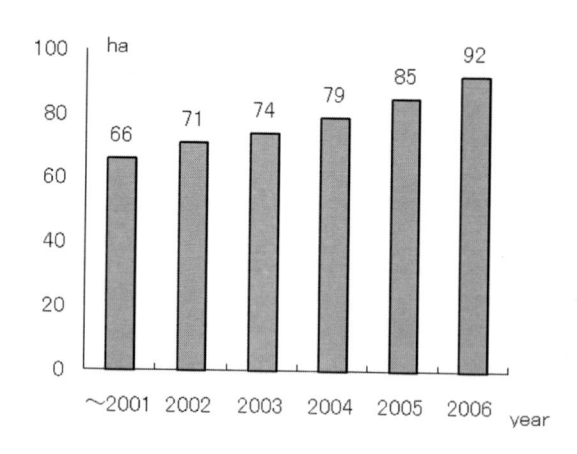

24.4.2 Price

The price of tea is increasing every year, with the market rate for dried tea leaves (old planted tea) moving from $0.61 kg^{-1} in 2002 to $3.1 in 2006. After the arrival of the Malaysian investors demand for tea increased and the price began to rise. As most tea buyers come to Laos from China, the local price is very much affected by the tea market in China, with the price in Phongsaly usually mirroring the rates on the Yunnan tea market. Compared with Yunnan tea, Phongsaly tea sells at a very low price: old planted tea in Phongsaly is the same as a price of Yunnan planting tea.

Investor-sponsored tea planting began in 2002 and bushes planted in that year gave their first harvest in 2005. Pu-erh tea is gaining popularity in China. Pu-erh tea made from old planted bushes is more valuable than new-planted tea leaves and the price for this tea is rising sharply in China. New species and leaves from newly planted tea bushes are not so attractive for Chinese companies and traders (Figs. 24.3 and 24.4).

Fig. 24.3 Tea price in Ban Khomen. Source: Authors' field survey (2007)

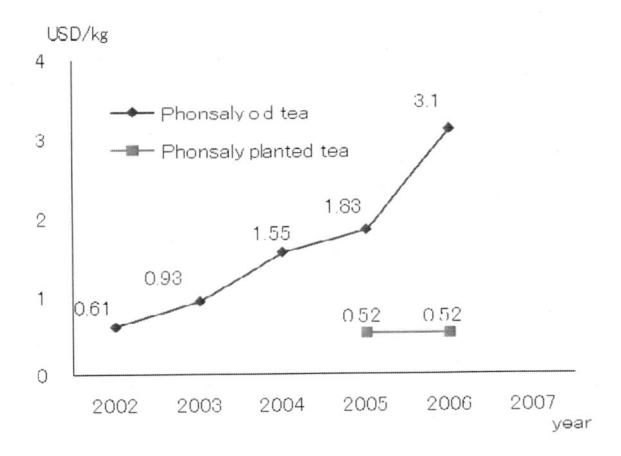

Fig. 24.4 Tea price in Yunnan. Source: Pu-erh Vista (2007)

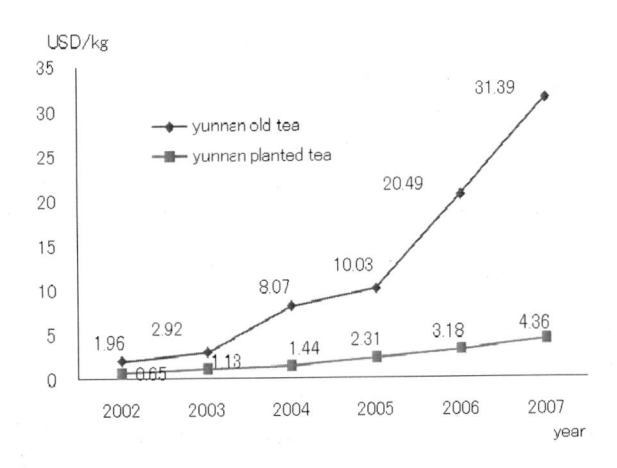

24.4.3 The Tea Trade

Before the tea projects were established local people use to take their tea to the local market but it was difficult to sell it there, given that the local market was full of tea and few outside buyers would attend (Fig. 24.5).

Since the tea projects were established local people do not need to take their tea products to the market as the Chinese company comes to Khomen village every week in the harvest season, and buys tea from all the villages based on the market

Fig. 24.5 Tea trade before tea projects

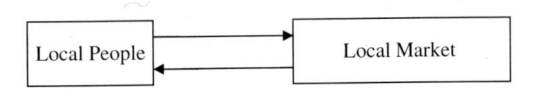

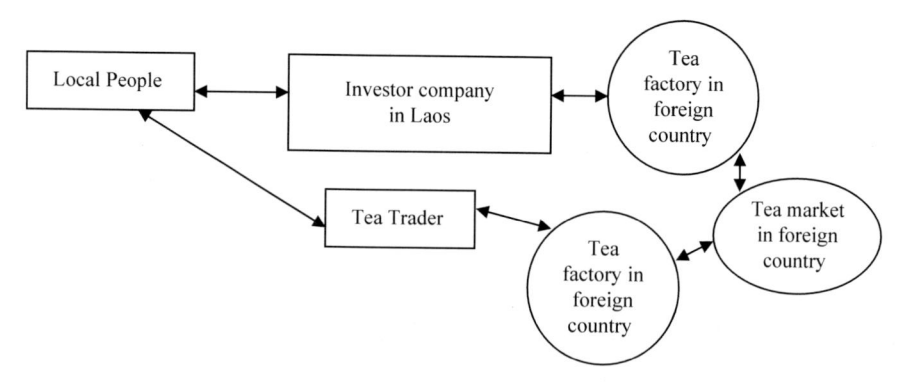

Fig. 24.6 New tea trade system

price. In addition, some independent tea traders come to buy Phongsaly tea and export it to China (Fig. 24.6).

24.4.4 Improving Local Livelihoods

Prior to the tea projects the lives of Ban Khomen people were tied to shifting cultivation with 3-year fallows. Now most of the villagers are engaged in tea planting and annual household income has risen to about $300. They can buy rice, and can afford to rebuild their traditional houses with new constructions.

In the 5 years that Phongsaly people have implemented these tea projects, living standards have improved. Most of those involved with the plantations have abandoned shifting cultivation and the area of land under slash-and-burn has decreased. At the same time there exist challenges to future tea development. The following section discusses how the existing tea development structure can be improved.

24.5 Key Challenges

While conducting this study it became clear that the local people and authorities would have great difficulty in improving the tea industry by themselves as:

1. There are no local tea experts with knowledge in tea marketing, tea science, tea ecology, tea culture etc;
2. Most Lao people do not habitually drink tea, so the tea market in Laos is limited. Access to tea markets in other countries is required;
3. The Phongsaly tea industry is in its early stages and therefore needs more investment. However, the local government lacks funds and the central government has transferred responsibility for such activities to the local authority.

4. A useful model or guidelines for agricultural commodity projects such as tea do not yet exist.

Considering the above four factors, it is recommended that Phongsaly looks to co-operate with Yunnan province at governmental level. Yunnan is not only next to Phongsaly Province but also has much experience in successfully developing the tea industry. The planting environments of Phongsaly and Yunnan are quite similar, so this experience should prove pertinent to Phongsaly.

24.5.1 Phongsaly Tea Development Model

The following steps are proposed to help the Phongsaly tea industry through its early stages, by creating a future tea development vision for a successful and sustainable industry (Fig. 24.7):

Step 1: Phongsaly government develops tea policy details in terms of investment incentives for foreign investors, foreign property protection, and training for government officials and coordinators;

Step 2: Phongsaly government makes a tea treaty with Yunnan, under which the Yunnan provincial agriculture department promotes tea investment in Phongsaly. After investment, exchange of experience, and agricultural trials, a fixed tea market is arranged with Yunnan province.

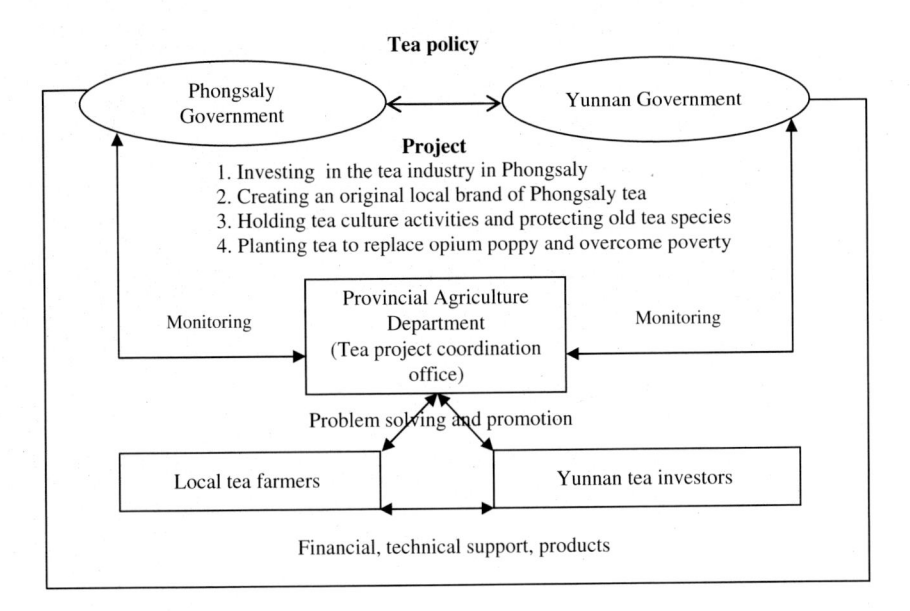

Fig. 24.7 Tea policy

Step 3: Phongsaly does not have its own tea brand, while Yunnan tea is sold under many brands that are well known on the Chinese market. This gives Yunnan tea a better price than Phongsaly tea, but Phongsaly can create its own tea brand with support from Yunnan investors.

Step 4: Cultural activities should be held to promote Phongsaly tea. For example, caravans could be held along the ancient tea road from Phongsaly.

24.5.2 Stakeholders

For the successful development of tea production and marketing, five organisations are proposed as key stakeholders in the industry in Phongsaly:

1 Phongsaly provincial government;
2 Yunnan provincial government;
3 Phongsaly Tea Project Coordination Office;
4 Yunnan tea investors;
5 farmers.

How these different stakeholders interact will have both economic and social consequences. The local Phongsaly and Yunnan authorities would be responsible for developing a policy for successful tea development and promotion. This policy should protect the rights of both investors and farmers while promoting Phongsaly's tea industry.

The Phongsaly provincial agriculture department will need to respond to the tea project. This office has the role of promoting the tea industry and coordinating between local farmers and Yunnan investors under the tea policy, reporting to both provincial governments on progress within the tea project. This office will also be the key body in drawing up contracts between investors and farmers and in organising activities to promote Phongsaly's tea industry.

24.6 Conclusions

The Phongsaly tea project has now been running for around 8 years. It is vital at this juncture that a focus be applied to making the crop sustainable for the future. If this can be achieved, then this project can attain the national government goals of eradicating shifting cultivation and opium production, and also of ending poverty in Phongsaly province.

Since the central government transferred autonomy to local authorities, provincial governments have not only had to maintain healthy administration, but are also responsible for securing their own sources of revenue. The tea project will both create revenue and also encourage people to settle in more concentrated populations in smaller same areas where they can grow tea, access social infrastructure, and find improved lifestyles.

If this project model is successful, it could be applied to other areas of Laos, which face similar problems to Phongsaly. The crop in question need not be tea, but could be another commodity which is suited to the local environment and which is sought after on neighbouring markets.

This study shows that it is difficult to resolve social problems in Laos independently: we must know how to co-operate with other countries to get their help and learn from them. In the case of Phongsaly and the tea industry, reaching an agreement with China is important as farmers and administrators in the province need to learn much in order to fulfil the potential of their situation. China has shown itself to be a key partner in helping the Lao PDR to develop in various fields, including the social, economic, environmental and political, and Lao people should be able to take advantage of this by promoting exchange with China.

Acknowledgement Heartfelt gratitude is extended to Mr. Sutanong, head of the forestry section of Phongsaly Province Lao PDR who actively assist us to collection data during in Phongsaly Province.

References

Department of Forestry in Lao PDR (DoF). (2005). Assessment of Forest Cover and Land Use 1992–2002. Vientiane.
Government of the Lao PDR (GoL). (2003). National Growth and Poverty Eradication Strategy. Vientiane, pp. 50–53.
Huang Guishu. (2005). Pu-er Tea Growing: A Broad Perspective. (In Chinese). Yunnan National Press, Kunming.
Lao National Front for Construction (LNFC). (2005). Ethnic Groups in the Lao PDR (In Lao). Vientiane, p. 223.
Lao Urban Planning and Research Institute. (2007). City Planning for the Lao PDR. Vientiane, pp. 20–21.
Ministry of Agriculture and Forestry (MAF). (2005). Agriculture and Forestry Strategy. Vientiane, pp. 18–30.
Phongsaly Agriculture and Forestry Office. (2007). Agriculture and Forestry Strategy 2006–2010. Phongsaly, pp. 1–2.
Phongsaly Planning and Investment Office. (2005). Socio-Economic Implementation 1997–2005 and Planning 2006–2010. Phongsaly, pp. 10–11.
Pu-erh Vista a Qinghai People. (2007). April 2007.04. p. 44.
Sanderwell. (1999). Forest Management in the Lao PDR.
The Yunnan Daily. (2006). 23.4.2006.
UNODC. (2002–2005). Lao PDR Opium Survey. Annual editions.

Chapter 25
Rehabilitation of Deserted Quarires in Lebanon to Initial Land Cover or Alternative Land Uses

T.M. Darwish, R. Stehouwer, C. Khater, I. Jomaa, D. Miller, J. Sloan, A. Shaban, and M. Hamze

Abstract Abandoned quarries in Lebanon represent not only deteriorated scenery but also a negative element leading to landscape fragmentation and ecosystem deterioration. Between 1996 and 2005 the number of quarries increased from 711 to 1,278 with a simultaneous increase of quarried land from 2,875 to 5,283 ha. Remote sensing data from 2005 showed that 21.5% of quarries were distributed on forest-land/arable land while 32.4% of quarries were detected on scrubland-grassland. Due to institutional weakness and the absence of a national policy, most Lebanese quarries have not been exploited following environmental concepts. To facilitate decision-making on rehabilitation options we propose two geospatial models based on bio and geophysical variables like precipitation, slope gradient, slope aspect, the availability of soil material, rock infiltration and soil texture to assess the probability of reclamation success. All attributes in the vegetation recovery model were assigned a weighted numeric score and separated into four classes of likely re-vegetation success. In addition to the above-mentioned parameters, water-harvesting potential was assessed based on catchment area above the quarry and rock permeability within the quarry. Therefore, deserted quarries were assessed for potential vegetation establishment and/or water harvesting. Results showed that potential re-vegetation success is strongly linked to terrain geomorphology and climatic conditions. Quarries suitable for water harvesting were spread all over the territory and can present useful additional water resources on the dry eastern mountain chain characterized with water scarcity. The good separation of the two sets of quarries suitable either for re-vegetation or water harvesting suggests the workability of the models, which need further field testing. But, limited national resources available for reclamation will be better targeted toward assessing those quarries where the

T.M. Darwish (✉)
National Council for Scientific Research-Remote Sensing Center, Mansourieh, Lebanon
e-mail: tdarwich@cnrs.edu.lb

likelihood of successful reclamation is most probable. The results of these models could be used to support decision-making concerning rehabilitation strategies towards restoration or possible alternative post-reclamation land uses.

Keywords Land degradation · Land cover change · GIS model · Quarries · Restoration · Water harvesting

25.1 Introduction

Quarrying activity is a necessary economic and financial activity serving the needs of social and urban development. Recent changes in the Lebanese governmental policy impose quarries rehabilitation. However, starting from the 1980s, a large number of quarries were abandoned in Lebanon and still left unmanaged. Decades of unregulated urban and industrial growth have left hundreds of quarry scars across the Lebanese territory. Official records estimated the number of quarries at 711 in 1996 (Handassah, 1996). Institutional weakness and the absence of a policy for integrated land resources management aggravated the deterioration of the mountain soil-forest ecosystem, especially in degraded quarries (Khawlie, 1998).

Moreover, sustained urban and industrial expansion is exerting increasing pressure on limited soil and water resources in the arid and subhumid Lebanese area (Eswaran and Reich, 1999; Darwish et al., 2004). Land degradation resulting from the mismanagement of quarried sites affects the soil-vegetation ecosystem and water balance in the watersheds leading to soil erosion and landslides. Most abandoned quarries in Lebanon leave lasting scars on the landscape and represent potential risks to natural ecosystems, particularly for the fragile Mediterranean forest and limited water resources (Atallah et al., 2003; Khater, 2004).

Climate change and recurrent droughts in the East Mediterranean complicate the restoration of quarries into initial land cover. The morphology of the southeast Mediterranean coast interacts with the atmospheric circulation features and strongly controls the spatial distribution of rainfall (Enzel et al., 2008). It creates a rainfall gradient promoting the dominance of forest on the western aspects of Mount Lebanon, while the eastern Lebanese and other east Mediterranean regions are characterized by a semi arid to arid climate and represent rare and degenerated forest and bare rocky lands. For these reasons, uncontrolled quarry expansion has altered natural landforms, and destroyed original vegetation and soil cover.

Abandoned and deserted quarries can negatively impact the ecological nature of landscapes (Mouflis et al., 2008), beside the pollution effects on forests health that can affect the natural recharge and deteriorate the groundwater quality (Cape, 2008). Reforestation in the highlands of Lebanon showed low performance and low survival rate for transplanted trees varying between 10 and 40% due to complex morphology and large human pressure (USAID/ECODIT, 2002). The spontaneous vegetation dynamics on deserted limestone quarries in Lebanon (Khater et al., 2003; Khater, 2004) and the suitability of quarries as landfill sites (El-Fadel et al., 2001) were evaluated. Besides the risks of groundwater pollution by seepage of toxic pollutants from landfill sites, harsh climatic conditions and human interaction reduce the possible re-vegetation success on poor substrates. Stehouwer et al. (2006) have assessed the use of bio-solids as amendment to restore the vegetation cover in surface coalmines post exploitation phase.

Overcoming strong barriers to forest establishment through enrichment planting was assessed on recently abandoned pastures (Zimmerman et al., 2000). It is important therefore to assess the likelihood of vegetation recovery based on the impact of different ecological factors to explore potential alternative uses of Lebanese abandoned quarries. Accordingly, the objective of this work was to involve east Mediterranean bio and geophysical variables into a spatial GIS model to assess the potential restoration success to previous land cover or water harvesting and suggest national course of action for ecosystem restoration efforts addressing the suitability of quarries in Lebanon for rehabilitation.

25.2 Materials and Methods

Satellite imagery of Landsat TM5 (30 m spatial resolution) from 1996 and 2005 as well as Ikonos (2 m resolution) from 2005 were used to locate the quarries in the Lebanese territory. The resulting spatial distribution of quarries was compared to the land cover map of Lebanon produced using 1998 IRS-1D (panchromatic 5 m spatial resolution). Satellite images were inspected visually on screen to identify quarry distribution and coverage patterns. This was followed by field surveys to support delineation of quarries. Distribution of quarries was examined with respect to individual thematic layers for which georeferenced data were available using the GIS facilities of ArcGIS 9.2. Thematic layers considered included landcover, soil, precipitation, geology and landform. Two models were developed to assess and rank quarries in Lebanon according to potential for (1) successful restoration to previous land use/re-vegetation, and (2) reclamation for water harvesting.

Parameters within these layers included in the models were assigned a numeric value based on their potential for future restoration of revegetation or reclamation for water harvesting. Both models assessed potentials associated with individual quarries based on a summation of the numeric values for each parameter included in the model. The distribution of numeric scores was then divided into ranked categories. Input parameters used in the model assessing potential for re-vegetation success included precipitation, slope gradient, slope aspect, soil material availability and soil texture (Table 25.1). Each of these parameters was assigned a numeric impact value (rating) ranging from 0 to 4. The quantity and quality of soil material adjacent to quarries was included in the vegetation model to evaluate the possibility of providing sufficient mineral substrate from neighbouring areas with deep soils possessing good physical-chemical properties for plant establishment and survival.

Because of the very strong correlation between elevation and precipitation, elevation was not explicitly included in the model. We also recognized interactive

Table 25.1 Parameter categorization and numeric values (rating) used to assess the potential for successful restoration by re-vegetation of quarries in Lebanon

Annual precipitation		Slope aspect (within precipitation classes)		Slope gradient (%)		Soil material availability		Soil texture	
Range (mm)	Rating	Category (mm year^{-1})	Rating	Range	Rating	Soil type	Rating	Class	Rating
>1,000	0	Flat >800 South >800 North 400–600	0	<8	1	Vertisols Luvisols Fluvisols Cambisols Arenosols	0	Loams	0
800–1,000	1	Flat 600–800 North 600–800	1	8–15	2	Sand beach	1	Sands	1
600–800	2	North <400 Flat 400–600 South 600–800 North >800	2	15–30	3	Calcisols Leptosols Regosols Andosols		Clays Silts	2
400–600	3	South 400–600 Flat <400	3	>30	4				
<400	4	South <400	4						

impacts among some parameters. For example, northern aspect was considered as a negative factor in the high precipitation (high altitude, lower temperature) areas due to decreased solar radiation and cooler temperatures. Northern aspect was classified positively in the lower rainfall zones (low altitude and high temperatures) due to decreased evaporative loss and moderated temperatures allowing for better vegetation establishment. Moreover, germination and early establishment in the field are favoured in shaded sites, which have milder environment and moister soil than open sites during low rainfall periods (Vieira and Scariot, 2006). Flat areas were also given higher ranking because decreased runoff and erosion and greater infiltration and soil water storage would favour establishment and growth of seedlings.

Because plant community composition in the country is strongly related to the "series of vegetation" which are more strongly associated with altitude and soil-parent material rather than with any other abiotic factor (Masri et al., 2006), soil type and rocks were introduced into the model. To assess the soil type and availability of soil material in the vicinity of degraded quarries, the updated soil map of Lebanon at 1:200,000 scale was used (Darwish et al., 2002). Soil taxonomic groups were categorized into two classes based on potential soil productivity and depth. Soil textural classes were separated into three groups on the basis of potential water holding capacity and susceptibility to compaction. On the basis of assigned value ranges, precipitation class was given the greatest weight in the model, slope factors intermediate weight, and soil factors the lowest weight because the first two parameters modify and sometimes predefine soil properties inherited from the parent material.

The other model prepared to assess the potential of quarries for water harvesting was developed using a GIS approach similar to the re-vegetation model. Parameters included in the water-harvesting model were slope, catchment area above the quarry, and rock permeability. Rock permeability at the quarry location was evaluated on the basis of the porosity associated with the types of basaltic, sandstone, marl and hard limestone bedrocks. These parameters and associated values are presented in Table 25.2. Catchment area categories were designed with respect to calculation of average area of catchment basins in Lebanon and the 4 quartiles calculated to define the class limits.

Precipitation was not included in the water-harvesting model because even in regions with low annual precipitation, water harvesting is feasible since much of the rainfall comes in intense, infrequent and short storms that generate significant runoff

Table 25.2 Parameter categorization and numeric values used to assess the potential redevelopment of quarries for water harvesting in Lebanon

Slope gradient		Rock infiltration		Catchment area (ha)	
Class(%)	Value	Class	Value	Class	Value
<8	3	Extreme	3	320–1,745	4
8–15	2	High	2	1,745–4,095	3
15–30	1	Medium	1	4,095–15,150	2
>30	0	Low	0	15,150–24,470	1

Table 25.3 Modelling the potential rehabilitation success of quarries in Lebanon

Re-vegetation success suitability for water harvesting	Very high (A)	High (B)	Medium (C)	Low (D)
High (1)	1-A	1-B	1-C	1-D
Medium (2)	2-A	2-B	2-C	2-D
Low (3)	3-A	3-B	3-C	3-D

and flash floods. Furthermore, it is precisely these regions of the country that would benefit most from water harvesting. By overlaying the output from re-vegetation and water harvesting models, a combined model was developed that could be used to facilitate prioritisation of quarry rehabilitation and to explore adequate decisions on the rehabilitation strategy to be tested and implemented. In the combined model, water-harvesting classes were assigned a numeric value and re-vegetation classes were assigned a letter value (Table 25.3). The resulting alphanumeric code thus contains an integrated assessment concerning both alternatives of potential rehabilitation for quarries. The approaches can reveal the potential for a given quarry for either or both options.

25.3 Results and Discussion

In less than 10 years, the number of quarries increased by 55.6% (from 711 in 1996 to 1,278 in 2005). The area of land affected by quarrying activities increased by 54.4% (from 2,875 to 5,283 ha) over the same period (1996–2005). Recent remote sensing data (2005) showed that 42.6% of these quarries are located on sparse grassland-bare rocky lands (Fig. 25.1), 21.5% on forestland-arable land while 32.4% of quarries were detected on scrubland-grassland and 3.2% of quarries were scattered inside urban zones.

The model for potential restoration success showed that relatively few quarries were assessed as having a very high (20 quarries over 175 ha) or high probability (144 quarries over 513 ha) of success as restoration to previous land cover (Fig. 25.2). Those sites are concentrated in the mountain sub humid area and western aspects of Mount Lebanon including the coastal area characterized by high winter rainfall followed by a long dry period. Some of them occurred on the eastern aspects of Mount Lebanon facing the semiarid Bekaa area and on the southern mountain range. These results reflect the difficulty in establishing lasting vegetative cover in regions characterized with low rainfall and high temperatures.

Quarries ranked as medium (603 quarries over 2,150 ha) and low (511 quarries over 2,428 ha) probability of success for restoration to previous land cover are scattered throughout the country reflecting how the complex morphology of the country and microclimate can affect the restoration success. Areas characterized by arid climate might require controlled intervention to sustain the vegetation

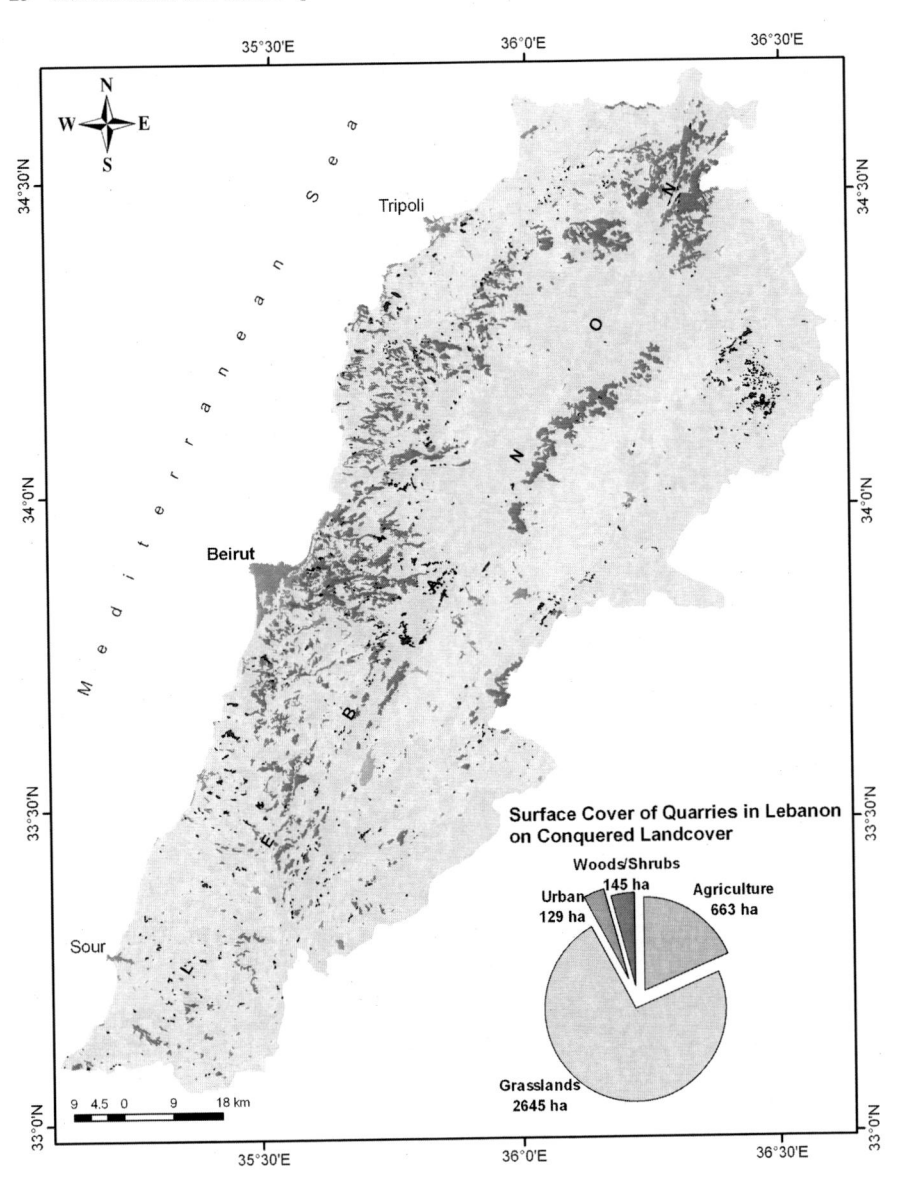

25 Rehabilitation of Deserted Quarires in Lebanon to Initial Land Cover

Fig. 25.1 Distribution of quarries on different land cover/use in Lebanon

recovery (Le Houérou, 2002). It is suggested that these quarries should be further assessed for alternative future land uses, as a large number of problems related to the rehabilitation process will be posed by the inaccessibility of the quarries and steep landform.

Most of the benefiting areas from higher rainfall in Lebanon tend to be located in steeply sloping areas where transporting soil material, spreading and incorporating

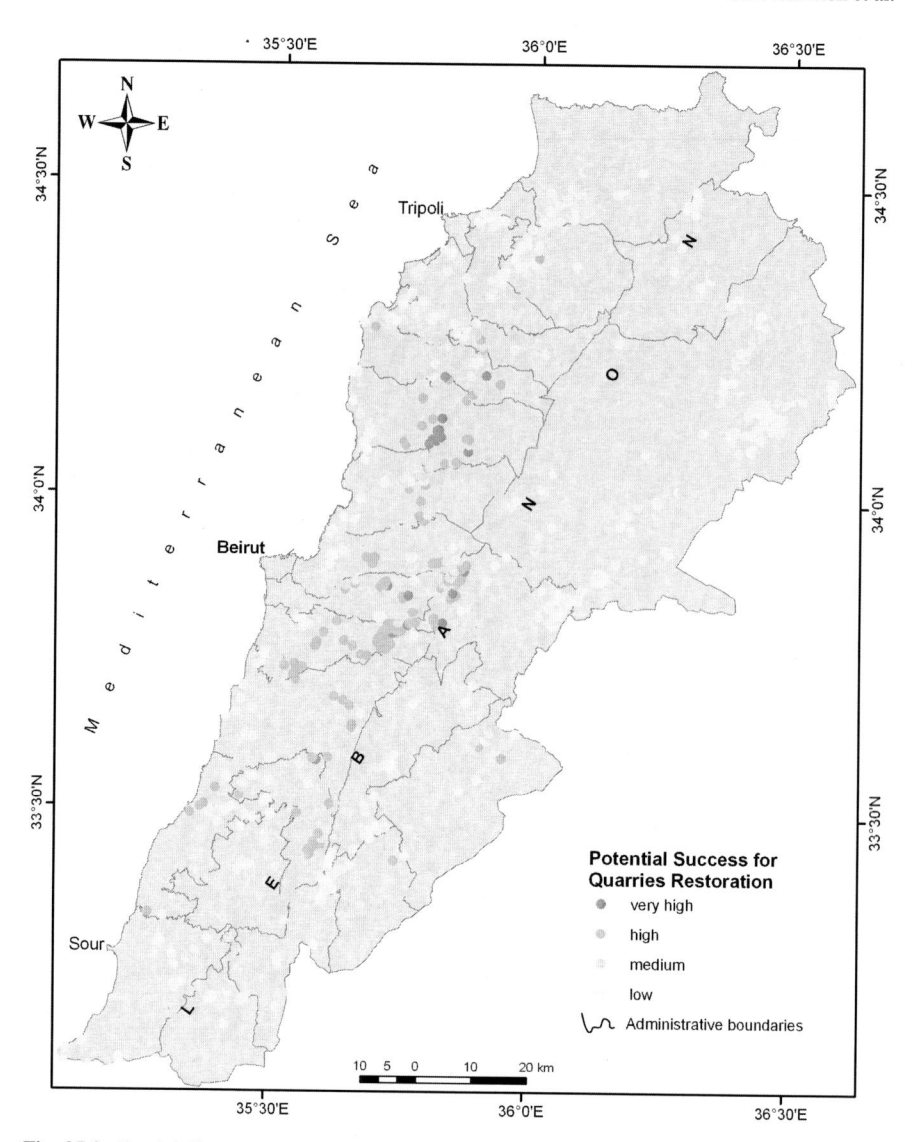

Fig. 25.2 Spatial distribution, number and area of quarries classified according to suitability for restoration to previous land cover

soil amendments, planting, and controlling erosion would be difficult. Such landscape specificity would require some modification of the observed shape (standing walls with a slope gradient over 90%) left after the exploitation by probable reshaping of deserted quarries by allowing controlled additional exploitation and blasting or transporting rock and soil material from close locations to bring the area into reclaimable shape.

In case policy is oriented to re-establish the pre disturbed landscape and land cover, it is necessary to consider management options to accelerate recovery and restore productivity, biodiversity and other values (Parrotta et al., 1997). These steps consist of recovering bare slopes of drylands in the North East Bekaa representing less challenging task than establishing initial vegetative cover on the abandoned quarries on western slopes of Mount Lebanon. However, several cost effective technical measures should be undertaken like re-mining the abandoned sites to bring them into updated exploitation criteria and shape including backfilling with inert materials, blasting the vertical walls to re-contour into milder slopes or terraces. Contouring and reshaping abandoned Lebanese quarried sites would permit surface placement of loose rocky and sandy material where annual and perennial vegetation could be re-established.

Output from the water harvesting assessment model produced a somewhat more even distribution of quarries within the three classes. A large number of quarries (616 units over an area of 2,129 ha) were ranked as having a high potential for redevelopment as water harvesting sites due to their position in the catchment and relatively low rock infiltration (Fig. 25.3). Water is an increasingly limiting resource in the country, and there is increasing competition for water among domestic, industrial, and agricultural users. Given the torrential nature of rainfall in Lebanon and its uneven distribution, many areas experience a severe shortage of fresh water in dry months.

Despite the fact that rainfall is more abundant in the western mountain chain, the country lacks large water harvesting structures. The eastern mountain chain having arid and semiarid climate showed large area of quarries suitable for water harvesting from recurrent flush floods, mainly in spring and fall. These water reservoirs will contribute to reducing the impact of devastating floods and provide additional water in this high water deficit area. A total of 442 quarries were found suitable for both water harvesting or ground water recharge purposes

Another alternative use that could be suggested for abandoned quarries in Lebanon is redevelopment into landfills. In fact, many are being illicitly used for this purpose. However, such practices could have devastating impact on groundwater quality if improperly managed. A key feature would be bedrock permeability and stability, factors included in the water-harvesting model. Thus, quarries ranking high for water harvesting could also represent potential sites for landfill development. However, other factors such as proximately to urban areas, transportation, and size would also need to be considered. By combining the outputs of the re-vegetation and water harvesting models, an overall assessment of quarry rehabilitation was generated (Fig. 25.4).

There is relatively little overlap of quarries given a high ranking by both of the assessment models. Thus the models were effective in segregating quarries into clear rehabilitation alternatives. Only three quarries were ranked in the highest categories for both re-vegetation and water harvesting. Of the 220 quarries ranked high for water harvesting, only 40 were also ranked high or very high for re-vegetation. Similarly, of the 124 quarries ranked high or very high for re-vegetation, only 40 were also ranked high for water harvesting.

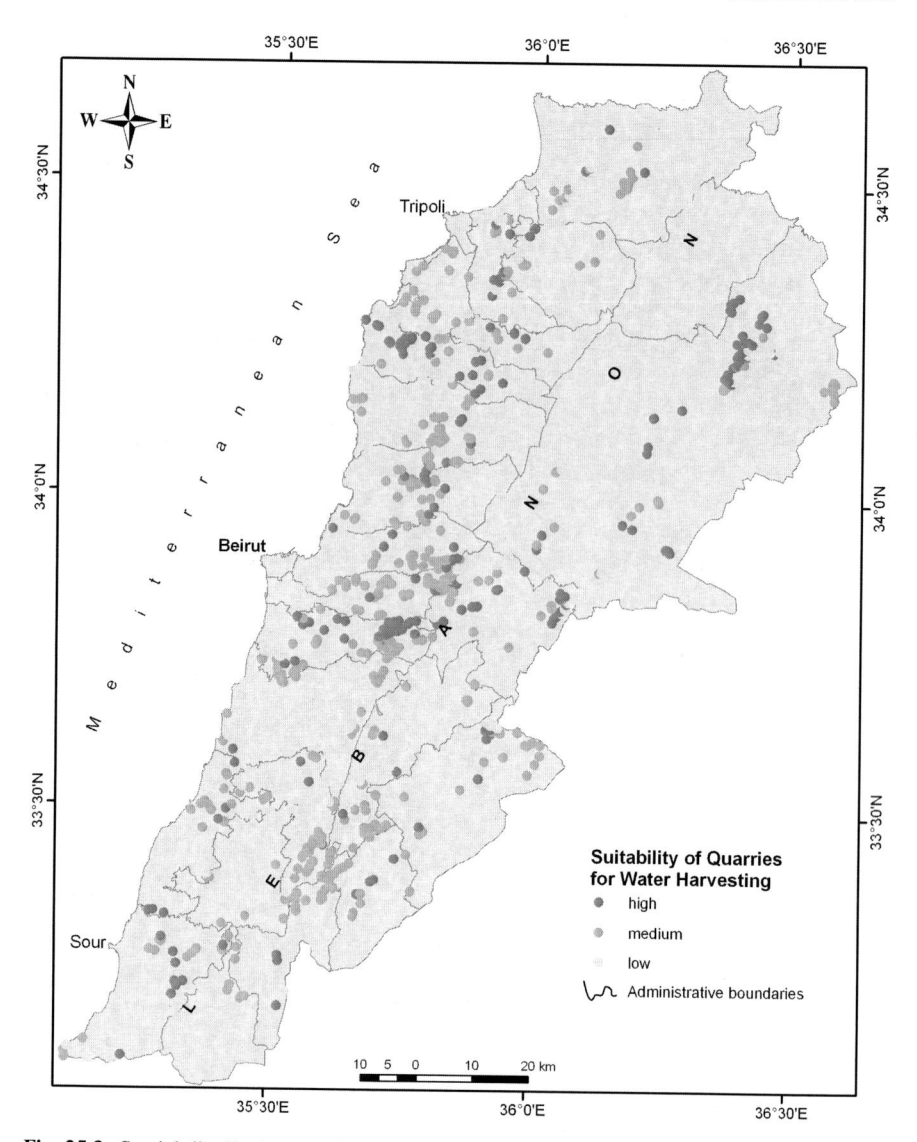

Fig. 25.3 Spatial distribution, number and area of quarries classified according to suitability for water harvesting

The combined model also demonstrates that quarries with a very high potential for re-vegetation (A) are mostly located in the Mount Lebanon mountain range (20 quarries), while those with high suitability for water harvesting (1) are more numerous (220) and are spread all over the territory.

Well-segregated quarries (high rankings in one model and low in the other) should be considered for rehabilitation following their high ranking. Quarries with

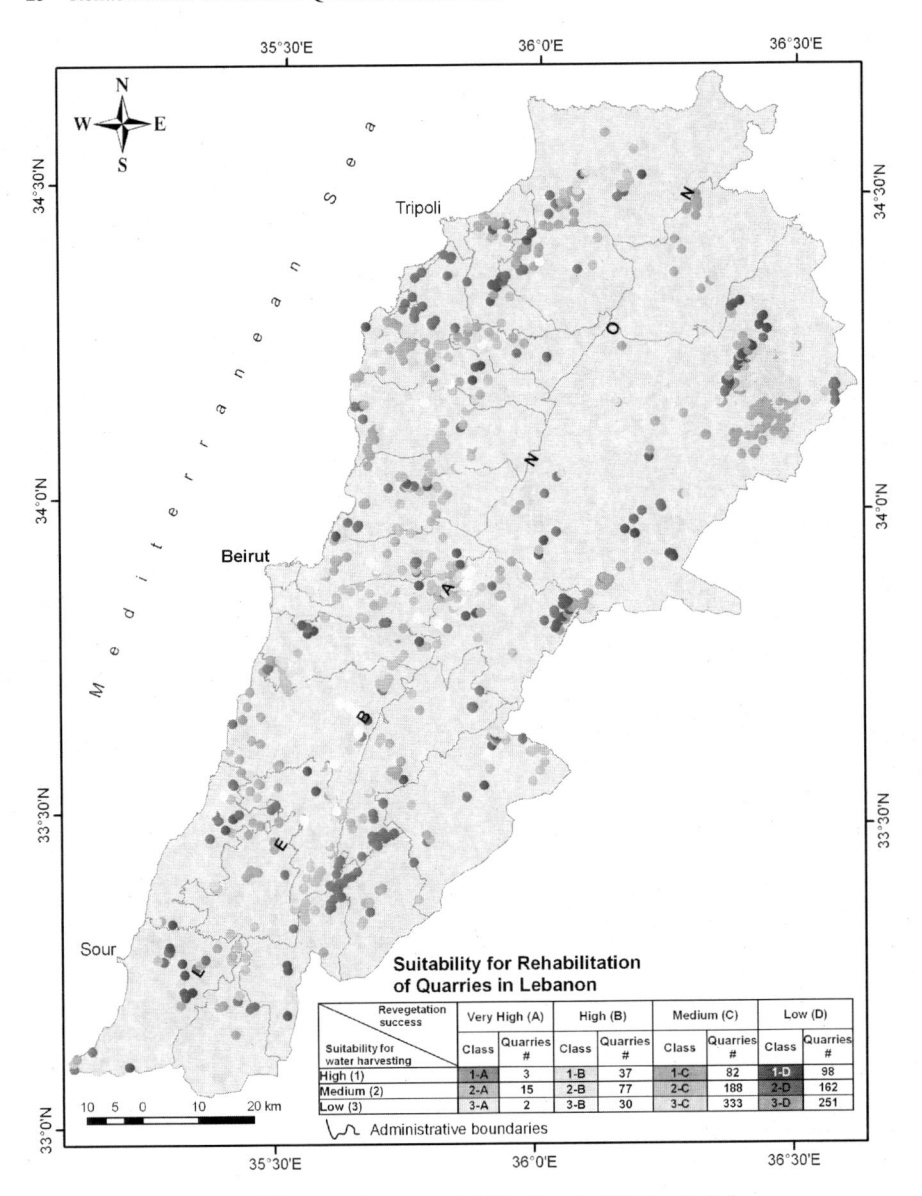

Fig. 25.4 Combined model output of quarry suitability for rehabilitation in Lebanon

high rankings in both models in all likelihood would be well suited for a combined approach of re-vegetation and development of reservoirs. This rehabilitation success model can be used as a tool to help with future prioritisation of quarry rehabilitation, locating quarries for rehabilitation, and to guide rehabilitation strategies. Although final decisions would clearly require extensive site investigations, this model can limit such investigations to those quarries with the highest probability of success.

25.4 Conclusion

The richness of agro-ecological zones in Lebanon derive from complex orography, variable climate and rich biodiversity. However, the chaotic and uncontrolled expansion of quarry sites and limited rehabilitation of quarries is affecting vegetation cover, causing forest fragmentation and loss of arable lands. Given the large number of quarries and limited availability of resources for rehabilitation, a national priority for Lebanon should be to direct resources to those sites with the greatest likelihood of successful rehabilitation. The GIS based models presented in this paper have ranked all quarries in Lebanon with respect to their potential for successful re-vegetation and their potential for redevelopment as water harvesting sites.

Both models produced a wide distribution of preferred uses for the quarries and effectively identified those that should be high priority candidates for rehabilitation. Detailed site investigations would be required to confirm the model output and to develop specific reclamation plans. However, use of these models can help targeting such investigations to sites with high potential for successful reclamation. Overlaying the output of the two models demonstrated a strong segregation of sites suitable for re-vegetation and those more suitable for water harvesting. Thus, these models could also be used to help direct reclamation strategies.

Prioritisation of quarries for reclamation should also consider the present environmental impact of the quarry and the potential for mitigation of that impact by reclamation. We are in the process of developing a third model that will assess potential environmental impact of existing quarries. The proposed model could also be used to limit future quarrying activity to sites with minimal potential environmental impact.

Overlaying this model with the combined model output presented in this paper will result in a further refinement of the tool to help direct national policy and limited resources toward those sites where the greatest benefits will be realized with less ecosystem damage. Assessing the environmental and health risk impact from current and future quarrying activities within a national policy for quarry exploitation and reclamation is essential to meet the needs of the growing population, while maintaining the sustainable use of natural resources for future generations. Legislation must consider not only licensing quarry exploitation but also their rehabilitation during the process of work. Monitoring the potential effect on vegetation cover and water resources is a continuous task serving the ecosystem conservation and public health.

Acknowledgment This work is a part of research supported by CNRS Lebanon and Penn State University within the frame of Fulbright Scholarship granted to the corresponding author in summer 2007.

References

Atallah, Th., Hajj, S., Rizk, H., Cherfane, A., El-Alia, R. and Delajudie, P. (2003). Soils of urban, industrial, traffic and mining. SUITMA, Nancy, Paris. 9–11 July, 2003.

Cape, J.-N. (2008). Interactions of forests with secondary air pollutants: Some challenges for future research. Environmental Pollution 155(3):391–397.

Darwish, T., Faour, Gh. and Khawlie, M. (2004). Assessing soil degradation by landuse-cover change in coastal Lebanon. Lebanese Science Journal 5(1):45–59.

Darwish, T., Khawlie, M., Jomaa, M., Awad, M., Abou Daher, M. and Zdruli, P. (2002). A survey to upgrade information for soil mapping and management in Lebanon. Options Méditerranéennes, Series A: Mediterranean Seminars. 50:57–71.

El-Fadel, M., Zeinati, M., El-Jisr, K. and Jamali, D. (2001). Industrial-waste management in developing countries: The case of Lebanon. Journal of Environmental Management 61:281–300.

Enzel, Y., Amit, R., Dayan, U., Crouvi, O., Kahana, R., Ziv, B. and Sharon, D. (2008). The climatic and physiographic controls of the eastern Mediterranean over the late Pleistocene climates in the southern Levant and its neighboring deserts. Global and Planetary Change 60(3–4): 165–192.

Eswaran H and Reich P (1999). Impacts of land degradation in the Mediterranean region. Bulgarian Journal of Agricultural Science, Plovdiv, Agricultural Academy of Bulgaria 1:14–23.

Handassah, D. (1996). A National Survey on Quarrying in Lebanon. Khatib & Alami, Beirut.

Khater, C. (2004). Dynamiques végétales post perturbations sur les carrières calcaires au Liban. Stratégies pour l'écologie de la restauration en régions méditerranéennes. Thèse de Doctorat. Université Montpellier II.

Khater, C., Martin, A. and Maillet, J. (2003). Spontaneous vegetation dynamics and restoration prospects for limestone quarries in Lebanon. Applied Vegetation Science 2:199–204.

Khawlie, M. (1998). An Environmental Perspective on Quarries for the Construction Industry in Lebanon. In Aggregate Resources- A Global Perspective. A. A. Balkema Publishers, Rotterdam, 387–395.

Le Houérou, H.-N. (2002). Man-made deserts: Desertization processes and threats. Arid Land Research and Management 16:1–36.

Masri, T., Khater, C., Masri, N. and Zeidan, Ch. (2006). Regeneration capability and economic losses after fire in Mediterranean forests-Lebanon. Lebanese Science Journal 1:37–48.

Mouflis, G.-D., Gitas, I.-Z., Iliadou, S. and Mitri, G. (2008). Assessment of the visual impact of marble quarry expansion (1984–2000) on the landscape of Thasos Island, NE Greece. Landscape and Urban Planning 86(1):92–102.

Parrotta, J.-A., Turbull, J.-W. and Jones, N. (1997). Catalysing native forest regeneration on degraded tropical lands. Forestry Ecology and management 99:1–7.

Stehouwer, R., Day, R. and Macneal, E. (2006). Nutrient and trace element leaching following mine reclamation with biosolids. Journal of Environmental Quality 35:1118–1126.

USAID/ECODIT. (2002). Lebanon environmental program assessment report. Final Report, 2002; LAG-I-00-99-00017-00.

Vieira D-L., M. and Scariot, A. (2006). Principles of natural regeneration of tropical dry forests for restoration. Restoration Ecology 14(1):11–20.

Zimmerman, J.-K., Pascarella, J.-B. and Aide, T.-M. (2000). Barriers to forest regeneration in an abandoned pasture in Puerto Rico. Restoration Ecology 8(4):350–360.

Chapter 26
The Impact of Land Use Change on Water Yield: The Case Study of Three Selected Urbanised and Newly Urbanised Catchments in Peninsular Malaysia

Mohd Suhaily Yusri Che Ngah and Ian Reid

Abstract Economic development and increasing population has brought significant changes in land cover and placed stress on water quantity within tropical river catchments, including those of Malaysia. There is a need to investigate the impact of land use changes on water yield in moderate-sized (c. 1,000 km^2) catchments, such as the Langat and Linggi, since these provide the water resources that cater for the rapid urbanization and industrialization that characterizes Malaysia. (On the contrary, most previous studies have been dealing with very small catchments i.e. <25 km^2, and frequently <1 km^2). Findings from our study provides information for local river managers and development planners and will assist them in minimizing the negative impacts of development on water resources, while promoting sensible planning within river basins especially in the newly developed catchments such as the Bernam. An analysis of land-cover in the Langat, Linggi and Bernam basins indicates that there has been a significant change from forest (primary and secondary selva) to agriculture, especially tree crops (rubber, oil palm), ranging from 7 to 15% in the three water catchments, and an increase in the urban area that ranges from 183 to 394% during the period 1984–2002. Despite this, the runoff coefficient shows no significant increase during 42 years of development. The coefficient lies between 22 and 48%. The outcome is not straightforward and counter intuitive when comparison is made with results from other experimental catchments in the tropics. This study suggests that up-scaling the findings of small catchment studies of forest removal is far from simple, especially in the wet tropics, where the impact of tree crops on water relations may be insufficiently distinguished from primary or secondary forests.

Keywords River catchment · Rainfall-runoff · Land use change · Water quantity · Water yield

---

M.S.Y.C. Ngah (✉)
Department of Geography, Faculty of Social Sciences and Humanities, Sultan Idris Education University, Tanjong Malim Perak 35900, Malaysia
e-mail: suhaily@upsi.edu.my

26.1 Introduction

Since its independence in 1957, development in Malaysia [i.e. deforestation, agriculture, urbanization and industrialization] has been focussed on the coastal areas, especially the West Coast of Peninsular Malaysia (Fig. 26.1). All the main drainage basins, namely, the Klang, Langat, and Linggi, have undergone rapid development since the 1970s, with a view to providing a catalyst for a development corridor and in order to achieve the Government's vision to achieve developed country status by the year 2020 (Vision 2020). Consequently, many of the West Coast river basins have experienced severe environmental problems, especially those related to water, such as pollution of water bodies, shortages of water and urban flooding (Jamaluddin, 2000).

Since all of these main river basins have already been developed, the Government has decided to expand development also to other river basins, which have a potential to support and accommodate the economic activities of the Kuala Lumpur area. The Bernam Valley, which is situated 73 km northwest of Kuala Lumpur, is one of those river basins that is now subject to extensive land development planning proposals by local governments and the corporate sector. The early stage of development in Bernam was started in 2000 and it is expected that various environmental problems, especially those involving water quantity will occur in Bernam, as they did in the Klang, Langat and Linggi catchments (Fig. 26.2). This is because part of the Bernam Valley is situated in a sensitive area of the Main Range, which supplies most of Peninsular Malaysia's water needs.

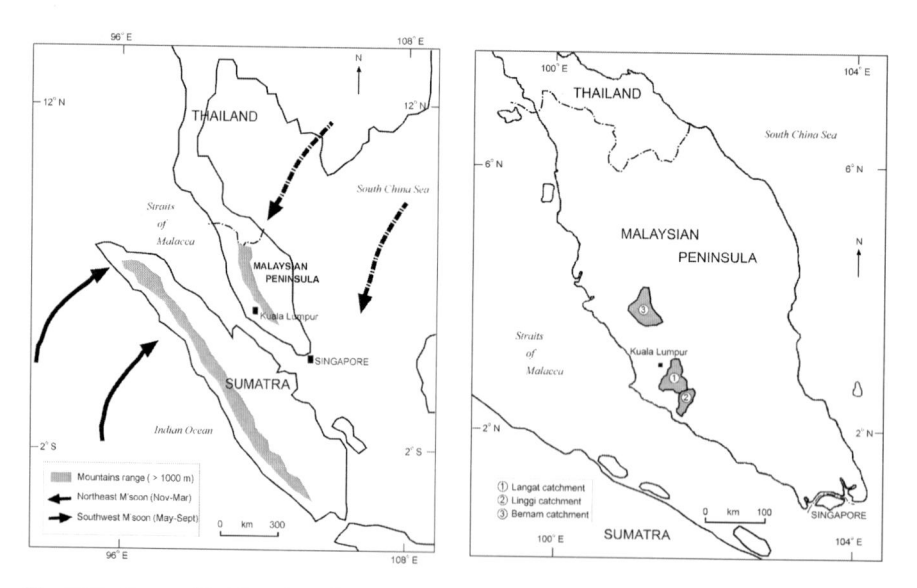

Fig. 26.1 Geographical location of study catchments in Peninsular Malaysia

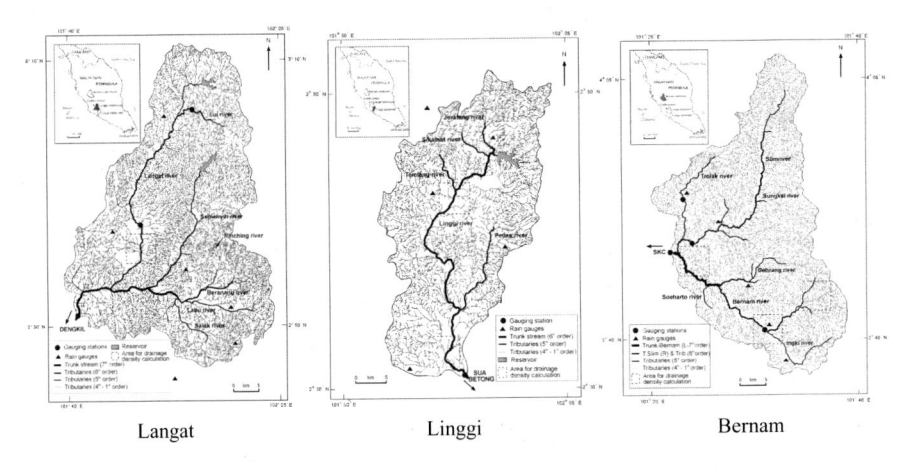

Fig. 26.2 Drainage network of the Langat, Linggi and Bernam catchments

To minimise the environmental degradation already experienced in the moderate-size Langat and Linggi basins, developers and managers need to be vigilant when dealing with land development and the need for environmental sustainability. In this context, the purpose of any development should "meet the needs of the present without compromising the ability of future generations to meet their own needs generated from environmental resources" (Brundtland Commission, 1987). Because land development in Malaysia has been driven by economic considerations, there is a need to study the hydrological impacts of rapid land use development so that lessons can be applied to new development areas. In this context, the Langat and Linggi catchments have been examined in order to provide an analogue for the Bernam. This is an important step in providing basic information to managers, so they can monitor and manage the new land development programme in the Bernam in a proper manner. The impact of economic development in Malaysia on the environmental quality of the river basin in general, and on water resources in particular, has not been studied so far in a significant way. The study that we report could help preventing further environmental deterioration by implementing appropriate policies in the future.

26.2 Material and Methods

Relations between key environmental factors and development processes in two developed catchments (the Langat River Basin – Selangor State and the Linggi River Basin – Negeri Sembilan state) have been studied and identified. It is important to understand hydrological relationships before, during and after land use changes have occurred. Managers can use these findings to attempt to predict potential impacts of land use change on the hydrology in the Bernam Valley and encourage the adoption of sustainable development practices. These "analogue" catchments were chosen because each of them has similarities with the Bernam in terms of climate, geology, river system, and rainfall.

Archived hydrological and land use data for the years 1960–2002 were analysed in order to assess whether they can be used as a basis for the development of transfer functions to predict what might happen in the Bernam Catchment. These would allow recognition of the possible impact of land development scenarios on rainfall-runoff relations, and they will be related to environmental sustainability issues over a time period since the 1960s. All of this information will allow development of a hydrological framework for river management in the Bernam Valley. All of these data sets were obtained from various Malaysian government agencies. It is important to convey that this study was dealing with imperfect archived datasets. The reasons include administrative instability following independence, lack of trained local engineers and technicians, and a lack of budget to repair and purchase new equipment. This caused existing stations not to be maintained, new data not to be collected and led to some data archives being lost.

Notwithstanding the data limitations, all available records were initially screened using various types of analysis. For example, time-series were scrutinized for the consistency of historical datasets of rainfall and runoff in order to identify anomalies in the sequence of observations; test of normality in the data frequency distributions (using the Kolgomorov-Smirnov test statistic) prior to rainfall-runoff analysis; double-mass analysis was also deployed in order to examine any peculiarities of rainfall and runoff data within the catchments; correlation analysis (Pearson, 2-tailed) was also widely used to establish the relations between rainfall gauges, indicating the strength and direction of covariation; ordinary least-squares regression (OLS), commonly used in hydrological studies, was also applied here to establish the rainfall-runoff relations and transfer functions; and student's t-test has been used to assess the significance of any differences in suspended concentration data during pre- and post-development periods in the same catchment and between catchments to help recognise land use change impacts in the Langat and Linggi basins. Finally the likely impact of such changes following new developments in the Bernam catchment was elaborated.

26.3 Results and Discussion

From this research, significant trends in land use change from forest (primary and secondary selva) to agriculture, especially tree crops (rubber, oil palm), ranging from 7 to 15% in the three water catchments, and an increase in the urban area that ranges from 183 to 394% during the period 1984–2002 throughout the three catchments have been documented. This confirms that the tropical countries have undergone very rapid changes involving deforestation, agricultural conversion and urbanisation. The major driving forces behind these activities are the expanding economic development and population growth. Many urban developments have been permitted on the floodplain by local authorities. Assessing the impact of land use change on hydrology, it is important to consider the impact of new land uses on rainfall-runoff intensity (Fig. 26.3).

Despite this, the runoff coefficient shows no significant increase during 42 years of development. The coefficient lies between 22 and 48%. The outcome is not straightforward and counter intuitive when comparison is made with results from other experimental catchments in the tropics, where there has been shown to be a significant increases (45–70%) in runoff when natural forests have been cleared as reported by Hsia (1987), Lal (1990), Malmer (1993), Abdul Rahim and Zulkifli (2004) and Bruijnzeel (1990, 2004). It is surmised that the fraction of rainfall leaving the drainage basin through the gauging station remains similar where tree crops are an important replacement for native forest and where urbanization covers less than 20% of the catchments (Figs. 26.4, 26.5 and 26.6).

However, there is a significant increase in sediment yield of 19% during the development period, which is functionally related to changes in land use cover, both in agricultural and urban areas. Biological oxygen demand (BOD_5) also shows a

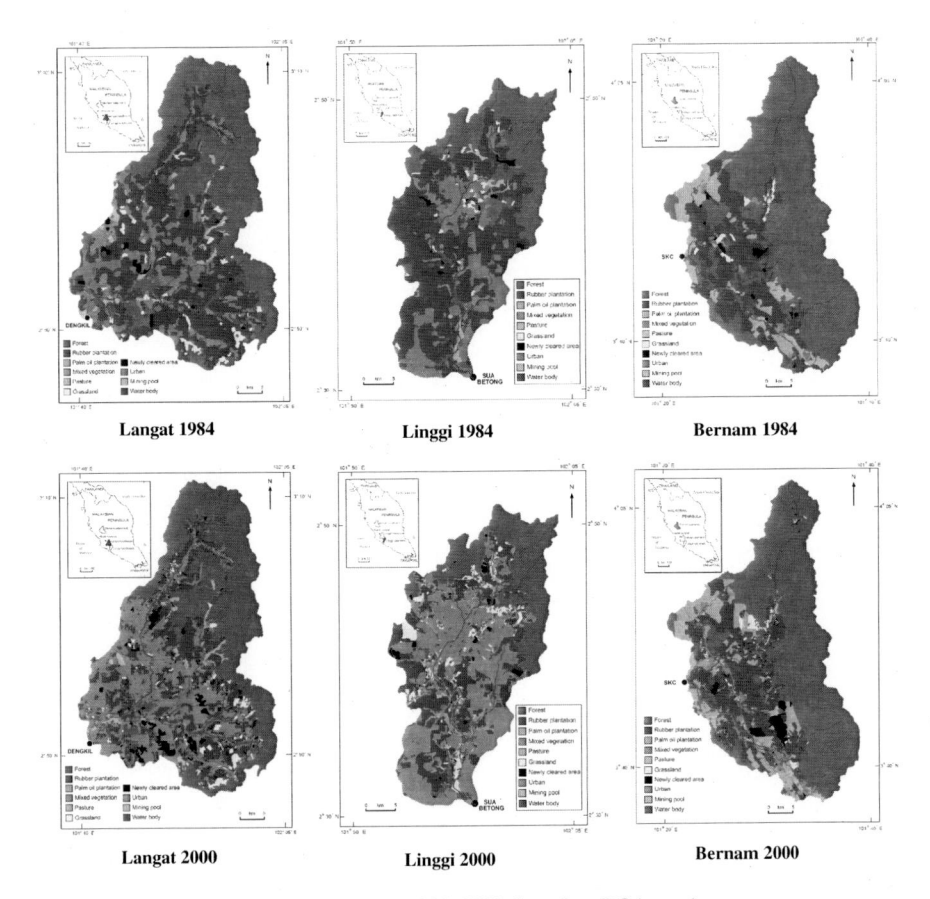

Fig. 26.3 All catchments – land use map 1984–2000 (based on DOA map)

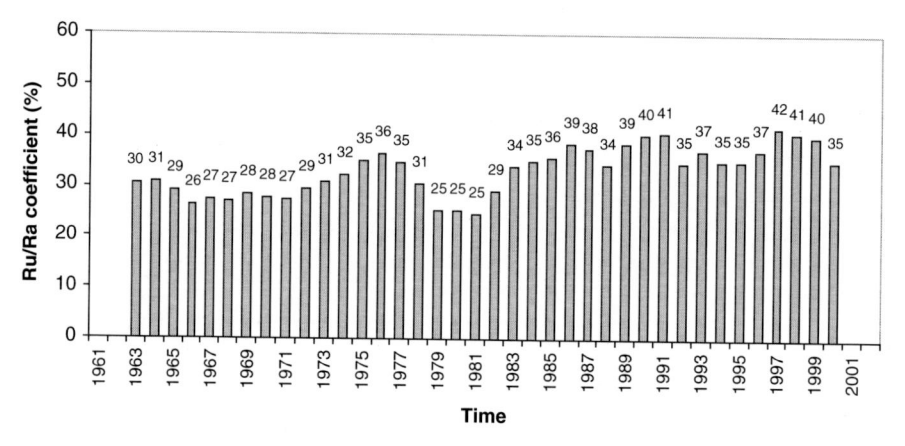

Fig. 26.4 Langat: rainfall-runoff coefficient for complete and partial monthly records – 5-year running mean (1961–2002)

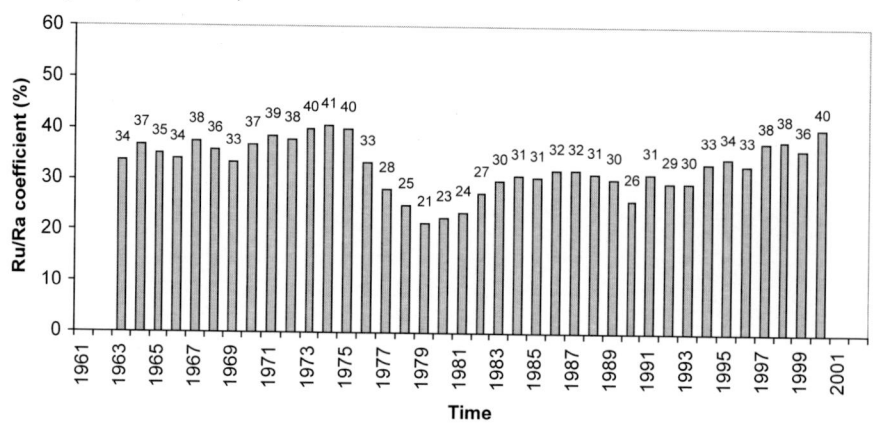

Fig. 26.5 Linggi: rainfall-runoff coefficient for complete and partial monthly records – 5-year running mean (1961–2002)

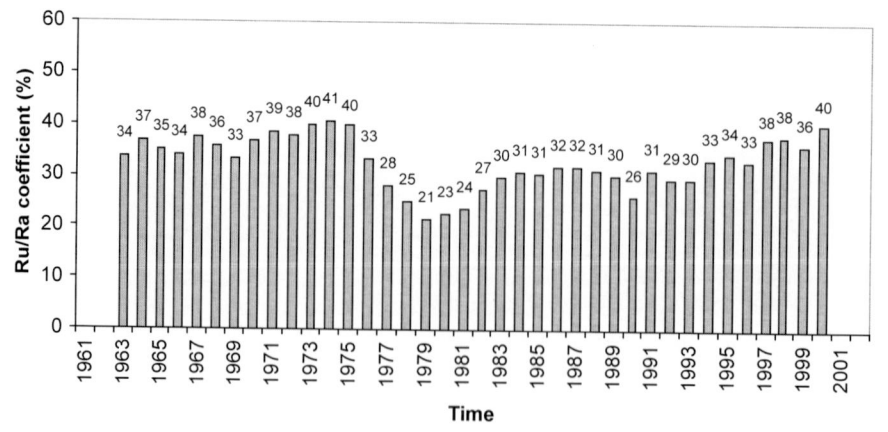

Fig. 26.6 Bernam: rainfall-runoff coefficient for complete and partial monthly records – 5-year of running mean (1961–2002)

significant increase where more domestic sewage and industrial waste discharges to the river as a function of higher population and industrial growth since 1990s. Water quality considerations dictate that the catchment must be properly protected and managed to ensure sustainable development.

26.4 Conclusions

The literature suggests that there is an impact on rainfall-runoff and water quality with urbanization expansion and development. This study shows that, in moderate-sized catchments of c. 1,000 km^2 in the wet tropics, the outcome is less straightforward. While water quality deteriorates, the fraction of rainfall leaving the drainage basin through the gauging station remains similar where tree crops are important replacement for native forest and where urbanization covers less than 20% of the catchments. Nevertheless, water quality considerations dictate that the catchment must be protected and managed to balance the needs of development against those of preservation of forestland. The initial intention to establish a predictive tool of land use impact on water relations i.e. a transfer function developed in analogue catchments that can be used to mitigate the effects of development has been abandoned in favour of assessing hydrological process-response in catchments as yet to be developed so that the findings can be used to inform land use management.

To conclude, an integrated study of land use, climatology and hydrology within river basins is essential to understanding the potential impact on the water quantity and water quality of an expanding population that is increasingly urbanised. This study provides information for the local river managers and development planners and could assist in minimizing negative impacts of development on water resources while promoting sustainable planning within river basins.

References

Abdul Hadi, H.S. and Abdul Samad, H. (2000). Modelling for integrated drainage basin management. In: Jamaluddin, M.J., Abdul Rahim, M.N., Abdul Hadi, H.S. and Ahmad Fariz, M. (eds.), Integrated Drainage Basin Management and Modelling. Centre for Graduate Studies, Universiti Kebangsaan Malaysia, Malaysia, pp. 164–190.

Abdul Rahim, N. and Zulkifli, Y. (2004). Hydrological impact of forestry and land use activities: Malaysian and Regional experience. In: Abdul Rahim, N. (ed.), Water: Forestry and Land Use Perspectives. IHP-VI/Technical document in Hydrology, No. 70. UNESCO, Paris, pp. 86–105.

Bruijnzeel, L.A. (1990). Hydrology of Moist Tropical Forests and Effects of Conversion: A State of Knowledge Review. UNESCO IHP Tropic Programme, Paris.

Bruijnzeel, L.A. (2004). Hydrological function of tropical forests: Not seeing the soil for the trees? Agriculture, Ecosystems and Environment 104:185–228.

Brundtland Commission. (1987). Report on the World Commission on Environment and Development. General Assembly Resolution 42/187 (United Nations). 9th Plenary Meeting 11 December 1987.

Hsia, Y.J. (1987). Changes in storm hydrographs after clearcutting a small hardwood basin in central Taiwan. Forest Ecology and Management 20:117–134.

Lal, R. (1993). Challenges in agriculture and forest hydrology in Humid Tropics. In: Bonell, M., Hufschmidt, M.M. and Gladwell, J.S. (eds.), Hydrology and Water Management in the Humid Tropics; Hydrological Research Issues and Strategies for Water Management. Cambridge University Press, UNESCO and Cambridge, Paris.

Malmer, A. (1993). Dynamic of hydrology and nutrient losses as response to establishment of forest plantation – a case study on tropical rainforest land in Sabah, Malaysia. Unpublished PhD thesis. Sweden: Swedish University of Agricultural Science.

Chapter 27
Reclamation of Land Disturbed by Shrimp Farming in Songkla Lake Basin, Southern Thailand

Charlchai Tanavud, Omthip Densrisereekul, and Thudchai Sansena

Abstract Based on remotely sensed data, it was revealed that, over the period of 24 years, from 1982 to 2006, areas devoted to shrimp farming in Songkla Lake Basin, southern Thailand, increased by a total of 7,401 ha, equivalent to an increase of 212.0%. The dramatically rise of the land under shrimp culture was attributed to culture technological advances, government subsidies and strong demand for shrimp in global markets. However, with the continued expansion of culture ponds, most farms were abandoned after 5–7 years of operation due to production losses stemming from problems with water quality and associated outbreaks of disease. Characterization of the properties of soil materials collected from an abandoned shrimp pond indicated that the shrimp farmed soils possessed several adverse physical and chemical impediments to the establishment of vegetation. The major physical limitations were largely associated with low saturated hydraulic conductivity and low oxygen diffusion rate whereas high salinity levels were the main chemical constraints. Methods to overcome these physical problems were investigated by mixing the farmed soils with locally available amendment materials such as gypsum and vermiculite. Findings revealed that the incorporation of these amendments into the farmed soils improved their physical properties through increased saturated hydraulic conductivity and oxygen diffusion rate values. The elevation of the saturated hydraulic conductivity values could also facilitate the leaching out of salt from the farmed soils resulting in reduced salinity levels. The applicability of this amelioration procedure under field conditions merits further research.

Keywords Shrimp farming · Shrimp farmed soils · Land reclamation · Gypsum · Vermiculite

C. Tanavud (✉)
Faculty of Natural Resources, Prince of Songkla University,
Hat Yai, Songkla, Thailand
e-mail: charlchai.t@psu.ac.th

27.1 Introduction

Songkla Lake Basin, located on the eastern coast of the southern Thai Peninsula, covers an area of approximately 8,463 km^2, of which 1,043 km^2 or 12.3% is open water in the form of three interconnected lakes (Fig. 27.1). The lake system, connecting to the Gulf of Thailand through a narrow channel outlet at the southern end of the lake, is subject to seasonal fluctuations in salinity. The economy of the basin is agricultural in nature. Rice is cultivated in the lowlands, rubber and mixed orchard in the foothills and terraces, and shrimps along the coastal shores (Tanavud et al., 2001). The Songkla Lake Basin, which was once the richest and most extensive rice growing area, was formerly described as the rice-bowl of southern Thailand.

Black tiger shrimp (*Penaeus monodon*) culture was introduced to the basin in the early 1980s as a mean of generating foreign exchange and enhancing employment opportunities in the basin's rural areas (Flaherty and Karnjanakesorn, 1995) (Fig. 27.2). Despite the large financial benefits generated, the production of farm-raised shrimp has given rise to deleterious environmental impacts such as soil salinization, water quality deterioration as a result of effluent disposal, and competition between agriculture and aquaculture for freshwater supplies (Flaherty et al., 2000) (Fig. 27.3).

The degradation of the coastal environment also contributed to shrimp disease outbreaks resulting in crop failures and pond abandonment (Dierberg and Kiattisimkul, 1996). These disease problems then led to the migration of the culture

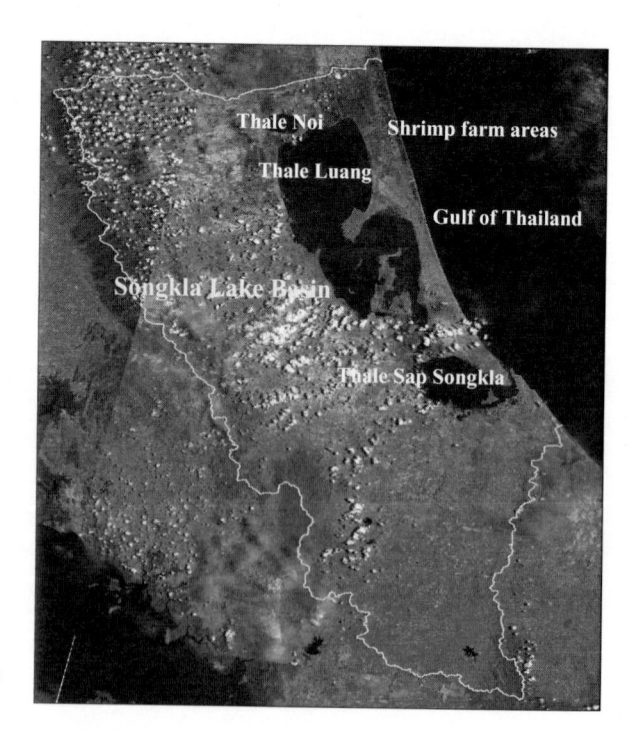

Fig. 27.1 Songkla lake basin from space

Fig. 27.2 The cultured pond

Fig. 27.3 Environmental impacts of farming operation

operations into rice-growing areas inland from the coast. At present, the exact area under shrimp farming in the basin is not known and hence the extent of the pond-induced problems is difficult to assess. In the majority of cases many of the ponds were abandoned after 5–7 years of operation as a result of environmental degradation and associated disease outbreaks.

In the early 2000s, white shrimp (*Penaeus vannamei*) was introduced into the basin after the tiger shrimp production declined due to slow growth rates and outbreaks of disease (Szuster, 2006). Despite this switch, it is anticipated that white shrimp operations will face the same problems with poor water quality and risks of disease as those experienced by the earlier tiger shrimp cultivations (Belton and Little, 2008).

The abandoned shrimp ponds are characterized by unstable, deformed landscapes, which are a potential hazard to the surrounding environment. Field observations indicate that natural establishment on these disturbed lands is negligible or slow even many years after cessation of farming, pointing to one or more growth limiting factors in the media. As crop production is inadequate to meet the requirements of the basin's population and concerns over environmental degradation are growing, reclamation of the disturbed lands has become an essential consideration.

The general objective of the reclamation is the return of disturbed lands to a condition approaching its original biological potential for crop production, amenity, and to an aesthetically pleasing condition (Bradshaw and Chadwick, 1980). Since lands disturbed by shrimp farming were originally in agricultural use, reclamation to agriculture is perhaps the most common land use objective. The identification of the limitations to growth in the rooting medium and the amelioration measures requirements have therefore become the essential fundamental prerequisite for successful land reclamation schemes. In general, soil physical limitations are the most difficult to ameliorate, and these often dictate production potential after the more readily altered chemical constraints are corrected. The aims of this study, therefore, were (i) to ascertain the areal extent of shrimp farming in Songkla Lake Basin using remotely sensed data and GIS, (ii) to define potential limitations to plant growth in the shrimp farmed soils by characterization of their major physical and chemical properties, and (iii) to evaluate the possibility of ameliorating the physical limitations of the farmed soils by means of mixing them with gypsum and vermiculite.

27.2 Materials and Methods

27.2.1 Areal Extent of Shrimp Farming

A Geographic Information Systems (GIS) was used in this study to compile spatially explicit data layers that described the basin's land use. All spatial analysis operations were performed using ArcGIS 9.0 software. To allow for comparisons, land use was determined for three time periods: 1982, 2000 and 2006. The 1982 land use coverage was generated by digitising from a paper map displaying 1982 land use at a scale of 1:50,000 prepared by the Department of Land Development. The 2000 and 2006 land use coverage were digitised from paper maps visually interpreted from 1:50,000 Landsat TM images acquired in 2000 and 2006, respectively. Ground truthing was also conducted to assist in the imagery classification and validate the final results. Following the preparation of land use coverage for the three dates, the areas of each category of land use for each period were calculated using ArcGIS.

27.2.2 Characterization of the Shrimp Farm Soils

Soil samples were taken from pond bottoms in an abandoned shrimp farm area in Songkla lake Basin. These soils are referred to as shrimp farmed soils in this study. According to the Department of Land Development (1973) using the USDA Soil Taxonomy system, the soils on which the cultured pond was built were identified as

Ranot series and classified as Typic Tropaqualfs. At each sampling location, three composite samples of disturbed farmed soils with an additional three samples of the undisturbed farmed soils were collected using a core sampler with an open ended metal cylinder of 5 cm internal diameter and 5 cm long (DIK-1621). Each composite sample was made up of three individual samples each taken randomly to a depth of 30 cm. The disturbed samples were air-dried, ground, passed through a 2 mm sieve and thoroughly mixed.

Particle size analysis was performed using the hydrometer method (Dirksen, 1999). Soil bulk density was determined by the core method. Plant available water was evaluated as the difference in water content held at a matric potential of −10 and −1,500 kPa (Rowell, 1994). Saturated hydraulic conductivity was determined in the undisturbed core samples using a Falling-Head Permeameter (DIK-4050). Oxygen diffusion rate (ODR) was measured in the core samples equilibrated at −10 kPa using an Oxygen Diffusion Rate Meter (DIK-5100). A pocket penetrometer (Eijkelkamp-06.03) with a probe diameter of 6.4 mm was used to measure mechanical resistance of the core samples at −10 kPa. Measurements of pH and electrical conductivity (EC) were made on a 1:5 soil/de-ionized water suspension. Soil organic matter content was measured using the Walkley-Black technique (Nelson and Sommers, 1982). Cation exchange capacity (CEC) was estimated by distillation of ammonia displaced by sodium (Thomas, 1982). Available phosphorus was analysed according to the Bray–2 method (Bray and Kurtz, 1945). Exchangeable potassium was extracted using ammonium acetate and determined by flame photometry (Thomas, 1982).

27.2.3 Physical Amelioration of the Shrimp Farmed Soils

Amelioration of physical limitations of the shrimp farmed soils was achieved by mixing them with locally available amendment materials such as gypsum ($CaSO_4.2H_2O$) and vermiculite $(Mg,Ca,K,Fe)_3(Si,Al)_4O_{10}(OH)_2.4H_2O$. In the case of the soil-gypsum mixture, the gypsum was added to the farmed soils at rates of 0, 12.5, 28.0 and 37.5 t ha^{-1}. Soil-vermiculite mixtures were also prepared in which the resulting vermiculite contents were 0, 12.5, 25.0 and 50.0% on a volume basis. After mixing, sufficient amounts of each mixture were transferred into a column constructed from a PVC tube (15 cm diameter and 100 cm in length) sealed at the bottom by a cap. To ensure free drainage, a hole at the bottom of each tube was made. During filling, the mixtures in the tube were packed by tamping the sides of the tube with a rubber hammer to give a final depth of 80 cm. The mixtures were maintained at constant soil water content, representing a matric potential of −10 kPa. After the 12 months period of the experimentation, three replicate undisturbed core samples were taken from the uppermost 30 cm of each mixture in the tube for saturated hydraulic conductivity and oxygen diffusion rate measurements using the methods described in the previous section. Analysis of variance (ANOVA) was used to test the effects of the inorganic amendments on physical properties of the mixtures. Values of soil properties that differed at $p < 0.05$ were considered significant (Fowler et al., 1998).

27.3 Results and Discussion

27.3.1 Areal Extent of the Shrimp Farming

Based on the analysis of satellite images, it was found that areas devoted to shrimp farming in 1982 covered an estimated 3,491 ha, equivalent to 0.47% of the total area of the basin (Table 27.1). At that time, the development of shrimp farming was limited to a relatively narrow band of coastal land, ensuring easy access to saline water. Over a period of 18 years, from 1982 to 2000, area under shrimp culture increased from 3,491 to 7,799 ha, equivalent to an increase of 123.4% (Table 27.1). In 2000, the ban of shrimp ponds along the coastline progressively widened as the coastal shore became increasing crowded with culture ponds. During the 6-year period from 2000 to 2006, shrimp cultivation areas further increased by 3,093 ha, representing an increase of 39.7% (Table 27.1). These new cultivation areas have emerged in freshwater areas inland from the coast as well as along the lakeshores. It is noteworthy that, over a period of 24 years, from 1982 to 2006, areas devoted to shrimp culture dramatically rose from 3,491 to 10,892 ha, representing an increase of 212.0%.

27.3.2 Characterization of the Shrimp Farmed Soils

The shrimp farmed soils had clay percentage of 69.19 and their textural class was clay (Table 27.2). The high bulk density in the farmed soils suggested that compaction had occurred, probably as a result of tractor wheel passage during pond construction. The air-filled porosity was below the critical value of 10%, indicating

Table 27.1 Estimates of land use in Songkla lake basin in 1982, 2000 and 2006

Land use categories	Area* (ha) 1982	2000	2006
Forest	146,568	86,225	78,052
Rubber	292,610	404,531	434,745
Rice field	208,599	164,209	141,200
Fruit orchards	21,412	126	2,766
Perennial crops	1,196	881	680
Shrimp farming	3,491	7,799	10,892
Mangrove forest	3,221	406	387
Fresh water swamp forest	24,821	18,682	12,917
Water bodies	1,334	2,632	2,762
Urban/built up area	30,990	54,568	56,213
Miscellaneous	7,725	1,908	1,353
Total	741,967	741,967	741,967

*Excludes lake surface.

27 Reclamation of Land Disturbed by Shrimp Farming

Table 27.2 Physical and chemical properties of the shrimp farmed soils

Properties of shrimp farmed soil	
Chemical properties	
pH	6.98
EC (mS cm^{-1})	13.74
Organic matter content (g kg^{-1})	4.50
CEC (cmol (+) kg^{-1})	21.95
Available phosphorus (mg kg^{-1})	19.51
Exchangeable potassium (cmol(+) kg^{-1})	1.54
Physical properties	
Texture	Clay
Sand (%)	0.27
Silt (%)	30.54
Clay (%)	69.19
Bulk density (g cm^{-3})	1.78
Particle density (g cm^{-3})	2.71
Air-filled porosity at field capacity (%)	9.95
Plant available water (%)	9.72
Saturated hydraulic conductivity (m s^{-1})	4.16×10^{-7}
Oxygen diffusion rate (g cm^{-2} min^{-1})	14.34×10^{-8}
Resistance to penetration (kPa)	120.63

that oxygen supply might become limiting to root growth (Landon, 1991). In addition, the farmed soils had low values of available water percentage (Table 27.2), apparently as a result of higher bulk density and low organic matter contents. The saturated hydraulic conductivity values fall within the low class according to Ghildyal and Tripathi (2005).

The predominance of fine-textured soils accounts for the low value of saturated conductivity, and possibly results in waterlogging conditions, which promote denitrification and runoff losses. If the diffusivity values of 20×10^{-8} g cm^{-2} min^{-1} is taken as a lower limit for plant growth as suggested by Stolzy and Latey (1964), then the dissuasive transport of gases through air-filled pores in the farmed soils would be restricted, affecting nutrient and water uptake, and eventually the growth of plant. Taylor and Burnett (1964) cited a mechanical resistance of about 2,500 kPa at –10 kPa was sufficient to prevent or markedly reduce root elongation. The mechanical resistance recorded in farmed soils at –10 kPa was 120.63 kPa. This resistance figure is well below 2,500 kPa and as such it could be assumed that mechanical impedance to root ramification would not be a limitation to plant growth.

The pH value for the farmed soils was in the optimum range for satisfactory growth (Table 27.2). The EC value was above the critical value of 4.0 ms cm^{-1}, indicating that plant grown on the farmed soils would be restricted by soluable salt (Rowell, 1994). The organic matter content was lower than published critical values of 5 g kg^{-1} (Department of Land Development, 1973), suggesting that the contribution to total nitrogen from organic sources could be expected to be extremely small. Phosphorus concentration in the farmed soils was higher than minimum threshold of

4 mg kg^{-1} (Landon, 1991), suggesting a probable sufficiency. The potassium levels in the farmed soils were found to be higher than the critical values of 0.15 cmol (+) kg^{-1}, and therefore deficiency was unlikely (Landon, 1991). Based on these findings, it is apparent that land from which shrimp has been farmed has been left in a condition that is far from ideal for agricultural purposes. Tanavud et al. (2001) obtained similar results working with the shrimp farmed soils collected from an abandoned shrimp farm area in Hat Yai District, Songkla Province.

27.3.3 Amelioration of Physical Limitations of Shrimp Farmed Soils

Under the conditions of this experiment, it was evident that the incorporation of gypsum into the shrimp farmed soils significantly increased saturated hydraulic conductivity value of the shrimp farmed soil-gypsum mixtures (Figs. 27.4 and 27.5 and Table 27.3), with a plateau being reached at 25.0 t ha^{-1}. Likewise, the addition of gypsum significantly elevated oxygen diffusion rate values of the mixtures, with a plateau being reached at 12.5 t ha^{-1}. With respect to vermiculite, values of saturated hydraulic conductivity and oxygen diffusion rate increased progressively with increasing quantities of vermiculite (Figs. 27.6 and 27.7 and Table 27.3). However, the levels of vermiculite applied were insufficient to produce a well-defined plateau for the farmed soil-vermiculite mixtures. Laboratory investigations indicated that the application of gypsum and vermiculite had a positive influence on the physical characteristics of the shrimp farmed soils. An additional benefit is a reduction in salinity levels resulting from improved leaching of salt brought about by enhanced saturated hydraulic conductivity values of the farmed soils. It is noteworthy that the effectiveness of incorporating gypsum in improving physical properties

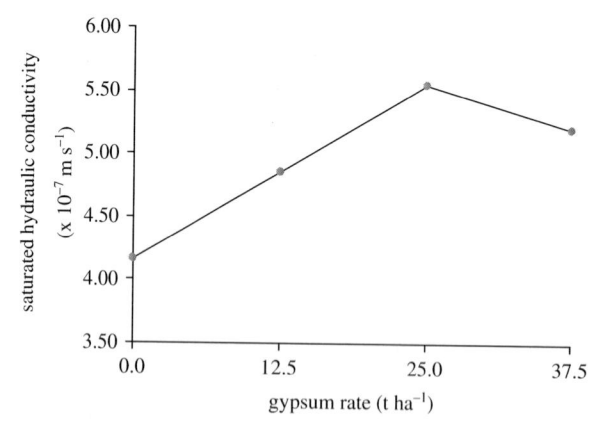

Fig. 27.4 Responses of saturated hydraulic conductivity values of the shrimp farmed soils to the addition of gypsum

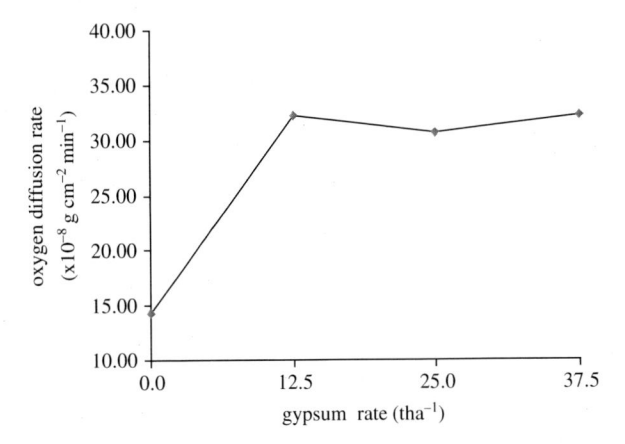

Fig. 27.5 Responses of oxygen diffusion rate values of the shrimp farmed soils to the addition of gypsum

Table 27.3 Effects of gypsum and vermiculite on saturated hydraulic conductivity and oxygen diffusion rate of the shrimp farmed soils. Numbers in brackets represent standard deviation

Physical properties	Gypsum rate (t ha^{-1})				Vermiculite (% by volume)			
	0.0	2.0	4.5	6.0	0.0	12.5	25.0	50.0
Saturated hydraulic conductivity ($\times 10^{-7}$ m s^{-1})	4.16a* (0.002)	4.86ab (0.001)	5.55b (0.007)	5.20b (0.003)	4.16w (0.002)	9.95x (0.007)	14.69y (0.007)	24.65z (0.003)
Oxygen diffusion rate ($\times 10^{-8}$ g cm^{-2} min^{-1})	14.34a (1.643)	32.27b (3.228)	30.66b (1.614)	32.27b (1.076)	14.34x (1.643)	41.96y (3.228)	47.33y (6.718)	105.42z (1.865)

*Means followed by the same letter within each parameter are not significantly different at 5% probability level.

of the farmed soils could be increased if more time is allowed for the removal of exchangeable sodium by calcium.

27.4 Conclusions

Characterization of the physical and chemical properties of the shrimp farmed soils has led to the conclusion that the physical problems of low saturated hydraulic conductivity and low oxygen diffusion rate values, and the major chemical constraint of high salinity level would pose limitations to the establishment of vegetation on

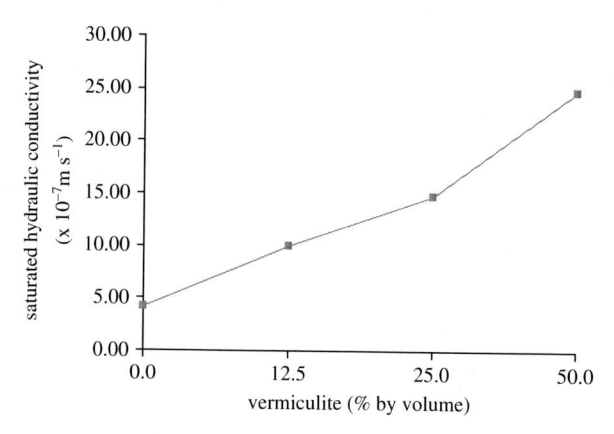

Fig. 27.6 Responses of saturated hydraulic conductivity values of the shrimp farmed soils to the addition of vermiculite

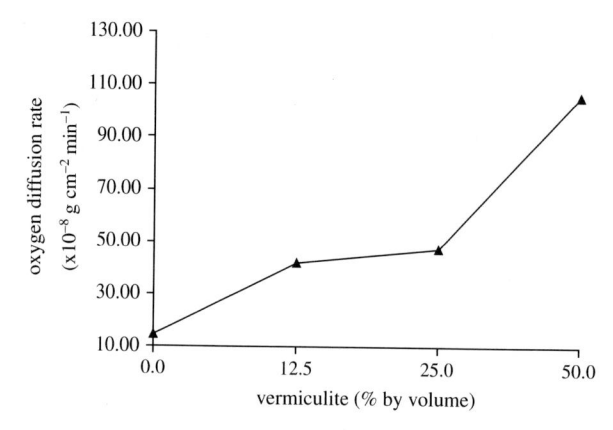

Fig. 27.7 Responses of oxygen diffusion rate values of the shrimp farmed soils to the addition of vermiculite

the farmed soils. To ensure successful establishment of plants on these farmed soils, these inherent limitations must be corrected. Laboratory studies have shown that the incorporation of the gypsum and vermiculite improved their physical properties of the farmed soils through increased values of the saturated hydraulic conductivity and oxygen diffusion rate. The elevation of the hydraulic conductivity values could also facilitate the leaching out of salt from the farmed soils resulting in reduced salinity levels. In alleviating physical impediment to growth, solutions reached were based solely on physical parameters measured in the laboratory. The ultimate indicator of soil physical conditions, however, is the plant, and thus there is a need for assessing the growth response of plants to the physical properties of the gypsum and vermiculite amended farmed soils in the glasshouse and in the field.

Acknowledgements This study was support in part by a grant from the Graduate School, Prince of Songkla University.

References

Belton, B. and Little, D. (2008). The development of aquaculture in central Thailand. Journal of Agrarian Change 8(1):123–143.

Bradshaw, A.D. and Chadwick, M.J. (1980). The Restoration of Land. Blackwell Scientific Publications, Oxford, 314 pp.

Bray, R.H. and Kurtz, L.T.. (1945). Determination of total, organic and available forms of phosphorus in soils. Soil Science 59:39–45.

Department of Land Development. (1973). Soil Map of Songkla Province. Provincial series no. 18. Ministry of Agriculture and Cooperatives, Bangkok.

Dierberg, F.E. and Kiattisimkul, W. (1996). Issues, impacts, and implications of shrimp aquaculture in Thailand. Environmental Management 20(5):649–666.

Dirksen, C. (1999). Soil Physics Measurements. Catena Verlag, Germany, 154 pp.

Flaherty, M. and Karnjanakesorn, C. (1995). Marine shrimp aquaculture and natural resource degradation in Thailand. Environmental Management 19(1):27–37.

Flaherty, M., Szuster, B. and Miller, P. (2000). Low salinity inland shrimp farming in Thailand. Ambio 29(3):174–179.

Fowler, J., Cohen, L. and Jarvis, P. (1998). Practical Statistics for Biology. John Wiley and Sons Ltd., West Sussex.

Ghildyal, B.P. and Tripathi, R.P. (2005). Soil Physics. New Age International, Ltd, New Delhi, India, 656 pp.

Landon, J.R. (1991). Booker Tropical Soil Manual. Longman Scientific and Technical, Essex, England, 474 pp.

Nelson, D.W. and Sommers, L.E. (1982). Total carbon, organic carbon, and organic matter. In: A.L. Page (ed.), Method of Soil Analysis. Part 2. Chemical and Microbiological Properties. 2nd edn. Agronomy Monograph No. 9. American Society of Agronomy Inc., Madison, Wisconsin, pp. 539–579.

Rowell, D.L. (1994). Soil Science. Methods and Applications. Longman Scientific and Technical, Essex, England, 350 pp.

Stolzy, L.H. and Latey, J.. (1964). Characterizing soil oxygen conditions with a platinum microelectrode. Advance Agronomy 16:249–279.

Szuster, B. (2006). Coastal shrimp farming in Thailand: Searching for sustainability. In: C.T. Hoanh, T.P. Tuong, J.W. Gowing and B. Hardy (eds.), Environment and Livelihoods in Tropical Coastal Zones. CAB International, Wallingford, pp. 86–98.

Tanavud, C., Yongchalermchai, C., Bennui, A. and Densrisereekul, O.. (2001). The expansion of inland shrimp farming and its environmental impacts in Songkla Lake Basin. Kasetsart Journal 35:326–343.

Taylor, H.M. and Burnett, E.. (1964). Influence of soil strength on the root growth habits of plants. Soil Science 98:174–180.

Thomas, G.W. (1982). Exchangeable cations. In: A.L. Page (ed.), Method of Soil Analysis. Part 2. Chemical and Microbiological Properties. Agronomy Monograph No. 9. American Society of Agronomy, Inc., Wisconsin, pp. 159–165.

Chapter 28
The Effect of Bio-solid and Tea Waste Applications on Erosion Ratio Index of Eroded Soils

Nutullah Ozdemir, Tugrul Yakupoglu, Elif Ozturk, and Orhan Dengiz

Abstract The aim of this study was to determine the effects of organic wastes such as bio-solid (BS) and tea waste (TW) on erosion ratio index (ER) of eroded soils in different levels (slight, moderate and severe). Soil samples used in this research were taken from surface soil (0–20 cm depth) located on agricultural land of Samsun province in the northern part of Turkey. These soil samples were treated with organic residues at four different levels (0, 2, 4, and 6% basis w/w) and each treatment was replicated three times in a split block design. All pot mixtures were incubated for 4 weeks under greenhouse conditions and later tomato plants were grown in all pots. When plant growth was in its 14th week, the experiment was ended. After greenhouse work, changes in ER values of eroded soils were determined. It was determined that the BS and TW treatments dramatically decreased ER values of the eroded soils compared to the control treatment. Statistical test results showed that erosion level ($p < 0.001$), type of residue ($p < 0.01$) and the application dose ($p < 0.001$) caused the changes of the ER values.

Keywords Erosion ratio · Organic wastes · Soil erodibility · Turkey

28.1 Introduction

There are a number of problems that threaten the life quality of mankind. One of the most severe recognized worldwide such threats is soil erosion. Losses of the most valuable natural resource-the soil-takes place almost imperceptibly and slowly affect the long-term productivity of the land (Mitra et al., 1998).

T. Yakupoglu (✉)
Department of Soil Science, Agricultural Faculty, Ondokuz Mayis University,
Samsun 55139, Turkey
e-mail: tugruly@omu.edu.tr

Almost 80% of Turkish soils are affected by erosion requiring radical precautions, especially on cultivated land. Hence it is of primary importance to know the major factors, which influence erosion at various degrees (Ozdemir and Askin, 2003). One of the said major factors is soil erodibility. Soil erodibility is a function of detachability and transportability, both of which are dependent on the physical, chemical and biological properties of the soil. The magnitude of erodibility is dependent on soil particle size and distribution, degree and stability of soil aggregation, and surface residue (Vanelslande et al., 1987; Choudhary et al., 1997). Treatment with organic conditioners to improve or maintain soil structure and aggregate stability may be one of the means of maintaining high water infiltration and low runoff and erosion (Ozturk et al., 2005). In addition, soil organic conditioning properties has been defined also by Brandsma et al. (1999) as the "upgrading of a poorly structured soil to one with suitably sized aggregates through proper tillage and subsequent stabilization of its tilth by the application of conditioning materials, naturally based".

Soil structure properties and the resistance to erosion, could be enhanced by adding appropriate organic matter to the affected soil (Turgut et al., 2004). Organic matter is a major stabilizing component that improves soil aggregation resulting in a higher total porosity (POR) and wider pore-size distribution (Zhang et al., 2005). Ozdemir (1993) examined the effect of organic residues on the soil structural stability and soil erodibility. The author indicated that additional organic matter was effective in decreasing the erosion ratio (ER) values of the soils.

The main objective of the present study was to determine the effects of different organic matter sources such as bio-solids (BS) and tea residues (TW) on the ER index of a soil eroded at different levels.

28.2 Materials and Methods

Soil samples used in this study were taken from a 0 to 20 cm depth of slight, moderate and severe eroded areas in Samsun-Turkey. This site is located in the Black Sea Region, Northern Anatolia (Latitude 41°C 19′N; longitude 36°C 02′W) as shown in Fig. 28.1. The climate is semi humid ($R_f = 47.21$) with monthly mean temperatures ranging from 6.6°C in February to 23°C in August. The annual mean temperature is 14.2°C and the annual mean precipitation is 670 mm (Anonymous, 2002). Soils were classified as Vertic Calciudolls according to the US Soil Taxonomy (USDA, 2003).

The bio-solids (BS) were obtained from the Bafra Municipality and tea residues (TW) from the Rize Tea Research Institute of the Ministry of Agriculture and Rural Affairs. The carbon nitrogen ratios of the BS and TW were 9.25 and 22.6, respectively. The soil samples were treated with these organic residues at four different levels (0, 2, 4 and 6% according to w/w) including the control treatments and each treatment was replicated three times in a split block design [(3 erosion

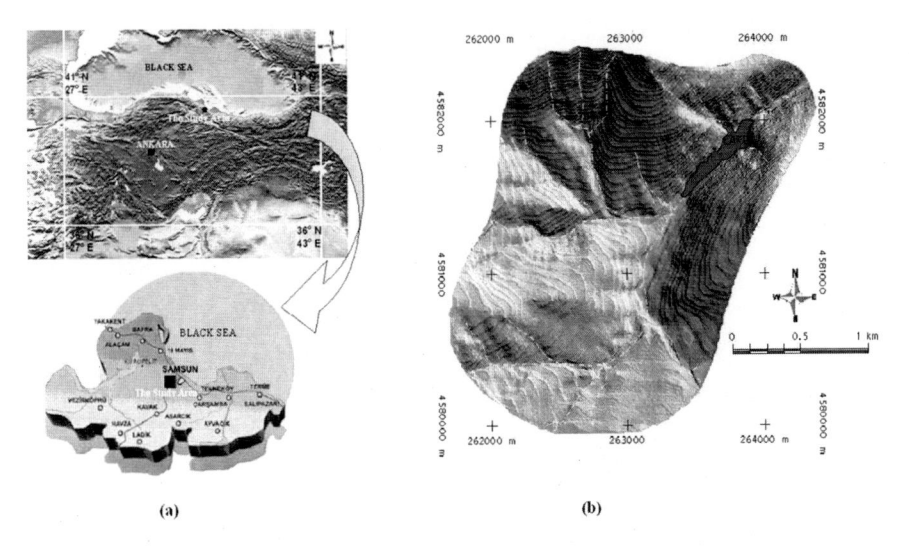

Fig. 28.1 Location (**a**) and physiographic (**b**) maps of the studied area

levels × 2 organic residues × 4 application doses) × 3 replicates]. All mixtures were transferred to plastic pots (height: 20 cm, diameter: 19 cm) and they were incubated at field capacity water contents at 20°C for 4 weeks. After the incubation period, tomato plants (cv. Tore F1) were grown in a greenhouse. After 112 days, the tomato plants were removed from the pots and the experimental work was accomplished. All soil samples including control were manually grinded to pass through a 2 mm sieve and the ER indices of soils were then determined. The effect of the plant roots on the ER indices was ignored and was determined by using Equation 1 developed by Lal (1988).

$$ER = ((DR \times ME)/C) \times 100) \tag{1}$$

In this equation, ER is the erosion ratio index (%), DR represents the dispersion ratio (%), ME is the moisture equivalent (%) and C is the clay content (%). Some physical and chemical soil properties were determined by the following methods: soil texture was determined by the hydrometer (Demiralay, 1993); soil organic matter content was measured by a modified wet digestion method (Nelson and Sommers, 1982); pH and EC values were determined in 1:2.5 soil:water suspensions measured by the pH and EC meters, respectively (Rowell, 1996); cation exchange capacity was determined by the ammonium acetate extraction method and $CaCO_3$ content was determined by Scheibler-calcimeter (Kacar, 1994).

The statistical assessment of the experimental data was undertaken using the SAS software package program (SAS Institute 1988).

28.3 Results and Discussions

28.3.1 Soil Properties

Some properties of the surface soil samples from the slightly, moderately and severely eroded area were determined as follows. The texture was fine with clay, silt and sand contents varying from 531 to 594 g kg^{-1}, 260–317 and 131–146 g kg^{-1}, respectively (Table 28.1). The soils were slightly alkaline having a pH value of 8.0. The free CaCO$_3$ content of the soils was determined to be between 16.6 and 21.9%. Cation exchange capacities were 37.4, 23.9 and 21.4 cmol kg^{-1} for slight, moderate and severe eroded soils, respectively. Soil organic matter content varied from 0.83 and 0.99%. EC values were found as 0.78, 0.65 and 0.64 ds m^{-1} for slight, moderate and severe eroded soils, respectively. The soils with ER ratio values greater than 10% were described as strongly erodible and soils with ER values less than 10% were described as weakly erodible (Tumsavas and Katkat, 2000). According to this assessment, slightly and moderately eroded soils were found to be weakly erodible and the severely eroded soil was found to be strongly erodible.

28.3.2 ER Index

Following the laboratory work, ER data were evaluated with the variance analysis test determining that the ER values were significantly influenced from the type of organic conditioner and application doses depending on the erosion levels. The mean of square values of the amendment materials (p <0.01) and their levels (p <0.001) were statistically significant. Meanwhile, interactions of the erosion level x conditioners, erosion level x doses, conditioners x doses and erosion level x conditioners x doses were significant (p <0.001).

Table 28.1 General properties of the experimental soils

Properties	Slight	Moderate	Severe
Clay, g kg^{-1}	594	561	531
Silt, g kg^{-1}	260	308	317
Sand, g kg^{-1}	146	131	152
Textural class	Clay	Clay	Clay
pH	8.0	8.1	8.1
EC, ds m^{-1}	0.78	0.65	0.64
OM, %	0.99	0.84	0.83
CEC, meq 100 g^{-1}	37.4	23.9	21.4
CaCO$_3$, %	16.6	19.4	21.9
ER, %	5.36	6.65	12.17

EC: electrical conductivity; OM: organic matter content; CEC: cation exchange capacity; ER: erosion ratio index

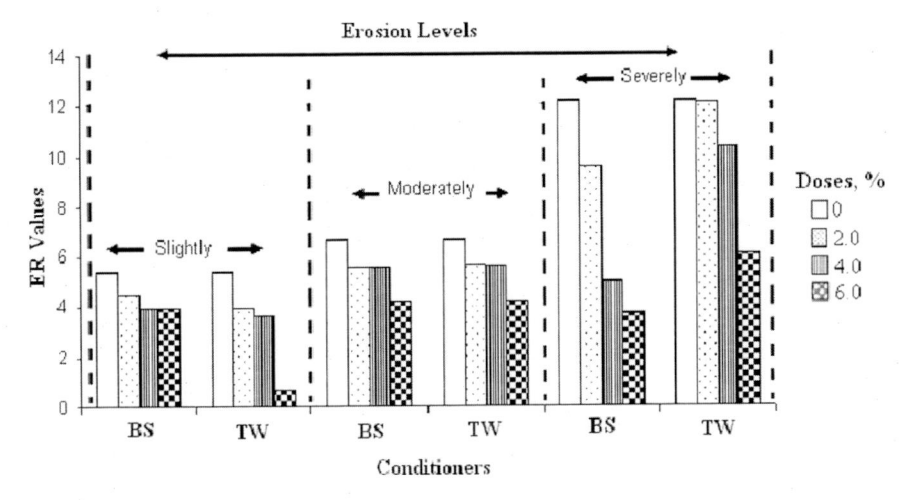

28 The Effect of Bio-solid and Tea Waste Applications on Erosion Ratio Index

Fig. 28.2 Medium ER values (%) obtained after the experiment

The ER values of the soils, determined after the greenhouse experiment in the mixtures of BS and TW with the soil samples are given in Fig. 28.2. Addition of the organic conditioners to the soil samples decreased the ER value ratio markedly depending on the type of organic matter conditioner, application dose and the erosion level of the soil. Generally, as the erosion level of the soils increased, the ER values increased too. The Mean ER value was 3.93% for slightly eroded soils while it was 5.50 and 8.99% for moderately and severely eroded soils, respectively (Fig. 28.2).

The $LSD_{0.05}$ (least standard deviation) results illustrating the effect of the applied organic residues on the ER index of the soils are given in Table 28.2. This table indicated that the application of organic wastes, their doses and the erosion level of the soils showed significant differences in terms of soil ER values (Table 28.2).

The organic residues, namely BS and TW decreased the ER value of the soils to 6.430 and 5.853, respectively. On the other hand, the effect of the BS and TW followed a descending order in relation to the effectiveness of the organic residues on the ER value. According to the application doses, the highest decrease in mean ER values compared with the control (8.059%) was found for 6% doses of organic conditioners as 3.816%. This may indicates that the organic residues had different resistances against decomposition. When taking the mean values into consideration, it is clear that organic residues decreased the ER values.

The texture of the soil along with the moisture equivalent was considered when the ER index was determined. A 10% threshold value was accepted for the ER index for determining soil erodibility. If the ER indices of the studied soils we determined to be greater than 10%, they would be described as erodible (Ngatunga et al., 1984).

This criterion is useful to determine the effects of the applied agricultural practices and the type of crop on soil structure. When the results from the present study

Table 28.2 LSD$_{0.05}$ test results of the ER data

	ER index
Erosion level	
Slight	3.933c
Moderate	5.503b
Severe	8.988a
LSD value for erosion level	0.452
Organic waste	
Bio-solid	6.430a
Tea waste	5.853b
LSD value for organic waste	0.099
Application dose, %	
0	8.059a
2	6.989b
4	5.702c
6	3.816d
LSD value for application dose	0.271

were discussed from this point of view, it is quite clear to see that the organic residues applied to the soils increased soil resistance reasonably well (Fig. 28.2) and the effects of the organic wastes on soil resistance varied depending on the degree of soil erosion accordingly. While the organic residues applied at the rate of 2 and 4% to the moderately eroded soils, decreased ER values to the level of erodibility of the non-applied slightly eroded soils. The 6% dose of TW applied to the severely eroded soils decreased the ER value of the soil to the level of the ER value of the non-applied soils in moderately eroded soil. These results were also supported by the work of Sonmez (1979), Ekwue (1990) and Haynes (2000).

28.4 Conclusions

The effect of organic residues, such as BS and TW, on a soil eroded at various levels, was studied in this work. It was determined that these soil conditioners have a positive effect on soil resistance against erodibility. The results can be summarized as: BS and TW treatments decreased ER values of eroded soils in the different levels. Effectiveness of the treatments varied depending on organic sources applied, its application dose and soil erodibility level. These results are of ample importance to countries suffering from severe erosion alike Turkey due to the low organic matter content of the regional soils and the magnitude of accelerated soil erosion. Organic residue applications have also an important role in their disposal as beneficial materials. Further studies are needed in the laboratory coupled with field experiments. They should determine the soil quality that is expected by the increasing amount of water stable aggregates and their significance in combating erosion.

References

Anonymous. (2002). Samsun Climatic Data (1974–2001) State Meteorology Services. Unpublished, Ankara, Turkey.

Brandsma, R.T., Fullen, M.A. and Hocking, T.J. (1999). Soil conditioner effects on soil structure and erosion. Journal of Soil and Water Conservation 54(2):485–489.

Choudhary, M.A., Lal, R. and Dick, W.A. (1997). Long-term tillage effects on runoff and soil erosion under simulated rainfall for a central Ohio soils. Soil and Tillage Research 42:175–184.

Demiralay, I. (1993). Toprak Fiziksel Analizleri (in Turkish). Erzurum

Ekwue, E.I. (1990). Organic matter effects on soil strength properties. Soil and Tillage Research 16:289–294.

Haynes, R.J. (2000). Interactions between soil organic matter and their effect on measured aggregate stability. Biology and Fertility of Soils 30:270–275.

Institute, S.A.S. (1988). SAS/STAT User's Guide, Release 6.03 Edition. SAS Institute, Cary, NC.

Kacar, B. (1994). Bitki ve Topragin Kimyasal Analizleri (in Turkish). Ankara

Lal, R. (1988). Soil Erosion Research Methods. Soil and Water Conservation Society, Ankeny, Iowa, pp. 141–153.

Mitra, B., Scott, H.D., Dixon, J.C. and McKimmey, J.M. (1998). Applications of fuzzy logic to the prediction of soil erosion in a large watershed. Geoderma 86:183–209.

Nelson, D.W. and Sommers, L.E. (1982). Total carbon, organic carbon and organic matter. In: Page A.L. (ed.), Methods of Soil Analysis, Part 2, Chemical and Microbiological Properties. Agronomy Monograph No 9. ASA Inc, Madison, WI, pp. 539–580.

Ngatunga, E.L.N., Lal, R. and Singer, M.J. (1984). Effects of surface management on runoff and soil erosion from some plot Milangano, Tanzania. Geoderma 33:1–12.

Ozdemir, N. (1993). Effects of admixturing organic residues on structure stability and erodibility of soils. Journal of Ataturk University Faculty of Agriculture 24(1):75–90.

Ozdemir, N. and Askin, T. (2003). Effects of parent material and land use on soil erodibility. Journal of Plant Nutrition and Soil Science 166:774–776.

Ozturk, H.S., Turkmen, C., Erdogan, E., Baskan, O., Dengiz, O. and Parlak, M. (2005). Effects of a soil conditioner on some physical and biological features of soils: Results from a greenhouse study. Bioresource Technology 96:1950–1954.

Rowell, D.L. (1996). Soil Science Methods and Applications. Wesley Longman Limited, Harlow.

Sonmez, K. (1979). Mus-Alparslan Devlet Uretme Ciftligi arazisinde yüzeyden alinan topraklarin strukturel dayanikliliği ve erozyona duyarliliği uzerine bir arastirma. Journal of Ataturk University Faculty of Agriculture 10(3–4):17–26.

Tumsavas, Z. and Katkat, A.V. (2000). A study on the determination of erodibilities of the sloping agricultural soils in the city of Bursa and vicinity against water erosion in laboratory conditions. Turkish Journal of Agriculture and Forestry 24:737–744.

Turgut, B., Olgun, M., Kucukozdemir, U., Karadas, K. and Gulseven, D. (2004). Effect of some organic residues on erodibility of soils. Paper Presented at the International Soil Congress (ISC) on Natural Resource Management Sustainable Development, Ataturk University, Erzurum, 7–14 June 2004.

USDA. (2003). Key to Soil Taxonomy. 9th edn. USDA, Natural Resources Conservation Services, Washington.

Vanelslande, A., Lal, R. and Gabriels, D. (1987). The erodibility of some Nigerian soils: A comparison of rainfall Simulator results with estimates obtained from the Wischmeier nomogram. Hydrological Processes 1:255–265.

Zhang B, Horn R and Hallett PD (2005). Mechanical resilience of degraded soil amended with organic matter. Soil Science Society of America Journal 69:864–871

Chapter 29
Modern and Ancient Knowledge of Conserving Soils in Socotra Island, Yemen

Dana Pietsch and Miranda Morris

Abstract At first view, soil erosion on Socotra Island, Yemen would seem to be a minor problem. This appraisal is based on the fact that on the one hand the island is poor in soil resources, and on the other hand research on the island has to date focussed on biodiversity. However, results of soil investigations on Socotra showed that in the Homhil Protected Area land degradation in terms of erosion due to soil structure deterioration and humus loss has increased drastically: within 3 years a loss of about 40 m^3 in a single gully head was estimated. Soil loss inevitably involves uprooting of trees and a decrease in soil fauna. Biodiversity is, of course, the most important argument for protecting the unique floral and faunal richness of the island, but what would terrestrial biodiversity be without soils? The present approach relies on a "down-to-earth" system of soil monitoring, based on both modern and ancient knowledge and oriented towards current environmental and political objectives. It should be understood as a first step towards conserving soils and vegetation in a Protected Area of this tropical island.

Keywords Ancient indigenous knowledge · Soil conservation · Biodiversity · Socotra island · Yemen

29.1 Introduction

The Island of Socotra is situated in the Arabian Sea (Fig. 29.1). It consists of a granite basement, limestone plateaus and basins, wadis and alluvial plains. Since large parts of the island are covered with Tertiary limestone (Geological Survey and Mineral Resources Board, 2003) and therefore partly covered with residual clays and coverloams, Cambisols are very common. More than 50% of the cambic soils

---

D. Pietsch (✉)
Institute of Geography, University of Tübingen, Tübingen 72070, Germany
e-mail: dana.pietsch@uni-tuebingen.de

Fig. 29.1 Location of Socotra island and Homhil

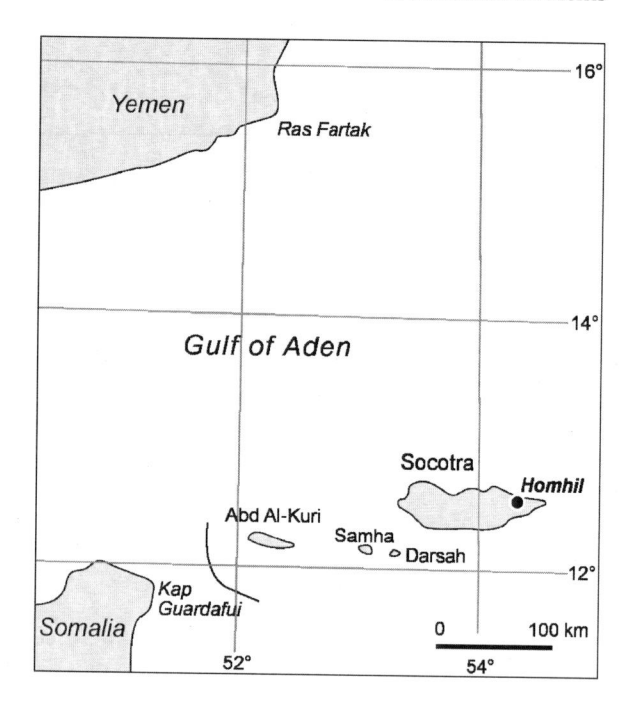

contain between 30 and 60% clay ($n = 30$). Furthermore, the coarse fragment content lies between 0 and 78%. In regions where erosion is most severe, Calcisols are often present. Calcisols in Socotra are silty (40–80% silt), but also clayic with clay contents of over 30%. Coarse fragment contents range from 0 to 63% (Pietsch, 2006; Pietsch and Kühn, 2009).

"Homhil" is situated in eastern Socotra and is one of the most extensively used intra-montane basins. It is characterised by limestone escarpments, a plain filled with red coverloam, and locally limited surface water. Vegetated with open woodland of *Boswellia* (frankincense), *Commiphora* and *Dracaena cinnabari* (the dragon's blood tree), it has been designated one of the island's Protected Areas in recognition of its rich biodiversity (Miller and Morris, 2004).

The people of Homhil are principally pastoralists. They also cultivate date palms and some are seasonal fisherman. A spring is the only permanent water-source of the area, though there are seasonal streams, and shallow wells also used for watering date-palms and, formerly, for the cultivation of finger-millet. Now a small number of kitchen gardens are cultivated (Miller and Morris, 2001).

From the socio-economic point of view an important fact is that from the late 1960s onwards the people of Homhil began moving down slope from their cave settlements in the mountain slopes to build more permanent stone dwellings around the valley floor (Morris, 2002). The reasons for this gradual move are various and complex, but one major effect was increased pressure on the valley floor grazing, which previously had been exploited only after rainfalls.

As people began to settle around the valley rim, the household herds – in general the productive animals, spent increasing amounts of time around the ten or so hamlets that developed over the next few decades. At the same time, aspirations changed, food security improved and alternative sources of livelihood became available. A consequence of these developments was the collapse of earlier systems of livestock management: controlled breeding, close herding and the slaughter of the majority of male offspring soon after birth in order to maximise milk production. The decline in the regular, seasonal transhumance of livestock between upland and lowland grazing meant that ever-greater numbers of animals spent more and more time around the permanent settlements. No longer managed as in the past, only disease and drought have prevented the numbers reaching uncontrollable proportions (Morris and Pietsch, 2008). Another important fact is that whereas cattle and sheep in general do not greatly disturb the soil surface when resting beneath trees and walking along the tracks, donkeys and goats are a prime cause of damage to land surfaces.

Disruption to economic and political systems against a background of land degradation has been investigated in other regions (Blaikie and Brookfield, 1987). There are also those involved in current research in this field that view land degradation as a "natural disaster" (Hudson and Ayala, 2006).

In the present case, land degradation in the form of soil erosion has arisen as a direct consequence of unsustainable land use. Land degradation of this kind leads to a decline in productivity over time and to a reduction in the actual and potential use of the land (Blaikie and Brookfield, 1987; Dalelo, 2001).

29.2 Materials and Methods

Soil data on Socotra obtained from 232 Puerckhauer drillings and analysis of 48 profiles, as well as from degradation mapping (see Fig. 29.4, Pietsch, 2006), combined with insights from traditional expertise in this field, were used to implement soil monitoring in Homhil. Knowledge of early Yemeni and Socotran expertise in soil management is based on the long-term experience of both authors. The soils were described by German Mapping Guidelines and FAO Guidelines (Ad-Hoc AG Boden, 2005; FAO, 2006) and classified after WRB (IUSS Working Group WRB, 2006). Soil analysis followed Blume et al. (2000).

To prevent badland development, a monitoring site was chosen whereon truncated Cambisols are distributed and endemic vegetation such as *Boswellia* (frankincense) is at risk of uprooting. This work is being carried out against a background of Yemeni Government policy that, over the last 10 years or so, has provided some tools for the "Conservation and Sustainable Use of Socotra Island" (EPA, 2000; EPC, 2000; FAO-UNCCD-UNDP, 2000).

29.3 Results and Discussion

29.3.1 Ancient Knowledge on Soil Conservation

In former times people developed many excellent strategies for conserving and exploiting soil resources, as is well known from the Yemen mainland, when as early as 3,600 BC, Yemenis started building terraces to conserve soils for agriculture (Wilkinson, 2003). People on Socotra also applied similar techniques: stones and rocks were used in different ways by the islanders to conserve soil (Morris, 2002). In the central Hagher mountains terrace systems were established (Fig. 29.2, Pietsch, 2003). Furthermore, large areas of the island are crossed by a network of low walling (Fig. 29.3).

Fig. 29.2 Old terrace systems in Wadi Di-Farhoh, Hagher mountains

Fig. 29.3 Walling in Diksam, western limestone plateau

Although to date the purposes and function of this extensive walling are not fully understood, in many areas they clearly have had an excellent side-effect: collecting soil and flood debris and supporting the succession of vegetation. Much of this early walling in Socotra is now buried beneath the shrubs and plants, which have become established on the accumulated sediments. Soil also collects around the stone-walled pens, byres and other livestock enclosures that proliferate on the island, and along the lines of rocks used to guide run-off to the rainwater catchments that are the main source of dry weather water.

29.3.2 Modern Knowledge of Conserving Soils

From the soil scientific point of view the main reasons for topsoil erosion in Socotra Island are humus and structure loss (Pietsch, 2006) mainly due to the increasing pressure of goats and feral donkeys. As a consequence, everywhere in the island, and especially after heavy rainfall and on sites which are more or less inclined in particular, surface water has been moving loose materials. One negative result is the truncation of soils due to gully erosion (Avni, 2005). As inclination increases, erosion by water is even more marked (Cerdá and García-Fayos, 1997). Among other visual indicators of land degradation are the uprooting of trees, a restricted rooting depth, crusting, a decrease in organic matter and in soil fauna (Stocking and Murnaghan, 2003).

Soil investigations show that in the Homhil Protected Area erosion is widespread, occurring in the form of gullies (LZ) and badlands (FB). Both forms were mapped, and are depicted in Fig. 29.4.

Measurement of the extent and size of the gullying and badlands (LZ1 and FB1) showed that erosion has increased drastically: in the period from 2003 to 2005 it was estimated that there had been a loss of about 40 m^3 soil within a single gully (Pietsch, 2007).

As mentioned earlier, Cambisols are common within the basin. The reference profile "Cambisol" (Hom9) is situated at the gully margin (LZ1, Fig. 29.4). Because of very slight inclination, the surface water is infiltrating the cambic soils optimally, because they have a loamy texture and a moderate skeleton content (Hom9, Table 29.1). At sites where Cambisols are shallow or even strongly calcified (crusting by calcification), soil erosion might have been happening over decades. Secondary calcification is the most common soil-forming process on Socotra (Pietsch and Kühn, 2009), and crusting due to loss of topsoil is one sign of human-induced soil degradation (Urban, 2002).

Since the lower part of the soils in Homhil basin consists mainly of petrocalcic or hypercalcic horizons (Cwck horizons), ongoing erosion will lead unavoidably to "plant-toxic" land surfaces with a low cation exchange capacity (CEC), as is seen in Table 29.1. In this regard, the reference profile is Calcisol Hom 14 at the badland margin (FB1, Fig. 29.4). The CEC in the Calcisol is at 5 cmol$_c$ kg^{-1}, i.e. very low.

Soil degradation pattern in Homhil basin

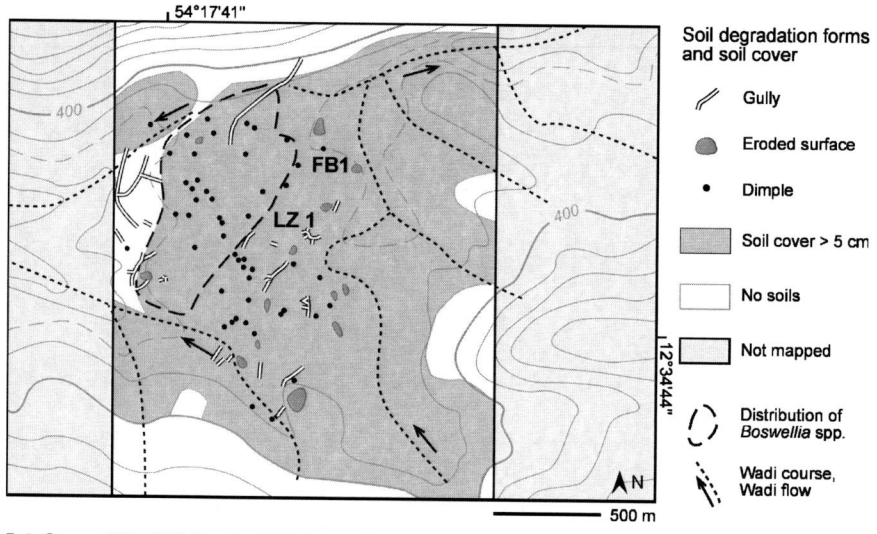

Data Source: SDB, GPS-Mapping 2004, Layout: D. Pietsch 2005

Fig. 29.4 Soil degradation pattern: gullies and badlands

Where one finds Hypercalcic Calcisols at the land surface, soil degradation has already reached the final stage (Pietsch, 2006) when only a very low nutrition budget is available for plant growing (Stocking and Murnaghan, 2003). The extent of land degradation in Homhil could be also appreciated by comparing the content of organic matter in the soil (SOM). Looking at these two reference profiles it is obvious that the already eroded Cambisol Hom9 (without Ah horizon) still carries 0.95% SOM in the upper horizon, but Calcisol Hom14 has lost more than 0.3% SOM due

Table 29.1 Soil physical and chemical data of a Cambisol (Hom9) and a Cambisol above a Calcisol (Hom14)

Horizon	Depth (cm)	Detritus (%)	Sand	Silt	Clay	SOM	CaCO$_3$	pH$_{H2O}$	CEC (cmol$_c$ kg^{-1})
Hom 9									
BCwk1	0–5	19.2	19.0	46.7	33.4	0.95	3.2	8.5	21.6
BCwk2	5–25	19.5	24.4	28.8	46.8	0.83	17.9	8.6	20.5
BCwk3	25–75+	23.1	24.0	29.3	46.6	0.50	27.5	8.6	13.4
Hom14									
BCwk1	0–5	10.7	15.4	46.9	37.8	0.59	17.6	8.2	38.3
BCwk2	5–30	25.5	15.3	35.8	48.9	0.31	21.1	8.2	13.9
Cwck	30–50	49.4	25.2	41.3	33.4	0.37	67.7	8.3	6.1
	50–85	48.8	27.7	38.4	33.8	0.36	63.0	8.6	4.7

to erosion. Its "topsoil" (BCwk1 and BCwk2 horizons) equates to the "topsoil" of Hom9 (BCwk3 horizon).

The main problem as a result of this at location LZ1 is the disturbance and uprooting of the endemic *Boswellia*, a thinning of the whole vegetation cover and consequent increasing topsoil erosion. Both are irreversible land degradation processes, or so called "natural disasters" (Hudson and Ayala, 2006).

29.4 Conserving Soils

Within the last few years, and in Less Developed Countries (LDC) especially, the call for soil conservation has suddenly become urgent (Anderson and Thampapillai, 1990). The Yemeni Government, like other LDCs, has an environmental policy, which takes soil conservation into consideration only indirectly (Republic of Yemen, 1995). However, "in the field" monitoring has shown that appropriate environmental policies can start from the bottom: during the "National Environmental Day Action in Homhil" on 20–22 February 2007, which was supported by the inhabitants of Homhil, the Socotra Conservation Development Program (SDCP), the Environmental Protection Authority (EPA) and the Royal Botanic Garden Edinburgh (RBGE), soil monitoring in Homhil basin started. The first step was to build stonewalls in a 100 m long gully with depths of 20–100 cm. The gully (LZ1, cf Fig. 29.4) is located in an area covered by Cambisols (cf Hom9). Sixteen walls were constructed, using wadi gravel, (Figs. 29.5, 29.6 and 29.7) to catch sediments transported along the gully by runoff during the rainy season (Pietsch, 2008).

In 2008 the walls proved to have remained stable and to have acted successfully as sediment traps: they remain perfectly intact (Figs. 29.6 and 29.7), except for two walls, which lie across a cattle path. The walls seem to function excellently as sediment traps, since within 1 year 10–20 cm depth of sediments had accumulated behind most of the sixteen stonewalls.

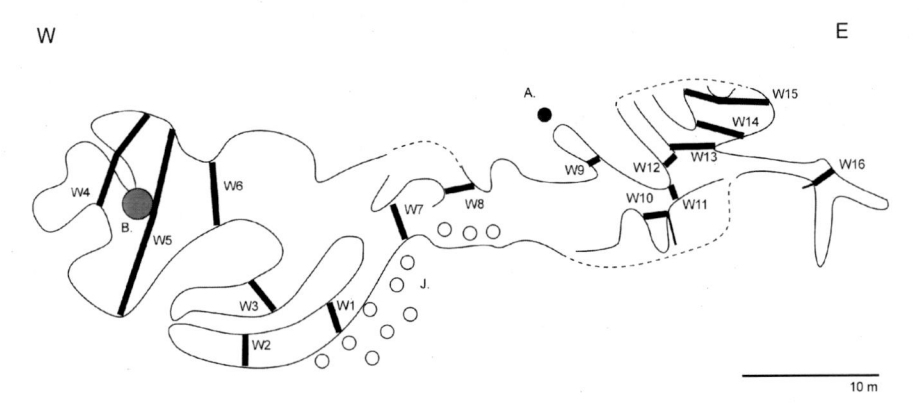

Fig. 29.5 Actual wall distribution

Fig. 29.6 Intact wall (W2)

Fig. 29.7 Sediment accumulation and succession of vegetation (W13 and W14)

Recently, *Asphodelus fistulosus* (Fig. 29.7) has become established on these loamy sediments along with other common plants such as *Tephrosia apollinea* and *Senna holosericea*. Furthermore, in some of these "dry walls", lizards, centipedes and ants have made their home (Fig. 29.8). It is interesting to compare the situation here with the Mediterranean region, where such walls are widely distributed and are well known to provide "eco-niches" for flora and fauna.

Further steps, such as planting young frankincense trees, will be carried out in autumn 2010, using ceramic plant rings produced by island women. The endemics *Boswellia elongata* and *Boswellia ameero* are available from nurseries in Hadiboh and Homhil (Fig. 29.9). They will need 2 years to become established and adapt to the arid climate and hot winds outside the nursery (Fig. 29.10). In preparation for

Fig. 29.8 Settling of soil fauna on walls (ants and lizards)

Fig. 29.9 Frankincense seedlings in the nursery (21.02.2008)

this planting, two different types of fence construction have been tested for feasibility as well as for their ability to withstand the summer monsoon winds in Homhil (Fig. 29.11).

This interdisciplinary and applied approach to combat land degradation, integrating social and natural science (Hudson and Alcántara-Ayala, 2006), is in line with the current environmental and political aims of Yemen, for example terracing or tree planting (FAO-UNCCD-UNDP, 2000). This system of soil monitoring should be understood as a first step towards conserving soils and vegetation in a Protected Area in Socotra.

Fig. 29.10 Frankincense seedlings outside the nursery (22.11.2008)

Fig. 29.11 Safety fences for plantings (04.03.2009)

29.5 Conclusions

Although in the past people did not have the soil data we possess today, they understood the vital importance of soil and practised soil conservation. Over time they even improved their soil-conservation techniques, and handed down this knowledge across the generations to modern times. It is thus possible to make use of Socotran skills in drystone wall construction as well as their insights into soil collection, and then combine this with topographical and soil data to devise "down-to-earth", basic soil monitoring programmes which are neither complicated nor expensive, but effective and very different from the often unsustainable soil monitoring practices seen elsewhere. In addition, the planting of trees on sediments accumulated behind walls imitates the quasi-natural eco-niche function of walls already recognised in many parts of Socotra and elsewhere in Yemen, as well as widely seen in the Mediterranean region.

The method of soil conservation discussed here is an important step towards achieving one of the environmental policies of Yemen: namely protecting the country's natural resources. This method is in tune with local customs and in sympathy with the local environment; it could be carried out by the islanders themselves without expensive machinery or other costly inputs; it causes no damage to the natural environment, is not unsightly, and its effects on the soil and the vegetation are clearly visible and readily understood.

The restoration of soils and the build-up of soil organic matter in this way is a sustainable system of combating soil erosion. Recently, the Yemeni government has made it a priority to try to manage its water crisis and land degradation due to pollution. Practical steps towards conserving soils, vegetation, and consequently water, such as those described above encourage local action undertaken by the beneficiaries themselves on a local rather than a national scale, and, most importantly, in accordance with the aims of the Conservation Zoning Plan as set out by the Government of Yemen.

Acknowledgements We would like to thank the *Socotra Conservation Development Programme* and the *Environmental Protection Authority* on Socotra, Yemen, for supporting this soil research. We also wish to thank the people of Homhil/Momi and the Hadiboh nursery for their help with wall construction, pottery production and plant monitoring.

References

Ad-Hoc-AG Boden. (2005). Bodenkundliche Kartieranleitung (KA5). BGR, Hannover.
Anderson, J.R. and Thampapillai, J. (1990). Soil Conservation in Developing Countries, Policy and Research Series of the World Bank 8. World Bank, Washington.
Avni, Y. (2005). Gully incision as a key factor in desertification in an arid environment, the Negev highlands, Israel. Catena 63:185–220.
Blaikie, P. and Brookfield, H. (1987). Land Degradation and Society. Methuen & co. Ltd, London.
Blume H.-P., Deller B., Furtmann K., Leschber R., Paetz A. and Wilke B.M. (eds.) (2000). Handbuch der Bodenuntersuchungen. Terminologie, Verfahrensvorschriften und Datenblätter. Physikalische, chemische, biologische Untersuchungsverfahren. Gesetzliches Regelwerk, Grundwerk. Berlin, Wien, Zürich.
Cerdá, A. and García-Fayos, P. (1997). The influence of slope angle on sediment, water and seed losses on badland landscapes. Geomorphology 18:77–90.
Dalelo, A. (2001). Natural Resource Degradation in Ethiopia. Flensburger Regionale Studien 11, Flensburg.
Environmental Protection Authority (EPA). (2000). Socotra Conservation Zoning Plan. Presidential Decree No. 275, YEM/96/623. Ministry of Planning and Development, Sana'a.
Environmental Protection Council (EPC). (2000). Socotra Archipelago Master Plan. YEM/B7 – 3000/IB/97/0787. Ministry of Planning and Development, Sana'a.
FAO. (2006). Guidelines for Soil Profile Descriptions. FAO, Rome.
FAO-UNCCD-UNDP. (2000). National Action Plan to Combat Desertification (Draft) Yemen. Ministry of Agriculture and Irrigation, Sana'a.
Geological Survey and Mineral Resources Board. (2003). Geological Map of Yemen. Socotra Island, 1:100.000. Geological Survey Department, Aden/Yemen.

Hudson, P.F. and Alcántara-Ayala, I. (2006). Ancient and modern perspectives on land degradation. Catena 65:102–106.

IUSS Working Group WRB. (2006). World Reference Base for Soil Resources (2007). World Soil Resources Reports No. 103. FAO, Rome.

Miller, A.G. and Morris, M.J. (2001). Conservation and Sustainable Use of the Biodiversity of the Soqotra Archipelago. Final Report. G.E.F., YEM/96/G32. Royal Botanic Garden, Edinburgh.

Miller, A. and Morris, M. (2004). Ethnoflora of Soqotra Archipelago. Royal Botanic Garden Edinburgh, Huddersfield.

Morris, M.J. (2002). Manual of Traditional Land Use in the Soqotra Archipelago. G.E.F. YEM/96/G32. Royal Botanic Garden, Edinburgh.

Morris, M. and Pietsch, D. (2008). Soil erosion in Homhil: Some background. Tayf 5:12–13.

Pietsch, D. (2003). Geo-ecological analysis of a sustainable development on Socotra Island. Dioscorida 2:6.

Pietsch, D. (2006). Böden der Trockentropen. Prozess- und Strukturindikatoren-gestützte Analyse geschichteter, polygenetischer und degradierter Böden der Insel Socotra (Jemen), Jemen-Studien, 17. Wiesbaden.

Pietsch, D. (2007). Socotra Final Soil Report Part 1. Homhil basin, Hadiboh coastal plain, Wadi Ayhaft. Tübingen.

Pietsch, D. (2008). National environmental day activities on Socotra in 2007: Soil conservation in Homhil. Tayf 5:10–11.

Pietsch, D. and Kühn, P. (2009). Soil developmental stages of Cambisols and Calcisols on Socotra island. Soil Science 174(5):292–302.

Republic of Yemen. (1995). EPL – Environmental Protection Law. No (26) 1995. Sana'a.

Stocking, M. and Murnaghan, N. (2003). Land Degradation Guidelines for Field Assessment (PLEC-Database Publication). Overseas Development Group, University of East Anglia, Norwich, UK.

Urban, B. (2002). Water and soil towards sustainable land use. In: S. Kunst, T. Kruse, A. Burmester (eds.), Sustainable Water and Soil Management. Springer Verlag, Heidelberg.

Wilkinson, T.J. (2003). Archaeological Landscapes of the Near East. University of Arizona Press, Tucson.

Part IV
Land Degradation and Mitigation in Europe

Chapter 30
Content of Heavy Metals in Albanian Soils and Determination of Spatial Structures Using GIS

Skender Belalla, Ilir Salillari, Adrian Doko, Fran Gjoka, and Majlinda Cenameri

Abstract This paper reports 8 years (2000–2007) of research results conducted in the soils of Albania aiming the establishment and the degree of pollution by heavy metals and their spatial distribution. One hundred and two soil samples (0–30 cm) from five representative soil types formed on different parent materials and wide physiographic regions of the country were sampled. Soil samples were analysed for total concentration of Pb, Ni, Cr and Co (Aqua Regia method) and for the main soil physical/chemical properties. Results showed that the total content of heavy metals varied widely between soil types as well as within the soil types. The highest total heavy metals concentrations were: 500, 761, 1,040 and 360 mg kg^{-1} for Pb, Ni, Cr and Co respectively. In many sites, concentration of heavy metals was higher than the limits permitted by the EU. Significant correlation between heavy metals and clay, pH and CEC were found. The thematic maps and the geostatistical analysis were performed using ArcGIS (v. 8.3.) and the extensions of Geostatistical and Spatial Analyst. Spatial distribution of heavy metals content in the studied areas was displayed via digitised maps. The thematic maps of heavy metals are expected to constitute an important tool, since they provide important information for the contamination levels and are vital for environmental risk assessments.

Keywords Total heavy metals · Spatial distribution · Permissible limits · Soil pollution · Albania

30.1 Introduction

The soils of Albania have been described as early as the 1930s (Zavalani, 1938) followed by other more profound studies (Veshi and Spaho, 1988; Zdruli, 1997; Zdruli et al., 2003). In terms of USDA Soil Taxonomy (Soil Survey Staff, 2006) dominant

---

I. Salillari (✉)
Centre of Agriculture Technology Transfer, Fushë Kruja, Albania
e-mail: ilirsalillari@gmail.com

soils are Inceptisols (35%) followed by Alfisols (17%), Mollisols (7%), Entisols (6%), Vertisols (2%) and pockets of Histosols. Large parts of the territory fells into a miscellaneous category that includes rocky areas, rock outcrops, bare lands, wetlands, saline areas, and urbanised zones. According to the World Reference Base for Soil Resources (WRB, 2006) classification system the soil distribution pattern in order of extensive coverage is as follows: Cambisols, Luvisols, Fluvisols, and Phaeozems, associated with Vertisols, Solonchaks, Gleysols, Arenosols, Histosols and Calcisols (Zdruli, 2005).

Until the 1990s surveys on heavy metals in Albania were concentrated especially in elements like Zn, Cu, Mn and Fe (so called "essential microelements") and problems related to soil fertility and plant nutrition (Gjoka, 1999). Other metals like Ni, Cr, Co and Pb have been totally ignored even though these elements are known for their toxic effects. The studies carried out over the last decades (Saraçi et al., 1995; Shallari, 1996; Gjoka et al., 2002; Sallaku, 2005) have been focused on the potential toxicity of heavy metals due to the industrial activities initiated after the 1960s and their potential waste deposits in the soil.

Soil and environmental pollution from heavy metals in general attracts major attention due to their negative effects on the environment. These elements have been studied intensively in many places around world due to the risks associated with them. There is concern that after they are accumulated in the soil, they could enter into the food chain and might be toxic for the surrounding organisms and the environment (Allowey, 1995). Additionally, special attention is given to the fact that remediation/amelioration of contaminated sites is a long and very costly process.

Monitoring heavy metals in agriculture and industrial areas of Albania has been and continues to be a main objective for various research institutions and governmental structures. Nevertheless, the real level of heavy metal contamination and their toxic effects in the Albanian soils are still unknown and detailed data are missing. Hence, there is an urgent need to overcome these shortcomings so that reference points could be established to determine the source and the level of heavy metal soil pollution. Our study is concentrated mainly with the determination of the content of heavy metals in different zones of the country, their spatial distribution and the evaluation of their pollution effects considering European Union regulations CEC (1986). All the data have been processed using a GIS system able to produce thematic maps and perform statistical analyses.

30.2 Materials and Methods

30.2.1 Study Area

The Republic of Albania is located in the Balkan Peninsula, between 39°38′ and 42°39′N, and 19°16′ and 21°40′ E longitude. The overall country's territory of 2,874,800 ha, includes 699,500 ha of agriculture land (24.4%), 1,062,770 ha of forests (36.9%), 414,517 ha of pastures and meadows (14.4%) and 698,013 ha (24.3%) for other land uses, such as urban and barren land, water bodies, etc.

Albania is a mountainous country (2/3 of the territory) and only 16% of its land is located at elevations of less than 100 m above the sea level. The western side of the country is characterised by Quaternary alluvial plains, the central part represents

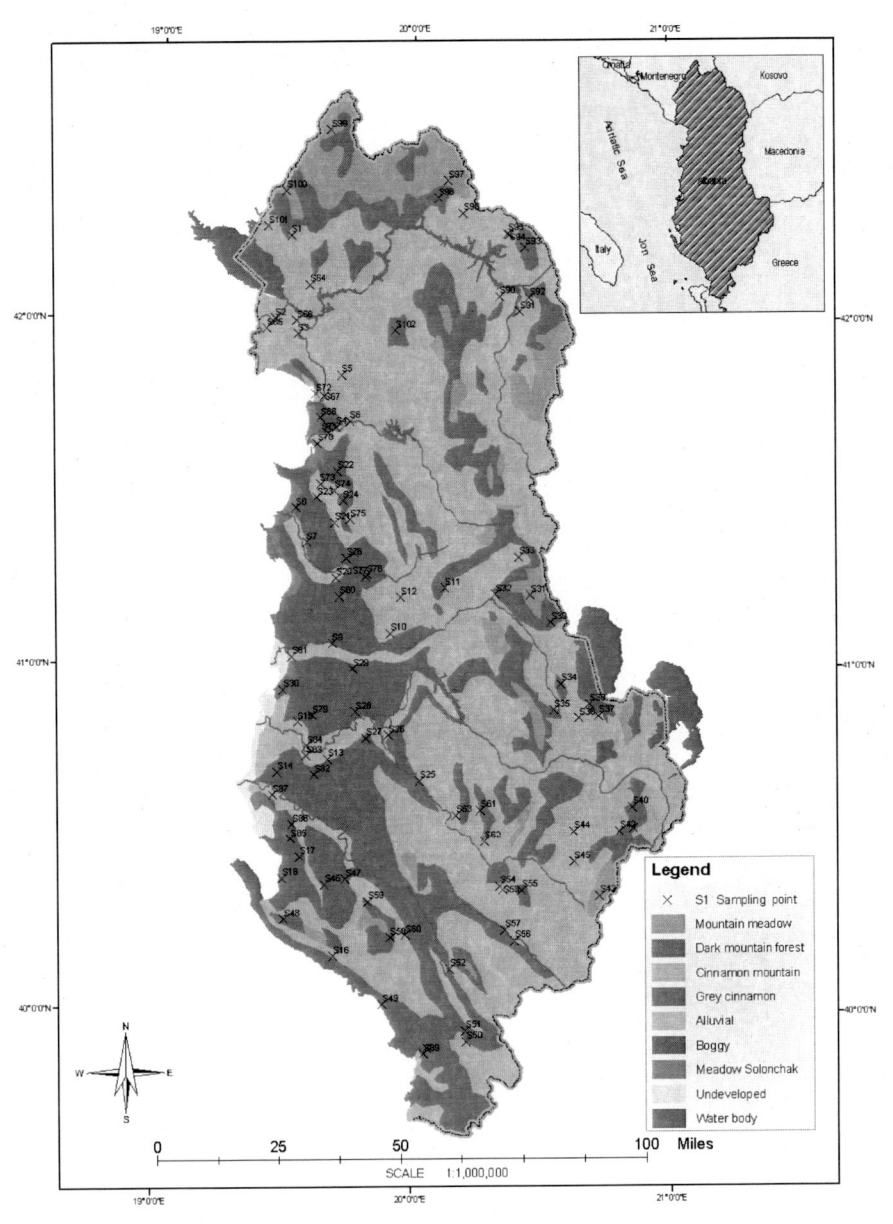

Fig. 30.1 Soil types according to the national soil classification systems and distribution of soil samples

mainly hilly Sedimentary landforms covered by fruit trees and olives and *macchia mediterranea* and high steep mountains (often sheltered by beech and pine trees) occupy the rest.

30.2.2 Sampling Sites

One hundred-two soil samples were collected from the topsoil (0–30 cm) representing the main soils type of Albania (Veshi and Spaho, 1988) i.e. mountain meadow, dark mountain forest, cinnamon mountain, grey cinnamon and alluvial soils. Each composite sample was prepared by mixing together of several sub samples taken from the soil over a surface with radius 2 m. Figure 30.1 shows sampling sites for each soil types.

30.2.3 Soil Analyses

After air-drying soil samples were passed through a plastic sieve with a 2 mm mesh and were analysed for basic soil characteristics by standards methods. Soil pH were measured in a 2.5:1 v/w 0.01 M $CaCl_2$ solution: soil ratio by pH-meter; particle size distribution by Pipette method; CEC by the Mehlich method; organic C and nitrogen the auto-analyser and for total Ni, Cr, Co and Pb concentrations after extraction with Aqua-Regia (AR) by atomic absorption (AA).

For exact determination of each location a Global Position System (GPS) was used, followed by Geographic Information System (GIS) technology for mapping and analyses. Kriging interpolation was applied to the evaluation of the undetermined points on the basis of the obtained semi-variable function model, hence to get a further accurate and direct description of the spatial distribution and variance of the content of heavy metal elements and their source components (Burrough and MCDonnell, 2000).

30.3 Analyses and Results

30.3.1 Soil Analyses

Soil analyses results (Table 30.1) showed great variability between soils i.e. strongly acid to strongly alkaline (pH 5.1–8.5), but most of them were neutral to slightly

Table 30.1 The range and the mean values of soil properties

Soil properties	Range	Mean	Median
pH	5.1–8.5	7.4	7.7
Clay	7.0–56.8	27.3	26.5
SOC	0.6–8.9	2.6	1.9
CEC	14.2–53.7	29.5	27.5

alkaline. Soil organic carbon (SOC) was generally high (0.6–8.9%), with only few samples of lower level. Cation exchange capacity was medium to very high (14.2–53.7 cmolc kg^{-1}). Soils were sandy loam to clay in texture (clay content 7.0–56.8%).

30.3.2 Statistical Summary of Total Content of Heavy Metals

Table 30.2 shows the range and the mean values of analysed metals. The total Ni varied from 20 to 761 mg kg^{-1} with a mean value of 254.3 mg kg^{-1}, Cr varied from 10 to 1,040 mg kg^{-1} with a mean value of 372.8 mg kg^{-1}, Co varied from 12 to 360 mg kg^{-1} with a mean value of 103.0 mg kg^{-1} and Pb varied from 17 to 500 mg kg^{-1} with a mean value of 170.5 mg kg^{-1}.

Table 30.2 Statistical summary of total content of heavy metals (mg kg^{-1})

Heavy metals	Mean	Median	Standard deviation	Variance	Range	Minimum	Maximum	EU limits
Ni	254.3	163.7	196.5	38629.3	741	20	761	75
Cr	372.8	345.5	306.8	94097.2	1,030	10	1,040	200
Co	103.0	82.9	78.9	6232.4	348	12	360	50*
Pb	170.5	102	133.5	17830.4	483	17	500	300

*Kabata-Pendias and Pendias (2001)

Descriptive statistical analysis made for total content of heavy metals showed that except Pb, average of Ni, Cr and Co exceeded the maximum permissible heavy metal concentration (Kabata-Pedias and Pendias, 2001) as well as those accepted for some European countries according to the guideline values of the 86/278/EEC Directive for agricultural soils CEC (1986).

Results showed that the total content of heavy metals for every soil type varies widely (Table 30.3). This manly reflects the different geologic conditions upon which the soils have formed.

Table 30.3 Minimal and maximal concentration of heavy metals according to national soil type classification

Heavy metals		Mountain meadow	Dark mountain forest	Cinnamon mountain	Grey cinnamon	Alluvial
Ni	Minimum	34	50	20	53	56
	Maximum	228	669	635	761	739
Cr	Minimum	23.4	55.6	12	18	10
	Maximum	300	850	1,040	1,000	909
Co	Minimum	34	42	38	12	13
	Maximum	55	226	360	326	266
Pb	Minimum	70	17	66	22	26
	Maximum	155	410	405	500	450

Table 30.4 Correlation between soil property and heavy metals

	Clay	pH	Org. C	CEC	Ni	Cr	Co	Pb
Clay	1							
pH	−0.2127	1						
Org. C	0.3533	−0.0282	1					
CEC	0.2244	−0.2503	0.3309	1				
Ni	0.0531	0.4408	0.1450	−0.4439	1			
Cr	−0.0626	0.4860	−0.0331	−0.4043	0.5501	1		
Co	0.0972	−0.0013	0.0938	−0.0087	0.2646	−0.0167	1	
Pb	0.0476	0.4151	0.3805	−0.3271	0.6864	0.5010	0.0283	1

Relationships between soil characteristics and heavy metal concentrations were also analysed (Table 30.4). Pb, Ni and Cr were significantly correlated with pH and Pb showed positive correlation with SOC. The same phenomenon was also observed between Pb with Ni and Cr. There were negative correlations though between Pb, Ni and Cr and CEC (Salillari et al., 1998).

30.3.3 Spatial Distribution

The highest concentration of Ni was located in the western, central and eastern parts of Albania (Fig. 30.2). The normal soil Ni content reported in literature (Kabata-Pedias and Pendias, 2001) varies from 1 to 100 mg kg^{-1}. Most of the analysed samples in this study were over this range except some of them in the south and north sides. High concentration of Ni in the central part of the country might be linked with the effect of human influence deriving form the metallurgical industrial complex established in Elbasan (Sallaku, 2005). The presence of ultramafic/ultrabasic rocks in the eastern part of Albania could play as well an important role in high concentration of Ni. Factors that influence high concentration of Ni in the western part remain still unknown at current research level.

Figure 30.3 shows spatial distribution structure of Cr. The high levels are located in the centre and southern parts of Albania. Based on this spatial distribution pattern it could be assumed that parent materials and anthropogenic factors played important roles in soil heavy metal concentrations. The main contributors on high concentration of Cr remain the parental material, although there are some areas in the centre of the country, with high concentrations of Cr, caused by human activities.

Spatial structure of Co is irregular compared with those of Ni and Cr (Fig. 30.4). The highest levels of Co were found in the north-western, western, central and southern parts of Albania. In the centre and in the north-western regions human activity is thought to contribute to increased Co levels resulting from industrial activities of the metallurgical complex of Elbasan and chemical-metallurgical plant

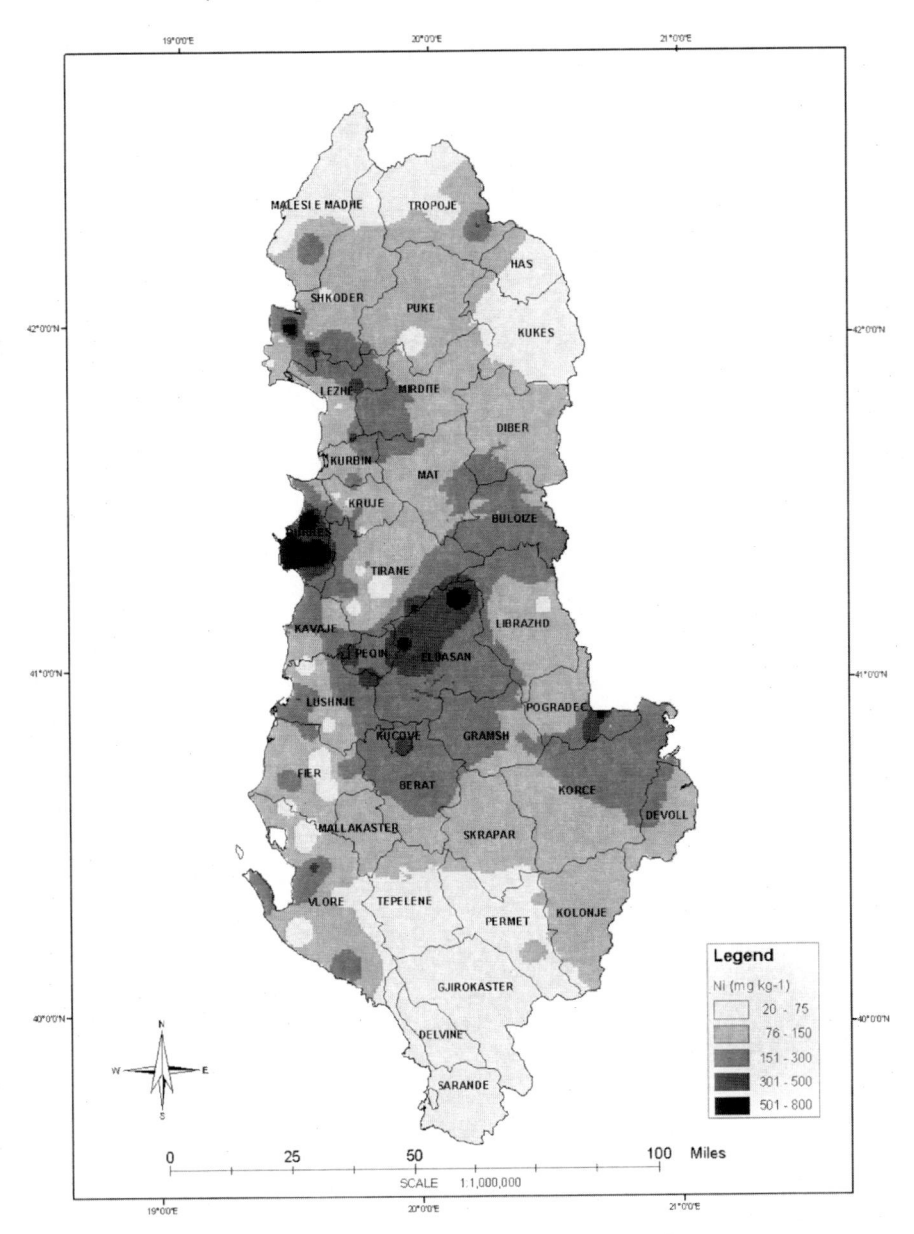

Fig. 30.2 Spatial variability of total concentration of Co in soil (mg kg^{-1})

of Laç respectively (Gjoka et al., 2002; Sallaku, 2005). Co content in sand and limestone rocks is low and subsequently Co concentrations in soils, formed from these rocks (Allowey, 1995) are usually low, which is the case in the southern part

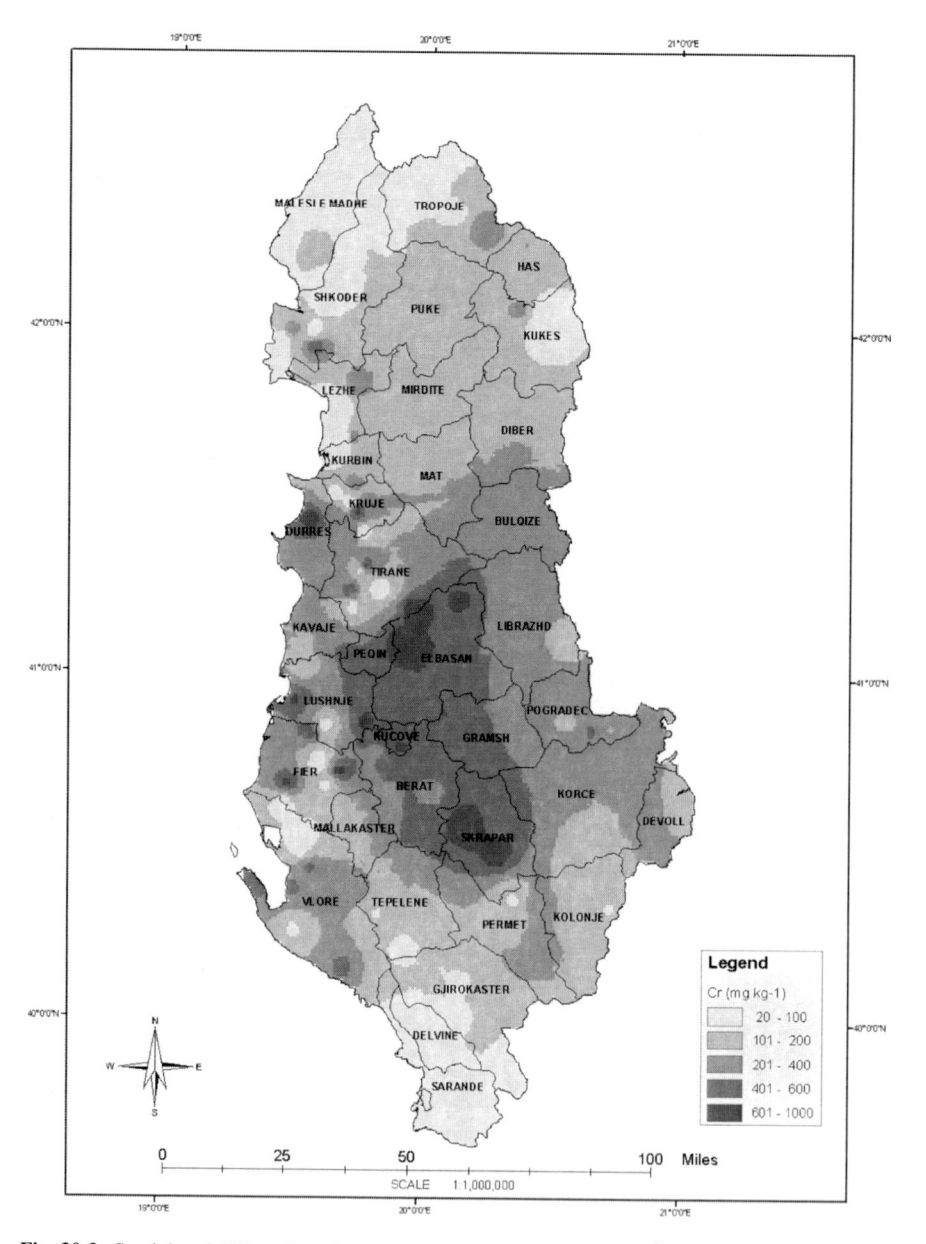

Fig. 30.3 Spatial variability of total concentration of Cr in soil (mg kg^{-1})

of Albania. Thus, a significant increase in soil Co content resulting from human activities could occur.

The highest levels of Pb were found mainly in the centre, western and some very small areas in the northwestern part of the country (Fig. 30.5). In the central part

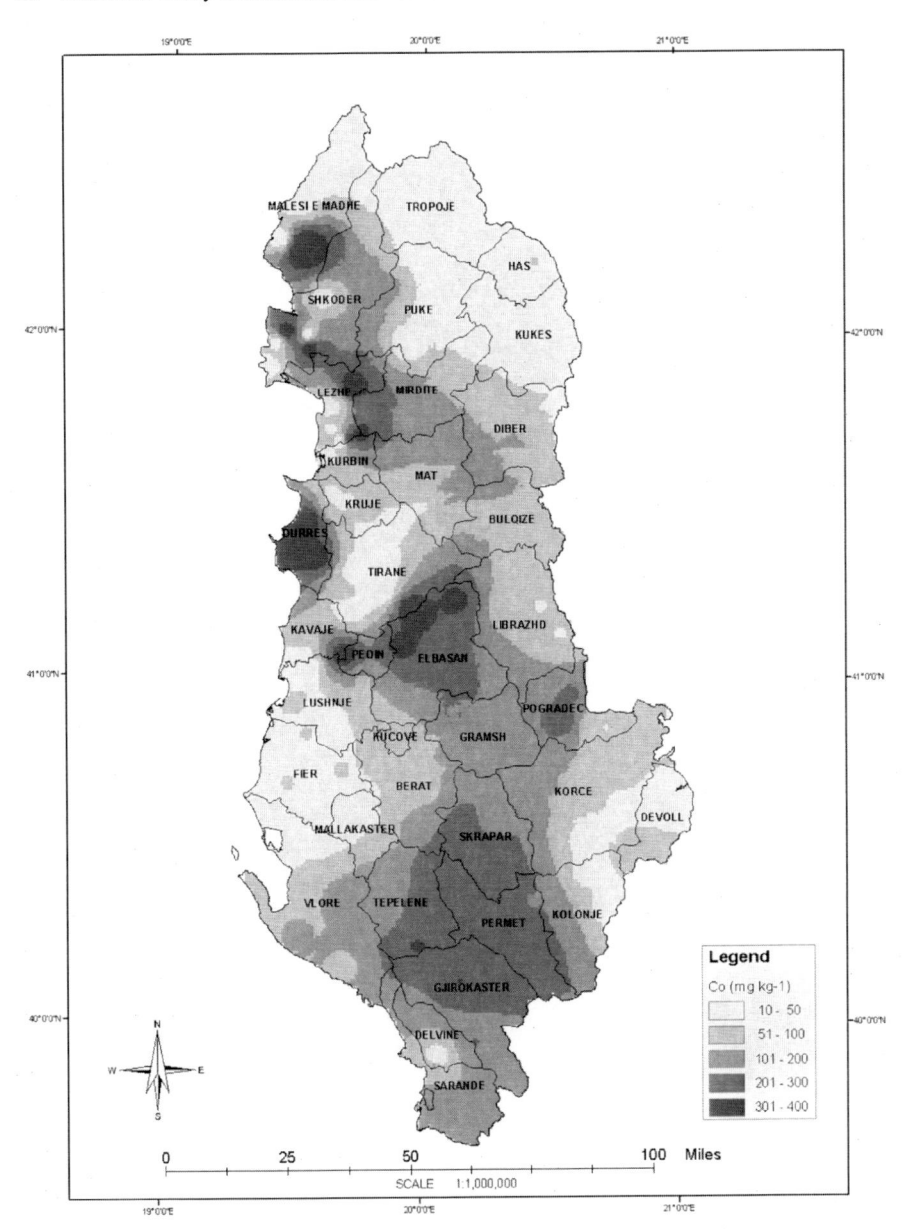

30 Content of Heavy Metals in Albanian Soils 397

Fig. 30.4 Spatial variability of total concentration of Co in soil (mg kg^{-1})

high concentration is referred to the metallurgical complex of Elbasan and petrol extraction industry of Kucova (Lushaj et al., 2002). The same as for the other elements mention above, factors that influence high concentration of Pb in the western part, remain unknown.

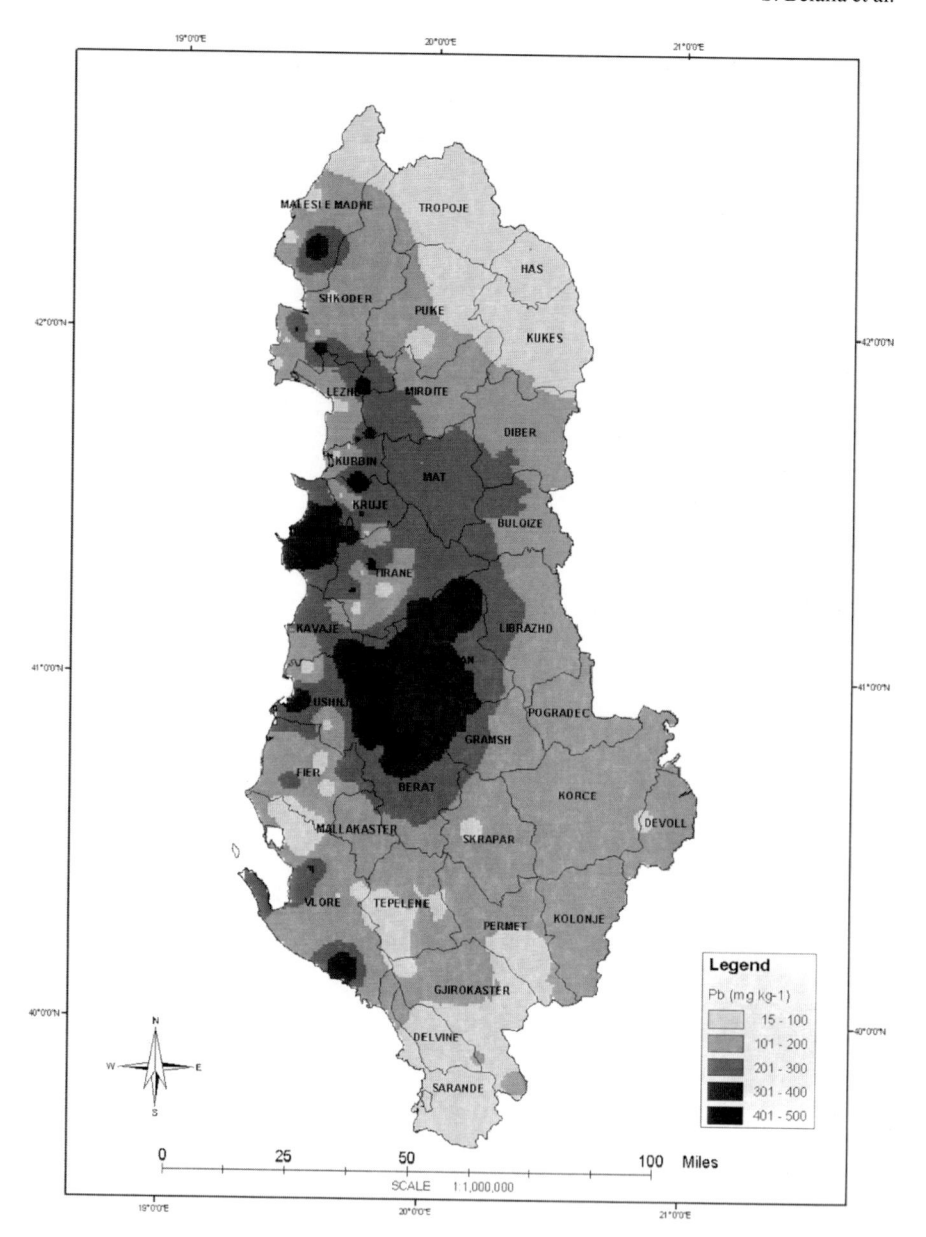

Fig. 30.5 Spatial variability of total concentration of Pb in soil (mg kg^{-1})

30.4 Conclusions

The following conclusions could be derived from this study:

1. Total content of Ni, in the western, central and eastern part, the content of Cr in the centre and southern part and the content of Co in the north-western, western, central and southern part of Albania, exceeded the maximum permissible heavy metal concentration levels established by the EU legislation. Total content of Pb in general is below maximum permissible, except for a small part in centre of the country.
2. Soil type does not influence heavy metals content or distribution. The results showed great variability of heavy metals content within each soil type.
3. The spatial distribution structure of Ni, Cr, Co and Pb are mainly controlled by parent materials and only in restricted areas by anthropogenic factors
4. A monitoring system should be established by the responsible governmental structures to control the trends and status of this form of soil contamination.

References

Allowey, B.J. (1995). Heavy Metals in Soils. Blackie, London, pp. 11–57.
Burrough, P.A. and MCDonnell, R.A. (2000). Principles of Geographical Information Systems. Oxford University Press, New York.
CEC. (1986). Council Directive of 12 June 1986 on the protection of the environment, and in particular of the soil, when sewage sludge is used in agriculture. Official Journal of the European Communities L181:6–12.
Gjoka, F. (1999). Available micronutrients in two Fluvisols of humid zone of northwest Albania. Bulletin of Agricultural Sciences Nr 4:27–34. Tiranë.
Gjoka, F. et al. (2002). Heavy metals in soils of vicinity of industrial site of Laç, Albania. Balkan Ecology 5:301–306.
IUSS Working Group WRB. (2006). World Reference Base for Soil Resources 2006. World Soil Resources Reports No. 103. FAO, Rome.
Kabata-Pedias, A. and Pendias, H. (2001). Trace Elements in Soils and Plants. 3rd edn. CRC Press, Boca Raton, Florida, 413 pp.
Lushaj, Sh. et al. (2002). Monitorimi i tokës dhe i ujrave. Anual report.
Salillari, A. et al. (1998). Eksperimentimi bujqësor. Universiteti Bujqësor i Tiranës, Tirana.
Sallaku, B. (2005). Concentracion and distribution of heavy metals in contaminated soils near the Metallurgical combine of Elbasani. Element Balance as a Tool for Sustainable Land Management, International Conference, Tiranë.
Saraçi, M. et al. (1995). Analytical investigation of soil from a tree plantation in South-East of Albania. Fresenius Environmental Bulletin 4:624–629.
Shallari, S. (1996). Diagnostikimi dhe evidentimi i ndotjes nga metalet e rënda nëtokat e bimësinë e serpentinave dhe zonave industriale.
Soil Survey Staff. (2006). Keys to Soil Taxonomy. 10th edn. USDA-Natural Resources Conservation Service, Washington, DC.
Veshi, L. and Spaho, Sh. (1988). Pedology. Agricultural University of Tirana, Tirana, Albania, 574 pp.
Zavalani, D. (1938). Die landwirtschaftlichen Verhaltnnisse Albaniens. Doctorial dissertation. Berlin, Germany

Zdruli, P. (1997). Benchmark Soils of Albania: Resource Assessment for Sustainable Land Use. USDA Natural Resources Conservation Service (NRCS), Washington DC and the International Fertilizer Development Centre (IFDC), Muscle Shoals, Alabama. 2 Volumes. 293 pp (PhD thesis).

Zdruli, P. (2005). Soil survey in Albania. In: R. Jones, B. Houskova, P. Bullock and L. Montanarella (eds.), Soil Resources of Europe. 2nd edn. European Soil Bureau Research Report No.9 EUR 20559 EN (2005). Office for Official Publications of the European Communities, Luxembourg, 420 pp.

Zdruli, P., Lushaj, Sh., Pezzuto, A., Fanelli, D., D'Amico, O., Filomeno, O., De Santis, S., Todorovic, M., Nerilli, E., Dedaj, K. and Seferi, B.. (2003). Preparing a georeferenced soil database for Albania at scale 1:250,000 using the European soil bureau manual of procedures 1.1. In: Zdruli P., Steduto P., and Kapur S. (eds.), OPTIONS Méditerranéennes. SERIE A: Mediterranean Seminars. Volume A 54. Selected papers of the 7th International Meeting on Soils with Mediterranean Type of Climate. Centre International de Hautes Etudes Agronomiques Mediterranéennes (CIHEAM), Paris, France.

Chapter 31
Radioisotopic Measurements (^{137}Cs and ^{210}Pb) to Assess Erosion and Sedimentation Processes: Case Study in Austria

L. Mabit, A. Klik, and A. Toloza

Abstract Twelve to seventeen percentage of the European soil is threatened by water erosion and around 13% of Austrian territory is affected. Only scarce information based on conventional assessment and measurements are available on erosion and sedimentation rates in Austria. The magnitude of sedimentation processes was evaluated in a small agricultural Austrian watershed using both nuclear techniques (^{137}Cs and ^{210}Pb) and conventional non-isotopic measurements in runoff erosion plots during the 1994–2006 periods. Using the erosion data provided by the plots (29.4 t ha^{-1} year^{-1} for the conventional tilled plot, 4.2 t ha^{-1} year^{-1} for the plot receiving conservation tillage and 2.7 t ha^{-1} year^{-1} for the plot receiving direct seeding treatment) and the ^{137}Cs soil profiles content and the conversion model mass balance 2 (MBM 2), a sedimentation rate of 13.2 t^{-1} ha^{-1} year^{-1} (value determined down slope of the runoff plot under direct seeding treatment) to 50.5 t^{-1} ha^{-1} year^{-1} (value determined in the lowest sedimentation area of the watershed under conventional tillage) was estimated. Under the experimental condition the conservation tillage and direct seeding system were effective in reducing the sedimentation magnitude by 65%. However, due to a high variability of the initial fallout inventory and a high γ-spectrometry measurement error, information provided by the ^{210}Pb method was not usable in the study area. The combined use of conventional erosion measurements and nuclear techniques appears to be a promising and complementary approach to evaluate sedimentation processes.

Keywords Water erosion · ^{137}Cs · ^{210}Pb · Runoff erosion plots · Sedimentation rates · Erosion rates

---

L. Mabit (✉)
Soil Science Unit, FAO/IAEA Agriculture & Biotechnology Laboratory, IAEA Laboratories Seibersdorf, Vienna A-1400, Austria
e-mail: l.mabit@iaea.org

31.1 Introduction

Degradation phenomena, such as erosion, desertification and salinization affect 65% of the soil in both developed and developing countries. In Europe, soil erosion by wind and water is the major threat to the resource soil and represents the main mechanisms of landscapes degradation. Indeed, around 115×10^6 ha of agricultural used land – around 12% of the total area of Europe – are highly affected by erosion processes (CEC, 2006).

On-site soil erosion affect farmers particularly through soil losses, removal, and reduction of soil quality; such processes are associated with fertility, productivity and yield decline. Sediment and associated potential pollutant can reach watercourses and contribute to the eutrophication of water resources, increase or create siltation problems and other major off-site impacts like mudslide and flow which can damage human infrastructure and habitation (Boardman and Poesen, 2006).

Influenced by the effects of climate change and global warming, water erosion risk is expected to increase in the EU by the year 2050 in about 80% of the agricultural areas, and this pressure will mainly take place in the areas where soil erosion is currently severe and Austria is part of this area (EEA, 1999, 2002). For the conservation, improvement and sustainable use of natural resources for food and agriculture there is a clear need of an integrated land and water management that should start firstly with an accurate spacio-temporal assessment of erosion and sedimentation magnitudes.

The Global Assessment of Human Induced Soil Degradation (GLASOD) survey carried out during the 1980s by the United Nations Environment Programme (UNEP) and the International Soil Reference and Information Centre (ISRIC) established that the severity of human induced degradation has been classified as severe for 37% of the Austrian territory (FAO, 2005). In 1999, about 4.4×10^4 ha agricultural land was classified highly erodible (Strauss and Klaghofer, 2006). Using long-term, remotely sensed Normalized Difference Vegetation Index (NDVI) and rainfall data Bai et al. (2008) evaluated that land degradation represents 28,291 km^2 equivalents to 34% of the Austrian territory.

However, no data on historical quantitative soil loss rates are available in Austria. Long-term annual rainfall (1961–1990) varies between 430 and 2,250 mm with an overall mean of 1,170 mm. This corresponds to theoretical R factors of 38–180 N h^{-1} (Strauss et al., 1995). However, intensive agricultural land use to which the Universal Soil Loss Equation (USLE) calculations are limited, is not accurate with annual rainfall above 1,500 mm.

Qualitative evidence of historical soil erosion exists in various forms. Especially in the wine-growing area situated in the eastern part of Austria the formation of large gullies on loess soils shows the impact of soil erosion processes. Results from field experiments mainly focussing on effects of different soil management practices can be used as indicator of the erosion severity. Klik (2003) investigated the effects of different tillage practices on runoff and soil loss at three sites in Lower Austria. Over 9 years he measured average soil loss rates of 5–39 t ha^{-1} year^{-1} for conventional plots with 2–6 t ha^{-1} year^{-1} for conservation tilled plots and 0.5–4 t ha^{-1} year^{-1} for no-till plots. Two-year experiments by Pollhammer (1997) at two sites in

Styria delivered soil loss rates of 8–72 t ha^{-1} year^{-1} for ploughed plots and 1–46 t ha^{-1} year^{-1} for chiselled plots. Soil erosion is mainly affected by extreme events. Strauss and Klaghofer (2004) mapped linear soil features within a 2.89 km^2 large experimental watershed after a 5-day period of heavy rain (115 mm) and recorded total amounts of soil transport by rilling of more than 730 t but only 17 t leaving the watershed. Soil erosion took place only in few fields with the highest soil loss recorded at almost 300 t ha^{-1}.

Traditional monitoring and modelling techniques (e.g. erosion pins, erosion modelling, e.g RUSLE, WEPP; sediment yields and reservoir silting measurements, runoff plots monitoring data under natural or artificial rainfall, sediment traps) for soil erosion/sedimentation require many parameters and of course years of measurements to integrate climatic inter-annual variability (Mabit et al., 2002a, b). Artificial radionuclide such as ^{137}Cs ($T_{1/2}$ = 30 years), geogenic radioisotope such as ^{210}Pb ($T_{1/2}$ = 22 years) and also more recently cosmogenic radioisotope such as ^{7}Be ($T_{1/2}$ = 53 days) have been used worldwide to assess medium and short term soil erosion and deposition processes (Mabit et al., 2008a; Ritchie and Ritchie, 2008).

Soil redistribution is commonly based on a comparison of the fallout radionuclides "FRNs" (^{137}Cs, ^{210}Pb and ^{7}Be) areal activity density in the landscape with a "stable" landscape position known as the "reference site", where neither erosion nor deposition has occurred (Mabit et al., 2008a). The erosion or sedimentation rates can be estimated using conversion models, which mathematically define the relationship between the increase or decrease in the FRNs inventory relative to the reference inventory (Walling et al., 2002). One of the main advantages of the use of the FRNs is that time-consuming, costly maintenance and installations required by non-isotopic and conventional methods can be avoided (Mabit et al., 2007, 2008b).

Innovative methodological approaches and protocols for the use of FRNs to obtain spatio-temporal patterns and magnitude of soil erosion/sedimentation were tested, compared and reported since the 1990s in many European countries such as Germany (Schimmack et al., 2002), France (Mabit et al., 1998), Spain (Navas and Walling, 1992), Italy (Porto et al., 2003), UK (Walling and Quine, 1991), Poland (Froehlich and Walling, 1992), Slovakia (Fulajtar, 2000), Slovenia (Mabit et al., 2009b). Such data from FRN test as soil tracer were not available in Austria until the present case study. The aim of the investigation in Mistelbach, a small Austrian agricultural watershed, was to measure soil redistribution rates using runoff plots and radioisotopics approaches (^{137}Cs and ^{210}Pb) and to assess the relative efficiency of soil conservation measures on soil losses.

31.2 Materials and Methods

31.2.1 Study Area, Soil Sampling and Laboratory Analyses

A field experiment was implemented from 1994 to 2006 in Mistelbach, a small agricultural watershed of 18 ha located 60 km north of Vienna, Austria, to investigate soil conservation impacts of three different tillage practices. Cereals growing on silty-loam gentle slopes dominate the landscape. The treatments included a conventional tilled plot (CVT), a conservation tillage plot (CST) with cover crops during winter and a plot under direct seeding (DS) with cover crops during winter. Each tillage practices tests were carried out on three 45 m² runoff plots with 14% slope to measure surface runoff and erosion processes (Fig. 31.1).

Runoff and associated sediment were collected from 1994 to 2006 using an automated erosion wheel (Klik et al., 2004). Shifting from conventional tillage to direct seeding the yearly soil loss was reduced significantly by a factor of 10. Indeed, over the 13 years of investigation, the annual average soil loss was 29.4 t ha^{-1} for the conventional tilled plot, 4.2 t ha^{-1} for the conservation tillage plot and 2.7 t ha^{-1} for the direct seeding treatment (Mabit et al., 2008c).

To establish the ^{137}Cs reference value 76 soil samples were collected along a multi-grid design (Mabit et al., 2008d) in an undisturbed small forest within the Mistelbach watershed (Soil newsletter). Eleven samples were randomly selected for analyses of total ^{210}Pb, ^{226}Ra and ^{210}Pb$_{ex}$. In the sedimentation area of the watershed (Fig. 31.1), two 0–100 cm composite samples of three cores were collected with an automatic soil column cylinder auger (Fig. 31.2). One soil profile was collected down slope of the field containing the runoff plots (Profile 1) and the other one was collected in the talweg runoff convergence located 15 m from the previous soil profile (Profile 2)

After pre-treatment (sieving at 2 mm after oven-drying for 48 h at 70°C) the gamma measurement of the soil samples, according to the recommendation provided by Shakhashiro and Mabit (2009) were completed by γ-spectrometry at the

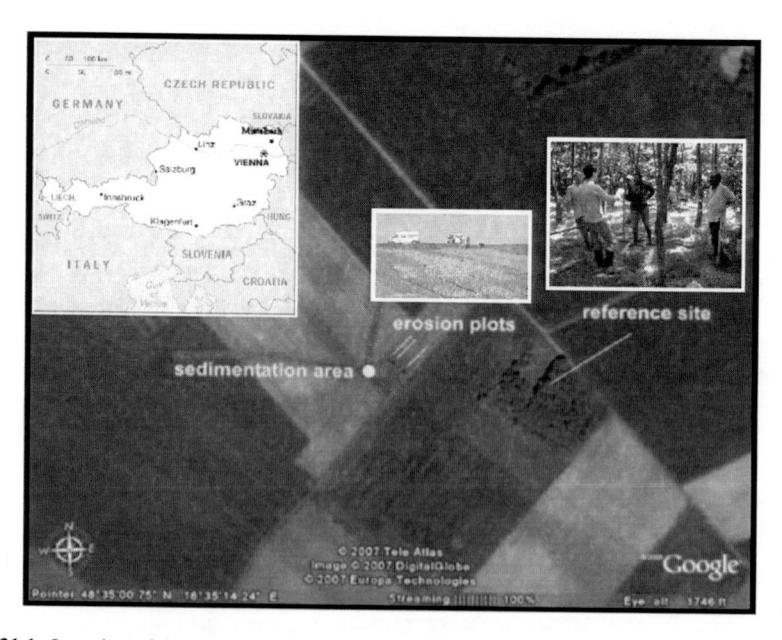

Fig. 31.1 Location of the study area (Adapted from Mabit et al., 2009a)

Fig. 31.2 Soil samples collection in the sedimentation area in Mistelbach watershed

IAEA Seibersdorf Laboratory (^{137}Cs) and at the CNESTEN in Morocco and at the CEA/CNRS in Gif-sur-Yvette in France (total ^{210}Pb, ^{226}Ra and ^{210}Pb$_{ex}$).

31.2.2 Conversion of FRNs Areal Activity into Soil Redistribution

For the soil profile 1, FRN areal activity was converted into soil redistribution using the Mass Balance Model 2 (MBM 2) (Walling et al., 2002) and linked with the runoff plots average erosion measurements. According to MBM 2, the erosion rate R can be estimated by solving numerically the following equation. For an eroding point ($A(t) < A_{ref}$), the change in the total ^{137}Cs inventory $A(t)$ with time can be represented as:

$$\frac{dA(t)}{dt} = (1 - \Gamma)I(t) - (\lambda + p\frac{R}{d})A(t) \quad (1)$$

$A(t)$ = cumulative ^{137}Cs areal activity (Bq m^{-2})
t = time since the onset of ^{137}Cs fallout (year)
R = erosion rate (kg m^{-2} year^{-1})
d = average plough depth represented as a cumulative mass depth (kg m^{-2})
λ = decay constant for ^{137}Cs (year^{-1})
$I(t)$ = annual ^{137}Cs deposition flux at time t (Bq m^{-2}) which depends on the ^{137}Cs reference inventory

P = particle size correction factor
Γ = proportion of the freshly deposited ^{137}Cs fallout removed by erosion before being mixed into the plough layer by cultivation.

For a depositional site ($A(t) > A_{ref}$), the sediment deposition rate R' can be estimated using the following equation:

$$R' = \frac{A(t) - A_{ref}}{\int_{t_0}^{t} C_d(t') e^{-\lambda(t-t')} dt'} \quad (2)$$

$C_d(t')$ is the ^{137}Cs concentration of deposited sediment assumed to be the weighted mean of the ^{137}Cs concentrations of the sediment mobilized originating from the entire eroded area represented by $C_e(t')$ from the upslope contributing area S (m^2). $C_d(t')$ can be calculated as following, where R is the erosion rate in the eroding zone and P' is the particle correction size:

$$C_d = \frac{1}{\int_S R dS} \int_S P' C_e(t') R dS \quad (3)$$

Depth redistribution activity of FRNs from both soil profiles (1 and 2) was also used to evaluate the sedimentation magnitudes. The following formula (Allison et al., 1998; Walling and He, 1999) was used:

$$R' = \frac{D_b - D_{pl}}{T_s} \quad (4)$$

T_s : time since the beginning of significant fallout and the collection of the sediment core (year)
D_b : plough depth (kg m^{-2})
D_{pl}: plough depth based on the ^{137}Cs depth profile distribution (kg m^{-2})
R' : sedimentation rate (kg m^{-2} year^{-1})

31.3 Results and Discussion

31.3.1 FRNs (^{137}Cs and ^{210}Pb) Baseline Level in the Reference Site and Test of the ^{210}Pb Methodology

Following an exponential decrease activity with depth, 90% of the ^{137}Cs is concentrated in the first 15 cm soil layer of the forest area. The ^{137}Cs areal activity of the seventy-six forested samples was 1,954 Bq m^{-2} with a coefficient of variation of 20%. This reliable value <2kBqm^{-2} ($n = 76$) demonstrates clearly a negligible impact of Chernobyl. It could therefore be concluded that the Chernobyl fallouts in the study area were small enough not to significantly influence the relationship between soil and ^{137}Cs losses.

Measurement error of total ^{210}Pb and ^{226}Ra is acceptable however and the bias concerning the value of ^{210}Pb$_{ex}$ can be as high as the measured values itself. Based on the laboratory γ-measurements there is no significant difference between the total ^{210}Pb and ^{226}Ra. Indeed, in most cases, the activity of total ^{210}Pb and ^{226}Ra from the reference site as well as from the two soil profiles collected in the sedimentation area are similar (Tables 31.1 and 31.2). The value of ^{210}Pb$_{ex}$ can sometimes even be negative due to the measurement uncertainty and the overlapping activity of total ^{210}Pb and ^{226}Ra. In that case the value of ^{210}Pb$_{ex}$ was reported as zero (Tables 31.1 and 31.2).

In summary, it was not possible to validate the ^{210}Pb method in the Mistelbach watershed, due to a very low level of 'unsupported' ^{210}Pb coupled with a high variability of the initial fallout inventory and a high measurement uncertainty (Tables 31.1 and 31.2). As mentioned by Mabit et al. (2009a) similar limitation like ^{210}Pb$_{ex}$ poorly enriched precipitations have been reported in several countries. With regard to gamma spectrometry and detection of total ^{210}Pb, a series of recommendations were also provided by the IAEA Seibersdorf Laboratory to improve analytical performance, in particular analytical quality assurance system to assess the accuracy and precision of the measurements (Shakhashiro and Mabit, 2009).

31.3.2 Assessment of Soil Deposition Rates Using ^{137}Cs Data

Based on the different soil layers areal activity, soil profile 1 (presence of ^{137}Cs till 40 cm) and profile 2 (presence of ^{137}Cs till 50 cm) revealed a radio-caesium activity typical of sedimentation processes significantly higher than the reference site of

Table 31.1 Results and error measurements of total ^{210}Pb, ^{226}Ra and ^{210}Pb$_{ex}$ from the reference site ($n = 11$)

^{210}Pb-total (Bq kg^{-1})	^{226}Ra (Bq kg^{-1})	^{210}Pb$_{ex}$ (Bq kg^{-1})
40.14 (6.76) †	35.36 (2.96) †	4.78 (7.38†; 154‡)
34.69 (6.08) †	33.38 (2.42) †	1.31 (6.54†; 499‡)
34.45 (5.18) †	35.63 (3.18) †	0 (n.a†; n.a‡)
35.52 (6.22) †	34.30 (3.18) †	1.22 (6.98†; 572‡)
42.60 (6.84) †	35.42 (3.36) †	7.18 (7.62†; 106‡)
42.25 (6.60) †	33.10 (2.90) †	9.15 (7.22†; 79‡)
39.69 (1.89) †	36.68 (0.41) †	3.01 (1.93†; 64‡)
40.06 (1.91) †	39.50 (0.41) †	0.56 (1.95†; 348‡)
43.00 (1.68) †	40.45 (0.37) †	2.55 (1.72†; 67‡)
34.02 (1.85) †	35.84 (0.40) †	0 (n.a†; n.a‡)
30.53 (1.91) †	39.10 (0.44) †	0 (n.a†; n.a‡)

n.a = non applicable; (†) measurement error in Bq kg^{-1} at 2σ; (‡) measurement error in % at 2σ.

Table 31.2 Results and error measurements of total ^{210}Pb, ^{226}Ra and ^{210}Pb$_{ex}$ in the sedimentation area

Soil samples	^{210}Pb$_{tot}$ (Bq kg^{-1})	^{226}Ra (Bq kg^{-1})	^{210}Pb$_{ex}$ (Bq kg^{-1})
P1 (0–5 cm)	44.16 (6.62) †	35.92 (3.08) †	8.24 (7.3†; 88‡)
P1 (5–10 cm)	41.45 (5.84) †	38.71 (3.00) †	2.74 (6.58†; 240‡)
P1 (10–15 cm)	36.79 (4.96) †	36.54 (3.14) †	0.25 (5.86†; 2344‡)
P1 (15–20 cm)	39.04 (5.70) †	36.99 (3.10) †	2.05 (6.5†; 317‡)
P1 (20–25 cm)	37.42 (4.88) †	37.03 (2.70) †	0.39 (5.56†; 1425‡)
P1 (25–30 cm)	37.41 (6.08) †	35.74 (2.96) †	1.66 (6.78†; 408‡)
P1 (30–35 cm)	32.19 (5.46) †	32.65 (2.88) †	0 (n.a†; n.a‡)
P1 (35–40 cm)	40.14 (6.62) †	36.83 (2.96) †	3.31 (7.24†; 218‡)
P1 (40–45 cm)	40.02 (5.54) †	37.58 (3.18) †	2.45 (6.4†; 261‡)
P1 (45–50 cm)	37.91 (6.42) †	34.14 (3.12) †	3.77 (7.14†; 189‡)
P1 (50–55 cm)	40.53 (6.30) †	37.17 (3.30) †	3.36 (7.1†; 211‡)
P1 (55–60 cm)	40.24 (6.04) †	35.99 (3.02) †	4.25 (6.74†; 158‡)
P1 (60–65 cm)	35.31 (4.58) †	35.32 (2.46) †	0 (n.a†; n.a‡)
P1 (65–70 cm)	45.27 (6.72) †	37.76 (3.08) †	7.51 (7.38†; 98‡)
P1 (70–75 cm)	37.17 (5.34) †	36.88 (2.78) †	0.29 (6.02†; 2075‡)
P1 (75–80 cm)	37.74 (5.08) †	36.10 (2.96) †	1.64 (5.88†; 358‡)
P1 (80–85 cm)	40.38 (5.84) †	36.02 (2.96) †	4.36 (6.54†; 150‡)
P1 (85–90 cm)	37.98 (4.90) †	36.60 (2.74) †	1.38 (5.6†; 405‡)
P1 (90–95 cm)	35.63 (5.84) †	36.92 (2.94) †	0 (n.a†; n.a‡)
P2 (0–10 cm)	40.25 (1.94) †	43.81 (0.44) †	0 (n.a†; n.a‡)
P2 (10–20 cm)	42.06 (1.66) †	38.37 (0.29) †	3.69 (1.69†; 46‡)
P2 (20–30 cm)	39.36 (2.18) †	38.03 (0.38) †	1.33 (2.21†; 166‡)
P2 (30–40 cm)	41.49 (2.12) †	39.15 (0.37) †	2.34 (2.15†; 92‡)
P2 (40–50 cm)	38.00 (2.19) †	38.86 (0.39) †	0 (n.a†; n.a‡)
P2 (50–60 cm)	40.14 (2.05) †	40.43 (0.37) †	0 (n.a†; n.a‡)
P2 (60–70 cm)	38.59 (2.09) †	38.94 (0.37) †	0 (n.a†; n.a‡)
P2 (70–80 cm)	41.64 (1.33) †	38.02 (0.23) †	3.62 (1.35†; 37‡)
P2 (80–90 cm)	39.38 (1.78) †	34.90 (0.30) †	4.48 (1.81†; 40‡)
P2 (90–100 cm)	37.56 (1.98) †	35.84 (0.33) †	1.72 (2.01†; 117‡)

n.a = non applicable; (†) measurement error in Bq kg^{-1} at 2σ; (‡) measurement error in % at 2σ.

respectively 2,836 and 4,776 Bq m^{-2} (Fig. 31.3). Using the information provided by the ^{137}Cs depth distribution, the sedimentation rate of profiles 1 and 2 was respectively estimated at 26.1 and at 50.5 t ha^{-1} year^{-1}. These results are in agreement with the landscape position of the selected profiles.

The information collected from profile 1 was compiled with the erosion plots results. As the sedimentation source of the deposited material is the eroded soils of the contiguous field containing the runoff plots located upstream, the MBM 2 was used to assess the sedimentation rates of the different treatments using the ^{137}Cs activity data derived from the runoff plots (685 Bq m^{-2} for CVT; 1,680 Bq m^{-2} for CST and 1,775 Bq m^{-2} for DS) and the ^{137}Cs information of the soil profile 1.

Profile 1 revealed a yearly sedimentation rate of 20.3, 13.5 and 13.2 t^{-1} ha^{-1} respectively for CVT, CST and DS. Comparing to the CVT, the DS conservation

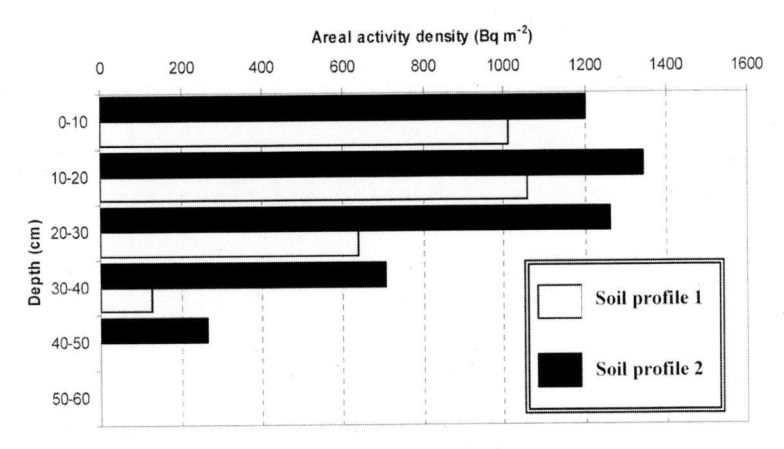

Fig. 31.3 ^{137}Cs areal activity density distribution in soil profiles 1 and 2

system was effective in reducing the sedimentation magnitude by 65%. In the conversion model the following parameters were used: bulk density = 1,380 kg m^{-3}; sampling year = 2007; tillage depth = 414 kg m^{-2} (0.3 m × 1,380 kg m^{-3}); year of initial tillage = 1954; particle size factor = 1; proportional factor = 1; relaxation depth = 4 kg m^{-2}.

Deposition rates provided by profile 1 located down slope the field using the ^{137}Cs depth distribution information (26.1 t ha^{-1} year^{-1}) and MBM 2 (20.3 t^{-1} ha^{-1} year^{-1}) are closely related to the erosion rates measured by the runoff plot under CVT (29.4 t ha^{-1} year^{-1}) in the upper part of the same experimental field.

Taking into account the average bulk density in the different soil layers (1.3 g cm^{-3}), and based on the fact that these annual average values obtained with ^{137}Cs method cover a 54 years period (1954–2007), it was possible to evaluate the soil layer deposited. The yearly sedimentation rate varies from 1 mm (13.2 t^{-1} ha^{-1} year^{-1}) to a maximum of 3.9 mm (50.5 t^{-1} ha^{-1} year^{-1}) corresponding to a respective total material accumulation over the last 54 years of 5.5 and 21 cm, respectively. This shows the magnitude of deposition that occurred in the selected watershed according to the topography and the different tillage treatment. The erosion rates under conventional tillage are in agreement with the sedimentation rates estimated down slope of the field by the ^{137}Cs depth distribution profile and MBM 2.

31.4 Conclusions

With the conjunctive use of isotopic and conventional erosion methodologies one can estimate soil loss and sedimentation rates as well as assess the effectiveness of soil conservation measures to reduce soil redistribution. A combined approach based on conventional runoff plots measurement and FRNs method (^{137}Cs and ^{210}Pb) was successfully implemented to assess sedimentation rates in the Mistelbach watershed in Austria. While the ^{137}Cs appeared to be a mature isotopic technique the ^{210}Pb was not applicable in the area under investigation due to very low concentrations of ^{210}Pb$_{ex}$ associated to a high uncertainty in the measurements.

There is always scope for further refinement of the FRN methodologies that previously had mostly focussed on soil degradation and redistribution at the hill slope, plot and field scale. New up-scale development to the watershed scale in a wide range of agricultural landscapes using interpolation tools and also considering sediment sources and budget, transfer residence and storage are needed to shift from use of these isotopic tracers (especially ^{137}Cs) as a research tool to a standardized decision support tool in the following years to protect natural resources quality in both developed and developing countries.

Acknowledgements This study was conducted in support of the FAO/IAEA Co-ordinated Research Project "Assess the effectiveness of soil conservation measures for sustainable watershed management using fallout radionuclides" (D1.50.08). The authors would like to thank Dr. Gudni Hardarson, Head of the Soil Science Unit and the Dr. Felipe Zapata (Soil and Water Management & Crop Nutrition Section) for suggestions to improve the manuscript.

References

Allison, M.A., Kuehl, S.A., Martin, T.C. and Hassan, A. (1998). Importance of flood-plain sedimentation for river sediment budgets and terrigenous input to the oceans: insights from the Brahmaputra-Jamuna River. Geology 26(2):175–178.

Bai, Z.G., Dent, D.L., Olsson, L. and Schaepman, M.E. (2008). Proxy global assessment of land degradation. Soil Use and Management 24:223–234.

Boardman J., Poesen J. (eds.) (2006). Soil Erosion in Europe. Wiley, Chichester.

Commission of the European Communities (CEC). (2006). Thematic Strategy for Soil Protection. Communication from the Commission to the Council, the European Parliament, the European Economic and Social Committee, and the Committees of the Region – COM (2006)231 final; Brussels 22.9.2006. 12 pp.

EEA – European Environmental Agency. (1999). Environment in the European Union at the turn of the century. Environmental Issue series, No. 2. Copenhagen: 448 pp.

EEA – European Environmental Agency. (2002). Down to earth: Soil degradation and sustainable development in Europe. A challenge for the 21st century. Environmental Issue series, No. 16. Copenhagen: 32 pp.

Food and Agriculture Organization of the United Nations. (2005). FAO/AGL. GLASOD. National Soil Degradation Maps. Soil Degradation Map of Austria. http://www.fao.org/landandwater/agll/glasod/glasodmaps.jsp?country=AUT&search=Display+map+%21

Froehlich, W. and Walling, D.E. (1992). The use of radionuclides in investigations of erosion and sediment delivery in the Polish Flysh Carpathians. IAHS publication 209:61–76.

Fulajtar, E. (2000). Assessment of soil erosion through the use of ^{137}Cs at Jaslovske Bohunice, Western Slovakia. Acta Geologica Hispanica 35(3–4):291–300.

Klik, A. (2003). Einfluss unterschiedlicher Bodenbearbeitung auf Oberflaechenabfluss, Bodenabtrag sowie auf Naehrstoff- und Pestizidaustraege. Oesterr. Wasserwirtschaft 55 (5–6):89–96.

Klik, A., Sokol, W. and Steindl, F. (2004). Automated erosion wheel: A new measuring device for field erosion plots. Journal of Soil and Water Conservation 59(3):116–121.

Mabit, L., Benmansour, M. and Walling, D.E. (2008a). Comparative advantages and limitations of Fallout radionuclides (^{137}Cs, ^{210}Pb and ^{7}Be) to assess soil erosion and sedimentation. Journal of Environmental Radioactivity 99(12):1799–1807.

Mabit, L., Bernard, C. and Laverdière, M.R. (2002a). Quantification of soil redistribution and sediment budget in a Canadian watershed from fallout caesium-137 (^{137}Cs) data. Canadian Journal of Soil Science 82(4):423–431.

Mabit, L., Bernard, C. and Laverdière, M.R. (2007). Assessment of erosion in the Boyer River watershed (Canada) using a GIS oriented sampling strategy and ^{137}Cs measurements. Catena 71(2):242–249.

Mabit, L., Bernard, C., Laverdière, M.R. and Wicherek, S. (1998). Spatialisation et cartographie des risques érosifs à l'échelle d'un bassin versant agricole par un radio-isotope (^{137}Cs). Étude et Gestion des sols 5(3):171–180.

Mabit, L., Bernard, C., Makhlouf, M. and Laverdière, M.R. (2008b). Spatial variability of erosion and soil organic matter content estimated from ^{137}Cs measurements and geostatistics. Geoderma 145(3–4):245–251.

Mabit, L., Klik, A., Benmansour, M., Toloza, A., Geisler, A. and Gerstmann, U.C. (2008c). Assessment of erosion and sedimentation rates using ^{137}Cs, ^{210}Pb$_{ex}$ and conventional erosion measurements within an Austrian watershed (Mistelbach). In: Laboratory Activities-Research, IAEA-Soils Newsletter 31(1):18–20.

Mabit, L., Klik, A., Benmansour, M., Toloza, A., Geisler, A. and Gerstmann, U.C. (2009a). Assessment of erosion and deposition rates within an Austrian agricultural watershed by combining ^{137}Cs, ^{210}Pb$_{ex}$ and conventional measurements. Geoderma 150(3–4):231–239.

Mabit, L., Klik, A., Toloza, A., Geisler, A. and Gerstmann, U.C. (2008d). Evaluation of the initial fallout of ^{137}Cs and characterisation of a reference site in the Mistelbach watershed (Austria). In: Laboratory Activities-Research, IAEA-Soils Newsletter 30(2):20.

Mabit, L., Laverdière, M.R. and Bernard, C. (2002b). L'érosion hydrique: méthodes et études de cas dans le nord de la France. Cahiers Agricultures 11(3):195–206.

Mabit, L., Zupanc, V., Martin, P. and Toloza, A. (2009b). Preliminary investigation using nuclear techniques to assess soil erosion in Slovenia. In: Laboratory Activities-Research, IAEA-Soils Newsletter 31(2):23–25.

Navas, A. and Walling, D.E. (1992). Using caesium-137 to assess sediment movement on slopes in a semiarid upland environment in Spain. IAHS publication 209:129–138.

Pollhammer, J. (1997). Die Auswirkung ausgewaehlter ackerbaulicher, pflanzenbaulicher und landtechnischer Maßnahmen auf den Bodenabtrag durch Wasser. Master Thesis. University of Natural Resources and Applied Life Sciences, Vienna.

Porto, P., Walling, D.E., Ferro, V. and Di Stefano, C. (2003). Validating erosion rate estimates provided by caesium-137 measurements for two small forested catchments in Calabria, Southern Italy. Land Degradation and Development 14:389–408.

Ritchie, J.C. and Ritchie, C.A. (2008). Bibliography of publications of Cs^{137} studies related to erosion and sediment deposition. http://www.ars.usda.gov/Main/docs.htm?docid=15237 .

Schimmack, W., Auerswald, K. and Bunzl, K. (2002). Estimation of soil erosion and deposition rates at an agricultural site in Bavaria, Germany, as derived from fallout radiocesium and plutonium as tracers. Naturwissenschaften 89(1):43–46.

Shakhashiro, A. and Mabit, L. (2009). Results of an IAEA inter-comparison exercise to assess ^{137}Cs and total ^{210}Pb analytical performance in soil. Applied Radiation and Isotopes 67(1):139–146.

Strauss, P., Auerswald, K., Blum, W.E.H. and Klaghofer, E. (1995). Erosivitaet von Niederschlaegen. Ein Vergleich Oesterreich–Bayern. Zeitschrift fuer Kulturtechnik und Landentwicklung 36:304–309.

Strauss, P. and Klaghofer, E. (2004). Scale considerations fort he estimation of processes and effects of soil erosion in Austria. In: Francaviglia R. (ed.), Agricultural Impacts on Soil Erosion and Biodiversity: Developing Indicators for Policy Analysis. March 2003. Proceedings OECD Expert Meeting. Proceedings from an OECD Expert Meeting, Rome, pp. 229–238.

Strauss, P. and Klaghofer, E. (2006). Status of Soil Erosion in Austria. Chap. 1.17. In: Boardman J., Poesen J. (eds.), Soil Erosion in Europe. Wiley, Chichester, pp. 205–212.

Walling, D.E. and He, Q. (1999). Improved models for estimating soil erosion rates from cesium-137 measurements. Journal of Environmental Quality 28:611–622.

Walling, D.E., He, Q. and Appleby, P.G. (2002). Conversion models for use in soil-erosion, soil-redistribution and sedimentation investigations. In: Zapata F. (ed.), Handbook for the Assessment of Soil Erosion and Sedimentation Using Environmental Radionuclides. Chap.7. Kluwer Dordrecht, The Netherlands, pp. 111–164.

Walling, D.E. and Quine, T.A. (1991). Recent rates of soil loss from areas of arable cultivation in the UK. IAHS publication 203:123–131.

Chapter 32
Development and Opportunities for Evaluation of Anthropogenic Soil Load by Risky Substances in the Czech Republic

Radim Vácha, Jan Skála, and Jarmila Čechmánková

Abstract This paper presents the evaluation of risky substance loads in the Czech agricultural land. The groups of potentially risky elements (REs) and persistent organic pollutants (POPs) are evaluated separately due to their characteristic differences and prevailing source of contamination. Risks deriving from increased RE and POPs soil loads could be controlled by the system of limit values. The limit values could be separated into two groups, the limits for REs and POPs contents in the soil and the limits regulating inputs of REs and POPs into the soil. The limit values system is based on the hierarchical limits from background values of REs and POPs in soils to the limits linked with particular risks and eventually to sanitary limits. The limit values regulating the entry of REs and POPs into the soils of the Czech Republic in case of fertilisers and additional soil substances are congruent with national Laws (Law No. 156/1998 of the Code of Law of the Czech Republic – subsequently amended) and sewage sludge (Directive from the Ministry of Environment No. 382/2001 of the Code of Law of the Czech Republic about conditions of modified sewage sludge application on agricultural land). The legislative regulation of the application of ponds and river sediments on agricultural land is under preparation.

Keywords Limit values system · Persistent organic pollutants · Potentially risky elements · Soil contamination

32.1 Introduction

Agricultural soil contamination is one of the most critical topics of soil degradation in Europe (EU Thematic Strategy for Soil Protection, COM 2006, 231). Increased soil loads by risky substances bring numerous problems to the environment, plant

R. Vácha (✉)
Research Institute for Soil and Water Conservation, Prague 15627, Czech Republic
e-mail: vacha@vumop.cz

production and food chains. The maintenance of suitable state of soil load by risky substances is thus at the best interest of every society. Nevertheless, the knowledge of risky substance background values, their inputs into soils, their behaviour and fate in the soil environment, and their transfer into the plants must support the evaluation of soil load by risky substances. From this viewpoint the approaches of individual countries not only around the world, but also in the European context are not unified and different methodologies may be used for the evaluation of soil load.

Long-term attention has been paid to soil contamination in the Czech Republic and monitoring of real state of soil load by a number of risky substances was previously reported (Sáňka, 1998). These activities led to the development of a unified system for the evaluation of soil contamination and helped draft legislative norms and the necessary amendments for soil protection in the country.

The potentially toxic compounds observed in Czech agricultural soils might be separated into two main groups of pollutants:

- Inorganic pollutants – potentially risky elements (REs), As, Be, Cd, Co, Cr, Hg, Cu, Mn, Ni, Pb, V, Zn, respectively Se and Tl
- Organic pollutants – persistent organic pollutants (POPs), a wide group of different organic substances, with linear or cyclic character. The current list of POPs observed in Czech regulations (Soil Protection Act) includes monocyclic and polycyclic hydrocarbons, PCBs, sum of DDT and petroleum hydrocarbons.

From the soil contamination assessment viewpoint it is necessary to depict load sources of risky substances that influence the behaviour of such risky substances into the soil (the mobility and bioavailability). Risky elements and POPs may be originated from:

- Natural sources – geochemical character of soil substrate (REs), volcanic activity (REs, POPs), natural fires (POPs) etc.
- Anthropogenic sources – industrial activities (REs, POPs), transport emissions (REs, POPs), the use of agrochemicals and biosolids in agriculture (REs, POPs), waste water production (REs, POPs) etc.

Increased inputs of potentially toxic compounds into the soils could result in soil contamination that may negatively influence:

- Ecosystem – soil functions, contamination of aquatic systems, plants, animals etc.
- Plant production – quantity and quality.
- Human health – via contamination of the food chain, dermal or inhalation intake etc.

There are a number of mandatory legislative limits established in the Czech Republic that control soil contamination from risky substances. The methodologies for their evaluation are described in this paper.

32.2 Materials and Methods

The establishment of limit threshold values was supported by various research activities and projects conducted in the country. The Research Institute for Soil and Water Conservation has been at the frontline of these efforts with its research that has been focusing on:

- *Monitoring* of the soil load by potentially risky elements and persistent organic pollutants in agricultural soils and protection of food chain. The Ministry of Agriculture supported such activity. Various authors provide interesting data on risky substance loads on food chains (Němeček et al., 1996; Podlešáková et al., 1998; Vácha et al., 2001).
- *Research* on the mobility of potentially risky elements and persistent organic pollutants in the soil and their inputs into the plants (Němeček et al., 2001; Podlešáková et al., 2002; Vácha et al., 2005; Vácha et al., 2008).
- *Research on the inputs* and the fate of risky substances into the soil by the application of soil additives (Vácha et al., 2005).

Monitoring of agricultural soil is subsequently realised in separate districts of the country. Soil samples are taken with a non-metallic tool from the topsoil humic horizons of the agricultural soils (inclusive of arable lands, pastures, and permanent grassland). Only in special cases, samples are taken from deeper soil horizons to be able to establish the sources of contamination, i.e. geogenic or anthropogenic. For each 25 km^2 area one soil sample is taken and its geographical coordinates are recorded through the Global Positioning System (GPS) equipment. Each sampling site is described in terms of pedologic characterisation (soil type, soil substrate, soil pH, etc).

The potentially risky element analyses are done at the Central laboratories of Research Institute for Soil and Water Conservation in Prague. The contents of As, Be, Cd, Co, Cr, Cu, Mn, Mo, Ni, Pb, V and Zn are assessed in the extract of agua regia by AAS. The content of Hg is analysed using the AMA. The mobile forms of potentially risky elements were analysed in the extract of 1 M NH_4NO_3. For better understanding the risky elements fractionation in the soil, sequential extraction procedure after Zeien and Brümmer (1989) are used in some cases. The soil additives are analysed and assessed by the same methods used in routine soils analyses. The contents of persistent organic pollutants (BTEX, polycyclic aromatic hydrocarbons – 12 substances, PCB_7, DDT, DDE, DDD, HCB, HCH and C_{10}–C_{40} and the content of polychlorinated dibenzodioxins and dibenzofuranes) are analysed in commercial laboratories. The methods are described by Podlešáková et al. (1998), (2000) and Vácha et al. (2008).

The contents of 13 risky elements and 26 persistent organic substances in some particular regions of the Czech Republic are assessed in relation to the possible food chain risks. The concentration levels are spatially shown in special maps using the inverse distance weighted interpolation technique (IDW interpolation).

32.3 Results and Discussion

The limit values system is a generally used tool for controlling the risks deriving from soil contamination and controlling the inputs of risky substances entering into the soil. The soil protection strategy provides the basis for establishing equilibriums between regulation requirements for preventing pollution from risky substances and the measures needed for remediation. We separate in two groups the soil contamination toxic compounds limits:

- The limit threshold values of risky substances in the soil.
- The limit threshold values regulating the inputs of risky substances entering into the soil.

32.3.1 Limit Threshold Values of Risky Substances in the Soil

The limit values of risky substances in the soil are derived in terms of the real state of soil load by risky substances reflecting natural and anthropogenic diffuse load. Limit values of this kind are usually specified as "background values" of risky substances (Podlešáková et. al., 1996; Němeček et al., 1996). The experimentally derived values that are focused on the target risk areas are derived from soil use and subsequently observed as environmental consequences i.e. quantity and quality reduction of agricultural production, reduction of soil microbial activity, etc.

One of the most effective and sophisticated limit values systems is the so-called hierarchical limit values system that should be able to register target risks deriving from soil contamination. This system is used in many European countries (Germany, Netherlands, Switzerland) as familiar system of "A, B and C limits" where:

> A – represents background values of risky substances in the soil. Generally, this limit value fulfils the principal of precaution.
> B – is focused on target risk. The limit can be targeted on the quality or quantity of plant production (this approach is used rarely and is determined rather for small allotment producers rather than for large scale agriculture) or on the decreasing of soil microbial activity and soil transformation functions etc.
> C – is used as remediation (decontamination) limit that is based on the risk assessment according to human health harm or environmental damage.

Hierarchical limits assessment differs in each country due to various soil substrate conditions, different research backgrounds and social and political conditions. In some countries (i.e. Great Britain, USA) the use of limit values system is focused on remediation needs (in the order of C limit level). A given limit value of risky substances delimits risky substances concentrations that could influence human health or the environment quality. After exceeding the risk limit the site-specific risk assessment on the field should be done and the results of risk assessment study determine the next approach (i.e. remediation or the land use change).

The structure of the proposed draft of the EU Soil Protection Act is based on similar philosophy and requires three steps at national level for all the member countries:

- The elaboration of soil contaminated sites register
- The realisation of risk assessment studies on contaminated sites
- The realisation and adoption of remediation approaches

The soil protection draft concept in the Czech Republic tends to resemble the prevalent hierarchical limit values system of Europe. The Directive of Ministry of Environment of the Czech Republic No. 13/1994 Coll. regulates the contents of REs and POPs in Czech agricultural soils. The limits of REs are determined for light texture soils and the other soils in the form of total content and the extract in 2 M HNO_3 (could method). The limit values for POPs are determined for the groups of monocyclic aromatic hydrocarbons, polycyclic aromatic hydrocarbons, chlorinated hydrocarbons including pesticides and petroleum hydrocarbons. All the limit values are defined like tolerable contents of risky substances but there is no relationship at any actual risk. However, it shows difficulties for the evaluation and interpretation of soil load by risky substances in many cases. This was one of the reasons for the proposal of Directive No. 13 amendment based on the principal of hierarchical limit values system (Sáňka et al., 2002).

Three levels of the limits were proposed:

Prevention limit – based on the background values of risky substances in Czech agricultural soils. Prevention limits were proposed for REs and POPs also. The exceeding of the limit shows increased anthropogenic soil load by risky substances. From the viewpoint of limit interpretation, it is prohibited to use the sludge or sediment for soil fertilization in the case of limit exceeding.

Indication limit – was derived experimentally and the exceeding limit indicates the risk of increased REs transfer from the soil into the plants. Indication limit was proposed for REs only. The limit values of REs are based on the comparison of their total contents and mobile fractions in the soil and soil pH in the case of Cd and Ni. The example of indication limit for Cd is presented in Table 32.1. The exceeding of critical REs contents in the plant will be assessed using statistical probability methods. Detailed assessment is recommended on special areas in the event of indication limit exceeding.

Decontamination limit – was not proposed yet.

The proposal of Directive No. 13/1994 Coll. amendment will be probably accepted in the new version of Soil protection act.

The following aspects complicate the assessment of indication limits for POPs:

- Technical and economical needs
- Relatively low mobility of POPs in the soil and low transfer into the plants, POPs decomposition and different toxicity of decomposition products.

Table 32.1 Proposed indication limit for Cd

Element	Soil	pH	Total content (mg kg^{-1})	1 M NH$_4$NO$_3$ (mg kg^{-1})
Cd		<4.0	0.7	–
	–	4.0–5.0	1.0	–
		5.0–6.5	1.5	–
	Light	>6.5	2.0	0.04
	Others	>6.5	2.0	0.1

- Risks comparison, for example the risk of food chain contamination and risk of endangered human health by dermal intake or inhalation of polluted soil particles from contaminated fields.

The transfer of POPs into the plants could be realised by different ways. Holoubek (2005) defined following transfer processes for polycyclic aromatic hydrocarbon (PAHs):

- Root intake from the soil solution (depending on the plant water regime and the content of lipid compounds in the root),
- Absorption of PAHs on root surface,
- Absorption of volatilised PAHs (from the soil) on the shoot,
- Absorption of PAHs on plants leaves (from emission fall-outs),
- Immediate synthesis of some PAHs by plants.

The transfer of POPs from the soil into the plant shoots is marginal in comparison with mobile REs and the emission contamination of plant surface (shoot) is usually

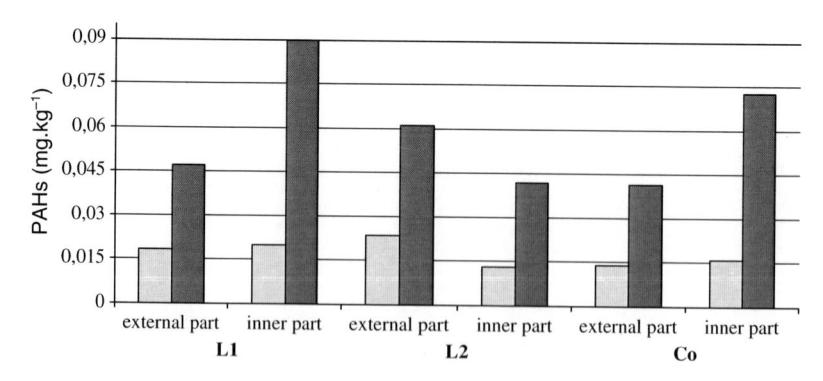

Fig. 32.1 Comparison of 2–3 nucleus PAHs and 4–6 nucleus PAHs contents in carrot roots (*external part* and *inner part*). L1 – loaded Fluvisol by PAHs, L2 – soil with sludge loaded by PAHs (1: 1), Co – control sample

prevailing. Nevertheless, the root plants (root vegetables) may be loaded by POPs significantly (Fig. 32.1) but the extent and way of contamination depends mainly on POPs properties, content and quality of soil organic matter (Vácha et al., 2008).

More individual risks must be accepted for POPs indication limit and the realisation of risk assessment is recommended for seriously contaminated sites (in order of C limit level usually). Despite these facts, Sáňka and Vácha (2006) proposed simplified indication limits for some POPs (Table 32.2). The limit values were determined as the lowest content of risky substances in the soil that could cause any health risk. The transfer of risky substance from the soil to human bodies by dermal, oral and dietary intake was accepted.

32.3.2 The Limit Threshold Values Regulating the Inputs of Risky Substances Entering into the Soil

The following documents are available in Czech legislation:

- The Act No. 156/1998 Coll., regulating the use of fertilizers in the agriculture.
- The Directive of the Ministry of Environment No. 382/2001 Coll., about the use of sludge on agricultural soils.
- The proposal of the Directive of Ministry of Environment about the use of extracted sediments.

The sludge, river or pond sediments seem to be the most important source of risky substances. The Directive for sludge application limits mainly the content of REs. Only seven congeners of polychlorinated biphenyls (PCB7) and the sum of

Table 32.2 The proposal of POPs limit values for Czech soils – indication values (Sáňka and Vácha, 2006)

POPs	Indication value (mg kg^{-1} of d.m.)
Benzo(a) pyrene	2.0
Sum PAHs[a]	30.0
Sum PCB[b]	1.0
DDT and metabolites	4.0
HCH (α, β, γ)	0.1
HCB	0.1
PCDDs/Fs[c]	20.0
Benzene	0.5
Ethylbenzene	5.0
Toluene	10.0
Xylene	10.0
Hydrocarbons C$_{10}$–C$_{40}$	500

[a]The sum of 16 individual PAHs (EPA).
[b]The sum of 7 PCB congeners (28 + 52 + 101 + 118 + 138 + 153 + 180).
[c]ng kg^{-1} I-TEQ PCDDs/Fs.

Table 32.3 Proposed POPs limit values in sludge ($\mu g\ kg^{-1}$, I-TEQ $ng\ kg^{-1}$* of d.m.)

Indicator	Σ PAU	Σ MAU	PCB$_6$	HCB	DDT	DDE	DDD	PCDD/F*
Proposed limit sludge	10,000	10,000	600	60	60	60	30	80
Primary proposal EU sludge	6,000	–	800	–	–	–	–	100
Background values in Czech soils	1,000	130	20	20	15	10	10	1

halogenated organically bound compounds (AOX) are limited from POPs group. Czech Directive is stricter in comparison with European Directive No. 86/278/EEC. The proposal of amendment of EU Directive (Working Document on Sludge, 2000) that introduced new criteria for REs and POPs was not accepted by the European Parliament. The proposal of Czech limits for some POPs in sludge was based (Vácha et al., 2006) on experimental data and the original proposal of EU Working Document on Sludge.

Proposed POPs limit values are presented in Table 32.3. The experimental data were originated from the monitoring of 46 wastewater factories and from the pot and field trials observing the influence of sludge application on soil and plant quality.

The absence of legislation for river and pond sediment application on agricultural soils brings many problems for sediment producers. The proposal of the national Directive is being developed at present. The principal structure of the proposal is similar with the Directive for sludge application; nevertheless more POPs groups are limited for sediment application (PCB$_7$, sum of PAHs, sum of BTEX-benzene, toluene, ethylbenzene and xylene, and hydrocarbons C_{10}–C_{40} for petroleum hydrocarbons indication).

In spite of relatively stricter limits for risky substances in the sludge, the inputs of risky substances into soils are much higher by sediment application (Table 32.4)

Table 32.4 The input of risky substances by the application of sediments and sludge

	Sediments The dose 750 t ha^{-1}/10 years RE input (g ha^{-1})	Sludge The dose 15 t ha^{-1}/10 years RE input (g.ha^{-1})
As	22,500	450
Be	3,750	
Cd	750	75
Co	30	
Cr	150,000	3,500
Cu	75,000	7,500
Hg	600	60
Ni	60,000	1,500
Pb	75,000	3,000
V	135,000	
Zn	225,000	37,500
PAHs (sum of 12 compounds)	750	–
PCB$_7$	150	9

because of higher doses of sediments in comparison with sludge (750 t ha^{-1} of sediments – dry matter, ones in 10 years maximally, compared to 5 t ha^{-1} of sludge – dry matter, ones in 3 years maximally). Legislation for controlling sediment application is seriously needed in the country due to huge amounts of extracted sediments. The current practice was based on the decisions of local authorities and the absence of legislative norms led to fundamental differences in decision-making process. This fact could negatively influenced soil quality.

32.4 Conclusions

It could be concluded that limit values system can protect agricultural soils from contamination by risky substances, but their assessment is complicated and the abovementioned concerns and questions must be solved first through experimental work and "limit adjustment", supported by juridical, political and economical acceptability and responsability.

Acknowledgments The article was prepared with the support of the Project of Ministry of Agriculture No. MZE0002704902

References

Czech Ministry of Agriculture. (1998). The act of Ministry of Agriculture for the management of fertilizers, soil and plant additives and agrochemical testing of soils.
Czech Ministry of Environment. (1994). The notice of the Ministry of Environment for the management of the soil protection, No. 13/1994 Coll (in Czech).
Czech Ministry of Environment. (2001). The notice of Ministry of Environment for the sludge application on agricultural soil, No. 382/2001 Coll (in Czech).
EEC Directive 86/278/. (1986). EU Sewage sludge directive No.86 /278/EEC, Brussels.
European Commission. (2006). Thematic Strategy for Soil Protection, COM. 231.
Holoubek, I. (2005). The chemistry of the environment IV. Polycyclic aromatic hydrocarbons (PAHs). http://recetox.muni.cz/index.php?id=23. Cited 2 Sept 2008.
Němeček, J., Podlešáková, E. and Vácha, R. (1996). Geochemical and anthropogenic soil load. Rostlinná Výroba 42(12):535–541.
Němeček, J., Podlešáková, E. and Vácha, R. (2001). Prediction of the transfer of trace elements from soils into plants. Rostlinná Výroba 47(10):425–432.
Podlešáková, E., Němeček, J. and Hálová, G. (1996). The proposal of limit values of soil contamination by potentially risk elements for CR. Rostlinná Výroba 42(3):119–125.
Podlešáková, E., Němeček, J. and Vácha, R. (2000). Zatížení zemědělských půd polychlorovanými dibenzo-p-dioxiny a dibenzofurany. Rostlinná Výroba 46(8):349–354.
Podlešáková, E., Němeček, J. and Vácha, R. (2002). Critical values of trace elements in soils from the viewpoint of the transfer pathway soil – plant. Rostlinná Výroba 48(5):193–202.
Podlešáková, E., Němeček, J., Vácha, R. and Pastuszková, M. (1998). Contamination of soils with persistent organic xenobiotic substances in the Czech Republic. Toxicological and Environmental Chemistry 66:91–103.
Sáňka, M. et al. (1998). Basal soil monitoring, monitoring of atmospheric deposition. Central Institute for Supervising and Testing, Brno (in Czech).
Sáňka, M., Němeček, J., Podlešáková, E., Vácha, R. and Beneš, S. (2002). The elaboration of limit values of concentrations of risky elements and organic persistent compounds in the soil and

their uptake by plants from the viewpoint of the protection of plant production quantity and quality. The report of the Ministry of Environment of the Czech Republic, 60 p (in Czech).
Sáňka, M. and Vácha, R., (2006). The evaluation of the limits for soils in normative regulations of CR and selected European countries. The Report of the Ministry of Environment of the Czech Republic, 38 p (in Czech).
Vácha, R., Čechmánková, J., Havelková, M., Horváthová, V. and Skála, J. (2008). The transfer of polycyclic aromatic hydrocarbons from the soil into selected plants. Chemicke Listy 11: 1003–1010.
Vácha, R., Podlešáková, E., Němeček, J. and Poláček, O. (2001). The state of load of agricultural soils by persistent organic pollutants. Chemické listy 95:590–593.
Vácha, R., Vysloužilová, M. and Horváthová, V. (2005). Polychlorinated dibenzo-p-dioxines and dibenzofurans in agricultural soils of Czech Republic. Plant Soil and Environment 51(10): 464–468.
Vácha, R., Vysloužilová, M., Horváthová, V. and Čechmánková, J. (2006). Recommended maximum contents of persistent organic pollutants in sewage sludge for application on agricultural soils. Plant Soil and Environment 52(8):362–367.
Working Document on Sludge. (2000). An EU-initiative to improve the present situation for sludge management, Brussels, ENV.E.3/LM, 19 p.
Zeien H. and Brűmmer G.W. (1989). Chemische Extraktionen zur Bestimmung von Schwermetallbildungformen in Bőden. Mitteilungen der Deutschen Bodenkundlichen Gesellschaft 59(1):505–510.

Chapter 33
Land Degradation in Greece

Sid. P. Theocharopoulos

Abstract This chapter represents the status of land degradation in Greece. As everywhere in the Mediterranean region, processes such as climatic change, extreme climatic events, and human actions accelerate land degradation and desertification and are reflected by the reduction of the capacity of the land to maintain its economic, ecological and productive functions. The main land degradation processes operating in the soil, water and biosphere system, cause soil degradation, water scarcity and decline of biodiversity. Greek soils are degradated by water and wind erosion, soil organic matter reduction, salinization, alkalization, fertility depletion, compaction, crusting, acidification, leaching, soil pollution and contamination, floods, landslides and sealing. Water scarcity and quality deterioration is exacerbated through continuing overexploitation of surface and ground water, poor management practices, pollution from point and diffuse sources and salinization through seawater intrusion. Land use change, wild fires, overgrazing, intensification of agriculture, and monoculture affect population dynamics and biodiversity while accelerating soil and water degradation processes and greenhouse gases emissions. The National Action Plan to Combat Desertification, the National Water Management plan, the Code of Good Agricultural Practices and the Soil Thematic Strategy in combination with international scientific cooperation and the creation of an harmonized monitoring system should provide to all stakeholders, especially to policy makers, the necessary knowledge, structure and measures to mitigate land degradation.

Keywords Land · Soil · Biodiversity · Degradation · Desertification · Climate change · Greece

---

S.P. Theocharopoulos (✉)
N.AG.RE.F. Soil Science Institute of Athens, Athens 14123, Greece
e-mail: sid_theo@nagref.gr

33.1 Introduction

Extensive soil and land degradation processes leading to desertification affect Greece. Many observed degradational processes are directly or indirectly induced by humans and accelerate climatic change and extreme climatic events, which may further exacerbate the impacts of land degradation. The aim of this chapter is to present the status of land degradation in Greece, to describe the natural processes in the different components of land, and the human activities that have direct or indirect impacts on resource degradation. Also, to highlight land conservation measures applied or needed to be applied to combat desertification and to propose extra measures and the necessary framework, which policy makers should develop and apply in order to reduce, stop or even reverse negative trends.

The total area of Greece, according to the last census (2001) is 13,195,740 ha, 30% of which, is cultivated land, 40% is pasture land, 22% forest land and the remaining 8% is water, buildings etc. Threats and degradation processes to soil and land as well as land protection policies and actions have been also presented previously by Theocharopoulos and Aggelides (1991), Davidson and Theocharopoulos (1992), Lyrintzis and Papanastasis (1995), Yassoglou (1987, 1999, 2005) Theocharopoulos and Panoras (2000) and Theocharopoulos (2007).

33.2 Land Degradation and Desertification Status

The United Nations Convention to Combat Desertification (UNCCD) defines "land" as "the terrestrial bio-productive system that comprises soil, vegetation, other biota, and the ecological and hydrological processes that operate within the system", while land degradation is defined as "the temporary or permanent lowering of the productive capacity of land". The same convention defines desertification as "land degradation in arid, semi-arid and dry sub-humid areas resulting from various factors, including climatic variations and human activities".

Land is a resource and also a complex living system. The rate of soil and land degradation processes as well as the land use and management practices, which operate in any point or ecosystem in Greece could be described by soil and land quality indicators, and supported by a proper monitoring system. A temporary or permanent lowering of the productive capacity of land is recognized throughout the country, but varies from place to place. Potential desertification and land degradation risks result from various factors, including climatic variations and human activities as presented in Fig. 33.1 prepared by the Greek National Committee to Combat Desertification (Yassoglou, 1999). Land degradation intensities are due to natural processes such as landforms shape and pattern, climate, extreme climatic events, but the human induced processes are those that in most cases accelerate the process. The final consequences are reflected by the inefficiency of the land to maintain its

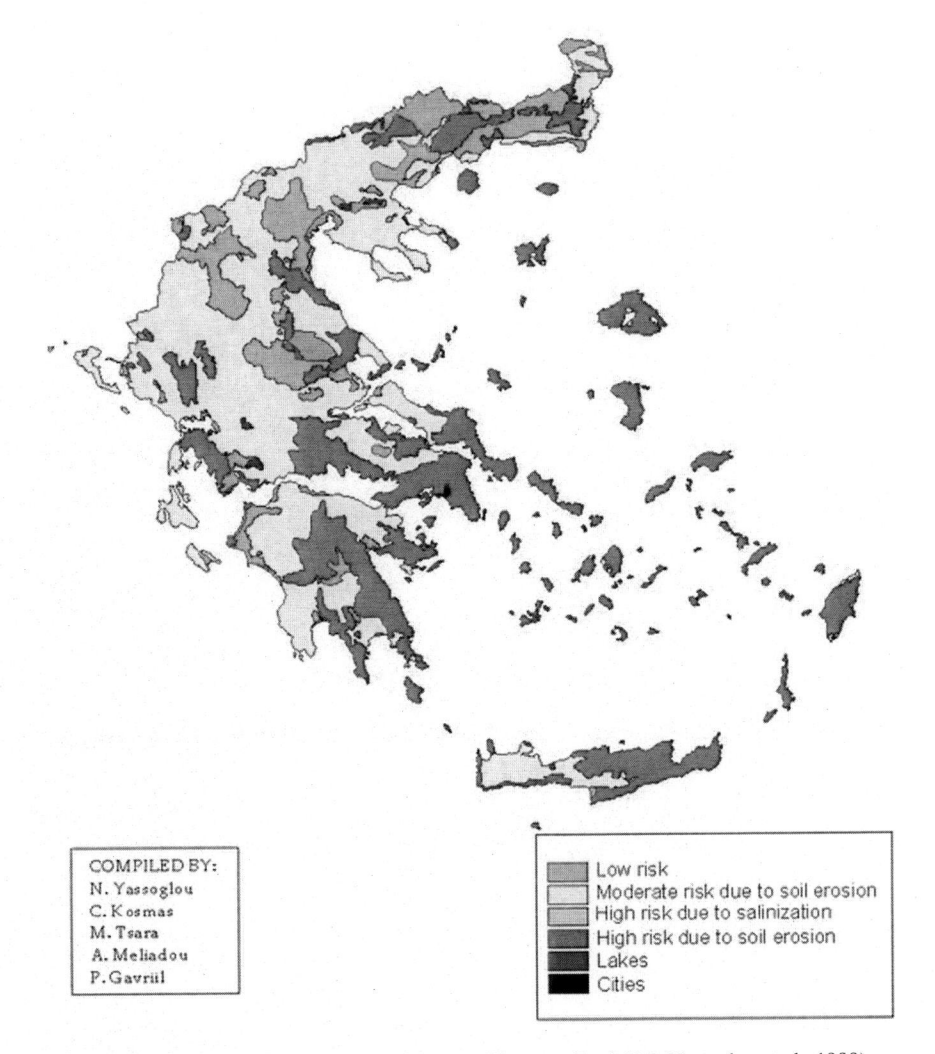

Fig. 33.1 Potential desertification map of Greece (Courtesy Prof. N.J. Yassoglou et al., 1999)

economic and ecological functions and an irreversible reduction capacity to produce goods and services. Based on soil, climatic and topographic characteristics land of potentially high quality covers 19%, land of moderate quality covers 18%, while the rest of the land (57%) is of low quality (CORINE, 1992).

The main human induced degradation processes in the country could be recorded as accelerated soil erosion, soil disturbance, removal of vegetative soil cover and/or hedgerows, abandonment of terraces, overstocking and overgrazing, poor and inappropriate crop management like burning of crop residues, wild fires etc.

33.2.1 Soil Degradation

Soil degradation is only one aspect of land degradation that describes the physical, chemical and biological degradation of the soil through the reduction in its ability to fulfil its functions related to productivity and the environment. Soil degradation process in Greece, is described by Theocharopoulos (2007), Kosmas et al. (2006), Yassoglou (1999), and others. The soils of Greece developed on the characteristic steep Mediterranean landscapes (Theocharopoulos, 2007) are mostly shallow without well-expressed pedomorphological horizons (Leptosols, Regosols), which mainly suffer from soil erosion, organic matter decline and nutrient depletion. The deep soils in the lower lying plains and the soils near the coast, where water is stagnant, or sea water intrusion occurs due to over pumping, are degraded by salinization and/or sodification. To this also contribute streams and groundwater, which contain significant amounts of wind-borne salt or salts from parent materials.

Water and wind erosion, loss of organic matter, salinisation, alkalisation, compaction, fertility depletion, crusting, decline of biodiversity, acidification, leaching, soil pollution and contamination, and sealing, seems to be the main degradation factors or processes occurring in Greece. The severity of each factor varies spatially and temporarily but no detailed information is available at present.

Soil erosion is the natural process of removal of soil by water or wind. Over millions of years, the deposition of these materials has built up fertile plains. Soil erosion is accelerated by inappropriate land management, by land use changes such as clearance of forest and grasslands followed by cropping that provides inadequate ground cover, inappropriate tillage, overgrazing, mining and earth moving, and poor maintenance of conservation measures such as terraces. Loss of topsoil means reduction of fertility, organic matter and losses of nutrients, reduction of water holding capacity, and decline of biodiversity, all of which create irreversible changes and reduce on-site soil quality. The eroded soil is usually deposited in downslope areas or could reach the seas. The off-site costs include damage to infrastructure and sedimentation of reservoirs, streams and estuaries, and losses of hydropower generation, that might in some cases be much greater than in situ losses of farm production. Widespread attempts to mitigate soil erosion in Greece during the past century have produced mixed results.

Measurements conducted in western Greece where flysh formations are dominant, and in a number of sites in eastern part of the country indicate that the sediment load transported to dams ranges between 1,200 and 2,000 t km^2 year^{-1} (Kosmas et al., 2006). Danalatos (1993) describes complete removal of the thick dark surface of soil horizon in the hilly Tertiary landscapes of central Greece at rates 1 cm year^{-1}, while Kosmas et al. (2006) mention erosion rates of 0–52 t km^2 year^{-1} in Viotia area, and from 15 to 252 t km^1 year^{-1} in vineyards in Attica. Using ^{137}Cs technique Theocharopoulos et al. (2003) estimated erosion rates in Mouriki catchment and Viotia area in the range of 3.54–95.78 t ha^{-1}year^{-1}, while the deposition rates ranged from 1.23 to 168.19 t ha^{1}year^{-1}. It is estimated that 8% of the hilly agricultural land in Greece has been abandoned in the last decades due to diminished productivity caused by soil erosion (Kosmas et al., 2006). According to Danalatos

(1993) analytical data in Thessaly have shown a reduction of soil organic matter, from 2.6 to 1.5% only during the last 6 decades.

Erosion by wind also occurs when the wind force is greater enough to detach bare soil structural units as mainly in the Aegean islands and Crete and carry soil particles away. This depends on wind speed, surface roughness, vegetative cover, soil moisture and erodibility, which are often aggravated by changes of land use, cropping and grazing regimes.

Wild fires in forests and bushes, plant cover removal or overgrazing, climatic change, rainfall pattern and intensity, soil structure deterioration, top soil treatment, current land use, soil cultivation patterns and stable burning are the main factors accelerating erosion, decreasing soil organic matter, and accelerating nutrients leaching (Blake et al., 2009; Theocharopoulos et al., 2003, 2004) with all other consequences to soil structure, erosion, fertility etc.

Agriculture intensification has led to increased and sometimes excessive application and leaching of fertilizers (Theocharopoulos et al., 1993). The use of excessive amounts of manure, sewage sludge and pesticides (Papadopoulou-Mourkidou, 1998; Lolas, 1998), has introduced soil pollution, contamination and decline of biodiversity. Haidouti et al. (1985) reports the presence of mercury in some Greek soils. Industrial development in Greece seems to have brought about, directly or indirectly large additions of wastes and pollutants, including heavy metals and acid rain.

The use of more powerful heavy machinery has led to compaction and loss of structure, and has, despite the crop production problems, indirectly accelerated soil erosion. Acidification, with the consequent plant toxicity, occurs because of soil carbonates leaching or improper use of acidifying fertilizers or industrial emissions. The need for housing, industry, infrastructure etc especially around big cities and near the coast has given rise to the removal or loss of high quality land from its natural function through its allocation for other uses.

The 2006 report on coastal zone management of Greece mention that 28.6% of the coastline is affected by erosion, while the total urbanised coastal area is estimated to be 1.31% of the total surface are. If this be combined with the fact that 70% of the coastline is rocky it is highlighted the increased degree of urbanisation in the coastal area.

33.2.2 Water Degradation, Scarcity and Quality Deterioration

Soil and water resources are in close contact in nature and interact between them. This is the reason these should treated together. Pollutants through water diffusion, mass or bypass flow either in solution or suspension reach the soil, surface and ground water. Also, soil processes like soil erosion, leaching, macropore flow, and mineralisation of humus affect water quality.

Surface water in Greece consists of rivers, lakes, and wetlands. In the northern part of the country there are the rivers Ebros, Nestos, Strimon and Axios with most of their catchment inside Greece. While other important rivers are Aliakmon,

Aheloos, and Pinios. Natural lakes are small apart from Prespes only part of which belongs to Greece. Groundwater aquifers throughout the nation are either carbonate rocks (karstic aqifers) or coarse-grained Neogene and Quaternary (porous aquifers) deposits.

The hydrologic regime corresponds to conditions characterized by the inadequate availability of water resources and the Mediterranean hydroclimatic conditions. In Eastern regions of the country, the islands of Aegean and Crete face a critical endemic shortage of water. This is enhanced by high agriculture water consumption especially by summer crops. Many areas in Greece do not satisfactorily cover their water demands and experience problems in water supply.

The main users of water resources are agriculture (85%), followed by urban uses and industry. Irrigated land (Theocharopoulos and Panoras, 2000) is estimated to be 1,250,000 ha; 600,000 ha of which are irrigated from surface water, while the rest 650,000 ha from pumped ground water. The spatial and temporal distribution of water demand is irregular while maximum demand is recorded during July and August and in the main plains or near the coast. Water management in agriculture is not rational and needs improvement especially to reduce water loss that start from the reservoirs to the field. Demands for potable use present also a spatial and temporal distribution due to tourism. The use of ground water resources has become particularly intensive in coastal areas due to intense urbanization, tourist development and irrigated land expansion (Daskalaki and Voudouris, 2007).

Lolas (1998) reports pollution from agrochemicals while Albanis (1992) reports herbicide losses through runoff from the agricultural land of Thessaloniki. Papadopoulou-Mourkidou (1998) reports that nitrates as well as atrazine, metalachlor and alachlor were detected in 78 out of 142 ground water samples in Northern Greece. Pesticide residues were also detected in Loudia River (Papadopoulou-Mourkidou, 1998). Mitsios et al. (2000a,b) detected heavy metals in soils and irrigation water in Thesaly as well as Borium in soils and irrigation water but not in high levels. Vizantinopoulos and Lolos (1994) studied the leaching and persistence of pesticides, while Lentza-Rizoy (1996) detected triazine in two areas. Miliadis and Aplada-Sarli (1995) report pesticides residues in surface and ground waters with seasonal fluctuation. Additional sources of water pollution come from the seawater intrusion due to over exploitation of coastal aquifers, the fertilizers used in agriculture and the disposal of non properly treated urban and industrial wastewater in torrents or in old pumping wells.

Overpumping from ground water is used to irrigate summer crops and satisfy tourism needs. This has created lowering of the ground water level while such form of water is most of the times of poor quality and enhances salinisation. Seawater intrusion is enhanced because the ground water level has lowered very much below the sea level in many places. Addition of nutrients through soil erosion, runoff and leaching could create eutrofication in many water reservoirs.

According to Xanthakis et al. (2009), from a total of 236 ground aquifers in Greece, 110 of them are threatened from degradation and pollution such as not to fulfil the qualitative and quantitative requirements of the 2000/60 EC guideline.

The major risks of surface and ground water uses in Greece could be summarised as follows:

1. Pollution from industrial and urban areas
2. Point pollution from non properly treated municipal and industrial wastes
3. Pollution from agrochemicals and fertilizers
4. Poor water use efficiency and high losses of water
5. Overexploitation and intensified drainage leading to lowering of ground water levels
6. Seawater intrusion in many coastal areas

The increased needs for water quantity and the protection of water quality necessitates optimum use of water resources, protection of water quality, and sustainable irrigation management systems. This could be achieved by using limited irrigation water supplies based on crop water requirements and optimum irrigation scheduling. The situation is becoming even worse under the systematic climate change. Improvement of soil infiltration, injection of water to ground and water recycling could be some adaptation remediation measures to face these problems.

Great efforts to save water are taking place in the country. In this context, water is being recycled (Panoras et al., 2000), and pressure is growing to allow transfer of agricultural water supplies to other uses. The EU directive 2087/92 is applied in order to minimize the detrimental effect of water misuse and pollution to the environment. The Ramsar agreement is also activated for the protection of wetlands. Despite the European Water Framework directive 2000/60 which has established a new legislation for sustainable management of water resources and protection of their relevant ecosystems, it seems that there is an urgent need for establishing the best institutional and policy practice for water management, along with the creation of a lasting network of institutional research policy for enhancing the productivity of water at national and local level (Mimikou, 2005).

33.2.3 Decline of Biodiversity

The biodiversity in the form of the variety of plants and animals in the protected areas of the country are well described in Earthtrends country report for Greece. As all other Mediterranean countries, Greece too is strongly affected by unsustainable development reflected by the loss of valuable forests, agricultural, coastal and marine ecosystems that were destroyed to accommodate recreational activities at sites that were important habitats for fauna and flora. It seems that wildfires, which occur almost every summer in Greece, cause the most severe damage to the ecosystems.

In addition to wildfires, monoculture, increased use of agrochemical in agriculture, unsustainable land, forest, soil and water management, have induced soil, water and air pollution and reduced biodiversity provoking changes in the population dynamics of flora and fauna and destroy the ecological balance of the natural ecosystems. According to Article 17 Report of the National Summary for Greece the frequency of pressures and threats in agriculture and forestry are 60 and 40% for natural habitat threats.

Frequent wildfires cause changes in the forest and the other ecosystems, and affect fauna and flora population dynamics as well as soils and downstream waters (Blake et al., 2009). Forest ecosystems play a crucial role in increasing water infiltration, preventing soil erosion and runoff, maintaining soil fertility, hosting most terrestrial biodiversity, and help sequester carbon in the soil thus should be protected. To face the problems of biodiversity decline especially after forest fires there are numerous areas in Greece that need urgent reforestation programmes as proposed by Trakolis et al. (2000) for the Voras Mountain. Biodiversity research is foreseen in two out of eleven national thematic priorities of the Strategic Development Plan for Research, Technology and Innovation under the 2007–2013 NSR Framework and the Greek National Biodiversity Strategy that match the EC communication on "Halting the loss of biodiversity by 2010 and beyond".

33.2.4 Land Use Change

The main human induced contributing factor to land degradation in Greece is land use change as it is shown in Table 33.1.

These changes, that in many areas are historical, have immediate consequences on food production, soil and freshwater resources, forest resources, biodiversity, climate and air quality. On the other side fertile soils were lost to host urban and industrial development. Fires and urbanization of the population of the rural areas are some other reasons of land use change. The use of genetically modified organisms and crops for food production has not been fully investigated and they are at the moment, contested scientifically and politically in the country. Land use change surely will be introduced to face climatic change through new crops that are tolerant to drought conditions, sown later in winter or early in summer in addition to crop redistribution to face the raising temperatures. To avoid further degradation

Table 33.1 Changes in the main land use categories in Greece for the period 1971–2001 (in 000 ha)

Year	Cultivated	Pasture	Forest	Horticulture	Orchard	Vines	Fallow	Total
1971	2,532.9	–	2,967.5	114.3	698.9	219.7		6,533.4
1981	2,423.8	5,252.2	2,951.1	122.1	840.0	186.4	506.4	12,282.0
1991	2,334.4		2,937.8	123.6	924.6	151.7		6,472.1
2001	2,213.2	5,219.1		118.9	997.7	134.3	454.5	9,137.7

Source: National Statistical Service of Greece (2002)

this has to be based on land capability and land suitability principles and requirements. In this context Xanthakis et al. (2009) estimated that if 75% of the Greek land cropped by cotton was shifted to winter wheat, then the irrigation water consumption decrease would be enough to cover the potable water needs for the whole population of Greece.

33.3 Actions to Mitigate Land Degradation

The national strategy for soil and water resources considering land use planning and climate change was presented, described and discussed in a Conference organized by the Ministry of Rural Development and Food (MRDF) in Athens, 25th February 2000 and presented by Missopolinos et al. (2000). The outcomes of this event have established the main priorities for sustainable land management as follows:

Improvement and broadening of scientific and technical knowledge on soil functions, scientific knowledge dissemination, selection of best soil uses in relation to soil functions, development of the proper tools to monitor, protect and improve soil quality, development of soil quality monitoring systems, improvement of the deteriorated soil functions, mapping and evaluating soil resources, monitoring and management, research, and enacting proper legislation.

Research projects to combat land degradation are carried out by the Soil Science and Forestry Institute of NAGREF and by relative units of many Greek Universities. All stakeholders, especially policy makers, should be informed and aware of the problems of land degradation and desertification.

Effective mitigation of land degradation and desertification should start from the prevention and establishment of early warning systems. The National Committee to Combat Desertification has produced the National Action Plan (Yassoglou, 1999) where all the factors and processes leading to land degradation and desertification are described in detail. Also, the general and specific measures and actions for agricultural practices, forest and pasture management and protection of biodiversity are presented along with the necessary socio-economic measures. A number of actions such as conservation agriculture with its three pillars of no till, mulching, and proper rotation as well as conservation tillage seems to be a good approach, to protect the soil and to maintain soil fertility.

The Greek Ministry of Rural Development and Food has also produced the Code of Good Agricultural Practice (CGAP). The "Greek Action Plan" for the mitigation of nitrates in water resources of vulnerable districts such as Thessaly, Kopais (Kallergis, 1997), is also implemented. It comprises a set of measures and practices targeting the protection of surface and groundwater aquifers from nitrate pollution of agricultural origin, through the rational management of inorganic fertilizers. On the other special attention is due to the use of sewage sludge is in accordance with EU Council Directive 86/278. Land protection is being implemented in the Framework of Reg (EU) 1257/99, which incorporates protection of the environment in the rural development areas. The 2000/60/EC Water Protection Directive has been into practice as well and will certainly contribute towards reversing land degradation trends.

Subsidies are offered to farmers who comply with good agricultural practices that are friendly to the environment. They include reduced nitrogen inputs, anti-erosion measures such as ploughing across slope, build or preserve terraces, extensification of animal husbandry, organic farming or organic husbandry, long time set aside land, protection of sloping landscapes etc. Through the integrated crop management systems, productive soils are protected from agrochemical residues and through water protection measures to conserve various lake ecosystems throughout the country.

The EU, through its new Common Agricultural Policy (CAP) and its Soil Thematic Strategy incorporates environmental issues in rural development and will contribute towards stopping and mitigating land degradation. A harmonized Land Degradation Monitoring System has to be established in Greece, based on new technologies with minimum sets of parameters/indicators, with quality control/assurance procedures for soil, water and biodiversity sampling, treatment and analysis and with traceability rules to be followed. Harmonized Guidelines for sustainable land use should be formulated too.

33.4 Conclusions

Land degradation is a serious problem in Greece. It needs proper actions and measures, education and public awareness. The country should consider the current threats and should update and implement its National Action Plan to combat desertification and land degradation. Harmonized guidelines for the sustainable use of land and protection of soil, water and the biodiversity should be formulated. A harmonized monitoring system for land degradation, based on remote sensing and expert based assessments with a minimum set of land quality indicators, needs to be developed. Finally, necessary legislation should be developed, approved and implemented.

Acknowledgements Thanks are expressed to Dr. M. Ntoula for reviewing this paper in the early stages. Many thanks to Professor N.I. Yassoglou for providing Fig. 33.1.

References

Albanis, T.A. (1992). Herbicide losses in runoff from the agricultural area of Thessaloniki in Thermaikos Gulf, N. Greece. Science of the Total Environment 114:59–71.
Blake, W.H., Theocharopoulos, S.P., Skoulikidis, N., Clark, P., Tountas, P., Hartley, R. and Amaxidis, Y. (2009). Impacts of wildfire on particulate phosphorus yields and bioavailability: Evidence from the burnt Evrotas River basin, Greece. Environmental Science and Technology in press.
CORINE. (1992). Soil erosion risk and important land resources. In the Southern Regions of the European Commission. EUR 13232.
Danalatos, N.G. (1993). Quantified Analysis of Selected Land Use Systems in the Larissa Region, Greece. Ph.D. Thesis. Agricultural University of Wageningen, Wageningen, 370p.

33 Land Degradation in Greece

Daskalaki, P. and Voudouris, K. (2007). Water resources of Greece. Environmental Geology 54(3):505–513.

Davidson, D.A. and Theocharopoulos, S.P. (1992). A survey in soil erosion in Viotia, Greece. In: Bell, M. and Boardman J. (eds.), Past and Present Soil Erosion. Oxbow Monograph 22. Oxbow Books, Oxford, pp. 149–154.

Haidouti, K., Skarlou, B. and Tsouloucha, F. (1985). Mercury content of some Greek soils. Geoderma 35:251–256.

Kallergis, Y. (1997). Vulnerable Zones to Agricultural Origin Nitrates Pollution. Department of Geography, University of Patras, Greece.

Kosmas, C., Danalatos, N. and Kosmopoulou, P. (2006). Soil erosion in Greece. In: J. Boardman and J. Poessen (eds.), Soil Erosion in Europe. John Wiley & Sons, Chichester, pp. 279–288.

Lentza-Rizoy, C. (1996). Determination of Triazine residues in water: Comparison between a gas chromatographic method and an enzyme-linked immunosorbent assay (ELISA). Bulletin of environmental contamination and toxicology 57:413–420.

Lolas, P. (1998). The fate of pesticides in the environment after their application. Proceedings of the 2nd Hellenic Plant Protection Meeting, Larissa, pp. 67–79, 5–7 May, 1998.

Lyrintzis, G. and Papanastasis, V. (1995). Human activities and their impact on land degradation-psilorites maintain in Crete. A historical perspective. Land degradation and rehanilitations 6:79–93.

Miliadis, G.E. and Aplada-Sarli, P. (1995). Detection of pesticides in surface and ground waters and seasonal concentrations. Agriculture-Animal Husbandry 4:38–42.

Mimikou, M.A. (2005). Water resources in Greece: Present and future. Global NEST Journal 7(3):313–322.

Missopolinos, N., Alifragis, D., Asimakopoulos, G., Grougou, S., Zalidis, G., Theocharopoulos, S., Mpalis, K., Panagiotopoulos, K., Stamatiadis, S., Sylleos, N., Tsantilas, C. and Haidouti, K. (2000). National Strategy for the soil Resources. MRDF (Gr).

Mitsios, I.K., Gatsios, F.A. and Floras, S.A. (2000a). Irrigation water evaluation, salinity and sodicity problems in soils of Magnesia region. Proceeding of the 2nd National Conference of Agricultural Engineering, pp. 305–319 (Gr,e).

Mitsios, I.K., Golia, E. and Christodoulou, E. (2000b). Determination of heavy metals in soils and irrigation water of Thessaly region. Proceeding of the 2nd National Conference of Agricultural Engineering, pp. 271–280 (Gr,e).

Panoras, A., Ilias, A., Skarakis, G. and Zdragas, A. (2000). Suitability of reclaimed municipal wastewater for sugar beet irrigation. Proceedings of the 5th International Congress on Environmental Pollution, Thessaloniki Greece, pp. 221–232 (Gr,e), August 28–September 1.

Papadopoulou-Mourkidou, E. (1998). Pollution of ground and surface waters from plant protection products in Greece and Europe. Proceedings of the 2nd Hellenic Plant Protection meeting, Larisa, pp. 153–168 (Gr,e), 5–7 May, 1998.

Theocharopoulos, S.P. (2007). Soils of Greece: Status, problems and solutions. In P. Zdruli, G. Liuzzi (eds.). Proceedings, Status of Mediterranean soil resources: Actions Needed to Support Their Sustainable Use. MEDCOASTLAND Conference, Tunis, pp. 145–164, 26–31 May 2007.

Theocharopoulos, S.P. and Aggelides, S., (1991). Current threats to soils and ecosystems in Greece. In J.M. Hodgson (ed.), Soil and Groundwater Research Report I, "Soil Survey -A Basis for European Soil Protection". Proceedings of the EEC Meeting of European Heads of Soil Survey. Commission of the European Communities, Silsoe, UK, pp. 157–162, 11–13 December 1989.

heocharopoulos, S.P., Christou, M., Touloucha, F., Tountas, P., Petrakis, P., Caramados, V., Kouloubis, P., Spyropoulou, I. and Ntoula, M. (2004). Soil erosion studies in central Greece based on rainfall simulator: Effect of soil type, rainfall intensity, soil slope and topsoil treatment. International Journal of Agricultura Mediterranea 134(3–4):165–177.

Theocharopoulos, S.P., Florou, H., Kalantzakos, H., Walling, D., Christou-Karayianni, M., Tountas, P. and Nikolaou, T. (2003). Soil erosion and redistribution rates in a cultivated catchment in central Greece, estimated using the Cs-137 technique. Soil and Tillage Research 69:153–162.

Theocharopoulos, S.P., Karayianni, S., Gatzogianni, P., Afentaki, A. and Aggelides, S. (1993). Nitrogen leaching from soils in the Kopais area of Greece. Soil Use and Management 9(2): 76–84.

Theocharopoulos, S.P. and Panoras, A. (2000). Protection and Management of Soil and Water Resources, Proceedings of the Mediterranean Conference for Agricultural Research Cooperation on "The Mediterranean nutritional model and cooperation to promote the international trade of Mediterranean agricultural products", Athens, Greece, pp. 248–268, 1–2 December 2000.

Trakolis, D., Platis, P. and Meliadis, A. (2000). Biodiversity and conservation actions on Mount Voras, Greece. Environmental Management 26(2):145–151.

Vizantinopoulos, S. and Lolos, P. (1994). Persistence and leaching of the herbicide Imazapyr in soil. Bulletin of environmental contamination and toxicology 52:404–410.

Xanthakis, E., Basilakos, Y., Mpoukis, Y., Tsiforos, Y., Smiris, M., Sofianos, X., Kontopoulos, G., Kamaras, G., and Mpotzios, A. (2009). Study of application of an irrigation water management model in Greek Agriculture. INASO (Gr).

Yassoglou, N.I. (1987). The production potential of soils Part II: Sensitivity of the soil systems in Southern Europe to degrading influxes. In: H. Barth and D. L'Hermite (eds.), Scientific Basis for Soil Protection in the European Community. Elsevier, London, pp. 87–122.

Yassoglou, N.J. (ed.). (1999). The Greek action plan for combating desertification. A document submitted to the Greek Government. (Gr, E).

Yassoglou, N. (2005). Soil Survey in Greece. European Soil Bureau, Research Report No 9, p. 159–168.

Chapter 34
Factors Influencing Soil Organic Carbon Stock Variations in Italy During the Last Three Decades

M. Fantappiè, G. L'Abate, and E.A.C. Costantini

Abstract Soils contain about three times the amount of carbon globally available in vegetation, and about twice the amount in the atmosphere. However, soil organic carbon (SOC) has been reduced in many areas, while an increase in atmospheric CO_2 has been detected. Recent research works have shown that it is likely that past changes in land use history and land management were the main reasons for the loss of carbon rather than higher temperatures and changes of precipitation resulting from climate change. The primary scope of this work was to estimate soil organic carbon stock (CS) variations in Italy during the last three decades and to relate them to land use changes. The study was also aimed at finding relationships between SOC and factors of pedogenesis, namely pedoclimate, morphology, lithology, and land use, but also at verifying the possible bias on SOC estimation caused by the use of data coming from different sources and laboratories. The soil database of Italy was the main source of information in this study. In the national soil database is stored information for 20,702 georeferentiated and dated observations (soil profiles and minipits) analysed for routine soil parameters. Although the observations were collected from different sources, soil description and analysis were similar, because all the sources made reference to the Soil Taxonomy and WRB classification systems, and soil analyses followed the Italian official methods. Besides horizon description and analysis, soil observations had a set of site information including topography, lithology, and land use. The SOC and bulk density referred to the first 50 cm, thus CS was calculated on the basis of the weighted percentage of SOC, rock fragments volume, and bulk density. A set of geographic attributes were considered to spatialize point information, in particular, DEM (100 m) and derived SOTER morphological classification, soil regions (reference scale 1:5,000,000) and soil systems lithological groups (reference scale 1:500,000), soil moisture and temperature regimes (raster maps of 1 km pixel size), land cover (CORINE project, reference scale 1:100,000) at three reference dates: years 1990 and 2000, and an original

---

E.A.C. Costantini (✉)
CRA-ABP, Research Centre for Agrobiology and Pedology, 50121 Florence, Italy
e-mail: edoardo.costantini@entecra.it

update to 2008, obtained with field point observations. The interpolation methodology used a multiple linear regression (MLR). CS was the target variable, while predictive variables were the geographic attributes. Basic statistical analysis was performed first, to find the predictive variables statistically related to CS and to verify the bias caused by different laboratories and surveys. After excluding the biased datasets, the best predictors were selected using a step-wise regression method with Akaike Information Criterion (AIC) as selection and stop criterion. The obtained MLR model made use of the following categorical attributes: (i) decade, (ii) land use, (iii) SOTER morphological class, (iv) soil region, (v) soil temperature regime, (vi) soil moisture regime, (vii) soil system lithology, (viii) soil temperature, (ix) soil aridity index (dry days per year), and, (x) elevation. The interaction between decade and land use variables was also considered in the model. Results indicated that CS was highly correlated with the kind of main type of land use (forest, meadow, arable land), soil moisture and temperature regimes, lithology, as well as morphological classes, and decreased notably in the second decade but slightly increased in the third one, passing form 3.32 Pg, to 2.74 Pg and 2.93 Pg respectively. The bias caused by the variables like "laboratory" and "survey source" could be as large as the 190%.

Keywords Carbon sequestration · Land use change · Factor of pedogenesis · Multiple regression

34.1 Introduction

Almost all European countries have ratified the Kyoto Protocol to reduce greenhouse gas (GHG) emissions for the period 2008–2012 by 6.5% compared to the 1990 level. Article 3.4 of the protocol indicates soil management as a carbon sequestering strategy to help achieve the emission reduction target (Morari et al., 2006). Sequestering carbon in soil is also beneficial to enhance soil quality: soil organic carbon (SOC) is a major indicator of soil quality and sustainability (Reeves, 1997). The communication of the European Commission "Towards a Thematic Strategy for Soil Protection" (COM 179, 2002; COM 231, 2006) as well as other documents (European Commission, 2008, 2009) points to soil organic matter decrease as one of the main European soil threats.

Both forestry and agricultural soils may be considered as carbon sinks according to the Kyoto Protocol. Agriculture and farming activities do approximately contribute 25% of the global GHG emissions. In Europe this figure is approximately 10%, excluding emissions due to land use change. Soils with high initial carbon contents are more prone to losses than soils with already low carbon content (Kätterer et al., 2004) assuming "high" SOC content values such as 2–3.4% and "low" SOC at <2%. Post and Kwon (2000) estimated that land use changes from arable cropping to grassland resulted in increases in soil carbon of 33 g C m^{-2} year^{-1}, although rainfall and the species sown in the new pastures could affect the rate substantially.

The flux exchange of CO_2 between soil and the atmosphere is also so large that it has been estimated at 10 times the flux of carbon dioxide from fossil fuels (Schils et al., 2008). If soil respiration, associated with decomposition and root activity, accounts for two thirds of carbon lost from terrestrial ecosystems (Luo and Zhou, 2006), recent research results (Kirk and Bellamy, 2008; Bouwman, 2001; Marland et al., 2003; West and Post, 2002; West and Marland, 2003; West et al., 2008) have shown that it is likely that past changes in land use history and land management were the dominant reasons for the soil carbon losses. Actually, land use changes, more than increased temperatures and changes of precipitation, resulted in an emission of nearly 2 Pg C year^{-1} during the 1990s at the terrestrial scale (Schimel et al., 2001; IPCC, 2001a, b). Costantini et al. (2007) pointed to the poorer organic matter content of Italian soils cultivated with row crops and/or vineyards and olive grooves, in comparison with vegetables, orchards and mixed cultivations, as well as the differences between irrigated crops compared with rainfed cultivations.

There are still many uncertainties and unanswered questions related to the issue of carbon sequestration, such as the relationships with the factors of pedogenesis, the size of sink and its accounting. Statistical analyses of spatially distributed soil samples provide information on changes in soil carbon pools when the measurements are taken at two points in time (Bellamy et al., 2005) or are from a chronosequence (simultaneous measurement at sites with different histories of change behind them, Covington (1981)), but such monitoring activity is absent in most European countries.

The only European region with "true" resampling data is England and Wales, where 40% of the original sites on a 5 × 5 km grid were resampled with an interval of 15–25 years (Bellamy et al., 2005; Bellamy, 2008). These authors reported on soil organic carbon changes in UK and Wales over the period 1978–2003. On the basis of data from the two samplings it was estimated that carbon was lost from soils across England and Wales over the survey period at a mean rate of 0.6% year^{-1} (relative to the existing soil carbon content in 1978). This estimate was based on the soil carbon content of the top 15 cm of soil. Converting this to carbon stocks, using a pedotransfer function to estimate bulk density, it was estimated that the soils of England and Wales were losing carbon at the rate of 4.44 Tg C year^{-1}. However, Smith et al. (2007a, b, c) and Smith (2008), using two soil carbon models, suggested that only 10–20% of the loss of carbon from soils in England and Wales reported by Bellamy et al. (2005) could be due to climate change. Moreover, recent studies have shown that it is likely that past changes in land use history and land management were the dominant reasons behind carbon losses rather than higher temperatures and changes of precipitation as result of the climate change (Kirk and Bellamy, 2008). Changes in bulk density over time, as well as precision and success rate of actual soil resampling, were acknowledged as more likely factors that dominated the observed changes of soil carbon.

In France, INRA has reported on measured carbon stocks in the top 0–30 cm layer. All data between 1970 and 2000 for different land uses have been pooled and used as an average value for 1990 stock of C (Arrouays et al., 2001, 2002a, b). The carbon stocks in the upper 30 cm of soils in France should vary from 15 to

40 Mg hm^{-2} in mid France, to 40–50 Mg hm^{-2} in the richer and more intensive cropping areas in the north and south–west, up to 70 Mg hm^{-2} in permanent grassland and forest, and >90 Mg hm^{-2} in more mountainous areas and wetlands (Arrouays et al., 2001; IFEN, 2007). The highest values are reported in organic soil at 350 Mg hm^{-2}. Soils that are under forest, grassland or pasture always have higher organic carbon stocks than identical soils under arable land. IFEN (2007) reported losses of carbon for soils in some regions and increases of soil carbon in other regions for agricultural soils in France.

The main difficulties with soil carbon monitoring are the large amount of work needed, and consequently high costs, plus the challenge to keep the study methods adequately similar between the monitoring periods. Combining modelling with monitoring can reduce the amount of work and the costs. Soil carbon stock (CS) estimation is also affected by many factors of uncertainty. For instance, depth of ploughing has changed over time. This change is hardly recognized in analysis of trends of the stock of organic C in soils (Schils et al., 2008). Increased temperatures may cause a not-linear carbon loss in combination with extreme drought, as reported in information derived from eddy-covariance studies across Europe in 2003 (Ciais et al., 2005; Reichstein et al., 2006). As bulk density and organic carbon are correlated, and as changes in bulk density may induce changes in the mineral mass of soil collected down to a given depth, it would be needed to have determined bulk density on all sites, with comparable methods, that is rarely the case in most databases.

Moreover, there are the sampling and laboratory biases. Although analysed with the same method, data coming from different laboratories and surveys could vary notably, for various reasons (Giandon, 2000; Ogle et al., 2006; Neff et al., 2002; Lal et al., 2001, 2008; Lal, 2008). In laboratory sources of bias are, for instance, sample handling and pre-treatments (exclusions of living roots, straws, intensity of grinding, etc.), which can be performed differently according to local protocols. In field sampling performed in different parts of the ploughed horizon can notably affect the SOC content, especially when, like in many parts of Italy, ploughing depth reaches 50 cm and more.

Also the particular time of sampling that could be soon after ploughing, or during crop vegetation, or after the harvest, can influence the bulk of the sample. The reference depth causes another important source of variability. In fact, as most soils are sampled at different depth, according to genetic horizons, the SOC content comes from the weighted averaging of the possible multiple analyzed sub-horizons within the reference depth (Franzluebbers, 2002).

Zdruli et al. (1999) made a first estimation of SOC content for Italy for the depth of 0–30 cm as part of the European Mediterranean SOC estimation at 1 × 1 km grid, on the basis of the European Soil Database. Jones et al. (2005) improved the previous estimates and provided a map of percentage SOC for the same depth. Other recent studies (Vitullo, 2006; Pilli et al., 2006) have estimated for instance the CS for forest soils in Italy. Some regional experiences have also been attempted to estimate CS in Emilia Romagna (Guermandi, 2005; Calzolari and Ungaro, 2005; Gardi, 2005), Piedmont (Petrella and Piazzi, 2005; Piazzi, 2006; Stolbovoy et al., 2006),

Lombardy (Solaro and Brenna, 2005; Cerli et al., 2009), Veneto (Dalla Valle, 2008; Garlato et al., 2009a), Trentino (Garlato et al., 2009b).

The present research work described in this paper was aimed at estimating CS variation in Italy during the last three decades and to relate it to land use changes. The study was also aimed at finding relationships between SOC and the factors of pedogenesis, namely pedoclimate, morphology, lithology, and land use, and at verifying the possible biases on SOC estimation caused by the use of data coming from different survey samplings, times and laboratories.

34.2 Materials and Methods

34.2.1 Methodological Approach

There are different methodological approaches to estimate soil carbon changes. A first distinction could be made between empirical methods that are based on sampling, and deterministic methods, based on theoretical models, derived from previous studies. A deterministic method is used in the procedures for estimating SOC changes under the Kyoto Protocol, in the International Panel on Climate Change report "Good Practice Guidance for LULUCF" (IPCC, 2003, 2007). Within empirical methods, a further division can be made on the basis of sources: data can come from either specific monitoring activities, or from existing databases. Using monitored data is possible to determine the sample design, to select the soil horizons to be studied, and to ensure the repeatability of sampling and laboratory measurements. In the case of data coming from existing different databases, the above parameters have to be checked before using the data itself.

Whichever the source of the data, they can be interpolated using pure statistical, geostatistical, or mixed approaches. In the statistical approach, data coming from more densely populated "external" datasets are combined with SOC measurements to obtain a statistical model of correlation, which is used to interpolate SOC content (Batjes, 2008; Geissen et al., 2009; Grimm et al., 2008; Hirmas et al., in press; Hoyos and Comerford, 2005; Meersmans et al., 2008; Nyssen et al., 2008). In the statistical approaches external datasets usually refer to the factors of pedogenesis (Jenny, 1941). Remote sensed data can also be added in the regression models (Gomez et al., 2008; Huang et al., 2007; Sankey et al., 2008; Vasques et al., 2008). In the pure geostatistical approach, both the SOC measurement and its localization are considered to obtain a spatial autocorrelation model, which is used for the spatialization. The geostatistical approach can be used to incorporate dense secondary information by means, for instance, of cokriging, multicollocated cokriging, or multicollocated cokriging with varying local mean (Castrignanò et al., 2009).

There are various mixed approaches available, but all of them consider a combination of target data autocorrelation and "external" effects (McBratney et al., 2003; Chai et al., 2008; Grunwald, 2009; Carrè et al., 2007; Simbahan et al., 2006). Hence, mixed approaches can add the geographical position as another factor of pedogenesis (SCORPAN model, McBratney et al., 2003). The external datasets can be combined with the spatial autocorrelation of residuals in different ways (e.g. regression kriging and kriging with external drift).

As previously stated, Italy is lacking a monitoring system of SOC, so we made use of the data collected in the national soil database, coming from different surveys and completed in different times. The inherited sample design was then random, with a great inhomogeneous spatial and temporal distribution of samples. Uniformity of soil horizons and repeatability of sampling and laboratory measurements were checked before performing the interpolation analysis.

Data stratification was made *ad posteriori*, attributing to the measured SOC content the "external" information coming from the different geographic attributes. The resulting table could then be used for basic statistic analysis, as well as to obtain the interpolation map. Therefore, our spatialization model can be considered a pure statistical approach, relating SOC to the soil forming factors.

Soil survey datasets were classed in 3 decades: between 01/01/1979 and 31/12/1988; between 01/01/1989 and 31/12/1998; between 01/01/1999 and 31/12/2008. The grouping was aimed at overlapping the times of land use/land cover databases, so that it could reflect the relevance of land use changes in SOC content and CS.

34.2.2 Data Sources and Data Preparation

The national soil database (Costantini et al., 2007) was the main source of information for SOC content and bulk density. The national soil database stores information of about 40,068 observations (soil profiles and minipits), 22,517 analyzed for routine and non-routine parameters, and 20,702 observations georeferentiated and dated (date of survey). The 20,702 observations were distributed rather unevenly in the last three decades (1979–1988: 1,676 observations; 1989–1998: 12,063 observations; 1999–2008: 6,963 observations). Although the observations were collected from different sources, soil description and classification were similar, because all the sources made reference to the Soil Survey Manual (USDA, Soil Survey Staff, 1983 and later versions), the Soil Taxonomy (Soil Survey Staff, 1975 and later versions) and the FAO-UNESCO soil classification (1974) and WRB (IUSS-ISRIC-FAO, 1998). Soil analyses always followed the Italian official methods (MIPAAF, 1992; Sequi and De Nobili, 2000). In particular, SOC content was determined using the Walkley-Black official procedure (1934). In this work, the values were converted to ISO (ISO14235) using the formula proposed by the ECALP project (Ecopedological Map of Alps, 2004–2006) of the European Soil Bureau (Garlato et al., 2009b):

$$SOC\_iso = 0.0763 + 1.0288\ SOC\_wb\ (R^2\ of\ 0.9763)$$

where SOC_iso is the estimation of SOC analysed with ISO (ISO14235) and SOC_wb is the SOC analysed with Walkley-Black.

We referred SOC and bulk density to the first 50 cm, which comprehend the plough layer, in agricultural soils, and the organic-mineral horizon (A horizon), in forest soils. In the elaboration, SOC of all A horizons with upper boundary within 50 cm from the mineral soil surface, and of any other type of soil horizon, except of O, Oh, Of, Oi and C, with lower boundary within 50 cm from the mineral soil surface, were expressed as percentage by weight (dag kg^{-1}). In the case of presence of more than one data of SOC content at the same location, for example in the case of more than one A horizon with upper boundary within 50 cm, one single data was obtained by weighted horizon thickness.

The database had also information about rock fragments content (daL m^{-3} of topsoil) and measured soil bulk density (Mg m^{-3}). However, only 37.5% of soil observations had measured bulk density, so the dataset was completed using a pedotransfer function, which related bulk density to the amount of clay, silt, OC, and CEC (Pellegrini et al., 2007). The CS was then calculated with the formula:

$$CS = D^* SOCcontent^* FEF^* BD$$

where CS is the carbon stock of topsoil (first 0.5 m from mineral soil surface) expressed as Mg hm^{-2}, D is the topsoil depth expressed as m, SOCcontent is the soil organic carbon content expressed as dag kg^{-1} of fine-earth fraction, FEF is the fine-earth fraction expressed as daL m^{-3}, BD is the bulk density expressed as Mg m^{-3}.

The national soil database is a geographical database, with geographical information such as the soil regions (Righini et al., 2001; Costantini et al., 2007) and soil systems of Italy (Costantini et al., 2003). The map of soil regions is the first informative level for the soil map of Italy and the tool for the soil correlation at the continental level. Soil region is a regionally restricted part of the soil cover characterized by a typical climate and parent material association, with reference scale 1:5,000,000. Soil regions were delineated according to the criteria of the Manual of Procedures Version 1.0 for the Georeferenced Soil Database of Europe (Finke et al., 1998). "Soil systems of Italy" is a national soil database with reference scale at 1:500,000. The geographical database contains information about physiography, morphogenetic processes, river drainage network, lithology, land cover, and land components of the soil systems. A "land component" is a specific combination of morphology, lithology, and land cover of the soil system, with indication of the dominant soil typological units (STU). All soil observations of the national soil database are related to the geography of soil systems by means of the STU to which they belong. Major landforms of the land systems follow the SOTER methodology (FAO, 1995).

In this study, the factor of pedogenesis relief was taken into account considering the SOTER morphological classes, which legend is summarized in Table 34.1. A SOTER morphological raster map of Italy was produced using the Digital Elevation Model of Italy at 100 m. SOTER morphological classes were further grouped in classes as follows:

Table 34.1 SOTER physiographical classification

Physiography and elevation (m a.s.l.)	Low hills (0–200)	Medium hills (200–300)	Medium hills (300–400)	High hills (400–600)	Low mountain (600–1,500)	High mountain (1,500–3,000)
Slope (%)						
0–2	LP1	LP1	LP2	LP2	LL1	LL2
2–8	LF1	LF2	LF2	LF3	RL1	RL2
8–15	SH1	SH2	SH2	SH3	SU1	SU2
15–30	SH1	SH2	SH2	SH3	SM1	SM2
30–60	TH1	TH2	TH2	TH3	TM1	TM2
>60	VH1	VH2	VH2	VH3	VM1	VM2

(a) LP1 and LF1. Levelled lowlands
(b) LF2, SH1, and SH2. Medium and low rolling hills
(c) LP2, LF3, and SH3. High rolling hills
(d) TH1, TH2, VH1 and VH2. Steep low hills
(e) TH3 and VH3. Steep high hills
(f) LL1, RL1, SU1, SM1, TM1, and VM1. Low mountain
(g) LL2, RL2, SU2, SM2, TM2, and VM2. High mountain

To account for the possible influence of soil parent material in SOC stocks, the lithological attributes of soil systems of Italy were grouped as follows:

(a) Marine sediments, Aeolian deposits, coastal and deltaic deposits, calcarenites and residual soil deposits;
(b) Alluvial and lacustrine deposits, clayey formations;
(c) Effusive and volcanoclastic formations, rudite, sandstone, metamorphic schist, clayey sandstone, marls and marly-pelitic turbidite;
(d) Lagoons and slope deposits;
(e) Calcareous and dolomitic rocks, intrusive and metamorphic non-schist rocks.

The influence of climate was taken into account by classifying the soil moisture and temperature regimes of the observation. The USDA Soil Taxonomy was the reference classification (Soil Survey Staff, 1999). The soil attribute was estimated using an original methodology based on the EPIC software (Costantini et al., 2002, 2005). The dry xeric soil moisture regime, postulated by Van Wambeke (1986), was also considered for a more detailed qualification of the driest pedoclimate in the Mediterranean environment. Maps of soil moisture and temperature regimes (pixel size 1 km) were produced by ordinary kriging of soil moisture and temperature regimes of the observations (L'Abate and Costantini, 2004). The raster maps were then transformed in vectors.

The control of land use on SOC was obtained considering the CORINE Land cover Maps of 1990 and 2000 (Sinanet, 2009), and an update to the year 2008 obtained with field point observations: 9,276 georeferenced point field information on land cover came from the LUCAS project (Land Use Land Cover Annual Survey, European Communities, 2003), and 65,536 from the SIN database (Sistema Informativo Agricolo Nazionale, 2009). A new, specific dataset was produced as revised CORINE land cover layer for the last decade (CORINE, 2009). CORINE polygons were not modified; only land cover attribution was corrected. Land cover classes were further grouped in three great classes: (i) arable land, (ii) forest, (iii) permanent meadow.

Beside the categorical predictive variables listed above, some continuous predictive variables were also considered: (i) the DEM of Italy at 100 m, and derived slope; (ii) the raster maps of soil temperature at 50 cm, and the soil aridity index (dry days per year) (Costantini and L'Abate, 2009; Costantini et al., 2009).

34.2.3 Data Selection

The data stored in the national soil database referred to soil samples collected in different surveys, various pedologists, and analysed in different laboratories. To check the presence of possible main biases in the SOC datasets, the values of the 5 datasets storing the largest amount of data (named A, B, C, D, and E) were compared to all the other datasets of the same soil region, analysed during the same decade and in the same land use class. Datasets that were significantly different from all the others were excluded from the successive elaborations. The significance of the differences between the means was tested with the t of Student statistic test.

34.2.4 Data Elaboration

The spatialization model considered the SOC content as dependent variable and the geographic attributes, as well as the decade of survey, as predictive variables. Geographic attributes were elevation, slope, soil region, soil system, lithology, soil moisture and temperature regimes, and land use at the date of the survey.

Basic statistic analysis was first performed to investigate the relationship between predictive and dependent variables. An analysis of variance was made to statistically compare the SOC content of samples, classed according to the different attributes, referring to the factors of pedogenesis.

A multiple linear regression analysis was then performed (MLRA) and the best predictors and combination of predictors were selected using a stepwise regression analysis with Akaike Information Criterion (AIC) as selection and stop criterion (Sakamoto and Akaike, 1978). As predictors were both categorical and continuous, values of the continuous were standardized using the formula $z = (x - \mu)/\sigma$.

On the basis of the MLRA model obtained, the categorical and continuous variables selected were used to obtain 3 maps of carbon stocks, one for each decade.

An estimation error analysis was also performed to derive the uncertainty of the prediction. A selection of biased data was then interpolated separately with the same method, to highlight the differences in the maps obtained with the spatial interpolation of biased and unbiased data.

34.3 Results and Discussion

34.3.1 Soil Organic Carbon and Factors of Pedogenesis

Taking into account the bulk of data, SOC content varies significantly according to soil temperature and moisture regimes (Figs. 34.1 and 34.2). The passage from the Mesic to the Thermic soil temperature regime comports a highly significant decrease of SOC of more than 0.35 dag kg^{-1}, meaning a relative lowering of more than 20%. Similarly, the passage between soil moisture regimes (SMR), from the Udic to the Ustic, Xeric, and dry Xeric, reveals a strong influence of soil humidity on the SOC content. As expected, the soils with a higher SOC content are located in the Udic soil moisture regime, while the passage to Ustic is underlined by a relative decrease of about 25%. A smaller, but always significant decrease marks the difference between the soil with Ustic and Xeric SMR, while the SOC content of soils with dry Xeric regime show the lowest values.

The morphological control on the SOC content is also evident (Fig. 34.3). The data evidence a clear increase sequence from plains to hills and mountains and with

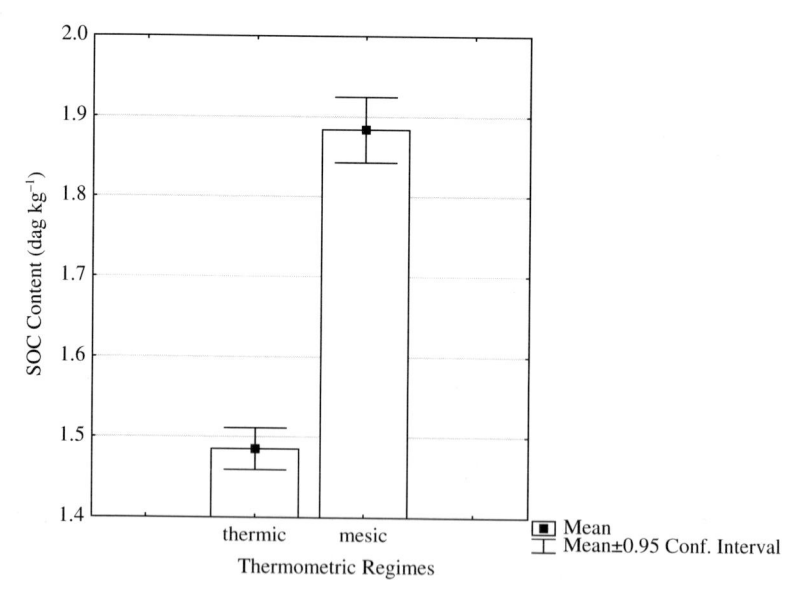

Fig. 34.1 Soil organic carbon content in the main soil temperature regimes of Italy. Differences between means are statistically different ($P < 0.01$)

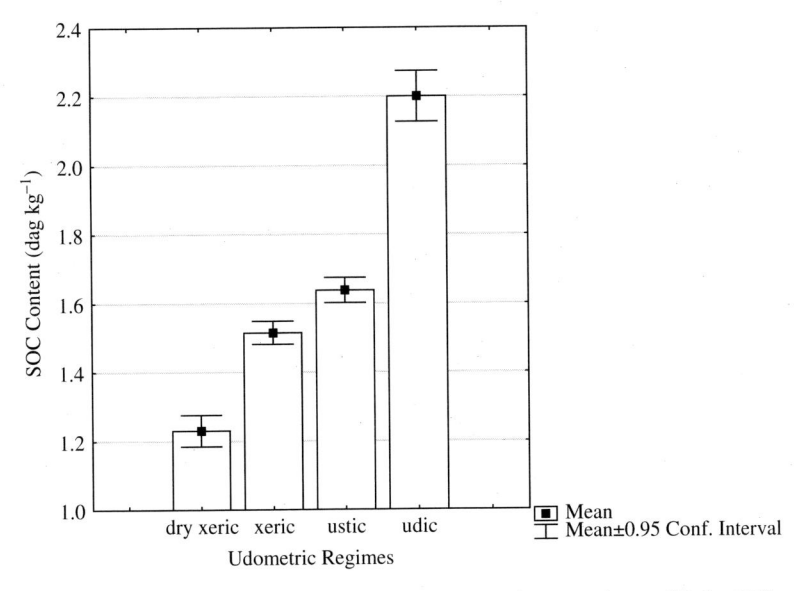

34 Factors Influencing Soil Organic Carbon Stock Variations

Fig. 34.2 Soil organic carbon content in the main soil moisture regimes of Italy. Differences between means are all statistically different ($P < 0.01$)

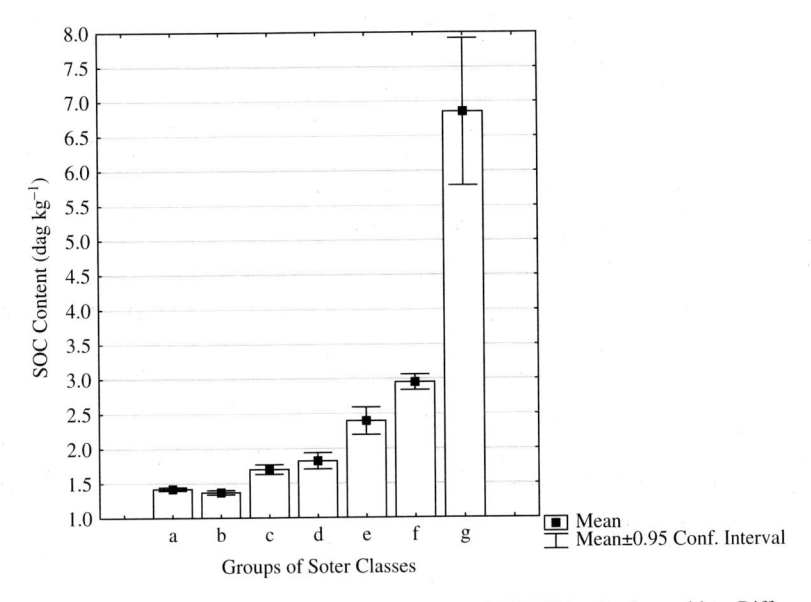

Fig. 34.3 Soil organic carbon content in the groups of SOTER's physiographies. Differences between means are all statistically different ($P < 0.01$, or $P < 0.05$, between *a* and *b* classes), except for the difference between *c* and *d* classes

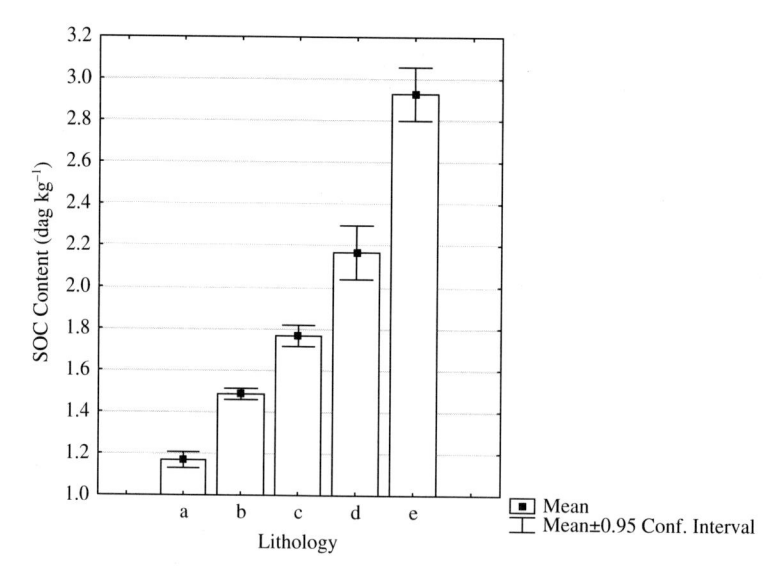

Fig. 34.4 Soil organic carbon content in the main lithological groups of Italy. Differences between means are all statistically different ($P < 0.01$)

the increase of steepness. The only exception is the passage from the first to the second class, that is, from levelled lowlands and rolling low hills, where there is a significant decrease. The interpretation is that the effect of morphology on SOC is mediated by the intensity of cultivation of arable lands, which decreases with elevation, where forests and meadows increase, and by climate, as the moister and colder climate enhances soil carbon sequestration. The inverse trend found at the passage from the first and second class can be explained considering that rolling low hills are, as a whole, intensively cultivated in Italy, and the slope of the cultivated fields may trigger soil water erosion.

Lithology influences significantly SOC content, although not as much as morphology (Fig. 34.4). Apart from lagoon and slope deposits, where the high SOC can be related with the presence of peat or organic matter rich deposits, the trend would point to a direct influence of the coherence and hardness of the substratum on SOC. In this case, the lower weathering rate of the rock would favour the organic matter accumulation in the first soil horizons. In addition, carbonate rocks evidence SOC enrichment. It is also reasonable to postulate an interaction with land use, as the harder the rock, the less intensive the agro-system, as well as with climate, in the passage from the lithological classes a and b, characterizing plains and hills, and c and e, typical of mountains.

The prominent and straightforward relationship between SOC and land use is evidenced in Fig. 34.5. The transition from arable land to permanent meadow is reflected with increase of SOC content that almost doubles, and triples in forests. However, if the relationship between SOC and land use is clear and simple, the influence of the soil forming factor time is not linear. The data reported in Fig. 34.6

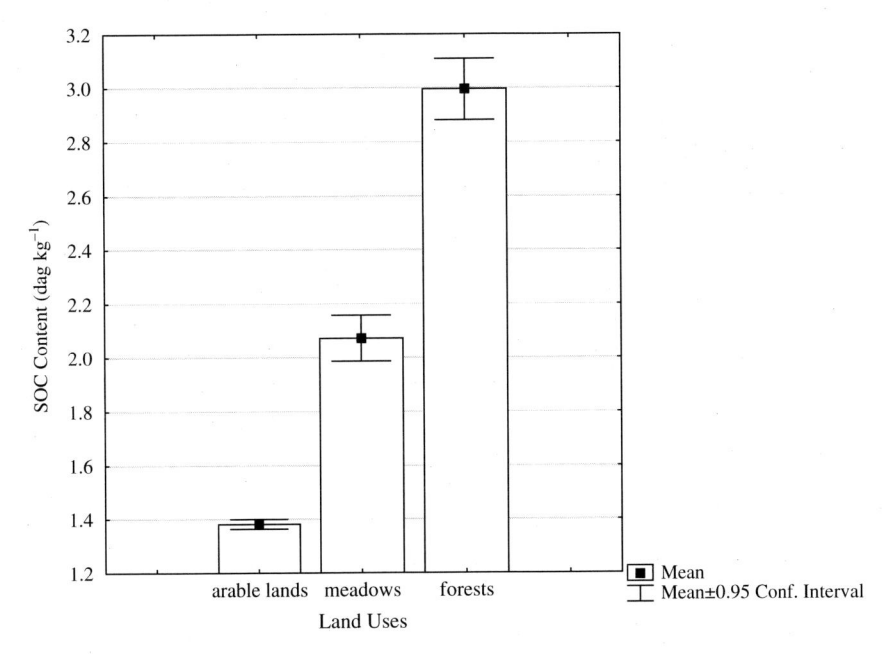

Fig. 34.5 Soil organic carbon content in the main land uses of Italy. Differences between means are all statistically different ($P < 0.01$)

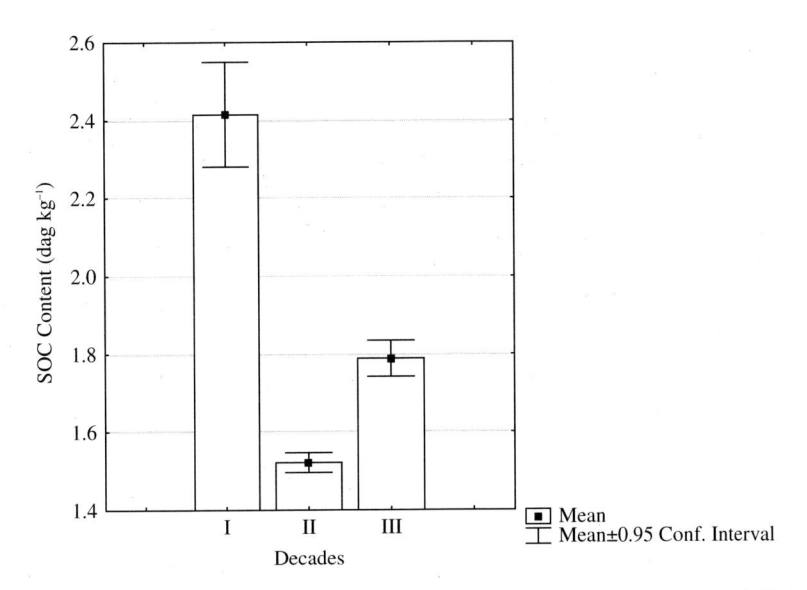

Fig. 34.6 Soil organic carbon content in the three decades considered (I: 1979–1988; II: 1989–1998; III: 1999–2008). Differences between means are all statistically different ($P < 0.01$)

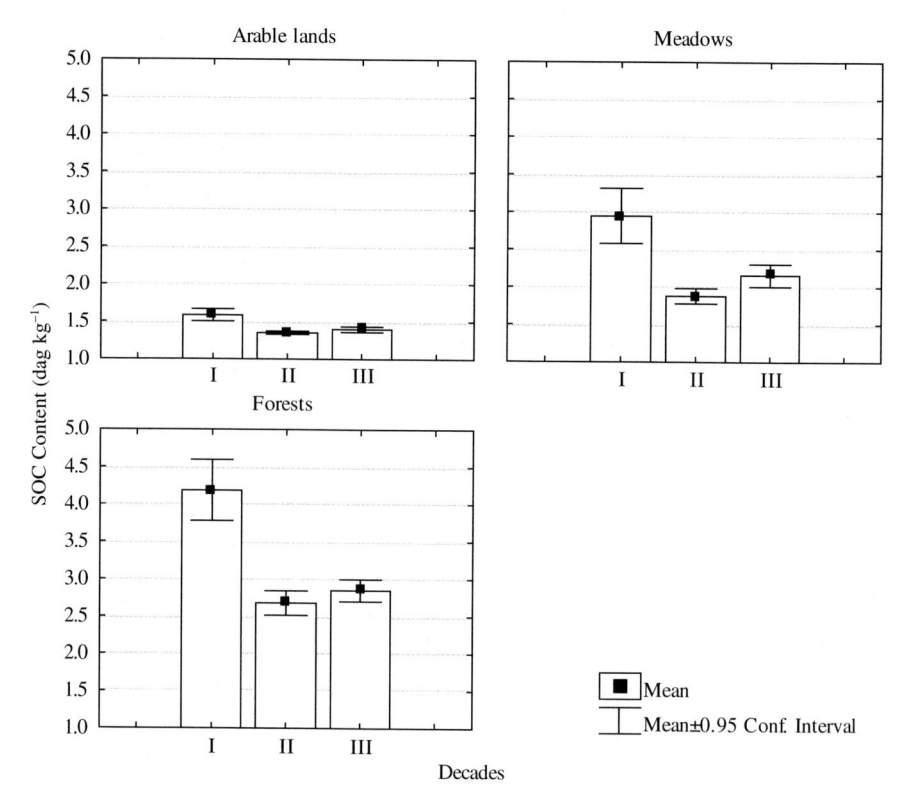

Fig. 34.7 Soil organic carbon content in the three decades considered and land uses. Differences between means are all statistically different ($P < 0.01$, or $P < 0.05$, between arable lands in decades II and III), except for the difference between forests in decades II and III

indicate a significant lowering of the overall mean in the nineties, with a certain recover in the last decade. The trend is common for the three land use classes considered (Fig. 34.7), although the differences between the second and the third decades become less significant.

34.3.2 Soil Carbon Stock Variations During the Last Three Decades

The MLRA model driven by stepwise regression is presented in Table 34.2. Among all the selected factors, the best predictive are land uses, decades, their interactions, SOTER morphological classes, and the continuous variables DEM, soil temperature and dry days. Almost all the soil regions are also highly predictive. Among the lithological groups, the best predictive is the e group (calcareous and dolomitic rocks, intrusive and metamorphic not-schist rocks).

Table 34.2 Multiple linear regression model adopted for the interpolations of carbon stock (Mg hm^{-2})

Predicting variables		Estimated coefficients	Std. error	t value	Pr(>|t|)	Significance level of P
	(Intercept)	86.134	4.33	19.862	<2e-16	***
Categorical						
Decade	II	−11.660	2.01	−5.777	7.71e-09	***
	III	−0.005	2.14	−0.002	0.998081	
Land use	Forests	47.881	3.63	13.169	<2e-16	***
	Meadows	14.696	4.10	3.58	0.000344	***
	18	20.421	2.88	7.076	1.54e-12	***
	34	21.503	4.12	5.214	1.87e-07	***
	35	23.141	7.48	3.092	0.001989	**
	37	16.690	4.55	3.663	0.000250	***
	56	8.624	3.72	2.313	0.020722	*
	59	8.635	3.01	2.867	0.004142	**
Soil region	60	−6.236	3.39	−1.836	0.066422	.
	61	−14.850	2.87	−5.164	2.45e-07	***
	62	−13.850	3.01	−4.589	4.49e-06	***
	64	−3.698	3.33	−1.109	0.267256	
	66	−5.433	4.11	−1.319	0.187204	
	67	24.654	7.05	3.494	0.000476	***
	72	−2.443	5.35	−0.456	0.648393	
	76	−11.632	4.00	−2.907	0.003656	**
	78	−0.175	3.04	−0.058	0.953938	
Soil systems lithology group	B	2.640	1.48	1.772	0.076350	.
	C	−0.645	1.55	−0.414	0.678809	
	D	1.604	2.16	0.739	0.459620	
	E	11.357	1.99	5.689	1.30e-08	***
Soil moisture regime	Udic	−4.833	3.71	−1.301	0.193141	
	Ustic	−5.116	2.83	−1.803	0.071382	.
	Xeric	2.944	2.14	1.376	0.168845	
Soil temperature regime	Thermic	3.198	1.49	2.133	0.032944	*
SOTER classes group	B	−9.981	1.22	−8.171	3.27e-16	***
	C	−15.448	2.45	−6.291	3.24e-10	***
	D	−11.734	1.86	−6.28	3.47e-10	***
	E	−12.437	2.72	−4.565	5.03e-06	***
	F	−4.515	3.29	−1.372	0.169998	
	G	25.360	7.95	3.189	0.001432	**
Continuous						
	Mean annual soil temp. at 50 cm	8.077	0.864018	9.349	<2e-16	***
	Soil aridity index	−8.478	1.177905	−7.198	6.34e-13	***
	Elevation	11.938	1.182095	10.099	<2e-16	***

Table 34.2 (continued)

Predicting variables		Estimated coefficients	Std. error	t value	Pr(>|t|)	Significance level of P
Interactions						
II decade	Land use forest	−16.316	4.16693	−3.916	9.05e-05	***
	Land use meadow	−3.728	4.398619	−0.848	0.396684	
III decade	Land use forest	−27.937	4.02584	−6.939	4.07e-12	***
	Land use meadow	−5.415	4.709544	−1.15	0.250171	

*** < 0.0001; ** <0.001; * <0.05; . <0.1.

The residual standard error is 56.12, with 17,824 degrees of freedom. Multiple R-Squared is 0.1643 and adjusted R-squared 0.1624. F-statistic is 87.62, with 40 and 17,824 degrees of freedom, P-value is < 2.2e-16. Therefore, although the F-statistic is very good, the multiple R-squared is quite low. This means that the high variability of the data cannot be well explained by the model, and a large amount of point variation remains unpredicted.

The bulk CS in Italy results 3.32 Pg in the eighties (107 Mg hm^{-2}), 2.74 Pg in the nineties (88 Mg hm^{-2}), and 2.93 Pg in the years 2000 (95 Mg hm^{-2}), (Figs. 34.8, 34.9 and 34.10). The distribution of estimation error is presented in Fig. 34.11. The RMSE were of 72.86 Mg hm^{-2} for the 1st decade, 44.78 for the 2nd decade and 65.37 for the 3rd decade. The variations between decades are reported in Figs. 34.12 and 34.13. The figures of the total budgets are intermediate between the 3.9 Pg postulated by the Natural Resources Conservation Service of the USDA (Schils et al., 2008) and the 2 Pg estimated by the European Soil Bureau (Stolbovoy et al., 2007a, 2007b; Schils et al., 2008).

The CS spatial distribution reveals larger amounts on the Alps, Apennines, and Sardinia, mainly coincident with forests, while the poorer areas are pretty well distributed all over the cultivated plains and hills of the country. It is interesting to note that many hilly lands of central and southern Italy, as well as in Sicily, are territories, which seem to be subjected to both negative and positive changes of CS over time. This could highlight a sensitivity of those soils to SOC modifications.

The trend during the last three decades shows an important decrease in the second decade, which can be probably related to the changes in land use and management, and their consequences on soil bulk density (Horn et al., 1995). Our data actually indicate a change in the distribution of the main land uses over the decades, which influences the calculation of the CS (Table 34.3).

The weight of bulk density on CS estimation in the three decades is highlighted in Fig. 34.14. We noticed that there is an average increase of soil bulk density with time, which is more evident in the third decade for arable lands, and in the second decade for meadows and forest. The outcome confirms what already was observed by many other authors on the enhanced risk of compaction for European soils, due

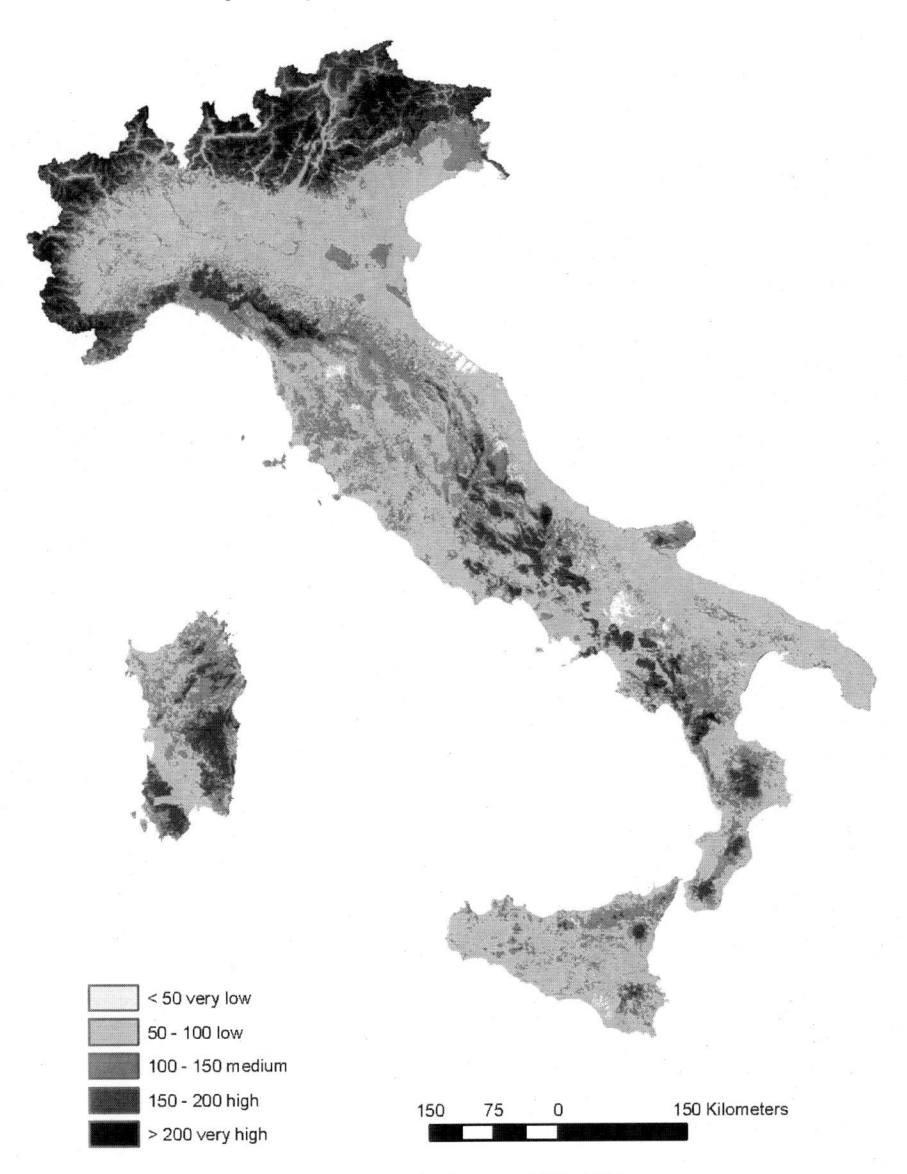

Fig. 34.8 Soil organic carbon stock of Italy in the years 1979–1988

Legend:
- < 50 very low
- 50 - 100 low
- 100 - 150 medium
- 150 - 200 high
- > 200 very high

to the steady increase in the diffusion of heavier tractors and machines (Słowińska-Jurkiewicz and Domazał, 1991; Alakukku, 1996; Bakken et al., 2009). On the other hand, the increase of soil bulk density in woodlands could be due to the reactivation of timber exploitation activities that occurred in the nineties, after about 20 years of silviculture decline (Vettraino et al., 2009).

It is also possible to observe a positive influence on CS of the European Union directives. As it is well known, during the nineties Italy, likewise many other

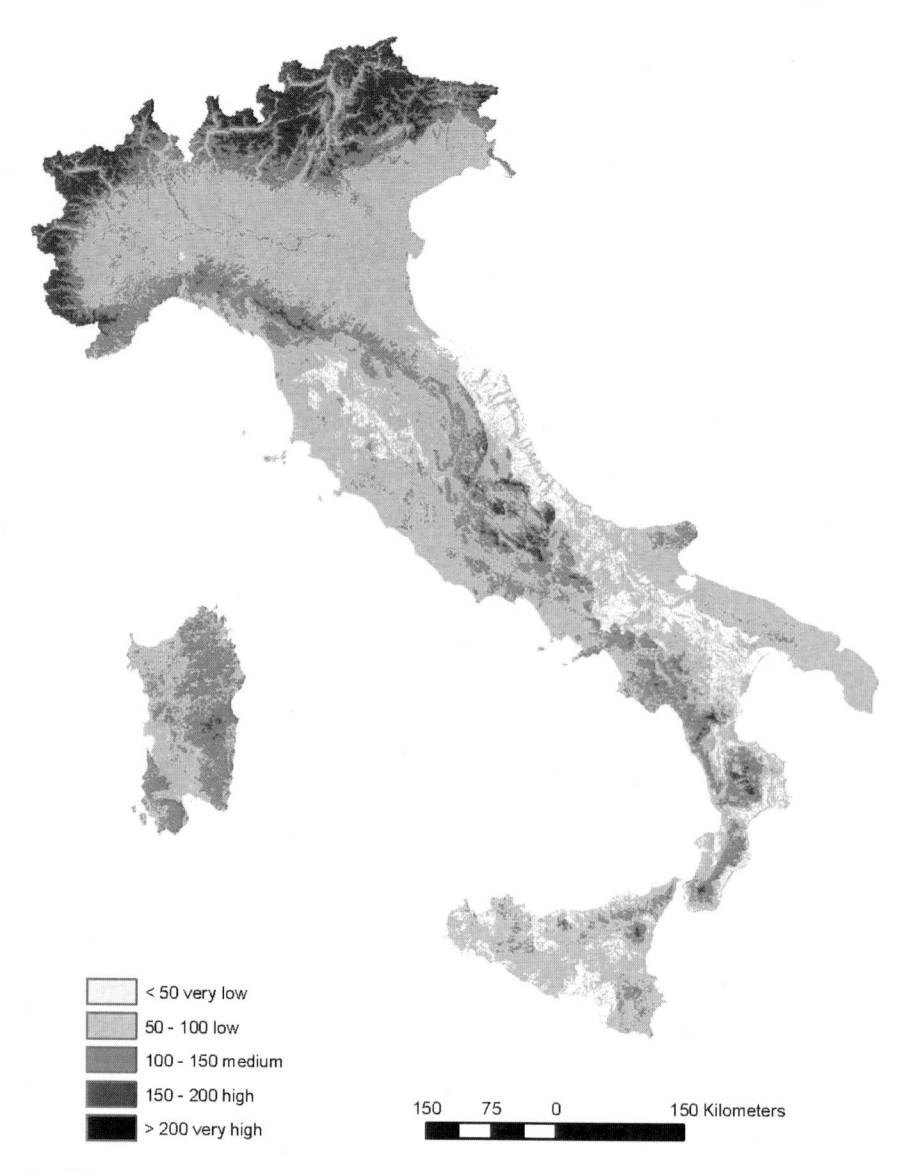

Fig. 34.9 Soil organic carbon stock of Italy in the years 1989–1998

European countries, adopted the so-called "agri-environmental measures" (Reg. CEE 2078/92). The EU applied agri-environmental measures which specifically supported designed farming practices, going beyond the baseline level of "good farming practices" which helped protect the environment and maintain the natural features of the countryside.

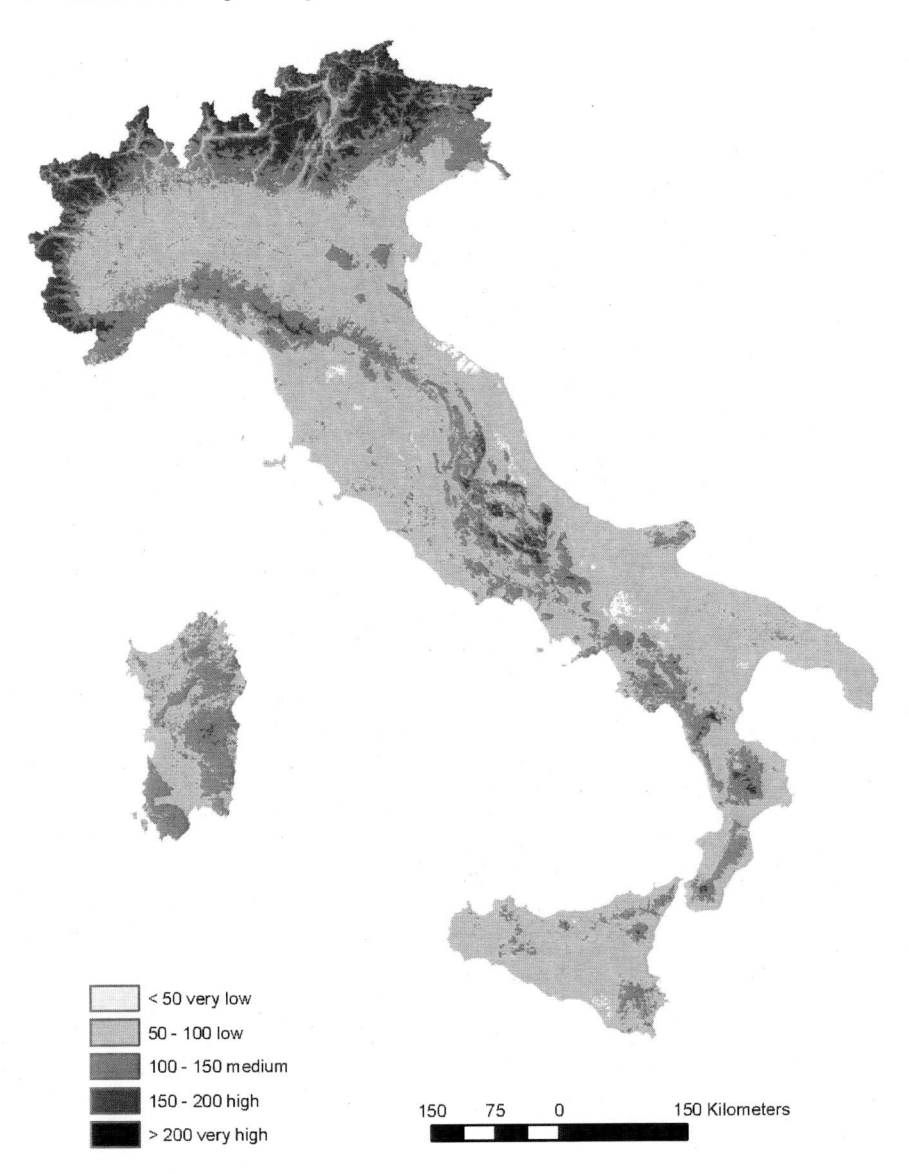

Fig. 34.10 Soil organic carbon stock of Italy in the years 1999–2008

34.3.3 Survey and Laboratory Biases

Some 2,937 values of SOC resulted biased in comparison to the others, representing 14.19% of the total (Table 34.4). They were rather randomly related to different surveys and soil regions, while resulting more frequent in the "arable land" in the second decade.

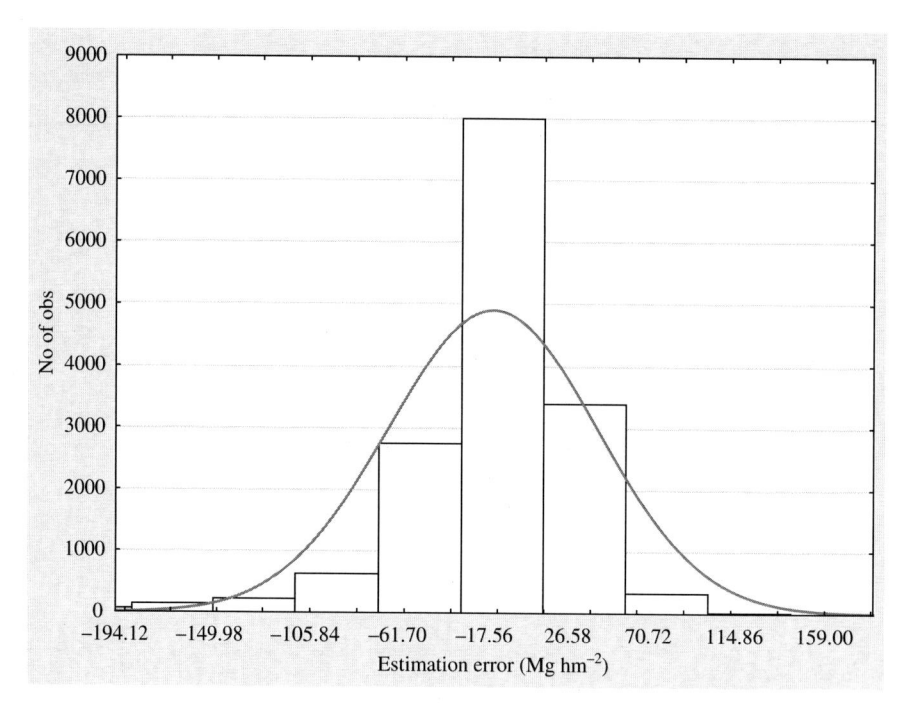

Fig. 34.11 Distribution of estimation error of soil carbon stock in Italy

A comparison between the exemplifying maps of CS made with biased (group C of the 3rd decade) and unbiased data shows a clear different estimation of CS (Fig. 34.15). The biased map gives an average lower CS estimation of 6.29 Mg hm^{-2}, and a range from –84 Mg hm^{-2} to +75 Mg hm^{-2}.

34.4 Conclusions

This study indicates that SOC content of Italian soils is rather low, on average, about 1.8 dag kg^{-1}. The outcome is consistent with what already estimated for the Mediterranean soils by Zdruli et al. (1999), showing that 74% of soils have less than 2% organic carbon. On the other hand, the comparison with the data reported for France (Arrouays et al., 2001; IFEN, 2007) indicates a slight larger SOC content of Italian soils. However, it must be considered that the reference depth was 50 in Italy and of 30 cm in France. Notwithstanding, our results are comparable and indicate an average CS content of 73 Mg hm^{-2} in arable lands, 95 in meadows, and 116 in forests.

The present research work does not consider the direct influence of climate or climate changes on SOC, but pedoclimate regimes instead. Additionally, soil moisture more than soil temperature regimes, result significantly related to SOC content. Therefore, it is probable that any change in rainfall amount and distribution, even

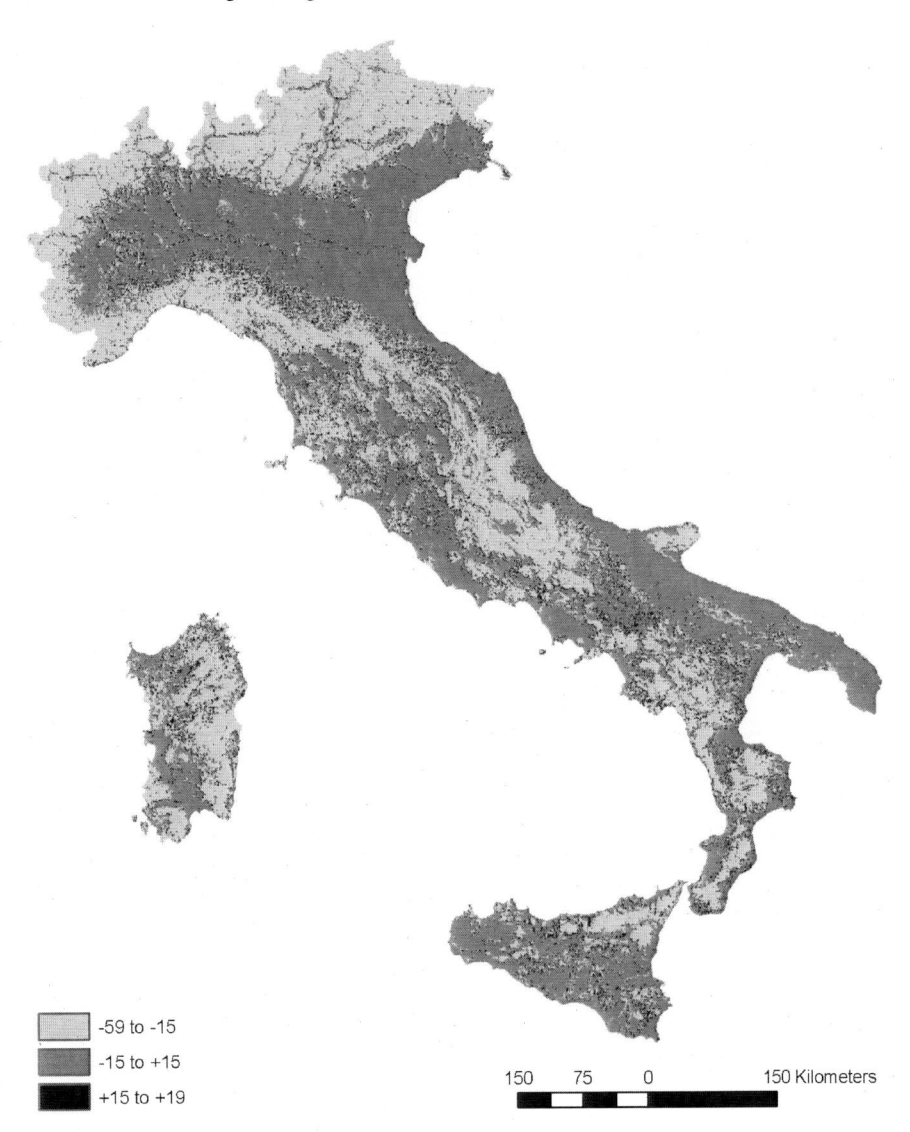

Fig. 34.12 Soil organic carbon stock variation in Italy between the first and the second studied decade

more than temperature, would affect SOC. Also, the physiographic position and lithology of the substratum are significantly related to SOC content, partly because of the interaction with climate and, most of all, land use. The class of land use in fact is by far the most important cause of SOC variation, pointing to the conservative role played by permanent meadows, and even more woodlands, in the Mediterranean environment.

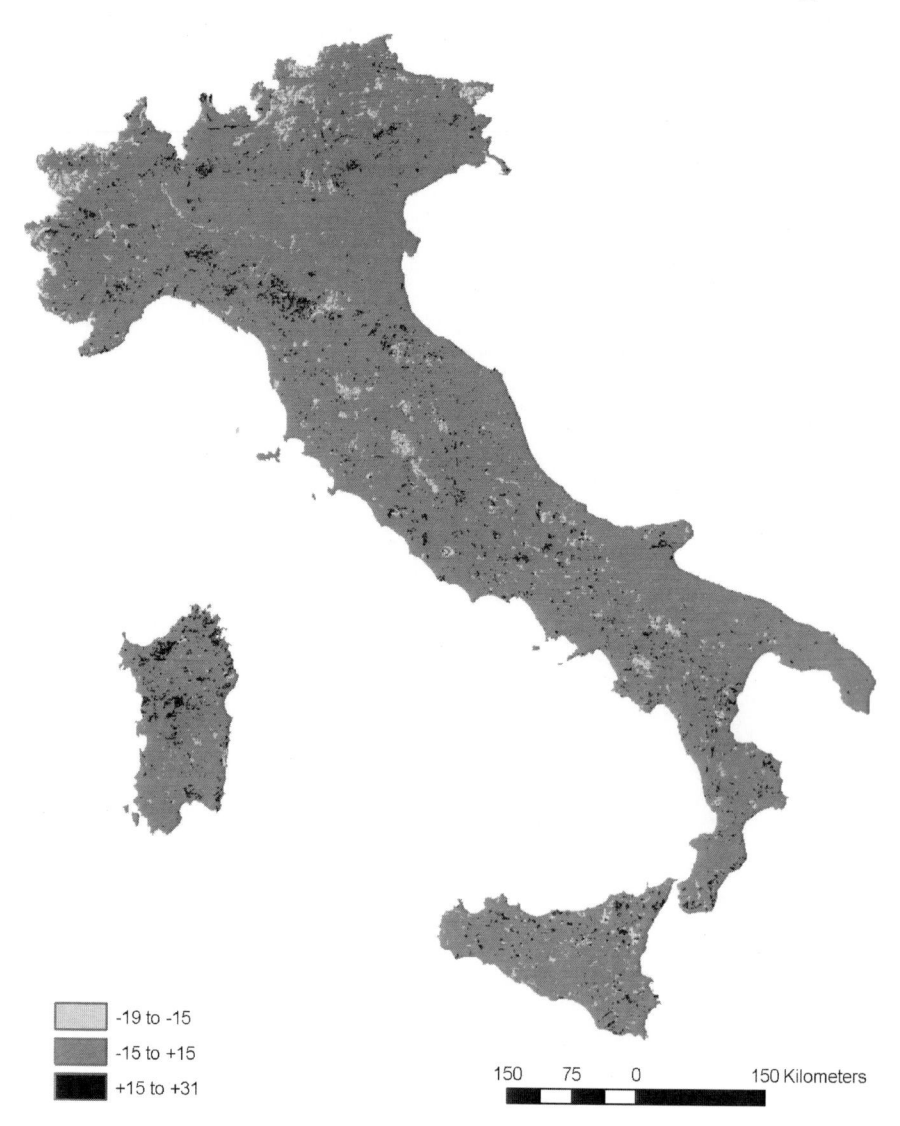

Fig. 34.13 Soil organic carbon stock variation in Italy between the second and the third studied decade

Our data highlight a significant change of the SOC content over the last three decades, which is not linear and apparently not related to major changes in main land uses. Other factors, like intensity of management, crop specialization, irrigation, adoption of conservation agriculture as a consequence of the European policies could have played an important role. In addition, we can not exclude the influence of the climate change occurred in Italy at the end of the eighties (Degobbis et al., 1995; Werner et al., 2000; Brunetti et al., 2004; Diodato and Mariani, 2007), which

34 Factors Influencing Soil Organic Carbon Stock Variations 457

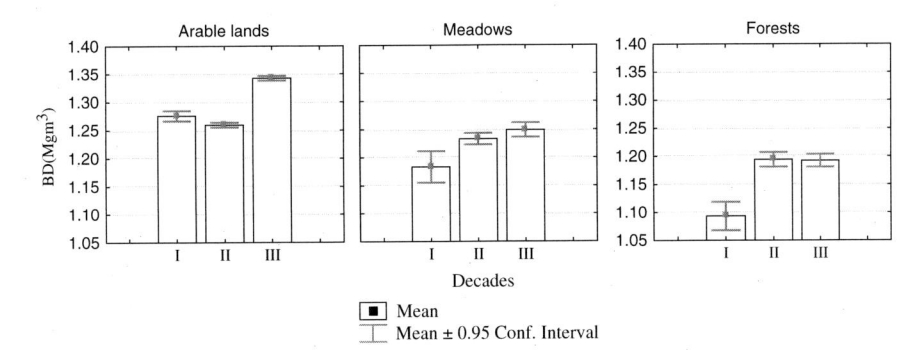

Fig. 34.14 Soil bulk density in the three considered decades and land uses. Differences between means are all statistically different ($P < 0.01$, or $P < 0.05$, between arable lands in decades I and II), except for the difference between meadows and forests in decades II and III

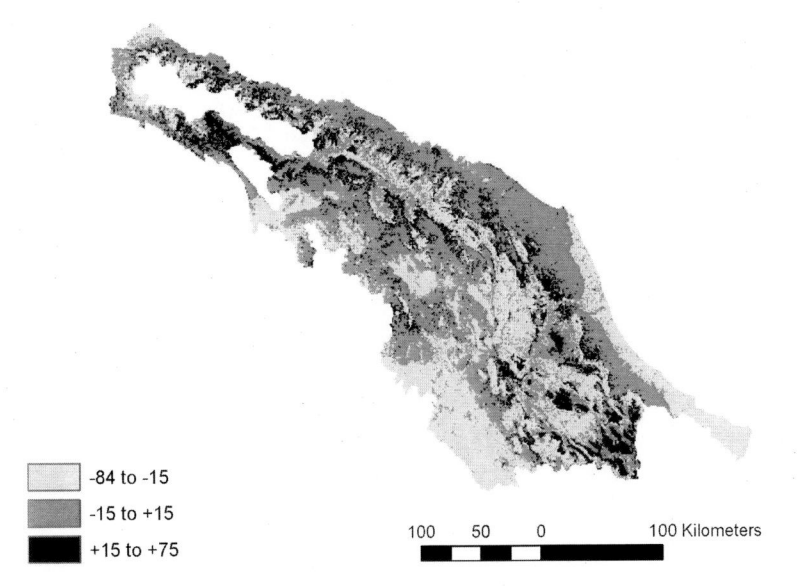

Fig. 34.15 Difference in the estimation of the soil organic carbon stock obtained with biased and unbiased data (exemplifying map)

Table 34.3 Main kind of land use of Italy, at the three reference times

Land cover	1990 (ha)	(%)	2000 (ha)	(%)	2008 (ha)	(%)
Arable lands	15,484.015	51.3	15,064.244	48.6	14,828.800	47.9
Forests	12,582.853	41.7	11,557.188	37.3	9,371.318	30.2
Meadows	494.125	1.6	1,883.553	6.1	3,158.724	10.2
Others	1,648.824	5.5	2,478.408	8.0	3,624.550	11.7
Total	30,209.817	100.0	30,983.393	100.0	30,983.393	100.0

increased average temperatures and augmented torrential regime of rainfall. All of these could have both directly and indirectly influenced soil erosion intensity and contributed to the observed SOC reduction.

CS of Italy is estimated to be at present about 2.9 Pg. The trend during the last three decades shows an important decrease in the second decade, followed by a slight increase in the third decade, mainly in arable lands. These results only partially correspond to what was found by some authors for European cultivated lands (Arrouays and Morvan, 2008), where the size of the soil organic carbon pool was estimated to be generally decreasing, while it seemed to be on increase in grasslands as well as in forests.

The observed average increase of soil bulk density of Italian soils during the last decade in arable lands, or in the nineties in permanents grasslands and forests, seems

Table 34.4 T-student test analysis for the independence of data coming from different survey sets, considered separately by soil region, decade and land use. Dashed rows indicate biased datasets

Survey set	Decade	Land use	Soil regions	Mean 1st group	Mean 2nd group	N 1st group	N 2nd group	Std. dev. 1st group	Std. dev. 2nd group	t-value	df	P (.) if <0.05	
A	2nd	Arable lands	16.4	1.569	2.378	96	14	1.190	1.077	−2.402	108	0.01799	(.)
			18.7–18.8	1.404	1.491	1,782	56	1.030	1.786	−0.600	1836	0.54829	
			34.2	1.723	1.326	93	36	0.751	0.926	2.518	127	0.01304	(.)
			37.1–37.3	1.663	1.387	56	18	1.091	0.752	0.995	72	0.32295	
			56.1	1.469	1.616	236	7	1.680	0.663	−0.231	241	0.81738	
			59.1–59.2–59.7	1.332	1.269	410	25	0.813	0.802	−0.375	433	0.70802	
			60.4	1.187	1.201	37	72	0.423	1.342	−0.062	107	0.95081	
			60.7	1.115	1.068	125	42	0.561	0.419	0.499	165	0.61872	
			61.1	1.106	1.304	432	86	0.591	0.902	−2.570	516	0.01046	(.)
			61.3	0.836	0.869	734	249	0.462	0.620	0.906	981	0.36510	
			62.1	1.103	0.999	409	61	0.515	1.453	1.065	468	0.28758	
			62.2	1.037	1.120	484	175	0.523	0.443	−1.876	657	0.06112	
			62.3	1.056	1.075	445	80	0.739	0.475	−0.227	523	0.82040	
			64.4	1.061	1.074	210	28	0.420	0.583	−0.148	236	0.88268	
			66.4	1.510	3.194	98	9	1.088	1.564	−4.273	105	0.00004	(.)
			72.2	1.468	0.887	133	12	0.678	0.397	2.918	143	0.00409	(.)
			76.1	0.885	0.918	224	24	0.467	0.488	0.328	246	0.74337	
			78.1	0.930	1.106	97	9	0.466	0.461	−1.085	104	0.28025	
			78.2	1.239	1.142	158	21	0.638	0.767	0.643	177	0.52082	
		Meadows	16.4–16.5	1.141	3.339	85	28	0.756	3.289	−5.763	111	0.00000	(.)
			59.1–59.2	1.288	2.876	58	67	0.825	1.658	6.616	123	0.00000	(.)
			59.7	1.668	2.172	75	16	1.072	1.028	−1.721	89	0.08878	
			60.4–60.7	1.300	1.368	9	15	1.031	0.350	−0.237	22	0.81511	
			61.1–61.3	0.935	1.105	181	34	0.806	0.674	1.158	213	0.24800	
			76.1	0.733	1.054	12	21	0.494	0.336	2.216	31	0.03415	(.)
			62.1–62.3	1.021	1.747	47	23	0.664	2.663	−1.772	68	0.08094	
			66.4–66.5	1.275	3.523	4	10	0.695	2.342	−1.847	12	0.08959	
B	1st	Arable	18.8–34.3–78.1	1.789	1.145	149	20	1.489	0.411	−1.918	167	0.05678	
		Meadows		1.895	2.126	78	2	1.082	0.588	−0.299	78	0.76556	
	2nd	Arable		1.487	1.425	1,392	106	1.216	1.427	−0.501	1496	0.61618	
		Meadows		1.818	1.019	512	6	1.304	0.619	1.497	516	0.13488	
	3rd	Arable		1.512	1.603	965	35	1.603	1.572	−0.332	998	0.73980	
		Meadows		1.672	4.089	165	10	1.329	5.619	4.075	173	0.00007	(.)
	1st 2nd 3rd	Forests		2.553	4.934	125	22	2.056	5.438	3.665	145	0.00035	(.)
				2.509	1.642	43	7	0.926	1.784	1.985	48	0.05287	
C	1st	Arable	16.4–56.1–	1.452	1.229	242	76	1.364	1.161	1.281	316	0.201006	

Table 34.4 (continued)

		Meadows	61.1–61.3–	5.413	4.337	55	9	3.850	4.158	0.769	62	0.444538	
		Forests	64.4–78.1–78.2	4.084	5.103	219	22	3.830	4.761	−1.162	239	0.246508	
		Arable		1.125	1.045	237	398	0.728	0.776	−1.281	633	0.200648	
	2nd	Meadows		5.656	2.151	54	64	3.562	2.537	−6.223	116	0.000000	(.)
		Forests		2.890	2.273	70	130	2.037	1.651	−2.316	198	0.021580	(.)
		Arable		1.465	1.161	652	773	2.281	0.790	−3.463	1423	0.000549	(.)
	3rd	Meadows		3.314	1.917	176	195	2.929	1.532	5.835	369	0.000000	(.)
		Forests		2.812	2.431	429	178	2.431	2.557	1.735	605	0.083281	
D	1st 2nd	Arable		1.979	0.926	53	75	3.601	0.449	−2.511	126	0.01331	(.)
		Forests	35.7–60.4–	2.080	4.373	80	28	1.380	4.310	4.210	106	0.00005	(.)
	1st 2nd	Arable	60.7–	1.318	0.975	508	412	1.350	0.794	−4.560	918	0.00001	(.)
		Forests	61.3–	1.948	2.284	368	69	1.442	1.320	1.800	435	0.07260	
	1st 2nd	Arable	64.4–78.2	1.149	1.007	104	538	0.809	0.586	−2.104	640	0.03574	(.)
		Forests		1.935	2.381	45	75	0.988	2.540	1.125	118	0.26281	
	1st 2nd 3rd	Meadows		1.777	1.622	122	92	1.739	1.523	−0.678	212	0.49823	
E	2nd	Arable	61.3–62.1–	1.155	0.981	44	73	0.575	1.336	−0.820	115	0.41375	
	3rd		72.2–72.3	1.068	0.982	295	96	0.551	0.409	−1.407	389	0.16034	

to have played a central role on CS temporal evolution. Land uses changes over the time modified the proportion of the conservative covers, thus affecting the CS. These results further stress the importance of soil management on the maintenance or increase of the national CS.

Finally, our study strongly suggest to carefully examine the bulk of data before to proceed with the elaboration of CS maps, as the values coming from different sources could be notably biased, even if samples were analysed with the same methodology. In the case of Italy, the CS estimations made using datasets that significantly deviated from the others could be as biased as the 190%.

References

Alakukku, L. (1996). Persistence of soil compaction due to high axle load traffic. II. Long-term effects on the properties of fine-textured and organic soils. Soil and Tillage Research 37(4): 223–238.

Arrouays, D., Balesdent, J., Germon, J.C., Jayet, P.A., Soussana, J.F. and Stengel, P. (2002a). Mitigation of the greenhouse effect. Increasing carbon stocks in French agricultural soils? Synthesis of an assessment report by the French Institute for Agricultural Research (INRA) on request of the French Ministry for Ecology and sustainable development. INRA, France, 36pp.

Arrouays, D., Balesdent, J., Gerom, J.C., Jayet, P.A., Soussana, J.F. and Stengel, P. (2002b). Contribution a la lutte contre l'effet de serre: Stocker du carbone dans les sols agricules de France? INRA, France, 332pp.

Arrouays, D., Deslais, W. and Badeau, V. (2001). The carbon content of topsoil and its geographical distribution in France. Soil Use and Management 17:7–11.

Arrouays, D. and Morvan, X. (2008). Inventory and monitoring report. ENVASSO Work package 2 report. JRC, Cranfield, UK.
Bakken, A.K., Brandsæter, L.O., Eltun, R., Hansen, S., Mangerud, K., Pommeresch, R. and Riley, H. (2009). Effect of tractor weight, depth of ploughing and wheel placement during ploughing in an organic cereal rotation on contrasting soils. Soil and Tillage Research 103(2):433–441. doi: 10.1016/j.still.2008.12.010.
Batjes, N.H. (2008). Mapping soil carbon stocks of Central Africa using SOTER. Geoderma, Elsevier, Amsterdam 146(1–2):58–65. doi: 10.1016/j.geoderma.2008.05.006.
Bellamy, P.H. (2008). UK losses of soil carbon – due to climate change? Report on the conference Climate change – can soil make a difference? Brussels, Thursday 12 June 2008. Full presentation. Available via DIALOG. http://ec.europa.eu/environment/soil/conf_en.htm Cited 13 August 2009.
Bellamy, P.H., Loveland, P.J., Bradley, R.I., Lark, R.M. and Kirk, G.J.D. (2005). Carbon losses from all soils across England and Wales 1978–2003. Nature 437:245–248.
Bouwman, A. (2001). Global Estimates of Gaseous Emissions from Agricultural Land. FAO, Rome, p. 106.
Brunetti, M., Maugeri, M., Monti, F. and Nanni, T. (2004). Changes in daily precipitation frequency and Distribution in Italy over the last 120 years. Journal of Geophysical Research D: Atmospheres 109 (5):D05102.
Calzolari, C. and Ungaro, F. (2005). La carta della dotazione in sostanza organica della pianura Emiliano Romagnola. Il suolo 34–36(1–3):29–32. Available via DIALOG. http://www.aip-suoli.it/editoria/bollettino/n1-3a05/n1-3a05_08.htm Cited 13 August 2009.
Carre, F., Mcbratney, A.B., Mayr, T. and Montanarella, L. (2007). Digital soil assessments: Beyond DSM. Geoderma, Elsevier, Amsterdam 142(1–2):69–79. doi: 10.1016/j.geoderma.2007.08.015.
Castrignanò, A., Costantini, E.A.C., Barbetti, R. and Sollitto, D. (2009). Accounting for extensive topographic and pedologic secondary information to improve soil mapping. Catena 77:28–38.
Cerli, C., Celi, L., Bosio, P., Motta, R. and Grassi, G. (2009). Effects of land use change on soil properties and carbon accumulation in the Ticino Pank (North Italy). In: G. Sartori (ed.), Studi Trentini di Scienze Naturali. Suoli degli ambienti alpini. volume 85, Museo tridentino doi scienze naturali, Trento, pp. 83–92.
Chai, X., Shen, C., Yuan, X. and Huang, Y. (2008). Spatial prediction of soil organic matter in the presence of different external trends with REML-EBLUP. Geoderma, Elsevier, Amsterdam 148(2):159–166.
Ciais, P., Reichstein, M., Viovy, N., Granier, A., Ogee, J. et al. (2005). Europe-wide reduction in primary productivity caused y heat and drought in 2002. Nature 437:529–533.
CORINE. (2009). COoRdination de l'INformation sur l'Environnement. http://www.eea.europa.eu/publications/COR0-landcover; http://www.clc2000.sinanet.apat.it/; http://stweb.sister.it/itaCorine/corine/progettocorine.htm Cited 13 August 2009.
Costantini, E.A.C., Barbetti, R. and L'Abate, G. (2007). Soils of Italy: Status, problems and solutions. In: P. Zdruli, G. Trisorio Liuzzi (eds.), Status of Mediterranean Soil Resources: Actions Needed to Support their Sustainable Use. Mediterranean Conference Proceedings, Tunis, Tunisia, IAM Bari (Italy), pp. 165–186.
Costantini, E.A.C., Barbetti, R. and L'Abate, G. (2009). The soil aridity index to asses the desertification risk. Advances in GeoEcology (in press).
Costantini, E.A.C., Castelli, F. and L'Abate, G. (2005). Use of the EPIC model to estimate soil moisture and temperature regimes for desertification risk in Italy. Advances in GeoEcology 251–263.
Costantini, E.A.C., Castelli, F., Lorenzoni, P. and Raimondi, S. (2002). Assessing soil moisture regimes with traditional and new methods. Soil Science Society of America Journal 66: 1889–1896.
Costantini, E.A.C. and L'Abate, G. (2009). The soil aridity index to asses the desertification risk. Advances in GeoEcology 40 CATENA VERLAG, 35447 Reiskirchen (in press).

Costantini, E.A.C., Magini, S. and Napoli, R. (2003). A Land System database of Italy. 4th European Congress on Regional Geoscientific Cartography and Information Systems. Proceedings 1:124–126. Available via DIALOG. http://www.regione.emilia-romagna.it/geologia/convegni/4th_congress/oral_4congr.htm#sess01 Cited 13 August 2009.

Covington, W.W. (1981). Changes in forest floor organic matter and nutrient content following clear cutting in northern hardwoods. Ecology 62:41–48.

Dalla Valle, E. (2008). Valutazione dello stock di carbonio e delle capacità fissative delle foreste assestate e dei boschi di neoformazione nella Regione Veneto. Available via DIALOG. http://paduaresearch.cab.unipd.it/1340/ Cited 13 August 2009.

Degobbis, D., Fonda-Umani, S., Franco, P., Malej, A., Precali, R. and Smodlaka, N. (1995). Changes in the northern Adriatic ecosystem and the hypertrophic appearance of gelatinous aggregates. Science of the Total Environment 165:43–58.

Diodato, N. and Mariani, L. (2007). Testing a climate erosive forcing model in the PO River Basin. Climate Research 33(2):195–205.

European Commission. (16 April 2002). Towards a Thematic Strategy for Soil Protection. COM 179, 16 April 2002.

European Commission. (2006). Thematic Strategy for Soil Protection, COM 231.

European Commission. (2008). Final Remarks. Report on the conference Climate change – can soil make a difference? Brussels, Thursday 12 June 2008.

European Commission. (2009). European research on climate change. Catalogue of FP6 projects. Available via DIALOG. http://ec.europa.eu/research/environment/pdf/european_research_climate_change_en.pdf. Cited 13 August 2009.

European Communities. (2003). The Lucas survey European statisticians monitor territory. Luxembourg p. 24. Available via DIALOG. http://circa.europa.eu/irc/dsis/landstat/info/data/studiesreports.htm Cited 13 August 2009.

Finke, P., Hartwich, R., Dudal, R., Ibanez, J., Jamagne, M., King, D., Montanarella, L. and Yassoglu, N. (1998). Georeferenced soil database for Europe. Manual of Procedures. Version 1.0. ESB-JRC-SAI. European Commission, EUR 18092 EN, p. 184.

Food and Agricultural Organisation. (1974). FAO-UNESCO soil map of the word; volume 1, Legend. UNESCO, Paris.

Food and Agricultural Organisation (FAO). (1995). Global and national soils terrain digital databases (SOTER) 74 Rev.1, FAO, Rome, Italy.

Food and Agricultural Organisation (FAO). (1998). World reference base for soil resources. World Soil Resources Reports, 84. Rome, Italy, p. 88.

Franzluebbers, A.J. (2002). Soil organic matter stratification ratio as an indicator of soil quality. Soil & Tillage Research 66:95–106. Elsevier Science B.V.

Gardi, C. (2005). Valutazione dello stock di carbonio nel suolo di prati stabili e seminativi avvicendati. Il suolo 34–36(1–3):46–49. Available via DIALOG. http://www.aip-suoli.it/editoria/bollettino/n1-3a05/n1-3a05_14.htm Cited 13 August 2009.

Garlato, A., Obber, S., Vinci, I., Mancabelli, A., Parisi, A. and Sartori, G. (2009b). La determinazione dello stock di carbonio nei suoli del Trentino a partire dalla banca dati della carta dei suoli alla scala 1:250.000. In: G. Sartori (ed.), Studi Trentini di Scienze Naturali. Suoli degli ambienti alpini. volume 85, Museo trdentino doi scienze naturali, Trento, pp. 157–160.

Garlato, A., Obber, S., Vinci, I., Sartori, G. and Manni, G. (2009a). Stock attuale di carbonio organico nei suoli di montagna del Veneto. In: G. Sartori (ed.), Studi Trentini di Scienze Naturali. Suoli degli ambienti alpini. volume 85, Museo tridentino doi scienze naturali, Trento, pp. 69–82.

Geissen, V., Sánchez-Hernández, R., Kampichler, C., Ramos-Reyesa, R., Sepulveda-Lozada, A., Ochoa-Goana, S., de Jonga, B.H.J., Huerta-Lwangaa, E. and Hernández-Daumasa, S. (2009). Effects of land-use change on some properties of tropical soils – An example from Southeast Mexico. Geoderma, Elsevier, Amsterdam 151(3–4):87–97. doi: 10.1016/j.geoderma.2009.03.011.

Giandon, P. (2000). Riproducibilità dei risultati delle analisi del terreno nei laboratori italiani. i risultati del confronto interlaboratorio gestito dalla società italiana dei laboratori pubblici agrochimici. ARPAV Centro Agroambientale, Italy.

Gomez, C., Viscarra Rossel, R.A. and McBratney, A.B. (2008). Soil organic carbon prediction by hyperspectral remote sensing and field vis-NIR spectroscopy: An Australian case study. Geoderma, Elsevier, Amsterdam 146(3–4):403–411. doi: 10.1016/j.geoderma.2008.06.011.

Grimm, R., Behrens, T., Märker, M. and Elsenbeer, H. (2008). Soil organic carbon concentrations and stocks on Barro Colorado Island – Digital soil mapping using Random Forests analysis. Geoderma, Elsevier, Amsterdam 146(1–2):102–113. doi: 10.1016/j.geoderma.2008.05.008.

Grunwald, S. (2009). Multi-criteria characterization of recent digital soil mapping and modeling approaches. Geoderma, Elsevier, Amsterdam 152(3–4):195–207. doi: 10.1016/j.geoderma.2009.06.003.

Guermandi, M. (2005). Protocollo di Kyoto. I suoli agricoli come "serbatoi" di anidride carbonica in Emilia-Romagna. ARPA Rivista N. 5 Settembre-Ottobre 2005.

Hirmas, D.R., Amrhein, C. and Graham, R.C. (in press). Spatial and process-based modeling of soil inorganic carbon storage in an arid piedmont. Geoderma, Elsevier, Amsterdam. doi: 10.1016/j.geoderma.2009.05.005.

Horn, R., Domzzal, H., Slowinska-Jurkiewicz, A. and van Ouwerkerk, C. (1995). The structure of the cultivated horizon of soil compacted by the wheels of agricultural tractors. Soil Compaction and the Environment. Soil and Tillage Research 35(1–2):23–36. doi: 10.1016/0167-1987(95)00479–C.

Hoyos, N.T. and Comerford, N.B. (2005). Land use and landscape effects on aggregate stability and total carbon of Andisols from the Colombian Andes. Geoderma, Elsevier, Amsterdam 129 (3–4):268–278.

Huang, X., Senthilkumara, S., Kravchenko, A., Thelena, K. and Qib, J. (2007). Total carbon mapping in glacial till soils using near-infrared spectroscopy, Landsat imagery and topographical information. Geoderma, Elsevier, Amsterdam 141(1–2):34–42. doi: 10.1016/j.geoderma.2007.04.023.

Institute Francais de l'environnement (IFEN). (2007). Le stock de carbone dans les sols agricoles diminue, 121. www.ifen.fr Cited 13 August 2009.

Intergovernamental Panel on Climate Change (IPCC). (2001a). Climate Change 2001: The Scientific Basis. Contribution of Working Group I to the Third Assessment Report of the Intergovernmental Panel on Climate Change [J.T. Houghton, Y. Ding, D.J. Griggs, M. Noguer, P.J. van der Linden, X. Dai, K. Maskell and C.A. Johnson (eds.)]. Cambridge University Press, Cambridge, p. 881.

Intergovernamental Panel on Climate Change (IPCC). (2001b). Climate Change 2001: Mitigation: Contribution of Working Group III to the Third Assessment Report of the Intergovernmental Panel on Climate Change [B. Metz, O. Davidson, R. Swart and J. Pan (eds.)], Cambridge University Press, p. 752.

Intergovernmental Panel on Climate Change (IPCC). (2003). Good Practice Guidance for Land Use, Land Use Change and Forestry [J. Penman, M. Gytarsky, T. Hiraishi, T. Krug, D. Kruger, R. Pipatti, L. Buendia, K. Miwa, T. Ngara, K. Tanabe and F. Wagner (eds.)], IPCC/OECD/IEA/IGES, Hayama, Japan.

Intergovernmental Panel on Climate Change (IPCC). (2007). Climate change 2007: Mitigation of Climate Change. Contribution of Working group III to the Fourth Assessment Report of the Intergovernmental Panel on Climate Change [B. Metz, O.R. Davidson, P.R. Bosch, R. Dave and L.A. Meyer (eds.)], Cambridge University Press, Cambridge, UK and New York, USA.

Jenny, H. (1941). Factors of Soil Formation – A System of Quantitative Pedology. McGraw-Hill, New York, USA, p. 281.

Jones, R.J.A., Hiederer, R., Rusco, E. and Montanarella, L. (2005). Estimating organic carbon in the soils of Europe for policy support. European Journal of Soil Science 56:655–671.

Kätterer, T., Andrén, O. and Person, J. (2004). The impact of altered management on long-term agricultural soil carbon stocks – a Swedish case study. Nutrient Cycling in Agroecosystems 70:179–187.

Kirk, G.J.D. and Bellamy, P.H. (2008). On the reasons for carbon losses from soils across England and Wales 1978–2003. Global Change Biology (in review).

Lal, R. (2008). The role of soil organic matter in the global carbon cycle. Report on the conference Climate change – can soil make a difference? Brussels, Thursday 12 June 2008. Full presentation. Available via DIALOG. http://ec.europa.eu/environment/soil/conf_en.htm. Cited 13 August 2009.

L'Abate, G. and Costantini, E.A.C. (2004). GIS Pedoclimatico d'Italia. Progetto PANDA. Istituto Sperimentale Studio e Difesa del Suolo, Centro Nazionale Cartografia Pedologica, Firenze, Italia. CD-Rom.

Lal, R., Kimble, J.M., Eswaran, H. and Stewart, B.A. (2008). Global Climate Change and Pedogenetic Carbonates. Lewis Publishers, Washington, DC. Available via DIALOG. http://ec.europa.eu/environment/soil/conf_en.htm. Cited 13 August 2009.

Lal, R., Kimble, J.M., Follet, R.F. and Stewart, B.A. (2001). Assessment Methods for Soil Carbon. Advances in Soil Science. Lewis Publishers, Washington, DC, p. 676. Available via DIALOG. http://books.google.it/books?id=kgiYYADtQx0C. Cited 13 August 2009.

Luo, Y. and Zhou, Z. (2006). Soil Respiration and the Environment. Academic/Elsevier, San Diego, p. 328.

Marland, G., West, T.O., Schlamadinger, B. and Canella, L. (2003). Managing soil organic carbon in agriculture: The net effect on greenhouse gas emissions. Tellus 55B:613–621.

McBratney, A.B., Mendonca Santos, M.L. and Minasny, B. (2003). On digital soil mapping. Geoderma, Elsevier, Amsterdam 117:3–52.

Meersmans, J., De Ridder, F., Canters, F., De Baets, S. and Van Molle, M. (2008). A multiple regression approach to assess the spatial distribution of Soil Organic Carbon (SOC) at the regional scale (Flanders, Belgium). Geoderma, Elsevier, Amsterdam 143(1–2):1–13. doi: 10.1016/j.geoderma.2007.08.025.

Ministero delle Politiche Agricole Alimentari e Forestali (MIPAAF). (1992). Decreto Ministeriale 11 maggio 1992 Approvazione dei Metodi ufficiali di analisi chimica del suolo. In: Suppl. ordinario alla Gazz. Uff., 25 maggio 1992, n. 121.

Morari, F., Lugato, E., Berti, A. and Giardini, L. (March 2006). Long-term effects of recommended management practices on soil carbon changes and sequestration in north-eastern Italy. Soil Use and Management 22:71–81. doi: 10.1111/j.1475–2743.2005.00006.

Neff, J.C., Townsend, A.R., Gleixner, G., Lehman, S.J., Turnbull, J. and Bowman, W.D. (2002). Variable effects of nitrogen additions on the stability and turnover of soil carbon Nature 419:915–917.

Nyssen, J., Temesgen, H., Lemenihd, M., Zenebe, A., Haregeweyn, N. and Haile, M. (2008). Spatial and temporal variation of soil organic carbon stocks in a lake retreat area of the Ethiopian Rift Valley. Geoderma, Elsevier, Amsterdam 146(1–2):261–268. doi: 10.1016/j.geoderma.2008.06.007.

Ogle, S.M., Breidt, F., Jay, W. and Paustian, K. (2006). Bias and variance in model results associated with spatial scaling of measurements for parameterization in regional assessments. Global Change Biology 12:516–523. doi: 10.1111/j.1365–2486.2006.01106.x.

Pellegrini, S., Vignozzi, N., Costantini, E.A.C. and L'Abate, G. (2007). A new pedotransfer function for estimating soil bulk density. In: C. Dazzi (ed.), Changing Soils in a Changing Wold: The Soils of Tomorrow. Book of Abstracts. 5th International Congress of European Society for Soil Conservation, Palermo, 25–30 June 2007. ISBN: 978–88–9572–09–2.

Petrella, F. and Piazzi, M. (2005). Il carbonio negli ecosistemi agrari e forestali del Piemonte: misure ed elaborazioni. Il suolo 34–36(1–3):33–34. Available via DIALOG. http://www.aipsuoli.it/editoria/bollettino/n1-3a05/n1-3a05_09.htm Cited 13 August 2009.

Piazzi, M. (2006). Le attività della Regione Piemonte nel settore dello studio del carbonio. Soil Indicators for the Soil Thematic Strategy Support/Indicatori e metodologie a supporto della strategia tematica per il suolo. Ispra, ITALY – 21–23 November 2006. Available via DIALOG. http://eusoils.jrc.ec.europa.eu/Events/Soil_Indicators/sessione_2/Documenti/Piemonte/piem_carbonio.doc, http://eusoils.jrc.ec.europa.eu/Events/Soil_Indicators/sessione_4/Documenti/Carbonio_piemonte/carbonio_t_250000.jpg

Pilli, R., Anfodillo, T. and Dalla Valle, E. (eds.)(5–8 Giugno 2006). Stima del Carbonio in foresta: metodologie ed aspetti normativi. Pubblicazione del Corso di Cultura in Ecologia, Atti del 42° corso, Università di Padova. San Vito di Cadore.

Post, W.M. and Kwon, K.C. (2000). Soil carbon sequestration and land-use change: Processes and potential. Global Change Biology 6:317–327.

Reeves, P.C. (1997). The development of pore-scale network models for the simulation of capillary pressure-saturation-interfacial area-relative permeability relationships in multi fluid porous media. PhD Thesis, Department of Civil Engineering and Operations Research, Princeton University, New Jersey, USA.

Reichstein, M., Ciais, P., Papale, D., Valentini, R., Running, S., Viovy, N., Cramer, W., Granier, A., Ogee, J., Allard, V. et al. (2006). Reduction of ecosystem productivity and respiration during the European summer 2003 climate anomaly: A joint flux tower, remote sensing and modelling analysis. Global Change Biology 12:1–18.

Righini, G., Costantini, E.A.C. and Sulli, L. (2001). La banca dati delle regioni pedologiche italiane. Bollettino della Società Italiana Scienza del Suolo 50(suppl):261–271.

Sakamoto, Y. and Akaike, H. (1978). Analysis of cross classified data by AIC. Annals of the Institute of Statistical Mathematics 30:185–197. Available via DIALOG. http://www.ism.ac.jp/editsec/aism/pdf/030_1_0185.pdf Cited 13 August 2009.

Sankey, J.B., Brown, D.J., Bernard, M.L. and Lawrence, R.L. (2008). Comparing local vs. global visible and near-infrared (VisNIR) diffuse reflectance spectroscopy (DRS) calibrations for the prediction of soil clay, organic C and inorganic C. Geoderma, Elsevier, Amsterdam 148(2):149–158. doi: 10.1016/j.geoderma.2008.09.019.

Schils, R., Kuikman, P., Liski, J., van Oijen, M., Smith, P., Webb, J., Alm, J., Somogyi, Z., van den Akker, J., Billett, M. et al. (2008). Review of existing information on the interrelations between soil and climate change. Climsoil, Final Report. 16 December 2008.

Schimel, D.S., House, J.I., Hibbard, K.A., Bousquet, P., Ciais, P., Peylin, P., Braswell, B.H., Apps, M.J., Baker, D., Bondeau, A. et al. (2001). Recent patterns and mechanisms of carbon exchange by terrestrial ecosystems. Nature 414:169–172.

Sequi, P. and De Nobili, M. (2000). Carbonio organico. In: Violante, P. (ed.), Metodi di analisi chimica del suolo. Collana di metodi analitici per l'agricoltura diretta da Paolo Sequi, Franco Angeli, Milano.

Simbahan, G.C., Dobermann, A., Goovaerts, P., Pinga, J. and Haddix, M.L. (2006). Fine-resolution mapping of soil organic carbon based on multivariate secondary data. Geoderma, Elsevier, Amsterdam 132(3–4):471–489. doi: 10.1016/j.geoderma.2005.07.001.

Sinanet. (2009). Rete del sistema Informativo Nazionale Ambientale. http://www.clc2000.sinanet.apat.it/. Cited 13 August 2009.

Sistema Informativo Agricolo Nazionale (SIN). (2009). http://www.sin.it. Cited 13 August 2009.

Słowińska-Jurkiewicz, A. and Domazał, H. (1991). The structure of the cultivated horizon of soil compacted by the wheels of agricultural tractors. Soil and Tillage Research 19(2–3):215–226.

Smith, P. (2008). The role of agricultural practices in keeping or increasing soil organic matter. Report on the conference Climate change – can soil make a difference? Brussels, Thursday 12 June 2008. Full presentation. Available via DIALOG. http://ec.europa.eu/environment/soil/conf_en.htm. Cited 13 August 2009.

Smith, P., Martino, D., Cai, Z., Gwary, D., Janzen, H.H., Kumar, P., McCarl, B.A., Ogle, S.M., O'Mara, F., Rice, C. et al. (2007a). Policy and technological constraints to implementation of greenhouse gas mitigation options in agriculture. Agriculture, Ecosystems and Environment 118:6–28.

Smith, P., Martino, D., Cai, Z., Gwary, D., Janzen, H.H., Kumar, P., McCarl, B.A., Ogle, S.M., O'Mara, F., Rice, C. et al. (2007b). Agriculture. In: B. Metz, O.R. Davidson, P.R. Bosch et al. (eds.), Climate Change 2007: Mitigation. Contribution of Working group III to the Fourth Assessment Report of the Intergovernmental Panel on Climate Change. Cambridge University Press, Cambridge, UK and New York, USA.

Smith, J., Smith, P., Wattenbach, M., Gottschalk, P., Romanenkov, V.A., Shevtsova, L.K., Sirotenko, O.D., Rukhovich, D.I., Koroleva, P.V., Romanenko, I.A. and Lisovoi, N.V. (2007c).

Projected changes in the organic carbon stocks of cropland mineral soils of European Russia and the Ukraine, 1990–2070. Global Change Biology 13:342–356.

Solaro, S. and Brenna, S. (2005). Il carbonio organico nei suoli e nelle foreste della Lombardia. Il suolo 34–36(1–3):24–28. Available via DIALOG. http://www.aip-suoli.it/editoria/bollettino/n1-3a05/n1-3a05_07.htm Cited 13 August 2009.

Stolbovoy, V., Filippi, N., Montanarella, L., Piazzi, M., Petrella, F., Gallego, J. and Selvaradjou, S. (2006). Validation of the EU soil sampling protocol to verify the changes of organic carbon stock in mineral soil (Piemonte region, Italy), EUR 22339 EN, p. 41. Available via DIALOG. http://eusoils.jrc.ec.europa.eu/ESDB_Archive/eusoils_docs/other/EUR22339EN.pdf Cited 13 August 2009.

Stolbovoy, V., Montanarella, L., Filippi, N., Jones, A., Gallego, J. and Grassi, G. (2007b). Soil sampling protocol to certify the changes of organic carbon stock in mineral soil of the european union. Institute for Environment and Sustainability. Version 2. EUR 21576 EN/2. p. 56. Office for Official Publications of the European Communities, Luxembourg. ISBN: 978-92-79-05379-5. Available via DIALOG. http://eusoils.jrc.ec.europa.eu/ESDB_Archive/eusoils_docs/other/EUR21576_2.pdf Cited 13 August 2009.

Stolbovoy, V., Montanarella, L. and Panagos, P. (eds.) (2007a). Carbon sink enhancement in soils of Europe: Data, modelling, verification. EUR 23037 EN. European Communities, 2007. Available via DIALOG. http://eusoils.jrc.ec.europa.eu/ESDB_Archive/eusoils_docs/other/EUR23037.pdf Cited 13 August 2009.

US Dept of Agriculture (USDA). (1992, 1983, 1972). Soil Conservation Service. Soil Survey Staff. National Soils Handbook. Washington, DC.

US Dept of Agriculture (USDA). (1999, 1975). Soil Conservation Service. Soil Survey Staff. Soil Taxonomy, USDA, National natural resources Conservation Service, Washington, DC, USA.

Van Wambeke, A. (1986). Newhall Simulation Model, a Basic Program for the IBM PC [Floppy Disk Computer File]. Dep of Agron, Cornell University, Ithaca, New York, USA.

Vasques, G.M., Grunwald, S. and Sickman, J.O. (2008). Comparison of multivariate methods for inferential modeling of soil carbon using visible/near-infrared spectra. Geoderma, Elsevier, Amsterdam 146(1–2):14–25. doi: 10.1016/j.geoderma.2008.04.007.

Vettraino, B., Carlino, M. and Rosati, S. (2009). La legna da ardere in Italia. Logistica, organizzazione e costi operativi. Progetto RES & RUE Dissemination. CEAR. Available via DIALOG. http://adiconsum.inforing.it/shared/documenti/doc2_56.pdf. Cited 13 August 2009.

Vitullo, M. (2006). Stime del carbonio in foresta: metodologie ed aspetti normativi. Gestione forestale sostenibile, lotta ai cambiamenti climatici e uso delle biomasse forestali: il progetto di ricerca del CISA. Progetto CISA. Porretta Terme, 7 luglio 2006. http://www.centrocisa.it/cisa2008/allegati/eventi/Vitullo_7luglio06.pdf Cited 13 August 2009.

Walkley, A. and Black, I.A. (1934). An examination of the Degtjareff method for determining organic carbon in soils: Effect of variations in digestion conditions and of inorganic soil constituents. Soil Science 63:251–263.

Werner, P.C., Gerstengarbe, F.W., Fraedrich, K. and Oesterle, K. (2000). Recent climate change in the North Atlantic/European sector. International Journal of Climatology 20(5):463–471.

West, T.O., Brandt, C.C., Wilson, B.S., Hellwinckel, C.M., Tyler, D.D., Marland, G., De La Torre Ugarte, D.G., Larson, J.A. and Nelson, G. (2008). Estimating Regional Changes in Soil Carbon with high spatial resolution. Soil Science Society of America Journal 72:285–294. doi: 10.2136/sssaj2007.0113.

West, T.O. and Marland, G. (2003). Net carbon flux from agriculture: Carbon emissions, carbon sequestration, crop yield, and land-use change. Biogeochemistry 63:73–83.

West, T.O. and Post, W.M. (2002). Soil organic carbon sequestration rates by tillage and crop rotation: A global data analysis. Soil Science Society of America Journal 66:1930–1946.

Zdruli, P., Jones, R. and Montanarella, L. (1999). Organic Matter in the Soils in Southern Europe, Expert Report prepared for DG XI.E.3 by the European Soil Bureau.

# Chapter 35
Monitoring Soil Salinisation as a Strategy for Preventing Land Degradation: A Case Study in Sicily, Italy

Giuseppina Crescimanno, Kenneth B. Marcum, Francesco Morga, and Carlo Reina

Abstract Water demand is increasing worldwide. In regions affected by water scarcity such as those located in the Mediterranean basin, water supplies are already degraded, or subjected to degradation processes, which worsen the water shortage. In Sicily, the increasing scarcity of good quality water is expanding irrigation with saline-sodic waters, thus enhancing the risk of secondary salinization and sodification. Adequate management practices are urgently needed for sustainable use of saline/sodic waters. This chapter illustrates how the Geonics EM-38 probe was used for monitoring salinization in a Sicilian area where irrigation with saline water is increasingly practiced, and the risk of salinisation and desertification is envisaged. Electrical conductivity of bulk soil (ECa) measurement grids were taken with the EM-38 in a Sicilian vineyard at several dates, both prior to, and following irrigations, with two distinct irrigation treatments of different salinities (0.6 and 1.6 dS m^{-1} respectively). Though both water sources contributed to field salinity, high salinity source had a more adverse effect. This investigation proved the usefulness of EM-38 for efficient, rapid monitoring of the progressive effects of different irrigation strategies on soil salinization.

Keywords Salinity control · Sodic water irrigation · Crop yield · Sicily

35.1 Introduction

Arid and semi-arid regions of the world are variably affected by soil salinization and sodification. Saline and sodic soils cover about 10% of the total arable lands and exist in over 100 countries. Of irrigated lands, about 344 millions of hectare (23%) are salty and another 560 millions of hectare (37%) are sodic. Saline and sodic soils, although affecting mostly arid and semi-arid regions, are not limited

G. Crescimanno (✉)
Dipartimento ITAF, Università di Palermo, 90128 Palermo, Italy
e-mail: gcrescim@unipa.it

to these regions. According to estimates, 10 millions of hectares of irrigated land are abandoned yearly as a consequence of the adverse effect of irrigation, mainly secondary salinization and sodification (Szabolcs, 1994). Furthermore, demand for fresh water is progressively increasing, due to rapid population growth and urbanization worldwide, resulting in problems associated with aquifer depletion, i.e. salt-water intrusion and use of salt-compromised waters for irrigation (Marcum, 2006). Salinization of water and soil represents a pre-condition for desertification, defined as "degradation of land in arid, semi-arid and dry sub-humid areas resulting mainly from adverse human impact" (UNEP, 1991).

Salinity may have negative direct effects on crop yield by (a) reducing the ability of plant roots to absorb water, due to increased soil osmotic potential, or (b) the direct effects of saline ions, resulting in either toxicity or nutrient imbalances. Threshold relationships between the soil electrical conductivity (EC) and crop yield have been empirically determined for several crops and can be used to evaluate the influence of saline irrigation water on agricultural production. Sodicity adversely impacts soil structure, particularly finer textured soils, due to deflocculation of clay aggregates. Direct effects include reduced soil hydraulic conductivity (Crescimanno et al., 1995), and therefore reduced leaching potential, resulting in accelerated soil salinization. Indirect effects include reduced soil macropores and oxygen exchange, resulting in anaerobic rhizosphere conditions. Root oxygen starvation typically results in reduction, or loss of plant salinity tolerance, due to reduced activity of ATP-driven ion partitioning/exclusion mechanisms at the root endodermis (Barrett-Lennard, 2003).

Salinity in irrigation water is defined as the total sum of dissolved inorganic ions and molecules. Soil salinity is generally measured by determining the electrical conductivity of the soil solution. EC measured in the saturated extract (EC_{sat}), a technique initially developed by USSL, is the most commonly used method of estimating soil salinity that uses reference water content (Rhoades et al., 1989). Though the method is labour-intensive and time consuming, it remains the standard benchmark for estimating soil salinity. Methods suitable for rapid assessment of soil salinity are necessary for survey and monitoring of large land areas susceptible to salinity degradation.

Application of electromagnetic induction sensor technology (EM) makes possible, after calibration, rapid surveys for determination of areas having the greatest hazard of salinisation. In these areas, detailed investigation is necessary to develop countermeasures and strategies suitable to control desertification. EM provides a measurement of the EC of the "bulk soil", or ECa (Hendrickx et al., 1992). As ECa is influenced not only by the chemical and physical properties of the soil solution, but also by those of the solid phase (soil texture, mineralogical composition of the soil), calibration is necessary for the individual soil monitored by EM, by taking an ECe grid sample of the field, thus allowing the conversion of ECa data into EC_{sat} (Corwin and Lesch, 2003; Lesch et al., 2005).

The use of saline-sodic waters for irrigation is prevalent in Sicily. Consider that in the last 30 years, no more than 300 million m^3 of good quality water have been available for irrigation, versus a requirement for 1,600 million m^3. As a result, increasing attention is being paid to the long-term hazards associated with prolonged application of these waters on Sicilian soils (Crescimanno, 2001b). A number of international and national projects have been developed since 1998 in Sicily with the objective of developing integrated approaches for sustainable management of irrigation (Crescimanno, 2001a; Crescimanno and Garofalo, 2005, 2006, 2007). This investigation is part of the "Evolution of cropping systems as affected by climate change" (CLIMESCO) project (2007–2009). CLIMESCO has the objective of developing management scenarios for optimizing the use of limited water resources while concurrently minimizing salinisation and the risk of desertification (Crescimanno and Marcum, 2009). This chapter illustrates results of a methodology based on EM-38 measurements for monitoring the soil salinisation risk in two different irrigation treatments in a vineyard located in western Sicily (Mazara del Vallo, Trapani) during the 2007 irrigation season.

35.2 Materials and Methods

35.2.1 Field and Irrigation Description

Apparent soil electrical conductivity (ECa) was measured in a vineyard (Fig. 35.1) in southwest Sicily, in a soil with silty clay textural class (44% clay, 46% silt, 10% sand). Soils like this are common in Sicily and have high shrink/swell potential, being particularly susceptible thus to salinization. Irrigation water derived from two sources (lake and well), having salinities of 1.6 and 0.6 ECw, respectively. The initial irrigation was made on 12 July 2007 with lake water throughout the field to establish uniform conditions. Second irrigation was a 2-day event, with lake water applied on the left (L) side of the field on 17 July, followed by well water on the right (R) side on 18 July. Third irrigation was again a 2-day event, with lake water applied on the L side of the field on 25 July, followed by well water on the R side on 26 July. Table 35.1 reports the irrigation schedule adopted. Irrigation amount was 15 mm depth per event. Reference evapotranspiration deficit (ETd) was on average 5–7 mm day^{-1} throughout the irrigation period, where ETd = ETo – measured precipitation (there was no measurable rainfall during the experimental period).

35.2.2 EM-38 Measurements

EM-38 measurements were taken using a grid pattern across the L and R treatments plots before the first irrigation, and subsequently after irrigation events. Figure 35.2 illustrates the plants rows located in the L and R treatments, and the sites along the rows where EM-38 measurements were taken. The first measurement was taken on 21 June, prior to irrigation. Two subsequent measurements were taken on 18 July, and again on 23 July, following the first two irrigation events (12 July and 17–18 July). The final measurement was taken on 27 July, following the third irrigation event. Table 35.1 reports the EM-38 measurement schedule adopted.

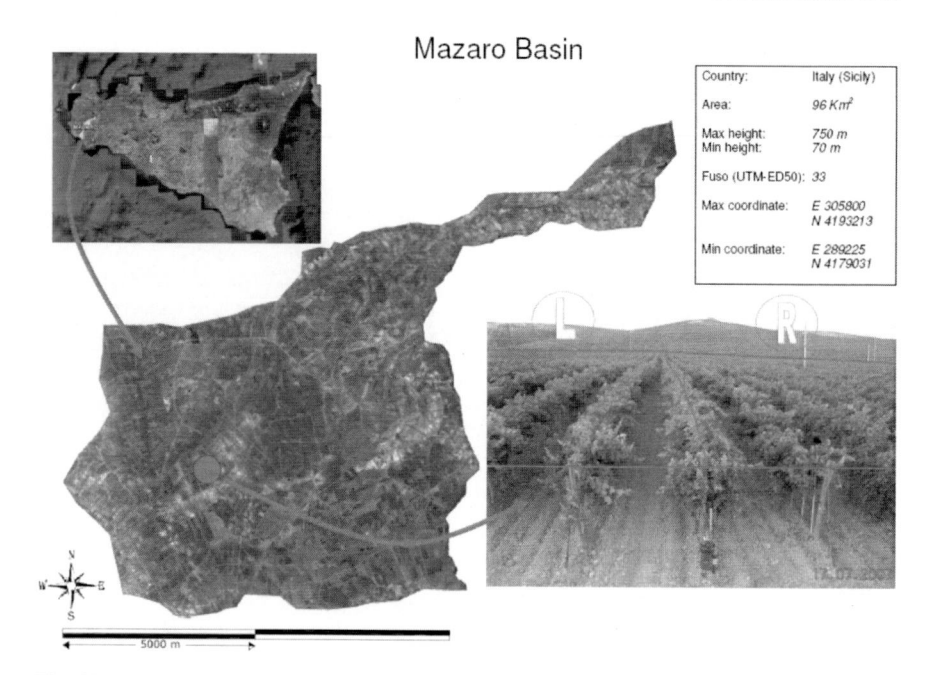

Fig. 35.1 Mazaro basin, Sicily (Italy), and location of the two treatments (Left, L, and Right, R) at the Foraci Farm

Table 35.1 Irrigation schedule adopted, measurement date, zonal mean EC_{sat} and the relative crop yield calculated according to Maas and Hoffman (1977)

Irrigation date	EM-38 measurement date	ZM_EC$_{sat}$ (dS m^{-1}) R	ZM_EC$_{sat}$ (dS m^{-1}) L	Relative crop yield (%) (Maas and Hoffman, 1977) R	Relative crop yield (%) (Maas and Hoffman, 1977) L
20 June 2007	–	–	–	–	–
–	21 June 2007	1.70	1.77	98.1	97.4
12 July 2007	–	–	–	–	–
–	18 July 2007	1.56	1.83	99.5	96.8
17–18 July 2007	–	–	–	–	–
–	23 July 2007	1.51	1.59	100	99.1
26 July 2007	–	–	–	–	–
–	27 July 2007	1.58	1.76	99.3	97.5

35.2.3 Instrumentation and Technique

ECa was monitored using a grid pattern, by an electromagnetic inductive meter (Geonics EM-38 probe, Geonics Limited, Mississauga, Ontario). The Geonics EM-38 sensor has an intercoil spacing of 1 m and operates at a frequency of 13.2 kHz, allowing measurement of ECa to effective depths of approximately 1

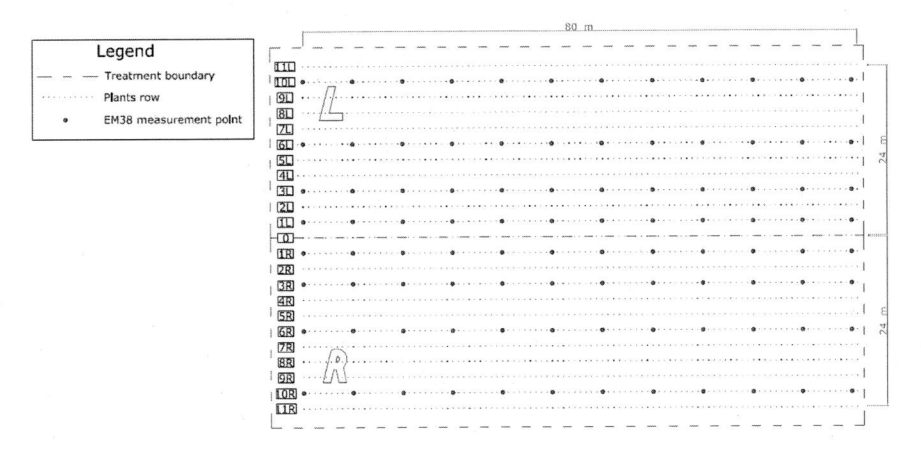

35 Monitoring Soil Salinisation as a Strategy for Preventing Land Degradation

Fig. 35.2 Left (L) and Right (R) treatments at the Foraci farm, plants rows and EM-38 measurement points

and 2 m when placed at the ground level in horizontal and vertical configuration respectively. An EM transmitter coil located in one end of the instrument induces circular eddy-current loops in the soil. The magnitude of these loops is directly proportional to the electrical conductivity (EC) of the soil in the vicinity of that loop. Each current loop generates a secondary electromagnetic field that is proportional to the value of the current flowing within the loop. The receiver coil intercepts a fraction of the secondary induced electromagnetic field from each loop, and the sum of these signals is amplified and formed into an output voltage that is linearly related to a depth-weighted soil EC.

For calibration of ECa data, soil samples were taken for measurements of the soil electrical conductivity, EC_{sat} (saturated extract) at representative points within the grid pattern, at 60 cm depth.

35.3 Results and Conclusions

Vertical EM-38 readings (EMv) were converted into EC_{sat} values according to the following calibration equation, obtained via data distributions taken in April 2007 and subsequently in June 2007:

$$EC_{sat} = 0.36 \cdot EMv + 1.10 \tag{35.1}$$

with R (correlation coefficient) $= 0.83$ (R significant at $P = 0.001$) and SEE (standard error of estimate) $= 0.16$ dS m^{-1}.

EC_{sat} values were elaborated with geostatistical techniques (Wackernagel, 2003) in order to create maps showing salinity distribution following irrigation events in both L and R treatments. EC_{sat} values were interpolated in a GIS environment, using a Gaussian semivariogram to express spatial correlation. Zonal statistics calculated

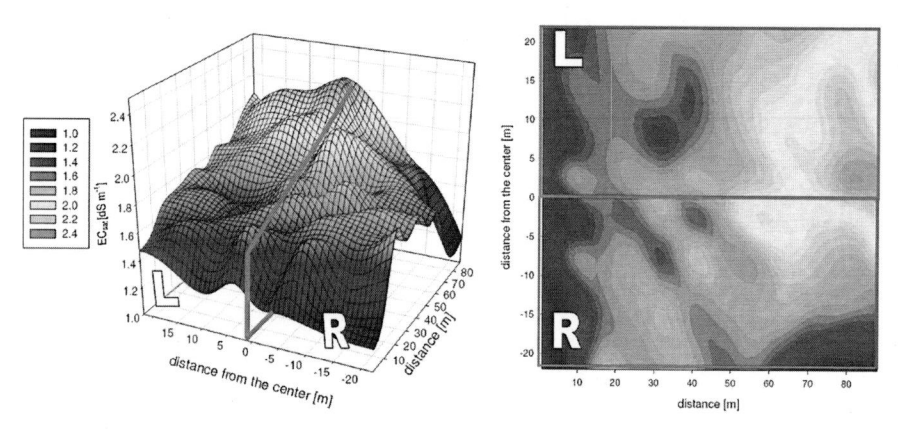

Fig. 35.3 Salinity distribution (3-D) and salinity map (2-D) on 21 June 2007

by ArcGIS® were used to determine zonal mean EC$_{sat}$ (ZM_EC$_{sat}$) for the L and R treatments, the value of which is reported in Table 35.1 and illustrated in Fig. 35.7.

First EM-38 survey (21 June) revealed the initial salinity distribution in the field prior to any irrigation (Fig. 35.3). The 3-D and 2-D maps show an increasing salinity gradient across the entire field, with EC$_{sat}$ values ranging from approximately 1 dS m^{-1} up to 2.4 dS m^{-1}. Higher salinity values occurred in the left part of the field, with ZM_EC$_{sat}$ equal to 1.77 dS m^{-1} in L, and 1.69 dS m^{-1} in R side, respectively.

Second EM-38 survey (18 July) describes the salinity distribution subsequent to irrigation of the whole field with lake water on 12 July, followed by irrigation of the left side only with lake water on 17 July. The 3-D and 2-D maps, illustrated in Fig. 35.4, reveal an increase in salinity on the L side, with ZM_EC$_{sat}$ = 1.82 dS m^{-1} (Fig. 35.7 and Table 35.1). However, salinity had a considerable spatial variability, and in some parts of the field salinity increased from 2 to 2.4 dS m^{-1}. The maps also show a decrease in salinity in the R side, with ZM_EC$_{sat}$ = 1.56 dS m^{-1}. Decrease in the value of ZM_EC$_{sat}$ compared to the value of 21 June indicates that some salt-leaching took place on the R side only, decreasing ZM_EC$_{sat}$ from 1.69 to 1.56 dS m^{-1}.

Third EM-38 reading (23 July) describes the salinity distribution following the initial irrigation of the R side with well water on 18 July, subsequent to irrigation of the L side with lake water on 17 July. Though this irrigation was expected to result in a further decrease in salinity on the R side only (Fig. 35.5), there was actually a decrease in salinity on both R and L sides, compared to the previous reading of 18 July (Fig. 35.4). This would indicate that this irrigation event resulted efficient in salt leaching on both R (ZM_EC$_{sat}$ decreased from 1.56 to 1.51) and L sides (ZM_EC$_{sat}$ decreased from 1.82 to 1.58). The decrease in ZM_EC$_{sat}$ measured on the L side could also be due to a significant salt redistribution from the L to the R side of the field resulting from slope effects, as this salinity reading was taken some days after irrigation events occurred.

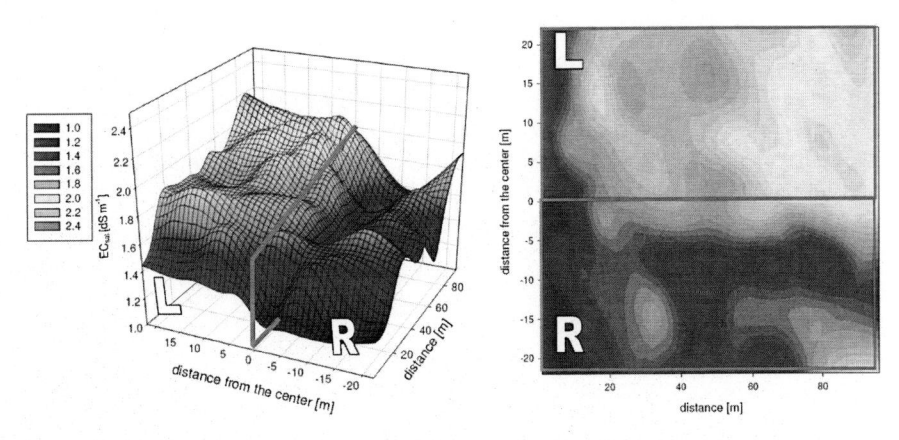

35 Monitoring Soil Salinisation as a Strategy for Preventing Land Degradation 473

Fig. 35.4 Salinity distribution (3-D) and salinity map (2-D) on 18 July 2007

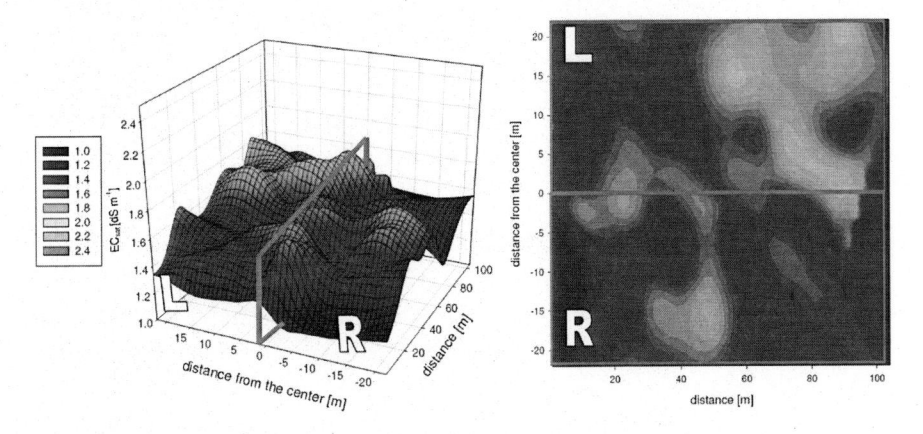

Fig. 35.5 Salinity distribution (3-D) and salinity map (2-D) on 23 July 2007

The final EM-38 reading (27 July) describes salinity distribution following irrigation of the whole field with lake (25 July, L side) and well (26 July, R side) water (Fig. 35.6). Left side shows an increase in salinity, in both maximum values (EC$_{sat}$ up to 2.4 dS m^{-1}), and ZM_EC$_{sat}$ (ZM_EC$_{sat}$ =1.75 dS m^{-1}, Table 35.1), and also in the relative field area affected, compared to the previous date, indicating a clear impact due to irrigation with the saline water source. Salinity is not uniformly distributed, with some areas undergoing maximum accumulation of salts. However, the R side also shows an increase in the ZM_EC$_{sat}$ (ZM_EC$_{sat}$ = 1.58 dS m^{-1}) compared to survey of 23 July (ZM_EC$_{sat}$ 1.51 dS m^{-1}), which could also be explained by redistribution of salts from the L to the R side, as previously observed in the 23 July survey.

Comparing ZM_EC$_{sat}$ values prior to any irrigation (Fig. 35.3) with values after all irrigations (Fig. 35.6) in the R and L sides, it can be seen that there was significant salt leaching in the R side, with a reduction of 6.5% in salinity from 21 June to

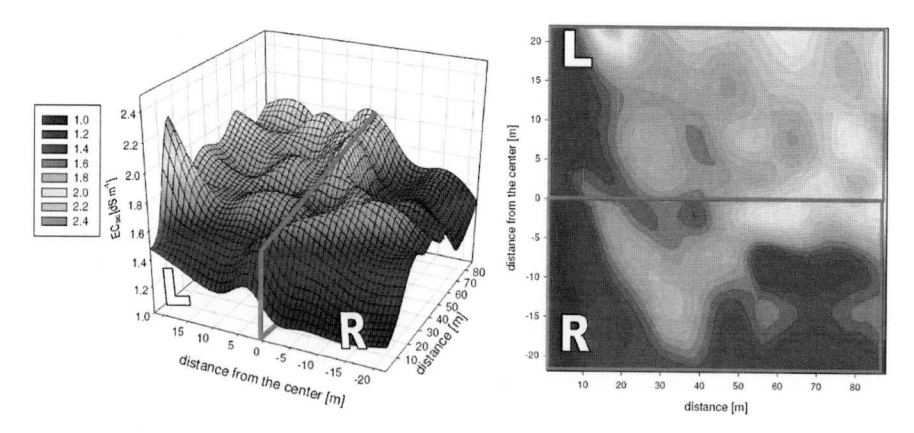

Fig. 35.6 Salinity distribution (3-D) and salinity map (2D) on 27 July 2007

Fig. 35.7 Zonal mean electrical conductivity (ZM_EC$_{sat}$) values for the R and L treatments.

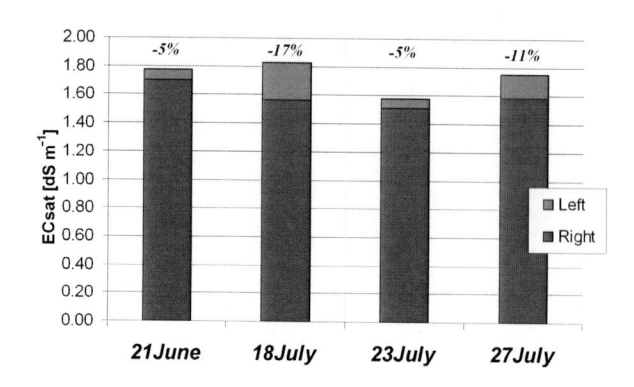

27 July 2007 (Fig. 35.7). This leaching might have been more significant if no flux from L to R had taken place during the irrigation season. In addition, an insignificant reduction in salinity (equal to 1.12%) was caused by irrigation in the L side from 21 June to 27 July 2007, perhaps also due to lateral water flux.

In conclusion, use of a moderate salinity water source (well water treatment) for irrigation proved to be an efficient practice, decreasing salinity in the R side of the field compared to salinity conditions observed in the L plot, following an annual irrigation season. ZM_EC$_{sat}$ values calculated during the irrigated season were converted into crop yield reduction according to the Maas and Hoffman equation (1977) (Table 35.1). Relative crop yield values calculated after irrigation events confirm more favourable salinity and crop conditions occurring in the R plot, relative to the L plot. Results indicate replacing lake water with well water irrigation to be a more sustainable practice for preventing soil salinisation in western Sicily.

This investigation also proved the usefulness of EM-38 for efficient, rapid monitoring of the progressive effects of different irrigation waters on soil salinisation, and for analysing spatial distribution of salinity before and after irrigation. Further investigation is under way using the same methodology in order to predict long-term salinity distribution changes, and to develop management scenarios aimed at reducing salinization and soil degradation.

Acknowledgments This research was supported by Ministero dell'Università e della Ricerca scientifica (MIUR, Italy) under the project: "Evoluzione dei sistemi colturali in seguito ai cambiamenti climatici" (CLIMESCO).

References

Barrett-Lennard, E.G. (2003). The interaction between waterlogging and salinity in higher plants: Causes, consequences, and implications. Plant Soil 253:35–54.

Corwin, D.L. and Lesch, S.M. (2003). Application of soil electrical conductivity to precision agriculture: Theory, principles, and guidelines. Agronomy Journal 95:455–471.

Crescimanno, G., Iovino, M. and Provenzano, G. (1995). Influence of salinity and sodicity on soil structural and hydraulic characteristics. Soil Science Society of America Journal 59:1701–1708.

Crescimanno, G. (2001a). An integrated approach for sustainable management of irrigated lands susceptible to degradation/desertification. Final Report ENV7-CT97–0681.

Crescimanno, G. (2001b). Irrigation practices affecting land degradation in Sicily. PhD Dissertation Thesis. Wageningen University, The Netherlands, ISBN 90-5808-426-4.

Crescimanno, G. and Garofalo, P. (2005). Application and evaluation of the SWAP model for simulating water and solute transport in a cracking clay soil. Soil Science Society of America Journal 69:1943–1954.

Crescimanno, G. and Garofalo, P. (2006). Management of irrigation with saline water in cracking clay soils. Soil Science Society of America Journal 70:1774–1787.

Crescimanno, G. and Garofalo, P. (2007). Irrigation strategies for optimal use of saline water in Mediterranean agriculture. International Conference on "Water Saving in Mediterranean Agriculture and Future Research needs". CIHEM, Bari, 14–17 February 2007.

Crescimanno, G. and K.B. Marcum (2009). Irrigation, salinization and desertification. Evolution of Cropping Systems as Affected by Climate Change. (CLIMESCO), Aracne (in press) (ISBN 978-88-548).

Hendrickx, J.M.H., Baerends, B., Raza, Z.I., Sadig, M. and Akram Chardhry, M. (1992). Soil salinity assessment by electromagnetic induction of irrigated land. Soil Science Society of America Journal 56:1933–1941.

Lesch, S.M., Corwin, D.L. and Robinson, D.A. (2005). Apparent soil electrical conductivity mapping as an agricultural management tool in arid zone soils. Computers and Electronics in Agriculture 46:351–378.

Maas, E.V. and Hoffman, G.J. (1977). Crop salt tolerance – current assessment. Journal of Irrigation and Drainage Division, ASCE 103:115–134.

Marcum, K.B (2006). Use of saline and non-potable water in the turfgrass industry: Constraints and developments. Agricultural Water Management 80:132–146.

Rhoades, J.D., Nahid, A., Manteghi, P., Shouse J. and Alves, W.J. (1989). Estimating soil salinity from saturated soil-paste electrical conductivity. Soil Science Society of America Journal 53:428–433.

Szabolcs, I. (1994). Prospects of soil salinity for the 21st century. 15th International Congress of Soil Science. Acapulco, Mexico.

UNEP. (1991). Status of desertification and implementation of the United Nations plan of action to combat desertification. UNEP, Nairobi.

Wackernagel, H. (2003). Multivariate Geostatistics: An Introduction with Application. Springer, New York, p. 387.

Chapter 36
Severe Environmental Constraints for Mediterranean Agriculture and New Options for Water and Soil Resources Management

N. Colonna, F. Lupia, and M. Iannetta

Abstract Mediterranean areas are characterized by strong and complex relationships between land use dynamics and land degradation process. The latter will probably worsen in the coming years because of climatic changes that are likely to cause a reduction of the quantity and quality of land resources. Understanding the relationships between land use dynamics and land degradation could allow evaluating the potential risks affecting a given area, and may provide means to implement suitable management policies for sustainable development especially in rural areas. This chapter describes the development of a methodology for the assessment of four typologies of risk in a given area (e.g. a river basin): land desertification, soil erosion, water degradation and soil sealing. This methodology could be useful for regional managers during the policy making process. The methodology was designed to provide an estimate of the risks for the current state and that for the year 2020 under various land resources management scenarios. The methodology was applied at the Ofanto river basin (South Italy). The approach we followed is expert based and makes use of GIS tools in order to produce easy to understand results and maps. It seems to be useful whenever decision makers have to deal with scarce data relative to large areas, and when different stakeholders are involved in a participative process.

Keywords GIS · Land degradation · Land systems · Mediterranean agriculture · Risk mapping

36.1 Introduction

Several research activities have been carried out in the drylands of the Mediterranean region during the last decade. This is an area characterized by a complex and

---

N. Colonna (✉)
Department of Biotechnologies, Agroindustry and Health Protection, ENEA, 00123 Rome, Italy
e-mail: nicola.colonna@enea.it

long history of land use (Thornes and Brandt, 1996). Beginning in the 1950s, large changes in the traditional land use occurred. The modernization and intensification of agriculture required increasing inputs in terms of fertilizers, water and heavy machinery (Margaris et al., 1996). The growing human pressure on the typically fragile Mediterranean agroecosystems brought about an increase of land degradation process.

The relationships between land use dynamics and land degradation are complex and many factors, both natural and non-natural, can have an impact on human activities. The cause-effect relationships between some natural and anthropic factors and specific degradation processes (e.g. soil erosion) have been studied for a long time and are well known (Wishmeier and Smith, 1978; Poesen, 1995). However, for more complex aspects (such as desertification) that involve physical, chemical, and biological processes in addition to the local socio-economic dimension, these relationships are still unclear and more research is needed (Iannetta et al., 2005). One of the more common approaches researchers are using to deal with this complexity is to develop models that can represent the reality. The definition of a model allows to clarify the conceptual perception of the process or the environmental system and to quantify the influence of the various factors that characterise it (Cross and Moscardini, 1985).

In a model, a real system is identified by its characteristics, its various components and interactions. The more these three elements are complex, the higher is the number of rules, functions and equations that must be implemented in the model itself. Moreover models can be divided into static and dynamic ones, the latter varying in time and typically also in space (Goodchild et al., 1993). The evolution of computer technology helped scientists to develop very complex models and their applications.

Since the 1970s, models have been largely used to support planning processes, to define future scenarios by projecting variables in time, to analyse and interpret the information deriving from monitoring campaigns on the status of the environment, to evaluate the effects of management policies in environmental issues and to evaluate the different forms of risk that affect the environment and landscape (Kersten and Gordon, 1999). The integration of models and data in a computer based system in order to pursue any of the applications above constitutes a Decision Support System (DSS) that can help a decision makers in the solution of unstructured problems (Scott Morton, 1971). Nowadays, a large number of DSSs has been developed and used by researchers in different fields of application and contexts. Since the 1980s several efforts have been made to develop specific models and DSS tools for the analysis and management of river basins (Gassman et al., 2009).

Many different tools have been developed, commercial or not (e.g. MIKE-SHE, SWAT) and are currently available to assess at catchment scale several different processes: runoff, erosion, sediment transport, water balance, climate change impacts and so on.

For example the Soil and Water Assessment Tool (SWAT) model is the results of 30 years of modelling efforts conducted by the USDA (United States

Department of Agriculture) that, over time, came to include different and specific models. Progressive enhancements during three decades introduced modifications that extended the model's capability to deal with a wide variety of watershed quality management problems (Gassman et al., 2009). Another example is Mike-She, a commercial software developed in 1982 under the name Système Hydrologique Européen (SHE). It is an integrated modelling framework to simulate all processes, and their interactions, involved at the basin level that could be tailored on the user needs and has been used extensively in several different contexts (Refshaard and Storm, 1995). All those models, and several others, are still evolving by integrating with other submodels or software applications (e.g. GIS tools) to increase their accuracy and their ability to show and map the simulation results.

DSSs have usually been developed in the context of high technology structures, such as private companies, universities or research institutes, where the personnel and technology to collect, catalogue and manage data are available (Rais et al., 1999). The software and tools that have been developed are sensitive to the quality of data and information and require highly skilled personnel to implement, manage and run them.

The risk is that these tools fail to address the user's needs and are not well accepted by the intended final users (policy makers) as has been reported (Mysiak, 2005). Some of the reasons underlying a low acceptance rate are: system complexity, high demand user interfaces, low transparency of the system's mode of operation and, last but not least, the difficulties of adaptation to the specific organisational and cultural context (Mysiak, 2005). Therefore, there is a need to reconsider the approaches to modelling and DSS development moving from what we may call a "model focused mind" to an "end-users needs" attitude.

With the aim to bridge the gap between the knowledge generated (results, maps) by DSSs with the processes underlying a specific phenomena and the practice of formulating policy to detect, prevent and resolve risks, one strategy could be make the tools more user friendly, communicative and transparent (Van Buuren et al., 2002) and to help local administrations and stakeholders to better clarify their needs to avoid a mismatch between requested and supplied functionality of the DSS.

Our work built on the large analytical experience that has been accumulated with modelling and attempted to reconsider the crucial role of the expert and to bring models closer to the decision makers and local stakeholders. Here, we describe our first field experience in an area characterized by large data scarcity and where local admistrators and stakeholders were not familiar with these kind of tools (Colonna et al., 2008).

In the framework of European IMAGE project (INTERREG III B) a methodology for land degradation risk assessment was defined, in order to evaluate both the current risks and, on the base of hypothetical scenarios, the expected risks for the year 2020. The methodology has been applied in the Ofanto river basin (South Italy), evaluating four main risk typologies: soil sealing, water degradation, soil erosion and land desertification.

36.2 Materials and Methods

36.2.1 Study Area

The Ofanto river basin has an extension of about 2,700 km^2 and its border cross the boundary of three different Italian Southern regions: Campania, Basilicata and Puglia. The main form of land use is forest and agriculture, especially arable land (Fig. 36.1). Large areas are cultivated with cereals (winter wheat) especially on the low hills, while grapes and olive trees occupy 7% of the total cultivated land particularly in the high hills and mountain areas. About 16% of land is occupied by pasture and grasses for sheep livestock. Intensive agriculture (irrigated horticulture) is mainly located in the lowlands along the riverbanks.

The pilot area is a representative site for the Mediterranean conditions. In particular, during the last 30 years the area under investigation has undergone relevant land use changes with evident phenomena of land abandonment, in the marginal areas, and agricultural intensifications of plain flat zones. As a preliminary investigation we compared land use maps of the Ofanto basin from 1960s until 2000.

The urban growth was very high in percentage but not relevant in terms of hectares. While in the lowlands we registered an intensification of land uses (from pastures to cereals, from cereals to horticulture) in the upper area we observed a less intensive land use due to land abandonment and to the consequent growth of spontaneous vegetation and expansion of semi-natural areas.

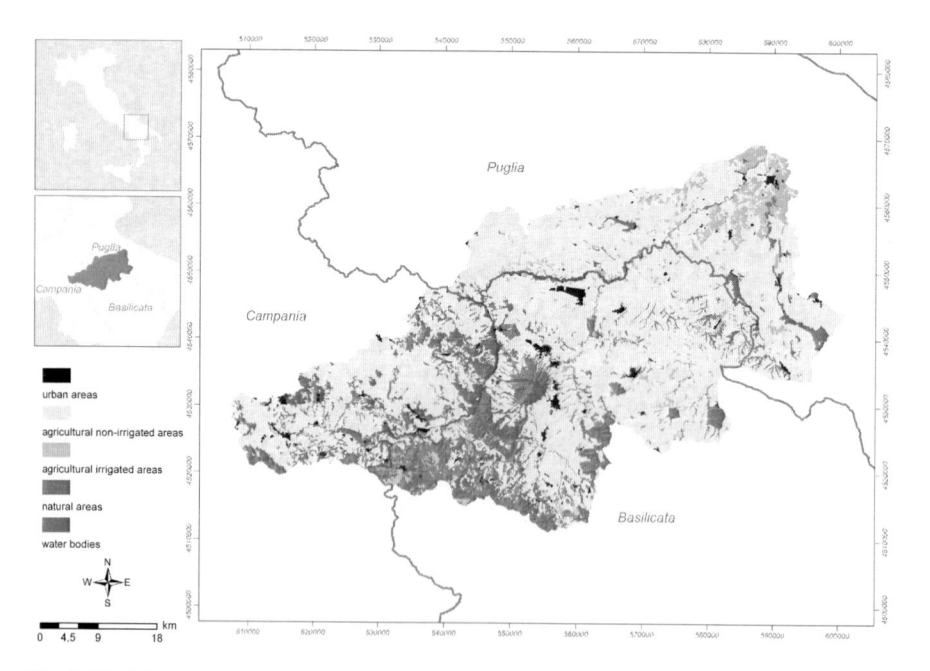

Fig. 36.1 Ofanto river basin: land cover map (2005)

The main driving forces of Mediterranean agriculture are the Common Agricultural Policy of the European Union and the consumer characteristics associated to socio-economic trends (Margaris et al., 1996). In our area, we observed that around 50% of land underwent a change in use, either intensification or an extensification in relation to land morphology, soil characteristics and water availability. Agriculture intensification can cause or promote soil erosion, soil compaction, and biochemical degradation of water quality, while agriculture extensification could be the first step towards abandonment, particularly in high hills or mountain area. Abandonment is considered to be a driving force to desertification because the next steps associated with it are very often the hydrogeological and fire risks.

As a consequence of the processes described above and taking into account the growing pressure of agricultural activities and the conflict for water exploitation between the different irrigated areas, the assessment of the main risks affecting the area are particularly valuable. They could represent a starting point for supporting policy makers throughout the decision making process for both allocation of water resources and the identification of sound conservative land resources management strategies.

36.2.2 The Approach for the Current Risk Evaluation

The aim of the work was to define a suitable methodology able to produce risk maps, for the Ofanto river basin, considering both the existing situation and the possible expected conditions generated by different scenarios. The methodology is therefore differentiated in current (M1) and the expected risk evaluation (M2).

The applied approach is expert-based, therefore qualitative, and the results obtained depend heavily on expert judgment. Notwithstanding it can be considered a good approach to derive information about land degradation for the territorial scale of the produced maps when the lack of quantitative data prevent the application, or make it difficult for analytical risks modelling. M1 methodology is described by the workflow of Fig. 36.2.

The core of M1 is an integrated analysis of the entities: Land Systems, Land Transition (1960–2000) and slope, which are interrelated through a three entries cross-matrix called EMIA (*Expert Matrix for Impact Evaluation*) used for the expert-based risks definition.

Land Systems (FAO, 1976), set up in scale 1:100,000, have been delineated detecting homogeneous geographic units in terms of environmental factors and agro-forestry resources, capable to influence their potential use and the possible degradation processes dynamics (Dent and Young, 1981). The result is a Land Systems map made up of different permanent environmental structures, linked to the integrated long-term action of climate, lithology, morphology, biotic community and permanent human changes (e.g. reclaimed land, terrace). The map of land systems (Fig. 36.3) was produced starting from the regional Pedological map, by

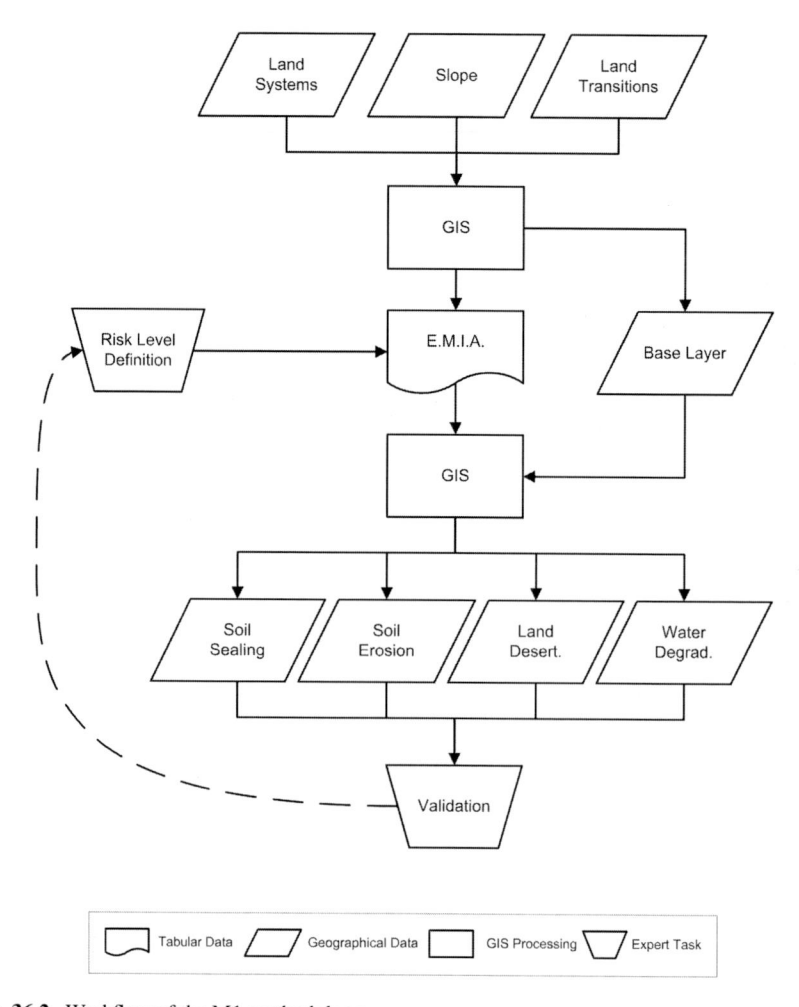

Fig. 36.2 Workflow of the M1 methodology

aggregating cartographical units at hierarchical levels of provinces and identifying physiographical areas, which are substantially homogeneous at regional scale.

The map could be considered a preliminary tool to analyse and evaluate natural resources of the rural space, the sustainability of agro-forestry and the degradation risk determined by the historical land use (FAO, 1995).

Land Transition map was produced by a land use change analysis using 1960 and 2000 land cover maps. A pre-elaboration was developed in order to harmonize map legends and produce a new one with a six items nomenclature: wood and shrub, grassland, agricultural and agro-forestry complex systems, permanent crops, annual crops and urban areas.

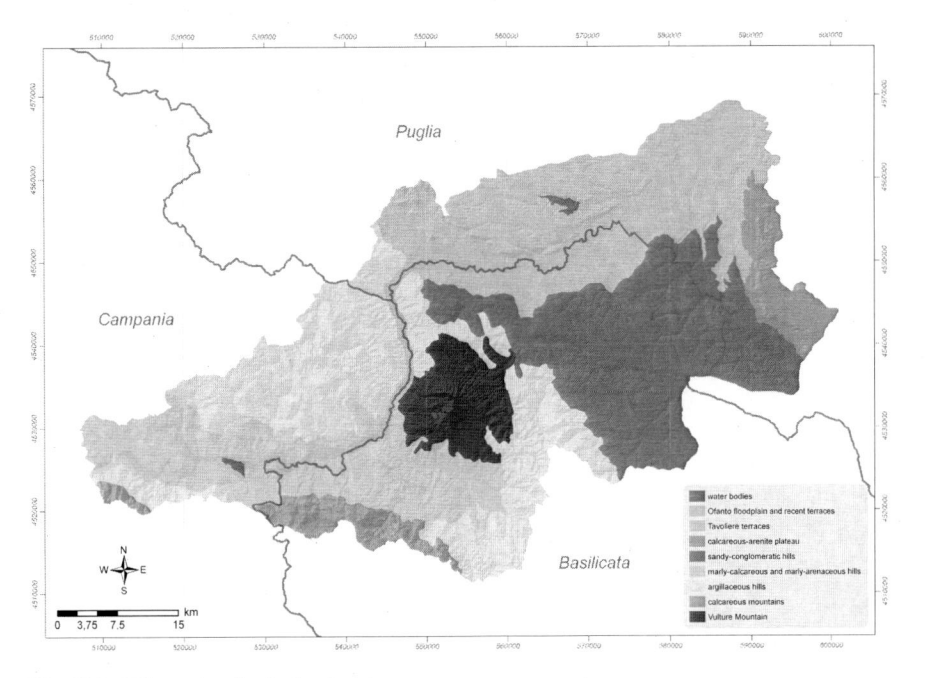

36 Severe Environmental Constraints for Mediterranean Agriculture 483

Fig. 36.3 Ofanto river basin: land systems map

Table 36.1 shows the main land cover dynamic typologies between 1960 and 2000 and in Table 36.2 different transition typologies are quantified for the same period.

The results showed some typical land use changes such as the progressive disappearance of pastures and grasslands in favour of the growth of woodland and shrubs as well as of cultivated area.

Table 36.1 Identification of main land cover dynamic typologies between 1960 and 2000

Land use transitions		Year 2000					
		Wood and shrubs	Grassland	Arable lands	Permanent crops	Complex agroforestry systems	Urban areas
Year 1960	Wood and shrubs	For_Pe	Def_Gr	Def_Ag	Def_Ag	Def_Ag	Urb_Tr
	Grassland	Gra_Fo	Gra_Pe	Gra_Ag	Gra_Ag	Gra_Ag	Urb_Tr
	Arable lands	Agr_Fo	Agr_Gr	Agr_Pe	Agr_In	Agr_In	Urb_Tr
	Permanent crops	Agr_Fo	Agr_Gr	Agr_Es	Agr_Pe	Agr_In	Urb_Tr
	Complex agroforestry systems	Agr_Fo	Agr_Gr	Agr_Es	Agr_Es	Agr_Pe	Urb_Tr
	Urban areas	–	–	–	–	–	Urb_Pe

Table 36.2 Hectares of land according to the different transitions typologies

	Sigla	Land transitions	Hectares	Basin percentage %
Intensification	Def_Gr	Deforestation for grazing	284.0	0.1
	Def_Ag	Deforestation for agriculture	7,863.3	2.9
	Gra_Ag	Grazing tilling	34,026.1	12.6
	Agr_In	Agricultural intensification	34,365.3	12.7
	Urb_Tr	Urban transformation	3,627.3	1.3
Persistence	For_Pe	Forest persistence	16,265.1	6.0
	Gra_Pe	Grazing persistence	3,084.7	1.1
	Agr_Pe	Agriculture persistence	116,541.7	43.0
	Urb_Pe	Urban persistence	700.5	0.3
Extensification	Gra_Fo	Grazing afforestation	10,777.7	4.0
	Agr_Fo	Agriculture afforestation	18,826.7	6.9
	Agr_Gr	Agriculture by grazing	3,909.9	1.4
	Agr_Es	Agriculture extensification	19,197.8	7.1

The last entity considered in the methodology is the "Slope" layer, produced processing a 20 m resolution DEM (*Digital Elevation Model*). For each land system 4 different slope intensity are considered (<8%, 8–25%, 25–50%, >50%).

The three maps was intersected using GIS to produce a base layer, it represents the grid of polygons used to render the risk maps for the investigated area.

EMIA matrix (Table 36.3) was processed by experts using a spreadsheet, associating the risk levels intensity for each Land Transition and for each Land System using the slope values as intensity modulator. The association procedure for risk level intensity is based on experts' judgment who have a comprehensive knowledge of the processes active inside each Land System of the study area and that is open to discuss it with local stakeholders. Four levels order the risk intensity: no, low, medium and high risk. As an example the erosion risk map of the pilot area (the Ofanto basin area within the Basilicata boundaries) is showed in Fig. 36.4.

EMIA matrix was realized for each risk type in order to subsequently create the relative risk maps by GIS processing. M1 final step is the risk map validation, through field campaigns, aimed at the confirmation of the risk intensity; this have been done for a consistent number of polygons representative of all the possible different conditions contained in the EMIA matrix. The validation may conduct, in case of inconsistency, to the redefinition of the intensity value defined by the experts in the EMIA matrix.

36 Severe Environmental Constraints for Mediterranean Agriculture

Table 36.3 Example of EMIA matrix to be filled in with the risk level intensity (0,1,2,3)

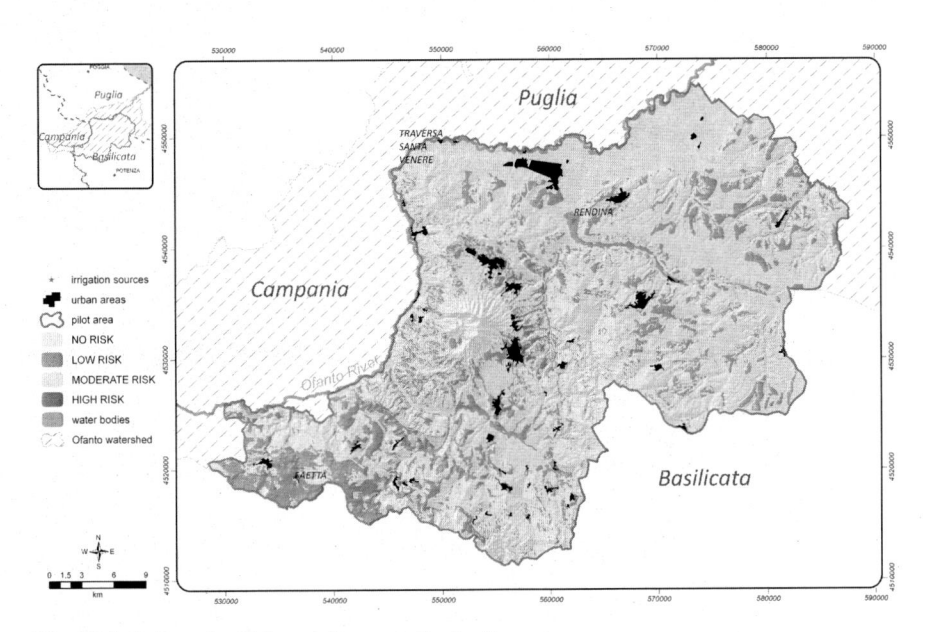

Fig. 36.4 Soil erosion risk map: "current situation"

36.2.3 The Approach for the Expected Risk Evaluation

M2 methodology has been developed in order to produce expected risk maps according to different foreseen scenarios; it is always an expert-based approach for risk evaluation. Two scenarios type have been considered: infrastructural and water saving. The former deals with the direct public intervention to restore dams and creating some new networks to increase water availability and enlarge irrigated areas, the latter focus on different actions to decrease water consumption by reducing network water losses, changing crops and irrigation methods.

Essentially M2 is based on a transformation of the current risk maps, by increasing or decreasing differentially the risk intensity at global scale (for example considering a Climate Change Scenario) or locally (for example for an Irrigation Network Expansion Scenario).

M2 is fully implemented in a GIS environment using geo-processing functions and different geographical data. The main inputs is represented by the current risk map that was obtained applying M1 (Fig. 36.5). M2 methodology is based on a scenario description consisting of a full statement of the involved phenomena with the possible effects on the landscape, like is shown in the Table 36.4 for the Erosion Risk.

Scenarios statements are used to define adequate rules for risk intensity transformation. The rules are implemented in the GIS environment through simple scripts in Visual Basic programming language and they drives the risk intensity transformation in a differential way considering the values of Land System, Land Transformation and slope coupled with additional layer (Table 36.4). The aim of the additional layer (for example an aridity risk map or a suitability irrigation map) is, besides the risk intensity modulation, the delineation of masks for applying the transformation.

The *scenario options* are all environmental, structural and technical conditions, which can determine the change in the current balance in the management of the Ofanto river basin system.

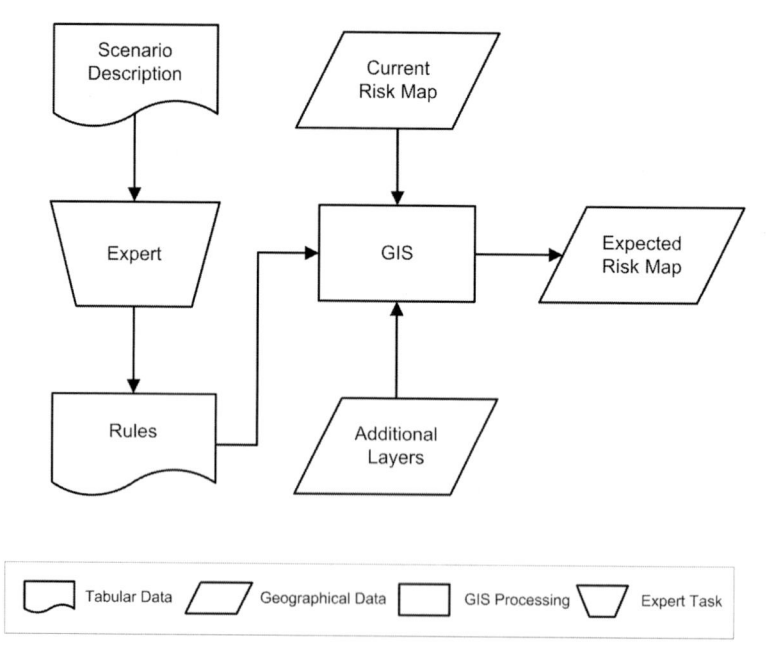

Fig. 36.5 Workflow of the M2 methodology

36 Severe Environmental Constraints for Mediterranean Agriculture

Table 36.4 Table summarizing assumptions and rules attributions to create one scenario for evaluating erosion risk in a business as usual scenario (example)

	Scenario no.1
Code	01 S_CC
Name	Business as usual but with climate change
Scenario description	Temperature rise, precipitation reduction and concentration, with higher intensity and short event duration (IPCC scenario)
Effects extension	Entire pilot area
Soil erosion risk	
Foreseen impacts	Decrease of soil coverage in rainfed areas Increase of soil erosion phenomena on medium and high slope zones
Impacts rules definition	The risk will increase in the zones where we observe an high value of aridity risk. In some areas the increased aridity will lead to a reduction of vegetation cover and the uncovered soil will be potentially affected by increased erosion phenomena when high intensity precipitation will occur.
Data required	- Ofanto current risks map - Ofanto aridity risk map

Rules description by VBA code	`'----------------------`
	`' scenario 1`
	`'----------------------`
	dim out as integer
	
	if [erosion_1] <> 999 then
	 if [AI]="H" or [AI]="VH" then
	  if ([LT] ="DbP" or [LT] ="PeA" or [LT] ="DsA" or [LT] ="DbA" or [LT] ="EsA" or [LT] ="InA" or [LT] ="EsP") and [SLOPE]=2 or ([SLOPE]=3 or [SLOPE]=4) then
	   out= [erosion_1] +1
	  else
	   out=[erosion_1]
	  end if
	 else
	  out=[erosion_1]
	 end if
	else
	 out=999
	end if

They can be human-controlled (and therefore can be planned and controlled) or, can depend on the environment (e.g. climate change) in which case cannot be planned and, to a certain extent, not even foreseen. In any case, the influence of all (human and non-human) factors on the model can be estimated and used in the decision making process.

A model building SW, called Environmental Model Builder (EMB, Iannetta et al., 2007) previously developed for another project and used as an engine to implement ad-hoc mathematical models (Compagnone et al., 2006) was implemented and tested to integrate data and maps, manage the matrix and the scenarios and implement the workflow showed in Figs. 36.2 and 36.5.

EMB is a Web Based tool, which allows performing simulations. It can be used independently by several users via a web browser interface and allows for multi-user sharing of the various models and maps.

This tool is addressed to various user types: expert researchers, that can create models with data and relationships, technical users that can change the input data, or change various options and analyse the results; and non-technical users that can displays the results and compare the various alternative scenarios.

36.3 Results and Discussion

By integrating all our workflow procedures (M1 and M2) via EMB we could produce scenario maps as the one showed in Fig. 36.6 about the evolution of soil erosion risk at the year 2020 in a climate change hypothesis without any change in rural policies.

Map interpretation by the use of 4 different colour levels is very easy and show that the areas with higher erosion risk are going to increase in the future according to the rules shared and implemented. The whole procedure, although based on specific maps that have been produced by complex analysis, is easy understandable.

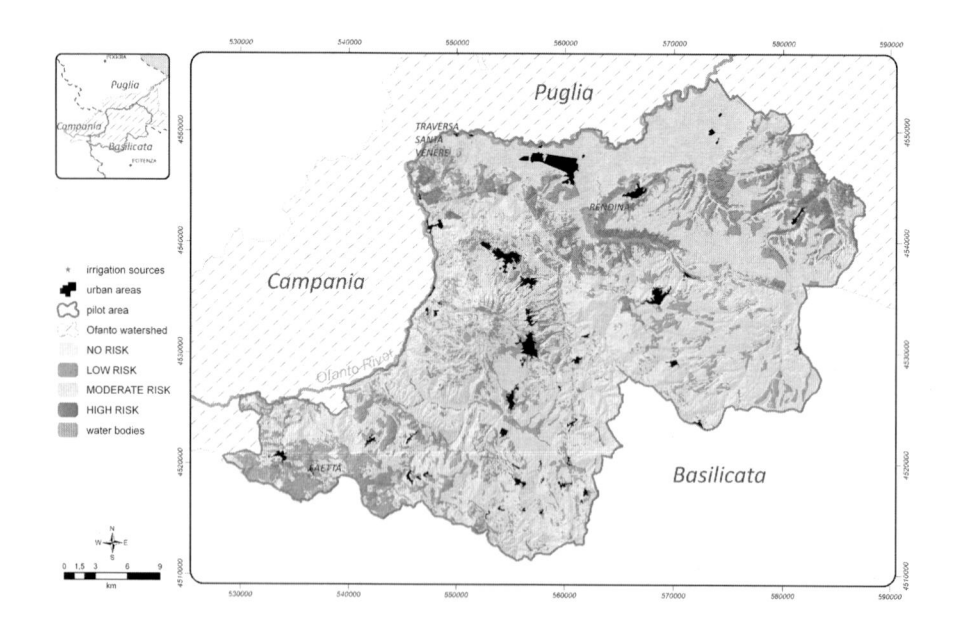

Fig. 36.6 Soil erosion risk map: scenario "climate change"

It can aid the social learning process among the stakeholders involved because it makes use of maps, and speaks about land use history, landscape characteristics and intuitive rules such as "what happens if" that the expert will translate in numbers and easy operational procedures (overlay, clip) via a GIS tool. Our analysis has not estimated quantitatively the risks by calculating the hectares associated to each risk level nevertheless the results could be used by the local administrator because they provide an early warning of what could happen in the near future.

The objective of this work was to define a simplified approach for risk identification and classification using the available data for the area under different scenarios to support the decision-making process. These approaches are often used for regional studies and are called "cognitive models" since rules are expert knowledge of phenomena (Debolini et al., 2008).

The methodology proposed is a useful tool for decision makers coping with issues related with land degradation in areas with data shortage and with low resolution information that limits the application of more reliable and accurate analytical risk evaluation.

In any case both part of the methodology (M1 and M2) produce results heavily dependent by expert opinion and by expert knowledge. In particular M2 methodology has to be considered a mere exercise if considering the hypothetical occurrence of some events and the high variability of conditions involved in a given scenario. Nevertheless sharing the results produced between the actors involved in the decision making process is a focal step for facilitating active participation and discussion about land degradation phenomena affecting a given area (Iannetta et al., 2008).

36.4 Conclusions

Expert-based methodologies, as the one presented here, could be considered a good starting point, for decision makers, to evaluate the existing and expected risk conditions in a given area and to define appropriate policies for conservative resources management or to decide some deeper analysis in circumscribed area.

The methodology is a valid alternative to any analytical model whenever it is required to realize first evaluation at broader scale in areas with data shortage and low accuracy/resolution information.

IMAGE project is now completing the full implementation of the described methodology by a development of a decision support system dedicated to the analysis of new options and scenarios for water and soil resources management to mitigate the degradation processes. The system is under test by local administrations and regional technical services in the framework of the River Basin and the Rural Development Plans.

The same tool (EMB) and methodological approach described in this chapter has been used for other prototype applications in different Italian regions: Sardinia and Basilicata. The Nurra Water Management (NWM) DSS has been developed for the sustainable management of water for civil and agricultural usage in the North of the Sardinia region (Barbieri et al., 2005).

The Ecological Network DSS has been implemented in the Basilicata region and concerns the simulation of the application of a range of possible agro-environmental measures and their impact on the various indicators of degradation. The results are displayed in terms of maps showing the result of the application of the various measures. The Ecological Network DSS (ENDSS) has been developed following the requirements of the Regional administrations, and provide a tool for assessing the impact of the application of the Priority Axe II Measures, indicated as *Improving environment and rural space*, for the *Rural Development Program (RDP) 2007 2010f the Basilicata Region* (Regione, 2009).

The implementation of the above described method and the further developments will provide the useful feedbacks to improve existing methodology and to test its real impacts of the cognitive approaches in addressing, at large scale, a very complex process such as desertification.

References

Barbieri, G., Ghiglieri, G. and Vernier, A. (2005). Aquifer vulnerability in the Alghero plain for integrated water resources management in NW Sardinia. 2nd International Workshop – AVR 05 Aquifer Vulnerability and Risk. Parma, Italy, September 2005.

Colonna, N., Lupia, F. and Iannetta, M. (2008). Expert knowledge-based methodology for land degradation risk evaluation. In: P. Rossi Pisa (ed.), ESA2008: Multifunctional Agriculture. 10th Congress of the European Society of Agronomy. Bologna, The Italian Journal of Agronomy 3(3 suppl).

Compagnone, L., Fattoruso, G. and Pace, G. (2006). Modellistica ambientale e sistemi di supporto alle decisioni per la lotta alla desertificazione. ENEA, Rome, p. 86.

Cross, M. and Moscardini, A.O. (1985). Learning the Art of Mathematical Modeling. Halsted Press, New York.

Debolini, M., Galli, M. and Bonari, E. (2008). Agro-environmental risk analysis at landscape scale: Limits for a sustainable land management. In: P. Rossi Pisa (ed.), ESA2008: Multifunctional Agriculture. 10th Congress of the European Society of Agronomy. Bologna, Italian Journal of Agronomy 3(3 suppl).

Dent, D. and Young, A. (1981). Soil Survey and Land Evaluation. G. Allen & Unwin, London.

FAO. (1976). A framework for land evaluation. FAO Soils Bulletin 32. FAO, Rome.

FAO. (1995). Planning for sustainable use of land resources. Toward a new approach. Land and Water Bulletin, 2. Rome

Gassman, P.W., Reyes, M.R., Green, C.H. and Aronold, J.G. (2009). The soil and water assessment tool: Historical development, applications and future research directions. In: J. Arnoldet al. (eds.), Soil and Water Assessment Tool (Swat) Global Applications. WASWC Special publications No. 4, Bangkok, p. 415.

Goodchild, M.F., Parks, B.O. and Steyaert, L.T. (eds.) (1993). Environmental Modeling with GIS. Oxford University Press, New York.

Kersten, G.E. and Gordon, L.O. (1999). DSS application areas. In: G.E. Kersten, M. Zbigniew and A.G.O. Yeh (eds.), Decision Support Systems for Sustainable Development a Resource Book of Methods and Applications. Springer, Berlin Heidelberg New York, p. 440.

Iannetta, M., Enne, G., Zucca, C., Colonna, N., Innamorato, F. and Di Gennaro, A. (2005). Il progetto Riade: i processi di degrado delle risorse naturali in Italia ed i possibili interventi di mitigazione. In: P. Gagliardo (ed.), "Lotta alla siccità e alla desertificazione", Geotema 25: 99–108.

Iannetta, M., Bizzi, S., Enne, G., Ghiglieri, G., Iocola, I. and Pace, G. (2007). EMB – Environmental Model Builder. IT Copyright 006415, 27 June 2007.

Iannetta, M., Lupia, F. and Colonna, N. (2008). Shared methods and tools to manage natural resources in the Mediterranean drylands. 2nd Conference on Desert Drylands and Desertification". Blaustein Institutes for Desert Research (BIDR) of Ben Gurion University of the Negev, 14–17 December, 2008.

Margaris, N.S., Koutsidou, E. and Giourga, Ch. (1996). Changes in Mediterranean Land-Use systems. In: J.B. Thornes and C.J. Brandt (eds.), Mediterranean Desertification and Land Use. Wiley, Chicester, pp. 29–42.

Mysiak, J. (2005). Decision support Systems for integrated water resources management value and success factors. Paper presented at the International Symposium on Environmental Software Systems. Sesimbra, Portugal, 24–27 May, 2005.

Poesen, J. (1995). Soil erosion in Mediterranean environments. In: R. Fantechi, D. Peter, P. Balabanis and J.L. Rubio (eds.), Desertification in an european context: Physical and socio-economic aspects. Report EUR15415 European Commission. Brussels, pp. 123–152

Refshaard, J.C. and Storm, B. (1995). MIKE SHE. In: V.P. Singh (ed.), Computer Models for Watershed Hydrology. Water Resources Publications, Highlands Ranch, CO, pp. 809–846.

Rais, M., Gamed, S., Craswell, E.T., Sajjapongse, A. and Bechstedt, H.D. (1999). Decision support system for sustainable land management. In: G.E. Kersten, M. Zbigniew and A.G.O. Yeh (eds.), Decision Support Systems for Sustainable Development a Resource Book of Methods and Applications. Springer, Berlin Heidelberg New York, p. 440.

Regione, B. (2009). Sistema Ecologico Funzionale Territoriale. Dipartimento Ambiente Territorio e politiche della sostenibilità, Potenza, p. 238.

Scott Morton, M.S. (1971). Management Decision Systems. Harvard Business School Press, Boston.

Thornes, J.B. and Brandt, C.J. (eds.). (1996). Mediterranean Desertification and Land Use. Wiley, Chicester.

Van Buuren, J., Engelen, G. and Van de Ven, K. (2002). The DSS WadBOS and EU Policies Implementation. Proceedings of the Conference Littoral 2002, The Changing Coast. Porto, Portugal, pp. 533–540

Wishmeier, W.H. and Smith, D.D. (1978). Predicting Rainfall Erosion Losses – A Guide to Conservation Planning. volume 537, USDA Agricultural Handbook, Washington, DC, p. 58.

Chapter 37
Assessment of Desertification in Semi-Arid Mediterranean Environments: The Case Study of Apulia Region (Southern Italy)

G. Ladisa, M. Todorovic, and G. Trisorio Liuzzi

Abstract This work focuses on the assessment of the areas threatened by desertification in the semi-arid Mediterranean environments. The presented approach represents a modification of the ESAs model (Environmental Sensitive Areas to Desertification; Kosmas et al., 1999) through a set of new indicators established to account for the regional-specific environmental characteristics as well as identifiable parameters relevant for land use planning and control measures. These supplementary indicators, comprehending socio-economic and environmental factors, were integrated in the ESAs model and, by using a GIS, were applied to Apulia region (Southern Italy), a typical representative of many Mediterranean areas affected by land degradation. The analyses include the elaboration of a whole set of indices on both regional and the administrative scales that constitute the principal territorial units for the management of natural resources. The results have demonstrated that the introduction of the new indices has improved substantially the overall evaluation of the desertification risk in the Apulia region. The proposed approach permits not only the identification and refinement of different degrees of vulnerability of an area to land degradation, but allows also the analyses of specific factors affecting desertification as well as their evaluation in terms of spatial and temporal distribution. Furthermore, the presented method is conceptually very simple and easy to implement from local to regional and national scale, and could be proposed as a standard methodology for the definition of priorities in implementation of strategies to mitigate desertification in the semi-arid Mediterranean environments.

Keywords Desertification risk · Sensitivity areas · Apulia region · Italy · Mediterranean environments

---

G. Ladisa (✉)
CIHEAM Mediterranean Agronomic Institute of Bari, Valenzano 70010, Italy
e-mail: ladisa@iamb.it

37.1 Introduction

The assessment of the state of land degradation and desertification represents the key phase in the studies aiming the identification of areas threatened by desertification and a consequent planning of mitigation and remediation measures. In these studies, the preliminary step is a careful selection of key-variables and indicators that should describe the actual state of the system and highlight the degradation changes and related effects in both time and spatial scales.

The explicit definition of the minimum threshold values signalling the conversion towards an irreversible state of land degradation represents a particularly complex issue. This is particularly the case in many semi-arid Mediterranean environments where the numerousness and variability of the factors influencing the process makes difficult to integrate and synthesize all of them in a unique value representing the threshold of degradation. Hence, it is necessary to assign to each triggering factor its own threshold value and to attribute to each one a different weight relevant to the role played in the desertification process in a space-specific and time-specific case (Hunsaker and Carpenter, 1990; Kosmas et al., 1999; Trisorio Liuzzi et al., 2004). Therefore, the solution must consist in a balanced merger of different aspects of environmental stress ensuring a straightforward link among indicators themselves and the state and the tendency of the system they are representing.

A minimum set of indicators (Minimum Data Set) is proposed for the Apulia region (Southern Italy), with the aim to identify the areas sensitive to desertification and subsequently to develop action plans to combat land degradation. The region represents a typical case for many Mediterranean areas, being seriously affected by land degradation due to both unfavourable climatic conditions and adverse

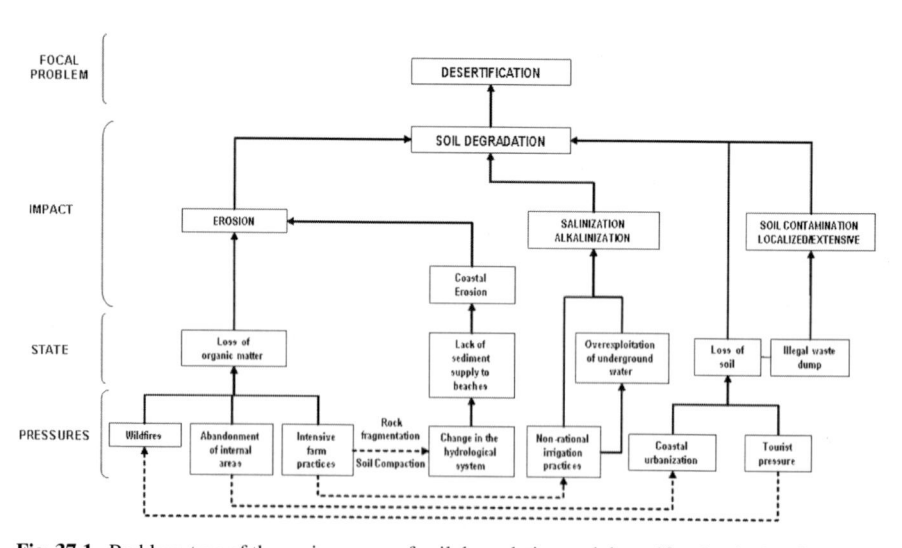

Fig. 37.1 Problem-tree of the main causes of soil degradation and desertification in Apulia region (Ladisa, 2007)

human activities triggering the desertification processes. Such factors include arid and semiarid climatic conditions, seasonal droughts, high rainfall variation and intensity, erodible soils, diversified landscapes, extensive man-induced deforestation and forest losses due to frequent wildfires, land abandonment and deterioration, unsustainable exploitation of water resources leading to environmental impairment, salinisation and exhaustion of aquifers, uncontrolled urbanization, concentration of economic activities in coastal areas, tourism pressure, etc (Fig. 37.1).

37.2 Reference Framework and Applied Methodology

The approach used for the identification of areas vulnerable to desertification represents a modification of the Environmental Sensitive Areas to Desertification (ESAs) model (Kosmas et al., 1999), formerly developed for the watershed scale, within the frame of the MEDALUS (Mediterranean Desertification and Land Use) project (Basso et al., 1999).

The ESAs model was based on four broad systems of indicators, within which the minimum data set selection has to be assessed, representative of the *soil* quality (texture, rock fragments, drainage, parent material, soil depth), the *climate* (rainfall, aridity, aspects), *vegetation* (plant cover, fire risk, erosion protection, resistance to aridity) and of the *management practices* (intensity of land use in rural zones, pastures and forest areas, managerial policies).

The indicators are grouped and combined into 4 quality layers, independently on the structure of the input layers (number of classes, etc.), allowing that the corresponding indices (the Soil Quality Index SQI, the Climate Quality Index CQI, the Vegetation Quality Index VQI and the Management Quality Index MQI) can be consistently compared among each other, no matter what could be the format of the input data/indicator (qualitative or quantitative; measured or estimated, etc.). In particular, the values of the Quality Indices for each elementary unit within a layer, are obtained as geometric average of the scores assigned (following the factorial scaling technique) to the single indicators according to the following formula:

$$Quality\_x_{ij} = \left(layer\_1_{ij}\right) \cdot \left(layer\_2_{ij}\right) \cdot \ldots \cdot \left(layer\_n_{ij}\right)^{1/n}$$

where i,j represent the "coordinates" (rows and columns) of a single elementary unit of a layer and n is the number of layers (equal to the number of indicators) used for determination of each quality layer.

The scores range from 1 (good conditions) to 2 (deteriorate conditions), whereas the "zero" value is assigned to the areas where the measure is not appropriate and/or those, which are not classified (e.g. water bodies, urban areas, etc.). It is possible in some cases, a non-linear variation of the function representing the variation of the indicators (scores) between the extreme values (Hunsaker and Carpenter, 1990; Kosmas et al., 1999).

The integration of four quality layers represents the synthetic Environmental Sensitive Areas Index (ESAI):

$$ESAI = (SQI^*CQI^*VQI^*MQI)^{1/4}$$

Such index identify three main classes of areas threatened by land degradation ("critic", "fragile" and "potentially affected"), that could be further differentiated in three subclasses (from low to medium and high sensitivity).

Applying the original ESAs model to the "regional (Apulia region) scale" (which is wider if compared to the ones of the pilot areas successfully tested previously (Kosmas et al., 1999; Basso et al., 1999), produced questionable results (Regione Puglia, 2000), as witnessed by clear evidences of the real conditions on the ground. The main reasons for this were in the simplifications caused by the lack of some input information, as well as the availability and increasing relevance of some other data non considered in the model.

Accordingly, the original approach was modified and applied on one of the Apulia provinces – the province of Bari (Ladisa, 2001; Ladisa and Trisorio Liuzzi, 2001; Ladisa et al., 2002; Trisorio Liuzzi et al., 2004; Trisorio Liuzzi et al., 2005) – introducing a new set of indicators, derived from the specific environmental context of the Apulia region, where the geographic, hydrological, geo-morphological, climatic and anthropic characteristics co-act in determining the complex of predisposing and triggering conditions to desertification. These supplementary indicators, introduced in the original ESAs approach are presented in Fig. 37.2.

Regarding the *climatic factors*, two additional new indicators were proposed for inclusion in the model. They both depend on the data availability and are related to the rainfall erosivity and have the scope to establish the erosion impact (that previously was simply derived either by a slope indicator included in the soil quality index (Kosmas et al., 1999) or by means of a Erosion Quality Index estimated from the Erosion Risk Map and based on the USLE equation).

Our first new climatic factor is the *Rainfall Erosivity Index (R)*, expressed through the factor R of the USLE-Universal Soil Loss Equation, and calculated according to the proposal of D'Asaro and Santoro (D'Asaro and Santoro, 1983) as:

$$R = 0.21 q^{-0.096} P^{2.3} NRD^{-2}$$

where q represents the altitude of meteorological stations (m), P is the average annual precipitation (mm) and *NRD* is the average number of rainy days during the year.

The second one is the *Modified Fournier's index (MFI)*, that, highly correlated to the R factor of USLE within the homogeneous climatic zones (Fournier, 1960; Ferro et al., 1999; Porto, 1994; Gabriels, 2000), is calculated according to Arnoldus (1980) as:

$$MFI = \sum \frac{p_i^2}{P}$$

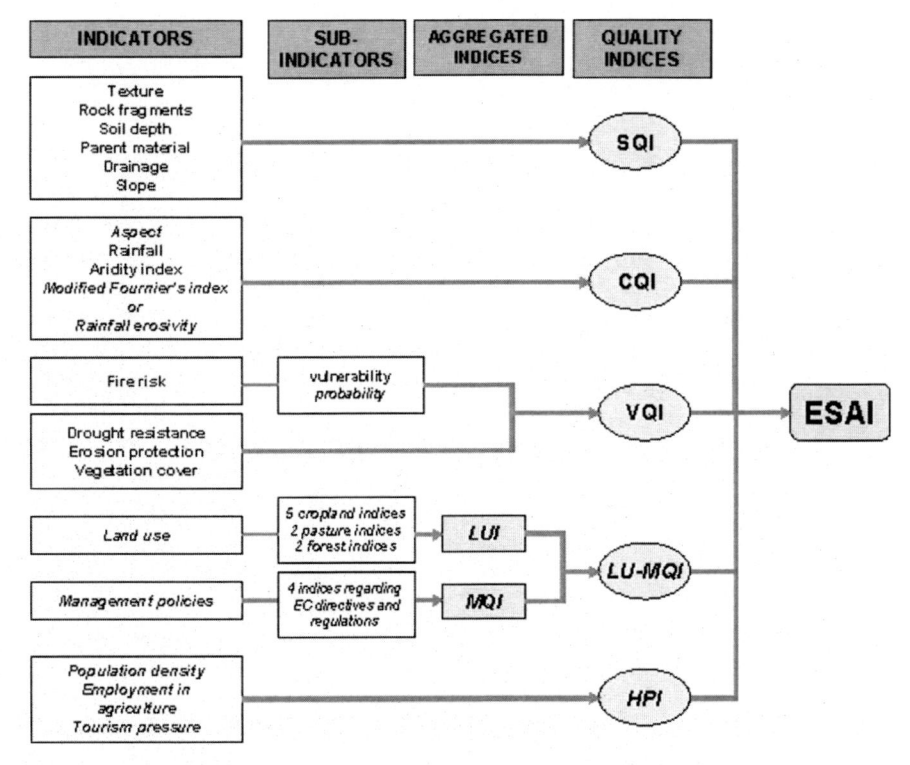

37 Assessment of Desertification in Semi-Arid Mediterranean Environments

Fig. 37.2 Indicators and quality indices used in the modified ESAs approach (in *italic* are written new and/or modified indicators and quality indices)

where p_i is the average rainfall of the month i ($i = 1$–12) and P is average annual precipitation.

These indices are used alternatively in the determination of Climate Quality Index as:

$$CQI = (BGI^*Aspect^*P^*R)^{1/4}$$

or

$$CQI = (BGI^*Aspect^*P^*MFI)^{1/4}$$

where:

BGI is the *Bagnouls-Gaussen Aridity Index* (Bagnouls and Gaussen, 1952) calculated as:

$$BGI = \sum_{i=1}^{12}(2t_i - p_i)k_i$$

where t is the average monthly air temperature, k is a coefficient indicating the number of months in which $2t>p$ and p_i is the average rainfall of the month i ($i = 1$–12).

The CQI was computed using alternatively both abovementioned indices. Since there is no significant difference between the two methods, due to the fact that they are highly correlated, in this study we show only the results of data elaboration with the Rainfall Erosivity Index. The similarity of results (Trisorio Liuzzi et al., 2005) demonstrates that Modified Fournier's Index and Rainfall Intensity Index may be applied alternatively for the definition of the Climate Quality Index, where the choice depends on the trustworthiness and availability of necessary input information. Slightly greater intensity of CQI, observed when the Rainfall Intensity Index was used, may be explained by the fact that this method takes into consideration the altitude of meteorological stations resulting in the higher impact to desertification of the high elevation areas.

The new Vegetation Quality Index takes into account, besides the vulnerability to burning (i.e. flammability and capacity of vegetation to recover after burning) of various species, already presented in the original approach, a new "probability of fire" index, developed by means of cluster analysis using the data available at municipality scale. Their multiplication gives a new aggregated index for fire risk, which, together with vegetation cover, drought resistance and capacity of protection from erosion, describes the Vegetation Quality Index.

The Land Use Index concept is explicated through the Crop Land Use Index (CLUI), the Pasture Land Use Index (PLUI) and the Forest Land Use Index (FLUI), in turn combined with management policies in order to obtain integrated land-use management quality index.

In particular, the Crop Land Use Quality Index (CLUI) is estimated as a geometric average of five new statistical indices on both provincial and municipal scale:

$$CLUI = (CLUI_1 * CLUI_2 * CLUI_3 * CLUI_4 * CLUI_5)^{1/5}$$

whose meaning is the following:

1. The *Intensity of land cultivation index (CLUI$_1$)* is defined as the ratio between the Utilized Agricultural Area (including arable land, permanent grassland, pastures, vegetables and orchards) and the total surface area. The lowest impact to desertification (score = 1) is assigned to the areas with the index ranging between 40 and 60% and the greatest impact (score = 2) to the areas where the index was either below 20% or greater than 80% (because a high percentage of cultivated area may indicate degradation risk due to intensification of agricultural practices, increased use of mechanization, fertilizers and pesticides etc., whereas a low percentage could display land abandonment). From the $CLUI_1$ it is possible to deduce the impact of farming activities on the environment, independently of the farm sizes and structures, applied agricultural practices, abandonment of marginal land and other phenomena correlated to either negative or positive effects on soil quality.

2. The *Intensity of irrigation index (CLUI₂)*, that represents the ratio between the Irrigated Area and the Agricultural Utilized one, can also highlight the effect of water withdrawal on land degradation in terms of seawater intrusion and salinization (two-thirds of irrigation water needs in the Apulia region comes from groundwater sources).

3. The *Use of mechanization index (CLUI₃)* links the compaction of superficial soil layers and the human induced activities (overgrazing, traffic of agricultural mechanization, etc.). The "proxy" indicator (unit of pressure: q/ha), based on the number of vehicles (tractors and hill-side combine harvester) present in the area, gives indications about their density, power and weight (directly proportional to the degree of modification of soil structure) and the number of travel-passes (related to the type of crop) (Fig. 37.3). In the following formula, *"0.5 q/kW"* represents the average weight of vehicles per kW, *"5 travelling"* corresponds to the principal agricultural activities where mechanization is used, and *"AS"* is agricultural surface area (in hectares):

$$q/ha = \frac{N°vehicles * kW^*_{avg}(0.5q/kW) * 5 traveling}{AS}$$

4. The *Use of nitrogen fertilizers index (CLUI₄)*, or the ratio between the applied nitrogen fertilizers (in q of N) and the agricultural surface (ha), is only referred

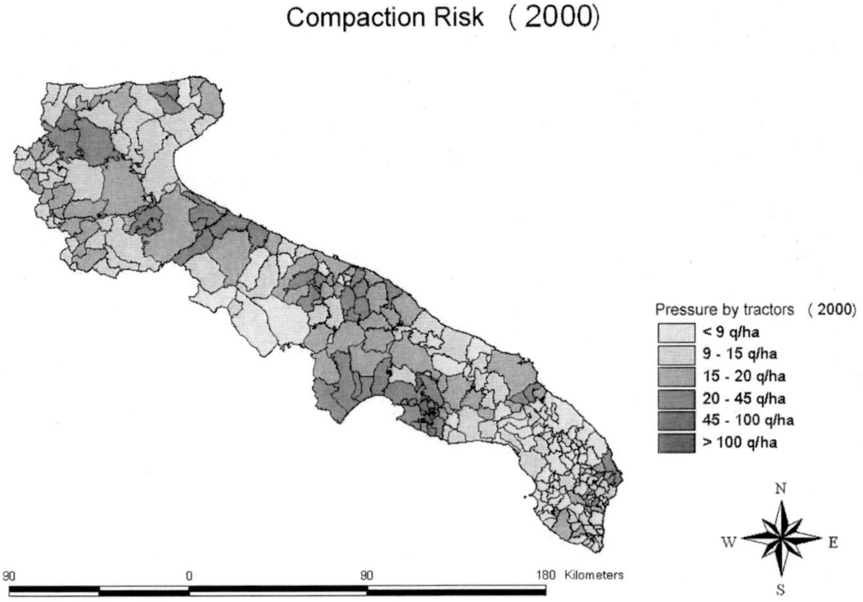

Fig. 37.3 Compaction risk related to number and power of tractors and hillside combined harvester in Apulia region (Blonda et al., 2007)

to the areas where fertilizers could be effectively applied (grassland and pastures are excluded). Just mineral fertilizers have been considered for the purpose of the application, representing the main cause of groundwater pollution.
5. The *Use of plant protection products index (CLUI₅)* defined as a ratio between the total quantity (in kg) of plant protection products (herbicides, fungicides, insecticides, etc.) sold for agricultural use and the surface area (ha) of agricultural land at which they could be applied (agricultural land excluding grassland and pastures).
6. The *Pasture Land Use Index (PLUI)*, globally describing the overall pressure of stocking rate, which includes both the surface area available for grazing and the number of animals existing in the total area, is obtained by the integration (by means of geometric average) of two "proxy" indices:

$$PLUI = (PLUI_1 * PLUI_2)^{1/2}$$

where:

1. the *Permanent Grassland and Pasture Index* $PLUI_1$ represents the percentage of permanent grassland and pasture land in respect to the potentially utilized agricultural land,
2. the *Intensity of grazing index* $PLUI_2$ provides information about the number of Adult Cattle Units (ACU) existing in one hectare of the potentially utilized agricultural land. It is computed by multiplying the number of cattle heads and the number of sheep and goat heads with the conversion coefficients, which are 0.85 and 0.1 respectively.

The *Forest Land Use Index (FLUI)* is computed integrating two new statistical indices:

$$FLUI = (FLUI_1 * FLUI_2)^{1/2}$$

where:

1. the *Forest Index (FLUI₁)* represents the degree of forestry cover through the ratio between the surface area covered by forest and total surface area,
2. the *Wood Harvesting Index (FLUI₂)*, defined as the ratio between harvested woody material in one administrative unit (in m³) and the total wood harvesting over the whole territory under consideration, is calculated from the statistical data (available for each of the five Apulia provinces).

The *Management Quality Index (MQI)* represents the aggregation of four indices as:

$$MQI = (DIR_1 * REG_1 * REG_2 * REG_3)^{1/4}$$

representing the following:

a. *DIR₁* regards to the Directive CEE 43/92 ("Habitat"), describing the percentage of protected land in respect to the total surface area and applied on the "municipality scale";
b. *REG₁* is related to the Regulation CEE 2078/92, regarding the low impact agro-environmental measures and applied on the "province – scale" as the ratio between the area considered by regulation and the utilized agricultural land;
c. *REG₂* regards to the Regulation CEE 2080/92, concerning the improvement of forestation and forestry practices and described at the "province – scale" through the percentage of agricultural land where the regulation is in use;
d. REG4 regards to the Regulation CEE 2092/91, concerning the development of organic agriculture and measured, at the "province – scale", by comparing the agricultural area converted to organic farming and total utilized agricultural land.

The effect of the management practices is globally estimated for each of three main agricultural land use types (cropland, pasture and forestry) as:

$$CLU\_MQI = (CLUI^*MQI)^{1/2}$$

$$PLU\_MQI = (PLUI^*MQI)^{1/2}$$

$$FLU\_MQI = (FLUI^*MQI)^{1/2}$$

where *CLU_MQI* is management quality index for cropland use, *PLU_MQI* is management quality index for pasture land use and *FLU_MQI* represents the management quality index for forest land use.

A new *Human Pressure Index (HPI)* has been proposed taking into account the population density, the resident population at the end of year, the employment in agricultural sector and the tourism pressure.

Very high and very low population density values, as well as excessive increases and decreases of resident population, are considered to have a negative influence on desertification processes.

The values of *resident population at the end of year* (measuring the population and migratory rates within an administrative unit) have been considered for three periods: 1980–1990, 1990–2000 and 2000–2005.

The employment in agricultural sector has been analysed because, if considered for a period of years, it could provide information about the shifting of workers from agriculture to other sectors as well as about the abandonment of agricultural land.

The tourism pressure represents a very important indicator for the coastal areas of the Apulia region, characterized with a significant increase of residents during the summer months (Fig. 37.4). This map shows that the tourist presence increases the population density by more than two times in the coastal areas, representing about 13% of the regional territory.

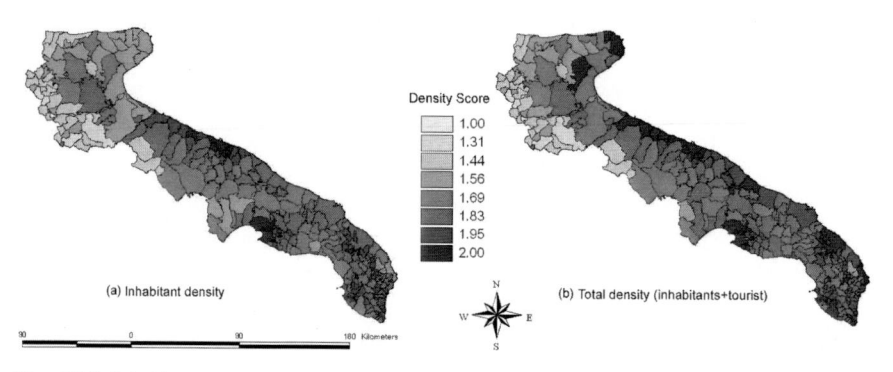

Fig. 37.4 Inhabitant density (**a**) and total density (**b**), considering tourist presence, of Apulia region in 2005

The Tourism Pressure Index (TPI) is estimated from the series of statistical data and referred to both the tourist presence and the tourism vocation of the area. A complex algorithm has been elaborated in order to make comparable the information related to both parameters. The algorithm attributes numerical values weighted for discrete classes of the presence of tourists in addition to a value assigned to the area particularly affected by tourism pressure and used as multiplication coefficient. In such a way is obtained a predictive indicator, which multiplies a percentage value determined from the vocation of the territory (and therefore its vulnerability) and a value of tourism pressure gained from the presence of tourists. The TPI is calculated as:

$$TPI = L^a * V$$

where L is the ratio between the presence of tourists and the number of residents, a is an exponent fixed to 3 as proposed by the literature (ANPA, 2000; Ladisa, 2001), and V represents the vocation to tourism estimated through the number of available beds at receptions over the surface area under consideration.

37.3 Description of the Study Area

The above-described approach was applied in Apulia Region that covers a surface area of approximately 19,500 km^2, subdivided in six Provinces (recently, a sixth one was established) on which the elaborations were based. Most of the region territory is flat to slightly sloping lowland except for the Gargano area, situated in the North-East, and Sub-Apennine part, located mainly in the North-West of the region.

A semi-arid Mediterranean type of climate with hot and dry summer and mild and rainy winter season characterise the region. The spatial interpolation of historical rainfall data shows that annual precipitation varies between 450 and 550 mm

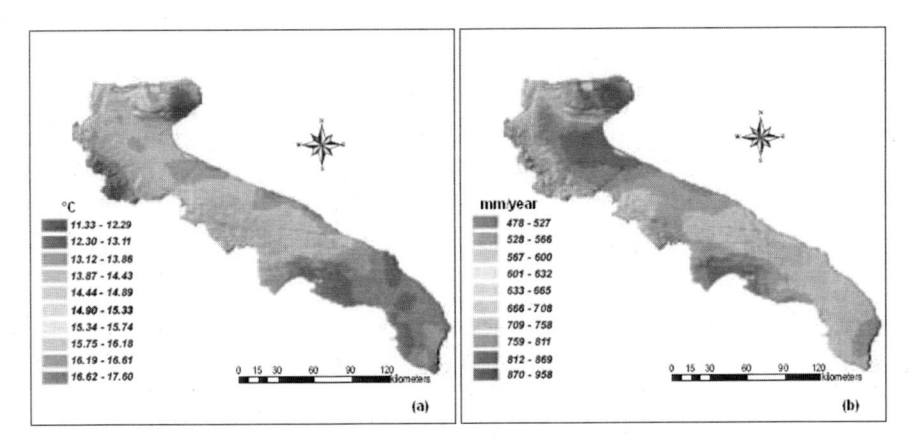

Fig. 37.5 Average annual temperature (**a**) and average annual precipitation (**b**) in Apulia region

in the greatest part of the region (Fig. 37.5). The lowest values, around 400 mm, are observed in the area of Tavoliere, in the province of Foggia, whereas the highest values of more than 900 mm year^{-1} referred to the Gargano area, in the North of the same province. The hydrological regimes are irregular, of torrential type with high flow rates during the rainy season and practically no water flow during summer.

The land use distribution in Apulia region is elaborated according to the CORINE Land Cover data (CORINE, 2000) as illustrated in Fig. 37.6. The greatest part of territory (81.4%) is allocated for agricultural use while forestry and semi-natural areas occupy about 13.3% of the region. Water bodies cover about 1.2% of territory including both natural lakes and artificial storage dams. Water availability is one of the main factors limiting agricultural productivity in the region.

Five physiographic units may be distinguished in the Apulia region (Fig. 37.7). Three of them (Sub-Appennino Dauno, Tavoliere delle Puglie and Gargano) are situated in the North from NW to NE respectively, the Murge covers predominantly the Central part of the region, and the peninsula of Salento is located in the South. The dominant soils are Cambisols, Luvisols and Vertisols formed mostly on cretaceous limestone, marl and clayey to sandy deposits.

Irrigation management is run by 6 reclamation consortia, which cover an administrative area of 1,743,591 ha, or about 90% of territory. The area equipped with the consortia water distribution networks, and therefore, potentially irrigated extends over 236,012 ha. However, in general, only one-third of it is effectively irrigated due to huge irrigation requirements and chronic shortage of water for irrigation and other uses.

The protected areas in Apulia region occupy about 238,535 ha, which represents 12.33% of territory (Table 37.1). These areas are mainly located in Foggia province (Foresta Umbra and Gargano National Park), in Bari province (Alta Murgia National Park) and in Taranto province (Gravine Joniche Park).

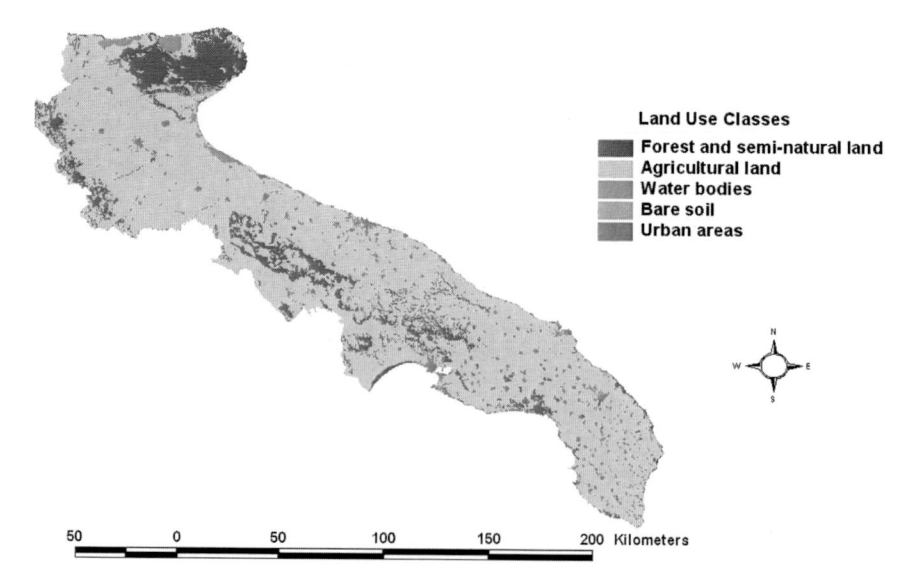

Fig. 37.6 Apulia region: main land use classes according the CORINE Land Cover data-catalogue (CORINE, 2000)

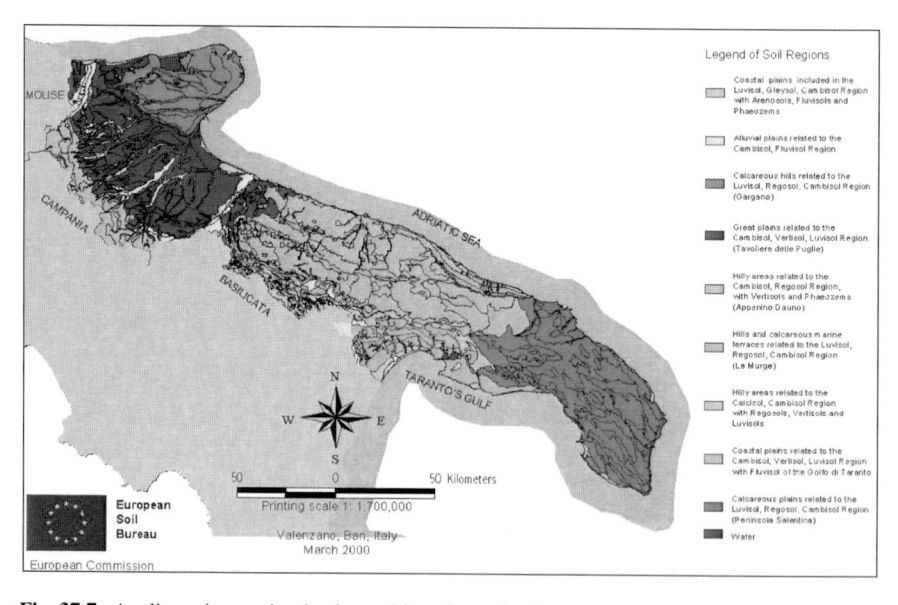

Fig. 37.7 Apulia region: main physiographic units and soil regions (Zdruli, 2000)

37.4 Results and Discussion

The data input, derived from statistical databases, already existing at municipal and provincial scales, as well as from specific regional projects (e.g. ACLA project on

37 Assessment of Desertification in Semi-Arid Mediterranean Environments

Table 37.1 Protected areas in Apulia Region – 2003 (Source: ISMEA elaboration on data from 5th updating of Natural Protected Areas Official List 2003 and from Apulia Region, Parks and Natural Reserves Office)

Protected area typology	Surface (ha)
National park	185,833.00
State natural reserve	9,906.33
Regional natural park	39,014.55
Oriented regional natural park	5,989.00
Municipality park	590.00
Marine natural protected area	20,347.00
Total regional (inland surface)	238,534.88
Protected areas surface/Region surface	12.33%

Agro-ecological characterization of Apulia region by means of potential productivity, Steduto and Todorovic, 2001), have been georeferenced and elaborated through a GIS (Todorovic and Steduto, 2003), and assembled in several thematic layers, each one representing one of five quality indices (the Soil Quality Index - SQI, the Climate Quality Index – CQI, the Vegetation Quality Index – VQI, the Land Use and Management Quality Index – LU_MQI and the Human Pressure Index – HPI) and their effects to desertification. These quality indices have been elaborated for the whole region characterizing the impact of each factor by means of several quality classes (low, medium, high and very high) as illustrated in Figs. 37.8, 37.9, 37.10, 37.11 and 37.12 for SQI, CQI, VQI, LU_MQI and HPI, respectively.

The data related to the Soil Quality Index (Fig. 37.8) indicate that the most of Apulia's territory is of medium soil quality (69.1%), almost one-fourth is of low quality and only 6.5% of territory is of high soil quality (Fig. 37.8b). The areas of high soil quality, corresponding to deep soils, are located mainly in the province of Foggia, close to the principal water courses (Ofanto, Celone, etc.) characterized by alluvial deposits. The low quality land, corresponding to shallow soils, is distributed primarily in the provinces of Bari and Brindisi characterized by the calcareous terraces well-known as "Le Murge".

Furthermore, it is important to underline that the SQI has been elaborated according to the original classification of parent material (developed for the Island of Lesvos by Kosmas et al., 1999), which results in a significant degree of homogeneity when applied to the Apulia region. Consequently, the future activities should be focused on the reclassification of parent material classes according to the erodibility as considered in USLE method.

The CQI characterizes almost half of the Apulia territory (48.8%) of medium quality, 42.4% of low quality and 8.7% of good quality (Fig. 37.9b). The good quality land by means of CQI is located almost exclusively in the province of Foggia due to favourable ratio between the precipitation amount and number of rainy days (rainfall intensity), low aridity index of Bagnouls-Gaussen and exposure (aspect). In particular, the CQI produces strong negative impact in the province of Taranto (77.2% is low quality land), and then, in the provinces of Lecce and Brindisi

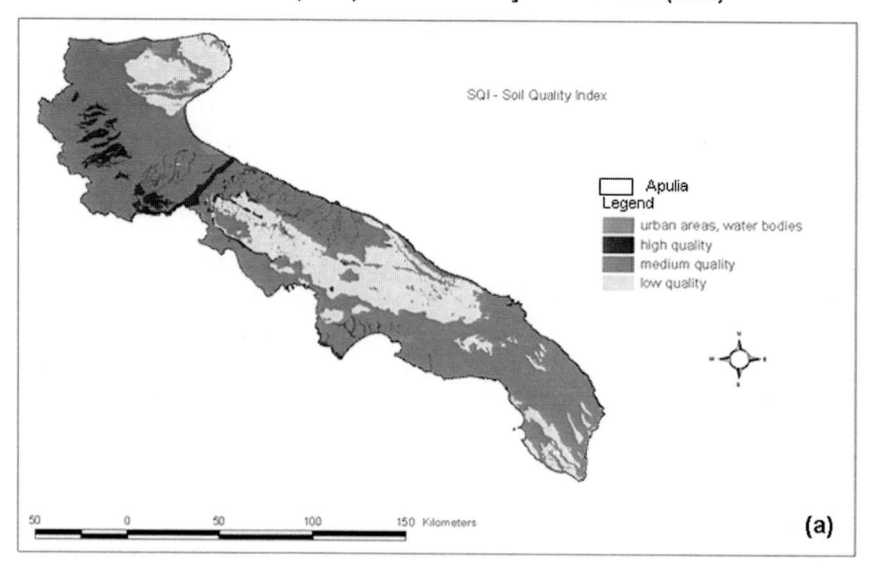

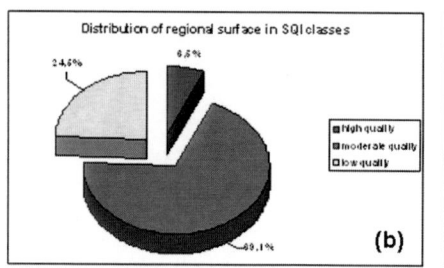

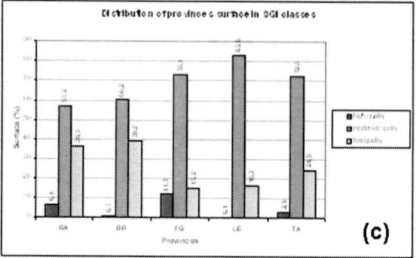

Fig. 37.8 Spatial distribution of SQI – Soil Quality Index (**a**), distribution of regional (**b**) and provincial (**c**) surface in SQI classes

(Fig. 37.9a, c) where the low quality land occupies more than 50% of territory (57.1 and 51.5%, respectively).

The Vegetation Quality Index shows that almost half of Apulia region is of low quality (46.1%), one-third (31.2%) is of medium quality and about 17.4% is of high quality (Fig. 37.10b). The low quality land by means of VQI is distributed predominantly in the province of Foggia (57.7%) and Taranto (50.2%) whereas in other provinces it occupies less than 50% of territory. Probably, such results originate from a large extension of annual crops (e.g. cereals) with low resistance to drought and presence of high fire risk areas especially in the Gargano relief and along the Gulf of Taranto.

The results of elaborations related to the impact of LU_MQI to desertification show that the greatest part of region (44.3%) can be classified as high quality land, about one-fourth (25.9%) as medium quality land, one-fifth (19.9%) as low quality land and about 4.6% as very high quality land. The impact of land use and

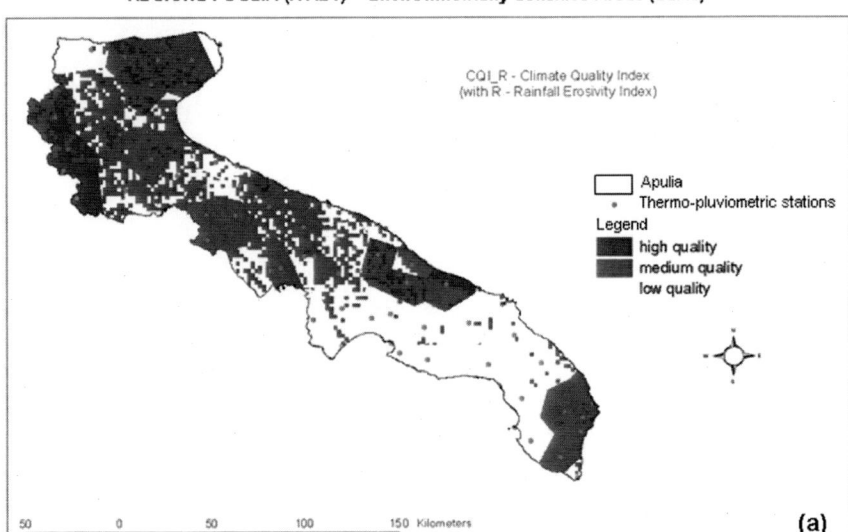

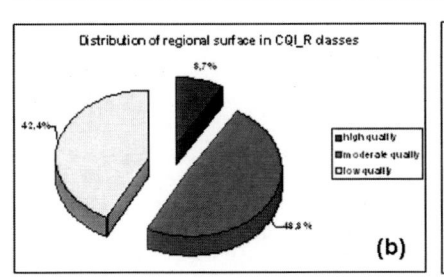

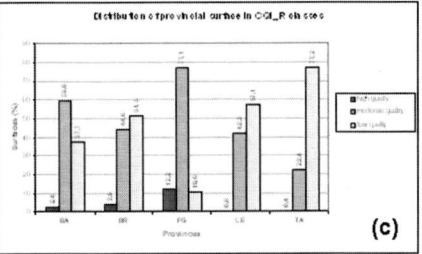

37 Assessment of Desertification in Semi-Arid Mediterranean Environments

Fig. 37.9 Spatial distribution of CQI – Climate Quality Index with R – Rainfall Erosivity Index (**a**), distribution of regional (**b**) and provincial (**c**) surface in CQI classes

management practices to desertification is particularly negative in the province of Foggia where more than half of territory (53%) is of low quality (Fig. 37.11a, c).

The Human Pressure Index depicts 9.4% of Apulia region as low-pressure territory, 29.8% as medium pressure land, 24% as high pressure and 36.8% as very high pressure areas. The results show that HPI has particularly great negative impact to desertification processes in the province of Foggia (Fig. 37.12a, c) where 64.5% of territory is classified as high-pressure area. This is mainly due to the important vocation of this province to tourism where many areas may be characterized with strong presence of non-resident population during the summer months and very limited resident population.

The proposed system of evaluation assigns the equal weight to each basic layer in the calculation of the quality indices (e.g. the resistance of vegetation to aridity has the same relevance as the vulnerability to fire, within the Vegetation Quality Index), and, simultaneously, each of the five quality indices has the same weight in the determination of the final Environmental Sensitive Areas Index (ESAI) independently of

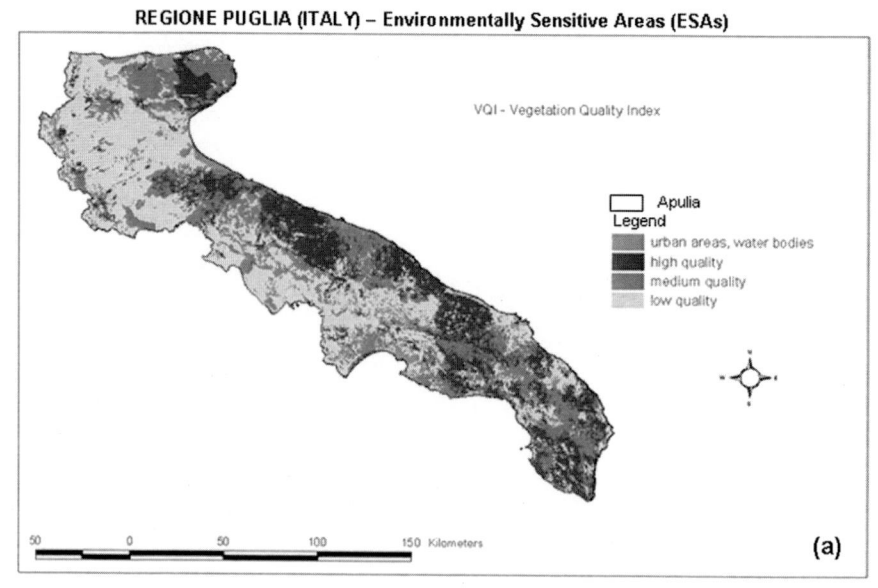

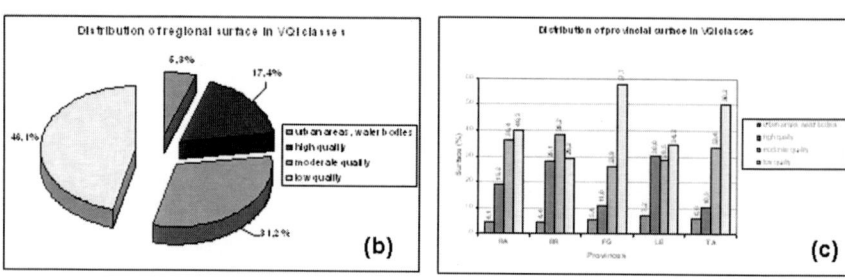

Fig. 37.10 Spatial distribution of VQI – Vegetation Quality Index (**a**), distribution of regional (**b**) and provincial (**c**) surface in VQI classes

the number of layers contributing to its definition. Consequently, the ESAI is not influenced by the number of the basic indicators (each of them represents one layer) which means that none of the main quality indices (soil, climate, vegetation, management and human pressure) is neither penalized nor advantaged by the fact to be constituted of different number of layers in respect to the other indices.

For the whole region, the integration of various indices necessary for the zoning of territory into several "sensitivity to desertification" classes (non-affected, potential, fragile, critical) has been done considering two scenarios:

  the first, including only physical aspects of territory by means of soil, climate and vegetation, and
  the second, embracing also the socio-economic factors, i.e. land use and management practices and human pressure.

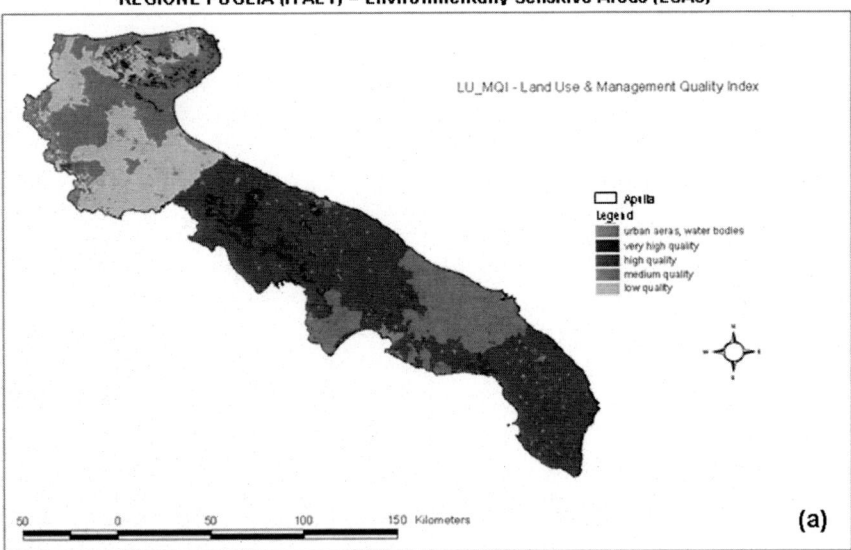

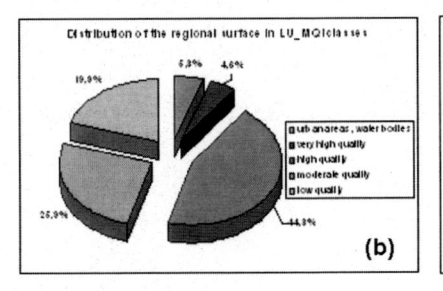

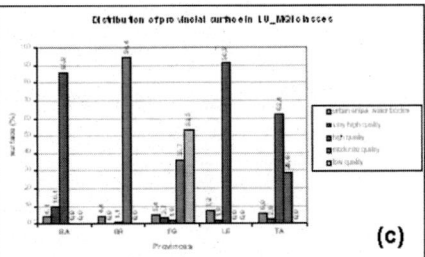

Fig. 37.11 Spatial distribution of LU_MQI – Land Use & Management Quality Index (**a**), distribution of regional (**b**) and provincial (**c**) surface in LU_MQI classes

Accordingly, the environmental areas sensitive to desertification have been obtained considering:

- firstly, the three primary indices, related to the soil, climate and vegetation databases, that is *SQI, CQI, VQI*, for deriving the Environmental Sensitive Areas Index (ESAI$_1$) as:

$$ESAI_1 = (SQI^*CQI^*VQI)^{1/3}$$

- secondly, including also the supplementary indices, related to the human-induced activities, that are *LU_MQI* and *HPI*, for deriving the final Environmental Sensitive Areas Index (*ESAI$_2$*) as:

$$ESAI_2 = (SQI^*CQI^*VQI^*LU\_MQI^*HPI)^{1/5}$$

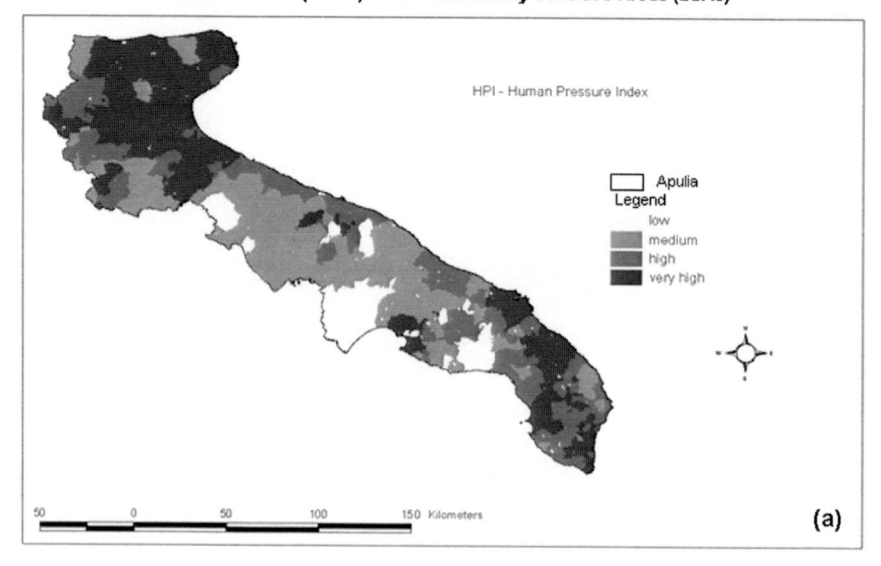

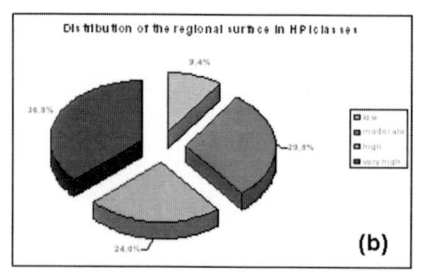

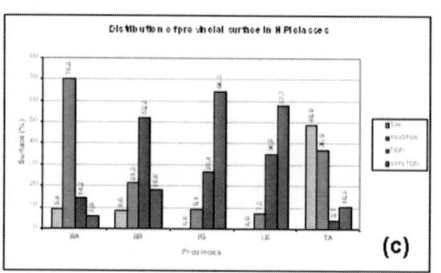

Fig. 37.12 Spatial distribution of HPI – human Pressure Index (**a**), distribution of regional (**b**) and provincial (**c**) surface in HPI classes

The mapping of these elaborations is given in Fig. 37.13, for the hypothesis that includes only physical aspects, and in Fig. 37.14 for the hypothesis with both physical and socio-economic indicators.

Furthermore, the overall results are elaborated for each province and the repartition into the different sensitivity classes is highlighted in Figs. 37.12c and 37.13c for both scenarios.

The results of desertification risk by means of physical factors (soil, climate and vegetation), presented in Fig. 37.13, show that more than half of the territory (51.7%) could be characterized as critical, 27.7% as fragile, 8% as potentially affected by desertification processes, 7% as non-affected land while 5.6% represents urban areas and water bodies. A particularly serious situation is observed in the province of Taranto, where 80% of territory is characterized as critical, followed

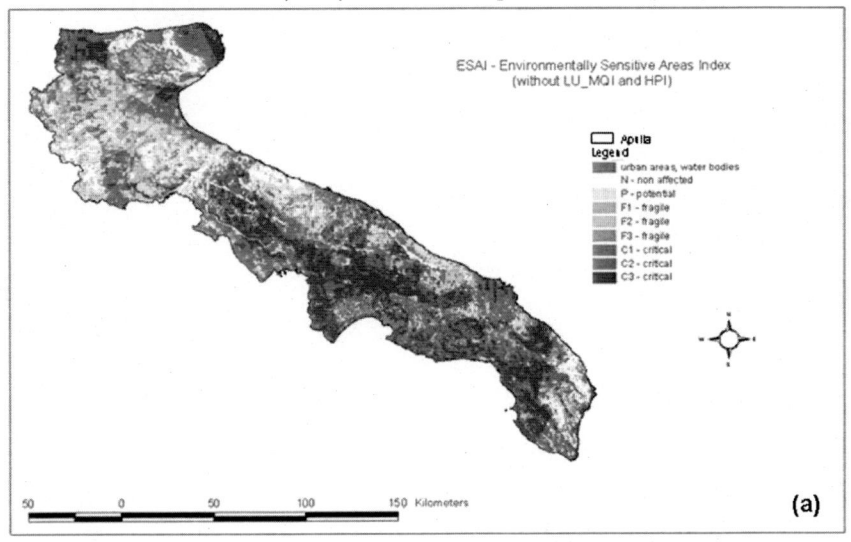

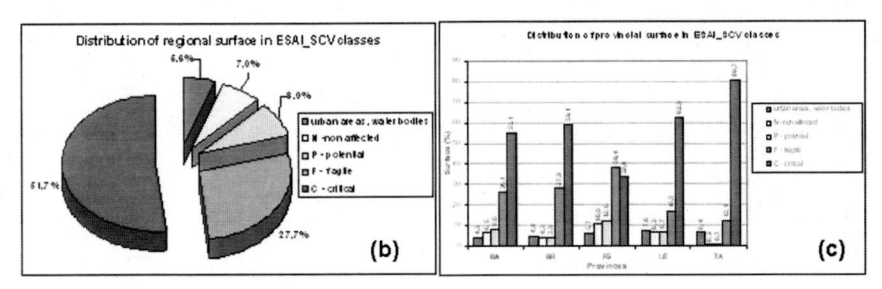

37 Assessment of Desertification in Semi-Arid Mediterranean Environments   511

Fig. 37.13 Spatial distribution of $ESAI_1$ – Environmentally Sensitive Areas Index (without LU_MQI and HPI) (**a**), distribution of regional (**b**) and provincial (**c**) surface in $ESAI1$ classes

by the provinces of Lecce, Brindisi and Bari, where the land areas critical to desertification occupy 63, 59 and 55% respectively. A relatively good situation is notified for the province of Foggia, where the critical to desertification land covers 33% of territory.

The inclusion of human-induced factors changes significantly the situation about desertification risk in the Apulia region by describing 80.1% of territory as critical, 12.9% as fragile, 1.2% as potentially affected and 0.2% as non-affected land (Fig. 37.14b). In particular, the results of elaboration per each province are totally different. In fact, the worst situation is indicated for the provinces of Foggia and Brindisi, both with 90% of land classified as critical to desertification, and then, for the provinces of Taranto and Lecce where the percentage of areas critical to desertification is 87 and 77% respectively (Fig. 37.14a, c). A more favourable situation is

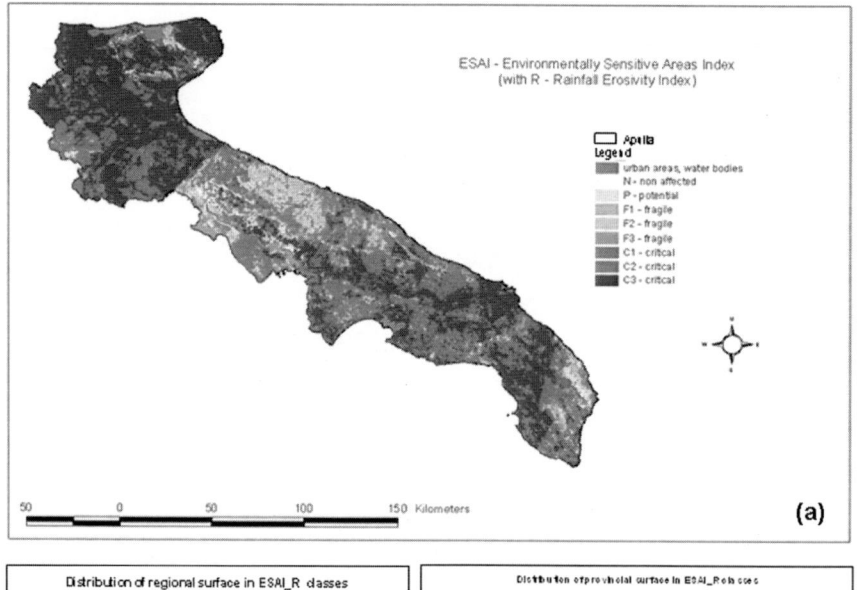

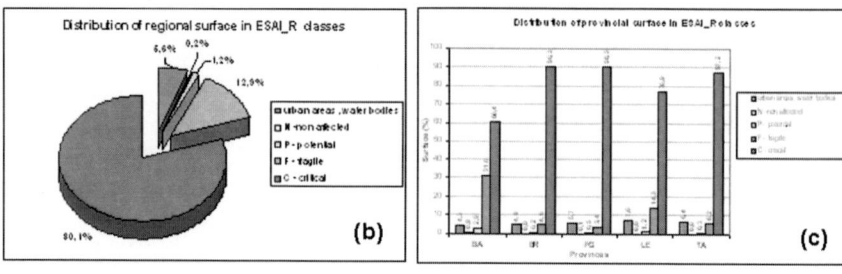

Fig. 37.14 Spatial distribution of $ESAI_2$ – Environmentally Sensitive Areas Index (with LU_MQI and HPI) (**a**), distribution of regional (**b**) and provincial (**c**) surface in $ESAI_2$ classes

observed for the province of Bari where 60% of territory is classified as critical to desertification processes.

These results are mainly due to lower human pressure impact in the province of Bari than in the other Apulian provinces (most of urban areas are regularly populated during the whole year), and, probably, due to better implementation of the EU regulations and directives related to the agronomical and forestry practices, protected areas and introduction of organic farming.

The introduction of the HPI and LU_MQI into the original approach has improved the overall evaluation of the desertification risk. Especially, it helps to distinguish between the areas with different management practices (poor and adequately applied) and those where the human pressure plays an important role. As an example, the difference between the ESAI indices, with and without consideration of HPI and $LU\&MQI$, is illustrated in Fig. 37.15.

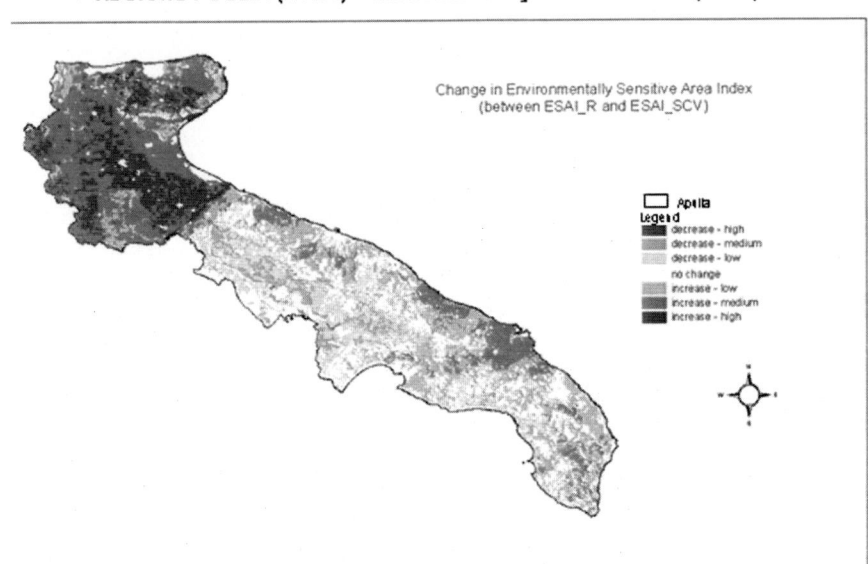

REGIONE PUGLIA (ITALY) – Environmentally Sensitive Areas (ESAs)

Fig. 37.15 Spatial distribution of variation between $ESAI_1$ and $ESAI_2$ for the whole region

A clear dissimilarity is observed between the provinces of Foggia and Brindisi and the other areas of Apulia (Fig. 37.15). While the provinces of Foggia and Brindisi are exposed to high human pressure and have medium-to-low level of quality of management and land use (which increases the overall desertification risk), such a risk is attenuated in the other areas, where the application of better (medium-to-high) management practices are supposed to be better effective, although the human pressure still remains high.

37.5 Conclusions

This study confirms high flexibility of the original ESAs approach that could be modified offering large opportunities for the evaluation of complex environmental conditions. The method is conceptually very simple, easy to be implemented from local to regional or national scale since much of the input information is already available in geo-referenced format (e.g. Soil Map of Europe, CORINE – Land Use and Land Cover database CORINE, 2000, different climatic database, etc.). Therefore, the presented methodology could be used as one of the guiding criteria for the definition of priorities in adoption of strategies to mitigate desertification not only in the Apulia region but also in other semi-arid areas of the Mediterranean region and beyond.

The introduction of the new indices, especially those with a notable degree of details (e.g. Land Use Intensity and Fire Risk) has improved the analysis of the range of the causes of the desertification process. In addition, the application of prevalently quantitative indicators, instead of the qualitative ones, as used in the original MEDALUS approach, has enhanced the definition of the areas vulnerable to desertification. The proposed modifications permit not only the identification and refinement of different degrees of sensitivity of an area subject to land degradation, but also allow the analyses of the factors affecting desertification and their evaluation in terms of spatial and temporal distribution.

It is worthwhile mentioning that the selection of indicators and the scale of application are "open" processes: some indicators may be excluded and some other may be added into the framework, in order either to adapt the model to the specific environmental conditions, or to improve the knowledge about some particular aspects of desertification.

Furthermore, the presented approach allows for the crossed analyses and elaborations focusing on the specific aspects of land degradation process. Such aspects could be particularly useful for the development of the strategies to combat desertification processes. They should be based on the local knowledge and specific features characterising each part of the territory. This would permit the identification of hot-spot areas and type of intervention, the allocation of financial resources, the commitment of local authorities, institutions and communities in the implementation process. In fact, in 2008, the regional authorities have applied the methodology presented in this study in a project dealing with the estimation of desertification risk at the regional scale following the guidelines of the EU Strategy for Soil Protection (COM, 2006 232 of 22 September 2006).

References

ANPA – Agenzia Nazionale per l'Ambiente – Centro Tematico Nazionale Conservazione della Natura. (2000). Selezione di indicatori ambientali per i temi relativi alla biosfera. RTI CTN_CON 1/2000. Roma, 167 pp.

Arnoldus, H.M.J. (1980). An approximation of the rainfall factor in the universal soil loss equation. In: M. De Boodt, D. Gabriels (eds.), Assessment of Erosion. John Wiley and Sons, Chichester.

Bagnouls, F. and Gaussen, H. (1952). L'indice xérothermique. Bulletin de l'Association de Géographes français. N° 222–223, janv.–févr., Paris.

Basso, F., Bellotti, A., Faretta, S., Ferrara, A., Mancino, G., Pisante, M., Quaranta, G. and Taberner, M. (1999). The agri basin. In: Kosmas C., Kirkby M., Geeson N. (eds.), The MEDALUS Project – Mediterranean Desertification and Land Use. Manual on key indicators of desertification and mapping Environmentally Sensitive Areas to desertification. EUR 18882.

Blonda, M., Ladisa, G. and Perrino, V.M. (2007). La desertificazione in Puglia: attività conoscitive avviate dall'ARPA Puglia. Poster presentato in occasione della II Conferenza Organizzativa "Stati Generali Arpa Puglia". Bari, 20 dicembre 2007.

CORINE. (2000). Soil erosion risk and important land resources in the southern regions of the European Community. EUR 13233 EN.

D'Asaro, F. and Santoro, M.. (1983). Aggressività della pioggia nello studio dell'erosione idrica del territorio siciliano. CNR – Progetto finalizzato "Conservazione del Suolo" – Sottoprogetto

"Dinamica dei versanti" pubblicazione 130; pubblicazione 164 dell'Istituto di Idraulica dell'Università di Palermo.

Ferro, V., Porto, P. and Yu, B. (February 1999). A comparative study of rainfall erosivity estimation for southern Italy and south-eastern Australia. Hydrological Sciences-Journal-des Sciences Hydrologiques 44(1):3–24.

Fournier, F. (1960). Climat et érosion: la relation entre l'érosion du sol par l'eau et les précipitations atmosphériques. Presses Universitaires de France, Paris, 201 pp.

Gabriels, D. (2000). Rain erosivity in Europe. En: Rubio J.L., Asins S., Andreu V., de Paz J.M. & Gimeno E. (eds.). ESSC III International Congress. Key Notes. Man and Soil at the Third Millennium. Valencia. pp. 31–43.

Hunsaker C.T. and Carpenter D.E. (eds.). (1990). Ecological Indicators for the Environmental Monitoring and Assessment Program. EPA 600/3-90/060. US Environmental Protection Agency, Office of Research and Development, Research Triangle Park, NC.

Kosmas, C., Ferrara, A., Briassouli, H. and Imeson, I. (1999). Methodology for mapping ESAs to desertification. In: Kosmas C., Kirkby M., Geeson N. (eds.) The MEDALUS Project – Mediterranean Desertification and Land Use. Manual on key indicators of desertification and mapping Environmentally Sensitive Areas to desertification. EUR 18882, pp. 31–47.

Ladisa, G. (2001). Criteri di quantificazione delle Aree Sensibili alla desertificazione in ambienti mediterranei. Tesi di dottorato di Ricerca, Università di Padova, Italy.

Ladisa, G. (2007). La desertificazione: priorità per la Puglia. In: La terra è una sola!. numero unico, Bari, pp. 2633, ottobre 2007.

Ladisa, G., Todorovic, M. and Trisorio Liuzzi, G. (2002). Characterization of areas sensitive to desertification in Sourthern Italy. In: Becciu G., Maione U., Majone Letho B., Monti R., Paoletti A., Paoletti M., Sanfilippo U. (eds.), Proceedings of the 2nd International Conference "New Trends in Water and Environmental Engineering for Safety and Life: Eco-compatible Solutions for Aquatic Environments". Capri (Italy), CDSU– Centro Studi Deflussi Urbani, Milano. Essestampa srl. Napoli. ISBN 88-900282-2X, June.

Ladisa, G. and Trisorio Liuzzi, G. (2001). Presentazione di alcuni indicatori sulla desertificazione. In: Biodiversità: Monitoraggio e indicatori ambientali. Seminario Nazionale del Centro tematico Nazionale Conservazione della Natura. ANPA – ARPA Valle d'Aosta. Saint Vincent (AO), 22–23 ottobre 2001.

Porto, P. (1994). Stima dell'aggressività della pioggia mediante l'Indice di Fournier. Un caso di studio. Annali dell'Accademia Italiana di Scienze Forestali XLIII:287–308.

REGIONE PUGLIA – Settore Programmazione Ufficio Informatico e Servizio Cartografico. (2000). Programma d'azione per la lotta alla siccità e alla desertificazione. Indicazione delle aree vulnerabili in Puglia. Bari, Italy.

Steduto, P. and Todorovic, M. (2001). Agro-ecological characterization of Apulia region: Methodologies and experiences. In: Zdruli, P., Steduto, P., Lacirignola, C., and Montanarella, L. (Eds.), Soil Resources of Southern and Eastern Mediterranean Countries. Options Méditerranéennes, Serie B: Studies and Research, Paris, N°34, pp. 143–158.

Todorovic, M. and Steduto, P. (2003). A GIS for irrigation management. In: Physics and Chemistry of the Earth, Special volume "Water for food and environment" Elsevier Science Ltd., Oxford, UK 28(4–5):163–174.

Trisorio Liuzzi, G., Ladisa, G. and Todorovic, M. (2004). Environmental Sensitive Areas to desertification model: Supplementary indicators accounting for socio-economic conditions of Southern Italy. In: Zdruli P., Trisorio Liuzzi G. (eds.) Workshop Proceedings on "Determining an income- product generating approach for soil conservation management, Marrakesh, Morocco. MEDCOASTLAND PROJECT – E.U. DG Research INCO-MED Programme – CIHEAM. MEDCOASTLAND publication 2. IAM- Bari, Italy. ISBN: 2-85352-311–X. 12–16 February.

Trisorio-Liuzzi, G., Ladisa, G. and Todorovic, M. (2005). Identification of areas sensitive to desertification in semi-arid Mediterranean environments: The case study of Apulia region (Southern Italy). In Hamdy A. (ed.), Water, Land and Food Security in Arid and Semi-Arid

Regions. International Conference Proceeding. Bari, Italy, IAM Bari, Italy, pp. 369–398 – ISBN2-85352-326-8, 6–11 September 2005.

Zdruli, P. (2000). Carta Ecopedologica della Puglia. In: Rusco, E.; Filippi, N.; Marchetti, M. and Montanarella, L. (eds.) Carta Ecopedologica d'Italia, scala 1:250,000. 2003. Relazione divulgativa. IES, CCR, CE, EUR 20774 IT, 45 pp.

Chapter 38
Spatial Variability of Light Morainic Soils

Michał Czajka, Stanisław Podsiadłowski, Alfred Stach, and Ryszard Walkowiak

Abstract The aim of the study was to estimate spatial variability of soil texture with particular consideration of soil susceptibility to wind erosion and it shows that in Central Poland this variability is high even within small fields. The fields' soil texture revealed a combined effect of the initial lithological variability of soil and the Aeolian selection of material. The latter embraces both the relatively permanent patterns of deflation and accumulation zones connected with the prevailing wind direction, and the pattern of permanent terrain barriers, such as the result of the last significant Aeolian episode. Wind erosion further increases the naturally high soil variability and makes it difficult for agriculture to progress in obtaining maximum yield from the practice of optimal organic manuring. The second aim was to show usefulness of the method, developed by Stach and Podsiadłowski, in the determination of deflation and accumulation zones on fields of up to 10 ha.

Keywords Wind erosion · Kriging · Deflation and accumulation zones · Poland

38.1 Introduction

The aim of the study was to estimate spatial variability of soil texture with particular consideration of soil susceptibility to wind erosion and it showed that in Central Poland this variability is high even within small fields. Wind erosion has become a major factor in the present soil forming process occurring in the lowland areas of Central Europe. It is caused by global changes in climate as well as the intensification of agriculture. New techniques of soil tillage reported in the 1960s resulted in the stimulation of the wind erosion process. As mechanization and effectiveness

---

M. Czajka (✉)
Department of Mathematical and Statistical Methods, Poznan University of Life Sciences, 60-637 Poznan, Poland
e-mail: michalczajka@gmail.com

of cultivation increased, the field area expanded, and roadside and field trees were removed, thus the menace of wind erosion developed.

Research on the influence of mechanical cultivation on wind erosion of light soils proved that in the Wielkopolska Region in Poland wind erosion processes generally take place on light soils with low mechanical strength of aggregate structure. This strength is mainly determined by the particle size distribution, particularly clay content. These soils are subject to wind erosion especially when the energy of wind is reinforced by the energy of tillage (Fig. 38.1). It happens especially during shallow tillage treatments at low soil moisture levels.

Results of a passive experiment established in Wielkopolska in 1986 indicate that the intensity of wind erosion amounts here from 5 to 20 t ha^{-1} year^{-1}. Soil removed from fields is deposited along roadside shelterbelts typical of the Wielkopolska landscape. The direct effect of wind erosion is that drainage ditches are covered by dust. A clear change in the size distribution of soil aggregates and humus content in the cultivated soil layer of an eroded field is the indirect effect of wind erosion. The intensity of pulverizing and wind erosion in Central Poland displays very high spatial variability, even within small, single fields. This is an effect of polyfractionality

Fig. 38.1 Pulverizing soil erosion on Brzeg Glogowski and Wierzenica sites

and diversity of morainic deposits that form the soil bedrock. Wind erosion further increases the naturally high soil variability. The high variability in texture and humus content, as well as derivative physical and chemical characteristics of soil make it difficult for agriculture achieve obtaining maximum yields from the practice of optimal organic manuring, Geostatistical methods were used to determine spatial variability of soil texture within the field under study.

38.2 Materials and Methods

The investigations described in this paper are a continuation of analyses made by Stach and Podsiadłowski (2001) on two fields. The first one with an area of 64 ha (the Wierzenica site) is located 14 km north–east of the center of Poznań (52°29'11"N, 17°24'06"E) near the village of Milno. It is an area of a flat ground moraine resulting from the Vistulian (Würm) Glaciation. The other, smaller, of 7 ha (Brzeg Głogowski) is located 10 km west of the Głogów (51°41'55"N, 15°54'48"E).

In order to verify the usefulness of this method on small fields, additional research was carried out on the Brzeg Głogowski field, and on a third smaller field (6.8 ha), located near the village of Bąblin, 32 km north–west from Poznan (52°40'50"N, 16°44'00"E). On its loamy sands and sandy loams there have developed soils lesivées (Typic Hapludalfs – Soil Survey Staff, 1975, Orthic Luvisols – FAO), which are common in the Polish Plain. Crop rotation practiced on the experimental fields is typical of loamy sand areas in Poland: potatoes or maize – grains (barley or rye) – lupin or rape – grains. The soil is cultivated using tractors with power ratings of up to 63 kW. The traditionally used pre-planting tillage causes breaking and pulverizing of aggregates in the topmost soil layer (0–3 cm), which has been shown to accelerate wind erosion (Fig. 38.1).

One of the reasons this plot of land was selected for analysis was its almost flat, little diversified relief, which suggests that the effect of water erosion on its soils is negligible. The Wierzenica field is surrounded on three sides by roadside trees with the following parameters: the north–eastern side: mean tree height $H = 17.5$ m, shelter porosity $P_0 = 14\%$; the south–eastern side: $H = 13$ m, $P_0 = 42\%$; and the south–western side: $H = 7.5$ m, $P_0 = 7\%$. There are also three tree clusters growing on the field itself. The Brzeg Głogowski field is sheltered on two sides by trees (Fig. 38.1) and was selected for the experiment because of its specific shape. It is approximately an isosceles triangle (Fig. 38.2), which north and south–east sides are sheltered by trees. The third side is open to the W–E wind, prevailing on the Polish Plains.

The Bąblin field is interesting with respect to wind erosion, because it is sheltered only to the south by low buildings and in its northern side has a shallow hollow, which should result in smaller soil erosion.

The Wierzenica and Bąblin fields, which are situated in a W–E oriented "corridor" between two patches of woodland, are particularly susceptible to wind erosion. Since 1986, measurements of Aeolian transport have been carried out on the Wierzenica field with the help of a dust trap. The mean intensity of this process

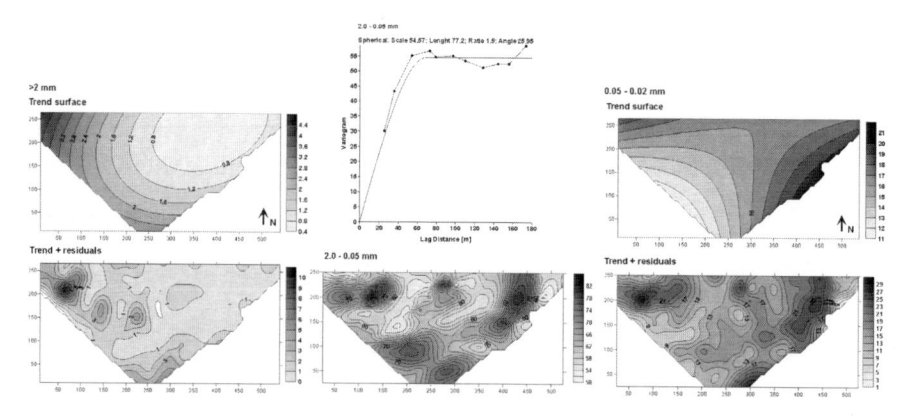

Fig. 38.2 Spatial variability of selected soil fractions in Brzeg Głogowski

was 8.9 ± 8.1 kg m^{-1} year^{-1} (± 1 SD). The minimum and maximum annual totals were 1 and 29 kg m^{-1} year^{-1}, respectively (Podsiadłowski, 1995). The process was usually most intense in early spring (March–April) during the pre-planting treatment and sowing periods (Hagen et al., 1999).

Wind erosion in the Wierzenica and Brzeg Głogowski fields were studied by Stach and Podsiadlowski (2001). Spatial variability of soils in the experimental fields was studied by analysing soil samples taken from a regular grid at 50 m intervals in the Wierzenica field and 25 m intervals in the Brzeg Głogowski and Bąblin fields. A detailed description of the sampling scheme and field and laboratory procedures is included in a study by Stach and Podsiadłowski (1998). The mechanical composition of fine particles was determined using sieves and the standard areometric method. The following fractions were distinguished: >2, 2 – 1, 1 – 0.1, 0.1 – 0.05, 0.05 – 0.02, 0.02 – 0.005, 0.005 – 0.002, and < 0.002 mm.

38.3 Results and Discussions

The analysis of spatial variations in soil texture resulting from wind erosion is very difficult in areas covered by deposits of direct glacial accumulation in the Polish Plain. The reason is the high random variability of soil lithology as a result of an uneven distribution of rock detritus in glacier ice, often at a very small scale. In this situation it is necessary to make use of geostatistical methods (Goovaerts, 1997; Oliver et al., 1989; Isaaks and Srivastava, 1989).

The random variability of lithology of soils in Wierzenica is most readily visible in the coarse fractions (>1 mm). In the spatial scale employed (interval = 50 m and scale = 900 m) they show no significant spatial autocorrelation. Systematic spatial variability in a range of 370–490 is displayed by 1–0.1, 0.05–0.02, 0.02–0.005 and 0.005–0.002 mm fractions in the ploughed-layer samples. In the <0.002 mm fraction from the ploughed-layer samples only "pure nugget" variance was found, as in the coarse fractions.

The most significant aspect from the point of view of the investigated effect of wind erosion on soil is spatial variability of the 0.1–0.05 mm fraction (very fine sand). In the case of soils with loamy sand and sandy loams in its composition, this is the fraction most susceptible to deflation (Fullen, 1985; Stach, 1995). Presumably, the lack of spatial autocorrelation is the result of the latest erosion episode caused either by high, variable winds or by fieldwork.

On the basis of the estimated semi-variance models, maps were made of the analysed textural parameters (interpolation by kriging). The analysis focused on spatial variability of particular fractions in the ploughed layer (Fig. 38.3).

Spatial variability of soil texture in the Wierzenica field may be explained much easier with the help of data from similar analyses carried out on a field in the vicinity of the villages of Brzeg Głogowski and Bąblin. A denser, 25 m interval, sampling grid was used there. In Brzeg Głogowski this made it possible to identify two types of spatial pattern: a quadratic trend embracing the entire field and autocorrelation up to 50–70 m. The character of this trend is different for specific fractions. It depends on the relative importance of two factors: (1) the type of sediment deposition (the site is located on the river terrace), and (2) the long-term pattern of wind erosion and accumulation. Structures in the range of 50–70 m are "belt and patch" in shape

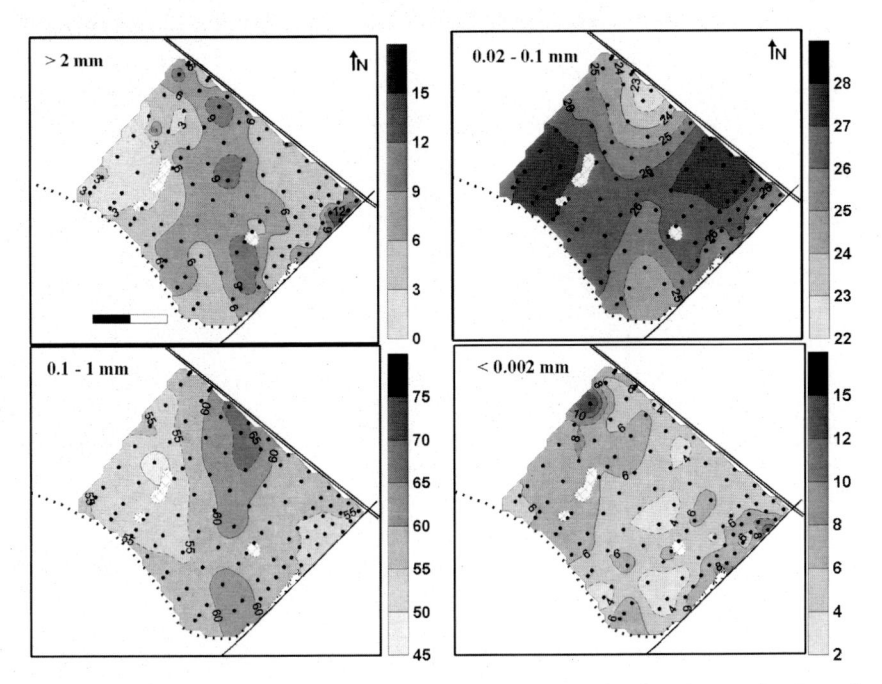

Fig. 38.3 Spatial variability in proportions (in %) of selected fractions in samples from the ploughed layer in Wierzenica

(Fig. 38.2). It is a trace of soil translocation at the time of the last episode of strong winds. This conclusion is confirmed by direct observations of the travel distance of soil material transported by saltation. In the case of the 2.0–0.05 mm fraction there was autocorrelation in the near East West direction. The semivariogram and map of this fraction were given in Fig. 38.2.

In the experimental field in Bąblin the application of geostatistical methods showed a lack of trend in spatial variation of the most important soil fractions. In turn, a rather weak spatial autocorrelation was detected (scale from 1.98 to 8.086) with a range from 71 m for dust to 167 m for loam. Semivariograms and maps of distribution for individual soil fractions are given in Fig. 38.4.

Based on the investigations carried out in 1988 in the field in Wierzenica, Podsiadłowski (1994) identified deflation and accumulation zones on the basis of humus content. Eight years later, on the same field Stach and Podsiadłowski (2001) applied a probe using a different method.

Soil eroded and transported by wind exhibits fast selection. The analysis of sediment samples collected several times just after strong wind episodes (Stach, 1995) makes it possible to describe the character of the sorting process in relation to transport type and distance (Fig. 38.5). Fraction class intervals in Fig. 34.4 are identical to those used for grain size analysis of samples taken from the Wierzenica field.

Erosion lag and material transported in traction over a distance of 10^{-1}–10^0 m are much coarser than the source soil. In the predominant fractions grain diameter is greater than 1 mm. There are practically no grains smaller than 0.05 mm. At a distance of 10^0–10^2 m homogeneous fine-grained sand is transported by saltation (especially fractions from 0.5 to 0.1 mm). The material in an air suspension rises and accumulates along shelterbelts and other barriers, and its grain diameter is smaller than 0.1 mm (fractions <0.02 mm make up close to 80% of the total).

Based on the above remarks, Stach and Podsiadłowski (2001) in order to identify deflation and accumulation zones in the Wierzenica field used standardised shares in the ploughed layer of fractions most susceptible to deflation and those constituting lag-deposits (Fullen, 1985; Jönsson, 1994; Lyles and Tatarko, 1986; Stach, 1995). It turned out that deflation and accumulation zones determined by this method (Fig. 38.6) are almost identical to zones identified by Podsiadłowski (1994).

An interesting issue is whether this method will also be applicable in case of small fields, such as those in Brzeg Głogowski and Bąblin. Thus analyses were conducted on standardised shares in the ploughed layer of fractions most susceptible to deflation (0.1–0.02 mm) and those constituting lag-deposits (>1 mm) (Stach, 1995). Kriging was used for this purpose. Two types of spatial pattern were detected: a quadratic trend embracing the entire field and autocorrelation up to 30 m in the NE–SW direction and 100 m in the NW–SE direction. Based on the semivariogram (linear semivariogram, the Nugget effect: 0.0249; slope of the straight line 2.09e-5; anisotropy ratio 2; anisotropy angle 110.7°) residuals were estimated, from which, following their addition to the trend, a map was created (Fig. 38.7), on which zones

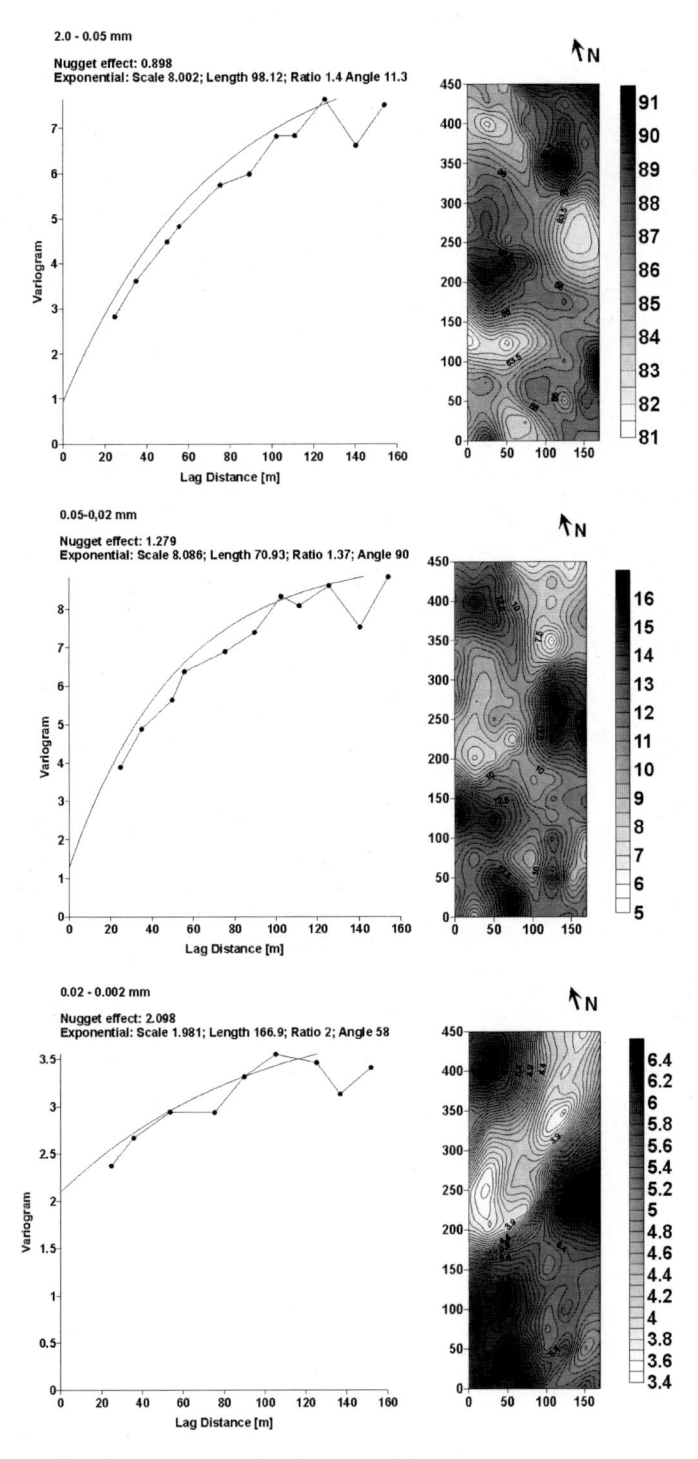

Fig. 38.4 Spatial variability of selected soil fractions in Bąblin

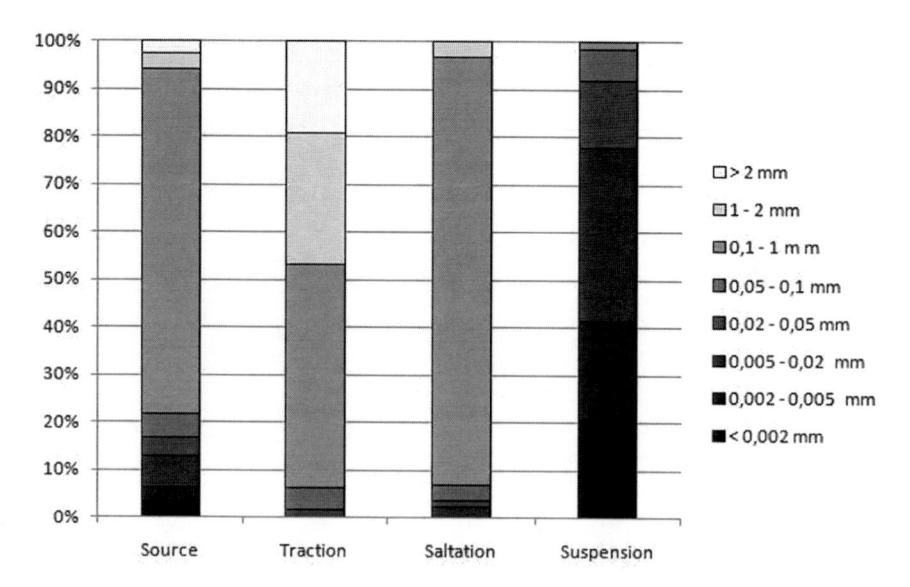

Fig. 38.5 Typical sorting processes of soil primary particles induced by wind erosion in Wielkopolska Region fields (Stach, 1995)

Fig. 38.6 Erosion (*dark*) and accumulation (*light*) zones (Stach and Podsiadłowski, 2001)

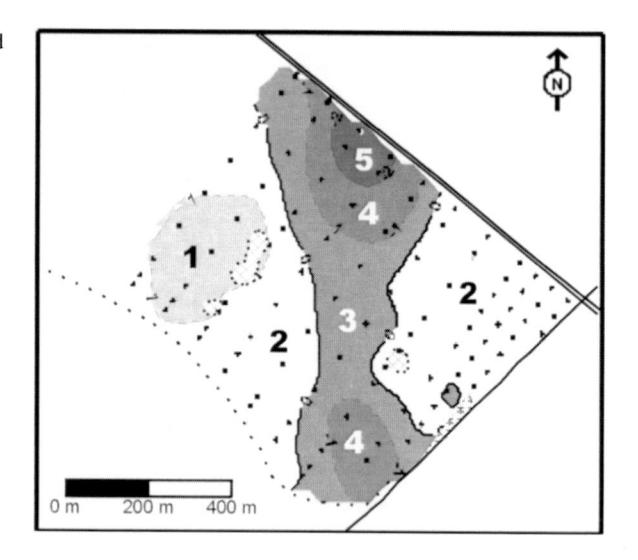

containing more 0.1–0.02 mm fractions are marked with a darker colour. The distribution of these zones is consistent with the direction of prevailing winds and tree planting belts limiting the field from the north and south–east.

In the Bąblin field soil variation is markedly smaller than that in Brzeg Głogowski, with only loamy sand found there. Despite that fact, the application of a slightly modified method by Stach and Podsiadłowski (2001) even in this small field

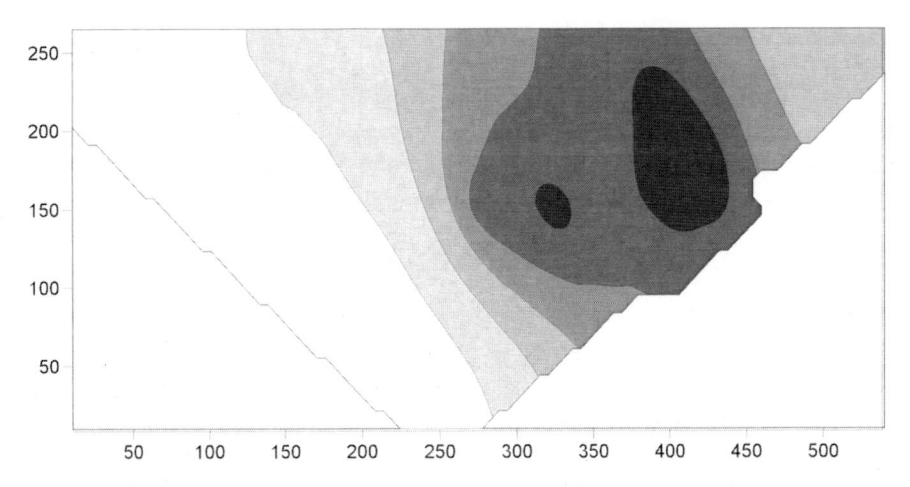

Fig. 38.7 Zones of more (*light*) and less (*dark*) eroded soil in the Brzeg Glogowski field

made it possible to identify erosion zones. Similarly as it is the case for the Brzeg Głogowski field, kriging was used to investigate the spatial distribution of deflation and accumulation zones. A quadratic trend was found. Based on a semivariogram, which range was 60–80 m, residuals were estimated. The sum of trend and residuals made it possible to create a synthetic picture of deflation zones in the Bąblin field (Fig. 38.8).

It may be seen that fragments of the field located in its northern and southern parts are less eroded. The southern part of the field, sheltered by buildings, and the northern part, located lower than the other sections of the field, were least eroded.

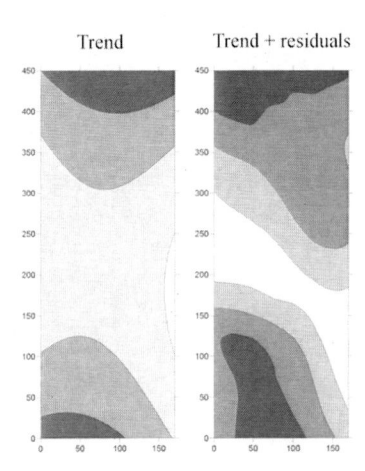

Fig. 38.8 Zones of more (*light*) and less (*dark*) eroded soil in the Bąblin field

38.4 Important Concluding Remarks

Analyses conducted on soil samples collected from experimental fields showed that even in a small field there is a rather considerable variation in grain size composition. Among other things, it is caused by long-term action of Aeolian erosion, especially intensive when there is no plant ground cover or when it is scarce. In such a situation it is necessary to search for tillage methods facilitating precision faming and at the same time potentially effectively preventing Aeolian erosion.

Investigations conducted on the effect of mechanical tillage of light soils on Aeolian erosion showed that a significant Aeolian erosion process occurs under the conditions found in the Wielkopolska region as a rule only on light soils, loamy sands, in the period corresponding to pre-sowing cultivation (Podsiadłowski, 1995). This is the case because:

- The aggregate structure of these soils exhibits poor strength,
- During traditional tillage the topsoil is subjected to the action of both tillage energy (kJ m^{-2}) and compaction energy generated by tractor wheels,
- Pre-sowing cultivation on light soils usually takes place at a relatively low moisture content of topsoil, which contributes to destruction and stimulates wind erosion.

As it is known, simplified tillage systems provide soil with Aeolian protection thanks to plant residue left on field surface. However, in case of soils with low natural porosity such systems cause a marked decrease in yields of traditionally cultivated crops, such as potatoes, sugar beets or barley.

It is advisable for light soils to consider introduction of integrated tillage. It consists in ploughing, post-ploughing tillage and sowing or planting during one tractor passage (Podsiadłowski, 2005). This is performed using a specially prepared machine aggregate (Fig. 38.9), consisting of a tractor, plough, land roller aggregate and a seeding or planting machine.

Integrated tillage is connected with the following advantages:

- A tractor travels only on soil having still a cohesive structure, which means that tractor wheels do not compact the soil having already an aggregate structure,
- Elimination of wheel tracks makes it possible to relatively arbitrarily modify total porosity of topsoil, according to requirements of crops and weather forecast, as well as limit the total tillage unit energy input and as a consequence – reduce fuel consumption;
- Limitation of tillage unit energy input and performance of tillage at stable (usually average) moisture content makes it possible to obtain relatively high strength of the formed aggregate structure, which reduces possible incidence of Aeolian erosion.

Fig. 38.9 Components of integrated tillage system: *1* – tractor, *2* – mouldboard plough, *3* – rollers, *4* – sowing machine

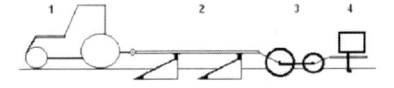

38.5 Conclusions

It is possible to sum up the study results as follows:

1. The small range of autocorrelation, 60–80 m, confirms very high spatial variability of light morainic soils.
2. Analysis of standardised shares of fractions most susceptible to deflation and those constituting lag-deposits in the ploughed layer may be useful in the detection of eroded zones.
3. It is advisable for light soils to consider introduction of integrated tillage.

Acknowledgments The study was conducted within the framework of grant no. R12 005 03. The development of a forecasting model for integrated anti-erosion tillage, financed by the Ministry of Science and Higher Education.

References

Fullen, M.A. (1985). Wind erosion of arable soils in East Shropshire (England) during spring 1983. Catena 12:111–120.
Goovaerts, P. (1997). Geostatistics for Natural Resources Evaluation. Oxford University Press, New York, pp. 1–483.
Hagen, L.J., Podsiadłowski, S. and Skorupski, D. (1999). Development of a tillage system to prevent soil pulverization and wind erosion. Scientific Papers of Agric. University of Poznan, Poland. Agriculture 1:15–27.
Isaaks, E.H. and Srivastava, R.M. (1989). Applied Geostatistics. Oxford University Press, New York.

Jönsson, P. (1994). Influence of shelter on soil sorting by wind erosion – a case study. Catena 22:35–47.
Lyles, L. and Tatarko, J. (1986). Wind erosion effects on soil texture and organic matter. Journal of Soil and Water Conservation 41:191–193.
Oliver, M., Webster, R. and Gerrard, J. (1989). Geostatistics in physical geography. Part I: Theory. Part II: Applications. Transactions Institution of British Geographers N.S. 14:259–269, 270–286.
Podsiadłowski, S. (1994). Próba oceny natężenia erozji eolicznej oparta na analizie zawartości próchnicy warstwy uprawnej gleby (sum.: An attempt to evaluate intensity of wind erosion process on basis of analysis of humus content in arable soil layer). Roczniki Akademii Rolniczej w Poznaniu 260:87–94 (in Polish).
Podsiadłowski, S. (1995). Rola uprawy mechanicznej w stymulacji procesu erozji eolicznej gleb lekkich (sum.: Influence of mechanical tillage on initiation of wind erosion on light soils). Roczniki Akademii Rolniczej w Poznaniu, Rozprawy Naukowe 264:1–47 (in Polish).
Podsiadłowski, S. (2005). Znaczenie zintegrowanej uprawy gleby w ograniczaniu deflacji gleb. Acta Agrophysica 5(1):111–120 (in Polish).
Soil Survey Staff. (1975). Soil Taxonomy. USDA, Washington, 754pp.
Stach, A. (1995). Procesy i osady eoliczne na polach środkowej Wielkopolski (sum.: Aeolian processes and deposits on the fields of central Wielkopolska). Studia z Geografii Fizyczne, Poznańskie Towarzystwo Przyjaciół Nauk, Sprawozdania Wydziału Matematyczno Przyrodniczego nr 109 za lata 1991–1994 1:145–153 (in Polish).
Stach, A. and Podsiadłowski, S. (1998). The effect of wind erosion on the spatial variability of cultivated soils in the Wielkopolska region (Poland). Proceedings of International Conference on Agricultural and Engineering. AgEng-Oslo 98 (CIGR), paper no: 98-C-089.
Stach, A. and Podsiadłowski, S. (2001). Pulverizing and wind erosion as influenced by spatial variability of soils texture. Questiones Geaграficae 22(1):67–78.

Chapter 39
Studding the Impacts of Technological Measures on the Biological Activity of Pluvial Eroded Soils

Geanina Bireescu, Costica Ailincai, Lucian Raus, and Lazar Bireescu

Abstract Pedo-biological indicators, together with physical and chemical indicators, are important elements in the complex task of determining soil quality and reducing negative effects of soil degradation. In this chapter we present the results of the soil biology research on degraded lands by pluvial erosion, disposed in agroterraces, located in North–Eastern part of Romania. The experiment was established in a sloping field in Podu-Iloaiei, Iasy county on a Cambic Chernozem at 0–20 cm depth to study the seasonal evolution (spring-summer) of the biotic potential (soil respiration and cellulolysae) without irrigation. Our data deriving from this experiment indicate that disposed antierosional measures such as agroterraces, have produced lower values of the biological activity on top slope, in control section and, also, in various fertilized variants. Much better results of tenaces were recorded on lower slopes. The effects of pluvial erosion have a higher negative influence on soil biological activity. The excessively dry summer season had a stressing and restricting effect on biological activity of the soil as well. The Indicator of Vital Activities Potential (IVAP) was correlated with soil physical and chemical indicators to establish possible relationships.

Keywords Agroterraces · Pluvial erosion · Romania · Soil biological activity · Technological elements

39.1 Introduction

The soil, the crucial element in the functioning of the biosphere (Rubio, 2008), is an open dynamic system, which realizes reversible changes of matter and energy with the surrounding environment (Mäder et al., 1997; Bireescu, 2001). It is a living space where several biological processes take place and contribute to transform the organic

G. Bireescu (✉)
Biological Research Institute, 700107 Iaşi, Romania
e-mail: bireescugeanina@yahoo.com

matter and also to ensure favourable conditions for plant nutrition. At the same time, the soil is a non-renewable resource and its conditions and well functioning influence food production and environmental efficiency of the natural ecosystems (Doran and Parkin, 1994; Dick, 1997; Doran and Zeiss, 2000; Gianfreda et al., 2005). The impacts of degraded processes have greater consequences on biological activity of the edaphically microorganisms and implicitly on soil fertility (Kubat et al., 2001; Bending et al., 2004).

Lal et al. (1999) suggest the existence of a strong correlation between soil quality and erosion. Thus, soil quality affects the rate of erosion and the erosion affects the quality of a soil. Lal describes the major soil quality effects of erosion that impact land productive capacity as follows: (1) decrease in rooting depth; (2) reduction in available water; (3) loss of soil organic matter; (4) loss of structure; (5) soil fertility problems; and (6) loss of soil biodiversity. The quantitative and qualitative evaluation of the biological activity of the soil resources ensures an estimation of the biological condition, evolutive edaphically processes and, also, points out the connection of the soil with biocenosis and the environment (McLaren, 1975; Rastin et al., 1988; Gianfreda and Bollag, 1996; Bireescu, 2001; Ştefanic et al., 2006). The complex study of the soil quality by physical, chemical and biological indicators, in an ecological context in different ecosystems contributes to the analysis and understanding of the causes and effects of the degradation processes (Fauci and Dick, 1994; Dilly and Blume 1998).

According to the *National Strategy and Action Programme to Combat Desertification, Land Degradation and Drought*, developed by the Institute for Soil Science and Agrochemistry in Bucharest (2001) and following the *National Strategic Plan of Romania* for the period 2007–2013, developed by Ministry of Agriculture, Forests and Rural Development in June 2006 it becomes evident that almost one third of the country is affected by various forms of soil degradation. The most important factors are the pluvial erosion and landslides, which affect about 7 million hectares. The soil organic matter (SOM) loss caused by the removal of the topsoil ranges between 45 and 90% of the total organic matter pool in the soil. At the country level, the total SOM losses amount are estimated at approximately 0.5 million tones year^{-1}. The areas with the highest rates of soil erosion are: Moldavian Plateau, Pericarpathian Hills between Trotuş and Olt, Transylvanian Plateau and Getic Piedmont.

The aim of this work was to highlight the spring-summer evolution of the biological activity (soil respiration, cellulosolysae and Indicator of Vital Activity Potential-IVAP) on degraded lands by surface pluvial erosion, and to test the efficiency of soil conservation measures such as terraces cultivated with winter wheat under organic, mineral and organo-mineral fertilization.

39.2 Materials and Methods

Recent studies (Farahbakhshazad et al., 2008) have shown that the complexities between soil management and biogeochemical cycles provide

Photo 39.1 The experimental site located at Podu Iloaiei in Moldavian Plateau in North Eastern part of Romania

opportunities for identifying and establishing the best management practices for specific agro-ecosystems. Consequently the National Research Programme in Romania is involved in long-term (40 years) research experience, which aimed the establishment of measures and technologies for sustainable use and improvement of degraded lands. Our research was conducted on a Cambic Chernozem WRB (2006) described by Munteanu and Florea (2001) and degraded by pluvial erosion. The experimental site (Photo 39.1) is located at Podu Iloaiei in the Moldavian Plateau in North Eastern part of the country, has a 16% slope and is cultivated with winter wheat. Before ploughing organic, mineral and organo-mineral fertilizers were added to the soil to boost soil micro organism activity.

Fractional mineral fertilizers ($N_{140}P_{140}$) were applied before sowing, during land preparation and in the beginning of spring during wheat vegetation. Organic fertilizer in the form of well-fermented manure (40 t ha^{-1}) was incorporated in the soil before ploughing. Other forms of manure (manure + $N_{70}P_{70}$) and vegetal residues as straws, maize stalks (6 t ha^{-1} + $N_{70}P_{70}$) and pea's creeping stalks (3 t ha^{-1} + $N_{70}P_{70}$) mixed with nitrogen and phosphorus mineral fertilizers have been also incorporated in the soil in different doses before ploughing. The vegetable residues were first minced with the disk. Terraces have been established in these degraded lands to control soil erosion as could be seen in Photo 39.1.

39.2.1 Eco-Pedological Research

The eco-pedological research pointed out a "constellation" of 20 main eco-pedological factors and determinants, included in the ecological specificity file. This file is characterized from a quantitative point of view by eight ecological size classes (zero-0, ... m-lack or minimum, I, II, III, IV, V, E_1-excesive with restricting effects on plants, E_2-excesive with toxic effects on plants) and from qualitative point of view through six ecological favourability classes (zero-0, ... m-lack or minimum, VL-Very Low, L-Low, M-Medium, H-High, VH-Very High). Such vales were based on specific published ecological criteria (Chirita, 1974; Bireescu et al., 2005, 2007).

39.2.2 Pedo-Biological Research

The research of soil biology analyse the role of the soil microflora in the processes of transformation of organic remains and, also, in the biological cycles of elements (Karlen et al., 1997, 2001; Pedro et al., 2002).

We determined experimentally the soil respiration and the cellulosolysae activity as biological processes of the soil. The determination of soil respiration was made with the respirometer, which replaces self-acting O_2 consumed in the process of soil respiration and collects disengaged CO_2. The determination of cellulosolysae was made after the methodology used by Unger (1960) and Vostrov and Petrova (1961) with the contribution to Ştefanic (1994a, b). The Indicator of the Vital Activity Potential (IVAP%) was developed under the new definition of soil fertility developed by Ştefanic (1994a, b).

$$IVAP\% = \frac{\sum_{k=1}^{2}(R,C)}{2}$$

where:

IVAP-Indicator of Vital Activity Potential;
R-soil respiration;
C-cellulosolysae.

This indicator was constituted by the method of numerical taxonomy, used in soil biology and, also, in soil chemistry by many scientists such as Verstraete and Voets (1974, 1977), Such et al. (1977), Misono (1977) and Teaci (1980).

39.3 Results and Discussions

The distribution of plant species depends on historical events and the ability to adapt to present environmental conditions (Schulze, 2005). Most agro-ecosystems are complex systems, within which climatic, soil and management factors intricately interact (Farahbakhshazad et al., 2008). Knowing the pedo-ecological background of the area can help us thus to understand local ecosystem functioning and enable to consider the best possible ways to implement sustainable land use and soil resource conservation programmes (Reintam et al., 2001; Kõlli et al., 2008).

Table 39.1 Ecological specificity file of ecopedotype (agro-ecosystem) from experimental field of Podu Iloaiei, Iasi in a Cambic Chernozem

Eco-pedological factors and determinants	Ecological size classes								Ecological favorability classes					
	0...m	I	II	III	IV	V	E_1	E_2	N...m	VL	L	M	H	VH
Total nitrogen content-Nt														
Available P content-P_2O_5														
Exchangeable K–K_2O														
Annual average temperature-T°C														
Annual average precipitations-Pmm														
Winds-W														
Summer precipitations-Pe														
Summer relative air humidity-Uer														
Edaphic volume-Ve														
Bioactive length period-BLP														
Alkality/Acidity-Alk														
Summer soil consistency-Con														
Soil Organic Matter content-SOM														
Soil texture-Tx														
Air porosity-PA														
Soil reaction-pH														
Base saturation-BS														
Pedo-biological activity-Bio														
Potential trophicity-PT														
Effective trophicity-ET														

Over the last decade Romania has experienced drought and flood periods that have become even more frequent, with negative impact on agricultural productivity, especially for wheat and corn, as well as for the flora and fauna species (National Strategic Plan of Romania, 2006). The year 2007 was excessively dry with only 487.9 mm annual average precipitations and 9.1°C annual average temperature.

Table 39.1 highlights the main 20 ecopedological factors and determinants, which act upon the soil quality. The majority of them are included in III and IV medium ecological size classes. In the lower ecological II class are included two stressful climatic factors, namely summer precipitation and summer relative air humidity and one pedo-ecological factor with lower value and un stressful namely alkality. In the high ecological V size class are included a growth factor i.e. the mobile phosphorus content, an ecologic factor condition of space and time or the bioactive length period and a pedologic determinant namely potential trophicity. In the ecological excessive E_1 class is included the hard soil consistency in the summer season, as a negative, limiting and stressful factor.

Referring to the favorability of biotope conditions for agricultural crops of ecological specificity file we note that most ecopedological factors and determinants with an essential role for the soil quality are included in medium (M) and high (H) ecological favorability classes. In very high (VH) ecological favorability class is included the high level of annual average temperature. Therefore analysis of ecological specificity file shows that the high trophic potential for this agro-ecosystem is stressed, limited and not fully exploited especially during the dry or rainy seasons. A possible explanation could be due to specific climatic conditions with prolonged drought in summer season starting from the Russian steppe all the way in northeastern area of Romania. Land degradation decreases infiltration, water holding capacity and transpiration, but enhances runoff and soil evaporation (Stroosnijder and Slegers, 2008). For these reasons we emphasize the lower level of summer rainfalls, the excessive drought and the lower level of air humidity in the summer season and hard soil consistency, which are included in the ecological very low (VL) favorability class for agricultural crops.

The study of soil biological activity analyses, along with physical and chemical indicators the quality of soil resources in different ecosystems and how the soil functions and interactions with climate and biocenoses are included in the ecosystem (Elliott, 1997; Seybold et al., 1997; Knoepp et al., 2000).

39.3.1 Biological Activity of the Degraded Soils During the Spring Season

In the spring season, in the control section there is the lowest value of soil respiration in agroterrace on the top slope. It observes a positive influence of manure and, also, the low energetically potential of straws, used as organic fertilizer, depending on variants of fertilization (chemical, organic or organic-mineral fertilization). Efficiency of applying plant remains depends on many factors but chemical composition, C/N ratio, period and depth of incorporation, and the grinding size of the plant remains have the greatest influence on supply of soil with nitrogen and organic matter. Besides, the surveys conducted in different conditions of climate and soil showed that decomposition of plant remains is up to five times slower if they are left on the soil surface rather than when they are incorporated into the soil (Ailincăi, 2007).

In the agroterrace on top slope the soil respiration is lower, comparatively with middle slope and especially with the lower slope (Fig. 39.1). These low values result as an effect of the stressing and negative impact of erosion that is more intense in top slope. In agroterrace on top slope there are the highest values of soil respiration in the case of mineral fertilization ($N_{70}P_{70}$). Actually, Kumar and Goh (2002), Shah et al. (2003) and Shafi et al. (2007) found that nitrogen content in the soil significantly increase due to the incorporation of crop residues.

Comparatively with agroterraces on the middle and lower slope, these values are still the lowest. In the agroterrace on the middle slope, soil respiration is higher in the control section, especially in the fertilized variants (Fig. 39.2). The highest values of soil respiration are in the agroterrace on the lower slope because, in this case, the impact of erosion is lower (Fig. 39.3). In the control section, soil respiration increases compared with the top slope. Also, in this agroterrace the highest positive influence of fertilization, which improves soil respiration, is in organic-mineral fertilization variant (manure 40 t ha^{-1} + $N_{70}P_{70}$).

Referring to cellulosolysae, the evolutive trend is similar with soil respiration. Thus, in the agroterrace on top slope there are the lower values of cellulosolythic potential (Fig. 39.4). We observed a positive influence of the manure and pea's creeping stalks. In conditions which encourage the use of lower doses of chemical fertilizers there is the need to provide alternative ways to maintain soil fertility

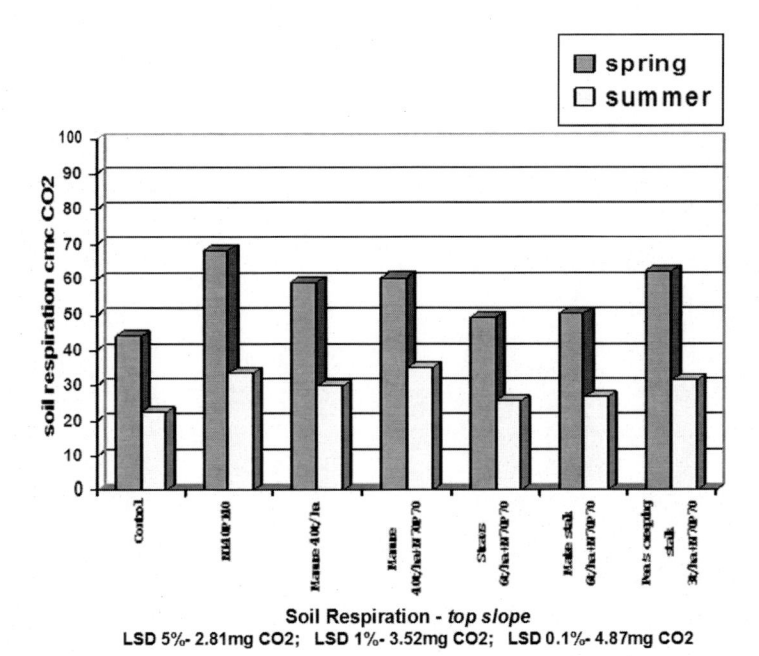

Fig. 39.1 Seasonal dynamics (spring–summer) of soil respiration (mg CO_2) to the top slope LSD-Limit Statistical Difference

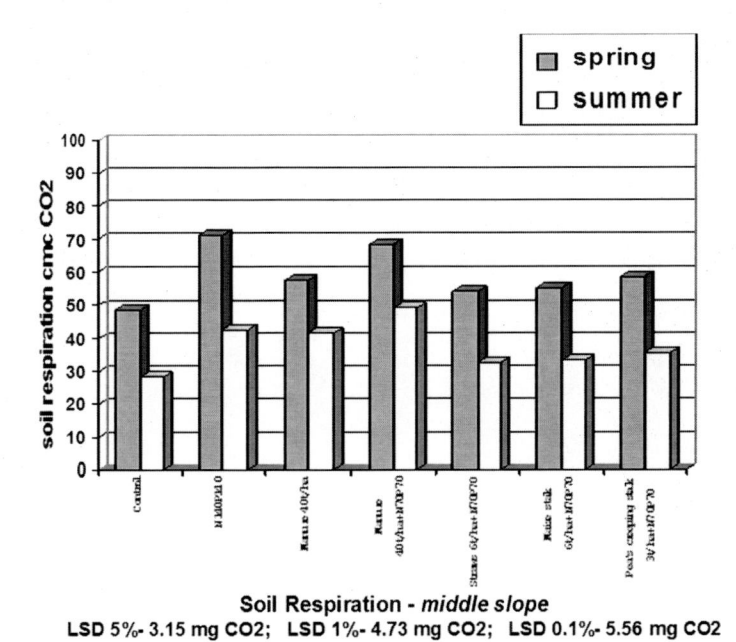

Fig. 39.2 Seasonal dynamics (spring–summer) of soil respiration (mg CO_2) to the middle slope LSD-Limit Statistical Difference

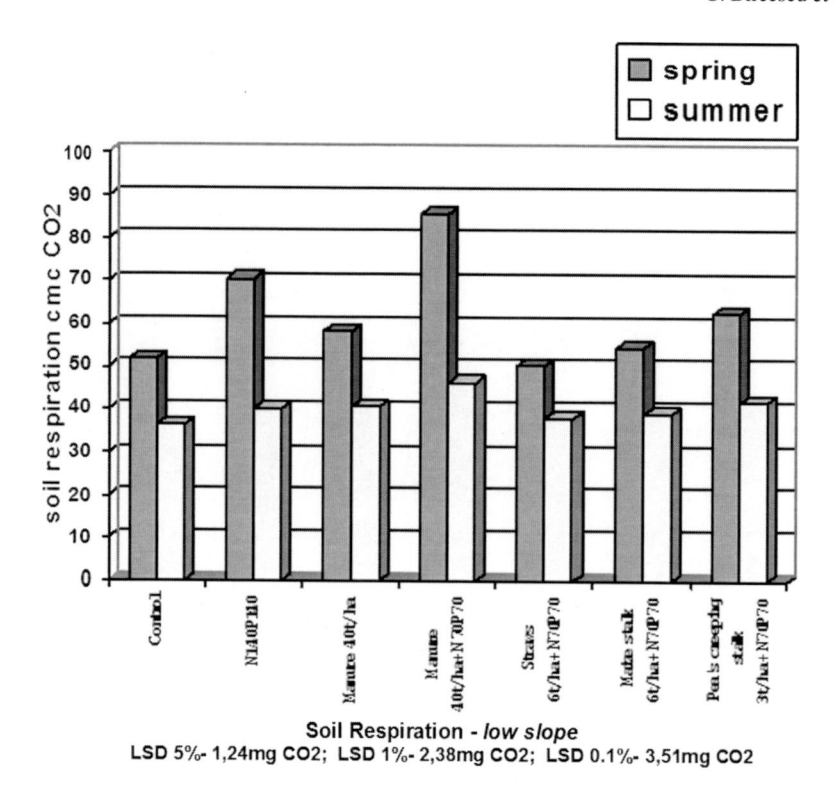

Fig. 39.3 Seasonal dynamics (spring–summer) of soil respiration (mg CO_2) to the low slope LSD-Limit Statistical Difference

and productivity at the most economical and practical means for preserving and recycling nutrients in the soil and maintaining a stable biomass and optimum conditions of its decomposition. Agro-ecosystem productivity with lower consumption is based on the internal circulation of biomass, part of which will be transformed for production and the difference to ensure maintaining the balance between production and decomposition of organic substances in the soil.

In the agroterrace on the middle slope and especially on the one on the lower slope, the negative impact of pluvial erosion is further diminished (Figs. 39.5 and 39.6). It observed superior values of cellulosolysae in control section and especially in the fertilized variants.

39.3.2 Biological Activity of the Degraded Soils in the Summer Season

During the summer, especially in the extremely dry period a strong stressing effect on the biological activity of the soil was observed (Figs. 39.1–39.6). In fact,

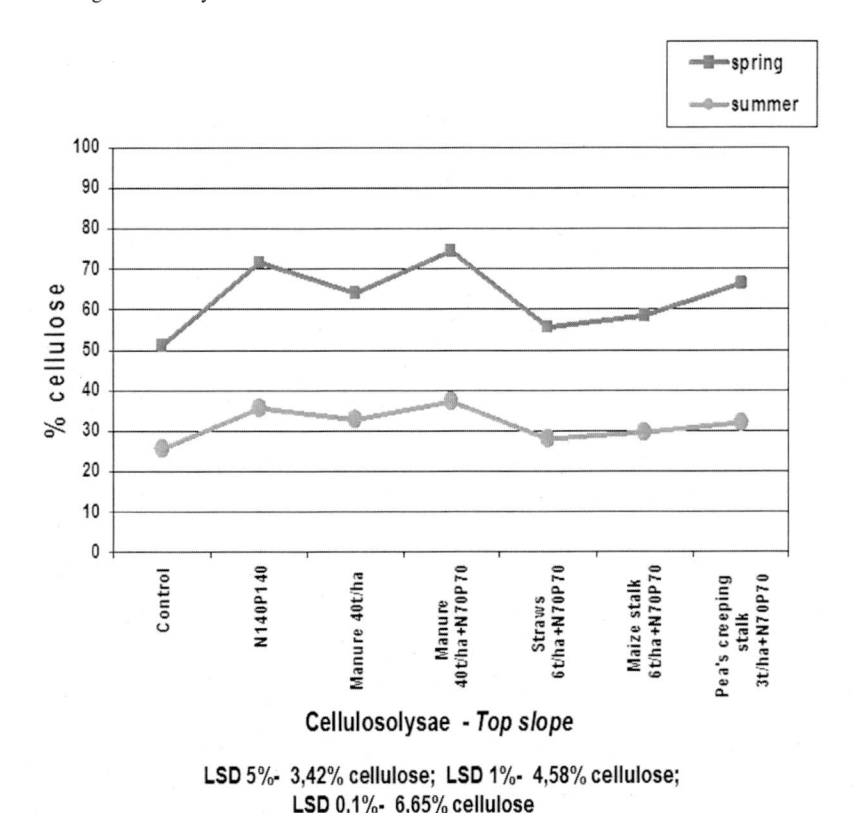

Fig. 39.4 Seasonal dynamics (spring–summer) of cellulosolysae (% cellulose) to the top slope LSD-Limit Statistical Difference

the prolonged low water availability produces changes in the biological activity of soils under stress conditions. The estimation of biological activity of the soil through pedo-biological indicators of fertility and soil quality, in the conditions of anthropogenic interventions (fertilization and agroterraces) in degraded agro-ecosystems shows the importance of the technological elements on the soil biological potential. Even in the stressing conditions produced by extreme drought, there are positive influences, as result of fertilization, especially by organic-mineral fertilization, in particular in the agroterrace on lower slope. The fertilization treatments maintain a C/N ratio that is highly conducive to microbial metabolism.

In the agroterrace on top slope, in the control section, there is the lowest value of soil respiration (22.44 cmc CO_2) because the extreme drought reduced approximately 50% soil respiration on the 0–20 cm depth. Instead at the agroterraces on lower slope there is the highest value of soil respiration (46.28 cmc CO_2) observed in the organic-mineral fertilized variant (manure 40 t ha^{-1} + $N_{70}P_{70}$).

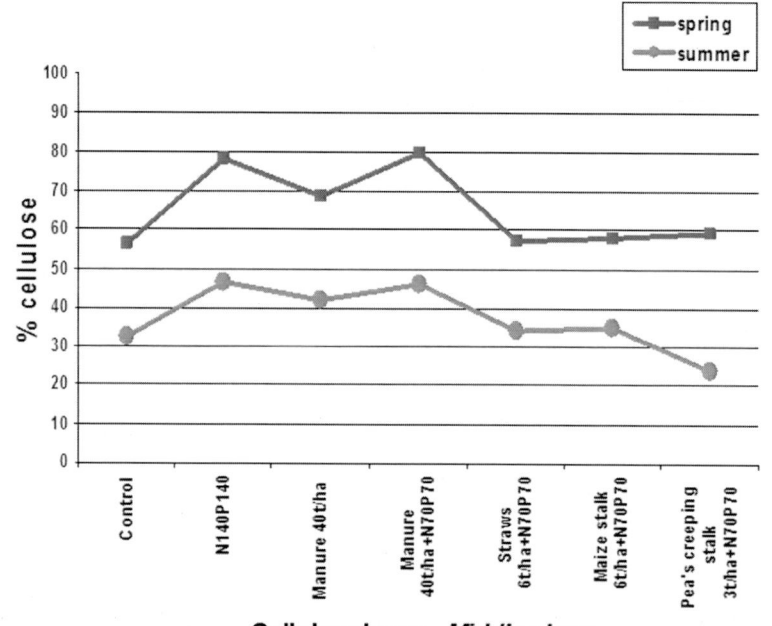

Fig. 39.5 Seasonal dynamics (spring–summer) of cellulosolysae (% cellulose) to the middle slope – LSD-Limit Statistical Difference

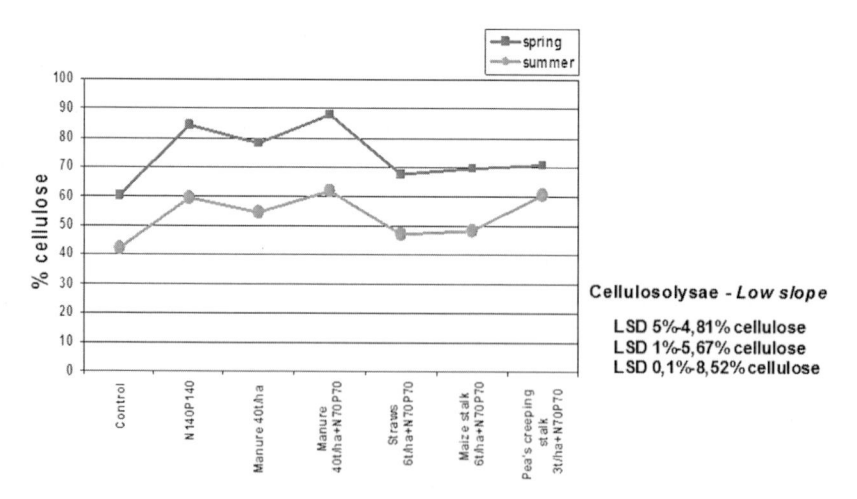

Fig. 39.6 Seasonal dynamics (spring–summer) of cellulosolysae (% cellulose) to the low slope LSD-Limit Statistical Difference

Referring to cellulosolysae there is a similar evolution. The negative impact of drought, corroborated with pluvial erosion it observed especially in the agroterraces on top slope, in control (25.56% cellulose, compared with 51.34% cellulose in the spring season). In the agroterraces on the lower slope, in the organic-mineral fertilized variant (manure 40 t ha^{-1} + N$_{70}$P$_{70}$) there is the highest value of cellulosolysae (61.87% cellulose).

39.3.3 The Indicator of Vital Activity Potential (IVAP%)

The complex and comprehensive characterization of biological activity and consequently of the fertility level of this soil as represented by the Indicator of Vital Activity Potential (IVAP%), shows a high negative impact of erosion in the agroterrace on top slope (Fig. 39.7). It observed the positive effect in the agroterrace on

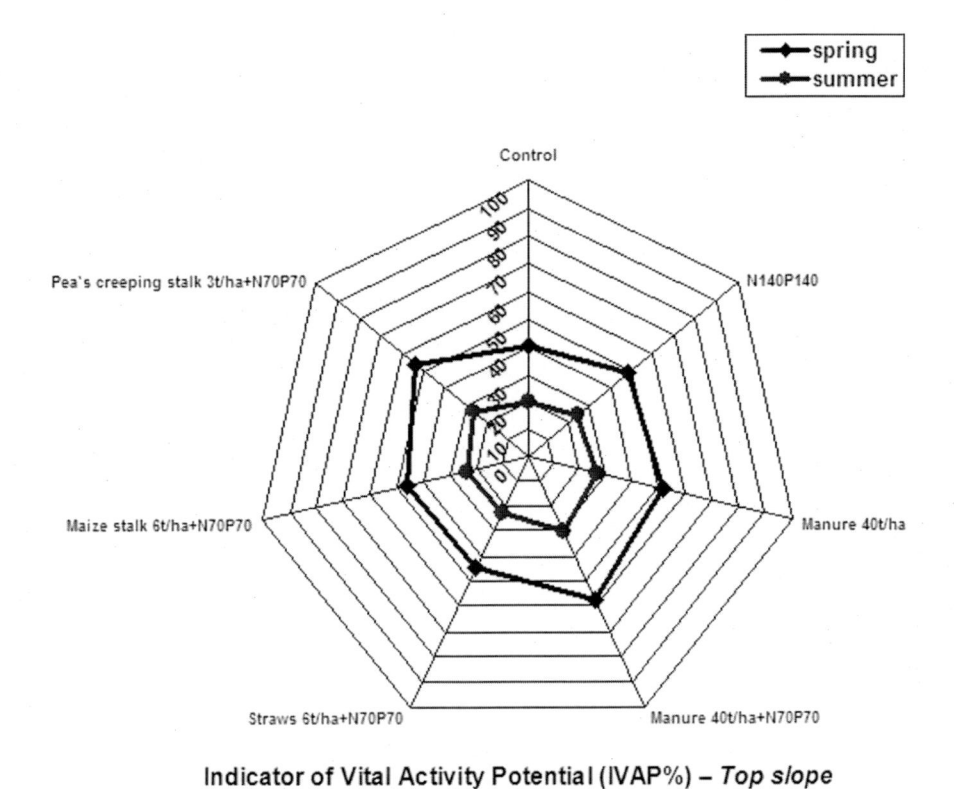

Indicator of Vital Activity Potential (IVAP%) – *Top slope*
LSD 5%- 2,17%; LSD 1%- 3,42%; LSD 0,1%- 6,51%

Fig. 39.7 Seasonal dynamics (spring–summer) of Indicator of Vital Activity Potential (%) to the top slope – LSD-Limit Statistical Difference

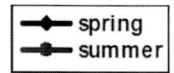

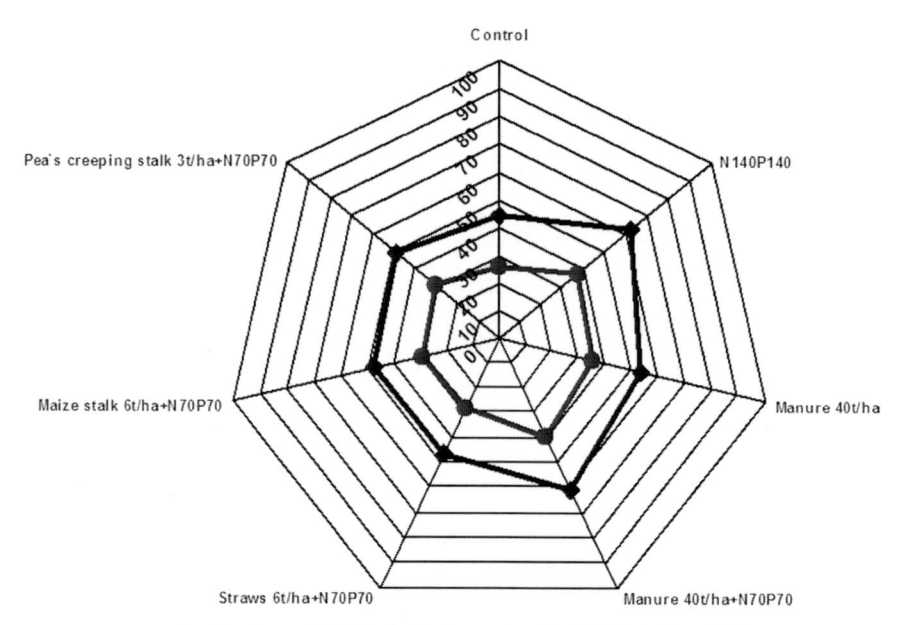

Indicator of Vital Activity potential (IVAP%) – Middle *slope*
LSD 5%- 1,18%; LSD 1%- 2,63%; LSD 0,1%- 3,05%

Fig. 39.8 Seasonal dynamics (spring–summer) of Indicator of Vital Activity Potential (%) to the middle slope – LSD-Limit Statistical Difference

middle and especially on the lower slope, where we added organic-mineral fertilization (manure 40 t ha^{-1} + N$_{70}$P$_{70}$ and pea's creeping stalks + N$_{70}$P$_{70}$) during the spring season (Figs. 39.8 and 39.9). The addition of manure and other organic fertilizers can supply the nutrient crop needs and help offset some loss of inherent fertility caused by soil erosion. In case of pluvial erosion, because the vitality of the surface horizon is very low it is important to bury the manure with a large diversity of microbes able to dissolve the soil nutrients.

Because the degraded soils are generally very poor in soil fertility, it is important to restore nutrients to the plants when the crops are deficient and not to the soil, which cannot store them efficiently (Roose, 2008). On the other hand, the vegetation cover, as a product of crop type and soil management, is the most important factor controlling the soil erosion rates (Ledermann et al., 2008). Furthermore, McVay et al. (1989), Kuo et al. (1997a) and Sainju et al. (2006), among others, showed that cover cropping provides additional residue that not only reduces soil erosion but also improves soil productivity by increasing soil organic carbon content.

Indicator of Vital Activity Potential (IVAP%) - *Low slope*
LSD 5%- 2,31%; LSD 1%- 3,01%; LSD 0,1%- 5,47%

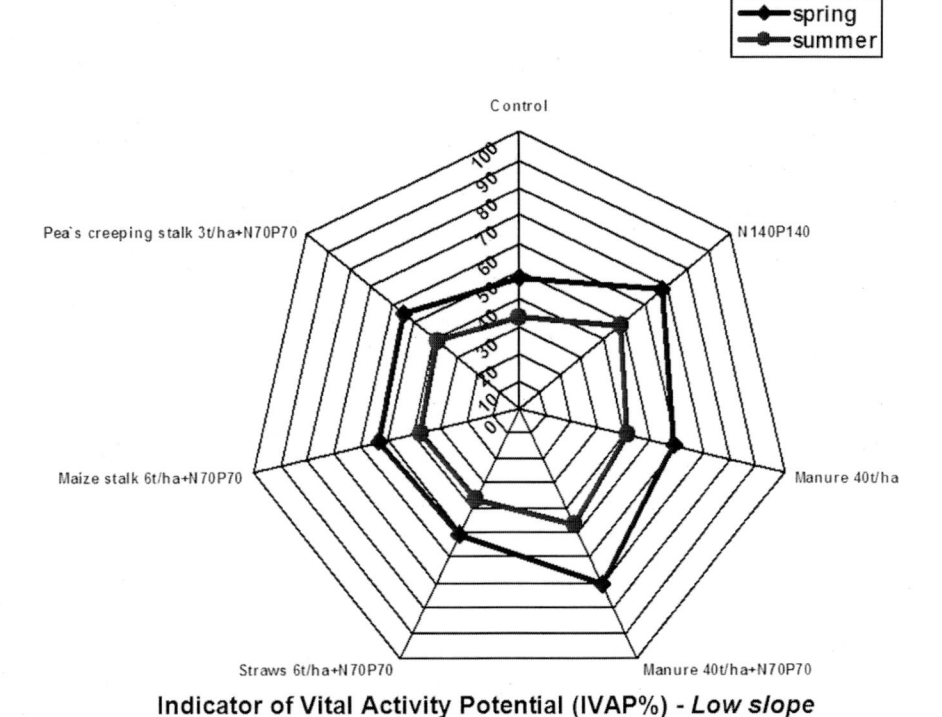

Fig. 39.9 Seasonal dynamics (spring–summer) of Indicator of Vital Activity Potential (%) to the low slope – LSD-Limit Statistical Difference

The values of IVAP show a decrease of the biological activity of the soil during summer season, depending on the dose of fertilization and altitude of the agroterrace. The lowest value (20.22%) was observed in the control section in the agroterrace on top slope, compared with 40.38% in the spring season. In the agroterrace on the lower slope, in the organic-mineral fertilized variant there is the highest value (46.36%). The value is still low, compared with the value obtained in the spring season (67.65%), in the same variant of fertilization, because of the response of the soil to a rapid change in environmental conditions (Antisari et al., 2008).

39.4 Conclusions

In the pluvial erosion degraded soils, the biological activity is dependent on the level of degradation and, also, on the anti-erosion measures and/or technological elements applied. Agroterraces are effective means for preventing water and fertility loss, because not only they conserve soil moisture, but also increase infiltration and decrease erosion. They also play a major role of improving soil structure and conserving soil nutrients.

The highest values of biological potential in our experiment were observed in organo-mineral fertilized variant (manure 40 t ha^{-1} + $N_{70}P_{70}$) on the lower slope, because the effect of pluvial erosion is less evident. Manuring alone will not restore the productivity of eroded soils. Combined farmyard manure and mineral fertilization had a significant effect on microbial biomass size and, in consequence, on biological activity. We observe, also, the important role of pea's creeping stalk as an organic and energetic matterial, especially in cellulosolysae processes. In hot and dry conditions, accelerated also by the global warming, the effect of the antierosional measures and the effect of the fertilization in agro-ecosystems take strategic importance to mitigate soil fertility and adaptation to climate change.

The study of the pedo-biological indicators of soil quality and fertility contributes to the elaboration of sustainable land management strategies to protect the environment following international standards. The agro-environment measures stipulated in the National Strategic Plan of Romania (2006) include combating soil erosion, contamination and other forms of degradation. The National Rural Development Plan relates to the development of sustainable agricultural systems to balance the effects of intensive exploitation of farmlands. The Romanian Rural Development Strategy is compliant with the objectives indicated in the 7th Framework Programme of the European Union, especially with those related to the promotion of natural diversity, water and soil protection, air pollution, the use of pesticides and mitigation of climate changes impacts.

References

Ailincăi, C. (2007). Agrotehnica Terenurilor Arabile. Editura "Ion Ionescu de la Brad", Iaşi, Romania.

Antisari, L.V., Vianello, G., Pontalti, F., Lorito, S. and Gherardi, M. (2008). Land use effects on organic matter in brown soils of the Emilian Apennines. In: C. Dazzi and E.A.C. Costantini (eds.), The Soils of Tomorrow-Soils Changing in a Changing World, Advances in Geoecology, Catena Verlag, Germany, 39:311–328.

Bending, G.D., Turner, M.K., Rayns, F., Marx, M.C. and Wood, M. (2004). Microbial and biochemical soil quality indicators and their potential for differentiating areas under contrasting agricultural management regimes. Soil Biology and Biochemistry 36:1785–1792.

Bireescu, G. (2001). Researches concerning vital and enzymatic processes in forestry and agricultural soils from Moldova. Ph.D Thesis, University of Agricultural Science and Veterinary Medicine Bucharest, Romania.

Bireescu, G., Bireescu, L., Breabăn, I., Lupaşcu, A. and Teodorescu, E., 2007. Ecopedological study over soil resources in forest ecosystems from Moldavian Plain. Lucr. şt. USAMV Iaşi, Romania, seria Horticultură, anul L. 1(50):863–870.

Bireescu, L., Bireescu, G., Lupaşcu, G., Secu, C. and Breabăn, I. (2005). Interpretarea ecologică a solului şi evaluarea impactului ecologic global în ecosisteme praticole situate peterenuri degradate din Podişul Bârladului. Lucr. Conf. Naţ. Şt. Sol., Timişoara, Romania, 2(34 B): 473–481.

Chiriţă, C. (1974). Ecopedologie cu baze de pedologie generală. Ed. Ceres, Bucureşti, Romania.

Dick, R.P. (1997). Soil enzyme activities as integrative indicators of soil health. In: C.E. Pankhurst, B.M. Doube and V.V.S.R. Gupta (eds.), Biological Indicators of Soil Health. CAB International, Wallingford, USA, pp. 121–156.
Dilly, O. and Blume, H.P. (1998). Indicators to assess sustainable land use with reference to soil microbiology. Advances in GeoEcology 31:29–36.
Doran, J.W. and Parkin, T.B. (1994). Defining and assessing soil quality. In: J.W. Doran (ed.), Defining Soil Quality for Sustainable Environment. SSSA Special Publication. Soil Science Society of America, Inc. and American Society of Agronomy, Inc., Madison, WI, pp. 3–23.
Doran, J.W. and Zeiss, M.R. (2000). Soil health and sustainability: Managing the biotic component of soil quality. Applied Soil Ecology 15:3–11.
Elliott, E.T. (1997). Rationale for developing bioindicators of soil health. In: C.E. Pankhurst, B.M. Doube and V.V.S.R. Gupta (eds.), Biological Indicators of Soil Health. CAB International, Wellington, pp. 49–78.
Farahbakhshazad, N., Dinnes, D.L., Li, C., Jaynes, D.B. and Salas, W. (2008). Modeling biogeochemical impacts of alternative management practices for a row-crop field in Iowa. Agriculture, Ecosystems and Environment 123:30–48.
Fauci, M.F. and Dick, R.P. (1994). Microbial biomass as an indicator of soil quality: Effects of long-term management and recent soil amendments. In: J.W. Doran, D.C. Coleman, D.F. Bezdicek and B.A. Stewart (eds.), Defining Soil Quality for a Sustainable Environment. SSSA Special Publication, 35, Minneapolis, Soil Science Society of America, 17:229–234.
Gianfreda, L. and Bollag, J.M. (1996). Influence of natural and anthropogenic factors on enzyme activity in soil. In: G. Stotzky and J.M. Bollag (eds.), Soil Biochemistry 9:123–194, New York.
Gianfreda, L., Rao, M.A., Piotrowska, A., Palumbo, G. and Colombo, C. (2005). Soil enzyme activities as affected by anthropogenic alterations: Intensive agricultural practices and organic pollution. Science of the Total Environment 341:265–279.
Government of Romania, Ministry of Agriculture Forests and Rural Development. (2006). National Strategic Plan of Rural Development, 2007–2013.
Karlen, D.L., Anrews, S.S. and Doran, J.W. (2001). Soil quality: Current concepts and applications. In: D.L. Sparks (ed.), Advances in Agronomy 74:1–40, Academic Press, San Diego, California.
Karlen, D.L., Mausbach, M.J., Doran, J.W., Cline, R.G., Harris, R.F. and Schuman, G.E. (1997). Soil quality: A Concept, Definition and a Framework for Evaluation. Soil Science Society of America Journal 61:4–10.
Knoepp, J.D., Coleman, D.C., Crosslez, D.A., Jr. and Clark, J.S. (2000). Biological indices of soil quality: An ecosystem case study of their use. Forest Ecology and Management 138:357–368.
Kõlli, R., Köster, T., Tõnutare, T. and Kauer, K. (2008). Influence of land use change on soil humus status and on soil cover environment protection ability. In: C. Dazzi and E.A.C. Costantini (eds.), The Soils of Tomorrow-Soils Changing in a Changing World, Advances in Geoecology, Catena Verlag, Germany, 39:27–36.
Kubat, J. et al. (2001). Relationships between soil productivity and soil quality. Symposium. Crop Science on the Verge of the 21st Century-Opportunities and Challenges. Prague, pp. 154–156.
Kumar, K. and Goh, K.M. (2002). Management practices of antecedent leguminous and non-leguminous crop residues in relation to winter wheat yield, nitrogen uptake, soil nitrogen mineralization and simple nitrogen balance. European Journal of Agronomy 16:295–308.
Kuo, S., Sainju, U.M. and Jellum, E.J. (1997a). Winter cover crop effects on soil organic carbon and carbohydrate. Soil Science Society of America Journal 61:145–152.
Lal, R., Mokma, D. and Lowery, B. (1999). Relation between soil quality and erosion. In: Soil Quality and Soil Erosion. Soil and Water Conserv. Soc., Ankeny, IA, pp. 237–258.
Ledermann, T., Herweg, K., Liniger, H., Schneider, F., Hurni, H. and Prasuhn, V. (2008). Erosion damage mapping: Assessing current soil erosion damage in Switzerland. In: C. Dazzi and E.A.C. Costantini (eds.), The Soils of Tomorrow-Soils Changing in a Changing World, Advances in Geoecology, Catena Verlag, Germany, 39:263–283.
Mäder, P., Pfiffner, L., Fliessbach, A., von-Lützow, M. and Munch, J.C. (1997). Soil ecology-Impact of organic and conventional agriculture on soil biota and its significance for soil fertility. Vth International Conference on Kyusei Nature Farming. Bangkok, Thailand, pp. 24–40.

McLaren, A.D. (1975). Soil as a system of humus and clay immobilized enzymes. Chemic Scripta 8:97–99.
McVay, K.A., Radcliffe, D.E. and Hargrove, W.L. (1989). Winter legume effects on soil properties and nitrogen fertilizer requirements. Soil Science Society of America Journal 53: 1856–1862.
Misono, S. (1977). Three phases distribution as a factor of soil fertility. Proceedings of the International Seminar on Soil Surviron and Fertility Management in Intensive Agriculture (SEFMIA). Tokyo, pp. 154–160.
Munteanu, I. and Florea, N. (2001). Present-day status of soil classification in Romania. In: E. Micheli, F.O. Nachtergaele, R.J.A. Jones and L. Montanarella (eds.), Soil Classification. Contributions to the International Symposium "Soil Classification 2001", 8–12 October 2001. Velence, Hungary.
Rastin, N., Rosenplänter, K. and Hüttermann, A. (1988). Seasonal variation of enzyme activity and their dependence on certain soil factors in a beech forest soil. Soil Biology and Biochemistry 20:637–642.
Reintam, L., Rooma, I. and Kull, A. (2001). Map of soils vulnerability and degradation in Estonia. In: D.E. Stott, R.H. Mohtar and G.C. Steinhardt (eds.), Sustaining the Global Farm. Purdue University and USDA-ARS NSERL, CD-ROM:1068–1074.
Research Institute for soil Science and Agrochemistry, Bucharest, Romania. (2001). National Strategy and Action Programme concerning Desertification, Land Degradation and Drought Prevent and Control – 27 November, 2001.
Roose, E. (2008). Soil erosion, conservation and restauration: A few lessons from 50 years of research in Africa. In: C. Dazzi and E.A.C. Costantini (eds.), The Soils of Tomorrow-Soils Changing in a Changing World, Advances in Geoecology, Catena Verlag, Germany, 39: 159–180.
Rubio, J.L. (2008). Soil, the crucial link. Foreword. In: C. Dazzi and E.A.C. Costantini (eds.), The Soils of Tomorrow-Soils Changing in a Changing World, Advances in Geoecology, Catena Verlag, Germany, 39:159–180.
Sainju, U.M., Singh, B.P., Whitehead, W.F. and Wang, S. (2006). Carbon supply and storage in tilled and nontilled soils as influenced by cover crops and nitrogen fertilization. Journal of Environmental Quality 35:1507–1517.
Schulze, E.D. (2005). Plant Ecology. Springer-Verlag, Berlin.
Seybold, C.A., Mausbach, J.M., Karlen, D.L. and Rogers, H.H. (1997). Quantification of soil quality. In: R. Lal, J.M. Kimble and S. Follet (eds.), Soil Processes and the Carbon Cycle. CRC Press, Washington, DC, USA.
Shafi, M., Bakht, J., Jan, M.T. and Shah, Z. (2007). Soil C and N dynamics and maize (Zea mays) yield as affected by cropping systems and residue management in North-western Pakistan. Soil and Tillage Research 94:520–529.
Shah, Z., Shah, S.H., Peoples, M.B., Schwenke, G.D. and Herriedge, D.F. (2003). Crop residue and fertilizer N effects on nitrogen fixation and yields of legume-cereal rotations and soil organic fertility. Field Crops Research 83:1–11.
Ştefanic, G. (1994a). Cuantificarea fertilităţii solului prin indici biologici. Lucr. Şt Conf. Naţ. Şt. Sol., Tulcea, S.N.R.S.S., 28A:45–55.
Ştefanic, G. (1994b). Biological definition, quantifyng method and agricultural interpretation of soil fertility. Romanian Agricultural Research 2:107–116.
Ştefanic, G., Sandoiu, D.I. and Gheorghita, N. (2006). Biologia Solurilor Agricole. Editura Elisavaros, Bucuresti, Romania.
Stroosnijder, L. and Slegers, M. (2008). Soil degradation and droughts in sub-Saharan Africa. In: C. Dazzi and E.A.C. Costantini (eds.), The Soils of Tomorrow-Soils Changing in a Changing World, Advances in Geoecology, Catena Verlag, Germany, 39:413–425.
Such, Y.S., Kyuma, K. and Kawaguchi, K., 1977. A method of capability evaluation for upland soil. 4 fertility evaluation and fertility classification. Soil Science and Plant Nutrition 23(3):275–286.
Teaci, D. (1980). Bonitatea Terenurilor Agricole. Editura Ceres, Bucureşti, Romania.

Unger, H. (1960). Der zellulosetest, eine Methode zur Ermittlung der Zellulolytischen Aktivität des Bodes in Feldversuchen. Zeitschrift für Pflanzenernährung, Düngung, Bodenkunde 91(1): 44–52.

Verstraete, W. and Voets, J.P. (1974). Impact in sugarbeet crops of some important pesticide treatment systems on the microbial and enzymatic constitution of the soil. Met. Fak. Landban., Gent. 39:1263–1277.

Verstraete, W. and Voets, J.P. (1977). Soil microbial and biochemical characteristics in reletion to soil management and fertility. Soil Biology and Biochemistry 9:253–258.

Vostrov, I.S. and Petrova, A.N. (1961). Opredelenie biologhiceskoi aktivnosti pocivîrazlicinîmi metodami. Mikrobiologiia 4:665–672.

WRB. (2006). World reference base for soil resources. World Soil Resources Reports No. 103. FAO, Rome.

Chapter 40
Achievements and Perspectives on the Improvement by Afforestation of Degraded Lands in Romania

Cristinel Constandache, Viorel Blujdea, and Sanda Nistor

Abstract One third of the territory of Romania is affected by land degradation (mainly water erosion and landslides). Climate change trends and impacts may double the areas affected. Since much of the country has lower forest coverage, urgent and ample measures of afforestations interventions are needed. The experience acquired on the improvement of degraded lands by forest plantations is very useful under the conditions where large areas of degraded lands (in the order of hundreds of thousands up to 1 millions of hectares) are not suitable to be used for agriculture. The country's goal in afforestation programmes is to increase the proportion of the forests from at 26.3% at present to around 40% by 2020.

Keywords Degraded lands · Erosion · Landslides · Afforestation · Forest plantations · Soil quality improvement

40.1 Introduction

In many part around Romania, as elsewhere in the world, there are large areas of degraded lands or lands vulnerable to degradation due to improper use for agriculture that could be otherwise used much better for afforestation (Giurgiu, 2004).

The processes of land degradation continue to affect rather large surfaces, with notable negative economic consequences. Loss of the fertile soil, substantial decrease of the agricultural production and agriculturally initiated erosion are common throughout the country. Other forms of degradation include mudflows with riverbed plugging and diminution of the draining capacity, associated with the

---

C. Constandache (✉)
Forest Research and Management Institute Bucuresti, Focsani Station, 620018 Focşani, Romania
e-mail: cicon66@yahoo.com

increase of flooding risks endangering thus a number of infrastructure objectives of strategic importance (railways, national roads, etc.).

Under natural geographical conditions of the country, where the slope grounds account up to 67% of the national territory, and backed by a complex number of natural factors, as well as due to the intense human interventions since the end of the nineteenth and beginning of the twentieth centuries, the vegetation cover and soils have been experiencing serious ecological imbalances. Consequently land degradation and the increasing frequency of torrential processes are creating some aspects typical for the semi-desert regions of the world. Under this picture, an overwhelming part of the forest areas encountered the reduction of the forest cover and deficient forests management, associated with heavy and uncontrolled grazing and improper agriculture use.

The main processes of land degradation in the country are the water erosion and landslides. Inventory of the actual state of the soil degradation processes according to the Institute of Research for Pedology and Agrochemistry in Bucharest, demonstrates that water erosion affects about 6.3 millions of hectares of agricultural land, out of which about 2.5 million are highly degraded lands, while the landslide affects about 0.7 millions of hectares (Dumitru et al., 2002). Thus, afforestation of the low productive degraded lands represents an efficient way to combating the degradation processes toward efficient use of lands unsuited for agricultural purposes. The forest vegetation is considered in general as having a particular role in preventing and stopping land degradation, while generate additional benefits (Untaru et al., 2006).

Research on degraded land amelioration and exploration of ways for its prevention and mitigation in Romania dates back as far as the second half the nineteenth century (Giurescu, 2004) Interventions by afforestation on degraded lands became more intense after 1930 when the "Law for ameliorating degraded lands" appeared. Total afforested area on degraded lands accounts presently to about 300,000 ha, according to the year 2000 data. Technical solutions for afforestation were based on long-term research and experimental result, starting in the year 1950, in cultures aged 10–45 years. These results created the basis for the elaboration of *Technical guidelines on compositions, schemes and technologies for forests regeneration and afforestation of degraded lands* (MAPPM, 2000).

Afforestation of degraded lands in Romania is labor and technology intensive activity, under strictly regulated ecological approach (i.e. only certain tree species may be planted under certain climates). The number of seedlings planted per area unit is rather high (i.e. 5,000 ha^{-1} in black locust; 6,700 ha^{-1} in oaks), coupled with intensive maintaining (3–4 yearly maintenances scheduled in the first 2 years since plantation, then 1–2 maintenances on year till 3–5 years old of plantation, according the species). Plantations are closely monitored for survival rate in first 3–5 years and gap filling is achieved in 2nd and 3rd year, if necessary.

The objectives of this paper were the assessment of previous interventions on improvement of degraded lands made in the past and to identify solutions for continuation of such operations in the current socio-economic and ecological context.

40.2 Materials and Methods

Surveys were conducted in forest plantations previously established on degraded lands, in permanent experimental plots, supplemented by temporary research plots in various locations. The permanent plots are long term monitored, as they have been established in 1981 (by C. Traci) in forest plantations established starting with 1954 (Traci and Untaru, 1986).

The analysis on the effect of afforestation interventions was conducted in two phases: comparing the situation before carrying out afforestation works and after a certain period of time since intervention was completed (the last few years from present).

For the reference situation, that before intervention, the records in early projects archived in present by the Forest Research and Management Institute (FRMI) were studied. A number of plots were selected for the study (upon the following criteria: type and land degradation intensity, type of land preparation techniques, afforestation assortments and schemes). Their time dynamic was surveyed in forest management plans. Observations included studying of environmental conditions, technologies for the plantations establishment, assessment of stands evolution (trees height, diameter, stand density, health status) and behaviour of plantations in relation to work techniques, the compositions of afforestation and planting schemes. Measurements on precipitation, liquid and solid runoff on forested degraded lands on various plots of the study areas was also made and the fixed mark method was used to determine the hydrological and antierosional efficiency. Specific plots physical and ecological conditions, as well as the geographical plot's locations are later on described under results and discussions.

A number of experiments on various sites were conducted with the scope to understand the effect of some particular type of interventions for land preparation (i.e. before planting, like land terracing) and afforestation species. This long run research achieved by the Forest Research and Management Institute (Focşani Research Station) resulting in significant support to establish the technologies for afforestation of degraded lands. The following land preparation works experiments were followed:

- Terraces 60–80 cm wide, supported by hurdles of oak poles and weaved willow twigs, placed each 3 m, from one beam to another, on lands with excessive eroded 30–40° slope (perimeters Bîrseşti and Ruget-Colacu in OS Exp/Experimental Forest District Vidra and perimeter Andreiaşu in OS/Forest District Focsani);
- Terraces 60–80 cm wide, supported by walls, placed each 2.5–3 m from one beam to another, on excessively eroded 25–40° slope (perimeters Bîrseşti and Valea Sarii – OS Exp. Vidra, Andreiaşu – OS Focsani);
- Terraces 50–60 cm wide, supported by root-suckers and sallow thorn branches, placed each 1.5–3 m from one beam to another, on lands with excessive erosion, with a 30–40° slope (perimeter Bîrseşti, OS Exp. Vidra);

- Vegetal corridors from root-suckers, sallow thorn and alder seedlings planted on narrow terraces (30–40 cm wide), on excessively eroded and gully lands, on 40–55° slopes (perimeter Bîrseşti, OS Exp. Vidra);
- On the planting procedures following procedures have been followed:
- Planting of Scots and black pine seedling grown into plastic bags of various dimensions, sizes and ages of seedlings;
- Various planting period (i.e. seasons) and composition of nutritive beds;
- Various planting techniques (i.e. holes of various dimensions, notches).

40.3 Results and Discussions

History of rehabilitation of land reflects a continuous process of experimentation and learning. Afforestation interventions in Romania have been carried out in two phases: before and after 1976 with the first national program for conservation and improvement of the forest fund (Programul national de conservare si dezvoltare a fondului forestier) and the publication of first technologically detailed national research on afforestation of degraded lands (Traci, 1976). During the *first phase*, before 1976, the most favourable degraded land sites were approached. The afforestation comprised assortments of pines (*Pinus nigra, Pinus silvestris*), in pure plantations and mixed with broadleaves species (*Cerasus avium, Acer sp., Fraxinus sp., etc.*) and black locust (*Robinia pseudoacacia*). Meantime, strongly and very strongly eroded lands, as well as the ones characterized by deep erosion have been afforested with sallow thorn (*Hippophae rhamnoides*). On very strongly eroded lands that were unstable and not suitable for the plantation of tree species, sallow thorn plantations have been extensively installed. After 10–15 years, the sallow thorn plantations got thicken, contributing thus to land consolidation, erosion reduction and soil amelioration. Therefore, the sallow thorn plantations and brushwood natural vegetation were gradually replaced with tree species (Bogdan and Untaru, 1972). Because of its ecological qualities the sallow thorn, it was one of the most appreciated and used species in the afforestation of excessively degraded lands. For this reason it was called "the salvation of degraded lands" (Haralamb, 1957) and used for afforestation in nearly all degraded lands categories, especially those with very advanced degradation status.

The *second phase*, included afforestation works on lands characterized by less favorable conditions for trees. On such land the afforestation assortments consisted in pines mixed with sallow thorn. Simultaneously, the substitution of the previous temporary plantations of sallow thorn bushes with other trees species occurred, in order to further reduce any remaining erosion and get benefit of shallow soil and dead organic matter accumulated. Besides the fact that it constituted the plantation core element, the sallow thorn equally contributed to successfully stabilize/consolidate the advanced degraded lands, in order to create the conditions for planting. Thus, the stems with branches of sallow thorn after cut were directly used as "vegetable reinforcement" to consolidate the terraces made on land with active erosion. Also, building of barrages across eroded ravine with local materials (i.e. soil or stone) and vegetable support (i.e. branches sea bucktorn) and longitudinally with fascines on the bed was used for their consolidation and stabilization. In all cases, stems and branches of sallow thorn entered in the growing phase in proportion of more than 40–50% (Traci and Untaru, 1986) resulting in the establishment of some really "anti-erosional barriers".

Among the land preparation and consolidation methods, the most successful were the terraces supported by sallow thorn branches and twigs, also less costly compared to terraces supported by hurdles. The cost is about 61% lower and the functioning period is 2–3 years longer (Traci and Untaru, 1986). Good results were achieved in Bîrsesti perimeter on the argillaceous marl and sandstone bedrocks, with the plantations made with pine seedlings raised in polyethylene bags. They ere planted on the terraces supported by root-suckers and sallow thorn branches, associated with works on ravine and gully consolidation.

To consolidate the torrential network, in most cases it was necessary to establish rapids and dams of concrete or stonework with cement. In the case of small gullies the rapids built in stone without concrete but in woody material (i.e. buckthorn twigs), earth and stone, which were particularly successful. In fact, protection of the slope stability by improving the hydrographic network by hydrotechnical works played a special part in the success of these afforestation programs. The seedlings grown in polyethylene bags perform as the best after planting, since the survival rate is higher than 90% and the increment was improved with 25–30% (Table 40.1).

The most important woody species used in Romania for the afforestation of degraded lands are: black pine *(Pinus nigra)*, Scots pine *(Pinus sylvestris)*, black locust *(Robinia pseudacacia)*, Norway maple *(Acer platanoides)*, sycamore maple *(Acer pseudoplatanus)*, cherry tree *(Prunus avium)*, ash *(Fraxinus excelsior)*, flowering ash *(Fraxinus ornus)* , mahaleb cherry *(Prunus mahaleb)*, gray and black alders *(Alnus spp.)*, sallow thorn *(Hippophae rhamnoides)*, oleaster *(Eleagnus angustifolia)*, e.a. Good result were obtained in the case of lands with advanced degree of degradation and on land where the sallow thorn that was removed after first plantation cycle of 20 years, when erosion stopped and the soil conditions improved for other tree species (usually Scots pine, black pine, forest cherry tree, ash and maple).

The types of forest plantations with a good evolution and high efficiency in both halting the erosion and stabilizing landslides on strongly to excessively eroded slopes were the following (Untaru et al., 2003; Constandache, 2003; Constandache et al., 2006):

- Mixed plantations of black or Scots pine with broadleaved trees (maple, ash, wild cherry tree, flowering ash, mahaleb cherry, red dogwood, etc.), on both strongly eroded and moderately fragmentised sliding land in both steppe and forests zone;
- Pine plantations associated with sallow thorn or sallow thorn in pure cultures, on strongly to excessively eroded lands in forests zone;

Table 40.1 Stand biometrical and ecological characteristics in Bîrseşti experimental perimeters

Land degradation type	Afforestation technique, scheme, composition	Tree species	Trees density (no/ha) at 10 years old	Mean stand height (m) at age ... 10	15	20	25	30	Mean stand diameter (cm) at age ... 10	15	20	25	30
R	Ta/3.0+Gr.o (Pi) 1.5/3.0	Pi	667		3.6	5.7	7.4	11		4.2	7.4	11	13.2
		Pi.n	351		2.8	5.1	6.7	10.8		3.6	7.2	11.2	13.5
	Cord. 0.33/3.0 (Ct.a)	An.a	386		6.8	7.9	9.8	16.6		6.9	7.8	9.9	18.6
	Gr.o (An.a) 1.5/1.0 m	Ct.a	4,561		1.7								
R	Ta/3.0+Gr.o (Pi) 1.5/3.0	Pi.n	1,654		2.8		6.4	10.5		3.1		10.9	11.9
	Cord. 0.33/3.0 (Ct.a)	Pi.n	455		3		6.6	10.9		4.5		10.2	11.4
E3	Tg 0.75/3+Gr.o (Pi) = 1.25/3.0 m;	Pi.n	630		2	3.3	5.1			2.1	4.8	7.7	
		Pi.n	2,195		3	3.9	5.5			2.3	4.9	7.8	
	Gr.o (Ct.a) = 1.0/3.0 m	Ct.a	3,100		1.8								
E3	Pp+Ta/3	Pi.n	1,783	2.2	2.8	4.0	6.2	8.3	2.1	4.8	7.4	9.3	11.2
	+Gr.o;1.25/3.0 m	Pi.n	1,102	3.0	3.3	4.1	7.5	10.9	2.3	5.1	7.8	10.3	12.7
	Tî/3 + Cord.0.33/3 (Ct.a)	Ct.a	9,100										
E3	Pp+Ta/3	Pi.n	805	3.0	3.4	5.5	8.1	10.5	2.2	5.5	8.6	10.3	12.0
	+Gr.o;1.25/3.0 m	Pi.n	910	3.1	4.3	6.0	8.8	11.5	2.3	5.4	8.8	10.4	12.0
	Tî/3 + Cord.1.0/3.0 (Ct.a)	Ct.a	8,600										

Tree species: Pi – Scots pine; Pi.n – black pine; An.a – gray alder; Ct.a – sallow thorn.
Erosion type: R – gully lands; E3 – excessively eroded lands.
Methods: Ta/3 – Terraces 50–60 cm wide, supported by root-suckers and sallow thorn branches, placed each 3 m; Tg 0.75/3 = Terraces 75 cm wide, supported by hurdles of oak poles and weaved willow twigs , placed each 3 m; Cord.0.33/3 = vegetal corridors from root-suckers, sallow thorn and alder seedlings planted on narrow terraces (Tî =30–40 cm wide); Pp = Planting operations with Scots and black pine seedling grown into plastic bags; Gr.o = Regular planting in holes of 30/30/30 cm.
Planting scheme – 1.25/3.0 m.

- Black locust plantations, on both eroded, gullied and sliding lands with sandy to loamy soils, from forest steppe up to the sessile oak sub-zone;
- Plantations with black and gray alder and willows, on lands with excess of water in forest zone.

Apart from amelioration of degraded lands with solely Scots pine and European black pine, good results are achieved also by mixing them with different broadleaved species planted or naturally regenerated. Pine species have provided important contribution to the recovery of degraded land with the largest vegetation representing similar conditions with those from the forest steppe (European black pine) and the mountain regions (Scots pine). This was mainly due to relatively fast growing and a strong rooting system, the densely canopy and the rich litter. In many situations, the two pine species intimately mix in plantations.

Stand growth is steady significant till 30 years old (Fig. 40.1), determined by both significant decadal growth in height and diameter. It is not clear at what age the growth starts to decline, as usually such stands are replaced after the first cycle with another cycle of same species or zonal ones (i.e. that correspond to local climate and soil). Nevertheless, the number of trees per area unit shows low density at age of 10, what suggest high loss under low survival rate, in the first 10 years of the plantations (Table 40.1). In general, pine species (*P. nigra, P. sylvestris*) showed lower survival rate compared to other species (i.e. sallow thorn).

Pines grown on degraded lands are generally even aged stands (homogenous age) showing a high degree of ecological fragility. In certain cases Scots pine or/and European black pine suffered serious damages because of the wind and snow (in forest zone) or from drying (in forest steppe). As well, they are very sensitive to disturbances by the biotic and abiotic factors and which worsen, in the case of illegal removal of trees, which often lead to cessation of their protective efficiency.

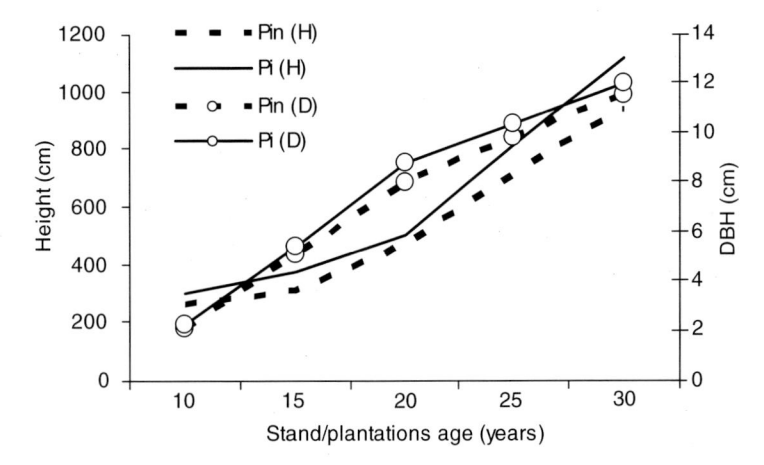

Fig. 40.1 Dynamic of stand biometrical characteristics in Bîrseşti perimeter

The observations and measurements made in various experimental plots established on degraded lands certify the functional and ameliorative efficiency of mixed stands of pines is comparable to broadleaved stands (Traci and Untaru, 1986). The litter decomposes more quickly, ensuring a better and faster improvement of the soil fertility. The fungi damage is frequent in pure pine stands and less virulent in mixed stands with pines with broadleaved trees, as their structure is more complex, leading to a high stability and resistance, and in case of mature plantations to the initiation of the natural regeneration process.

Former research showed that the stability of stands from degraded lands, as well as other stands, is in relation with the number of species, the tending (silvicultural) operations and the environmental conditions (Constandache, 2003). On very strong eroded lands the number of species is further reduced, being limited by the ecological requirements of forest species. Therefore, the diversity and thus stability (i.e. tolerance to damages) of these stands is reduced. In these conditions, the sallow thorn cultures have demonstrated a significant anti-erosional efficiency. Having a large suckering capacity, the brushwood is denser, so it entirely covers the ground, while the rich root system contributes to effective soil stabilization. The research results shown that mixtures of Scots and black pine with sallow thorn (see Photo 40.1) under identical conditions, increase growth of pines with 20–30% higher compared to pure plantations of pine, as a result of the enrichment in soil with nitrogen and on symbiosis with actinomicets (Untaru et al., 2008).

This way, the unproductive or very low productive lands (Photo 40.2) were reintegrated into the economic chain, as degradation process almost halted on much of the soil surface. The soil has been improved and the disagreeable landscape

Photo 40.1 Mixtures of Scots and black pine with sallow thorn (Bîrsesti perimeter, photo C. Constandache, 2008)

Photo 40.2 Historical photograph of unproductive lands in Romania (Valea Sării – Scaune experimental area, photo E. Costin, 1954)

Photo 40.3 Same photo 40.2 area after afforestation interventions (photo E. Untaru, 1997)

(shown in Photo 40.2) lacking its protective vegetation shield has been replaced by a beautiful landscape with forest plantations (Photo 40.3). It was demonstrated as well the possibility to achieve a proper natural and socio-economic framework by improving the land productive potential and landscape restoration, thus achieving aesthetic-sanitary, recreational and climatic functions.

Besides of soil protection positive impacts, these interventions on degraded lands generated also a series of important direct economic effects, represented by the wood, timber, honey, resin, berries and other income generating products.

Research emphasized the fact that forest vegetation has a determinant importance in rehabilitating degraded lands by reducing the volume of the surface water flows and also by creating favorable conditions for water storage, infiltration, accumulation of organic substances, etc. Forest vegetation also guarantees land stability as consequence of the water regime regularization, amelioration of the temperature regime and mechanic consolidation of the shallow soils horizons by root systems. In 25–28 years old anti-erozional plantations the specific erosion[1] is quantitatively very low (Table 40.2).

Previous research that the average muddiness of the discharged waters dropped down to 84 g/l in the case of very strong to excessively eroded, not forested lands, to 7–13 g/l, in the case of forested degraded lands, while in a mature beech stand, the medium turbidity was 2.6 g/l (Untaru et al., 2006). Accordingly, the average specific erosion reduced from 57.5 t ha^{-1} year^{-1} (between 55 and 60 t ha^{-1} year^{-1}) on lands with active erosion and no vegetation to 0.41 t ha^{-1} year^{-1} (between 0.15 and 0.75 t ha^{-1} year^{-1}) in forest plantations with pines as main tree species aged between 12 and 20 years. In the case of mature beech forests erosion rates were only 0.12 t ha^{-1} year^{-1}. After 25 years the forest plantations, black and Scots pine performed better with an average specific erosion rate below 0.1 t ha^{-1} year^{-1}. The strong reduction of the runoff and soil erosion due to the presence of forest vegetation (from 50 to less than 1 t ha^{-1} year^{-1}) has lead to the restoring of soil formation on lands with ground stones (especially due to the increase capacity for water retaining and stocking).

The research conducted in Bârseşti area (Constandache, 2003; Untaru et al., 2006) regarding the runoff and soil erosion have shown that on very strongly and excessively eroded forested lands, after 10–20 years since the installation of the forest cultures, the runoff load were reduced by 4–10 times, as compared to lands with active erosion and almost lacking vegetation (Table 40.3).

These results confirm the forest as a very efficient vegetation type in protecting the soils against erosion. In almost all situations, by establishing the protective forest cultures, after the degradation processes have stopped, a general improvement of the site and environmental conditions have been noticed. The positive impact of protective forest cultures on the improvement of vegetation and implicitly on the soil is also associated with climate improvement, by alleviating the temperature extremes and reduction of solar radiation. This leads especially on moderately to strongly eroded soils to natural establishment of shelter planted species of oak, sessile oak, beech, wild cherry, ash, etc. (Constandache, 2003).

In the case of lands with advanced degree of degradation stopping erosion would require in the first place the establishment of conditions for ecological reconstruction and essentially for the resumption of pedogenesis. This is possible after establishing forest cover that even in a period of 5–10 years after tree planting soil formation may be accelerated by the increase of soil organic matter and reduced effects of torrential rainfalls.

---

[1] Specific erosion – the amount of eroded matter in runoff on the parcel land area after certain rain intensity

Table 40.2 The main characteristics of the research plots and average values of runoffs and erosion, on degraded lands from the experimental Bîrseşti perimeter, during the period of May until October 2005, 2007 and 2008

Erosion type	Exposition, slope (grades)	Vegetation description (2003)	Average runoff (%) 2005	2007	2008	Specific erosion (average, to matter/hectare) 2005	2007	2008
E2	V (30)	Stand of *Pinus sylvestris* (Pi) and *Pinus nigra* (Pi.n) mixture, 28 years old, full canopy cover, with following characteristics: Pi average height (Hm) =8.1 m; diameter (Dm) =10.2 cm; Pi.n. Hm=7.9 m, Dm=10.4 cm	1.2	0.8	0.8	0.177	0.073	0.053
E3	V (20)	*Pinus nigra* stand, 28 years old, canopy cover of 0.6, Hm = 8.1 m, Dm = 9.6 cm	2.0	1.7	2.3	0.209	0.088	0.040
E3	V (27)	*Pinus nigra* stand, 25 years old, canopy cover of 0.9, Hm = 8.1 m, Dm = 9.5 cm	2.2	1.6	1.6	0.278	0.110	0.110
E3	E (11)	Stand of *Pinus sylvestris* (Pi) and *Pinus nigra* (Pi.n) mixture, 28 years old, full canopy cover, with follwing features: Pi: Hm = 10.4 m, Dm = 10.8 cm; Pi.n: Hm = 10.3 m, Dm = 10.7 cm, sallow thorn undercovered	2.0	1.4	2.0	0.220	0.079	0.066
E3	E (11)	Stand of *Pinus sylvestris* with sallow thorn, 28 years old, and consistency 0.9, with follwing characteristics: Pi: Hm = 10.8 m, Dm = 11.1 cm; Pi.n: Hm = 10.5 m, Dm = 10.8 cm, sallow thorn undercovered	1.7	1.0	1.4	0.147	0.022	0.037

Degradation type: E2 – strong erosion; E3 – excessively eroded lands.

Table 40.3 The average amount of eroded matter in Caciu catchement, determined by surveys on fixed benchmarks, for the period of 1st of June until 30th of September 2006

Slope (grades)	Eroded layer (mm)	Eroded volume ($m^3\ ha^{-1}$)	Weight of eroded matter ($t\ ha^{-1}$)
1. Strongly to excesivelly degraded slopes, with sallow thorn on 30% of area			
35–40	3	30	45
40–45	5	50	75
2. Active gullying, with no vegetation			
30–35	7	70	105
40–45	8	80	120
3. Ravine billows with outer bedrock (marls with plaster stone)			
40–45	12	120	180

40.4 Outlook and Conclusions

One third of the territory of Romania is affected by land degradation, mainly erosion and landslides. Climate change impacts may further aggravate the situation, thus ample and necessary conservation measures are needed. Among them we mention afforestation as an anti-erosional approach to combat soil erosion and promote sustainable development in rural areas. At present, forests cover only 7% of the plain areas, 27%, in hills while in mountain the figure is around about 60%.

Under extreme site conditions (such as very strongly eroded lands, ravened lands, riverbanks and gradients of the torrential hydrographic network, and landslides with strong division into fragments), the installation of the forest vegetation represented the only way of mitigation and improvement of the degraded lands to balance and rehabilitate the ecological and social-economic functions of these affected areas.

The National Strategy and Action Programme Concerning Desertification, Land Degradation and Drought – Prevent and Control (2001), provides the specific measures and actions toward the prevention and combating desertification, drought and land degradation in the country as far as afforestation actions are concerned. These measures include:

- Endorsement of agroforestry and a targeted balance share between agricultural and forestry ecosystems in the support of sustainability;
- Gradual restriction of the arable interventions on lands with slope over 12% showing soil degradation problems;
- Conversion of agricultural use to afforestation;
- Establishment of protective forests belts in affected areas;
- Integrity assurance and sustainable development of forests.

The intensity of the land degradation problems in the country makes such measures necessary and urgent. But afforestation of the degraded lands must be based on proven scientific results and by introducing compositions and mixture schemes that are ecologically compatible (Ciortuz and Păcurar, 2004) with local conditions. Besides the main forestry species, introduction of fruit tree species and other productive species for food and accessory products is quite important to increase forest economic productivity.

Because of the climatic conditions in such degraded areas, especially on sunny slopes with steppe features, is necessary to foresee suitable technologies for better hydrological and anti-erosion management of the land. In order to control the run-off on the slopes, simple terraces (width of 75–80 cm) are, by far, the most appropriate solution. On land with low slope it is recommended to plough the soil in strips that are further improved by establishment of large range of woody species well adapted to water shortage and sunstrokes. Another important condition for successful forest plantations is to keep the maintenance works at appropriate/optimum times, preventing the weeds to consume soil water and nutritive substances. Additionally, an important issue is to safeguard and protect the forests against the biotic and abiotic damaging factors.

Future actions needed to create new forests by afforesting degraded lands and establishing forest protection shelterbelts foresee an annual rate of 20–30 thousand hectares of afforestation in the next periods, gradually reaching 40–50 thousand hectares annually for the period 2010–2020 (Giurgiu, 2004). The goal of the national strategic objective is to increase the proportion of the forests up to 40% throughout the country's territory (compared to 26.3% at present).

References

Bogdan, N. and Untaru, E. (1972). Substitution of Sallow Thorn Cultures on the Degraded Lands of Vrancea. Ceres Publishing House, Bucharest, pp. 81–155.

Ciortuz, I. and Păcurar, V. (2004). Forest Ameliorations. Lux Libris Publishing House, Braşov, 231p.

Constandache, C. (2003). The Aspect Concerning at Installation of Condition and Functional Efficiency of Forest Protection Cultures Installated in the Degraded Lands from Vrancea County, ANALE ICAS – Seria I. Volume 46, Silvica Publishing House, Bucharest, pp. 432–433.

Constandache, C., Nistor, S. and Ivan, V. (2006). Afforestation of Degraded Lands Bad for Agriculture in South-East of Romania, ANALELE ICAS – Seria I. Volume 49, Silvica Publishing House, Bucharest, pp. 187–204.

Dumitru, M. et al. (2002). Monitoring of Lands and Soils in Romania, in Academician Constantin Chiriţă in Memoriam. Ceres Publishing House, Bucharest, pp. 215–231.

Giurescu, C.C. (2004). Istoria pădurii româneşti din cele mai vechi timpuri până astăzi. Orion Publishing House, Bucharest, 398p.

Giurgiu, V. (2004). Forest Science. Volume III B, Sustainable management of Romanias' forests, Academia Româna Publishing House, Bucharest, 320p.

Haralamb, At. (1957). Cultivation of Forest Species. Agro-silvica Publishing House, Bucharest, 755p.

Traci, C. (1976). Types of Forest Cultures for Eroded Lands Afforestation in Sessile Oak and Beech, ICAS Seria II-a. 41p.

Traci, C. and Untaru, E. (1986). Development and Improvement and Consolidation Effect of Forest Plantations on Degraded Lands in Experimental Areas, ICAS, Seria II-a. Propagandă Tehnică Agricolă Publishing House, Bucharest, 70p.

Untaru, E., Constandache, C., Ivan, V. and Munteanu, Fl. (2003). Achievements and Perspectives in Improving and Use by Forestation of Degraded Lands in Vrancea, ANALE ICAS – Seria I. Volume 46, Silvica Publishing House, Bucharest, pp. 363–375.

Untaru, E., Constandache, C. and Nistor, S. (2006). Afforestation of Degraded Lands and Forestall Floods, SILVOLOGIE. Volume V, Academia Româna Publishing House, Bucharest, 285p.

Untaru, E., Constandache, C. and Roşu, C. (2008). The Effects of Forest Plantations Installed on Eroded and Sliding Lands, Related to their Evolution in the Time, SILVOLOGIE. Volume VI, Academia Româna Publishing House, Bucharest, pp. 137–168.

MADR. (2000). National Strategy and Action Programme concerning Desertification, Land Degradation and Drought-Prevent and Control. Bucharest (2001).

MAPPM. (2000). Norme tehnice privind compozitii, scheme si tehnologii de regenerare a padurilor si de impadurire a terenurilor degradate, Ministerul Apelor, Padurilor si Protectiei Mediului, Romania.

Chapter 41
Investigating Soils for Agri-Environmental Protection in an Arid Region of Spain

C. Castaneda, S. Mendez, J. Herrero, and J. Betran

Abstract The saline wetlands of the Monegros Desert, in the central Ebro Basin (NE Spain), host valuable biodiversity and pedodiversity. A part of this area has been proposed for inclusion in the European Union Natura 2000 network. However, agricultural intensification is changing the area as more land is consolidated for new irrigation or is plowed to earn CAP (Common Agricultural Policy) subsidies. Soil mapping is needed to assist in the delimitation of natural habitats and to make conservation compatible with agriculture. The methodology presented here to characterize agri-environmental areas takes into account current agricultural and environmental practices. We examined the opinions of farmers and agricultural and environmental officers concerning a new agri-environmental measure, which could be proposed for inclusion in the Rural Development section of the CAP. The measure would save agriculture inputs in unproductive areas and also comply with nature conservation objectives. A GIS database was built for selecting the farming plots suitable for new agricultural practices favoring biodiversity and pedodiversity. At the local scale, we used remote sensing and pedodiversity criteria for selecting low production areas to be prospected. The opinion poll resulted in a positive response and confirmed the interest of farmers in having detailed maps of those soil features that limit crop production. The soil survey reveals soil salinity, and high contents of gypsum and calcium carbonate, as significant features to map low production areas for making agriculture compatible with habitat conservation.

Keywords Monegros · Spain · Salinity · Pedodiversity

C. Castaneda (✉)
Soils and Irrigation Department (associated with CSIC), AgriFood Research and Technology Centre of Aragon (CITA), 50059 Zaragoza, Spain
e-mail: ccastanneda@aragon.es

41.1 Introduction

The Monegros saline wetlands, known locally as "saladas", are part of a unique landscape where scattered gentle depressions are surrounded by dry-farmed or recently irrigated areas. The arid climate plus the gypsiferous and calcareous rocks are determinant of the genesis and evolution of *saladas*. More than a hundred depressions containing ephemeral brines and halophylous vegetation have been inventoried. These wetlands are of scientific and environmental value as natural habitats for endemic microbes (Casamayor et al., 2005), plants (Braun-Blanquet and Bolòs, 1958; Pedrocchi, 1998; Domínguez et al., 2006) and animals (Melic and Blasco-Zumeta, 1999; Baltanás, 2001). Half of the agricultural area has been proposed for inclusion in the European Union's Natura 2000 network. The other half will be irrigated.

The soils are shallow, calcareous or gypseous, with a low organic matter content; whitish spots occur in relatively higher topographic positions. The soils are deeper and saline in the wetlands and have a dark upper horizon; their salinity largely exceeds thresholds for crop production. Despite the low returns imposed by natural limits, dry farming often is practiced in areas of shallow soils with low water-holding capacity, even though evapotranspiration much greater than precipitation results in no yield. Due to scarce and very irregular precipitation from year to year, the only feasible crops are winter cereals, which, in many years, remain unprofitable. Earning CAP subsidies compels farmers to expand the plowed surface, even into saline depressions of remarkable environmental value hosting endemic and endangered species. Moreover, land consolidation, pipelines, pumping stations, roads, or other works associated with the new irrigated areas, have increased the degradation or disappearance of habitats (Castañeda and Herrero, 2008; Herrero and Castañeda, 2009) leading to a decrease in the extent of natural vegetation.

Maps are needed to identify soil features that limit agricultural production and to pinpoint pedodiversity spots. The agri-environmental rules can be adapted to better allocate subsidies, without increasing the CAP budget. The new allocation rules ought to encourage farmers to manage their land in a way that saves on labor and farm inputs, as well as address nature conservation issues related to biodiversity (van der Horst, 2007) and pedodiversity.

This chapter presents an integrated methodology aimed at identifying low production areas where plowing-exemption rules should be implemented as a means for promoting habitat protection on a local scale. The methodology takes into account rural practices and seeks the agreement of local farmers.

41.2 Materials and Methods

As a starting point, information was available from environmental and agricultural Geographical Information System (GIS) sources. The environmental data were the wetland inventory (Castañeda and Herrero, 2008), the 1:5,000 scale maps of halophilous vegetation (Domínguez et al., 2006), and the Special Protection Area (SPA) for birds (European Union, 1979). The agricultural data were the Spanish Farming Land Geographic Information System (SIGPAC), and the 2005–2006 alphanumerical data from the GIS of Herbaceous Crops of the Spanish Ministry of Agriculture, Fisheries and Food.

In order to select agri-environmental plots for soil prospecting, a new GIS database was created using the most detailed SIGPAC delimitations as the basic geographic unit, i.e. the plots declared by farmers for CAP subsidies. The above-mentioned environmental and agricultural data were superimposed onto the farming plots' GIS coverage, together with an image of the Quickbird satellite acquired on July 11, 2007.

Using GIS, we first identified the agri-environmental plots adjacent to the saladas (Fig. 41.1) because of their sensitivity to biodiversity conservation. For this purpose we drew a buffer line 200 m from the edge of the saladas. When a farming plot was only partially included within the 200-m limit, the entire plot was computed. All these farming plots were extracted from the database furnished by the Government of Aragon. Four of the accessible saladas (Agustín, Gramenosa, Guallar and Salineta) were selected as representative of dry-farming, imminent irrigation, and Natura 2000 areas (Fig. 41.1).

We contacted 26 people from the list of CAP subsidy recipients in the municipality of Bujaraloz (Mendez, 2006). Only 16 agreed to be interviewed. Most of them were land managers, representing a total of 34 subsidy recipients. A poll was carried out in order to learn their opinions and concerns about a new agricultural measure to be applied under the Natura 2000 conservation policy. The local officers in charge of environment and agriculture were also interviewed.

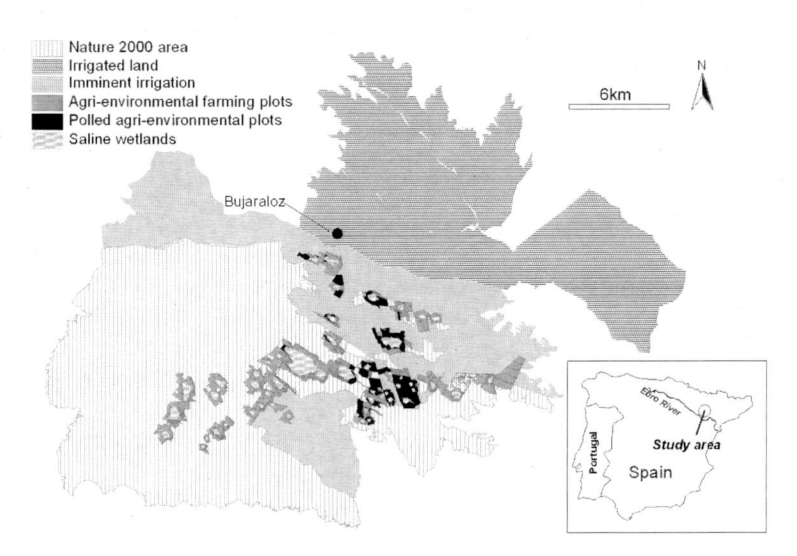

Fig. 41.1 Location of the study area and the selected farming plots. Most of the area not included in Natura 2000 is, or will be, equipped for irrigation with Pyrenean water

We delimited low production enclaves in the agri-environmental plots selected by means of a visual analysis of the Quickbird image. The 432 RGB composition showed a number of white patches close to the wetlands, and some prominent patches in nearby additional plots. A subsequent field survey was carried out to decide on the prospecting sites using geomorphologic and soil surface color criteria (white, dark, and reddish colors), allowing for sampling points of contrasting colour to be selected.

The soil prospecting was based on auger holes in the four sites selected. The most contrasting color patches were found around Agustín. The prospecting there was then intensified by describing the soil surface and opening pits for pedon study. Crop development at several stages was also recorded, as was the yield from three randomly selected 0.25 m^2 areas around the pit location.

A total of 117 soil samples were taken for laboratory analysis. The results shown here are for (i) electrical conductivity in saturation extracts (ECe) and 1:5 soil to water ratio extracts (EC1:5), according to US Salinity Laboratory Staff (1954); (ii) calcium carbonate equivalent (CCE) by titration of excess HCl, a method validated by the Laboratory of the Aragón Government as yielding results similar to the Bernad calcimeter; and (iii) gypsum by thermogravimetry (Artieda et al., 2006).

41.3 Results and Discussion

The farming plot database contains 1,264 plots located in the wetland environment, with a total surface area of 5,747 ha, five times the wetland surface extent. SPA for birds covers 81% of the surface area of the selected agri-environmental plots, and 19% is included in the near future irrigation. The poll was carried out over 936 ha, i.e. 111 farming plots, or 16% of the selected agri-environmental surface area extent. The size distribution of the polled plots was similar to that of the plots of the entire area studied.

The opinions gathered from the interviewees were attributed to the number of hectares they were responsible for. The subsidy recipients were direct cultivators of only 50% of the polled surface area. Three respondents managed around 80% of the polled area, indicative of the disappearance of small farming enterprises. Hence, the decisions made by few people will impact large areas. From a total of 27 agriculture/environment ideas discussed in the interviews, we concluded that 37% of those polled agreed with a new ploughing exemption measure, 30% agreed but strictly in exchange for legal economic compensation, and 16% were distrustful about future changes of the CAP. Environment officers would prefer a measure adapted to the existing agri-environmental program, while agricultural officers were uncertain about the technical and economical feasibility of its application, although they approved of new agri-environmental measures, especially non-ploughing. The poll confirmed the farmers' interest in the production of detailed maps of soil features limiting crop production. Farmers had a personal stake in the soil and crop surveys due to the potential source of supplementary income that could result from any habitat protection efforts (Fig. 41.2).

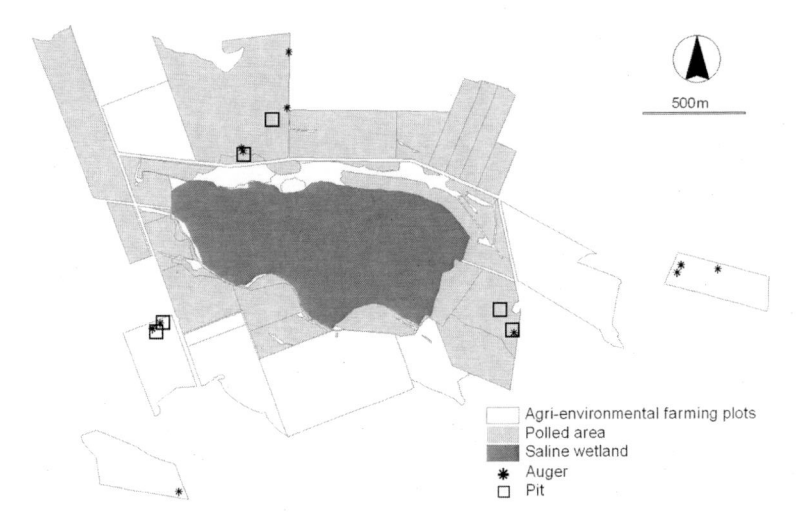

Fig. 41.2 Farming plots in the Agustín site, and points sampled for soil characterization

Most of the farming plots studied are located on gentle slopes at the edges of the depressions. Supplemental field inspections showed halophilous vegetation in some patches with stunted barley or durum wheat. The white patches drawn from the Quickbird image occupy 13.4% of the 711 ha of agri-environmental farming plots, this percentage being different at each site: 20.7% in Agustín, 10.6% in Guallar, 8.4% in Pito, and 10.3% in Salineta.

The grain production measured in 2008 ranges from 0 to 639 kg ha^{-1}. These figures translated into agricultural production equivalent to a zero yield, because harvesting becomes unprofitable. Bad agricultural years are frequent in this area, characterized by a mean yield of 900 kg ha^{-1} (McAneney and Arrúe, 1993), strongly conditioned by rainfall during cereal growth and by the application of herbicides.

The soil sampling depth ranged from 30 to 150 cm (Table 41.1). Most of the white patches are in the top positions with shallow lithic or paralithic contacts, as shown by a more frequent occurrence of the shallow auger holes in these patches. Table 41.1 shows the ranges of ECe, gypsum, and CCE for the samples at each site studied. For all samples, ECe ranges from 0.7 to 30.3 dS m^{-1} with 80% of the samples having ECe <8 dS m^{-1}. CCE ranges from 2.7 to 58.6% (w/w), with 66% of samples having >20% CCE, and 18% having >40% CCE.

Gypsum content determinations give results from 0.5 to 94.7%, with 51% of samples having >40% gypsum and 39% with >60% gypsum. For the six samples giving results <2%, a qualitative test of gypsum would be needed for confirmation (Artieda et al., 2006). The six samples come from the dark patches. Five of them are neither saline nor saturated in calcium sulfate (EC1:5 < 1.5 dS m^{-1}); the other one is very saline (EC1:5 = 3.66 dS m^{-1}, and ECe = 10.43 dS m^{-1}). The gypsum content in these six samples is assumed to be insignificant. All soil samples from the Salineta site have higher CCE than gypsum.

Table 41.1 Soil sampling sites, sampling techniques used, and ranges of: depth, ECe, calcium carbonate equivalent (CCE), and gypsum content

Sampling site	Number of pits (P) and auger holes (A)	Number of soil samples	Depth (cm)	ECe (dS m^{-1})	CCE (% w/w)	Gypsum (% w/w)
Agustín	6 P	20	80–150	3.0–13.4	2.7–46.7	1.6–94.7
	12 A	47	50–125	2.0–20.3	6.9–41.4	*in.–89.3
Guallar	7 A	20	30–110	3.3–19.5	7.1–49.8	9.9–94.6
Pito	5 A	14	25–100	0.9–30.3	5.1–41.9	1.9–90.4
Salineta	5 A	16	50–125	0.7–15.4	22.6–58.6	1.9–19.5

*in. = insignificant.

As expected, the colour of the soil surface detected by means of remote sensing is related to the gypsum and CCE content in the upper horizon. Figure 41.3 shows the difference between white and dark patches. The sum of gypsum and CCE at the upper horizon is about 100% in most white patches, indicative of the general lack of clay minerals or other non-soluble components in the soil, especially at the depth of roots. Dark patches, frequently in gentle hollows where the organic matter, clay, and moisture accumulate, have <50% gypsum plus CCE, allowing for better growth of the cereal in dry years.

Provided that gypsum is ubiquitous in the landscape, soil samples with ECe ≤ 2.25 dS m^{-1} cannot be deemed saline. Aridity is the key constraint for life in Monegros. Main plant stresses are related to (1) lithic or paralithic contacts at shallow depths, (2) occurrence of horizons with low clay content near the surface, due to high gypsum and/or calcium carbonate contents, and (3) saline horizons, i.e. occurrence of salts more soluble than gypsum somewhere in the soil profile. These soil features were considered the blueprint for easy criteria to identify low crop production areas. Their future mapping and the analysis of their distribution in the soil profile will provide basic knowledge on favouring pedodiversity and biodiversity.

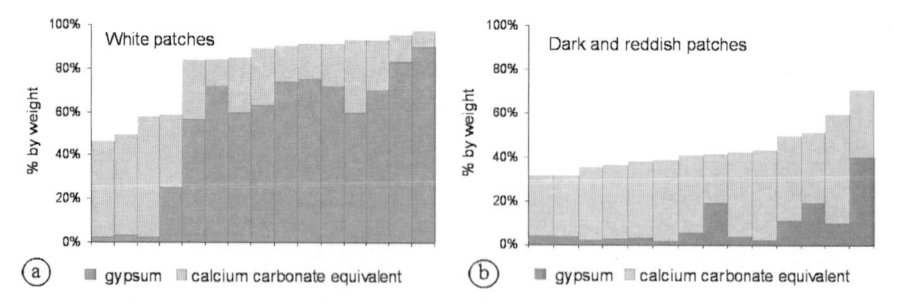

Fig. 41.3 Gypsum and calcium carbonate equivalent in auger soil samples of the upper horizons (0–25 cm) in the white (**a**) and the dark and reddish (**b**) patches

41.4 Conclusions

The considerable amount of information from different database sources, processed by means of GIS tools, enabled the recognition of farming plots with agri-environmental interest and the selection of the farmers for the opinion poll.

The low production criteria extracted from the analyses of satellite images and from the soil survey (presented here in a test area) are complementary, and the results are coincident. The soil prospection reveals soil salinity and the high gypsum and calcium carbonate content as significant pedodiversity criteria. These soil features condition production in the dry-farmed area, and their mapping can help to reconcile agriculture with habitat conservation. We consider it worthwhile to extend these criteria to the remaining farming plots.

The GIS database created and the consultative approach will be advantageous for the systematic selection of plots for agri-environmental purposes. A measure for ploughing exemption – without a decrease in CAP subsidies to farmers – in low-production areas with environmental value would combat the degradation of biodiversity and soils. Farmers would be rewarded by saving on agricultural inputs, with no increase in the CAP budget.

Acknowledgments This article is a result of projects AGL2006-01283 and GALC006-2008.

References

Artieda, O., Herrero, J. and Drohan, P.J. (2006). A refinement of the differential water loss method for gypsum determination in soils. Soil Science Society of America Journal 70: 1932–1935.

Baltanás, A. (2001). *Candelacypris* n. gen. (Crustacea, Ostracoda): A new genus from Iberian saline lakes, with a redescription of *Eucypris aragonica* Brehm & Margalef, 1948. Bulletin de la Société des Naturalistes Luxembourgeois 101:183–192.

Braun-Blanquet, B. and Bolòs, O. (1958). Les groupements végétaux du bassin moyen de l'Ebre et leur dynamisme. Anales de la Estación Experimental de Aula Dei 5:1–266.

Casamayor, E., Castañeda, C., Pena, A., Vich, M.A. and Herrero, J. (2005). Monegros: riqueza escondida bajo la sal del desierto. Investigación y Ciencia 349:38–39.

Castañeda, C. and Herrero, J. (2008). Assessing the degradation of saline wetlands in an arid agricultural region in Spain. CATENA 72:205–213.

Domínguez, M., Conesa, J.A., Pedrol, J. and Castañeda, C. (2006). Una base de datos georreferenciados de la vegetación asociada a las saladas de Monegros. In: M.T. Camacho, J.A. Cañete and J.J. Lara (eds.), El acceso a la información espacial y las nuevas tecnologías geográficas. University of Granada, Spain, Actas del XII Congreso Nacional de Tecnologías de la Información Geográfica.September 2006.

European Union. (1979). Council Directive 79/409/EEC of 2 April 1979 on the conservation of wild birds. Official Journal of the European Communities, Luxembourg 103:1–18.

Herrero, J. and Castañeda, C. (2009). Delineation and functional status monitoring in small saline wetlands of NE Spain. Journal of Environmental Management 90:2212–2218. doi: 10.1016/j.envman.2007.06.026

McAneney, K.J. and Arrúe, J.L. (1993). A wheat fallow rotation in northeastern Spain: Water balance-yield considerations. Agronomie 13:481–490.

Melic, A. and Blasco-Zumeta, J. (eds.). (1999). Manifiesto Científico por los Monegros. Boletín de la Sociedad Entomológica Aragonesa 24:1–266.

Mendez, S. (2006). Intégration d'indices agro-environnementaux auprès des acteurs locaux de las saladas du sud de Monegros. Project of Master Professionnel d'Ingénierie en Écologie et en Gestion de la Biodiversité. Université des Sciences et Techniques de Montpellier, Francia.

Pedrocchi, C. (1998). Ecología de los Monegros. Instituto de Estudios Altoaragoneses, Huesca.

US Salinity Laboratory Staff. (1954). Diagnosis and improvement of saline and alkali soils. In: Agriculture Handbook. volume 60, US Gov Print Office, Washington, DC.

Van der Horst, D. (2007). Assessing the efficiency gains of improved spatial targeting of policy interventions; the example of an agri-environmental scheme. Journal of Environmental Management 85:1076–1087.

Chapter 42
Risk Assessment in Soils Developed on Metamorphic and Igneous Rocks Using Heavy Metal Sequential Extraction Procedure

S. Martínez-Martínez, Angel Faz Cano, Gerhard Gerold, J.A. Acosta, and R. Ortiz

Abstract Six areas developed on metamorphic and igneous parent materials were selected in Murcia Province (SE Spain), two from diabases, two from mica schist and two from andesites. Three of them potentially polluted by anthropogenic sources and the other three no polluted. One profile and 6 surface samples were collected from each one in order to identify the Pb and Zn concentrations bounded to different soil phases using sequential extraction and to verify if the existent differences in Pb and Zn concentrations from soils developed on the same parent material are the consequence of the anthropogenic activities, or natural geochemical contributions. The results showed that the soils that present higher total Pb and Zn concentrations that come from anthropogenic sources registered higher percentages of these heavy metals in the most labile phases (exchangeable and organic matter bounded heavy metals). In addition, the carbonate phase is also an important sink of metals, however in soils developed on metamorphic and igneous rocks this constituent is very low due to organic and exchangeable phases that retain these metals. This behaviour increases the risk of mobility to trophic chain to plants and microorganisms that may absorb these metals.

Keywords Sequential extraction procedure · Metamorphic and igneous rock · Heavy metals · Spain

42.1 Introduction

The degree of heavy metals pollution in soils and their mobility depends on the soil retention capacity, especially on physicochemical properties and characteristics

A. Faz Cano (✉)
Sustainable Use, Management, and Reclamation of Soil and Water Research Group, Department of Agriculture Science and Technology, Technical University of Cartagena, 30203 Cartagena, Spain
e-mail: angel.fazcano@upct.es

(mineralogy, grain size, organic matter) affecting soil particle surfaces and also on the chemical properties of the metal (Moral Robles et al., 2005). When a metal reaches soil surface, depending on soil properties and characteristics, can be bound to different soil constituents which include water-soluble, exchangeable organic associated, carbonate associated phases, and could be bound and occluded in oxides and secondary clay minerals phase, and residual (RES) within the primary mineral lattice phase (Li and Shuman, 1996). Water soluble and exchangeable fractions are considered readily mobile and bioavailable, whereas other metal fractions, especially a residual fraction, are considered immobile and tightly bound and may not be ready to be released under natural conditions. Sequential extraction procedures are usually used in order to determine in what phase heavy metals are bounded; moreover, this analytical procedure could provide valuable information for predicting metal availability to plants, metal movement in the soil profile and transformation between different forms in soils in the long term (McGrath and Cegarra, 1992; Anderson et al., 2000). Although, in the literature, there are several sequential extraction methods (Meguellati et al., 1983; Forstner, 1985; Shuman, 1985; Beckett, 1989; Ahnstrom and Parker, 1999), the most frequently used sequential extraction method was proposed by Tessier et al. (1979), which provides information about forms of trace metals associated with soil and sediment components, such as clay minerals, hydrous ferric and organic matter. This was the method selected for this study. The degree of metal association with different geochemical forms strongly depends upon physico-chemical properties and characteristics of the soils, such as pH, calcium carbonate and organic matter content (Kabata-Pendias and Pendias, 2001; Alvarez et al., 2006).

The interest of carrying out the sequential extraction was due to some soils evolved from previous rock types that presented higher total heavy metals content than other soils derived from the same type of parent material. This analytical procedure scheme may be useful to distinguish between anthropogenic and geochemical sources of the metal species (Gouws and Coetzee, 1997) and moreover, this method will provide information on the potential metal bioavailability and mobility of soil-bound metals (Stone and Marsalek, 1996). At the present time, it is widely recognized that distribution, mobility and bioavailability of heavy metals in the environment depends not only on their total concentration but also on the association form the solid phase to which they are bound. There is a concern to know the soil metal bioavailability and toxicity to plants, animals and man and the efficiency of the soil as a sink for metals and the potential capacity of a metal to be mobilized from the soil.

Natural and anthropogenic environmental changes greatly influence the behaviour of metallic pollutants as the association form in which they occur can be changed. Such external influences can include pH, temperature, redox potential, organic matter decomposition, leaching and ion exchange processes and microbial activity. The metal content bound to carbonates is sensitive to pH changes and could become mobilised when pH is lowered; the metal fraction bound to Fe–Mn oxides and organic matter can be mobilised with increasing reducing or oxidising conditions in the environment, and finally, the metal fraction associated with the residual fraction (e.g. silicate) can only be mobilised as a result of weathering, which can only cause long-term effects.

This survey presents the results from six natural sites, – two metamorphic areas and four ones developed on igneous parent material, in pursue of the following goals: (1) to know the heavy metals concentrations retained in different soil phases (e.g. carbonates, Fe and Mn oxides, organic matter), (2) to verify if the existent differences in heavy metals concentrations from soils developed on the same parent material occur as a consequence of the anthropogenic activities, or due to geochemical contribution.

42.2 Materials and Methods

42.2.1 Soil Sampling

Six areas were selected in the Murcia Province (SE Spain), two sites were predominated soils derived from diabases: (1a-SAN) from "Cerro Volcánico de Santomera" and (1b-OFI) form "Cabezo Mingote"; two areas developed on andesites: (2a.-CVE) from "Cabezo Ventura" and (2b.-ISL) from "Isla del Ciervo"; and another two areas evolved on mica schists: (3a.-MIC) from "Cerro Metamórfico de Cala Reona" and (3b.-GRE) from "Cerro Metamórfico de Calnegre" (Fig. 42.1).

All soils are influenced by the Mediterranean climate characterised by low rainfall and high average summer temperatures, where the nature of the parent material and the topography have a strong influence on soil genesis and evolution. According

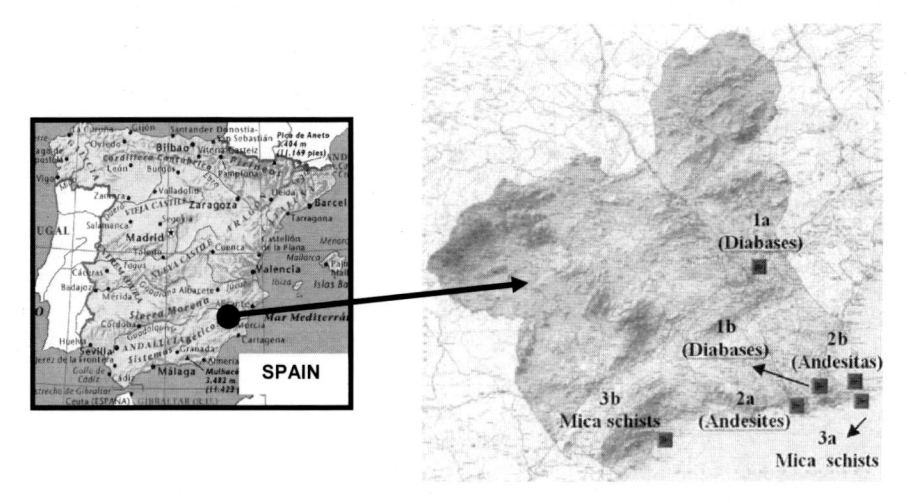

Fig. 42.1 Location maps of selected areas: *1a* (SAN) – "Cerro Volcánico de Santomera" (diabases), *1b* (OFI) – "Cabezo Mingote" (diabases), *2a* (CVE) – "Cabezo Ventura" (andesites), *2b* (ISL) – "Isla del Ciervo" (andesites), *3a* (MIC) – "Cerro Metamórfico de Cala Reona" (mica schists) y *3b* (GRE) – "Cerro Metamórfico de Calnegre" (mica schists)

with U.S.D.A. Soil Taxonomy (2006) the soils present an Aridic soil moisture regime and a Thermic soil temperature regime.

One representative soil profile and 6 representative surface samples (0–15 cm) were taken in each area. The soils were macro-morphologically described following the FAO-ISRIC (2006) guidebook.

42.2.2 Analytical Methods

All soil samples were air-dried and sieved to pass a 2-mm mesh prior to the determination of: pH measured in H_2O and 1 m KCl using a 1:1 soil/solution ratio (Peech, 1965), electrical conductivity (Bower and Wilcox, 1965), total organic carbon (Duchaufour, 1970), equivalent calcium carbonate (Bernard calcimeter), cation exchange capacity (Chapman, 1965), and particle-size analysis (Robinson's pipette method). Total heavy metals content of bulk soil samples was determined by acid digestion using HNO_3/HF.

The sequential extraction scheme was developed from that of Tessier et al. (1979) modified by Li et al. (1995). The extraction was carried out progressively on an initial weight of 2,000 g of soil samples and reference materials.

The reagents and operationally defined chemical fractions were as follows:

Fraction 1: exchangeable. Sample extracted with 16 ml of 0.5 m magnesium chloride ($MgCl_2$) at pH 7.0 for 20 min, with continuous agitation, at room temperature.

Fraction 2: bound to carbonate and specifically adsorbed. Residue from Fraction 1 leached for 5 h with 16 ml of 1 m sodium acetate (NaOAc; adjusted to pH 5.0 with acetic acid, HOAc) at room temperature. Continuous agitation was maintained during the extraction.

Fraction 3: bound to Fe–Mn-oxides. Residue from Fraction 2 was extracted with 40 ml of 0.04 M hydroxylammonium hydrochloride (NH_2 OH HCl) in 25% (v/v) HOAc for 6 h. The extraction was performed at 96°C with occasional agitation. After extraction, the extract solutions were diluted to 50 ml with DIW.

Fraction 4: bound to organic matter and sulphide. To the residue from Fraction 3, 6 ml of 0.02 m HNO_3 and 10 ml of 30% hydrogen peroxide (H_2O_2; adjusted to pH 2.0 with HNO_3) were added. The sample was heated progressively to 85°C, and maintained at this temperature for 2 h with occasional agitation. A second 6 ml aliquot of 30% H_2O_2 (adjusted to pH 2.0 with HNO_3) was then added, and the mixture was heated again at 85°C for 3 h with intermittent agitation. After cooling, 10 ml of 3.2 m ammonium acetate (NH_4OAc) in 20% (v/v) HNO_3 were added, the tubes were then continuously agitated for 30 min. the finally, extract solutions were diluted to 25 ml with DIW.

Fraction 5: residual phase. Residue from Fraction 4 was digested with 4 ml concentrated HNO_3 (65% w/w) and 2 ml hydrofluoric acid (HF, 40% w/w) using the following heating regime:

(a) Starting to 20°C → 30 min to 90°C → 11 h to 175°C → 8 h to 20°C.
(b) Starting to 20°C → 6½ h to 150°C.
(c) Starting to 20°C → 2 h to 185°C.

Before starting with the (c) step, 2 ml HNO_3 (65% w/w) and 15 ml of DIW are added into each vessel. Finally, extract solutions were diluted to 50 ml with DIW.

When continuous agitation was required, samples were shaken on a mechanical shaker at 160 rpm. Heating of the samples was operated using an aluminium-heating block. Following each extraction, the mixtures were centrifuged at 2,000 rpm for 20 min at room temperature. Prior to the start of next extraction step, samples were shaken with 16 ml of DIW for 5 min, and the wash solution was discarded after 20 min centrifuging.

In order to ensure one uniform matrix for ICP-OES analysis, the different extraction matrices were digested with concentrated HNO_3. To the aliquots of extractant solutions from fractions 1–4 (Fraction 1: 10 ml, Fraction 2: 10 ml, Fraction 3: 20 ml, Fraction 4: 20 ml), 1 ml HNO_3 (concentrated, 70% w/w) was added, and then the solutions were heated to dryness on an aluminium heating block at 140°C. Care should be taken when heating the solution from extraction fractions 3 and 4 because a violent reaction of HNO_3 with NH_2 OH HCl or NH_4OAc may occur. After adding HNO_3, these extract solutions were left overnight on a heating block at 90°C, and the temperature rose slowly to 140°C and heated to dryness. The residue in the tubes was then leached with 2 ml of 5 m HCl, and made-up to final volume of 20 ml with DIW.

To determine the accuracy of the results, two certified reference materials were used: SO-3 y SO-4 (Canadian Certified Reference Material Project, Canada; CCRMP). Additionally, to estimate the standard deviation of the method for each fraction, three replicates were carried out. Finally, in order to find the percentage of recovery, total concentration of the bulk samples was obtained using the same procedure used in the Fraction 5.

42.3 Results and Discussion

42.3.1 Soil Properties and Characteristics

Soil profiles taken in each site were classified as follows: Lithic Haploxeroll (areas 2a, 2b, 3b), Lithic Torriorthent (areas 1a, 1b), and Aridic Haploxeroll (3a).

The mean of physical and chemical properties and characteristics of each studied area are showed in Table 42.1. pH values are alkaline for every selected area, with a mean values ranging from 8.4 and 6.6, showing a high buffer capacity of these soils.

Table 42.1 Mean values of physical and physico-chemical characteristics of the studied soils

Areas	pH H$_2$O	pH KCl	E.C. (μS cm^{-1})	C.E.C. (cmol kg^{-1})	Percentage of O.C.	Percentage of CaCO$_3$	Percentage of clay	Percentage of silt	Percentage of sand
ISL	7.0	6.2	163.5	20.6	1.7	0.8	3.3	37.0	59.7
CVE	6.6	5.8	412.7	26.3	2.52	5.5	23.02	31.49	45.49
SAN	7.7	7.0	161.2	10.3	1.17	13.2	11.47	21.41	67.11
OFI	8.2	7.2	159.3	15.5	1.05	3.6	16.40	27.80	55.79
MIC	8.2	7.3	250.9	5.3	0.59	0.6	7.32	22.67	70.02
GRE	8.4	6.9	93.4	8.6	0.82	2.3	7.12	23.57	69.31

It is clearly appreciated how the content of organic carbon influences the pH$_{water}$ values, since the samples that presented lower contents of O.C. are those that registered at lower acidity. It indicated that when the OM decomposition takes place, organic acids are generated and as consequence pH decreases lightly (Brady Nyle, 2000). In addition, the organic carbon percentage is associated with CEC, where higher percentages of OC correspond to high cation exchange capacity (Porta et al., 1999). The soils studied from different parent material are not saline since they presented lower values than 1 mS cm^{-1}. The majority of the studied soils have coarse textures. The clay and silt contents are less than 30% and sand percent is more than 50%.

42.3.2 Risk Assessment Using Sequential Extraction Procedure

Table 42.2 shows the results of both single total extraction and the summation of the five fractions from sequential extraction procedure. In addition the percentage of recovery is also shown. For lead, the recovery percentage range from 70.7 to 119.7%, while for zinc range from 76.5 to 118.2. These results indicated that the sequential extraction procedure used has a high recovery percentage for these two metals and represents the total concentration present to the soil.

Table 42.2 Pb and Zn concentrations obtained from the total extraction, sum of the 5 fractions and percentage recoveries obtained from the sequential extraction

Sample	Pb Σ 5 fractions (mg kg^{-1})	Pb Total extraction (mg kg^{-1})	R (%)	Zn Σ 5 fractions (mg kg^{-1})	Zn Total extraction (mg kg^{-1})	R (%)
ISL-1	245.2	224.4	109.3	201.0	203.8	98.6
ISL-4	113.1	102.9	109.9	138.1	121.6	113.6
ISL-7	263.4	251.0	104.9	142.3	148.8	95.6
ISL-8	303.6	288.1	105.4	239.3	256.2	93.4
ISL-11	143.8	127.3	112.9	134.5	134.2	100.2
ISL-12	164.8	149.7	110.1	123.7	142.4	86.9
HOR A	178.9	158.0	113.2	160.5	135.8	118.2

42 Risk Assessment in Soils Developed on Metamorphic and Igneous Rocks

Table 42.2 (continued)

	Pb			Zn		
Sample	Σ 5 fractions (mg kg^{-1})	Total extraction (mg kg^{-1})	R (%)	Σ 5 fractions (mg kg^{-1})	Total extraction (mg kg^{-1})	R (%)
CVE-2	807.1	821.4	98.3	358.2	385.8	92.8
CVE-3	549.4	573.4	95.8	322.3	363.0	88.8
CVE-5	600.3	628.3	95.5	296.1	338.4	87.5
CVE-6	429.6	434.1	99.0	255.3	256.1	99.7
CVE-8	882.4	869.9	101.4	351.5	363.3	96.8
CVE-11	880.2	885.5	99.4	463.1	499.6	92.7
HOR A	416.8	401.2	103.9	199.9	191.5	104.4
SAN-1	27.9	35.2	79.4	58.1	57.0	102.0
SAN-3	19.5	20.1	97.0	42.7	38.7	110.2
SAN-4	16.6	18.7	88.4	39.3	37.5	104.9
SAN-5	34.5	36.9	93.5	48.4	51.4	94.2
SAN-6	45.3	50.1	90.3	58.5	54.4	107.5
SAN-8	34.8	40.3	86.5	49.5	56.3	87.9
HOR A/R	32.8	46.4	70.7	47.7	42.8	111.4
OFI-5	451.7	392.2	115.2	438.0	396.1	110.6
OFI-6	319.6	288.6	110.7	259.0	267.8	96.7
OFI-8	243.1	212.9	114.2	291.6	269.7	108.1
OFI-9	362.8	311.0	116.7	287.2	275.5	104.3
OFI-11	325.6	290.3	112.2	405.6	348.3	116.5
OFI-12	195.7	167.4	116.9	225.1	219.1	102.7
HOR A	76.4	75.1	101.7	214.8	219.6	97.8
MIC-1	414.5	437.4	94.8	796.3	801.3	99.4
MIC-4	633.0	604.4	104.7	629.3	784.1	80.3
MIC-7	331.5	277.0	119.7	503.0	470.8	106.9
MIC-8	334.3	320.2	104.4	562.1	499.6	112.5
MIC-9	680.9	672.1	101.3	661.0	637.7	103.7
MIC-11	341.7	340.7	100.3	341.8	293.0	116.7
HOR A1	291.5	301.1	96.8	344.8	367.1	93.9
HOR A2	299.2	279.8	106.9	266.9	325.6	82.0
HOR A/C	244.8	268.1	91.3	319.1	371.0	86.0
HOR C	78.6	90.3	87.0	203.7	266.4	76.5
GRE-1	49.0	42.2	116.2	127.3	96.2	115.2
GRE-2	77.1	65.4	117.7	106.6	100.8	105.8
GRE-8	81.9	78.7	104.2	170.3	157.2	108.4
GRE-9	65.8	59.7	110.3	114.9	109.1	105.3
GRE-10	84.2	79.3	106.1	127.1	113.8	111.7
GRE-11	76.0	72.7	104.6	111.8	114.5	97.7
HOR A	50.9	48.8	104.4	103.3	105.1	98.3

Note – Σ concentrations from 5 fractions: exchangeable, carbonate, iron/manganese oxide, and organic bounds, and residual fraction; R: percentage of recuperation of the method used.

Regarding total Pb and Zn single extraction, three areas presented a significant accumulation of these metals (Table 42.2) as documented in the soils derived from mica schists in Cala Reona (3a), of andesites at the "Cabezo Ventura" (2a) and on diabases at the "Cabezo Mingote" (1b). Mean lead concentrations are 359, 659 and 248 mg kg^{-1}, respectively and the mean zinc concentrations are 481, 342 and 285 mg kg^{-1}, respectively. These high concentrations are probably due to the influence of the mining activity from La Unión for 2,500 years and the industrial development in the surroundings of Cartagena city, located less than 2 km from these sites. Studies carried out in this area have shown as well high levels of heavy metal concentrations (Clemente et al., 2002; Moreno et al., 2002).

Metal mines are one of the major industries producing metal containing dusts and slags (Porter and Bleiwas, 2002); although they can vary considerably in composition, these wastes usually contain high concentrations of hazardous metals. Oppositely, the other three areas showed lower concentration for both metals, and mean values range between 35 and 185 mg kg^{-1} for lead and between 163 and 48 mg kg^{-1} for zinc. Thus, they are not considered as polluted areas from anthropogenic sources. Instead their concentrations are due to natural parent material contribution.

The lead and zinc retained in the five phases from sequential extraction procedure can be observed in the Figs. 42.2 and 42.3, respectively. Important differences are obtained in the soils that showed the highest total Pb and Zn concentrations, soils developed from andesites (2a-CVE), diabases (1b-OFI) and mica schists (3a-MIC), presenting higher percentages of exchangeable (1st step) and adsorbed in organic matter (4th step) heavy metals.

It is especially remarkable the percentage of lead and zinc bound to organic matter in "Cabezo Ventura", with a value up to 16% of the total concentration, that it is due to high percentage of OC that content this soil (2.52%). In addition, in spite of low carbonate content of these soils, the percentage of lead and zinc bound to carbonates is also high, with a value close to 10 and 6% of the total concentration respectively, showing the high accumulation of lead and zinc in the absorbed-carbonate phase. Cabral reported similar results as well as Lefebvre (1998) in soils contaminated with high Pb concentration. For carbonate bound fraction, the zinc can be retained as Zn carbonate (Kiekens, 1995; Adriano, 2001), or adsorbed Zn in existed carbonates (McLean and Bledsoe, 1992).

It is expected that if the soil would have more amount of carbonates, lead and zinc will bound to this phase before that others. These results indicate that when the lead and zinc comes from anthropogenic sources and reach soils developed from siliceous parent material they are bound to organic fraction or more labile fractions due to the fact that there are not enough carbonates to bind them. These distribution patterns represent a risk for the ecosystems since plants or microorganisms could uptake these metals and introduce them to the trophic chain.

However, in the six areas the metals are mostly bound to the most stable phase, residual fraction, where the metals are retained in the crystalline net of the minerals indicating that high concentrations of these metals in soils develop from metamorphic and igneous rocks come from parent material.

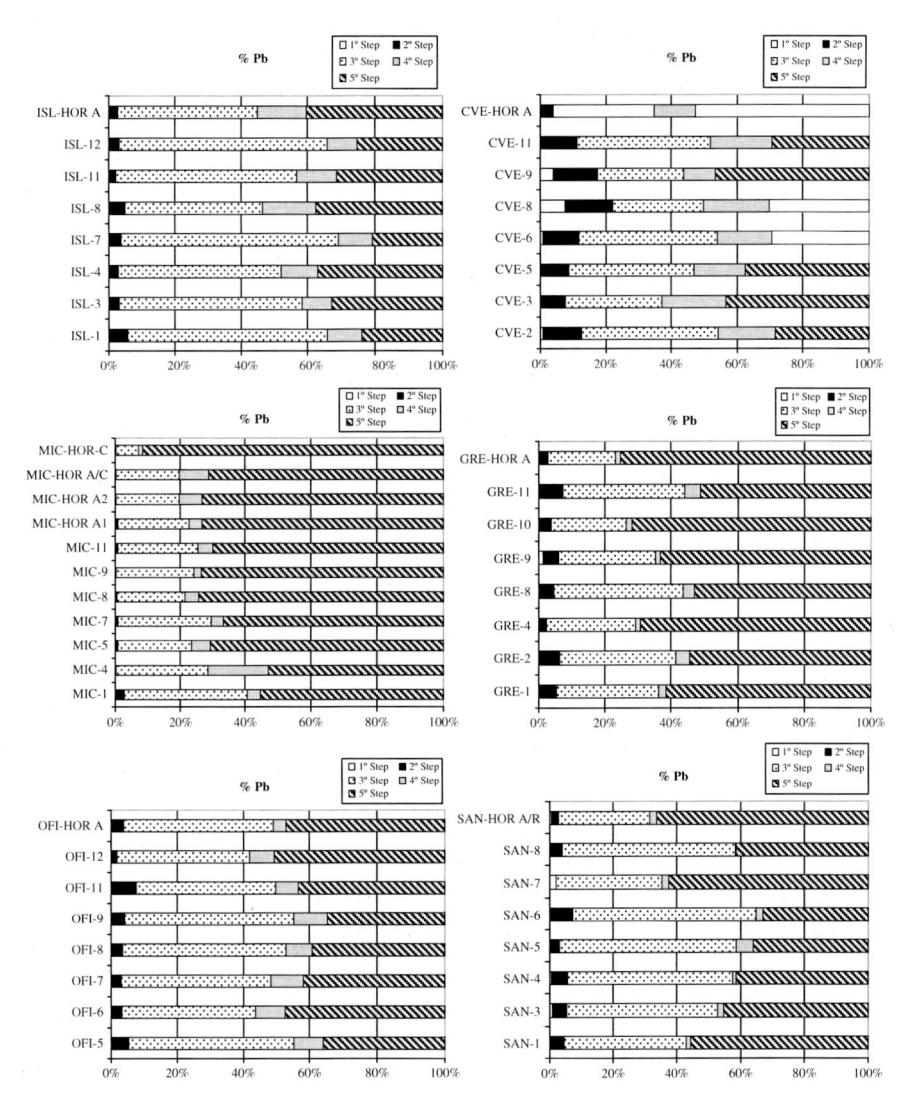

Fig. 42.2 Percentages of Pb easily exchangeable (1st step), bound to carbonates (2nd step), bound to Fe or Mn oxides (3rd step), bound to organic matter or as sulphides (4th step) and that presents in the residual fraction (5th step) from different soils

42.4 Conclusions

Results showed that the soils that present higher total Pb and Zn concentrations registered higher percentages of these heavy metals in the most labile phases, exchangeable and organic bound. Since the origin of Pb and Zn in these three areas could be anthropogenic, environmental risks may be generated due to Pb and Zn that might be absorbed by the plants and microorganisms affecting hence the trophic

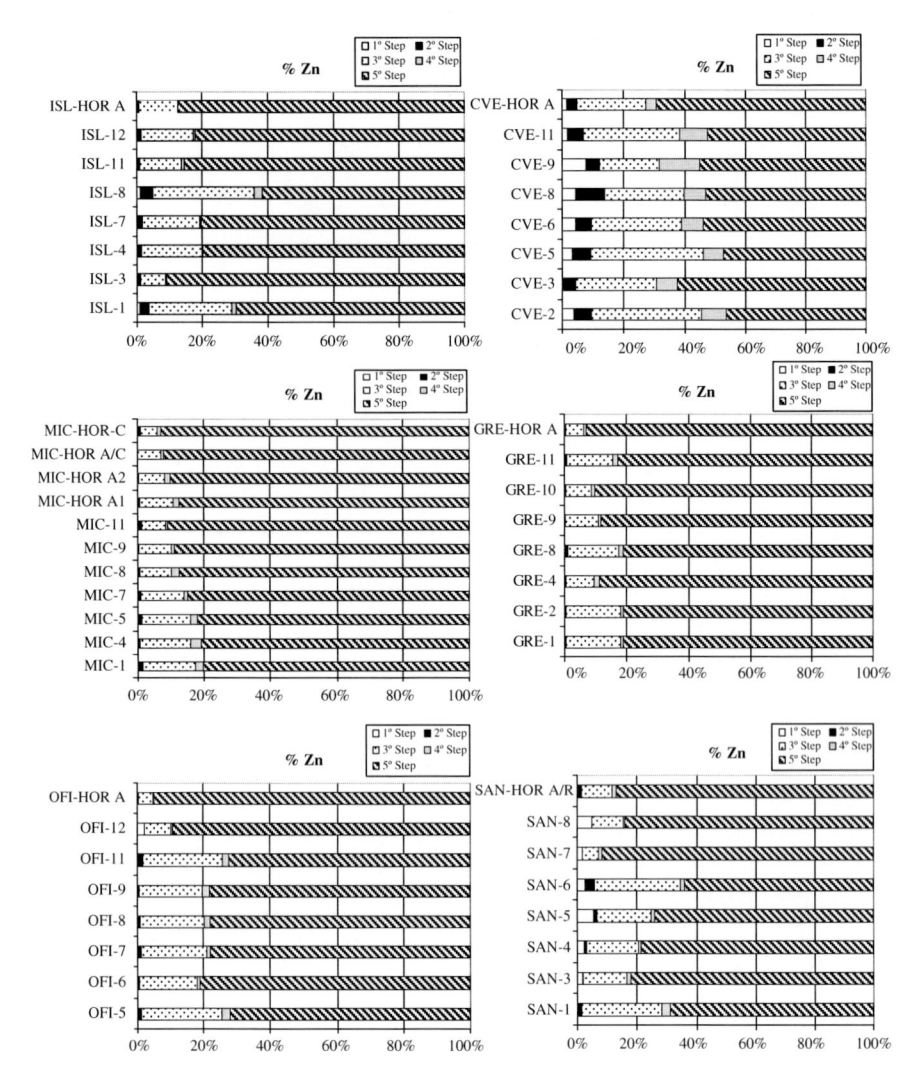

Fig. 42.3 Percentages of Zn easily exchangeable (1st step), bound to carbonates (2nd step), bound to Fe or Mn oxides (3rd step), bound to organic matter or as sulphides (4th step) and that presents in the residual fraction (5th step) from different soils

chain. The soil in the immediate vicinity of plant roots (rhizosphere) has chemical, physical and biological properties that are substantially different from those of the bulk soil. The exudation of organic substances is a mechanism by which plant roots can mobilize metal ions. For this reason, another complementary study in these areas should be carried out to establish possible potential environmental risks.

Acknowledgments The "Caja de Ahorros del Mediterráneo (CAM)" is acknowledged for its financial support. The authors would like to thank staff members of the Department of Landscape Ecology of George August University of Göttingen for the technical and other support rendered.

References

Adriano, D.C. (2001). Trace Elements in the Terrestrial Environment. Biogeochemistry, Bioavailability and Risk of Metals. 2nd edn. Springer-Verlag, New York, 880pp.

Ahnstrom, Z.S. and Parker, D.R. (1999). Development and assessment of a sequential extraction procedure for the fractionation of soil cadmium. Soil Science Society of America Journal 63:1650–1658.

Alvarez, J.M., Lopez-Valdivia, L.M., Novillo, J., Obrador, A. and Rico, M.I. (2006). Comparison of EDTA and sequential extraction tests for phytoavailability prediction of manganese and zinc in agricultural alkaline soils. Geoderma 132:450–463.

Anderson, P., Davidson, C.M., Duncan, A.L., Littlejohn, D., Ure, A.M. and Garden, L.M. (2000). Column leaching and sorption experiments to assess the mobility of potentially toxic elements to assess the mobility of potentially toxic elements in industrially contaminated land. Journal of Environmental Monitoring 2:234–239.

Beckett, P.H.T. (1989). The use of extractants in studies on trace metals in soils, sewage sludges, and sludge-treated soils. Advances in Soil Science 9:143–176.

Bower, C.A. and Wilcox, L.V. (1965). Soluble salts. In: C.A. Black (ed.), Methods or Soil Analysis. Volume 2, American Society of Agronomy, Madison, Wisconsin, USA, pp. 933–940.

Brady Nyle, C. (2000). The Nature and Properties of Soils. 8th edn. Macmillan Publishing, CO., IND, New York.

Cabral, A.R. and Lefebvre, G. (1998). Use of sequential extraction in the study of heavy metal retention by silty soils. Water, Air, and Soil Pollution 102:329–344.

Chapman, H.D. (1965). Cation Exchange capacity. In: C.A. Black (ed.), Methods of Soil Analysis. Volume 2, American Society of Agronomy, Madison, Wisconsin, USA, pp. 891–900.

Clemente, R., Walker, D.J., Roig, A. and Bernal, M.P. (2002). The effect of organic matter addition on the bioavailability of heavy metals in a contaminated semiarid soil from Murcia (Spain). In: A. Faz, R. Ortiz and A.R. Mermut (eds.), Sustainable Use and Management of Soil in Arid and Semiarid Regions. Volume II, pp. 425–426.

Duchaufour, P. (1970). Précis de Pedologie. Masson y Cie (Ed), Paris, 481p.

FAO-ISRIC. (2006). Guidelines for Soil Description. 4th edn. Food and Agriculture Organization of the United Nations, Rome, 97pp.

Forstner, V. (1985). Chemical forms and reactivities of metals in sediments. En: R. Lechsber, R.A. Davis and P. LHermitte (eds.), Chemical Methods for Assessing Bioavailable Metals in Sludges. Elsevier, London, pp. 1–30.

Gouws, K. and Coetzee, P.P. (1997). Determination and partitioning of heavy metals in sediments of the Vaal dam system by sequential extraction. Water SA 23:217–226.

Kabata-Pendias, A. and Pendias, H. (2001). Trace Elements in Soils and Plants. CRC Press, Boca Raton, Fla, 413pp.

Kiekens, L. (1995). Zinc. In: B.J. Alloway (ed.), Heavy Metals in Soils. 2nd edn. Blackie Academic and Professional, Glasgow, pp. 284–305.

Li, X., Coles, B.J., Ramsey, M.H. and Thornton, I. (1995). Sequential extraction of soils for multi-element analysis by ICP-AES. Chemical Geology 124:109–123.

Li, Z. and Shuman, L.M. (1996). Redistribution of forms of zinc, cadmium, and nickel in soils treated with EDTA. Journal Science of Total Environment 191:95–107.

McGrath, S.P. and Cegarra, J. (1992). Chemical extractability of heavy metals during and after long-term applications of sewage sludge to soil. Journal of Soil Science 43:313–321.

McLean, J.E. and Bledsoe, B.E. (1992). Behaviour of Metals in Soils. EPA/540/S-92/018, United States Environmental Protection Agency (USEPA). Washington, DC, 25pp.

Meguellati, N., Robbe, D., Marchandise, P. and Astruc, M. (1983). A new chemical procedure in the fractionation of heavy metals in sediments – Interpretation. Proceedings of the International Conference on Heavy Metals in the Environment. C.E.P. Consultants Publ., Heidelberg, pp. 1090–1093.

Moral Robles, F.J., Romero Diaz, A. and Garcia Fernandez, G. (2005). Erosión eólica en el área minera de Cartagena-La Unión, Sureste de España. Primeros resultados. En: R. Jiménez

Ballesta y A.M. Álvarez (eds.), Resúmenes del II Simposio Nacional sobre control de la degradación de suelos. Universidad Autónoma De Madrid, Madrid, pp. 747–752.

Moreno, J., Faz, A., Arnaldos, R. and Perez, J.A. (2002). Pb, Cu and zn in soils of mining zones in La Union (SE Spain) and their accumulation in Lettuce. In: A. Faz, R. Ortiz and A. Mermut (eds.), Sustainable Use and Management of Soil in Arid and Semiarid Regions. Volume II, pp. 504–505.

Peech, M. (1965). Hydrogen-ion activity. In: C.A. Black (ed.), Methods for Soil Analysis. Volume 2, American Society of Agronomy, Madison, WI, USA, pp. 914–916.

Porta, J., López-Acevedo, M. and Roquero, C. (1999). Edafología para la agricultura y el medio ambiente. 2nd edn. Ediciones Mundi-Prensa, Madrid.

Porter, K.E. and Bleiwas, D.I. (2002). Physical aspects of waste storage from a hypothetical open pit porphyry copper operation. USGS Open File Report 03–143.

Shuman, L.M. (1985). Fractionation method for soil microelements. Soil Science 140:11–22.

Stone, M. and Marsalek, J. (1996). Trace metal composition and speciation in street sediments: Saultste. Marie, Canada. Water, Air, and Soil Pollution 87:149–169.

Tessier, A., Campbell, P.G.C. and Bisson, M. (1979). Sequential extraction procedure for the speciation of particulate trace metals. Analytical Chemistry 51:844–851.

U.S.D.A. (2006). Keys to Soil Taxonomy. 10th edn. United States Department of Agriculture, Natural Resources Conservation Service. U.S. Government Printing Office, Washington, DC.

Chapter 43
Assessing the Impact of Fodder Maize Cultivation on Soil Erosion in the UK

Mokhtar Jaafar and Des E. Walling

Abstract The short-term gross and net soil erosion in a bare maize stubble field during a period of heavy rainfall in the winter of 2004–2005 was estimated using ^{7}Be measurements, with the longer-term erosion rate, estimated using Cs-137 measurements and compared with the longer-term soil erosion rate for the field, estimated using ^{137}Cs measurements. The results show that the gross and net soil erosion occurring in the field under maize stubble were considerably greater than the longer-term gross and net soil erosion rates for the field. Both sets of measurements demonstrate that the field is characterized by a high sediment delivery ratio, emphasizing that a large proportion of the sediment mobilised by erosion was transported beyond the field and towards the stream network. Depending on the weather conditions, late harvesting, the associated compaction of the soil by heavy machinery and leaving the stubble field bare over the ensuing winter can result in a substantial increase in erosion relative to the longer-term erosion rate under more traditional land use. The need to implement measures to reduce soil erosion associated with maize cultivation in England, such as the Code of Good Agricultural Practice and the Agri-Environment Schemes is clearly demonstrated.

Keywords Soil erosion · Maize · Radionuclide technique · ^{7}Be measurements · ^{137}Cs measurements

43.1 Introduction

Maize has become an important fodder crop in England in recent years. Its cultivation expanded rapidly between 1990 and 2004. According to available land use statistics (Defra) the area under fodder maize in England was 33,300 ha, and this

M. Jaafar (✉)
Faculty of Social Sciences and Humanities, School of Social, Development and Environmental Studies, Universiti Kebangsaan Malaysia, 43600 Bangi, Selangor, Malaysia
e-mail: mokhtar@eoc.ukm.my

more than trebled to reach 107,400 ha in 2004. This major increase in the area under fodder maize cultivation reflects changes in both agricultural policy and livestock husbandry. The EU Common Agricultural Policy (CAP) provided a major stimulus to the growing of fodder maize through subsidies and the use of fodder maize for silage production.

This rapid expansion of the area devoted to maize cultivation has been implicated as a cause of increased soil erosion and of sediment problems in adjacent streams and rivers. Poor land management and environmentally unfriendly attitudes among maize growers and harvest contractors have been identified as an important cause of the increased soil erosion and off-site sediment delivery. In this context, the key problems associated with fodder maize cultivation relate to its relatively long growing season. This can delay harvesting into the early autumn, when soils are wet and unsuitable for ploughing, with the result that the harvested field is frequently left bare during the ensuing winter, which is commonly a period of heavy rainfall.

Furthermore, the use of heavy machinery for harvesting under wet conditions will compact the soil and destroy the soil structure, resulting in reduced infiltration and increased surface runoff (Clements and Lavender, 2004). The harvesting machinery also frequently travels up and down the slope, creating wheelings, thereby further increasing the potential for erosion by concentrated flow.

Considering these important environmental problems that can occur as a result of maize cultivation, and the major expansion of the area cultivated for maize in England in recent years, there is a need to obtain reliable information on rates of soil loss from stubble fields that are left bare after harvest and through the ensuing winter and to assess the magnitude of the increase in soil erosion over more traditional land use practices. In this contribution, we describe an attempt to document the short-term soil erosion occurring in a bare maize stubble field, by using measurements of the fallout radionuclide beryllium-7 (^{7}Be), and to compare this with an estimate of the longer-time erosion rate in the same field obtained using caesium-137 (^{137}Cs) measurements. By virtue of its short-life (53 days) and near continuous fallout, ^{7}Be measurements can be used to document the erosion and soil redistribution associated with individual storm events and short-periods of heavy rainfall (see Walling et al., 1999; Sepúlveda et al., 2008).

In contrast, longer-lived ^{137}Cs (half-life 30.2 years), which originated as fallout in the late 1950s and the 1960s, as a result of the testing of nuclear weapons, provides a means of estimating longer-term erosion rates over the period extending from the commencement of fallout in the mid 1950s to the present (Walling et al., 1999; Zapata, 2002).

The study site was a 5.6 ha maize field located at Little Landsite Farm, in Devon, UK. A typical brown earth of the Crediton Series developed on Permian strata underlies the field, and the slope gradient ranges between 2.2° and 6.4°. Maize grown in the field during the preceding summer was harvested in late October 2004 and the field was left bare without any surface protection over the subsequent winter period.

43.2 Materials and Methods

The period extending from December to early January was very wet and a total of 85.4 mm of rainfall was recorded at a nearby rainfall measuring station between the 15th and 31st of December, 2004. A further 30.3 mm of rainfall was recorded between January 1st and 10th, 2005 (Fig. 43.1). A total of more than 115 mm of rainfall therefore fell within a period of about 3 weeks and there was evidence of significant soil erosion in both the study field and other fields with bare maize stubble in the local area. Soil sampling for both radionuclides was conducted on 11th January 2005, when there was clear evidence of both sheet erosion and the development of small rills within the study field, as a result of the preceding rainfall. The period prior to December 2004 had been relatively dry, with little opportunity for significant erosion. The soil sampling was undertaken along two downslope transects separated by a distance of ca. 20 m. The cores were collected at an interval of ca. 10 m along the transects.

Soil cores for ^{137}Cs analysis were collected using a percussion corer fitted with a 70 mm internal diameter metal core tube. The core tube was driven into soil to a depth of more than 30 cm to ensure that the core included all the ^{137}Cs contained in the soil profile. At locations where deposition could be expected to have occurred, cores were collected to a depth ca. 50 cm. A separate purpose-built manual corer was used to collect the shallow soil cores required for analysis of their ^{7}Be content. These cores were 15 cm in diameter and penetrated to a depth 30 mm. In addition, soil cores for ^{137}Cs and ^{7}Be analysis were also collected from an adjacent uncultivated field with minimum gradient that was selected to provide a reference site, which had experienced no erosion either during the preceding winter period or the preceding 50 years.

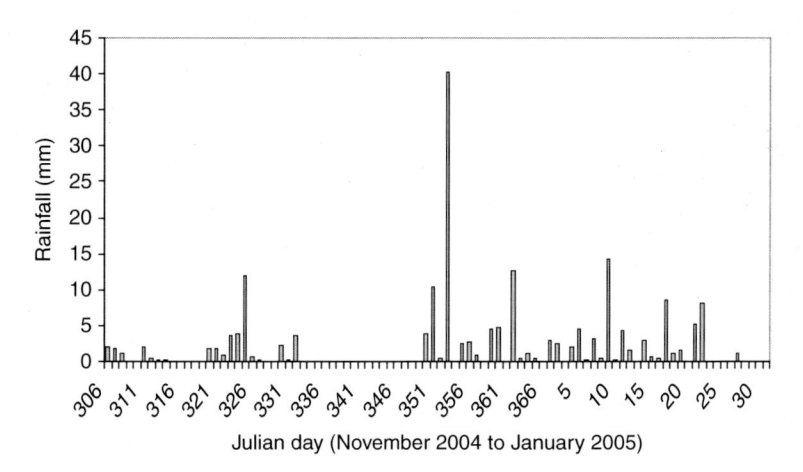

Fig. 43.1 The daily rainfall recorded at Hemyock during the period November 2004–January 2005

The cores collected from this field were used to provide values for the ^{137}Cs and ^{7}Be reference inventory, which could be compared with the inventories associated with the sampling points in the study field. In addition, a further core was collected from the reference site for sectioning into 5 mm depth increments. These depth incremental samples were required to characterize the depth distribution of ^{7}Be in the soil. The information was needed as an input to the conversion model used to estimate soil redistribution from the ^{7}Be measurements for the individual bulk cores collected from the study field.

All soil samples were dried, gently disaggregated and sieved to <2 mm prior to determination of their ^{137}Cs and ^{7}Be activity by gamma spectrometry, using an HPGe coaxial gamma detector. Count times of at least 7 h (25,200 s) were employed, in order to provide an effective compromise between the precision of the associated measurements and the need to complete the measurements within a relatively short period, in view of the short half-life of ^{7}Be. The precision of the ^{7}Be and ^{137}Cs measurements was generally ± ca. 10% at the 95% level of confidence. The values of mass activity density provided by the gamma spectrometry measurements were converted to estimates of the areal activity density (Bq m^{-2}) for the individual sampling points, based on the known surface area of the core samples.

Comparing the measured inventory values with the reference inventories and using a conversion model to derive the estimates of soil redistribution estimated soil redistribution rates at the sampling points. The profile distribution model described by Walling et al. (1999) and Schuller et al. (2006) was used to convert the ^{7}Be measurements into estimates of soil redistribution rates, and mass balance model III described by Walling et al. (2002) was used to derive the estimates of soil redistribution rate from the ^{137}Cs measurements. In both cases, the conversion models were applied using the conversion model software developed by the Department of Geography at the University of Exeter, UK.

43.3 Results and Discussion

The measurements of areal activity density obtained from the cores analysed for ^{7}Be showed a clear difference between the mean inventory for the study field (150 Bq m^{-2}) and the local reference inventory (259 Bq m^{-2}). This difference provides clear evidence of significant erosion and net soil loss from the study field during the period of heavy rainfall that preceded sample collection. In the case of ^{137}Cs, the difference between the mean inventory for the sampling points and the reference inventory was less, with equivalent values of 1,966.6 and 2,427.8 Bq m^{-2}, respectively. Again, however, the results provide clear evidence of significant net soil loss from the field over the past ca. 50 years.

Figures 43.2 and 43.3 present more detailed results for both the ^{7}Be and ^{137}Cs measurements, obtained for the individual transects. The estimates of soil redistribution presented for the individual sampling points show that there is evidence of both,

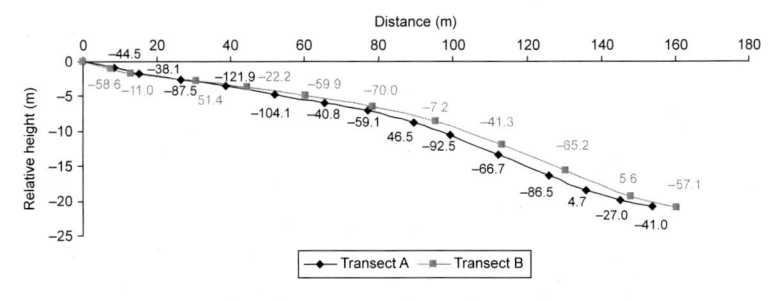

43 Assessing the Impact of Fodder Maize Cultivation on Soil Erosion in the UK 585

Fig. 43.2 Estimates of soil erosion and deposition (t ha^{-1}) associated with the period of heavy rainfall in late December 2004 and early January 2005, for the two transects in the study field, derived from the ^{7}Be measurements

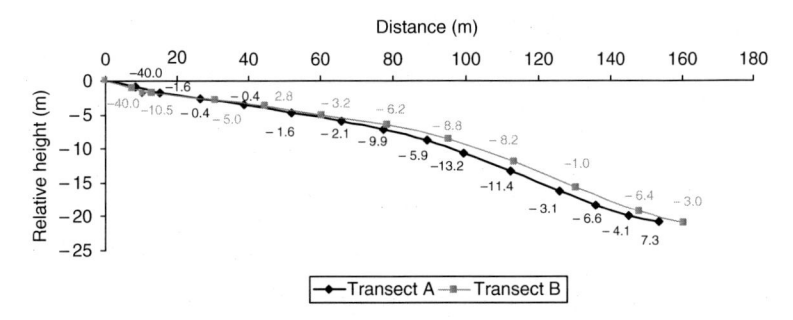

Fig. 43.3 Estimates of the mean annual rates of soil redistribution (t ha^{-1} year^{-1}) for the sampling points along the two transects in the study field, derived from the ^{137}Cs measurements

erosion and deposition along each transect. In the case of the ^{7}Be measurements (Fig. 43.2), the soil redistribution rates derived using the conversion model provide an estimate of the gross erosion rate from the field of 46.7 t ha^{-1} and a net erosion rate of 42.3 t ha^{-1}. These values relate only to the period of heavy rainfall occurring in December and early January 2005 and must be considered high for the local area. As such they emphasize the potential impact of maize cultivation in increasing erosion rates.

There is some evidence to suggest that erosion rates associated with transect A are higher than those associated with transect B and this is likely to reflect contrast in slope steepness between the two transects. The slope gradient in the middle part of transect A ranges from 4° to 6° whereas the equivalent value for transect B is only 3° to 4°. The estimates of gross and net erosion based on the ^{7}Be measurements indicate that only a relatively small proportion of sediment mobilized by erosion is redeposited further down the slope and within the field. Most of the mobilized sediment (i.e. ca. 90%) is removed from the field, resulting in the high estimate of net soil loss and a high potential for sediment delivery to the stream network.

In the case of the estimates of longer-term mean annual soil redistribution rates provided by the ^{137}Cs measurements and presented in Fig. 43.3, the mean annual gross erosion rate estimated for the study field is ca. 7.2 t ha^{-1} year^{-1}, and the mean annual net erosion rate for the field is 6.4 t ha^{-1} year^{-1}. This result again indicates that ca. 90% of the soil mobilized by erosion over the past ca. 50 years was transported out of the field. The high erosion rates documented at the top of the field are likely to reflect the effects of tillage translocation at the head of the slope.

A comparison of the estimates of gross and net erosion generated from the ^{7}Be and ^{137}Cs measurements provides a basis for assessing the impact of maize cultivation in increasing winter erosion rates from the study field. In comparing the two sets of estimates, it must be recognized that the values based on the ^{7}Be measurements relate only to the period of heavy rainfall occurring in December 2004 and early January 2005. They are likely to underestimates the annual rate of soil redistribution under maize cultivation. Equally, it is important to recognize that the estimates of mean annual soil redistribution provided by the ^{137}Cs measurements will include the periods when the study field was cultivated for maize. As such, the erosion estimates based on the ^{7}Be measurements are likely to underestimates the annual erosion rate for the field when cultivated for maize and the estimates based on the ^{137}Cs measurements are likely to overestimate the longer-term mean annual erosion rate associated with land use other than maize cultivation.

Any assessment of the increase in erosion rates caused by maize cultivation based on the present results must therefore be seen as a minimum estimate of the increase. It can be suggested that the cultivation of maize in the study field increases both gross and net erosion rates by at least 6 times, as compared to the longer-term erosion rate under other land use. This value is similar to that provided by Walling et al. (1999), who used an equivalent approach, involving ^{7}Be and ^{137}Cs measurements, to show that erosion rates under maize in another field in Devon associated with a period of heavy winter rainfall were about 5 times greater than the longer-term mean annual erosion rate. It is also important to note that the results from the study field indicate that the sediment delivery ratio is high and approximately 90% for both timescales.

This value is again similar to that of ca. 80% reported by Walling et al. (1999) for another maize field in Devon. This high value emphasizes the potential for high connectivity between the slopes and the stream network, although it is possible that a significant proportion of the sediment leaving the study field may be trapped and stored before reaching the stream network.

The result presented above have important implications for any broader assessment of the impact of the expansion of maize cultivation in England on soil erosion, diffuse source pollution and sediment-related problems in rivers and streams. A growing awareness of the magnitude of the problem has prompted the UK government to develop guidelines related to environmentally friendly practices to be implemented by maize growers. These include the Code of Good Agricultural Practice (CoGAP) and the Agri-Environment Schemes (AES), which require farmers to follow the environmental standards of sustainable and environmentally friendly practices, as incorporated into the Common Agricultural Policy (CAP). Generally, maize growers need to be sensitive to both the potential on-site and off-site impacts of maize cultivation on soil erosion and sediment transfer to watercourses and to protect the land from environmental damage by promoting sustainable farming practices.

43.4 Conclusions

The study reported has used ^{7}Be and ^{137}Cs measurements to compare soil redistribution rates in the study field during a period of winter rainfall, when the field was left bare and compacted after the maize harvest, with the longer-term average erosion rate in the field which is representative of more traditional land use. The results highlight the potential magnitude of the increase in soil erosion and sediment delivery caused by maize cultivation and, more particularly, the practice of leaving bare compacted stubble fields exposed to winter rainfall. For the study field, the period of heavy winter rain following the maize harvest in December 2004 and early January 2005 was associated with erosion rates at least 6 times, and possibly as much as an order of magnitude, greater than the longer-term mean annual erosion rate.

The lack of deposition within the field, as reflected by its high sediment delivery ratio, emphasizes the potential sensitivity the local area to increased sediment inputs to rivers and streams, which can seriously degrade aquatic habitats and the ecological status of the receiving river. It is important that improved land management practices should be promoted to reduce the potential impact of fodder maize cultivation on both on-site soil loss and off-site sediment transfer to watercourses. The recent development of maize strains requiring a shorter growing season and thus permitting an earlier harvest should also prove helpful in this regard.

Acknowledgment The work reported in this contribution was undertaken whilst the first author was a postgraduate student in the Department of Geography at the University of Exeter supported by a Scholarship from the Malaysian Government. The support and assistance of Mr. Jim Grapes with sample processing and analysis and the help of the landowner in permitting access to the study field and the collection of the soil samples are gratefully acknowledged.

References

Clements, R.O. and Lavender, R.H. (2004). Measurement of surface water runoff from maize stubbles in the Parrett Catchment area (Somerset): Winter 2003/04. Report to FWAG, 13 July.

Defra. http://farmstats.defra.gov.uk/cs/farmstats_data/DATA/nuts_data/nuts_query.asp

Schuller, P., Iroume, A., Walling, D.E., Mancilla, H.B., Castillo, A. and Trumper, R.E. (2006). Use of beryllium-7 to document soil redistribution following forest harvesting operations. Journal of Environmental Quality 35(5):1756–1763.

Sepúlveda, A., Schuller, P. and Walling, D.E. (2008). Use of ^{7}Be to document erosion associated with a short period of extreme rainfall. Journal of Environmental Radioactivity. doi: 10.1016-j.jenvrad.2007.06.010 (in press).

Walling, D.E., He, Q. and Appleby, P.G. (2002). Conversion models for use in soil-erosion, soil redistribution and sedimentation investigations. In F. Zapata (ed.), Handbook for the

Assessment of Soil Erosion and Sedimentation Using Environmental Radionuclides. Kluwer Academic Publishers, Dordrecht, pp. 111–162.

Walling, D.E., He, Q. and Blake, W. (1999). Use of ^{7}Be and ^{137}Cs measurements to document short- and medium-term rates of water-induced soil erosion on agricultural land. Water Resources Research 35(12):3865–3874.

Zapata, F. (ed.). (2002). Handbook for the Assessment of Soil Erosion and Sedimentation Using Environmental Radionuclides. Kluwer Academia Publisher, Dordecht, 219p.

Part V
Land Degradation and Mitigation in the Americas

Chapter 44
Evolution and Human Land Management During the Holocene in Southern Altiplano Desert, Argentina (26°S)

Pablo Tchilinguirian and Daniel Olivera

Abstract The aim of this chapter is to present a stratigraphic and geomorphic analysis of the Holocene historical paleowetland records and to study the factors that involve high Andean wetland degradations related to human settlements in the Southern Altiplano of Argentina. In order to obtain the records and to test the hypothesis we proposed a cross-disciplinary study with methods and techniques from archaeological, geological and biological sciences, but in an adequate and integrated approach. This paleohydrological study may influence on decisions that have an impact on the current regional economy, for instance in the case of ill used technology and therefore degrading the wetlands or the possibility of reinforcing land reactivation of former productive areas.

Keywords Holocene · Paleowetland · Ancient human settlement · Climate change · Argentina

44.1 Introduction

In the Southern Altiplano (26°S) of Argentina, Holocene and actual wetlands deposits (peat and organic soils) are preserved in numerous non-glacial valleys. Paleowetland and peat environments are very important for the actual economy due to the fodder and permanent water supply in this arid region (Annual precipitation (P) < 127 mm year^{-1} and Ev: 550 mm year^{-1}).

Numerous contributions in the Puna de Atacama of northwestern Argentina discuss the role of the wet environment in cultural societies. Yacobaccio and Morales (2005) inferred periods of higher humidity through diatoms analyses from fluvial

D. Olivera (✉)
CONICET-INAPL-UBA, CP 1426, Buenos Aires, Argentina
e-mail: deolivera@gmail.com

terrace paleowetlands of lower Holocene and indicated that hunter-gatherer populations used these wet environments. Rodríguez et al. (2006) showed the relationship between the environment and archaeological locations from late Holocene based on archaeo-botanical information (Nuñez et al., 2002) indicated that the collapse of human occupation is associated with the drying of the lakes under very arid conditions. In more recent times, Inca and Belen occupations (*ca.* 1,100–500 year BP) used water resources from Altiplano wetlands to make intensive and extensive agricultural production.

Paleoenviromental studies have determined significant climatic changes in the Altiplano deserts during the Holocene (Fernández et al., 1991; Baied and Wheeler, 1993; Kulemeyer et al., 1999; Thompson et al., 2000; Grosjean et al., 2003; La Torre et al., 2003; Servant and Servant-Vildary, 2003; Maldonado et al., 2005; Liu et al., 2005; Olivera et al., 2006). Several studies have proved impacts on water availability associated with climatic changes during the last few centuries (Liu et al., 2005; Valero-Garces Blas et al., 2003). Therefore the question arose if climatic conditions and/or human use of water resources might have played a major role in the wetland's degradations and in the human behaviour of Holocene societies.

We present in this chapter a stratigraphic and geomorphic analysis of Holocene and historical paleowetland records and we study the factors that involve high Andean wetland degradations related to human settlements.

44.2 Geomorphic and Climatic Setting

The South Altiplano of Argentina is a high crustal region (3,300–4,000 m) situated between the fold and thrust belt of the eastern Cordillera (5,000 m) and the active volcanic arc of the western Cordillera (6,000 m). It covers 2,000,000 km^2, and extends from 24 to 26°S latitude and 66 to 71°W longitude. There are many saline aquifers in the Altiplano, which contain lacustrine, paleowetlands and alluvial fan quaternary deposits. During the Mesozoic Era, crustal and lithosphere scale extension produced a graven system partially coincident with the modern topographic closed saline aquifers. Graven north–south mountain system developed across the Altiplano (Calalaste and Laguna Blanca ranges) and divided the Altiplano plains fluvial valleys.

The Austral Altiplano is dominated by tropical summer moisture in the east and, in lower proportions, by winter precipitations (cold fronts from the Pacific) in the west. Average annual rainfall is <127 mm year^{-1} at 3,300 m (Antofagasta station, 3,400 m a.s.l.) and increases at higher levels (150–300 mm year^{-1}). This shows that the geographical sector is an extremely dry region.

Although the climate is desertic, wetlands formed by organic and hydromorphic soil could be found only in the drainage system. High wetlands of Austral Puna of Argentina are ecosystems where the greatest primary production, biomass and water of the Puna desert can be found, and they play a fundamental role in the behaviour of the fauna (Caziani and Derlindati, 1999). Around 99% of the wetlands water comes from a permanent regime of springs with a scarce seasonal (10%) and annual (15%) flow variation. The Punilla River and its tributaries (Las Pitas, Ilanco, Miriguaca, Mojones, and Curuto streams) are perennial rivers and only spread over 1% of the drainage system. The remaining streams are ephemeral.

44.3 Methods

Detailed field studies of the Holocene deposits were performed on outcrops exposed over four valleys (Curuto, Mojones, Las Pitas, Miriguaca) and two salars (Colorada and Carachipampa). Lithofacies and depositional environments were determined on all recovered units, using visual core descriptions and standard sedimentologic facies analysis techniques (Miall, 1978). Emphasis was placed on the identification of vertical and horizontal facies changes as indicated by variations in grain size, sedimentary structures, biogenic, pedogenic components and the abundance of organic material. Organic and inorganic carbon content was used with the sedimentary facies analysis to qualitatively determine palaeowater depths and paleohydrological conditions. Radiocarbon (AMS) dates (26) were obtained from bulk organic matter or plant debris. The AMS analyses were undertaken at the Center for Applied Isotope Studies of the University of Georgia (12 samples), The National Science Foundation – University of Arizona (10 samples) and Laboratorio de Radiocarbono y Tritio (University of La Plata, Argentina) (4 samples).

^{2}H, ^{18}O and ^{3}H were analysed in Los Colorados River (26.031°S, –67.448°W, 3,421 m) and Las Pitas (26.028°S, 67.343°W, 3,581 m) in order to determine the origin of the recharge, altitude and age. The stable isotope analyses were carried out at the INGEIS laboratories following Coleman's techniques (Coleman et al., 1982), and Panarello and Parica's techniques (Panarello and Parica, 1984) for ^{2}H and ^{18}O respectively.

The soil profiles in ancient archaeological farms were described following the terminology of Soil Taxonomy. Soil sampling was undertaken at fixed depth intervals (20 cm) in each of the 10 profiles and the total depth of each profile was recorded.

Finally, we also analysed a group of archaeological sites and data that ranged between 4,500 year BP to pre Hispanic times, to establish several relationships between paleoecological proxies and archaeological record (Table 44.1).

Table 44.1 Radiocarbon paleopeat data from the study area

Lab. No.	Depth (m)	No. Profile-core	Material	^{14}C corrected dates ±1	δ^{13}C (‰)	Laboratory
Laguna Colorada section: –26.028824°S, –67.448208°W, 3,420 m						
1	1.2	13/99	Peat paleosoil	1,620 ± 70	s/d	LATYR
2	1.3	13/99	Peat paleosoil	1,600 ±60	s/d	LATYR
3	1.5	13/99	Peat paleosoil	2,270± 60	s/d	LATYR

Table 44.1 (continued)

Lab. No.	Depth (m)	No. Profile-core	Material	^{14}C corrected dates ±1	$\delta^{13}C$ (‰)	Laboratory
8785	1.65	12/99	Peat paleosoil	2,890 ± 40	−5.14	CASI-UG
8786	2.47	12/99	Peat paleosoil	3,430 ± 40	−17.26	CASI-UG
8787	3.25	12/99	Peat paleosoil	3,910 ± 40	−12.87	CASI-UG
Miriguaca section: 25.998471°S, 67.388867°W, 3,482 m						
8793	0.4	3/99	Peat paleosoil	1,560 ± 40	−24.2	CASI-UG
8792	1.8	3/99	Peat paleosoil	3,060 ± 40	−15.77	CASI-UG
Curuto section: 25.911644°S, 67.347936°W, 3,682 m						
8794	0.46	100/07	Peat paleosoil	1,610 ± 60	−24.84	CASI-UG
8795	1.55	100/07	Peat paleosoil	2,280 ± 60	−24.23	CASI-UG
8796	2.8	100/07	Peat paleosoil	2,880 ± 70	−26.45	CASI-UG
Las Pitas section: 26.015196°S, 67.325651°W, 3,645 m						
AA78534	0	55/04	Peat paleosoil	152 ± 43	−27.1	NSF-UA
15108	0.8	53/04	Peat paleosoil	200± 35	−26.4	CASI-UG
AA78537	0.8	55/04	Peat paleosoil	202 ± 38	−26.5	NSF-UA
AA78546	0	53/04	Peat paleosoil	115 ± 37	−27.4	NSF-UA
AA78533	0.2	51/04	Peat paleosoil	3,620 ± 48	−23.4	NSF-UA
AA78536	0.6	51/04	Peat paleosoil	3,917 ± 44	−23.3	NSF-UA
Mojones section: 25.856719°S, 67.425160°W, 3,727 m						
AA78538	0.2	96/05	Peat paleosoil	241 ± 38	−28.0	NSF-UA
AA78535	0.6	96/05	Peat paleosoil	305 ± 43	−26.7	NSF-UA
AA78539	0.5	200/07	Peat paleosoil	1,936 ± 41	−26.9	NSF-UA
Carachipampa section: 26.416337°S, 67.481082°W, 3,013 m						
AA78540	0.40	101/05	Peat paleosoil	644 ± 43	−24.9	NSF-UA
AA78541	1.04	101/05	Peat paleosoil	1,905 ± 41	−26.7	NSF-UA
Ilanco section: 26.077644°S, 67.279309°W, 3,878 m						
15107	0.20	85/04	Peat paleosoil	695 ± 30	−25.8	CASI-UG
AA78547	0.75	67/04	Peat paleosoil	2,558 ± 45	−25.5	NSF-UA
Confluencia section: 26.070°S, 76.41°W, 3,351 m						
4	1.08	2/98	Peat paleosoil	4,110 ±180	s/d	LATYR
5	1.24	2/98	Peat paleosoil	4,560 ± 60	s/d	LATYR

CASI-UG: Center for Applied Isotope Studies of the University of Georgia, USA.
NAF-UA: The National Science Foundation – University of Arizona, USA.
LATYR: Tritio laboratory, University of La Plata, Argentina.
s/d: without data.

44.4 Results

The wetland's geological record of the Holocene age is composed by a variation of sedimentary facies, which were deposited under different wet conditions. On the one hand, there are sediments formed by organic paleosoils (peat) and hydromorphic paleosoils, diatomite and stratified gravels with organic material. These sediments would have been deposited in environments with lower groundwater levels, high productivity, and in permanent bodies of water. On the other hand, there are sediments, which belong to ephemeral fluvial systems and eolic environments that had buried valleys and *salars*. In this case, the sediments would indicate higher levels of aridity conditions.

The geomorphology of the Puna wetlands also indicates changes in the environmental conditions. Organic terraces in ephemeral regime streams and old inactive floodplains can be found. These geological forms would indicate changes in the base level and depth of the phreatic layer. These evidences would allow proposing the following hypothesis: 1 – the fluvial systems had inconstant paleohydrologic conditions along the Holocene in the Puna desert, and 2 – fluvial wetlands had changes in geographic extension (wetlands transgression vs. wetland degradation-regression).

The data shows that paleohydrologic reconstructions based on sedimentological and palaeopedological records indicate abrupt hydrological fluctuations during the last 4,500 years (Fig. 44.1). Despite the differences due to local geological and geomorphologic patterns, the records show similar general trends that subdivide the late Holocene into six main paleohydrological phases: (1) wet phase *ca.* 4,500–1,600 ^{14}C year BP, (2) dry phase *ca.* 1,600–600 ^{14}C year BP, (3) wet phase *ca.* 600 ^{14}C year BP, (4) dry *ca.* 600–300 ^{14}C year BP, (5) wet *ca.* 300–150 ^{14}C year BP, (6) dry *ca.* 150 year – actual.

The wet phases are related to: (a) High water level, (b) peat downstream transgression, (c) hydromorphic and organic soils development, and (d) perennial fluvial systems. The wet phase that took place between 4,500 and 1,600 ^{14}C year BP is the most important of the late Holocene and there are registers of it in every studied location. There are traces of the wet phase of the 600 in Carachipampa and Ilanco stream, and of the one between 300 and 150 ^{14}C year BP in Los Mojones, Mirihuaca, Las Pitas streams and upper Ilanco catchment.

Between 1,600–600 ^{14}C year BP, 600–300 ^{14}C year BP, and after 150 ^{14}C year BP wetland sedimentation was interrupted by erosion phases when water level decreased and the wetlands began to reduce the surface area (wetland regression and segmentation, Figs. 44.1 and 44.2).

The O (δ^{18}O: –6.6 ± 0.2) and H (δ^2H: –50 ± 1) isotope studies in the modern wetlands seem to be the only supported by small groundwater reservoirs, which respond to high mountain (>4,500 m) precipitation. ^{3}H data (UT: 0.0 ± 0.6) indicates long periods of time (>60 year) in which the water remains in this arid environment.

44.5 Cultural Process and Land Management

The cultural settlement in the southern Puna of Argentina was complex (Olivera and Vigliani, 2000/2002; Olivera et al., 2006). By 3,000 year BP there is evidence of the introduction of domesticated plants. High fluvial valleys were used for hunting

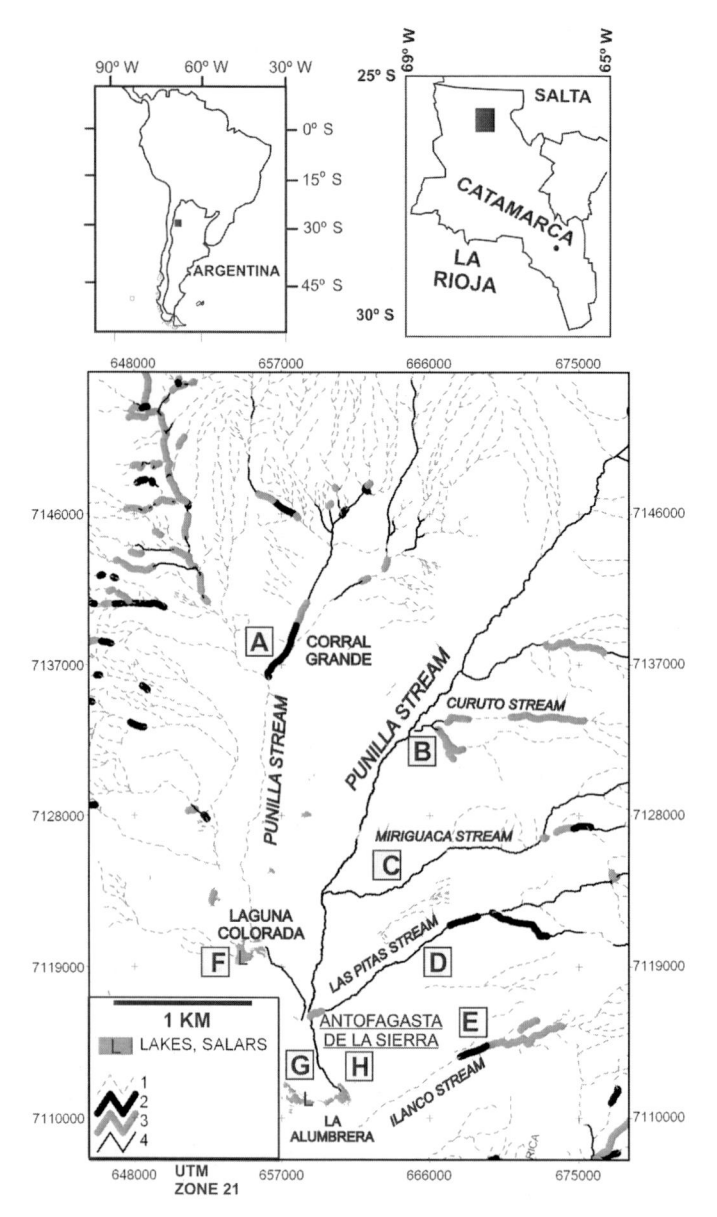

Fig. 44.1 Peat degradation evolution in lagoon systems (**a**) and rivers (**b**). (**a**) (*1*) Pleistocene Alluvial terrace, (*2*) paleocliff, (*3*) Holocene lacustrine terrace, (*4*) marginal lacustrine wetlands, (*5*) lagoon, *P*: organic sediments and paleosoils, *D*: diatomite sediments, *S*: Sands playa deposits. (**b**): (*1*) Late Pleistocene alluvial terrace, (*2*) Lower Holocene alluvial terrace, (*3*) Holocene peatland

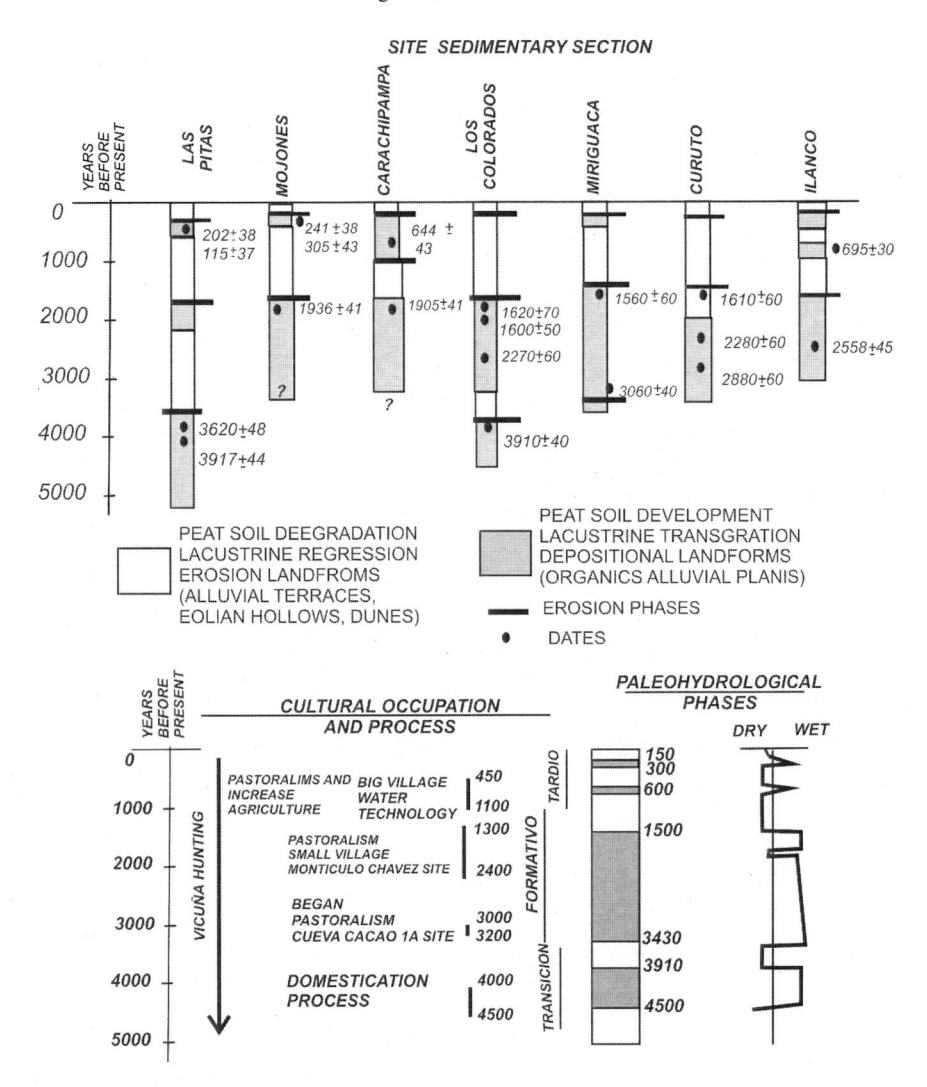

Fig. 44.2 Map of the Punilla river system showing location site and geographic features. References: 1 – Ephemeridal streams, 2 – Complete peat degradation, 3 – Peat in degradation, 4 – Peat development in permanent stream, *A*: Mojones section, *B*: Curuto section, *C*: Miriguaca section, *D*: Las Pitas section, *E*: Ilanco section, *F*: Los Colorados saline section, *G*: Bajo del Coypar and Coyparcito site and *H*: Casa Montículos Chávez site

(vicuña) and, probably, also to shepherd (llama). At 2,500 year BP, a small village appeared at the bottom of the basin (Casa Chavez Montículos, near Antofagasta de la Sierra village), with evidence of agriculture and cattle – although the latter seems to be more important (Fig. 44.3). New evidences under study suggest that another similar village (Las Escondidas) could have settled near Mirihuaca River (Escola et al., 2007). About 2,000 year BP the middle basins of lateral valleys (Curuto,

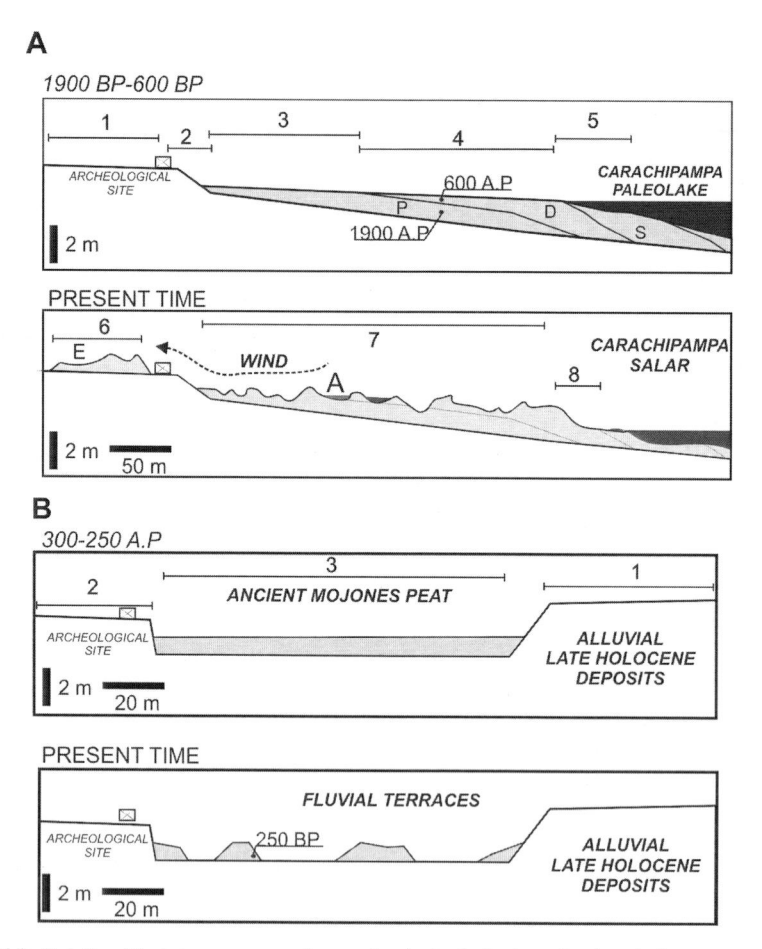

Fig. 44.3 Relationship between proxy data and paleohydrologic and cultural changes (Dates in ^{14}C year BP)

Miriguaca, Las Pitas streams) were populated, bringing an increase in agriculture, which was associated with archaeological materials from other regions (pre-Puna and lower valleys). The amount of water used for irrigation was considerably lower than the water available in the wetland systems at the time. We therefore speculate that there was not an impact on the wetland ecosystem.

From 1,100 year BP the village moved downstream (Bajo del Coypar and La Alumbrera). New technology for watering was introduced and cultural components changed (e. g., Belén pottery). The lower sector of Bajo del Coypar (Fig. 44.4a, b) was expanded and used for intensive and extensive agricultural production (surface area: 540 ha). Between 1,100 and 450 year BP new agricultural settlements appeared in the Antofagasta catchment at Miriguaca (35 ha), Curuto (8 ha), Mojones (Corral Grande site: 9 ha), Punta Calalaste (2.3 ha) and Campo Cortaderas (5.3 ha).

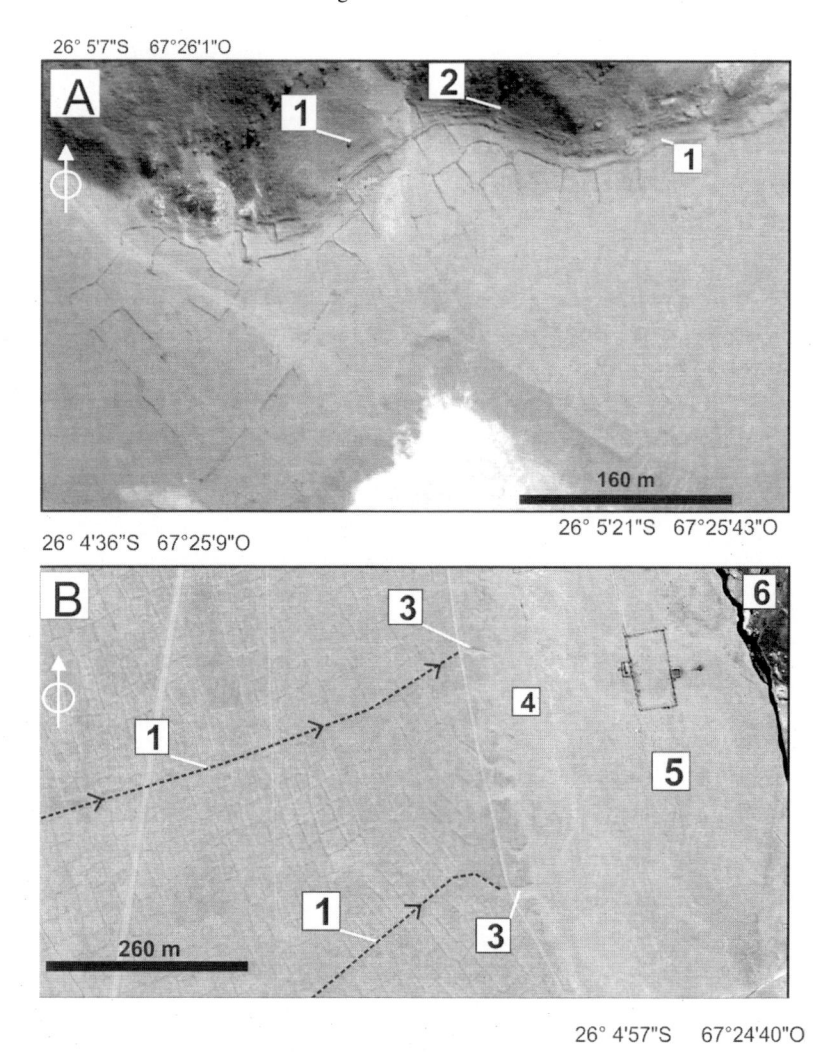

Fig. 44.4 Agricultural fields at Bajo del Coypar, Antofagasta de la Sierra (Austral Argentine Puna). (**a**) channel irrigations and terrace land development during Inca times; (**b**) piedmont agricultural sector whose development began during previous Belén times. *1* – Archaeological channels; *2* – agricultural terraces; *3* – gully erosion induced by irrigation; *4* – fan sedimentation by irrigation; *5* – land without agricultural activity; *6* – actual wetland

At *ca.* 450 years BP, the Incas coming introduced new improvements in the watering system (more sophisticated irrigation channels and terrace land developments, Fig. 44.4a) and reorganised the space with new settlements in Bajo del Coypar. But, later on during the Hispanic times, the agricultural system of Bajo del Coypar was abandoned (Fig. 44.3). At present, only a little part of the land (10 ha) is used to cultivate alfalfa for pasture. This land does not show severe salt accumulation, although an increase of 1,000 mho cm^{-1} in water conductivity was recorded.

Chemical analysis of cultivated land at present shows a greater content of organic material (0.6–0.5%) and phosphorous (6.4–14 ppm) than the land with no cultivation. The same occurs with potassium, sodium and soluble Ca + Mg, where a considerable increase in organic material and phosphorous takes place (op. cit. 2000).

There is an increase in the superficial horizon conductivity (0–20 cm) in regard to sub superficial horizon (20–40 cm) or uncultivated areas (300 mho cm^{-1}). This phenomenon affected some specific sections of the 540 ha of the cropland and did not have an impact on the crops performance, which were able to tolerate salinity (corn and fodder). All the above further proves that the land is far from becoming exhausted and its potential is improved by reasonable cultivation.

Crop fields started to be built and expanded during a mainly arid climatic phase. (1,600–600 year BP; 600–300 year BP) and the wet conditions lasted a short period of time (*ca.* 600 BP; 300–150 year BP). These data could suggest that the society decided to incorporate technology and working efforts (building of terraces, irrigation channels and dams) to avoid the decrease of resources, instead of moving to more suitable locations or reducing the number of members. This would be evident in the construction of long irrigation channels in Corral Grande 1 (4 km), Miriguaca (5 km), Bajo del Coypar, sector II (3 km), and the expansion of La Alumbrera urban site at the bottom of the basin.

Therefore, the sustenance conditions could not be worse not even at present time, which leads to think that the water storage basin, properly handled and associated to optimal crop criteria (species rotation, alternate soil rest, manure) would have assured the exploitation of most of the sections, total or partially.

These data are also supported by the cultivation of 1 ha within the archaeological fields, which is being carried out by one of the area's inhabitants (Mrs. Santos Claudia), who uses traditional techniques (Fig. 44.5). This parcel of land served as an experimental control in order to gather valuable data on the cultivation technique and the productivity of the farms. As a result, the following was established: 1 – Characteristics in the preparation and maintenance of the land, sowing and harvesting; 2 – Amount of man/hours used. 3 – Irrigation schedules and amount of water used; 4 – Productivity (and efficiency). This experience has been carried out during 6 years (2002/2007); two interruptions took place during summer (January and March) and the sheep were confined during winter (July) for pasturing; the average summer harvest is estimated in 5,000/5,500 kg ha^{-1}.

In contrast with other agrarian societies in arid and semi arid locations, human settlement did not produce salinisation and abandonment due to high land performance. The good quality of the soil, water and low levels of evapotraspiration (lower than in mesothermal valleys) allowed the soil not to be degraded as elsewhere. Likewise, the shortage of precipitations and associated erosive processes also helped to preserve the soil properties.

Fig. 44.5 View of experimental cultivated area at Bajo del Coypar archaeological site (Antofagasta de la Sierra, Austral Argentine Puna)

44.6 Conclusions

The analysis of the records leads us to the conclusion that, facing similar environmental challenges and changes (i.e., aridisation processes) the answers of human societies do not necessarily show equifinality. The complexity of the relationships between human society and the environment offers a variability that has to be analysed in each case with extreme care.

An important element is that in the Puna of Antofagasta de la Sierra there were six palehydrological phases of environments that indicated changes in humidity in late Holocene (Fig. 44.3).

The hydrologic cycle of these endorheic lakes (Antofagasta Lake and Laguna Colorada Salar) and pluvial wetlands is a direct response to effective moisture (precipitation–evaporation) fluctuations (Stoertz and Ericksen, 1974; Grosjean, 1994), and therefore, climate variability played a major role in sedimentological facies deposition and lakes level fluctuations (Fig. 44.1). Consequently, climate forcing has controlled the depositional history of both lakes during the Holocene. Therefore, the values in precipitation and temperatures, speculated from reconstruction of past atmospheric circulation, are considered as the main climatic forcing responsible for pluvial or dry phase creation (Markgraf et al., 1992) and wetlands and lakes transgression-degradation.

The chronological differences indicate that all the environments and landscapes did not have exactly the same palaeoenvironmental behaviour. On the other hand, some of them have more evolution than others in the sense of availability of water and biomass. This is due to the different water volume that they received according to their particular basin geomorphology (catchment surface over 4,000 m). This situation must have influenced in some way the settlement and subsistence of human groups.

We stress that the cultural responses to the paleoenvironmental changes involve economical and social decisions. The challenges of the environment (wetland degradation, reduction of water availability) may imply the introduction of more sophisticated technology (channel irrigation systems and land terrace levelling) that could allow its development in the long term. An example of this process is the development of Bajo del Coypar extensive agricultural system and the socio-political development of La Alumbrera (demographic increase and more energetic investment) during a dry phase (1,600–600 ^{14}C year BP).

We must be cautious in the sense that a complex and specialized cultivation system through irrigation in a desert area could be considered to be highly vulnerable and particularly sensitive to small, short cycle climate variations, typical of the micro-scale climate cycles in Puna de Atacama. Therefore, those periods with greater continued drought frequency, with an approximant duration of 4 or 5 years, could have had greater impact on the Puna than on other regions, specially taking into account that they would have affected the agricultural production as well as the natural forage productivity of the wetland. As a consequence, the large population, which was settled in the basin bed area at the time could have partially migrated to other less affected areas.

The soil quality deterioration due to salt is very unlikely since no significant variations were recorded either in salt levels or in cultivation potential. Furthermore, we must recall the high productivity of the experimental parcel and the present 10 ha of fields, which have been cultivated for decades. Therefore, according to the above-mentioned points, the important decrease in land productivity does not seem to have been the sole direct cause for the abandonment of the agricultural system.

Our opinion is that the possibilities for reaction of these societies to general climatic and environmental changes are not restricted only to one chance but cover a wide range of variability. In front of significant changes in the environment, human populations react with changes in their ways of relating to the environment in order to survive.

The economic changes may be intimately related to their technology, mobility strategies, socio-political organization and ideological system. The degree and form in which these different aspects are affected in each case, however, is variable and is mediated by decision making in each society. This implies that the new cultural and environmental phases may be more or less successful, involving its stability in time. In high risk, non-predictable environments such as high deserts, it is usually proposed that sociocultural changes have the main goal of diminishing the number of members. We think that society and environment are a mutually related equation but the results are not unique.

These paleohydrological study may influence decisions that have an impact on the current regional economy, for instance not using technology in the right way and therefore degrading the wetlands (depth of irrigation channels, unnecessary irrigation and saltiness due to water evaporation) or the possibility of reinforcing land reactivation of former productive areas (as in the case of the new alfalfa farmers in the town of Antofagasta de la Sierra).

Acknowledgments We would like to thank: 1 – The group of Antofagasta de la Sierra Archaeological Project for their important and unconditional support in the field and laboratory work, 2 – The people of Antofagasta de la Sierra for the hospitality and friendliness, 3 – Carlos Aschero group research (University of Tucumán, Argentina) for using existing ^{14}C data to reconstruct Holocene fluvial activity at Miriguaca section. 4 – The editors, especially Dr. Pandi Zdruli, for the invitation to participate in this book, 5 – the National Council of Scientifics and Techniques Researches (CONICET, Argentina), the National Agency of Scientifics and Techniques Promotion (ANPCYT, Argentina) and University of Buenos Aires (UBACYT Program, Argentina) which provides the researches with financial support.

References

Baied, C.A. and Wheeler, J.C. (1993). Evolution of High Andean Puna ecosystem environment, climate and culture change over the last 12,000 years in the Central Andes. Mountain Research and Development 13:145–156.

Caziani, S.M. and Derlindati, D.J. (1999). Humedales Altoandinos del Noroeste de Argentina. Su contribución a la biodiversidad regional. In: A.I. Malbares (ed.), Tópicos sobre Humedales Subtropicales y Templados de Sudamérica. MAB, Montevideo, Uruguay, pp. 1–15.

Coleman, M.L., Sheperd, T.J., Durham, J.J., Rouse, J.E. and Moore, F.R. (1982). A rapid and precise technique for reduction of water with Zinc for Hydrogen isotope analysis. Analytical Chemistry 54:993–995.

Escola, P., López Campeny, S., Martel, A., Romano, A. and Hocsman, S.. (2007). (MS). Prospecciones en un sector de quebrada de Antofagasta de la Sierra (Catamarca). Paper to the XVI Congreso Nacional de Arqueología Argentina. Facultad de Humanidades y Ciencias Sociales. Universidad Nacional de Jujuy, San Salvador de Jujuy.

Fernández, J., Markgraf, V., Panarello, H., Albero, M., Angiolini, F., Valencia, S. and Arriaga, M. (1991). Late Pleistocene-early Holocene Environment and climates, fauna, and human occupation in the Argentine Altiplano. Geoarchelogy 6 and 3:251–272.

Grosjean, M. (1994). Paleohydrology of the Laguna Lejía (north Chilean Altiplano) and climatic implications for late-glacial times. Paleogeography, Paleoclimatology, Paleoecology 109:89–100.

Grosjean, M., Cartajena, I., Geyh, M.A. and Núñez, L. (2003). From proxy data to paleoclimate interpretation: The mid-Holocene paradox of the Atacama Desert, northern Chile. Paleogeography, Paleoclimatology, Paleoecology 194:247–258.

Kulemeyer, J.A., Lupo, L.C., Kulemeyer, J.J. and Laguna, L.R. (1999). Desarrollo paleoecológico durante as ocupaciones humanas del precerámico del norte de la Puna Argentina. In: Beiträge zur quartären Landschftsentwicklung Sudamerikas. volume 65, Festschrift zum, Bamberg, pp. 233–255.

La Torre, C., Betancourt, J.L., Rylander, K.A., Quade, J. and Matthei, O. (2003). A vegetation history from the arid prepuna of northern Chile (22–23°S) over the last 13,500 years. Paleogeography, Paleoclimatology, Palaeoecology 194:223–246.

Liu, K.B., Reese, C.A. and Thompson, L.G. (2005). Ice-core pollen record of climatic changes in the central Andes during the last 400 year. Quaternary Research 64(II):272–278.

Maldonado, A., Betancourt, J.L., Latorre, C. and Villagrán, C. (2005). Pollen analyses from a 50,000 year rodent midden series in the southern Atacama Desert (25° 30)́. Journal of Quaternary Science 20(5):493–507.

Markgraf, V., Dodson, J.R., Kershaw, A.P., McGlone, M.S. and Nicholls, N. (1992). Evolution of Late Pleistocene and Holocene climates in the circum-South Pacific land areas. Climate Dynamics 6:193–211.

Miall, A.D. (1978). Lithofacies types and vertical profile models in braided river deposition: A summary. In A.D. Miall (ed.), Fluvial Sedimentology. Memoir 5, Canadian Society of Petroleum Geologists, Calgary, pp. 597–604.

Núñez, L., Grosjean, M. and Cartajena, I. (1999). Un ecorefugio oportunístico en la Puna de Atacama durante eventos áridos del Holoceno Medio. Estudios Atacameños 17:125–174.

Olivera, D., Tchilinguirian, P. and de Aguirre, M.J. (2006). Cultural and environmental evolution in the meridional sector of the Puna of Atacama during the Holocene. Change in the Andes: Origins of Social Complexity Pastoralism and Agriculture. Acts of the 14th UISPP Congress. BAR, UK, pp. 1–7.

Olivera, D. and Vigliani, S. (2000/2002). Proceso cultural, uso del espacio y producción agrícola en la Puna meridional argentina. CUADERNOS del Instituto Nacional de Antropología y Pensamiento Latinoamericano 19:459–481.

Panarello, H.O. and Parica, C.A. (1984). Isótopos del oxígeno en hidrogeología e hidrología. Primeros valores en aguas de lluvia de Buenos Aires. Asociación Geológica Argentina, Revista XXXIX(1–2):3–11.

Rodríguez, M.F., Rúgolo de Agasar, Z. and Aschero, C. (2006). El uso de las plantas en unidades domésticas del sitio arqueológico Punta de la Peña 4. Puna Meridional Argentina Revista Chungará 38(2):257–271.

Servant, M. and Servant-Vildary, S. (2003). Holocene precipitation and atmospheric changes inferred from river paleowetlands in the Bolivian Andes. Palaeogeography, Paleaeoclimatolgy, Palaeoecology 194:187–206.

Stoertz, G.E. and Ericksen, G.E. (1974). Geology of salars in Northern Chile. Geological Survey Professional Paper 811, Washington, DC.

Thompson, L.G., Mosley-Thompson, E. and Henderson, K.A. (2000). Ice-core palaeoclimate records in tropical South America since the Last Glacial Maximum. Journal of Quaternary Science 15:377–394.

Valero-Garces Blas, L., Delgado-Huertas, A., Navas, A., Edwards, L., Schwalb, A. and Ratto, N. (2003). Patterns of regional hydrological variability in central southern Altiplano (18–26°S) lakes during the last 500 years. Palaeogeography, Palaeoclimatology, Palaeoecology 194: 319–338.

Yacobaccio, H.D. and Morales, M. (2005). Mid-Holocene environment and human occupation of the Puna (Susques, Argentina). Quaternary International 132(1):5–14.

Chapter 45
Metal Pollution by Gold Mining Activities in the Sunchulli Mining District of Apolobamba (Bolivia)

T. Teran, Angel Faz Cano, M.A. Munoz, J.A. Acosta, S. Martinez-Martinez, and R. Millan

Abstract The Sunchullí district is located inside of the National Integrated Management Area of Apolobamba in Bolivia where intense gold mining activities have been carried out since ancient times to the present but with very poor gold extraction and mineral processing technology where mercury is still being used in the amalgamation processes. The aim of this work thus was to evaluate the heavy metal pollution by gold mining activities in the area. Soil samples were taken and Hg, Pb, Cu, Zn and Cd were analysed. Total DTPA and water extractable metals were also determined along with soil organic carbon, total nitrogen, calcium carbonates, exchangeable bases (Na, Mg, K), pH, electric conductivity and texture. The results show that the mining activities do not increase heavy metals levels, except for Hg that presents high concentrations surpassing the Holland reference levels. Also, the decreased TOC and TN in the zone located near the mining operation site could give rise to the extinction of the plant cover affecting biodiversity, soil stability and accelerating soil erosion processes.

Keywords Gold mining · Heavy metal pollution · Chemical analyses · Sunchulli district · Bolivia

45.1 Introduction

Since historical times metal mining activities in Bolivia have been some of the most important causes of environmental pollution. This is also the case of the National Integrated Management Area of Apolobamba (ANMIN of Apolobamba) in La Paz, Bolivia, where intense gold mining activities have been carried out for a long time

---

A. Faz Cano (✉)
Sustainable Use, Management, and Reclamation of Soil and Water Research Group, Department of Agriculture Science and Technology, Technical University of Cartagena, 30203 Cartagena, Spain
e-mail: angel.fazcano@upct.es

until the present, but with very little gold extraction and very primitive mineral processing technology.

Mercury is still being used in the amalgam processes of the mineral concentration. In the final phase, after being obtained the gold amalgam, this is burned outdoors to recover the gold. Data obtained in field show that the amount of Hg still being used varies from 0.25 to 6 kg month^{-1} of Hg per mine (Ramírez and Terán, 2002). Depending upon the mining company, Hg is not always recovered and 5–45% of the total Hg used can be released into the environment: soil, water and atmosphere (Malm et al., 1990; Maurice-Bourgoin et al., 1999).

Mercury passes directly to the atmosphere by the open air burning of the amalgam. It can also be released from the gold mines, and may provoke the pollution of aquatic ecosystems (Malm et al., 1990, 1995; Nriagu et al., 1992; Pfeiffer et al., 1993). Due to its high toxicity, mercury contamination is one of the most critical environmental problems in the area. Mercury interaction with the ecosystem depends on the ionic species, pH, redox potential of water, organic matter and climatic factors (Gochfeld, 2003). Hg is considered little dangerous in metallic form and in salts, but in organic compounds its action is very toxic for human health. Inhabitants of this area consume fish from lakes and rivers and use the waters for the livestock, domestic use, and irrigation. Fish and other animals may accumulate heavy metals in its tissues (Roldán, 1992).

The aim of this work was to evaluate the heavy metal pollution by gold mining activities in Sunchullí area. This area is a representative of the mining district in Apolobamba where mining activity is mainly related to gold extraction despite that the extracted mineral volume and the technology used are very obsolete compared to present technologies.

45.2 Description of the Study Area

The ANMIN of Apolobamba is located in the western part of La Paz (Fig. 45.1), in the Andean region. This area is bordering with Peru and ANMI Madidi. It is located between the coordinates 14°40′ and 15°10′ south latitude, 68°30′ and 69°20′ west longitude and covers an area of 483,743 ha. The area belongs to the Altoandina, Puna and Mountainous Humid Forest subregions of "Yungas". It has an altitude between 800 and 6,200 m.a.s.l (Parks Watch, 2007). The gold mining district selected for our study was Sunchullí that is located in the central part of ANMIN Apolobamba (Fig. 45.1).

Sunchullí water effluents end up to the Amazonian basin and hence these high Hg values could contribute to the environmental degradation of a much larger area. For example, total mercury concentrations measured in surface waters of the upper Beni basin varied, during the dry season, from 2.24 to 2.57 ng l^{-1} in the Madeira River at Porto Velho, and in the range 9.49–10.86 ng l^{-1} at its confluence with the Amazon (Laurence et al., 2000).

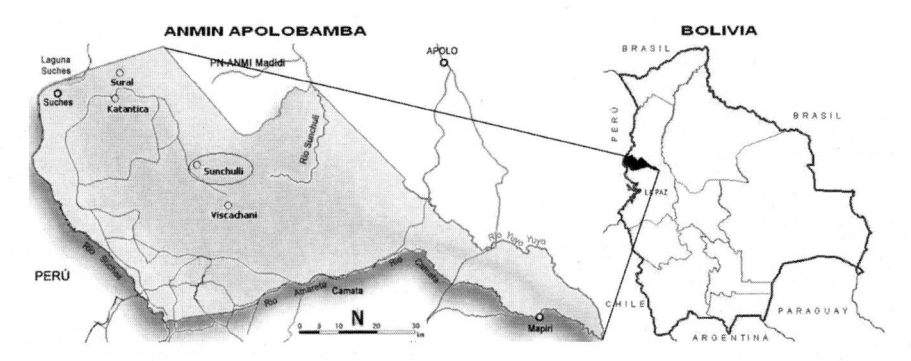

Fig. 45.1 Sunchullí location Map in ANMI Apolobamba

45.3 Materials and Methods

45.3.1 Sampling Procedures

Two representative zones were distinguished and sampled. The zone near the mining operation site was considered as affected by mineral extraction processes, while the far away zones represented the non-affected ones. In each zone, 3 plots (I, II and III) of 5 × 5 m were established. In each plot, 3 soil sampling points were selected in a random manner and analysed separately. In each sampling point, two samples were taken, one at the surface, from 0 to 5 cm depth (topsoil), and the other between 5 and 15 cm (subsurface). In addition a representative soil profile was taken for each zone. All samples were stored in polyethylene bags and stored at 4°C (Parker and Bloom, 2005).

45.3.2 Analysis

Samples were oven dried at 30°C, then sieved to a size less than 2 mm and then were grinded. Total organic carbon (TOC) (Anne, 1945) and total nitrogen (TN) were determined applying the method described by Duchaufour (Duchaufour, 1970). Cation exchange capacity (CEC) was obtained according to Chapman method (Chapman, 1965) while exchangeable bases (Na, Mg, K) were determined using Pratt (Pratt, 1965). pH in water and KCl were measured according to Peech method (Peech, 1965) and the electrical conductivity (EC) was determined through Bower and Wilcox method; the calcium carbonate was determined by the volumetric method of Bernard. The values of texture were obtained according to F.A.O.-I.S.R.I.C. (1990) system, based on the particle size percentages, determined with Robinson pipette method combined with sieving.

Total, DTPA and water extractable mercury concentration were determined using an advanced Mercury analyser (AMA-254, LECO Company) with a detection limit of 0.5 $\mu g\ kg^{-1}$ (Sierra et al., 2008), taking into account three replicates of each sample. AMA-254 is based on the thermal decomposition of the sample and the collection of the Hg vapour on a gold amalgamator. The analysis is performed on solid samples without any further preparation. Samples are initially dried at 125°C and then thermally decomposed at 550°C. Mercury vapour is then trapped on the gold amalgamator. Certified reference materials (CRM) were used to determine the accuracy and precision of the measurements and validate the applied methods. They were obtained from the Community Bureau of Reference (BCR). These certified reference materials consist of SRM 2709 (San Joaquin agricultural soil, 1.40 ± 0.08 mg kg^{-1} of Hg), BCR–CRM 62 (olive leaves, 0.28 ± 0.02 mg kg^{-1} of Hg).

Total Pb, Cu, Zn, and Cd were analyzed after nitric-perchloric acid digestion method. DTPA extractable metals were determined according to Lindsay and Norwell (1978) and Norwell (1984) methods, in order to evaluate heavy metals bioavailability. Water extractable metals were obtained by Ernst (1996) technique, with distilled water addition and agitation at room temperature and pressure, and filtered (pore diameter 0.45 μ). Measurements were carried out using an atomic absorption spectrometer AAnalyst 800 (PerkinElmer), with flame mix of air-acetylene atomization. Calibration of standard solutions was made by stock solution 1,000 mg l^{-1} Panreac Química.

45.4 Results

45.4.1 Physical and Chemical Parameters

Table 45.1 shows the summary of the physical and chemical parameters of Sunchullí soils.

Calcium carbonate is not present in Table 45.1, due to the fact that values were below the detection level.

45.4.1.1 Soils from the Plots

pH values in both, affected and non affected zones, were strongly acid (5.1–5.5) and very strongly acid (4.5–5.0) (Soil Survey Division Staff, 1993). This means that these soils are depleted from their bases, and it is also corroborated by the absence of calcium carbonate (Cobertera, 1993). pH values of the topsoil samples are slightly higher than subsurface soils. The difference between pH (H$_2$O) and pH (KCl) indicated that an acidity reserve in the soils exists. These soils were not saline (Soil Survey Division Staff, 1993).

TOC and TN concentrations of the soils from the far away plots were twice higher in comparison to the concentrations in soils from plots near these mining activities. TOC and TN values were higher in topsoil samples, what could be explained due to the vegetation contribution (the vegetable cover is 60% approximately). In both zone, C/N values showed a possible equilibrium between mineralization and humification favouring soil fertility (Cobertera, 1993).

Table 45.1 Physical and chemical parameters of Sunchullí soils (Mean: Standard Deviation, SD; minimum, min; maximum, max)

Samples	pH$_{H2O}$	pH$_{KCl}$	EC (dS m^{-1})	OC (%)	TN (%)	C/N	CEC (Cmol$_{(+)}$ kg^{-1})	Na$^+$ (Cmol$_{(+)}$ kg^{-1})	K$^+$ (Cmol$_{(+)}$ kg^{-1})	Mg^{+2} (Cmol$_{(+)}$ kg^{-1})	Clay (%)	Silt (%)	Sand (%)
Zone far away from the mining operation site (n = 9)
Topsoil[a]	5.33	4.42	0.31	12.33	1.05	11.72	41.92	0.35	1.55	2.77	17.80	36.41	45.80
SD	0.41	0.25	0.17	2.64	0.18	1.40	4.07	0.05	0.73	0.57	3.02	9.44	8.87
Min	4.81	4.13	0.09	8.78	0.79	8.44	35.40	0.29	0.55	2.29	14.57	24.60	33.59
Max	6.14	4.96	0.67	17.27	1.44	13.18	48.20	0.45	2.55	4.17	23.55	48.43	58.35
Subsurface[b]	4.99	4.03	0.17	7.90	0.71	10.93	37.11	0.32	0.42	1.38	15.39	35.27	49.34
SD	0.19	0.12	0.08	3.36	0.24	1.81	4.50	0.13	0.25	0.59	2.48	7.99	7.04
Min	4.58	3.81	0.09	4.20	0.51	6.80	30.47	0.00	0.00	0.00	12.14	26.20	38.32
Max	5.21	4.18	0.33	15.92	1.30	12.23	45.25	0.46	0.93	1.92	19.83	46.31	55.60
Zone near the mining operation site (n = 9)
Topsoil[a]	5.02	3.99	0.96	5.55	0.53	10.45	22.00	0.29	0.86	1.47	12.77	45.15	42.08
SD	0.47	0.32	1.51	2.00	0.16	1.05	5.45	0.18	1.02	1.27	2.80	4.26	5.43
Min	4.35	3.64	0.04	3.18	0.33	8.45	15.15	0.12	0.16	0.38	8.77	36.56	34.43
Max	5.56	4.70	4.07	9.95	0.86	11.88	32.84	0.68	3.32	4.42	16.87	50.97	52.13
Subsurface[b]	4.81	3.77	1.33	4.85	0.48	10.18	21.19	0.26	0.39	0.69	12.67	45.08	42.25
SD	0.41	0.13	2.33	1.49	0.10	1.89	4.69	0.10	0.25	0.19	1.58	4.96	6.19
Min	4.29	3.60	0.03	2.79	0.27	6.29	12.77	0.14	0.14	0.46	10.64	35.80	33.55
Max	5.29	4.00	5.88	7.13	0.59	12.07	28.14	0.44	0.96	1.07	14.68	52.10	53.30

(continued)

Table 45.1 (Continued)

Samples	pH$_{H2O}$	pH$_{KCl}$	EC (dS m^{-1})	OC (%)	TN (%)	C/N	CEC (Cmol(+) kg^{-1})	Na$^+$ (Cmol(+) kg^{-1})	K$^+$ (Cmol(+) kg^{-1})	Mg^{+2} (Cmol(+) kg^{-1})	Clay (%)	Silt (%)	Sand (%)
Profile far away from mining operation site
A1
(0–16 cm)	4.94	3.98	0.13	3.71	0.51	7.35	39.57	0.33	2.56	3.08	18.34	47.42	34.24
A2
(16–24 cm)	5.07	3.89	0.06	2.07	0.29	7.21	32.45	0.35	0.51	1.53	14.53	45.95	39.51
C/R
(24–43 cm)	5.16	3.83	0.07	1.23	0.12	6.94	30.67	0.43	0.37	1.57	13.69	49.70	36.61
Profile in zone near the mining operation site
A1
(0–10 cm)	4.49	3.82	0.12	2.77	0.55	5.01	22.59	0.21	0.79	0.61	14.30	50.77	34.93
A2
(10–26 cm)	4.66	3.86	0.05	0.83	0.44	4.88	16.79	0.25	0.43	0.65	13.58	49.58	36.64
C/R
(26–64 cm)	5.84	4.27	0.02	0.67	0.31	4.18	15.35	0.12	1.42	0.30	5.23	46.30	48.48

[a] 0–5 cm
[b] 5–15 cm

However this ratio between the TOC and the TN is lower for the affected zone samples due to the most important nitrogen source in the soil and mineralization of organic matter contributed by the scarce plant cover (Calvo et al., 1992). In the future, the decreased TOC and TN in the zone located near the mining operation site could give rise to the extinction of the plant cover affecting the soil stability and promoting soil erosion processes.

CEC showed high values, especially in the far away plots, what can be related to the TOC higher concentrations, the clay content and its composition by mainly vermiculites (Honorato, 2000). There are not important CEC differences between topsoils and subsurface ones. Furthermore the availability of nutrients (exchangeable bases Na, Mg, K) is normal for these soil types, although there is a slight Mg deficiency (Cobertera, 1993).

Soils had either sandy or silt sandy soil texture (F.A.O.-I.S.R.I.C., 1990), which makes the soil moderately permeable, promoting the metals mobility.

45.4.1.2 Profiles

pH values of the soil profiles were strongly acid (5.1–5.5) and very strongly acid (4.5–5.0) (Soil Survey Division Staff, 1993) and showed a light increment with depth. These soils were non-saline (Soil Survey Division Staff, 1993) and did not present calcium carbonate in both topsoil and subsurface horizons.

TOC and TN concentrations decreased with depth; which is related to the higher inputs of organic matter due to the contribution of vegetation at the soil surface.

C/N ratio decreased with depth, what showed a tendency of the humus mineralization in deeper horizons (Cobertera, 1993). CEC shows high values (Cobertera, 1993), especially in the zone located far away from the mining activities; this could be related to the high TOC concentration and also to the composition of the clays. There are not important differences of CEC between topsoil and subsurface.

45.4.2 Heavy Metals

In order to compare the results, Belgium reference levels for Cd (0.8 mg kg^{-1}), Cu (17.0 mg kg^{-1}), Pb (40.0 mg kg^{-1}) and Zn (62.0 mg kg^{-1}) are considered (BWRHABTGG, 1995), while in the case of Hg the Holland reference levels are considered (0.3 mg kg^{-1}) (NMHPPE, 1994).

45.4.2.1 Soils from the Plots

In the zone located near the mining operations site, Cu, Pb and Hg concentrations were higher than in the far away zone; on the contrary, Zn and Cd values were lower (Table 45.2 and Fig. 45.2), what is related to the background levels of the parent material.

Table 45.2 Heavy metals in Sunchulli soils (Mean; Standard Deviation, SD; minimum, min; maximum, max)

Samples	Total metals (mg kg^{-1})					DTPA extractable metals (mg kg^{-1})					Water extractable metals (mg kg^{-1})				
	Cu	Zn	Cd	Pb	Hg	Cu	Zn	Cd	Pb	Hg	Cu	Zn	Cd	Pb	Hg
Zone far away from the mining operation site ($n = 9$)															
Topsoil[a]	14.43	154.72	0.53	16.49	0.85	2.17	24.78	0.05	4.45	0.01	0.04	0.17	1.0E-04	4.2E-03	1.2E-03
SD	2.78	49.98	0.85	2.26	0.13	0.30	5.73	0.01	1.11	4.6E-03	0.06	0.13	9.0E-05	3.4E-03	1.1E-03
Subsurface[b]	15.89	107.07	1.09	21.99	0.37	1.69	8.56	0.01	4.24	0.01	0.05	0.19	1.5E-04	6.5E-03	2.4E-04
SD	2.80	33.33	2.54	8.01	0.11	0.33	2.84	0.01	1.54	4.6E-03	0.06	0.09	1.6E-04	5.0E-03	8.7E-05
Zone near the mining operation sites ($n = 9$)															
Topsoil[a]	21.43	77.17	0.13	33.02	0.87	2.18	7.39	0.03	10.64	0.02	0.03	0.22	2.0E-04	4.5E-03	1.5E-03
SD	4.15	26.37	0.06	28.17	0.73	0.79	8.87	0.03	20.80	0.01	0.03	0.18	1.6E-04	4.2E-03	1.2E-03
Subsurface[b]	24.28	81.23	0.09	25.21	0.32	1.80	2.28	0.02	3.72	0.01	0.04	0.12	1.6E-04	5.1E-03	3.7E-04
SD	4.38	33.82	0.04	3.76	0.23	0.22	1.29	0.01	1.91	3.5E-03	0.05	0.17	7.9E-05	5.4E-03	8.7E-05
Profile far away from mining operation site															
A1 (0–16 cm)	11.26	59.04	0.14	17.94	0.18	2.05	6.12	0.00	5.58	0.004	3.9E-01	4.9E-01	<DL	4.8E-02	6.2E-04
A2 (16–24 cm)	10.01	68.40	0.10	27.48	0.05	2.77	2.84	<DL	11.66	0.003	2.3E-01	4.2E-01	<DL	3.9E-02	1.0E-04
C/R (24–43 cm)	3.54	59.95	0.11	15.63	0.03	0.99	1.42	<DL	4.18	0.002	1.5E-01	2.8E-01	<DL	1.3E-01	5.4E-05

Table 45.2 (Continued)

Samples	Total metals (mg kg^{-1})					DTPA extractable metals (mg kg^{-1})					Water extractable metals (mg kg^{-1})				
	Cu	Zn	Cd	Pb	Hg	Cu	Zn	Cd	Pb	Hg	Cu	Zn	Cd	Pb	Hg
Profile in zone near the mining operation site															
A1 (0–10 cm)	19.24	67.25	0.18	21.52	0.48	2.44	5.18	0.01	2.75	0.004	3.2E-03	2.7E-01	<DL	5.4E-03	9.1E-04
A2 (10–26 cm)	19.25	61.63	0.22	21.14	0.23	2.15	3.81	0.02	2.85	0.003	1.1E-03	2.3E-01	<DL	2.4E-03	1.2E-04
C/R (26–64 cm)	19.87	69.75	0.22	23.15	0.01	2.59	2.15	0.01	2.61	0.002	<DL	2.0E-01	<DL	<DL	6.0E-05

<DL Cd: Below the detection level (0.002 µg kg^{-1}).
[a]0–5 cm
[b]5–15 cm

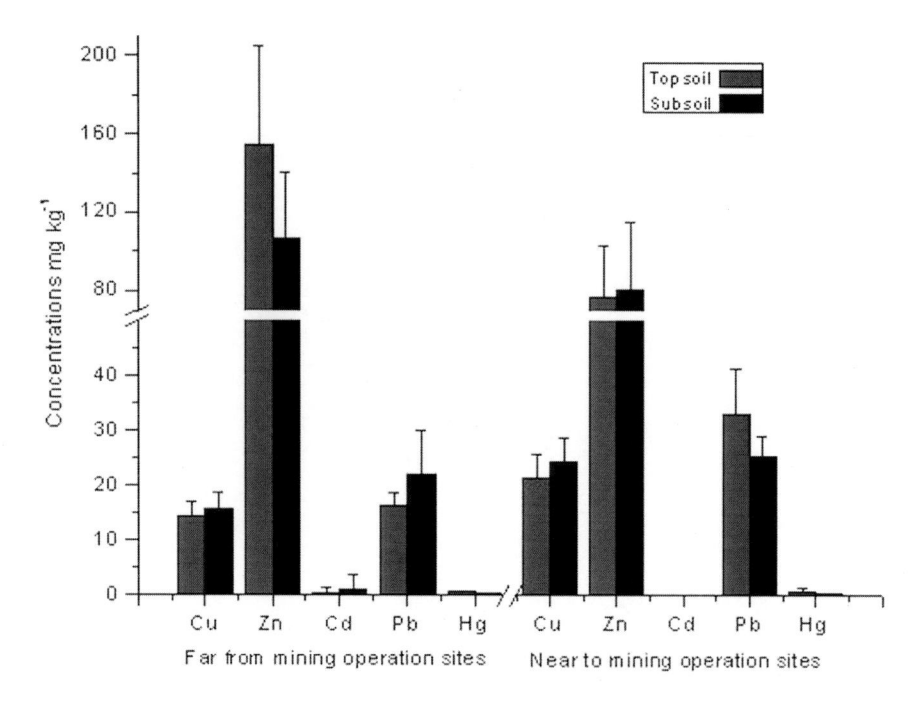

Fig. 45.2 Total metals in zones near and far from mining operation sites in Sunchullí

Cd and Pb values in both zones did not surpass the reference levels mentioned. The following values were recorded: for total Zn (154.7, 107.1 mg kg^{-1} and 77.2, 81.2 mg kg^{-1}, values for topsoil and subsurface in the areas located away and near the mining operation, respectively), total Cu (21.4 and 24.3 mg kg^{-1}, values for topsoil and subsurface respectively in the area located near the mining operation) and total Hg (0.8, 0.4 mg kg^{-1} and 0.9, 0.3 mg kg^{-1}, values for topsoil and subsurface in the areas located away and near the mining operation, respectively) values surpass those levels. pH values (strongly acid and very strongly acid) and texture (sandy or silt sandy) could permit the mobility of some metals, especially Zn (Adams and Sander, 1984).

The normal Hg concentrations in soils are in the average of 0.01–0.03 mg kg^{-1} (Senesi et al., 1999), however in places where Hg is extracted, the concentrations are higher. This is the case the Azogue Valley in Almeria (Spain) where an average Hg concentration of 357.30 mg kg^{-1} was found (Mendoza et al., 2005); also in the Callao (Venezuela) the Hg levels were in the range from 0.05 to 20.06 mg kg^{-1} (Carrasquero and Adams, 2002). In our case, although the Hg values were not so high it is possible that its presence in both zones could be caused by the deposition of the Hg^0 from the evaporation carried out after the amalgamation (Carmouze et al., 2001). This explains the difference between topsoil and subsurface concentrations (0.8 and 0.9 mg kg^{-1}, total Hg values for topsoil in both areas, away and near the mining operation, respectively).

These results clearly show that the mining activities do not increase heavy metals (Cu, Cd, Pb and Zn) levels; however Hg presents high concentrations surpassing the Holland reference levels.

45.4.2.2 Profiles

All the metals concentrations from the horizons of the profile taken in the zone far away from mining operations site did not surpass the reference levels mentioned.

The profile taken near the mining operation site presented total concentrations of Cu (A1: 19.2 mg kg^{-1}, A2: 19.2 mg kg^{-1} and C/R: 19.9 mg kg^{-1}) and Zn (A1: 67.2 mg kg^{-1}, A2: 61.6 mg kg^{-1} and C/R: 69.8 mg kg^{-1}) that surpassed the reference levels; on the other hand Cd and Pb were below those reference values. Total Hg value in the surface horizon (0.48 mg kg^{-1}) was higher than in the subsurface horizons; Hg concentration in C/R horizon was almost null (0.01 mg kg^{-1}), what suggests that this element is not a constituent of the rock materials. As it has already been mentioned for the soils of the plots, these values can be increased by the deposition of the Hg^0.

DTPA and water extractable metals were lower than total metals in all cases as we could expect and the Cd values were lower than the detection limit. In all samples, DTPA metals were below 15% (Cu), 16% (Zn), 10% (Cd), 10% (Pb) and 2.5% (Hg) related to total metals; while water extractable metals were below 0.4% of the total concentrations. Hg bio-available values were also low; this is very important due to the relationship between this aspect and the Hg entrance into the trophic chain.

45.5 Conclusions

Soils were strongly acid and very strongly acid but and non-saline, in both zones. TOC and TN concentrations of the far away zone were twice higher in comparison to these concentrations in the zone near the mining activities. In this last zone, the extinction of the plant cover could affect the soil stability and initiate soil erosion processes. TOC and TN concentrations in the studied profiles decreased with depth: This could be related to the increase of organic matter in surface horizon due to the contribution by plants.

C/N values in plots showed a possible equilibrium between mineralization and humification, while in the a few profiles there is a tendency of the increased humus mineralization with depth as this ratio decreases from surface to deeper horizons.

CEC shows high values especially in the zone located far away from the mining activities; this could be related to the high TOC concentrations and the composition of the clay. Heavy metal results show that the mining activities do not increase heavy metals (Cu, Cd, Pb and Zn) levels, except for Hg that presents high concentrations surpassing the Holland reference levels. It occurs due to the deposition of the Hg^0 from the evaporation carried out after the amalgamation.

Acknowledgments This work was financed by Education and Science Ministry of Spain with Formation of Personal Investigator (FPI) Grant (Project CTM2006-02812: Natural resources and biodiversity assessment in Apolobamba (Bolivia): Risk assessment by mining activities and remediation of affected areas) and Secretary of Exterior Action and Relation with the Union European of Autonomous Community of Murcia Region.

References

Adams, T.M. and Sander, J.R. (1984). The effect of incubation of soil solution displaced from for soils treated with zinc, copper or nickel loaded sewage sludge. In: P. L'Hermite and H. Ott (eds.), Processing and Use of Sewage Sludge. Reidel, Dordrecht.

Anne, P. (1945). Sur le dosage rapide du carbone organique des sols. Annales Agronomique 2: 161–172.

BWRHABTGG. (1995). Besluit van de Vlaamse Regering Houdende Achtergrondwaarden. Bodernsaneringsnomen en Toepassingen van Gereinigde Grond. Ministry of Environment and Employment Brussels. Belgium.

Calvo, R., Macias, F. and Riveiro, A. (1992). Aptitud Agronómica de los suelos de la provincia de La Coruña. Diputación Provincial de La Coruña, La Coruña, pp. 88–100.

Carmouze, J.P., Lucotte, M. and Boudou, A. (2001). Mercury in Amazon. Importance of Human and Environment, Health Hazards. Synthesis and Recommendations. IRD edn. IRD, Paris, Francia, pp. 1–37.

Carrasquero, A. and Adams, M. (2002). Comparación de métodos para el análisis de mercurio en suelos procedentes de El Callao, Estado Bolívar, Venezuela. Asociación Interciencia. INCI. [online]. 27(4): 191–194.

Chapman, H.D. (1965). Cation exchange capacity. In: C.A. Black (ed.), Methods of Soil Analysis. Volume 2, American Society of Agronomy, Madison, Wisconsin, USA, pp. 891–900.

Cobertera, E. (1993). Edafología Aplicada: Suelos, Producción Agraria, Planificación Territorial e Impactos Ambientales. Ed Cátedra, Spain.

Duchaufour, Ph. (1970). Precis de Pedologie. Masson, Paris, 481 pp.

Ernst, W.H.O. (1996). Bioavailability of heavy metals and decontamination of soils by plants. Applied Geochemistry 11:163–167.

F.A.O.-I.S.R.I.C. (1990). Guidelines for Soil Description. 3rd edn. Soil Resources Managemente ando Conservation Service Land and Water Development Division, Rome, 69p.

Gochfeld, M. (2003). Cases of mercury exposure bioavailability and absorption. Toxicology and Environmental Safety 56:174–179.

Honorato, R. (2000). Manual de Edafología. Ediciones Universidad Católica de Chile, Santiago, Chile.

Laurence, M., Quiroga, I., Chincheros, J. and Courau, P. (2000). Mercury distribution in waters and fishes of the upper Madeira rivers and mercury exposure in riparian Amazonian populations. Science of the Total Environment 260:73–86.

Lindsay, W. and Norwell, W. (1978). Development of a DTPA soil test for Zn. Fe. Mn. and Cu. Soil science Society of America Journal 42:421–428.

Malm, O., Branches, F.J.P. and Akagi, H. (1995). Mercury and methylmercury in fish and human hair from the Tapajós river basin. Brazil. The Science of the Total Environment 175:141–150.

Malm, O., Pfeiffer, W.C., Souza, C.M.M. and Reuther, R. (1990). Mercury pollution due to gold mining in the Madeira river basin, Brazil. AMBIO 19(1):11–15.

Maurice-Bourgoin, L., Quiroga, I., Guyot, J.L. and Malm, O. (1999). Mercury pollution in the upper Beni River basin, Bolivia. AMBIO 28(4):302–306.

Mendoza, J.L., Navarro, A., Viladevall, M. and Doménech, L.M. (2005). Caracterización y tratamiento térmico de suelos contaminados por mercurio. In: J.A. López-Geta, J.C. Rubio y M. Martín Machuca (eds.), VI Simposio del Agua en Andalucía. IGME. pp. 1077–1088.

NMHPPE. (1994). Netherlands Ministry of Housing. Physical Planning and Environment. Leidschendam, Holland.

Norwell, W.A. (1984). Comparison of chelating agents as extractans for metals in diverse soil material. Soil science Society of America Journal 48:1285–1292.

Nriagu, J.O., Pfeiffer, W.C., Malm, O., Souza, C.M.M. and Mierle, G. (1992). Mercury pollution in Brazil. Nature (London) 356:389.

Parker, J. and Bloom, N. (2005). Preservation and storage techniques for low-level aqueous mercury speciation. Science of the Total Environment 337:253–263.

Parks Watch. (2007). Fortaleciendo parques para proteger la Biodiversidad 2007. In: www.parkswatch.org

Peech, M. (1965). Hydrogen-ion activity. In: C.A. Black (ed.), Methods or Soil Analyses. Volume 2, American Society of Agronomy, Madison, Wisconsin, USA, pp. 914–916.

Pfeiffer, W.C., Lacerda, L.D., Salomons, W. and Malm, O. (1993). Environmental fate of mercury from gold-mining in the Brazilian Amazon. Environmental Review 1:26–37.

Pratt, M. (1965). Potassium and sodium. In: C.A. Black (ed.), Methods of Soil Analysis. Volume 2, American Society of Agronomy, Madison, Wisconsin, USA, pp. 1022–1033.

Ramírez, V. and Terán, N. (2002). Informe de trabajo de campo: Inventariación de actividades mineras e impactos ecológicos y socioeconómicos en el Área Natural de Manejo Integrado Apolobamba. La Paz. Conservación Internacional/CEPF.

Roldán, G. (1992). Fundamentos de Limnología Tropical. Ed. Universidad de Antioquia, Medellín, 529p.

Senesi, G.S., Baldassare, G., Senesi, N. and Radina, B. (1999). Trace elements inputs into soils by anthropogenic activities and implications for human health. Chemosphere 39:343–377.

Sierra, M.J., Millán, R. and Esteban, E. (2008). Potential use of solanum melongena in agricultural areas with high mercury background concentrations. Food and chemical toxicology 46: 2143–2149.

Soil Survey Division Staff. (1993). Soil Survey Manual. USDA Handbook 18. U.S. Government Printing Office, Washington, DC, 435 pp.

Chapter 46
Areas Degradated by Extraction of Clay and Revegetated with *Acacia mangium* and *Eucalyptus camaldulensis*: Using Soil Fauna as Indicator of Rehabilitation in an Area of Brazil

Cristiane Figueira da Silva, Eliane Maria Ribeiro da Silva, William Robertson Duarte da Oliveira, Maria Elizabeth Fernandes Correia, and Marco Antonio Martins

Abstract Biological indicators of soil quality can be utilised to evaluate the degree of recovery of a degraded area. The aim of this work was to evaluate the edaphic fauna, in areas degraded by the extraction of clay and revegetated with *Acacia mangium and Eucalyptus camaldulensis* both in monoculture and intercropped systems. Samples of soil and litter were collected in August 2006, utilising a square wooden probe. The samples were placed in a battery of Berlese-Tullgren-type extractors, and after 15 days, the extracted organisms were collected in glass tubes containing a 50% alcohol solution. The specimens were transferred to a 70% alcohol solution in the laboratory, and the organisms were screened with the help of a binocular loupe and separated into large taxonomic groups. Results showed that the type of plant cover influenced the soil fauna density as well as diversity. The acacia-eucalyptus intercropping was the system that gave better distribution of organisms within each group (equability) and consequently greater diversity, compared to monocultures and to the area with spontaneous vegetation.

Keywords Biological indicators · Arboreal legumes · Acacia · Eucalyptus · Revegetation · Mineralisation

46.1 Introduction

The ceramic industry of the state of Rio de Janeiro, Brazil is composed of more than 300 businesses spread throughout the state. The main producing site is located at

E.M.R. da Silva (✉)
Embrapa Agrobiologia, Seropédica, CEP 23890-000, RJ, Brazil
e-mail: eliane@cnpab.embrapa.br

Campos dos Goytacazes which, according to estimates of the syndicate of ceramists, consists of 120 ceramics producing 5 million pieces per day, generating 4,500 direct jobs and 15,000 indirect jobs (Valicheski et al., 2006).

Despite being of high socio-economic importance, this activity generates a number of environmental threats, due to the removal of soil. It is estimated that 7,000 m^3 of soil are removed daily (Schiavo, 2005), which impacts water, soil and landscape overall.

Brazil's legislation legally requires that areas degraded by the extraction of clay have to be recovered. One process that has been economically viable for their recovery is revegetation with species of the family Leguminosae (Franco et al., 1995). Studies have shown that these species develop well in soils of low natural fertility and in degraded areas where the surface layers have been removed, either by the erosive processes or mineral exploration (Franco and Faria, 1997) since they favour natural succession.

One of the characteristics that make these legumes suitable for such a process is their ability to engage in symbiotic relationships with bacteria known as *rhizobium*. Symbiosis allows atmospheric nitrogen to be converted and transferred to the plant in forms that can be assimilated, through the process referred to as biological nitrogen fixation (BNF) (Franco et al., 1995). Besides, many legumes that form nodules are also symbiotic with mycorrhizal fungi. The association with these fungi could generate additional nutritional benefits, such as greater absorption of nutrients. It could bring non-nutritional benefits as well, such as improved soil aggregation (Rillig and Steinberg, 2002).

The planting of eucalyptus, intercropped with legumes, favours the recovery of sites used for clay extraction, as it could also supply the energy demand for the production of ceramics. According to Borlinil et al. (2005), the fuel most utilized by the red ceramic industries is firewood, with a mean consumption of 0.4 m^3 h^{-1} in a density of 0.4 t m^{-3}.

Therefore, considering the great importance of eucalyptus for the ceramic industry in the Norte Fluminense region and the added value of legume species for the recovery of degraded areas, their intercropping results offer remarkable economic and environmental benefits.

The environmental benefits provided by the recovery of a degraded area could be evaluated using indicators of soil quality (Kennedy and Parpendick, 1995). These indicators are based on physical, chemical and biological characteristics, which can be measured to monitor changes in soil quality (Santana and Bahia Filho, 1999). As the soil fauna and litter composition show great diversity and rapid capacity for reproduction, they are excellent bio-indicators. Moreover, their properties or functions indicate and determine the status of soil degradation level (Knoepp et al., 2000). Thus, the study of the community of edaphic fauna can help in determining the influence of the quantity and mainly the quality of vegetal material added to the soil (Correia and Andrade, 1999).

46.2 Materials and Methods

The experiment was set up in August 2002, in a clay extraction pit, belonging to the ceramic company Stilbe Ltda., in Poço Gordo, in the municipality of Campos dos Goytacazes (Rio de Janeiro-Brazil) (Schiavo, 2005). The area was revegetated with *Acacia mangium* and *Eucalyptus camaldulensis* in monoculture and intercropped, utilizing a spacing of 2 × 3 m.

Prior to the implementation of the experiment, the area was ploughed twice and tilled once, and Araxa rock phosphate was applied in the pits at a dose of 100 mg kg^{-1} soil. The legume (*Acacia mangium*) was inoculated with arbuscular mycorrhizal fungi (*Glomus macrocarpum*, *Glomus etunicatum* and *Entrophospora colombiana*) and a specific strain of rhizobio, Br 5401. Two lines of plants of *E. camaldulensis* were placed between the treatments, which constituted the borders. The experimental plot consisted of 16 plants.

The experimental design utilized was a randomised block designed with four treatments and three repetitions. The following treatments were used: monocultures of *A. mangium* and *E. camaldulensis*; intercropping of *A. mangium* × *E. camaldulensis*; and control treatment, represented by the area of the pit with spontaneous vegetation (grasses).

To evaluate the community of soil fauna, two soil samples were collected per plot, utilizing a 25-cm^2 wooden probe, which delimited the area sampled. Each sample was separated into sub-samples of litter and topsoil (0–5 cm), except for the control treatment where there was no litter. The sampling was carried out in August 2006 (dry period). To separate the fauna from the soil, the samples were processed in a battery of Berlese-Tullgren type extractors (Garay, 1989), which were modified.

After 15 days, the extracted organisms were collected in glass tubes containing 50% alcohol solution. Samples were screened in the laboratory with the help of a binocular loupe, and the organisms were separated into large taxonomic groups. After the screening and the determination of the total number of individuals, the means ± SE were calculated and the number of individuals m^{-2} for each treatment was estimated.

Statistical analysis of the means for the taxonomic groups was carried out using Tukey's test at 5% probability. Diversity was calculated based on the Shannon index using the formula:

$$H = -\sum p_i . \log_2 p_i$$

where:

$p_i = n_i/N$;
n_i = importance value for each species or group;
N = total of importance values.

Equability was determined by the Pielou index using the formula:

$$e = H/\log_2 R,$$

where:

H = Shannon index;
R = number of taxonomic species or groups.

The vertical distribution of the organisms was determined using the chi-squared test (X^2). Cluster analysis was performed using Ward's linkage method, and measure of distance by Pearson's r correlation. These tests and analyses were carried out with the aid of the statistical programmes SAEG-5.0 (System of Statistical Analyses and Genetics – University Federal de Viçosa) and ESTAT – 1.0 (System for Statistical Analysis – University Estadual Paulista, FCAV, Campos de Jaboticabal).

46.3 Results and Discussion

The mean richness (or number of groups) and total organisms found in the plantings and in the area with spontaneous vegetation are given in Table 46.1.

According to Stork and Eggleton (1992), both total richness and mean richness of species can be considered as a general measure of biodiversity. In the plantings, there was greater mean richness of taxonomic groups in relation to the area with spontaneous vegetation (Fig. 46.1).

Rovedder et al. (2004) pointed out that the presence of arboreal species creates propitious conditions for colonisation by these organisms, through the deposition of litter on the soil, which acts as a source of food and shelter, easing the variations of soil temperature and moisture and creating hence a more favourable microclimate. Consequently, the available food resources, as well as the structure of the microhabitat produced by the plantings, are positive factors that allow colonisation by various

Table 46.1 Density, total richness, mean richness, and Shannon and Pielou indices of the soil fauna with different plant covers. Intercropping = *A. mangium* × *E. camaldulensis*; Control = area with spontaneous vegetation

Treatment	Density (ind. m^{-2} ± SE)	Mean richness[a]	Total richness	Shannon index	Pielou index
Acacia	2,029 ± 301	15 a	20	2.24	0.51
Eucalyptus	1,120 ± 454	9 ab	14	2.39	0.63
Intercropping	1,333 ± 534	11 a	18	2.75	0.66
Control	203 ± 89	3 b	5	1.41	0.60

[a] Means followed by the same letter in the column did not differ according to Tukey's test at 5%.

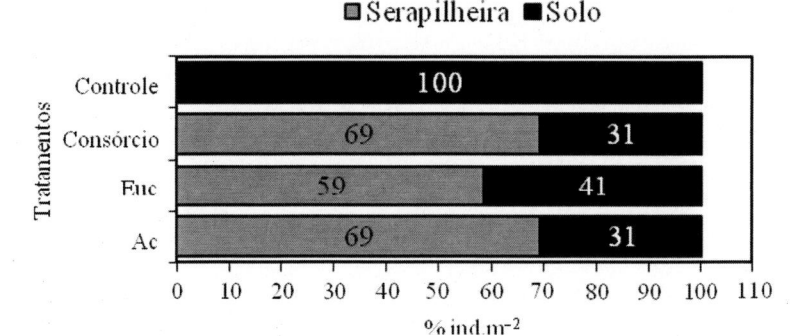

46 Soil and Forest Rehabilitation in an Area of Brazil

☐ Serapilheira ■ Solo

Tratamentos	Serapilheira	Solo
Controle		100
Consórcio	69	31
Euc	59	41
Ac	69	31

% ind.m^{-2}

Fig. 46.1 Vertical distribution (soil-litter) of edaphic fauna in different plantings and in the area with spontaneous vegetation (control). *A. mangium* (Ac); *E. camaldulensis* (Euc)

species of soil fauna with different survival strategies (Nunes et al., 2008). In addition, the presence of an arboreal legume creates favourable conditions for the fauna as well, since it is expected that the litter deposited has a greater level of N and a lower C:N ratio, which favours soil fauna (Dias et al., 2007).

In laboratory experiment with diplopods Correia (2003) observed a consumption of litter from legumes greater than that from eucalyptus, mainly due to the lower C:N ratio of legumes. On the other hand, in the area with spontaneous vegetation, the little plant cover and intensity of erosive processes, impede the establishment of a bio diverse ecosystem (Table 46.1).

Although there were no significant differences between plantings with respect to mean richness, the Shannon index showed that acacia-eucalyptus intercropping had a greater diversity of organisms. In addition, statistics showed that there is a transition between plantings and the area with spontaneous vegetation, where eucalyptus although showing a greater number of organisms, did not differ statistically from the area with spontaneous vegetation (Table 46.1).

In relation to abundance of organisms, the monoculture of acacia showed greater density in relation to other treatments. However, this planting showed the least diversity and equability, with exception of the area with spontaneous vegetation (Table 46.1). This could be explained by the fact that the high density of fauna found in this planting could have a reduced diversity, because the greater the density of organisms in a particular cover, the greater the chance is that some group predominating and thereby reducing equability, since species diversity is associated with the relation between number of species (species richness) and distribution of the number of individuals among the species (equability).

The fact that intercropping showed greater diversity (Table 46.1), could be related to the decomposing organic material being more diversified and meeting the nutritional and microclimatic requirements of a wider spectrum of organisms. That could indicate more favourable conditions for the development of edaphic fauna. According to Correia and Andrade (1999), the more diverse is the plant cover, the more the heterogeneity of the litter will be and, consequently, the greater the diversity of the soil fauna as well. On the other hand, the lower diversity found in the monoculture of eucalyptus followed by that of acacia (Table 46.1), could be attributed to the low heterogeneity of the food resources for these communities.

Although studies of natural regeneration under mono-specific plantings of eucalyptus have indicated that this crop does not reduce the diversity of species underground (Silva Junior et al., 1995), other studies show that the quantity and diversity of animal species that are found in a given forest ecosystem, depend on the number of available niches in the habitat. Therefore, a monoculture system, whether eucalyptus or any other species, is appreciably less capable of supporting a high diversity of fauna (Lima, 1993).

An analysis of the vertical distribution of the organisms showed the importance of the litter for the fauna, since it was observed that for all the plantings, the organisms had a preference for this compartment in relation to the soil (Fig. 46.1 and Table 46.2). According to Takeda (1995), the litter can serve as a habitat and food, contributing to an increase in abundance and diversity of all groups of soil organisms. The greater population of fauna in the litter is possibly related to the fact that this compartment is richer than soil in labile C and N.

Cluster analysis separated at a distance of 100% the area with spontaneous vegetation from those with plantings (Fig. 46.2). Among the plantings, there were two groups formed, one of which grouped the intercropping and monoculture of acacia, at a distance of about 15%. The other consisted of only the monoculture of eucalyptus, separated from the above group by a distance of 30%.

The relative proximity between the plantings (Fig. 46.2) suggests that the populations share some regulatory factors (Dias et al., 2007), in contrast to the control treatment. One of the main factors that could have contributed to the separation between the plantings and the control is the greater layer of organic residues present in the plantings, which provide a greater supply of food and better conditions of temperature and moisture in the soil, thereby resulting, in greater abundance and richness of the soil fauna. The modifications in the environmental conditions and in the soil-litter interface of areas under tree canopies exert positive effects on the biological activity of the soil (Frank and Furtado, 2001; Dias et al., 2007).

Table 46.2 Chi-squared analysis (X^2) of vertical distribution (soil-litter) of edaphic fauna in different treatments

Treatment	Chi-squared (X^2)
Eucalyptus	9.81*
Acacia	84,11**
Intercropping	38.21**
Control	–

Significant difference at 5% (∗) and at 0.1% (∗∗).

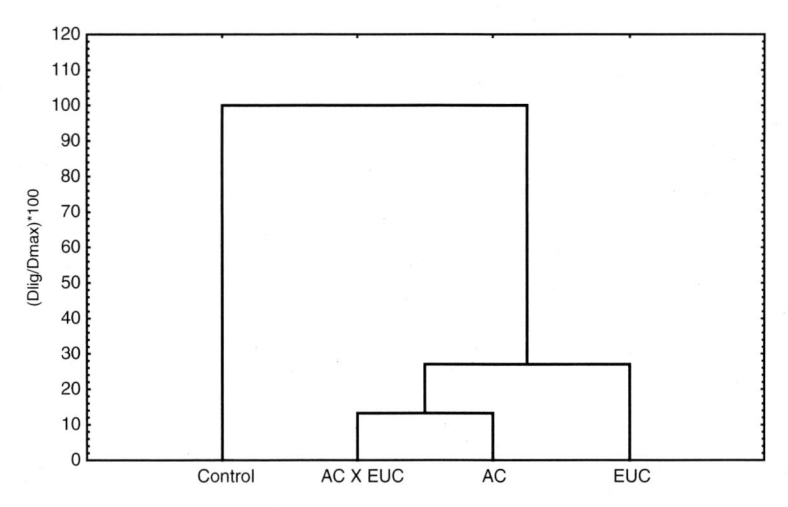

Fig. 46.2 Dendrogram of treatments studied for density of different taxonomic groups. Measure of distance: complement of Pearson's correlation coefficient *r*; linkage method: Ward's. Intercropping = *A. mangium* × *E. camaldulensis*

46.4 Conclusion

Based on the data obtained, it could be stated that the revegetation of areas degraded by the extraction of clay contributed to the reestablishment of the soil fauna. Both density and diversity were influenced by the type of plant cover, where intercropping was the system that showed the best distribution of organisms within each group (equability) and consequently the best diversity.

References

Borlinil, M.C., Sales, H.F., Vieira, C.M.F., Conte, R.A., Pinatti, D.G. and Monteiroi, S.N. (July/September 2005). Cinza da lenha para aplicação em cerâmica vermelha Parte I: características da cinza. Cerâmica 51(319):192–196. São Paulo.
Correia, M.E.F. and Andrade, A.G. (1999). Formação de serapilheira e ciclagem de nutrientes. In: G.A. Santos and F.A.O. Camargo (eds.), Fundamentos da Matéria Orgânica do solo: Ecossistemas Tropicais e Subtropicais. 1. Genesis, Porto Alegre, pp. 197–225.
Correia, A.A.D. (2003). Distribuição preferência alimentar e transformação de serapilheira por diplópodes em sistemas florestais. Tese de Doutorado, Curso de Pós-Graduação em Ciência do Solo, UFRRJ, Seropédica.
Dias, P.F., Souto, S.M., Corrêia, M.E.F., Rodrigues, K.M. and Franco, A.A. (2007). Efeito de leguminosas arbóreas sobre a macrofauna do solo em pastagem de Brachiaria brizantha cv. Marandu. Pesquisa Agropecuária Tropical 37:38–44.
Franco, A.A. and Faria, S.M. (1997). The contribution of N2-fixing tree legumes to land reclamation and sustainability in the tropics. Soil Biology and Biochemistry 29:897–903.

Franco, A.A., Dias, L.E., Faria, S.M., Campello, E.F.C. and Silva, E.M.R. (1995). Uso de leguminosas florestais noduladas e micorrizadas como agentes de recuperação e manutenção da vida do solo: um modelo tecnológico. In: F. Esteves (ed.), Oecologia Brasiliensis: Estrutura, Funcionamento e Manejo de Ecossistemas. UFRJ, Rio de Janeiro, pp. 459–467.

Frank, I.L. and Furtado, S.C. (2001). Sistemas Silvipastoris: Fundamentos e Aplicabilidade. Embrapa Acre, Rio Branco, 51p. (Série Documentos, 74).

Garay, I. (1989). Relations entre l'hétérogéinéité des litières et l'organization des peuplements d'arthropodes édaphiques. École Normale Supérieure, Paris, 192p. (Publications du Laboratoire de zoologie no. 35).

Kennedy, A.C. and Parpendick, R.I. (1995). Microbial characteristics of quality soil. Journal of Soil and Water Conservation 50:243–248.

Knoepp, J.D., Coleman, D.C., Crossey, D.A., Jr. and Clark, J.S. (2000). Biological indices of soil quality: An ecosystem case study of their use. Forest Ecology and Management 138:357–368.

Lima, W.P. (1993). Impacto Ambiental do Eucalipto. 2nd edn. Editora da Universidade de São Paulo, São Paulo, 301p.

Nunes, L.A.P.L., Araújo Filho, J.A. and Menezes, R.I.Q. (2008). Recolonização da fauna edáfica em áreas de Caatinga submetidas a queimadas. Caatinga 21:214–220.

Rillig, M.C. and Steinberg, P.D. (2002). Glomalin production by arbuscular mycorrhizal fungus: A mechanism of habitat modification? Soil Biology & Biochemistry 34:1371–1374.

Rovedder, A.P., Antoniolli, Z.I., Spagnollo, E. and Venturini, S.F. (2004). Fauna edáfica em solo suscetível à arenização na região sudeste do Rio Grande de Sul. Revista de Ciências Agroveterinárias 3:87–96.

Santana, D.P. and Bahia Filho, A.F.C. (1999). Indicadores da qualidade do solo, CDROM dos Anais do XXVII Congresso Brasileiro de Ciência do Solo. Brasília-DF.

Schiavo, J.A. (2005). Revegetação de áreas degradadas pela extração de argila, com espécies micorrizadas de Acacia mangium, sesbania virgata e Eucalyptus camaldulensis. Tese (Doutorado em Produção Vegetal) – campos dos Goytacazes – RJ, Universidade Estadual do Norte Fluminense, 117p.

Silva Júnior, M.C., Scarano, F.R. and Cardel, F.S. (1995). Regeneration of an Atlantic forest formation in the understorey of a Eucaliptus grandis plantation in south-eastern Brazil. Journal of Tropical Ecology 11:147–152.

Stork, N.E. and Eggleton, P. (1992). Invertebrates as determinants and indicators of soil quality. American Journal of Alternative Agriculture 7:2–6.

Takeda, H. (1995). Templates for the organization of collembolan communities. In: C.A. Edwards, T. Abe and B.R. Striganova (eds.), Structure and Function of Soil Communities. Kyoto University, Kyoto, pp. 5–20.

Valicheski, R.R., Marciano, C.R. and Ponciano, N.J. (2006). Viabilidade econômica da reutilização de áreas de extração de argila em Campos dos Goytacazes-RJ. www.ebape.fgv.br/radma/doc/GEM/GEM-031.pdf . Accessed on 22 May 2006.

Chapter 47
Conservation Tillage in Potato Rotations in Eastern Canada

M.R. Carter, R.D. Peters, and J.B. Sanderson

Abstract Potato (*Solanum tuberosum* L.) farming systems are often associated with soil degradation due to their excessive use of tillage and production of low levels of crop residue in the potato year. Results from a 12-year study, initiated on a fine sandy loam (Orthic Podzol) in Prince Edward Island (eastern Canada), were evaluated to assess the use of conservation tillage (CT), compared to conventional tillage, in 2-year (barley-potato) and 3-year (barley – red clover – potato) rotations. The CT strategy was to shift the primary tillage event for the potato phase from the autumn to spring, and to reduce the degree and depth of tillage. Mulches were used on all plots after potato harvest to provide soil cover over the cool season. Potato yield, soil cover, soil organic matter and structure, and soil-borne diseases were used as indicators to assess the feasibility of the cropping systems. Marketable potato yield was similar between the two tillage systems. The CT system provided relatively high surface residue levels after potato planting, compared to the bare soil surface in the conventional tillage system. Soil organic carbon and soil structural stability were significantly increased at the 0–10 cm soil depth in the CT, compared to the conventional system. Soil-borne diseases of potato were significantly reduced in 3-year rotations compared to 2-year rotations, but were mainly unaffected by tillage practice. Overall, use of CT in 3-year potato systems has the potential to maintain crop productivity and protect the soil resource.

Keywords Conservation tillage for potato · Soil residue cover · Soil organic matter · Soil structure · Soil-borne diseases

---

R.D. Peters (✉)
Agriculture and Agri-Food Canada, Crops and Livestock Research Centre, Charlottetown, PE C1A 4N6, Canada
e-mail: rick.peters@agr.gc.ca

47.1 Introduction

In Prince Edward Island, in eastern Canada, a significant proportion (up to 40%) of the agricultural land is committed to potato production on a rotation basis, and the dominant form of primary tillage is autumn mouldboard ploughing (Carter and Sanderson, 2001), which raises concerns for both energy use (Sijtsma et al., 1998) and soil erosion (Kachanoski and Carter, 1999). Soil degradation concerns have precipitated experimentation with conservation tillage (CT) systems (Carter and Sanderson, 2001). Specific objectives for CT in potato rotations can include reduction in energy use; prevention of poor soil structure; improvement in soil biological properties; and protection against soil loss due to tillage and water erosion (Carter and Sanderson, 2001; Riley, 2006).

Successful use of CT in potato rotations is dependent on soil type and climate. Long-term studies in Norway on imperfectly drained loams indicate that CT did not adversely influence cereal yields but did significantly reduce potato yield, compared to conventional ploughed systems (Riley, 2006). In short-term studies on a range of sites in Prince Edward Island, replacement of the mouldboard with a chisel plough within the potato phase had no adverse effect on potato yield and quality in both 2-and 3-year potato rotations (Holmstrom et al., 1999; Carter and Sanderson, 2001).

A study was initiated in 1994 to assess the feasibility and long-term effects of using CT in combination with crop residue mulches (after the potato harvest) on potato production in 2- and 3-year potato rotations on a fine sandy loam. CT in the potato phase involved replacing the conventional autumn mouldboard ploughing with a herbicide treatment followed by reduced tillage in the spring. It was considered that certain criteria must be met before CT could be considered an acceptable management system to combat soil loss and degradation. In addition to the soil conserving aspect of the system, the need to maintain potato productivity and quality is required. Further, a low soil-borne disease pressure to prevent a decline in tuber quality and marketable yield must also characterize the soil conserving measures. Thus, the following indicators were utilized to assess the soil conserving system: maintenance of potato productivity and quality; enhanced surface residue levels after potato planting; increased organic matter and structural stability at the soil surface; and low levels of soil-borne potato tuber disease. We hypothesized that the 3-year CT rotation system would meet the above criteria, while the 2-year CT rotation system would be marginal. Thus, the overall objective of the study was to assess the influence of CT on these indicators.

47.2 Materials and Methods

47.2.1 Experimental Site and Treatments

The study was conducted at the Harrington Research Farm in central Prince Edward Island on a Charlottetown fine sandy loam, an Orthic Podzol (FAO) [clay (<2 mm), silt (2–50 mm), sand (50–2,000 mm), and gravel (2–80 mm) content of 165, 280, 534, and 21 g kg^{-1}, respectively]. Annual precipitation at the site ranges from 800 to 1,100 mm, while the soil moisture regime and soil temperature is humid to perhumid and cool boreal, respectively.

The replicated experiment was initiated in 1994 to assess CT in both 2-and 3-year potato rotations. The experimental design was a split-plot with five main plots and six replicates, which allowed each phase of the rotation to be present each year. Main plot size was 23 × 5 m. The 2-year rotation was barley (*Hordeum vulgare* L.) and potato (cv. "Russet Burbank"), while the 3-year rotation was barley (undersown with red clover, *Trifolium pratense* L.), red clover and potato (cv. "Russet Burbank"). In the potato year only, the main plots were split to provide conventional and CT sub-plots.

A conceptual overview of the rotations indicating time of crop growth and agronomic practices is given in Fig. 47.1. Carter and Sanderson (2001) provide more detail on management practices. Peters et al. (2003, 2004, 2005) describe studies on soil-borne diseases in the experiment; while Carter et al. (2009a, b, c) provide

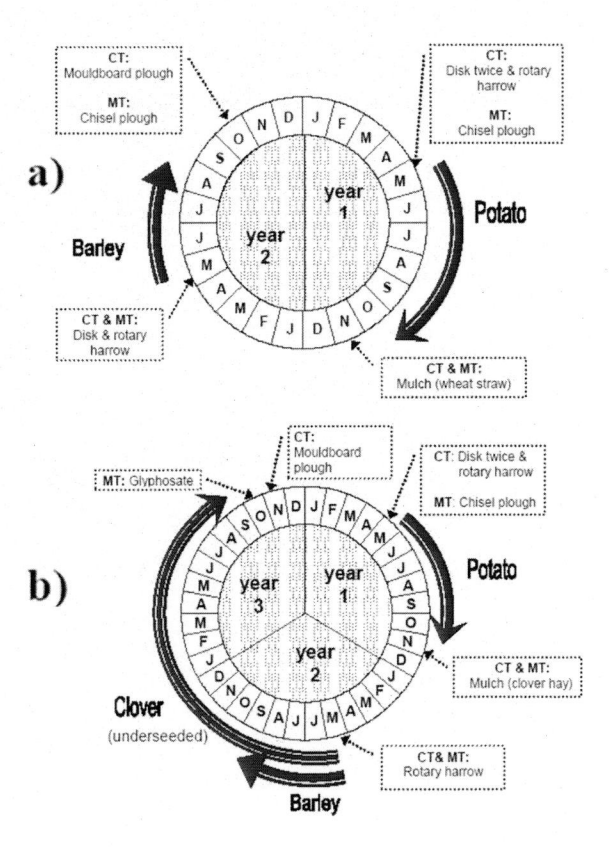

Fig. 47.1 Conceptual outline of tillage and management practices for conventional (CT) and conservation (MT) tillage in the (**a**) 2-year and (**b**) 3-year potato rotation (after Carter et al., 2009c). Note: CT refers to conventional tillage and MT to conservation tillage

results on potato productivity and quality, soil cover, soil organic matter, and soil biological properties.

In the CT treatment, the red clover was sprayed with glyphosate [N-(phosphonomethyl) glycine] at 2 l (356 g l^{-1} a.i.) ha^{-1} in the autumn prior to the potato year and the residue left as a soil cover over the cool season (November–April). For the potato phase, the conventional tillage consisted of mouldboard ploughing (20 cm deep), in the autumn (October) prior to the potato year, followed by two or more passes (10 cm deep) with a disc and harrow in the spring (May). CT consisted of one pass (15 cm deep) of a chisel plough with wide (36 cm) sweeps in the spring. During the potato growing season, both conventional and CT treatments were subject to the same in-row cultivation for ridging (hilling), fertilizer and pesticide applications, and harvesting operations.

Mulch (barley straw for the 2-year rotation and red clover hay for the 3-year rotation) was applied to the soil surface immediately after the potato harvest, in both tillage treatments, to provide winter cover. The specific rate (4 Mg ha^{-1}) of mulch corresponds to the optimal rate for reduction of cool season soil erosion (Edwards et al., 1995).

47.2.2 Methodology for System Indicators

47.2.2.1 Potato Yields and Quality

Yields were estimated by harvesting one row (23 m) per sub-plot. As a measure of yield quality for the processing industry, tubers were graded to obtain marketable yield (i.e. total tuber yield minus yield of tubers ≤ 90 g and misshapen tubers).

47.2.2.2 Surface Crop Residue Cover

Soil cover was measured after potato planting but before hilling each year and was estimated using a tape-transect method (Hartwig and Laflen, 1978), by counting the pieces of plant residue under each mark (2.5 cm spacing) on a 30.5 m tape.

47.2.2.3 Soil Organic Matter and Structure

Soil samples were obtained in 1999, 2003 and 2006 from the 3-year rotation. At each sampling time, three soil cores (8 cm inside diameter × 8 cm long) were obtained from 0 to 10 cm soil depth of each plot in the potato phase (within the potato ridge) of each rotation. Soil from each core was removed and air-dried. Soil organic C was determined by dry combustion using a LECO CNS 1000 analyser, while wet sieving of whole soil was used to determine the Mean Weight Diameter (MWD) of water stable aggregates.

47.2.2.4 Soil-Borne Disease

A number of potato diseases caused by soil-borne pathogens were assessed from 2000 to 2008 in the potato phase of the 2- and 3-year rotation. Specific sampling and experimental details are given in Peters et al. (2004). Potato disease assessment was conducted on both non-tuber and tuber tissues. Of the four potato rows in each plot, one internal row was used for destructive sampling and assessment of potato disease development on non-tuber tissues and a second internal row was harvested for assessment of tuber disease. Six plants were randomly harvested from each plot at the end of August and rated for percent necrosis of stem tissue caused by *Rhizoctonia solani* Kühn. Following harvest, 50 tubers from each experimental plot were rated visually for percent of tuber surface covered with sclerotia of *R. solani* (black scurf), as well as percent of tuber surface covered with lesions caused by *Helminthosporium solani* Durieu & Mont. (silver scurf), *Fusarium* spp. (dry rot), and *Streptomyces scabiei* (Thaxter) Lambert & Loria (common scab). In addition, tuber tissue samples were excised from the margins of diseased areas, surface-sterilized, and then plated onto potato dextrose agar in small Petri dishes. Petri dishes were incubated in the dark at 22°C for 7–10 days after which they were scored for the presence of the pathogens of interest.

47.2.3 Statistics

Analysis of variance was used to determine treatment differences (Genstat 5 Committee, 1987).

47.3 Results and Discussion

47.3.1 Potato Yield and Quality

Moving the time of tillage from the fall to the spring and reducing the degree and intensity of tillage did not adversely influence marketable potato yield (Fig. 47.2). Potato marketable yield over the 14-year duration of the study, as reported in Carter and Sanderson (2001) and Carter et al. (2009c), indicated that CT, compared to conventional tillage, was yield neutral. Tillage differences in marketable yield were evident only for two of the 14 years in favour of the CT system (Carter and Sanderson, 2001; Carter et al., 2009c). Comparison of other tuber quality parameters (e.g. number of misshapen tubers separated at tuber sizing and percentage of small tubers) did not show any main differences between the tillage treatments (data not shown).

Marketable potato yield showed clear differences between rotations, for six of the 14 years, in favour of the clover-based 3-year rotation (Carter and Sanderson, 2001; Carter et al., 2009c). However, these yield differences, between the 2-and

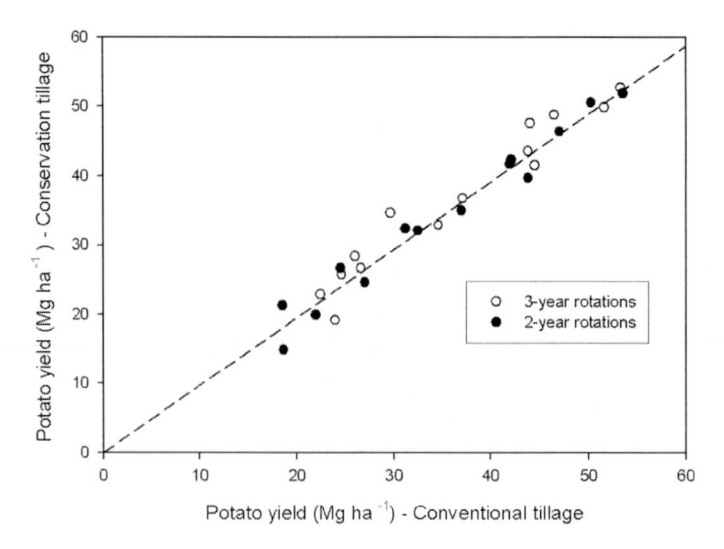

Fig. 47.2 Comparison of marketable potato yield for 2-year and 3-year potato rotations, under conventional and conservation tillage, over a 14-year period (1994–2007) (after Carter et al., 2009a)

3-year rotation, were only evident after 4 years indicating the need for at least one to two cycles of the rotation before yield and tuber quality (i.e. marketable yield) are influenced by rotation treatment (Carter and Sanderson, 2001). The 14-year average for marketable yield was 27 Mg ha^{-1} for the 2-year rotation and 30 Mg ha^{-1} for the 3-year rotation. Overall, tillage differences did not greatly influence potato yield or yield parameters. Most of the potato yield parameters were influenced only by crop rotation.

47.3.2 Surface Crop Residue Cover

Residue levels at the soil surface in the spring after potato planting reflected the differences in tillage system (data not shown). Mouldboard plough systems subject to tillage inversion and subsequent intensive spring tillage had no measurable residue on the soil surface. The CT treatments, however, for both crop rotations contained relatively high levels of surface residue. The 14-year mean (and standard deviation) for residue cover in the CT treatments was 31 ± 11.5% with a range of 22–60%. Thus, shifting the time of tillage from the fall to the spring and the use of mulches after the potato harvest presented a practical form of CT for potato production (Carter et al., 2009c). The soil was continually covered using live (crop canopy), mulch, or dead (residue) cover. In the case of the latter, the surface cover met the commonly recognized indicator or target (i.e., 30%) for CT even when measured after crop emergence just prior to hilling.

47.3.3 Soil Organic Matter and Structure

The CT system (data given for the 3-year rotation only) showed an increased mass of SOC at the 0–10 cm soil depth, at each of the three sampling times, compared to the conventional tillage system (Table 47.1). The MWD, an indicator of soil structural stability, was also increased under CT in 1999 and 2003. Over time, SOC also increased under both tillage systems probably due to the input of the crop residue mulches applied after potato harvest. Increases in SOC and soil structural stability in the surface soil has been shown to have benefits for soil water-holding capacity for potato systems on the same soil type as the present study (Carter, 2007). In combination with the increased soil surface cover, the enhanced levels of SOC in the surface soil would improve the soil conservation potential and help resist the soil degradation so often associated with intensive potato farming systems.

Companion studies have shown that increased SOC at the soil surface is also associated with positive effects on soil biota and soil biological properties (Carter et al., 2009a, 2009b). For example, the soil densities of Collembola (springtails) and Acari (mites), and the beneficial bacterial-feeding nematodes were significantly enhanced under the 3-year CT system, compared to the conventional system.

47.3.4 Soil-Borne Diseases

Over the course of the study (2000–2008), stem canker (Table 47.2) and black scurf (Table 47.3) caused by *R. solani* were often significantly reduced under the 3-year rotation, regardless of tillage regime. Increasing the rotation duration from 2 to 3 years was the most important factor, compared to tillage type, in the reduction of Rhizoctonia stem canker and black scurf. These results corroborate those found in previous studies, which have documented the benefits of crop rotation for control of diseases caused by *R. solani* (Scholte, 1987; Johnston et al., 1994; Ball et al., 2005).

Table 47.1 Comparison of soil organic carbon (SOC) and mean weight diameter (MWD) for water-stable aggregates, at the 0–10 cm soil depth in the potato phase, over time for the 3-year rotation on a fine sandy loam Podzol under conservation and conventional tillage

Tillage system	1999	2003	2006
SOC (g m^{-2})			
Conventional tillage	2,091	2,108	2,351
Conservation tillage	2,299	2,316	2,563
Significance level	0.001	0.050	0.002
MWD (mm)			
Conventional tillage	1.00	1.20	1.26
Conservation tillage	1.20	1.60	1.27
Significance level	0.001	0.05	0.929

Table 47.2 Mean stem canker severity (percent of stem tissue covered with canker) in Rhizoctonia-infected potatoes harvested from field plots managed with 2- and 3-year crop rotations and conventional (C) and conservation tillage (CT) systems from 2000 to 2008[a]

Rotation[c]	Tillage[b] 2000 C	2000 CT	2001 C	2001 CT	2002 C	2002 CT	2003 C	2003 CT	2004 C	2004 CT	2005 C	2005 CT	2006 C	2006 CT	2007 C	2007 CT	2008 C	2008 CT
2-year	19.1	28.6	34.0	32.3	54.2	42.2	44.2	42.9	24.2	21.9	37.0	30.0	36.3	48.1	26.9	31.2	35.2	38.8
3-year	25.0	17.3	23.3	20.5	24.0	22.5	30.3	28.2	23.4	16.0	33.2	24.8	29.9	37.0	30.2	32.8	30.7	35.3
LSD ($P = 0.05$)[d]	8.8		11.6		10.9		9.1		ns		6.7		8.0		ns		5.0	
Standard error of the mean[e]	2.3		3.8		3.6		4.2		3.1		2.1		3.7		2.8		2.1	
Significance level (R, rotation)	0.162		0.036		<0.001		0.005		0.376		0.121		0.022		0.171		0.030	
Significance level (T, tillage)	0.770		0.539		0.083		0.575		0.096		0.002		0.004		0.168		0.028	
Significance level (R H T)	0.012		0.870		0.167		0.911		0.351		0.722		0.367		0.740		0.777	

[a] Data for 2000–2002 taken from Peters et al. (2004); data for 2003 taken from Carter et al. (2009b).
[b] C = conventional tillage; CT = conservation tillage.
[c] Rotations of 2 years (barley-potato) or 3 years (barley underseeded with clover-clover-potato).
[d] LSD = least significant difference; ns = not significant.
[e] number of observations = 36, error degree freedom = 15.

Table 47.3 Mean black scurf severity (percent of tuber tissue covered with sclerotia) in Rhizoctonia-infected potatoes harvested from field plots managed with 2- and 3-year crop rotations and conventional (C) and conservation tillage (CT) systems from 2000 to 2008[a]

Rotation[c]	Tillage[b]																	
	2000		2001		2002		2003		2004		2005		2006		2007		2008	
	C	CT	C	CT	C	CT	C	CT	C	CT	C	CT	C	CT	C	CT	C	CT
2-year	3.7	3.6	2.9	2.7	3.3	2.9	2.9	3.1	0.5	1.2	1.0	0.7	1.1	1.9	0.2	0.4	0.1	0.2
3-year	1.4	1.8	2.2	1.6	0.3	0.3	0.8	0.5	0.6	0.8	0.5	0.3	0.3	0.5	0.2	0.3	0.1	0.1
LSD ($P = 0.05$)[d]	1.2		1.0		2.5		1.1		ns		ns		1.4		ns		ns	
Standard error of the mean[e]	0.4		0.3		0.8		0.4		0.3		0.3		0.5		0.2		0.1	
Significance level (R, rotation)	0.001		0.002		0.043		<0.001		0.558		0.138		0.007		0.672		0.133	
Significance level (T, tillage)	0.817		0.425		0.752		0.907		0.145		0.320		0.277		0.147		0.336	
Significance level (R H T)	0.570		0.642		0.808		0.563		0.427		0.724		0.556		0.650		0.365	

[a] Data for 2000–2002 taken from Peters et al. (2004); data for 2003 taken from Carter et al. (2009b).
[b] C = conventional tillage; CT = conservation tillage.
[c] Rotations of 2 years (barley-potato) or 3 years (barley underseeded with clover-clover-potato).
[d] LSD = least significant difference; ns = not significant
[e] number of observations = 6, error degree freedom = 15.

Potato dry rot and common scab were generally found at very low frequency and severity, precluding determination of treatment differences (data not shown). *Fusarium* spp. commonly infects tubers via wounds created during harvest and handling operations. Since care was taken to avoid tuber bruising during harvest, conditions conducive to the development of dry rot were generally absent. Similarly, the potato cultivar employed in this study ("Russet Burbank") has some resistance to common scab.

However, parallel studies comparing conventional tillage and CT in grower fields in Prince Edward Island have sometimes found a correlation between increased soil residues and the severity of common scab in susceptible potato cultivars (Peters et al., unpublished; data not shown). More research is required to elucidate this relationship and determine its impact on the adoption of CT as a standard practice in potato agriculture in Prince Edward Island.

The effects of tillage practice and rotation length on the incidence and severity of silver scurf varied from year to year, with no clear trend emerging (data not shown). However, for all diseases assessed, the lowest disease severity was commonly found in tubers from plants grown in 3-year rotations with CT. Previous research conducted on this field site suggests that this disease suppression is at least partly due to the activity (antibiosis abilities) of beneficial endophytic and root zone bacteria (endo- and exoroot) stimulated in soils managed with CT (Peters et al., 2003, 2005).

47.4 Conclusions

Major attributes of CT for potato systems in humid regions, subject to high precipitation and an extensive cool season, would be the continuum of soil cover over the relatively long winter period; the use of mulches and residue management; and the in depth reduction and intensity of tillage. Based on the criteria used to assess the soil conservation practices in this long-term study, CT produced similar yields as the conventional system, but had the added benefits of fewer tillage inputs and improved soil conservation potential. The 2-year, compared to the 3-year potato rotation, significantly decreased the marketable potato tuber yield. Diseases of potato caused by *Rhizoctonia solani* were suppressed in plots managed with a 3-year rotation, including those managed with CT. In addition, CT was not shown to acerbate other tuber blemish diseases. Increases in SOC and soil structural stability to water at the surface depth, under CT, would be potentially beneficial for both protection of the soil resource from erosive forces and enhancing soil water holding capacity.

The long-term results indicate that CT in a 3-year rotation had a positive effect on potato growth and quality, and appears to be a viable form of management for intensive potato production systems in Atlantic Canada.

References

Ball, B.C., Bingham, I., Rees, R.M., Watson, C.A. and Litterick, A. (2005). The role of crop rotations in determining soil structure and crop growth conditions. Canadian Journal of Soil Science 85:557–577.

Carter, M.R. (2007). Long-term influence of compost on available water capacity of a fine sandy loam in a potato rotation. Canadian Journal of Soil Science 87:535–539.

Carter, M.R. and Sanderson, J.B. (2001). Influence of conservation tillage and rotation length on potato productivity, tuber disease and soil quality parameters on a fine sandy loam in eastern Canada. Soil Tillage Research 63:1–13.

Carter, M.R., Noronha, C., Peters, R.D. and Kimpinski, J. (2009a). Influence of conservation tillage and crop rotation on the resilience of an intensive long-term potato cropping system: Restoration of soil biological properties after the potato phase. Agriculture, Ecosystems and Environment (in press).

Carter, M.R., Peters, R.D., Noronha, C. and Kimpinski, J. (2009b). Influence of 10 years conservation tillage on some biological properties of a fine sandy loam in the potato phase of two crop rotations in Atlantic Canada. Canadian Journal of Soil Science (in press).

Carter, M.R., Sanderson, J.B. and Peters, R.D. (2009c). Long-term conservation tillage in potato rotations in Atlantic Canada: Potato productivity, tuber quality and nutrient content. Canadian Journal of Plant Science 89:273–280.

Edwards, L.M., Burney, J. and DeHaan, R. (1995). Researching the effects of mulching on cool-period soil erosion control in Prince Edward Island, Canada. Journal of Soil Water Conservation 50:184–187.

Genstat 5 Committee. (1987). Genstat 5 Reference Manual. Oxford University Press, Oxford.

Hartwig, R.O. and Laflen, J.M. (1978). A meterstick method for measuring crop residue cover. Journal of Soil Water Conservation 32:90–91.

Holmstrom, D.A., DeHaan, R., Sanderson, J.B. and MacLeod, J.A. (1999). Residue management for potato rotation in Prince Edward Island. Journal of Soil Water Conservation 54:445–448.

Johnston, H.W., Celetti, M.J., Kimpinski, J. and Platt, H.W. (1994). Fungal pathogens and *Pratylenchus penetrans* associated with preceding crops of clovers, winter wheat, and annual ryegrass and their influence on succeeding potato crops on Prince Edward Island. American Potato Journal 71:797–808.

Kachanoski, R.G. and Carter, M.R. (1999). Landscape position and soil redistribution under three soil types and land use practices in Prince Edward Island. Soil Tillage Research 51:211–217.

Peters, R.D., Sturz, A.V., Carter, M.R. and Sanderson, J.B. (2003). Developing disease-suppressive soils through crop rotation and tillage management practices. Soil Tillage Research 72: 181–192.

Peters, R.D., Sturz, A.V., Carter, M.R. and Sanderson, J.B. (2004). Influence of crop rotation and conservation tillage practices on the severity of soil-borne potato diseases in temperate humid agriculture. Canadian Journal of Soil Science 84:397–402.

Peters, R.D., Sturz, A.V., Carter, M.R. and Sanderson, J.B. (2005). Crop rotation can confer resistance to potatoes from *Phytophthora erythroseptica* attack. Canadian Journal of Plant Science 85:523–183.

Riley, H. (2006). Recent yield results and trends over time with conservation tillage on morainic loam soil in southeast Norway. Acta Agriculturae Scandinavica, B 56:117–128.

Scholte, K. (1987). The effect of crop rotation and granular nematicides on the incidence of *Rhizoctonia solani* in potato. Potato Research 30:187–199.

Sijtsma, C.H., Campbell, A.J., McLaughlin, N.B. and Carter, M.R. (1998). Comparative tillage costs for crop rotations utilizing minimum tillage on a farm scale. Soil Tillage Research 49: 223–231.

Chapter 48
An Assessment of Soil Erosion Costs in Mexico

Helena Cotler and Sergio Martínez-Trinidad

Abstract Soil erosion presents a critical environmental hazard to Mexico as it threatens the sustainability of agricultural production. Although several conservation programmes addressing this issue have been developed over the past four decades, implementation has failed because, regardless its relevance, soil erosion has not yet been integrated into the political agenda. In order to ascertain the magnitude of the environmental risk, we assessed the cost of soil erosion in terms of productivity and nutrient loss. However, during the analysis of 140 publications from various sources, we came upon major stumbling blocks; i.e., a wide variation of results due to significantly different evaluation methods. Case in point: runoff plots reflected the lowest rate of soil erosion (0.1–5.9 mm), whereas RUSLE results were significantly higher (0.4–32.8 mm). In financial terms, following two scenarios suggested by the Mexican Ministry of Agriculture soil erosion costs may vary from US$ 39.7 to US$ 79.4 ha^{-1} and total nutrient loss could top at US$ 19 million – amounting to 85% of the government agricultural subsidies for rural families. These numbers stress the severity of soil erosion in Mexico as an obstacle to the financial sustainability of rural families and an expression of inadequate governmental support.

Keywords Soil erosion · Environmental assessment · Economic costs · Nutrient loss

48.1 Introduction

The impact of water soil erosion on agricultural lands and the implications of on-site and off-site damages have been widely documented throughout the world, including long-term productivity loss of degraded soils (Pimentel et al., 1993; Stocking, 2003;

H. Cotler (✉)
Instituto Nacional de Ecología, Coyoacán, México D.F., CP 04530, México
e-mail: hcotler@ine.gob.mx

Tengberg et al., 1998); soil and nutrient loss (Martinez-Casasnovas and Ramos, 2006; Maass et al., 1988); and a wide range of environmental consequences from the loss of ecosystem services provided by the soil (Barrios, 2007; Hodson and Dixon, 1988).

The implications of soil erosion are far-reaching and encompass the social, political, economic and environmental spheres (Gamini, 2003). Researchers acknowledge that soil degradation in Mexico has become an environmental, social and financial dilemma (Maass and García-Oliva, 1990; Cotler et al., 2007) that often triggers political conflict, as witnessed by the phenomenon of rural migration prompted by reduced yield (SIACON, 2008) and the ever-diminishing income in rural households (Scherr and Yadav, 1996). Farming is still the leading landscape-transforming force, with nearly 11% of the Mexican territory currently exploited in agricultural activities that provide direct sustenance to 5.8 million individuals and indirectly benefit about 25% of the population, mostly in rural areas (CONAPO, 2008). Nevertheless, agricultural output hardly impacts the national GDP and consequently fails to improve rural conditions, where poverty remains a critical issue (Vélez et al., 2007) often exacerbated by the increasing cost of crop management, as farmers require larger amounts of mineral fertilizer to maintain soil fertility.

Despite the environmental and social repercussions, soil conservation has not yet found its rightful place in the Mexican political agenda. Governmental policies in this area have not been successful. After nearly 40 years in operation, the federal agency in charge of soil and water conservation programmes has been able to restore merely 2% of the total surface affected by erosion (Vázquez, 1986). Furthermore, conservation policies are quite diluted even among the limited amount of government programmes spearheaded by two loosely connected federal agencies, the Ministry for the Environment and the Ministry of Agriculture (SEMARNAT and SAGARPA) – each with a meagre 3% of their total federal allotments earmarked for soil conservation projects (Cotler et al., 2007). This clearly reflects the scarce interest vested in the soil, the environmental services it provides and the repercussions of its deterioration on communal welfare.

To change this conservational paradigm and bring home the hazardous extent of soil erosion to the Mexican society, it is imperative to consider off-site and on-site erosion costs – financially and otherwise. The assessment of soil erosion costs in other countries has shined a glaring light on this issue and sparked significant management decisions (Cohen et al., 2006; Martinez-Casasnova and Ramos, 2006; Pimentel et al., 1995). Our aim here is to provide an insight of the effects of soil erosion on the Mexican economy from the perspective of loss of productivity and nutrients, and its consequences to public policy.

48.2 Soil Erosion Research in Mexico

Soil should be envisioned as a dynamic, multifunctional system that provides vital services to ecosystems and human activities. Using different methodologies and scales, with high seasonal variability, several authors have published disparate erosion data affecting 40–98% of the Mexican territory (Estrada and Ortiz, 1982; García, 1983). Some even postulate that soil loss has been underestimated over the past four decades (Maass and García-Oliva, 1990). However, the most recent report (SEMARNAT-Colegio de Posgraduados, 2002) acknowledges that 45% of the national territory (888,968.75 km^2) shows some degree of soil loss, mainly through water erosion and nutrient depletion, especially where agricultural activities account for 77% of soil degradation.

In their review of soil erosion in Mexico from 1940 and 1988, Maass and García-Oliva (1990) found a wide variation in methodologies used to identify water erosion of soil – ranging from satellite imaging to mathematical models and pebbles-, and a notable scarcity of publications supported by laboratory results (under 2%), which points at a prevailing insufficiency of proper empirical data.

Furthermore, as most studies span short periods and a very few are intended as a long-term soil erosion research, they provide limited and incomplete information of yield over time and under different production systems. Research has mostly focused on two subjects: erosion-causing conditions and amount of soil loss. This means emphasis has been placed on a single soil function related to productivity. While most studies revolve around the plot and later progress to the state- and nationwide levels, all of them reflect an obvious lack of homogenous and comparative methodologies.

Soil erosion has both on- and off-site effects, with major consequences of the former being reduced soil productivity and nutrient loss. While agricultural output is directly contingent on productivity and indirectly dependent on interactions with the rest of the economy (Alfsen et al., 1996), the impact of erosion on productivity (and the environment) remains largely underrepresented in available research data (Stroosnijder, 2005).

Most researchers have made half-hearted attempts to determine the financial costs. In an estimation of productivity loss, Magulis (1992) suggested that the effect of on-farm erosion (in terms of soybean, maize, sorghum and wheat yield) could top at a billion dollars, while McIntire (1994) established that the cost of soil erosion in the Mexican maize production for 1988 amounted to between 2.7 and 12.3 of GDP. However, government programs addressing soil degradation allow for an economic loss below these figures.

48.3 Materials and Methods

Soil erosion information was obtained through a meticulous revision of existing reports, which included theses from leading Mexican universities (1946–2006); abstracts of the Mexican Soil Congress (1969–2000); articles from Earth and Soil Science Journals (~1980–2006); and internal government documentation (Table 48.1). In all, we compiled a total of 140 studies on soil erosion spanning from 1960 to 2006.

Table 48.1 Reviewed documentation on Mexican soil erosion (Martínez, 2007)

Bibliography	Reviewed term
Theses	
Universidad Autónoma de Chapingo	1946–2006
Colegio de Posgraduados	1970–2006
Universidad Nacional Autónoma de México	1990–2006
Universidad Autónoma de México	1990–2006
Congress Abstracts	
Congreso Nacional de la Ciencia del Suelo	1969–2000
Governmental documentation	
Comisión Nacional del Agua	1980–2006
Instituto Nacional sobre Recursos Bióticos	1980–2006
Journals	
Terra	1980–2006
Agrociencia	1990–2006
Investigaciones Geográficas	1990–2006
Applied Soil Ecology	1996–2004
Advances in Soil Science	1985–1992
European Journal of Soil Science	1994–2004
Catena	1980–2006
Geomorphology	1988–2006
Geoderma	1980–2006
Soil Science Society of America	1988–2005
Soil Science	1996–2006
Soil Technology	1988–2006
Soil and Tillage Research	1986–2006

Despite starting as early as the 1960s, the studies on soil erosion peaked three decades later (Fig. 48.1). Nevertheless, the subject has increasingly lost impetus among research centres ever since.

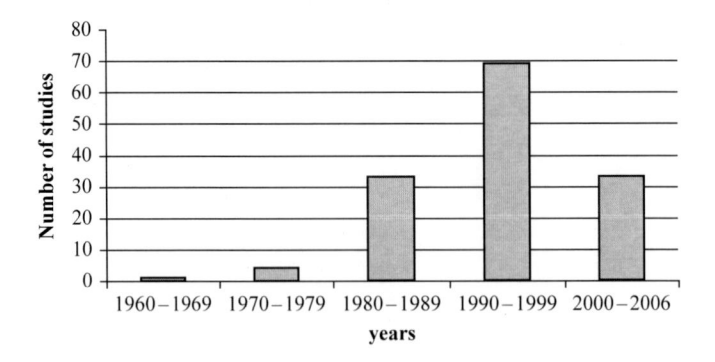

Fig. 48.1 Erosion studies in Mexico by decades

48 An Assessment of Soil Erosion Costs in Mexico

Soil erosion occurs at different scales, both spatial and temporal. Research efforts have prompted the development of several quantitative and qualitative methodologies involving various spatial references. Studies implemented in past decades have been based on punctuated analyses, with 45% focused on quantitative soil loss as determined by runoff plots.

Hill-slope evaluations by perched stones and pebbles have been implemented nationwide, while soil loss evaluations in the plot and watershed levels have used the RUSLE model (Fig. 48.2).

On-site effects remain at the core of research, especially in the plot level (nearly 68% of studies), while hill-slope, watershed and regional evaluations are mostly RUSLE-based, although some also use the perched stones and pebbles methodologies (Fig. 48.3).

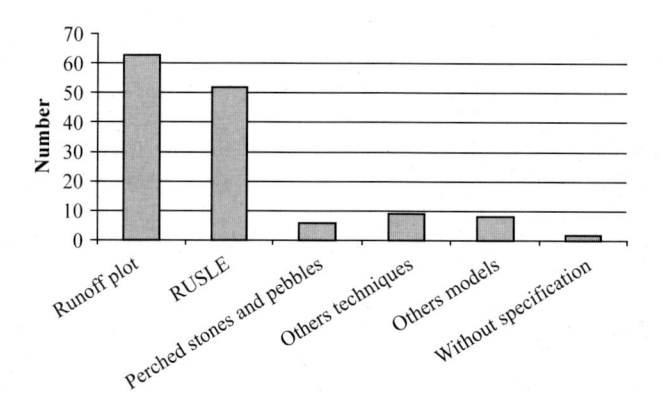

Fig. 48.2 Erosion-measuring methodologies

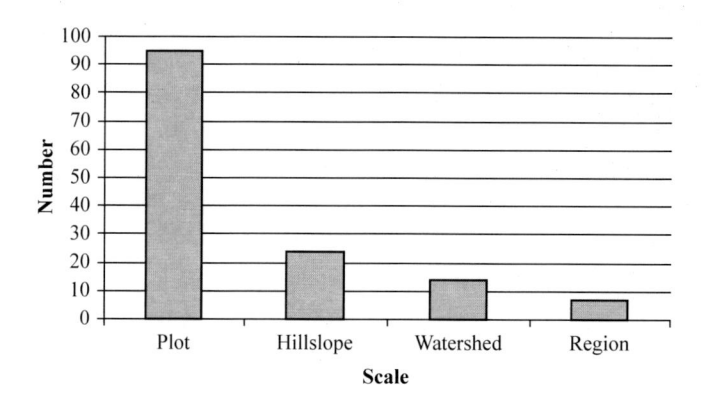

Fig. 48.3 Spatial scale of the soil erosion studies

48.3.1 Economic Assessment of Productivity Loss

Data was extrapolated to bordering areas with similarities of climate, soil type, land uses and slope. This generalization was initially done using ArcView 3.1 (Fig. 48.4), and later by digital coverage crossings on ArcGIS 8.1.

The relationship between yield and soil loss is not merely site- or soil-specific, but also dependent on the farming system. Therefore, data reflecting this ratio varies widely. Some researchers suggest that experimental yield reduction is 4% in the first 10 cm of topsoil (Bakker et al., 2004), while the results supported by findings of the Service of Soil Conservation (SCS, 1977) identify a yield reduction of 10% from the loss of 10 cm of topsoil.

Meanwhile the Mexican Ministry of Agriculture (SAGARPA, 2007) reports that 1 cm of soil loss decreases the yield by 150–300 kg ha^{-1}. This presents with two possibilities: a *conservative scenario*, which sets productivity at minus 150 kg cm^{-1} of soil loss and a *critical scenario*, with a production loss of 300 kg cm^{-1} of eroded soil.

The economic analysis was mostly based on rain-fed maize cropping, as it covers the largest surface in Mexico. According to SAGARPA (2008) the average maize yield is currently 2.95 t ha^{-1}, with an average cost of US$ 304 t^{-1} (SNIIM, 2009).

The documentation analysis review for the present report was focused on: (a) identification of soil loss through all the revised documents; (b) assessment of soil loss through decreasing yield of maize crops; and (c) determination of the economic cost of soil loss for both scenarios (*conservative* and *critical*).

Fig. 48.4 Distribution of extrapolated soil erosion data

Table 48.2 Fertilizer and nutrient costs

Fertilizer	Fertilizer cost (US $ t^{-1})[a]	Nutrient	Nutrient cost (US $ kg^{-1})
Urea	606.8	Nitrogen	1.31
Simple superphosphate	292.14	Phosphorus	1.46
Potassium chloride	755.38	Potassium	1.25

[a]Cost of fertilizer as of January 2009; USD exchange rated adjusted for April 14, 2009.

48.3.2 Economic Assessment of Nutrient Loss

The estimation of the economic effects from nutrient loss was based on the amount and cost of fertilizers needed for replacement. Fertilizers most frequently used in Mexico include urea, simple superphosphate and potassium chloride (Ávila, 2001), whose costs are displayed in Table 48.2. The financial cost of nutrient loss was established by: (a) identification of average nutrient loss in all reviewed documents; and (b) cost of fertilizers.

48.4 Results

Erosion data were mostly focused on the three main soil types in Mexico (Phaeozems, Regososl and Vertisols) and spans a total surface area of 27,733.7 km^2, including sub-humid to temperate climates and rain-fed crops on slopes under 5°. We found that the wide variations of soil erosion results are dependent upon evaluation methodologies. For instance, runoff plots reflect the lowest rate of soil erosion (0.1–5.9 mm of soil depth) whereas RUSLE results tend to be higher (0.4–32.8 mm of soil depth).

Based on the *conservative scenario*, where 1 cm of soil loss results in 150 kg of reduced productivity, total cost amounts to US$ 42,448,284.6 US with a value of $39.7 US ha^{-1}, while in the *critical scenario* of minus 300 kg cm^{-1} of soil loss, cost increases to US$ 84,896,599.6 US or $89.4 US ha^{-1}. Meanwhile, nitrogen, phosphorus and potassium losses amount to US$ 6.43, US$ 9.85 and US$ 7.19 ha^{-1}, respectively. Therefore, total nutrient loss for the entire Mexican territory is worth US$ 23.5 ha^{-1} or about US$ 19,704,055.

Although these amounts represent the annual economic cost caused by soil erosion based on the existing data for the country, the numbers could easily change, as soil erosion is a dynamic process. Consequently, the continuance of the current farming system over the following decade could well increase the cost of the *conservative scenario* to US$ 424,482,846 with the *critical scenario* topping at US$ 848,965,996.

It should be noted that, while the reviewed area spans 27,733.7 km^2 – about 1.4% of the total Mexican territory-, the latest study on soil degradation points out that 45% of the nation's surface shows some form of soil degradation, which suggests that the total cost could increase to US$ 1,123,631,062.9 for the *conservative scenario* and US$ 2,247,262,930.6 in the *critical scenario*.

48.5 Conclusions

Soil erosion is one of the most critical environmental hazards in Mexico. Despite having been studied for several decades, the current knowledge of soil erosion in our country is still quite rudimentary, resulting in limited and deficient public policy programmes (Cotler et al., 2007). A detailed analysis of documented studies unveils significant methodological differences, which prevent comparisons. As Lal (1994) has stated, it is indeed compulsory to transcend basic research in order to generate accurate and reliable data using standardized methods.

Scale is critical to soil erosion research because it influences our understanding of the elements that need to be accounted for in model development. Appreciating the scale at which soil erosion occurs is also important in a policy context because it defines the institutions and types of public policies required (Cotler and Ortega, 2006). Government can play significant and multiple roles in controlling erosion. Stroosnijder (2005) even proposed that policymakers should: (i) resort to moral persuasion combined with regulations; (ii) define cultural, supporting or structural measures – regardless of cost; or (iii) offer subsidies and other incentives.

So far, institutional response to this environmental risk has been twofold. On the one hand, by proposing programmes to support certain soil conservation actions – mostly structural measures to control sedimentary flows, which are both costly and difficult to replicate. On the other side, granting various types of economic incentives, although the amount currently offered as a direct subsidy (known as PROCAMPO: http://www.procampo.gob.mx/artman/publish/article_183.asp) is US$ 73 ha^{-1}, the net on-farm loss through soil erosion often expends up to 85% of that allowance. This underlines not only the severity of the problem in Mexico, but also the insufficiency of the government financial support which, consequently, contributes to drive the rural income to ever-diminishing levels.

Most Mexican soil erosion analyses aim at determining on-site effects focusing on runoff plots. However, design and enforcement of public policies for soil preservation should integrate both the on-site effects and the watershed scale, as the latter affects a larger population. Indeed, future studies must consider additional soil functionalities, including carbon retention, infiltration or detoxification. In summary, soil erosion studies should transcend from the productive to the environmental context, allowing for a fairer appreciation of the environmental services and the cost of soil loss. Although the inaccuracy inherent to available data prevented us from reaching an unambiguous conclusion, our review underlines the importance of introducing the economic factor of soil erosion into the formula of proposed policies to mitigate soil erosion and improve the conditions and income of the rural poor.

References

Alfsen, K.H., De Franco, M., Glomsrod, S. and Johnsen, T. (1996). The cost of soil erosion in Nicaragua. Ecological Economics 16:129–145.
Avila, J.S. (2001). El mercado de los fertilizantes en México, situación actual y perspectivas. Problemas del desarrollo 32(17):189–207.
Bakker, M.M., Govers, G. and Rounsevell, M.D.A. (2004). The crop productivity-erosion relationship: An analysis based on experimental work. Catena 57(1):55–76.
Barrios, E. (2007). Soil biota, ecosystem services and land productivity. Ecological Economics 64:269–285.
Cohen, M., Brown, M. and Sheperd, K. (2006). Estimating the environmental costs of soil erosion at multiple scales in Kenya using synthesis. Agriculture, Ecosystems and Environment 114:249–269.
Comisión Nacional de Población-CONAPO. (2008). La situación demográfica de México 2008. Mexico.
Cotler, H. and Ortega, M.P. (2006). Effects of land use on soil erosion in a tropical dry forest ecosystem, Chamela watershed, Mexico. Catena 65:107–117.
Cotler, H., Sotelo, E., Dominguez, J., Zorrilla, M., Cortina, S. and Quiñones, L. (2007). La conservación de suelos: Un asunto de interés público. Gaceta Ecológica 83:1–71. Instituto Nacional de Ecología, México.
Estrada, J. and Ortiz, S.C. (1982). Plano de erosión hídrica del suelo en México. Geografía agrícola 3:23–27.
Gamini, J. (2003). Soil erosion in developing countries: A socio-economic appraisal. Journal of Environmental Management 68:343–353.
García, L.R. (1983). Diagnóstico sobre el estado actual de la erosión en México. Terra 1(1):11–14.
Hodson, G. and Dixon, J.A. (1988). Measuring economic losses due to sediment pollution: Logging versus tourism and fisheries. Tropical Coastal Area Management 7:5–8.
Lal, R. (1994). Soil erosion by wind and water: Problems and prospects. In: R. Lal (ed.), Soil Erosion Research Methods. Soil Water Conservation Society, St. Lucie Press, Ankeny, IA.
Maass, J.M. and García-Oliva, F. (1990). La investigación sobre erosión de suelos en México: Un análisis de la literatura existente. Ciencia 41(3):209–228.
Maass, J.M., Jordan, C.F. and Sarukhan, J. (1988). Soil erosion and nutrient losses in seasonal tropical agroecosystem under various management techniques. The Journal of Applied Ecology 25:595–607.
Magulis, S. (1992). Back-of-the-envelope estimates of environmental damage costs in Mexico. Working Papers, World Bank, Washington, DC.
Martinez, S. (2007). Estimación de la valoración económica de la erosión de suelos. INE working paper http://www.ine.gob.mx/dgioece/cuencas/proyectos.html Cited 28 May 2009.
Martínez-Casasnova, J.A. and Ramos, M.C. (2006). The cost of soil erosion in vineyard fields in the Penedes-Anoia Region (NE Spain). Catena 68:194–199.
McIntire, J. (1994). A review of the soil conservation sector in Mexico. In: E. Lutz, S. Pagiola and C. Reiche (eds.), Economic and Institutional Analyses of Soil Conservation Projects in Central America and the Caribbean. World Bank Environment Paper 8, Washington, DC.
Pimentel, D., Allen, J., Beers, A. et al. (1993). Soil erosion and agricultural productivity. In: D. Pimentel (ed.), World Soil Erosion and Conservation. Cambridge University Press, Cambridge, MA.
Pimentel, D., Harvey, C., Resosudarmo, P. et al. (1995). Environmental and economic costs of soil erosion and conservation benefits. Science 267:1117–1123.
SAGARPA. (2007). Informe de diagnóstico de la degradación de suelos e impacto de los programas de conservación de suelos en México. http://www.sagarpa.gob.mx/agricultura/info/sust/suelo/at-degrada.pdf Cited 14 May 2009.
SAGARPA. (2008). Resumen Nacional por productos agrícolas. Servicio de información y estadística agroalimentaria y pesquera (SIAP) http://www.siap.sagarpa.gob.mx/

Scherr, S. and Yadav, S. (1996). Land degradation in the developing World: Implications for food, agriculture and the environment to 2020. IFPRI Discussion Paper no. 14, Washington DC.

SEMARNAT-Colegio de Posgraduados. (2002). Evaluación de la degradación del suelo causada por el hombre en la República Mexicana, escala 1:250,000. Memoria Nacional, SEMARNAT-Colegio de Posgraduados, México.

Service of Soil Conservation – SSC. (1977). Midwest Technical Service Center. TSC Advisory Soils L1-13, July 14.

SIACON. (2008). Sistema de Información Agropecuaria de Consulta. SAGARPA

Sistema Nacional de Información e Integración de Mercados – SNIIM. (2009). Precios de mercado: Granos básicos e insumos agrícolas. Sistema Nacional de Información e Integración de Mercados, Secretaría de Economía. http://www.economia-sniim.gob.mx> Available at http://www.oeidrus-tamaulipas.gob.mx/cd_anuario_06/TS.html Cited 14 May 2009.

Stocking, M. (2003). Tropical soils and food security: The next 50 years. Science 302:1356–1359.

Stroosnijder, L. (2005). Measurement of erosion: Is it possible?. Catena 64(2–3):162–173.

Tengberg, A., Stocking, M. and Dechen, S. (1998). Soil erosion and crop productivity research in South America. In: H.P. Blume, H. Eger, E. Fleischhauer, A. Hebel, C. Reij and K.G. Steiner (eds.), Towards Sustainable Land Use, Advances in Geoecology. volume I, Catena Verlag, Reiskirchen, Germany, pp. 355–362.

Vazquez, A.V. (1986). La erosión y la conservación del suelo en México (realidades y perspectivas). Terra 4(2):158–172.

Vélez, F., Meléndez, A. and García, H. (2007). Algunas consideraciones de política agropecuaria. In: P. Cotler (ed.), Políticas públicas para un crecimiento incluyente. Universidad Iberoamericana, México.

Chapter 49
Predicting Winter Wheat Yield Loss from Soil Compaction in the Central Great Plains of the United States

Joseph G. Benjamin and Maysoon M. Mikha

Abstract Adoption of methods to minimize the effects of soil compaction on crop production by farmers has been slow. Often farmers do not equate degradation of soil physical properties with reduction in crop yield. The objective of this study was to determine the potential yield loss caused by degradation of soil physical quality due to compaction. Soil conditions and winter wheat (*Triticum aestivum* L.) yields were observed on the Alternative Crops Rotation study at Akron, Colorado in 1996 and 1997. Changes in soil physical properties were determined by observing changes in the soil Least Limiting Water Range (LLWR), which includes limitations of water holding capacity, soil strength and soil aeration, on crop production. Grain yield decreased approximately 1,000 kg ha^{-1} per 0.1 unit decrease in LLWR, showing that soil compaction can cause serious yield reductions if not managed properly. Soil compression curves were developed to help predict the amount of soil compaction, and subsequent yield loss, to be expected with wheel traffic at various tire pressures and soil moisture conditions. Methods such as controlled wheel traffic or the use of low-pressure tires should be used to reduce soil compaction and maintain soil productivity.

Keywords Winter wheat · Yield loss · Soil compaction · Central Great Plains · USA

49.1 Introduction

The use of no-till cropping systems and better residue management in the Central Great Plains has led to water savings that allow increased cropping intensity and more diversity of crop species (Anderson et al., 1999). However, because no tillage is done to loosen the soil, concerns arise that the long-term effects of no tillage could

---

J.G. Benjamin (✉)
USDA-ARS, Central Great Plains Research Station, Akron, CO 80720, USA
e-mail: joseph.benjamin@ars.usda.gov

result in increased soil compaction and possible degradation of the soil physical environment for crop production.

Soil compaction is an unavoidable consequence of using machinery for crop production in modern mechanized agriculture. Some tractors used in field operations such as tillage, planting, and harvesting may weigh over 20,000 kg and provide over 300 kW of power. The use of such equipment has the potential to compact soils, making them less productive due to poor soil physical conditions.

Soil compaction changes the arrangement of soil solids and soil pores. It can affect the soil physical environment in three ways. First, compaction can affect water holding capacity by changing the soil pore size distribution (Horton et al., 1994). Secondly, compaction may alter soil aeration by decreasing O_2 diffusion into the soil and CO_2 diffusion out of the soil (Stepniewski et al., 1994). A third soil limitation to crop production is high soil strength (Guerif, 1994). Soil compaction can push the particles of the soil together such that the penetration resistance of the soil increases high enough to limit root proliferation. Each of these factors interacts with the others to determine limiting conditions to crop production. A method called the Least Limiting Water Range (LLWR) has been developed to account for each of these limiting soil physical conditions and provides a method to evaluate changes in soil potential productivity caused by compaction (Letey, 1985; Benjamin et al., 2003).

When compaction occurs, few options exist in no till systems for alleviating that. Natural processes of freeze-thaw and wetting-drying will reduce the effects of compaction, but the processes may take a long time and require a change in vegetation to be most effective (Benjamin et al., 2007).

Over the last few decades, much work has been focused on predicting the amount of compactive force transferred to the soil from tractors with various weight and power. Gupta and Raper (1994) summarize various research results for predicting soil compaction under vehicles with various size, power, and wheel configurations. They conclude that these models are adequate for many experimental conditions, but more research is needed to make the models acceptable for everyday use. Raper (2005) lists several machinery and soil management options for minimizing soil compaction. These recommendations include minimizing traffic when the soil is moist, applying controlled wheel traffic systems to contain the traffic to specific areas, reducing axle loads by minimizing the size of the vehicle for the specific field activity, and using the minimal inflation pressure for the size of the tire. Adoption of these practices by farmers is dependant on the realization of the yield loss and subsequent loss of income from soil compaction. Knowledge of the response of soil to compactive forces and the response of crops to changes in the physical properties of the soil will help convince farmers to use management techniques to minimize the effects of soil compaction on winter wheat yield under no-till systems in the central Great Plains of the United States.

49.2 Materials and Methods

This study was conducted at the Central Great Plains Research Station near Akron, Colorado on a Weld loam (fine, smectitic, mesic, Aridic Argiustolls). All data presented in this paper were collected from the ongoing Alternative Crops Research (ACR) study. The experiment consists of three replications of several rotations of crops suited for dryland crop production in the central Great Plains. Each phase of each rotation occurs each year. Crops included in the rotations are wheat, abbreviated W, corn (*Zea mays* L.), abbreviated C, and proso millet *(Panicum miliaceum* L.), abbreviated M, with or without various intensities of fallow (F). More detail about the experimental design and crop management techniques can be found in Anderson et al. (1999) and Bowman et al. (1999). We selected the wheat plots from the WF, WCF, WCM rotations in the experiment.

To construct the LLWR for a particular soil, knowledge of field capacity, wilting point, air-filled porosity and soil strength are needed for the range of bulk densities likely to occur in the field. In this paper we have defined field capacity as the water content at −33 kPa water potential, the wilting point as the water content at −1,500 kPa water potential, the aeration limitation as 10% air-filled porosity, and the strength limitation as 2 MPa cone penetrometer resistance. These criteria have also been used by da Silva et al. (1994) and Betz et al. (1998).

Soil cores (75 mm diam. by 75 mm tall) were collected with a Giddings[1] hydraulic soil probe. Cores were taken immediately after wheat harvest in July. The cores were placed in individual moisture desorption cells and the 33 kPa (field capacity) water content was determined. Bulk density was determined on the same cores. Disturbed soil samples were used to determine 1,500 kPa (wilting point) water content. Measurements of cone penetrometer resistance and corresponding water content and bulk density were taken in the field. More detail in sampling procedures can be found in Benjamin et al. (2003).

Winter wheat yields from 1996 and 1997 were plotted against the corresponding LLWR. The yield data were separated into wheat yields following a fallow period under no-till management and wheat yields either directly following millet or wheat yields under sweep tillage management.

A series of compaction tests were run on disturbed soil samples to determine the response of the Weld soil to compactive pressure. An automatic soil compactor (ELE International) was used to compact the soil. The amount of energy was varied by changing the number of blows each sample received or by changing the weight of the tamper and drop height of the tamper. The machine turns the sample such that the entire surface of the soil in the mold is covered by overlapping tamper blows. Triplicate samples were prepared at each compaction energy level. The standard

---

[1]Mention of trade names in for reference only. It does not imply a recommendation of this equipment over similar makes or models.

Proctor density test (Spangler and Handy, 1982) was run by compacting soil into a 101 mm diameter by 116 mm thick cylindrical mold with a 2.5 kg tamper dropped from a height of 305 mm. Soil was added in three layers and each layer received 25 blows by the hammer. A modified Proctor test was also used that compacts soil into a 152 mm diameter by 116 thick cylindrical mold by dropping a 4.5 kg tamper from a 457 mm height. Soil was added in five layers and each layer received 25 blows. Modifications of the standard Proctor test were made to compact soil with less energy. Instead of 25 tamper blows per layer as in the standard Proctor test, tests were run with 7 blows per layer and 14 blows per layer. Compaction energy was calculated by

$$E = Mhbg \tag{49.1}$$

Where:

E is the compaction energy (J),
M is the mass of the tamper (kg),
h is the drop height (m),
b is the number of tamping blows (–) and
g is the gravitational constant (9.8 m s^{-2}).

The energy applied to the soil sample was expressed as an equivalent pressure by

$$P = EV^{-1} \tag{49.2}$$

Where:

P is the compaction pressure (kPa) and
V is the sample volume (m^3).

The summary of compaction levels is shown in Table 49.1. Compaction at each energy level was applied to soils at 0.05, 0.10, 0.15, 0.20 and 0.25 g g^{-1} water content. After compaction the sample was trimmed and the bulk density and ending soil water content were determined.

Table 49.1 Compaction energy applied to determine compaction characteristics of a Weld loam

Mold volume	Number of blows	Applied energy	Pressure
m^{-3}	–	J	kPa
9.11 10^{-4}	21	157	172
9.11 10^{-4}	42	313	344
9.11 10^{-4}	75	560	614
2.10 10^{-3}	125	2,512	1,193

49.3 Results

The Least Limiting Water Range (LLWR) has been used as a method to combine limitations of the soil physical environment for crop production. The LLWR can be thought of as the range of water contents, at a given bulk density, where none of these soil physical properties are limiting to crop production. Plots of −33 kPa water content vs. bulk density, −1,500 kPa water content vs. bulk density, water content and bulk density which gives 2 MPa cone penetrometer resistance, and water content and bulk density which gives 10% air-filled pore space were made and the LLWR was determined (Fig. 49.1). The range of water contents where none of these properties are limiting is shown in the cross-hatched zone. For instance, the LLWR at a bulk density of 1.2 Mg m^{-3} would be between 0.23 and 0.38 volumetric water content, resulting in a LLWR of 0.15. The LLWR is smaller as bulk density increases. The LLWR at a bulk density of 1.6 Mg m^{-3} would be between 0.25 and 0.29 water content, resulting in a LLWR of 0.04.

Wheel traffic effects on soil bulk density, and the corresponding effect on LLWR, are dependent on compaction pressure and the water content of the soil when trafficked. The effects of compaction pressure and soil water content for a Weld loam are shown in Fig. 49.2. For a compaction pressure of 172 kPa, the range of bulk density would be 1.4–1.54 Mg m^{-3} depending on the water content of the soil at compaction. For higher compaction pressures, the bulk density increases. For a compaction pressure of 614 kPa, the range of bulk density would be 1.5–1.7 Mg m^{-3}. The optimum water content for compaction decreases with increasing compaction pressure. The optimum water content for compaction at 172 kPa is about 0.20 g g^{-1}. The optimum water content for compaction at 614 kPa is 0.15 g g^{-1}.

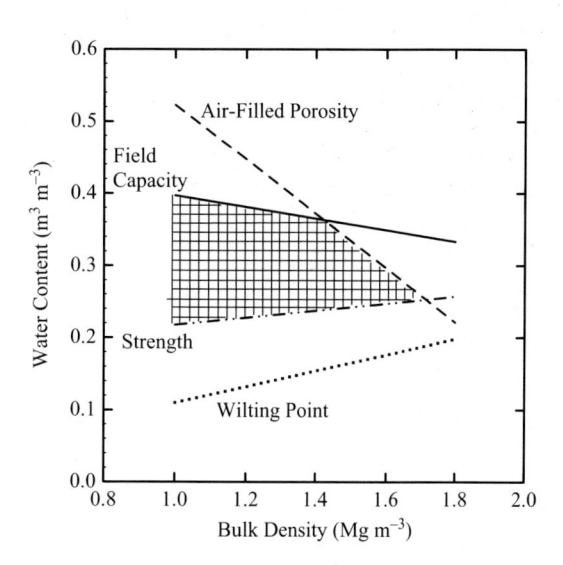

Fig. 49.1 Determination of the Least Limiting Water Range (LLWR) for a Weld loam at Akron, Colorado

Fig. 49.2 Response of a Weld loam to compactive pressure. Each *curve* corresponds to a compactive pressure calculated from the energy used to pack the soil with an automated soil compactor

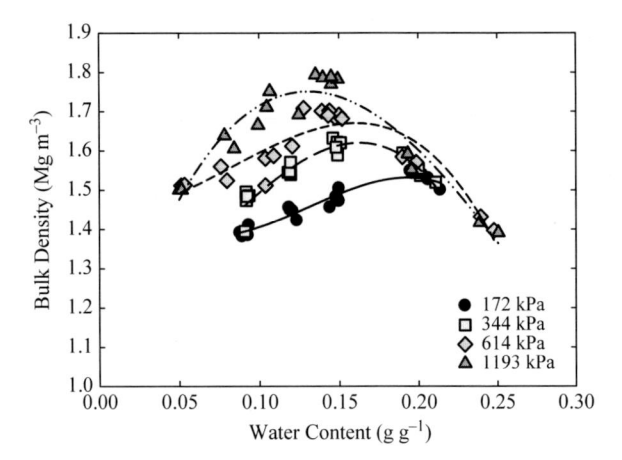

There was a linear response of wheat yield to LLWR at this site (Fig. 49.3). The squares indicate wheat yields for wheat after fallow in no-till cropping systems at Akron, Colorado in 1996 and 1997. The circles are wheat yields for no-till wheat grown either after millet or wheat grown after fallow in a sweep tillage system. Even though overall wheat yields were lower in cropping systems with no fallow period or with tillage, wheat response to LLWR was similar in all cropping systems. In each cropping system wheat yields declined approximately 1,000 kg ha^{-1} for each 0.1 decline in LLWR.

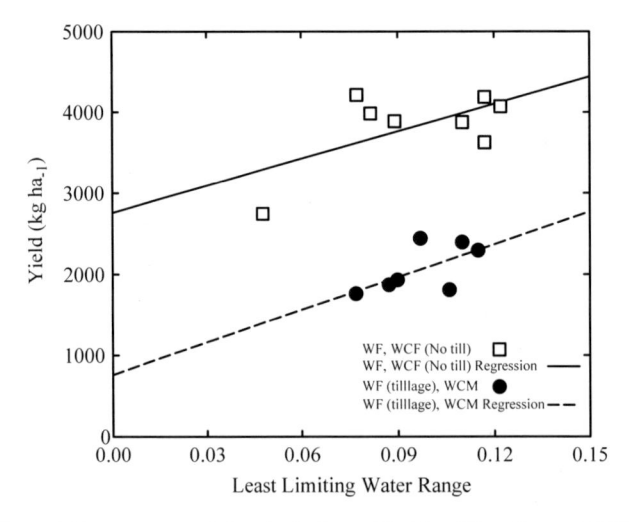

Fig. 49.3 Wheat yield response to change in the Least Limiting Water Range (LLWR) for a Weld loam at Akron, Colorado. WF indicates the Wheat-Fallow rotation, WCF indicates the Wheat-Corn-Fallow rotation, and WCM indicates the Wheat-Corn-Millet rotation

49.4 Discussion

Uncontrolled traffic patterns are common in many commercial agricultural fields. Some fields may be covered many times by implements in the course of a crop year (Kuipers and van de Zande, 1994). Farmers can use information on the compaction characteristics of the soil and the response of the crop to soil physical conditions to make better decisions on management of their fields. Compaction information may help them determine the effects of machinery operations on soil compaction and subsequent effects on potential wheat yield.

For instance, farmers must often decide when the water content of the soil in a field is suitable for field operations. If a farmer were to traffic this soil with a water content of 0.10 g g^{-1} water content with an implement that provides 172 kPa pressure, the farmer could expect the soil to compact to a bulk density of about 1.4 Mg m^{-3} (Fig. 49.2). If rainfall or irrigation was to occur such that the water content increased to 0.2 g g^{-1} and the field was trafficked with the same implement, the farmer could expect the soil to compact to a bulk density of about 1.54 Mg m^{-3}, increasing the amount of compaction. If the entire surface of the soil were covered with wheel tracks the difference in LLWR would be the change of LLWR from 0.13 to 0.08 (from Fig. 49.1), a decrease of 0.05. A decrease in LLWR of 0.05 would result in a winter wheat yield loss of about 500 kg ha^{-1} (from Fig. 49.3). Information such as this can point out to the farmer the risk involved when trafficking the soil when it is too wet.

Farmers often have decisions to make on the size of machinery used and the compactive pressure the selected implement will have on the soil. An implement that provides 172 kPa compaction pressure on a soil with a water content of 0.15 g g^{-1} will compact the soil to a bulk density of about 1.5 Mg m^{-3}, whereas an implement that provides 344 kPa compaction pressure on the same soil under the same conditions will compact the soil to a bulk density of about 1.6 Mg m^{-3} (Fig. 49.2). The change in LLWR would be from 0.09 to 0.04 (Fig. 49.1) and a winter wheat yield loss of about 500 kg ha^{-1} (Fig. 49.3). Farmers can use this information to make decisions on the size and weights of machines for field operations.

Sometimes field operations on soil that is too wet or using relatively large machines for farming is unavoidable. Devising a controlled wheel traffic pattern on the field helps limit the damage caused by compaction to the entire field. The goal of a controlled wheel traffic system is to create poorer conditions, as noted in the above examples, on part of the field but preserve more optimal conditions on the area between the wheel tracks. Showing the direct influence of wheel traffic on the soil physical condition and the subsequent affects on productivity may provide incentive for farmers to devise such controlled wheel traffic systems for their operations.

49.5 Conclusions

Soil compaction has the potential to severely limit crop production. The primary method to avoid compaction is to not traffic the soil when the soil is wet, as that is when it is most susceptible to compaction. Many soils are most easily compacted at water contents slightly less than field capacity. A second method to minimize compaction is to use implements with low tire pressures. By using lower tire inflation pressure, the weight of the implement is spread over a larger surface area and decreases the amount of force applied to the soil and compaction energy. An analysis of the compaction characteristics of soil, the response of a specific crop to soil compaction, and the proposed machinery used for crop production may help the farmer to devise machinery systems that will minimize adverse effects of soil compaction on crop production.

References

Anderson, R.A., Bowman, R.L., Nielsen, D.C., Vigil, M.F., Aiken, R.M. and Benjamin, J.G. (1999). Alternative crop rotations for the central Great Plains. Journal of Production Agriculture 12:95–99.

Benjamin, J.G., Nielsen, D.C. and Vigil, M.F. (2003). Quantifying effects of soil conditions on plant growth and crop production. Geoderma 116:137–148.

Benjamin, J.G., Mikha, M., Nielsen, D.C., Vigil, M.F., Calderon, F. and Henry, W.R. (2007). Cropping intensity effects on physical properties of an no-till silt loam. Soil Science Society of America Journal 71:1160–1165.

Betz, C.L., Allmaras, R.R., Copeland, S.M. and Randall, G.W. (1998). Least limiting water range: Traffic and long-term tillage influences in a Webster soil. Soil Science Society of America Journal 62:1384–1393.

Bowman, R.A., Vigil, M.F., Nielsen, D.C. and Anderson, R.L. (1999). Soil organic matter changes in intensively cropped dryland systems. Soil Science Society of America Journal 63:186–191.

da Silva, A.P., Kay, B.D. and Perfect, E. (1994). Characterization of the least limiting water range of soils. Soil Science Society of America Journal 58:1775–1781.

Guerif, J. (1994). Effects of compaction on soil strength parameters. In: B.D. Soane and C. van Ouwerkrek (eds.), Soil Compaction in Crop Production. Elsevier, Amsterdam, pp. 191–214.

Gupta, S.C. and Raper, R.L. (1994). Prediction of soil compaction under vehicles. In: B.D. Soane and C. van Ouwerkrek (eds.), Soil Compaction in Crop Production. Elsevier, Amsterdam, pp. 71–90.

Horton, R., Ankeny, M.D. and Allmaras, R.R. (1994). Effects of compaction on soil hydraulic properties. In: B.D. Soane and C. van Ouwerkrek (eds.), Soil Compaction in Crop Production. Elsevier, Amsterdam, pp. 141–165.

Kuipers, H. and van de Zande, J.C. (1994). Quantification of traffic systems in crop production. In: B.D. Soane and C. van Ouwerkrek (eds.), Soil Compaction in Crop Production. Elsevier, Amsterdam, pp. 417–445.

Letey, J. (1985). Relationship between soil physical properties and crop production. Advances in Soil Science 1:277–294.

Raper, R.L. (2005). Agricultural traffic impacts on soil. Journal of Terramechanics 42:259–280.

Spangler, M.G. and Handy, R.L. (1982). Soil density and compaction. In: Soil Engineering. 4th edn. Harper and Row, New York, pp. 170–201.

Stepniewski, W., Glinski, J. and Ball, B.C. (1994). Effects of compaction on soil aeration properties. In: B.D. Soane and C. van Ouwerkrek (eds.), Soil Compaction in Crop Production. Elsevier, Amsterdam, pp. 167–189.

Index

A
Acacia, 117, 264–265, 550–551, 619–625
Acid–sulphate, 262
Afar region, 97–107
Afforestation, 208, 484, 547–559
Agroforestry, 8, 40, 42, 74–76, 115, 159, 268, 275, 281, 483, 558
Agroterraces, 534, 537, 539, 541–542
Albania, 389–399
Algeria, 81–94
Ancient human settlement, 592, 601
Ancient indigenous knowledge, 6, 116
Apulia region, 493–514
Arboreal legumes, 623
Argentina, 45, 591–603

B
Bare Soil Index, 172, 174, 176–178
Baringo, 111–126
7Be measurements, 215–216, 582, 584–586
Biodiversity, 6, 8, 10, 16, 19–20, 64, 76–77, 82, 112, 120–121, 148, 157, 341, 344, 375–376, 426–427, 429–432, 530, 561–563, 566, 622
Biological indicators, 530, 537, 542, 620
Biomass, 5, 16, 41, 73, 75–76, 84, 122, 125, 220–230, 298, 300, 302–304, 536, 542, 592, 601
  dynamics, 220–222, 226–227
Bolivia, 45, 605–615

C
Carbon sequestration, 46, 70, 72–74, 76–77, 297–304, 437, 446
Carbon stocks, 72, 76, 298–299, 301–304, 437–438, 443
Central Great Plains, 649–656
Chemical analyses, 310–311, 314, 600
Chemical techniques, 308–309, 311, 316

China, 44, 116, 132, 207–216, 219–230
China PR, 323–324, 326, 328
Climate change, 6–7, 9–10, 16–18, 37, 45, 69–71, 73, 136, 268, 290, 298, 334, 402, 429, 431, 437, 439, 454, 456, 469, 478, 486–488, 542, 558
Climate variability, 67–68, 601
Compaction, 649–656
Conservation agriculture (CA), 40, 42–43, 431, 456
Conservation tillage for potato, 627–636
Crop suitability, 179–191
Crop yield, 152, 157–158, 263–264, 467–468, 470, 474, 649
^{137}Cs, 401–410
^{137}Cs measurements, 212, 216, 582, 584–587

D
Deflation and accumulation zones, 522, 525
Degradation
  land, 3–12, 37–46, 49–64, 97–107, 163–168, 219–230, 285–295, 423–432, 467–475
  soil, 5–6, 15–21, 25–34, 38, 75–76, 138, 253–256, 263, 379–380, 402, 410, 426–427, 548, 558, 620, 628, 633, 640–641, 646
Degraded lands, 8, 76, 113, 174, 236, 530–531, 547–559
Desertification, 131–144, 179–191, 424–425, 493–514
  risk, 81–94
  sensitivity, 82, 86, 93–94, 132, 134–135, 137–140, 144
Detailed soil survey, 238
DPSIR (drivers-pressures-state-impacts-responses), 49–64, 132

Drought, 6–7, 40, 70–72, 78, 86–87, 105–106, 119, 136, 174–175, 180, 254, 308, 377, 430, 438, 498, 506, 530, 533–534, 537–539, 558, 602
Drought-resistant shrubs, 78, 180
Drylands, 4–5, 8, 49–64, 73–74, 98, 112, 120, 179, 297–304, 341, 477

E
Economic costs, 644–645
Enclosures, 111–126, 379
Environmental assessment, 639–646
Environmental degradation, 38, 70, 171–178, 349, 357–358, 606
Erosion
  rates, 21, 43, 215–216, 409, 426, 540, 556, 582, 585–587
  ratio, 367–372
ESR (exchangeable sodium ratio), 259
Ethiopia, 97–106, 116, 122, 196
EU, 11, 17–21, 26–29, 31, 33–34, 393, 402, 417, 420, 431–432, 452, 512, 514, 582
Eucalyptus, 210, 212, 264–265, 619–625

F
Factor of pedogenesis, 441
Food security, 39, 41, 115, 119, 260, 287–288, 295, 377
Forest plantations, 549, 551, 555–556, 559
FRN (fallout radionuclides), 43–45, 208–209, 211, 403, 405, 410
Future perspectives, 45–46, 125–126, 559, 566, 646

G
Geographical Information System (GIS), 51–53, 81–94, 132, 134–135, 187, 238, 335, 337, 344, 358, 389–399, 562–563
  model, 335
Gold mining, 605–615
Grain yield, 159, 195, 197, 202, 261, 264, 649
Grassland, 112, 122, 208, 210–211, 213, 216, 221–230, 297–299, 338, 415, 426, 436, 438, 482–483, 498, 500
Grazing, 97–107, 113–117, 120–121, 123–125, 138, 141, 150, 158, 172, 210, 484, 500, 548
Greece, 423–432
Gypsum, 260–264, 358–359, 362–364, 564–567

H
Heavy metals, 389–399, 569–578, 611
  pollution, 605–615
Holocene, 591–603
Hydraulic conductivity, 256, 258, 263, 274, 359, 361–364, 468

I
IAIE, 38
Income generating, 6, 10, 124, 142, 323, 555
Indices, 53, 88, 140, 177, 220, 224, 294, 298, 301–302, 369, 371, 495, 497–498, 500, 505, 507–509, 512, 514, 622
Indo-Gangetic plain, 257
Invasive shrubs, 122, 163
Iran, 285–295
Italy, 435–459, 467–475, 493–514

K
Kazakhstan, 297–304
Kenya, 111–126
Kriging, 51–52, 392, 440, 442, 521–522, 525

L
Land
  conflicts, 98, 105–106
  conservation, 147–160, 424
  cover, 61–62, 64, 150–152, 154–155, 160, 181, 188, 268–270, 272, 301, 333–344, 347, 435–441, 443, 483, 503
  change, 221, 227–228, 272, 334
  degradation, 3–12, 37–46, 49–64, 97–107, 163–168, 219–230, 285–295, 423–432, 467–475
    assessment, 5, 7, 49–64, 101, 104
    evaluation, 180–181, 184, 187, 191
    policies, 125, 164, 220
    reclamation, 221, 228, 230, 358
  systems, 52–57, 481, 483–484, 486, 598
  use change, 11, 70, 220–222, 228, 273–274, 276, 281, 291, 298, 430–431, 436–437, 439–440, 480, 482–483
    and land degradation, 347–353, 431–432
  use policy, 325
Landslides, 4, 16–17, 25–27, 30–31, 34, 290, 334, 530, 548, 551, 558
Lao PDR, 323–331
Legislation, 16–19, 28, 291, 294–295, 344, 420–421, 429, 431–432, 620
Limit values system, 416–417, 421

M

Maize, 150, 154–155, 270, 272–273, 276–277, 281, 291, 519, 531, 581–587, 641, 644
Mapping, 51–52, 54, 58–59, 63, 84, 99, 165, 174, 181, 187, 236, 251, 269, 300, 302, 377, 392, 431, 510, 566
MEDALUS concept, 131–144
Mediterranean
  agriculture, 477–490
  environments, 493–514
Metamorphic and igneous rock, 569–578
Mineralisation, 75, 278, 280, 427
Mitigation, 5–7, 10, 70, 72, 76, 134, 140, 191, 344, 431, 494, 542, 548, 558
Monegros, 562, 566
Morocco, 131–144, 180, 405
Moving Standard Deviation Index (MSDI), 174, 176
Multiple regression, 439–440, 443, 448–449
Multi-scale assessment, 53–54
Multi-temporal remote sensing, 219–230

N

Natural disaster, 285–295
Nodulation, 198–200
Nomadic pastoral, 98, 100–101, 105, 125
Nuclear and isotopic techniques, 38–42, 45–46
Nutrient(s), 16, 43, 121–122, 153, 196, 236, 246, 255, 262, 426–428, 536, 540, 542, 611, 620, 640
  loss, 236, 268, 640–641, 645

O

Ordos, 219–230
Organic wastes, 371–372
Overgrazing, 97–107

P

Paleowetland, 592
^{210}Pb, 43, 207–213, 216, 401–410
Pedodiversity, 562, 566–567
Permissible limits, 393, 399
Persistent organic pollutants, 414–415
Pluvial erosion, 530–531, 536, 539–542
Poland, 517–527
Policies, 9, 10–11, 17–18, 27, 34, 45, 63, 164, 220–222, 228–230, 254, 289, 294–295, 323, 349, 381, 385, 424, 456, 478, 488–489, 495, 498, 640, 646
Possible solutions, 6, 10, 126, 293, 364, 548
Potentially risky elements, 414–415
Poverty alleviation, 295, 324

Q

Quarries, 334–344

R

Radionuclide technique, 43, 403, 583
Rainfall-runoff, 350–353
Rainforest conversion, 280–281
Reforestation, 8, 148, 334, 430
Rehabilitation, 8, 45, 112–115, 117, 119–122, 125–126, 148, 290, 333–344, 550, 619–625
Remediation, 45, 390, 416–417, 429, 494
Remote sensing, 26, 51–52, 58, 132, 167–168, 219–230, 238, 255, 297–304, 338, 432, 566
Research, 8–9, 16–17, 20–21, 39–43, 46, 73–74, 76–77, 106, 117, 125, 149–150, 166, 180, 182, 196, 257, 259, 268, 274–275, 324–325, 350, 377, 390, 394, 410, 415–416, 429–431, 437, 439, 454, 477–479, 518–519, 531–534, 547–550, 554, 556–557, 628–629, 636, 640–643, 646, 650–651
Restoration, 17, 112, 114–116, 120, 122, 180, 210, 334–336, 338, 340, 555
Revegetation, 113–114, 116, 333, 335, 620, 625
Risk
  cartography, 81–94
  mapping, 510
River
  catchment, 271, 347–353
  discharge simulation, 268, 270–278, 280–281
Romania, 29, 31–32, 530–531, 533, 542, 547–559
Runoff erosion plots, 408

S

Salinity, 11, 46, 71, 138, 182, 190–191, 236, 238, 245–246, 248–249, 251, 255, 257, 262–263, 288, 292, 294, 356, 362–364, 468, 471–474, 562, 564
  control, 263, 288, 467
  development, 246–249, 251
Salinization, 17, 27–28, 31, 135–136, 148, 263, 292, 356, 402, 426, 468–469, 499
Sand encroachment, 82, 171, 178, 180, 309
Sedimentation rates, 43, 45, 403, 408–409
Semi-arid rangelands, 111–126
Sensitivity areas, 82, 94
Sequential extraction procedure, 415, 569–578
Sheet erosion, 154, 240, 246–247, 583

Shrimp
  farmed soils, 358–364
  farming, 355–364
Sicily, 450, 467–475
SLWM (sustainable land and water management), 38, 44, 78
Socotra Island, 375–385
Sodic saline soil, 259
sodic water irrigation, 468–469
soil
  awareness, 16, 17, 20–21
  biological activity, 529–542
  -borne diseases, 629–630, 633–636
  compaction, 19, 28–29, 136, 481, 649–656
  conservation, 8, 11, 20, 40, 75, 138, 288, 378, 381, 384–385, 403–404, 409–410, 530, 633, 636, 640, 644, 646
    measures, 6, 43–46, 114, 207–216, 403, 410, 530
  contamination, 17, 19, 399, 414, 416
  erodibility, 368, 371–372
  erosion, 6–7, 10–11, 18, 21, 28–29, 33, 37–38, 40–41, 43–45, 75, 85–86, 113–114, 119, 121–122, 141, 153, 207–216, 221, 228, 236, 246–247, 251, 254, 268, 270, 274–275, 280–281, 287–291, 334, 367, 372, 377, 379, 381, 402–403, 425–428, 430, 478–479, 481, 485, 487–488, 518–519, 530–531, 540, 542, 556, 558, 581–587, 611, 615, 628, 630, 639–646
  fixation, 319
  organic matter, 8–9, 11, 18, 25–27, 42, 44, 73–75, 77, 211, 254, 258, 359, 369–370, 385, 419, 427, 436, 530, 630, 633
    decline, 28, 34
    pollution, 19, 390, 426, 427
  quality, 6, 11, 18–20, 37, 44, 75, 83, 85–89, 134–135, 138, 158, 207–216, 372, 402, 421, 426, 436, 495–496, 498, 505–506, 530, 533, 537, 602, 620
    improvement, 547–559
  residue cover, 630, 632
  series, 238–240, 250
  structure, 18, 64, 120–121, 123, 254, 258, 368, 371, 427, 468, 499, 542, 582, 628
  survey, 84, 135, 180, 187, 238, 440, 567
  thematic strategy, 16, 17, 20, 21, 27, 432
  threats, 17, 21, 27, 29–31, 34, 436
Solute nutrient output, 280–281
Souss river basin, 131–144

South Africa, 163–168
South Aurès, 81–94
Spain, 4, 29, 31–32, 182, 403, 561–567, 571–572, 614
Spatial distribution, 32, 303, 334–335, 340, 342, 390, 392, 394–398, 450, 474, 506–513, 525
Sudan, 116, 171–178
Sulawesi-Indonesia, 267–281
Sunchulli district, 605–616
Suspended sediment output, 268, 272
Sustainable land management, 8–12, 43, 106, 117, 138, 143, 542
Sustainable land use, 149, 159, 164, 254, 432, 532

T
Tea industry, 323–331
Techniques, 37–46, 147–160, 307–321
Technological elements, 537, 541
Total heavy metals, 570, 572
Trade, 149, 327–328
Tropical catchment, 268
Tunisia, 179–191

U
UNCCD (United Nations Convention to Combat Desertification), 5–6, 10–11, 49, 112, 132, 308, 377, 383, 424
USA, 11, 27, 41–42, 75, 271, 287, 289, 416

V
Vermiculite, 358–359, 362–364

W
Water
  erosion, 21, 136–137, 171, 178, 190, 208–209, 213, 254–255, 291, 321, 402, 446, 519, 548, 628, 641
  harvesting, 114, 117, 121, 141, 143, 295, 335, 337–338, 341–342, 344
  management, 7, 38–40, 44, 46, 69–78, 138, 160, 208, 248–249, 286, 295, 402, 428–429
  quantity, 348, 353, 429
  yield, 347–353
Watershed management, 10–11, 285–295
Wind erosion, 82, 177–178, 180, 183, 190–191, 208, 210–211, 215, 254–255, 426, 517–521, 524, 526
Winter wheat, 431, 480, 530–531, 649–656

Y
Yemen, 375–385
Yield loss, 158, 649–656